ELECTRONIC DEVICES
Electron-Flow Version

Second Edition

ELECTRONIC DEVICES
Electron-Flow Version

THOMAS L. FLOYD

Prentice Hall
Englewood Cliffs, New Jersey   Columbus, Ohio

Cover Photo: Copyright © Superstock
Editor: Dave Garza
Developmental Editor: Carol Hinklin Robison
Production Editor: Rex Davidson
Text Designer: Anne D. Flanagan
Cover Designer: Brian Deep
Production Manager: Patricia A. Tonneman
Marketing Manager: Debbie Yarnell
Illustrations: Rolin Graphics Inc.

This book was set in Times Roman by The Clarinda Company and was printed and bound by R. R. Donnelley & Sons Company. The cover was printed by Phoenix Color Corp.

Library of Congress Cataloging-in-Publication Data

Floyd, Thomas L.
     Electronic devices : electron-flow version / Thomas L. Floyd. — 2nd ed.
          p.   cm.
     Includes bibliographical references and index.
     ISBN 0–13–363599–6
     1. Electronic apparatus and appliances.   2. Solid state electronics.   I. Title.
TK7870.F52   1996
621.3815—dc20   95–15275
                                                                                CIP

©1996 by Prentice-Hall, Inc.
A Simon and Schuster Company
Englewood Cliffs, New Jersey 07632

First Edition copyright ©1992 by Macmillan Publishing Company.

Printed in the United States of America
10   9   8   7   6   5   4   3   2   1

ISBN: 0-13-363599-6

Prentice-Hall International (UK) Limited, *London*
Prentice-Hall of Australia Pty. Limited, *Sydney*
Prentice-Hall Canada Inc., *Toronto*
Prentice-Hall Hispanoamericana, S.A., *Mexico*
Prentice-Hall of India Private Limited, *New Delhi*
Prentice-Hall of Japan, Inc., *Tokyo*
Simon & Schuster Asia Pte. Ltd., *Singapore*
Editora Prentice-Hall do Brasil, Ltda., *Rio de Janeiro*

ONCE AGAIN, TO SHEILA, WITH LOVE.

PREFACE

Welcome to the Second Edition of *Electronic Devices: Electron-Flow Version.* This full-color edition provides a thorough, comprehensive, and practical coverage of electronic devices, circuits, and applications in a student-friendly format. The extensive troubleshooting coverage, the system applications, and the expanded use of data sheets provide an important link between theory and the real world.

This book consists of two basic topic areas. Chapters 1 through 11 cover discrete devices and circuits, while Chapters 12 through 18 deal primarily with linear integrated circuits, including an extensive coverage of operational amplifiers.

Overview

Discrete Devices and Circuits   Basic semiconductor theory, *pn* junctions, and the diode are introduced in Chapter 1. Diode rectifiers and other diode applications are covered in Chapter 2. Special-purpose diodes including the zener, varactor, and optical devices are introduced in Chapter 3. Bipolar junction transistors (BJTs), small-signal amplifiers, and power amplifiers are covered in Chapters 4 through 7. Chapters 8 and 9 cover field-effect transistors (FETs) and FET amplifiers. Amplifier frequency response is covered in Chapter 10, and thyristors and other devices are introduced in Chapter 11.

Linear Integrated Circuits   Operational amplifiers are introduced in Chapter 12, and their frequency response and stability characteristics are covered in Chapter 13. Basic op-amp circuit configurations are studied in Chapters 14 and 15. Active filters are covered in Chapter 16 followed by oscillators (both integrated and discrete) in Chapter 17. Finally, Chapter 18 provides a coverage of voltage regulators.

Organization

Generally this Second Edition retains the basic organization of the First Edition, with only three notable changes that have been incorporated to improve the sequence of topics.

BJT power amplifiers (Chapter 7) now immediately follows the sequence of BJT topics in Chapters 4, 5, and 6. A new chapter (Chapter 15) has been added to expand the coverage of basic linear integrated circuits. Active filters (Chapter 16) now precedes the coverage of oscillators in Chapter 17.

Features

This text includes many features that make it an effective teaching and learning tool. The most significant features (not necessarily in order of importance) are listed below, with those that are new to this edition indicated by an asterisk.

☐ Balanced treatment of both discrete and integrated circuits

☐ Full-color format*

☐ Extensive use of examples

☐ Related exercise in each numbered example

☐ Extensive use of color illustrations

☐ Clear writing style

☐ Standard component values

☐ Two-page chapter openers with system application preview

☐ Section openers

☐ Performance-based section objectives*

☐ Section reviews

☐ Answers to section reviews and related exercises at end of chapters

☐ Mathematics kept to a minimum

☐ Illustrated summaries at key points in several chapters*

☐ System application section in all chapters except Chapter 1 (revised and improved from Third Edition)

☐ Unique circuit board color graphics in all system applications

☐ Extensive troubleshooting coverage (expanded from Third Edition)

☐ Data sheets used throughout text*

☐ End-of-chapter summaries

☐ End-of-chapter glossaries

☐ End-of-chapter multiple-choice self-tests

☐ End-of-chapter formula lists

☐ Basic end-of-chapter problems

☐ Data sheet end-of-chapter problems*

☐ Troubleshooting end-of-chapter problems (expanded from Third Edition)

☐ Advanced end-of-chapter problems*

The comprehensive ancillary package for this edition includes the following:

☐ Instructor's Solution Manual*   (ISBN: 0-13-397738-2)

☐ *Laboratory Exercises for Electronic Devices* by Dave Buchla*   (ISBN: 0-02-316374-7)

☐ *Experiments in Electronic Devices* by Howard Berlin   (ISBN: 0-13-399544-5)

☐ Study Guide prepared by Osama Maarouf   (ISBN: 0-13-398454-0)

☐ Test Item File   (ISBN: 0-13-399510-0)

☐ Test Manager Version 2.0   (ISBN: 0-13-399528-3)

☐ Transparencies and Transparency Masters   (ISBN: 0-13-399536-4)

☐ PSpice/Electronics Workbench® Data Disk created by John Borris   (ISBN: 0-13-399551-8)

☐ Electronics Workbench® available to adopters* (contact your Prentice-Hall representative for details)

☐ Video Library available to adopters* (contact Prentice-Hall representative for details)

☐ Electronic Bulletin Board* available through America OnLine® (contact your Prentice-Hall representative for details)

In addition to these specific features, many areas have been strengthened and improved by the inclusion of new or rewritten material and/or enhanced illustrations.

Chapter Pedagogy

Chapter Opener   Each chapter begins with a two-page spread as shown in Figure P–1. The left page provides the chapter title, a section listing, chapter objectives, and a brief introduction. The right page presents a brief description of the system application along with a block diagram and a section of circuit board art.

Section Opener and Section Review   Each section within a chapter begins with an introduction that provides a general section overview followed by a list of performance-based objectives. Each section ends with a set of review questions or problems that focus

List of performance-based chapter objectives.

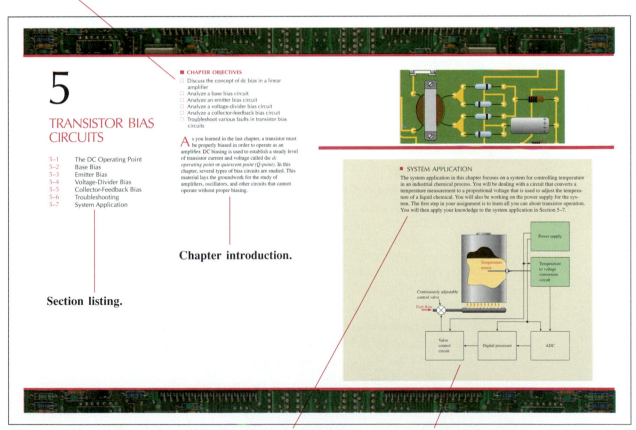

Chapter introduction.

Section listing.

Introduction to
the system application.

System block diagram.

FIGURE P–1
A typical chapter opener.

FIGURE P–2

A typical section opener and section review.

Section review questions end section.

Introductory paragraph begins each section.

Performance-based section objectives.

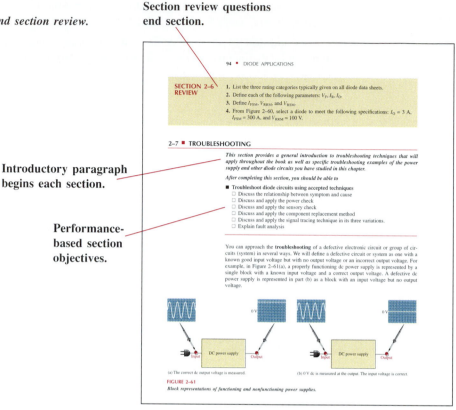

on the main concepts that were covered. Answers to the reviews are given at the end of the chapter. Figure P–2 illustrates these features.

Examples and Related Exercises   Worked out examples are used extensively to illustrate and clarify topics covered. At the end of each example and within the example box is a related exercise, which reinforces or expands on the example. Some related exercises require the student to repeat the procedure illustrated in the example with a different set of values or conditions. Others focus on a more limited part of the example or encourage further thought beyond the procedure shown in the example. Figure P–3 illustrates a typical example with related exercise. Answers to related exercises can be found at the end of the chapter.

Illustrated Summaries   Topics are summarized at appropriate points in certain chapters by an illustrated summary that pulls together important concepts, symbols, circuits, and formulas that are significant in the development of the topic. A typical illustrated summary is shown in Figure P–4.

System Application   The last section of each chapter (except Chapter 1) presents a system application of devices and circuits covered in the chapter. A series of activities with a practical slant are provided to give the student a simulated "on-the-job" experience. These activities include relating a schematic to a graphical representation of a printed circuit board, analysis of a circuit, development of test procedures, setting up a test bench, troubleshooting circuit boards and writing a final report. A typical system application sec-

FIGURE P–3

A typical example and related exercise.

Each example is
enclosed in a box.

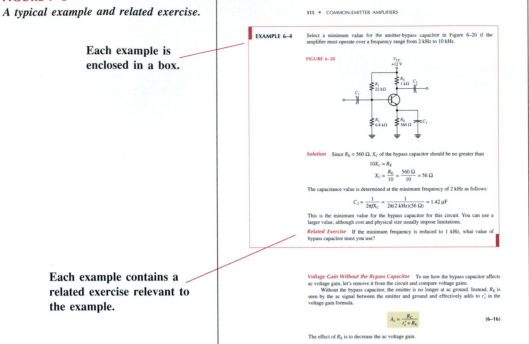

Each example contains a
related exercise relevant to
the example.

FIGURE P–4

A typical illustrated summary.

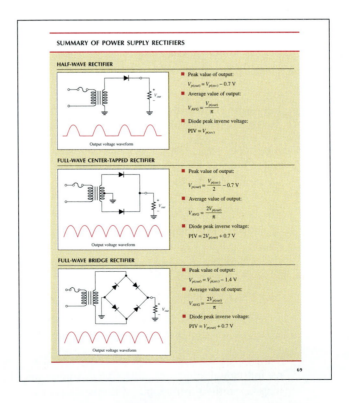

A series of activities is provided, which simulate "on-the-job" experiences.

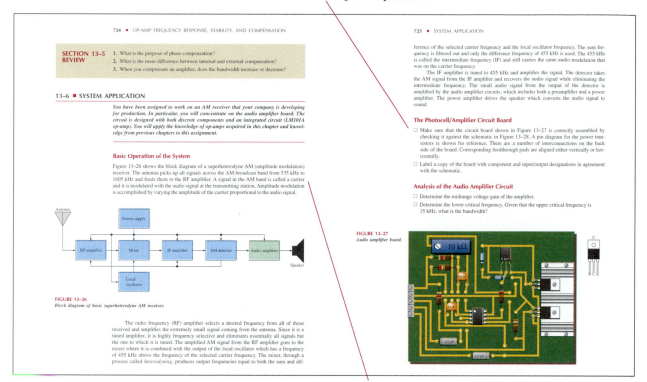

An overall introduction to the system application is provided before a particular pc board circuit is focused on.

FIGURE P–5

Portion of a typical system application section.

tion is illustrated in Figure P–5. Results of system application activities are available in the Instructor's Edition.

Chapter End Matter A chapter summary, a glossary, a list of key formulas, and a multiple-choice self-test follow the text of the chapter. Terms that appear boldface in the text are defined in the glossary.

All chapters have a sectionalized set of basic problems with answers to the odd-numbered ones provided at the end of the book. Additionally, many chapters include a set of data sheet problems, a set of troubleshooting problems, and/or a set of advanced problems.

Answers to section reviews and related exercises are the last items in each chapter.

Suggestions for Use

Electronic Devices: Electron-Flow Version, Second Edition, can be used to accommodate a variety of scheduling and program requirements.

Two-Term Course Chapters 1 through 11 can be covered in the first term. Depending on individual situations, selective coverage may be necessary. For example, Chapter 11 may be omitted if the topic of thyristors is covered in a later industrial electronics course.

Chapters 12 through 18 can be covered in the second term. Again, selective coverage may be necessary.

One-Term Course By omitting certain topics and/or by maintaining a rigorous schedule, this text can be used in one-term courses which cover only discrete devices and circuits or only linear integrated circuits. Also, a significant reduction in the coverage of discrete devices in Chapters 1 through 11 and a limited coverage of integrated circuits (only op-amps, for example) will allow use in a one-term course.

To the Student

You should realize that knowledge and skills are not obtained without effort. Hard work is required to properly prepare yourself for any career, and electronics is no exception. Much reading, thinking, and doing are required to make the most effective use of this book. Don't expect every concept or procedure to become immediately clear. Some topics may take several readings, working many problems, and help from your instructor before you really understand them.

Work through each example step-by-step and then do the related exercise. Work through each section review and check your answers at the end of the chapter. If you don't understand an example or if you can't answer a review question, go back into the section until you can and only then move on to the next section.

The multiple-choice self-tests at the end of each chapter are a good way to check your overall comprehension and retention of the material covered in a chapter. You should do the self-test before working the end-of-chapter problems and check your answers at the end of the book.

The problems at the end of each chapter provide exercises with varying degrees of difficulty. Many chapters offer a set of advanced problems for those of you who require additional challenge. Working through a problem gives you a level of insight and understanding that reading or classroom lectures alone do not provide. Never think that you fully understand a concept or procedure by simply watching or listening to someone else. In the final analysis, you must do it yourself and you must do it to the best of your ability.

A Brief History

Before you begin your study of electronic devices and circuits, let's briefly look back at the beginnings of electronics and some of the important developments that have led to the electronics technology that we have today. It is always good to have a sense of the history of your career field. The names of many of the early pioneers in electricity and electromagnetics still live on in terms of familiar units and quantities. Names such as Ohm, Ampere, Volta, Farad, Henry, Coulomb, Oersted, and Hertz are some of the better known examples. More widely known names such as Franklin and Edison are also very significant in the history of electricity and electronics because of their tremendous contributions.

The Beginning of Electronics The early experiments in electronics involved electric currents in glass vacuum tubes. One of the first to conduct such experiments was a German named Heinrich Geissler (1814–1879). Geissler removed most of the air from a glass tube and found that the tube glowed when there was an electric current through it. Around 1878, British scientist Sir William Crookes (1832–1919) experimented with tubes similar to those of Geissler. In his experiments, Crookes found that the current in the vacuum tubes seemed to consist of particles.

Thomas Edison (1847–1931), experimenting with the carbon-filament light bulb he had invented, made another important finding. He inserted a small metal plate in the bulb. When the plate was positively charged, there was a current from the filament to the plate. This device was the first thermionic diode. Edison patented it but never used it.

The electron was discovered in the 1890s. The French physicist Jean Baptiste Perrin (1870–1942) demonstrated that the current in a vacuum tube consists of the movement of negatively charged particles in a given direction. Some of the properties of these particles were measured by Sir Joseph Thomson (1856–1940), a British physicist, in experiments he performed between 1895 and 1897. These negatively charged particles later became known as electrons. The charge on the electron was accurately measured by an American physicist, Robert A. Millikan (1868–1953), in 1909. As a result of these discoveries, electrons could be controlled, and the electronic age was ushered in.

Putting the Electron to Work

A vacuum tube that allowed electrical current in only one direction was constructed in 1904 by British scientist John A. Fleming. The tube was used to detect electromagnetic waves. Called the Fleming valve, it was the forerunner of the more recent vacuum diode tubes. Major progress in electronics, however, awaited the development of a device that could boost, or amplify, a weak electromagnetic wave or radio signal. This device was the audion, patented in 1907 by Lee deForest, an American. It was a triode vacuum tube capable of amplifying small electrical ac signals.

Two other Americans, Harold Arnold and Irving Langmuir, made great improvements in the triode vacuum tube between 1912 and 1914. About the same time, deForest and Edwin Armstrong, an electrical engineer, used the triode tube in an oscillator circuit. In 1914, the triode was incorporated in the telephone system and made the transcontinental telephone network possible. The tetrode tube was invented in 1916 by Walter Schottky, a German. The tetrode, along with the pentode (invented in 1926 by Dutch engineer Tellegen), greatly improved the triode. The first television picture tube, called the kinescope, was developed in the 1920s by Vladimir Sworykin, an American researcher.

During World War II, several types of microwave tubes were developed that made possible modern microwave radar and other communications systems. In 1939, the magnetron was invented in Britain by Henry Boot and John Randall. In the same year, the klystron microwave tube was developed by two Americans, Russell Varian and his brother Sigurd Varian. The traveling-wave tube (TWT) was invented in 1943 by Rudolf Komphner, an Austrian-American.

Solid-State Electronics

The crystal detectors used in early radios were the forerunners of modern solid-state devices. However, the era of solid-state electronics began with the invention of the transistor in 1947 at Bell Labs. The inventors were Walter Brattain, John Bardeen, and William Shockley, shown in Figure P–6.

In the early 1960s, the integrated circuit (IC) was developed. It incorporated many transistors and other components on a single small *chip* of semiconductor material. Integrated circuit technology has been continuously developed and improved, allowing increasingly more complex circuits to be built on smaller chips.

Around 1965, the first *integrated general-purpose operational amplifier* was introduced. This low-cost, highly versatile device incorporated nine transistors and twelve resistors in a small package. It proved to have many advantages over comparable discrete

FIGURE P–6

Nobel Prize winners Drs. John Bardeen, William Shockley, and Walter Brattain, shown left to right, with apparatus used in their first investigations that led to the invention of the transistor. The trio received the 1956 Nobel Physics award for their invention of the transistor, which was announced by Bell Laboratories in 1948. (Courtesy of Bell Laboratories)

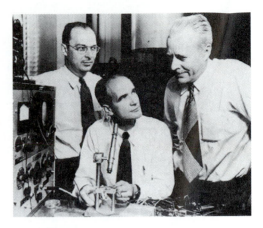

component circuits in terms of reliability and performance. Since this introduction, the IC operational amplifier has become a basic building block for a wide variety of linear systems.

Acknowledgments

This Second Edition of *Electronic Devices: Electron-Flow Version* was made possible by the efforts and talents of many people. I want to express my appreciation to Carol Robison, Dave Garza, Rex Davidson, and Pat Tonneman of Prentice-Hall. Also, many thanks to Lois Porter for another thorough job in editing the manuscript, to Dave Buchla for his valuable input, and to Gary Snyder for an excellent job of checking the manuscript for accuracy.

I also wish to thank the many contributors to *Electronic Devices: Electron-Flow Version*. Without their input, an effective revision would be next to impossible. The following are those individuals who have provided many excellent suggestions, which have brought *Electronic Devices* to its current state:

Don Arney, Indiana Vocational Technical College;
Dave Bechtal, DeVry–Columbus;
Howard Berlin, Delaware Technical and Community College;
Frank Brattain, Indiana Vocational Technical College;
William Burns, Greenville Technical College;
Jeff Carpenter, Columbus State Community College;
Ron Coash, AMTECH Institute;
Darrell Cole, Southern Arkansas University–Tech;
Mark Coleman, Lamar University–Orange;
Ian Davis, Miami Dade Community College;
Jim Davis, Muskingum Technical College;
William Doi, Jr.; Electronics Institute;
Michael Dotson, Southern Illinois University;
Pamela Estep, Electronics Institute;
Ray Fleming, Edison Community College;
George Fredericks, Northeast State Technical Community College;
Kevin Gray, DeVry–Columbus;
Roger Hack, Indiana University–Purdue University;

Byron Hall, ITT–Dayton;
Chuck Kenny, Delaware Technical and Community College;
Ed Kerly, Delaware Technical and Community College–Terry;
Robert Koval, Salem Community College;
Don LaFavour, Columbus State Community College;
James Lovvorn, RETS Electronic Institute;
Mario Marsico, Central Carolina College;
Gary McDowell, Itawamba Community College;
Wally McIntyre, DeVry–Chicago;
William Mead, Wisconsin School of Electronics;
Mike Merchant;
Vic Michaels, Pennsylvania College of Technology;
Rick Miller, Ferris State University;
Maurice Nadeau, Central Maine Vocational Technical Institute;
Larry Oliver, Columbus State Community College;
Mark Porubsky, Milwaukee Area Technical Community College;
Thomas Ratliff, Central Carolina Community College;
Robert Reaves, Durham Technical Community College;
Fred Schuneman, Pierce College;
Mohammed Taher, DeVry–Lombard;
James Walker, Greenville Technical Institute.

We depend on you, so keep up the good work.

Tom Floyd

CONTENTS

ELECTRONIC DEVICES
Electron-Flow Version

1

INTRODUCTION TO SEMICONDUCTORS

■ CHAPTER OBJECTIVES

☐ Discuss the basic structure of atoms
☐ Discuss semiconductors, conductors, and insulators and how they basically differ
☐ Discuss covalent bonding in silicon
☐ Describe how current is produced in a semiconductor
☐ Describe the properties of *n*-type and *p*-type semiconductors
☐ Describe a *pn* junction and how it is formed
☐ Discuss the bias of a *pn* junction
☐ Analyze the current-voltage (*I-V*) characteristic curve of a *pn* junction
☐ Discuss the operation of diodes and explain the three diode models

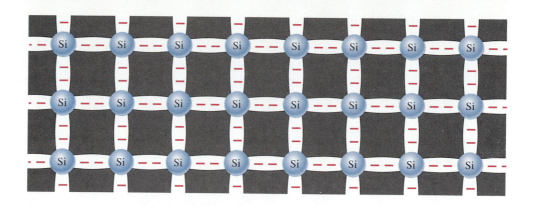

Electronic devices such as diodes, transistors, and integrated circuits are made of a semiconductor material. To understand how these devices work, you need a basic knowledge of the structure of atoms and the interaction of atomic particles. An important concept introduced in this chapter is that of the *pn* junction that is formed when two different types of semiconductor material are joined. The *pn* junction is fundamental to the operation of devices such as the diode and certain types of transistors. Also, the function of the *pn* junction is an essential factor in making electronic circuits operate properly.

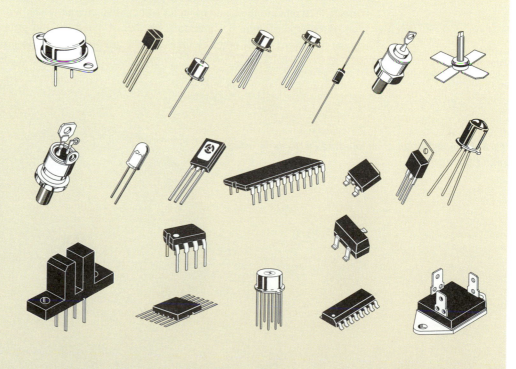

1–1 ■ ATOMIC STRUCTURE

All matter is made of atoms; and all atoms are made of electrons, protons, and neutrons. In this section, you will learn about the structure of the atom, electron orbits and shells, valence electrons, ions, and two semiconductor materials—silicon and germanium. Semiconductor material is important because the configuration of certain electrons in an atom is the key factor in determining how a given material conducts electrical current.

After completing this section, you should be able to

■ **Discuss the basic structure of atoms**
  □ Define *nucleus, proton, neutron,* and *electron*
  □ Define *atomic weight* and *atomic number*
  □ Define *shell*
  □ Explain what a valence electron is
  □ Describe ionization

An **atom** is the smallest particle of an element that retains the characteristics of that element. Each of the known 109 elements has atoms that are different from the atoms of all other elements. This gives each element a unique atomic structure. According to the classical Bohr model, atoms have a planetary type of structure that consists of a central nucleus surrounded by orbiting electrons, as illustrated in Figure 1–1. The **nucleus** consists of positively charged particles called **protons** and uncharged particles called **neutrons. Electrons** are the basic particles of negative charge.

Each type of atom has a certain number of electrons and protons that distinguishes it from the atoms of all other elements. For example, the simplest atom is that of hydro-

FIGURE 1–1
The Bohr model of an atom showing electrons in orbits around the nucleus.

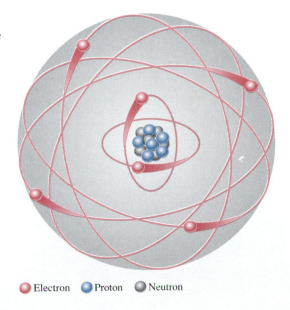

● Electron ● Proton ● Neutron

gen, which has one proton and one electron, as shown in Figure 1–2(a). As another example, the helium atom, shown in Figure 1–2(b), has two protons and two neutrons in the nucleus and two electrons orbiting the nucleus.

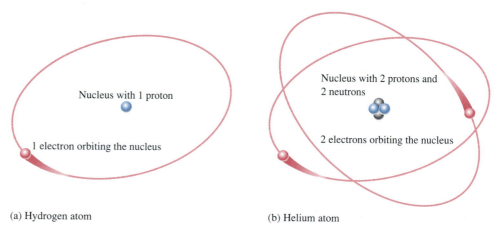

(a) Hydrogen atom   (b) Helium atom

FIGURE 1–2
The two simplest atoms, hydrogen and helium.

Atomic Number and Weight

All elements are arranged in the periodic table of the elements in order according to their **atomic number,** which equals the number of electrons in an electrically balanced (neutral) atom. The elements can also be arranged by their **atomic weight,** which is slightly greater than the number of protons plus neutrons in the nucleus. For example, hydrogen has an atomic number of 1 and an atomic weight of 1.0079. The atomic number of helium is 2, and its atomic weight is 4.00260. In their normal (or neutral) state, all atoms of a given element have the same number of electrons as protons; the positive charges cancel the negative charges, and the atom has a net charge of zero.

Electron Shells and Orbits

Electrons orbit the nucleus of an atom at certain distances from the nucleus. Electrons near the nucleus have less energy than those in more distant orbits. It is known that only discrete (separate and distinct) values of electron energies exist within atomic structures. Therefore, electrons must orbit only at discrete distances from the nucleus.

Energy Levels   Each discrete distance (**orbit**) from the nucleus corresponds to a certain energy level. In an atom, the orbits are grouped into energy bands known as **shells.** A given atom has a fixed number of shells. Each shell has a fixed maximum number of electrons at permissible energy levels (orbits). The differences in energy levels within a shell are much smaller than the difference in energy between shells. The shells are designated *K, L, M, N,* and so on, with *K* being closest to the nucleus. This energy band concept is illustrated in Figure 1–3, which shows the *K* and *L* shells. Additional shells may exist in other types of atoms, depending on the element.

FIGURE 1–3

Energy levels increase as the distance from the nucleus increases.

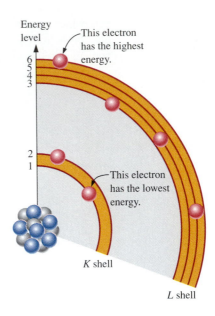

Valence Electrons

Electrons that are in orbits farther from the nucleus have higher energy and are less tightly bound to the atom than those closer to the nucleus. This is because the force of attraction between the positively charged nucleus and the negatively charged electron decreases with increasing distance from the nucleus. Electrons with the highest energy levels exist in the outermost shell of an atom and are relatively loosely bound to the atom. This outermost shell is known as the **valence** shell and electrons in this shell are called *valence electrons*. These valence electrons contribute to chemical reactions and bonding within the structure of a material and determine its electrical properties.

Ionization

When an atom absorbs energy from a heat source or from light, for example, the energy levels of the electrons are raised. When an electron gains energy, it moves to an orbit farther from the nucleus. Since the valence electrons possess more energy and are more loosely bound to the atom than inner electrons, they can jump to higher orbits within the valence shell more easily when external energy is absorbed.

If a valence electron acquires a sufficient amount of energy, it can actually escape from the outer shell and the atom's influence. The departure of a valence electron leaves a previously neutral atom with an excess of positive charge (more protons than electrons). The process of losing a valence electron is known as **ionization** and the resulting positively charged atom is called a *positive ion*. For example, the chemical symbol for hydrogen is H. When a neutral hydrogen atom loses its valence electron and becomes a positive ion, it is designated H^+. The escaped valence electron is called a **free electron.** When a free electron falls into the outer shell of a neutral hydrogen atom, the atom becomes negatively charged (more electrons than protons) and is called a *negative ion,* designated H^-.

The Number of Electrons in Each Shell

The maximum number of electrons (N_e) that can exist in each shell of an atom is determined by the formula,

$$N_e = 2n^2 \qquad (1-1)$$

where n is the number of the shell. The innermost (K) shell is number 1, the L shell is number 2, the M shell is number 3, and so on. The maximum number of electrons that can exist in the innermost shell (shell 1) is

$$N_e = 2n^2 = 2(1)^2 = 2$$

The maximum number of electrons that can exist in the second shell is

$$N_e = 2n^2 = 2(2)^2 = 2(4) = 8$$

The maximum number of electrons that can exist in the third shell is

$$N_e = 2n^2 = 2(3)^2 = 2(9) = 18$$

The maximum number of electrons that can exist in the fourth shell is

$$N_e = 2n^2 = 2(4)^2 = 2(16) = 32$$

All shells in a given atom must be completely filled with electrons except the outer (valence) shell.

SECTION 1–1 REVIEW	**1.** Describe an atom.
	2. What is an electron?
	3. What is a valence electron?
	4. What is a free electron?
	5. How are ions formed?

1–2 ■ SEMICONDUCTORS, CONDUCTORS, AND INSULATORS

In terms of their electrical properties, materials can be classified into three groups: conductors, semiconductors, and insulators. In this section, we will examine the properties of semiconductors and compare them to conductors and insulators.

After completing this section, you should be able to

■ Discuss semiconductors, conductors, and insulators and how they basically differ
  ☐ Define the core of an atom
  ☐ Describe the atomic structure of copper, silicon, germanium, and carbon
  ☐ List the four best conductors
  ☐ List four semiconductors
  ☐ Discuss the difference between conductors and semiconductors
  ☐ Discuss the difference between silicon and germanium semiconductors
  ☐ Explain why silicon is much more widely used than germanium

All materials are made up of atoms. These atoms contribute to the electrical properties of a material, including its ability to conduct electrical current.

For purposes of discussing electrical properties, an atom can be represented by the valence shell and a **core** that consists of all the inner shells and the nucleus. This concept is illustrated in Figure 1–4 for a carbon atom. Carbon is used in many types of resistors. Notice that the carbon atom has four electrons in the valence shell and two electrons in the inner shell (*K*). The nucleus consists of six protons and six neutrons so the +6 indicates the positive charge of the six protons. The simplified representation shows the four valence electrons and a core with a net charge of +4 (+6 for the nucleus and −2 for the two inner shell electrons).

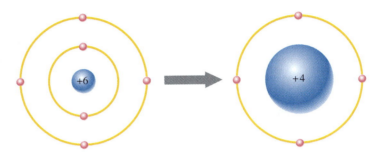

FIGURE 1–4

Diagram of a carbon atom and its simplified form, which consists of the valence shell and the core.

Conductors

A **conductor** is a material that easily conducts electrical current. The best conductors are single-element materials, such as copper, silver, gold, and aluminum, which are characterized by atoms with only one valence electron very loosely bound to the atom. These loosely bound valence electrons can easily break away from their atoms and become free electrons. Therefore, a conductive material has many free electrons that, when moving in a net direction, make up the **current.**

Insulators

An **insulator** is a material that does not conduct electrical current under normal conditions. Most good insulators are compounds rather than single-element materials. Valence electrons are tightly bound to the atoms; therefore, there are very few free electrons in an insulator.

Semiconductors

A **semiconductor** is a material that is between conductors and insulators in its ability to conduct electrical current. A semiconductor in its pure (intrinsic) state is neither a good conductor nor a good insulator. The most common single-element semiconductors are **silicon, germanium,** and **carbon.** Compound semiconductors such as gallium arsenide are also commonly used. The single-element semiconductors are characterized by atoms with four valence electrons.

Energy Bands

Recall that the valence shell of an atom represents a band of energy levels and the valence electrons are confined to that band. If an electron acquires enough additional energy from an external source, it leaves the valence shell and becomes a free electron and exists in what is known as the *conduction band*.

The difference in energy between the valence band and the conduction band is called the *energy gap*. This is the amount of energy that a valence electron must have in order to jump from the valence band to the conduction band. Once in the conduction band, the electron is free to move throughout the material and is not tied to any given atom.

Figure 1–5 shows energy diagrams for the three types of electrical materials. Notice in part (a) that insulators have a very wide energy gap. Valence electrons do not jump into the conduction band except under breakdown conditions where extremely high voltages are applied across the material. As you can see in part (b), semiconductors have a much narrower energy gap. This gap permits some valence electrons to jump into the conduction band and become free electrons. By contrast, as part (c) illustrates, the energy bands in conductors overlap. In a conductive material there is always a large number of free electrons.

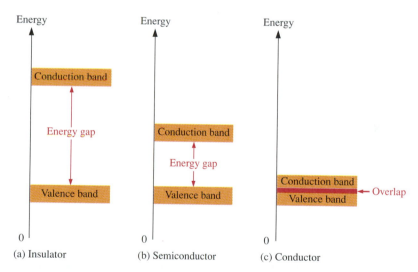

FIGURE 1–5
Energy diagrams for the three types of materials.

Comparison of a Semiconductor Atom to a Conductor Atom

Now let's examine some basic reasons why silicon is a semiconductor and copper is a conductor. Diagrams of the silicon atom and the copper atom are shown in Figure 1–6. Notice that the core of the silicon atom has a net charge of +4 (14 protons − 10 electrons) and the core of the copper atom has a net charge of +1 (29 protons − 28 electrons).

The valence electron in the copper atom "feels" an attractive force of +1 compared to a valence electron in the silicon atom which "feels" an attractive force of +4. So, there is four times more force trying to hold a valence electron to the atom in copper than in

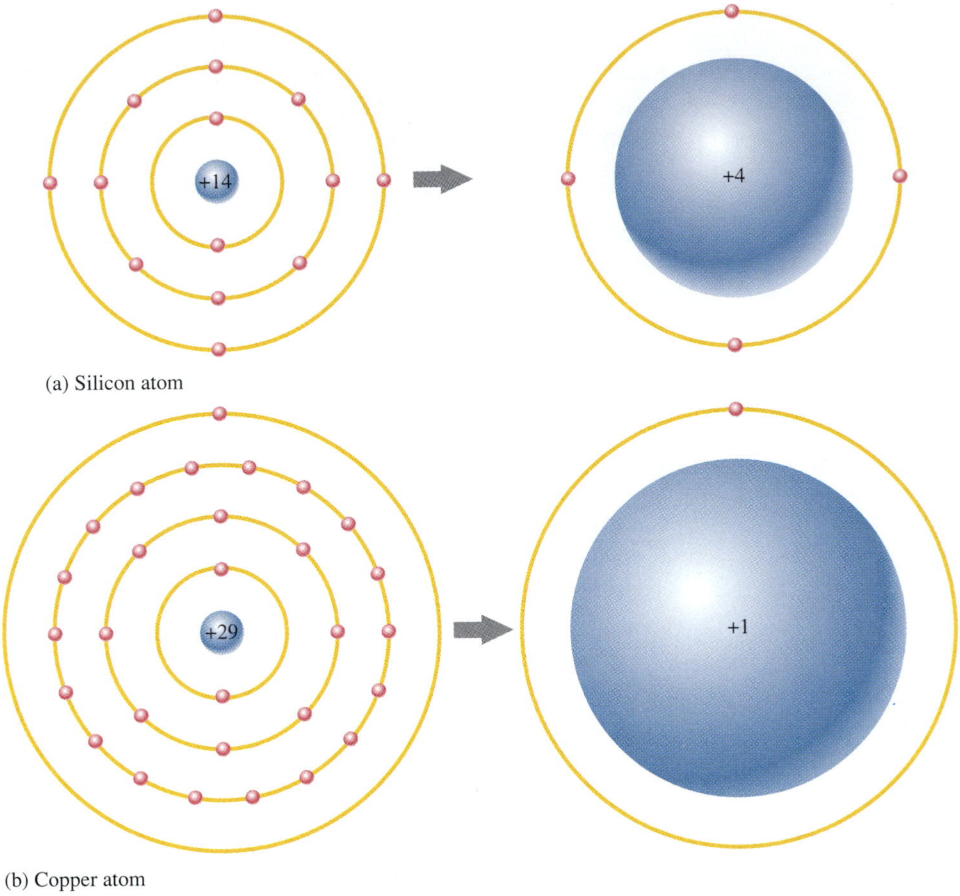

(a) Silicon atom

(b) Copper atom

FIGURE 1–6
Diagrams of the copper and silicon atoms.

silicon. Also, the copper's valence electron is in the fourth shell, which is a greater distance from its nucleus than silicon's valence electron in the third shell. Recall that electrons farthest from the nucleus have the most energy.

Therefore, in copper, the valence electron has less force holding it to the atom than does the valence electron in silicon. Also, the valence electron in copper has more energy than the valence electron in silicon. This means that it is easier for valence electrons in copper to acquire enough additional energy to escape from their atoms and become free electrons in the conduction band than it is in silicon. In fact, large numbers of valence electrons in copper already have sufficient energy to be free electrons as indicated by the overlapping of the valence band and the conduction band.

Silicon and Germanium

The atomic structures of silicon and germanium are shown in Figure 1–7. Silicon is the most widely used material in diodes, transistors, integrated circuits, and other semiconductor devices. Notice that both silicon and germanium have the characteristic four valence electrons.

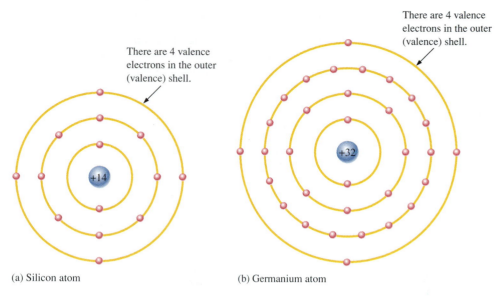

There are 4 valence electrons in the outer (valence) shell.

There are 4 valence electrons in the outer (valence) shell.

(a) Silicon atom   (b) Germanium atom

FIGURE 1–7
Diagrams of the silicon and germanium atoms.

The valence electrons in germanium are in the fourth shell while the ones in silicon are in the third shell, closer to the nucleus. This means that the germanium valence electrons are at higher energy levels than those in silicon and, therefore, require a smaller amount of additional energy to escape from the atom. This property makes germanium more unstable at high temperatures, and this is a basic reason why silicon is the most widely used semiconductor material.

SECTION 1–2 REVIEW

1. What is the basic difference between conductors and insulators?
2. How do semiconductors differ from conductors and insulators?
3. How many valence electrons does a conductor such as copper have?
4. How many valence electrons does a semiconductor have?
5. Name three of the best conductive materials.
6. What is the most widely used semiconductor material?
7. Why does a semiconductor have fewer free electrons than a conductor?

1–3 ■ COVALENT BONDS

When certain atoms combine to form a solid material, they arrange themselves in a fixed pattern called a crystal. The atoms within the crystal structure are held together by covalent bonds, which are created by the interaction of the valence electrons of the atoms. Silicon is a crystalline material.

After completing this section, you should be able to

■ **Discuss covalent bonding in silicon**
   ☐ Define a covalent bond
   ☐ Explain what a covalent bond consists of
   ☐ Explain how a silicon crystal is formed

Figure 1–8 shows how each silicon atom positions itself with four adjacent atoms to form a silicon **crystal.** A silicon atom with its four valence electrons shares an electron with each of its four neighbors. This effectively creates eight valence electrons

FIGURE 1–8
Covalent bonds in silicon.

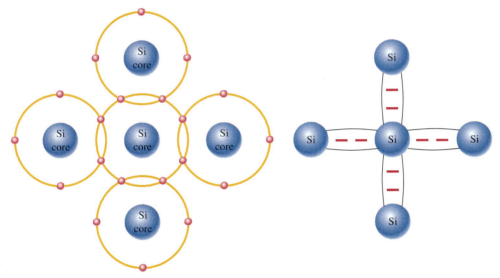

(a) The center atom shares an electron with each of the four surrounding atoms creating a covalent bond with each. The surrounding atoms are in turn bonded to other atoms, and so on.

(b) Bonding diagram. The red negative signs represent the shared valence electrons.

FIGURE 1–9
Covalent bonds form a crystal structure.

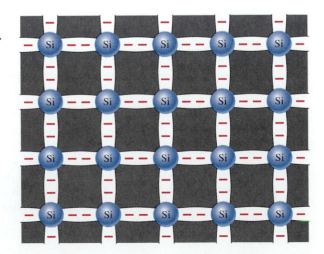

for each atom and produces a state of chemical stability. Also, this sharing of valence electrons produces the **covalent** bonds that hold the atoms together; each shared electron is attracted equally by two adjacent atoms which share it. Covalent bonding in an intrinsic silicon crystal is shown in Figure 1–9. An **intrinsic** crystal is one that has no impurities. Covalent bonding for germanium is similar because it also has four valence electrons.

1–4 ■ CONDUCTION IN SEMICONDUCTORS

How a material conducts electrical current is important in understanding how electronic devices operate. You can't really understand the operation of a device such as a diode or transistor without knowing something about the basic current phenomenon. In this section, you will see how conduction occurs in semiconductor material.

After completing this section, you should be able to

■ **Describe how current is produced in a semiconductor**
   □ Define *conduction electron*
   □ Define *hole*
   □ Explain what an electron-hole pair is
   □ Discuss recombination
   □ Explain the difference between electron current and hole current

As you have learned, the electrons of an atom can exist only within prescribed energy bands. Each shell around the nucleus corresponds to a certain energy band and is separated from adjacent shells by energy gaps, in which no electrons can exist. This is shown in Figure 1–10 for an unexcited (no external energy such as heat) silicon atom. This condition occurs *only* at a temperature of absolute 0° K.

Conduction Electrons and Holes

An intrinsic (pure) silicon crystal at room temperature derives heat (thermal) energy from the surrounding air, causing some valence electrons to gain sufficient energy to jump the gap from the valence band into the conduction band, becoming free electrons not bound to any one atom but free to drift. Free electrons are also called **conduction electrons.** This is illustrated in the energy diagram of Figure 1–11(a) and in the bonding diagram of Figure 1–11(b).

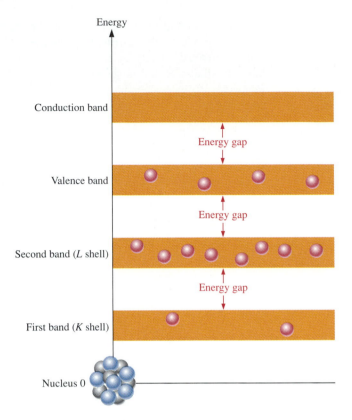

FIGURE 1–10

Energy band diagram for unexcited silicon atom. There are no electrons in the conduction band.

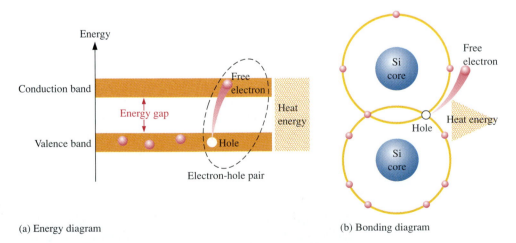

(a) Energy diagram

(b) Bonding diagram

FIGURE 1–11

Creation of an electron-hole pair in an excited silicon atom. An electron in the conduction band is a free electron.

When an electron jumps to the conduction band, a vacancy is left in the valence band. This vacancy is called a **hole.** For every electron raised to the conduction band by external energy, there is one hole left in the valence band, creating what is called an **electron-hole pair. Recombination** occurs when a conduction-band electron loses energy and falls back into a hole in the valence band.

To summarize, a piece of intrinsic silicon at room temperature has, at any instant, a number of conduction-band (free) electrons that are unattached to any atom and are essentially drifting randomly throughout the material. There is also an equal number of holes in the valence band created when these electrons jump into the conduction band. This is illustrated in Figure 1–12.

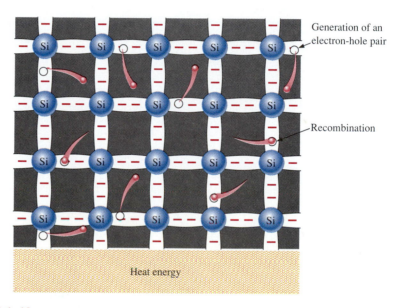

FIGURE 1–12

Electron-hole pairs in a silicon crystal. Free electrons are being generated continuously while some recombine with holes.

Electron and Hole Current

When a voltage is applied across a piece of intrinsic silicon, as shown in Figure 1–13, the thermally generated free electrons in the conduction band, which are free to move randomly in the crystal structure, are now easily attracted toward the positive end. This movement of free electrons is one type of current in a semiconductor material and is called *electron current*.

Another type of current occurs at the valence level, where the holes created by the free electrons exist. Electrons remaining in the valence band are still attached to their atoms and are not free to move randomly in the crystal structure as are the free electrons. However, a valence electron can move into a nearby hole, with little change in its energy level, thus leaving another hole where it came from. Effectively the hole has moved from one place to another in the crystal structure, as illustrated in Figure 1–14. This is called *hole current*.

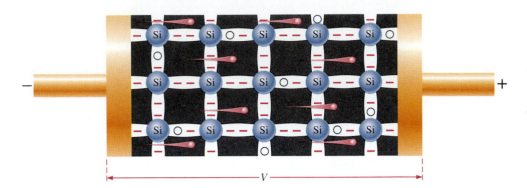

FIGURE 1–13

Electron current in intrinsic silicon is produced by the movement of thermally generated free electrons.

Valence electron moves into 4th hole and leaves a 5th hole.

Valence electron moves into 2nd hole and leaves a 3rd hole.

Free electron leaves hole in valence shell.

Valence electron moves into 5th hole and leaves a 6th hole.

Valence electron moves into 3rd hole and leaves a 4th hole.

Valence electron moves into 1st hole and leaves a 2nd hole.

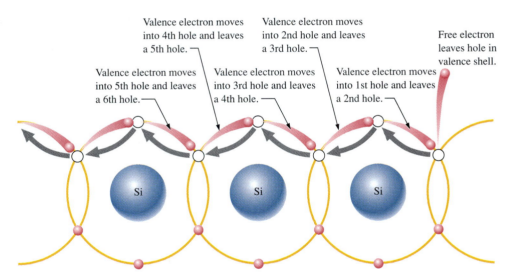

As a valence electron moves left to right to fill a hole while leaving another hole behind, a hole has effectively moved from right to left.

FIGURE 1–14

Hole current in intrinsic silicon.

SECTION 1–4 REVIEW

1. Are free electrons in the valence band or in the conduction band?
2. Which electrons are responsible for current in a material?
3. What is a hole?
4. At what energy level does hole current occur?

1–5 ■ *N*-TYPE AND *P*-TYPE SEMICONDUCTORS

Semiconductor materials do not conduct current well and are of little value in their intrinsic state. This is because of the limited number of free electrons in the conduction band and holes in the valence band. Intrinsic silicon (or germanium) must be modified by increasing the free electrons and holes to increase its conductivity and make it useful in electronic devices. This is done by adding impurities to the intrinsic material as you will learn in this section. Two types of extrinsic (impure) semiconductor materials, n-type and p-type, are the key building blocks for all types of electronic devices.

After completing this section, you should be able to

■ Describe the properties of *n*-type and *p*-type semiconductors
   ☐ Define *doping*
   ☐ Explain how *n*-type semiconductors are formed
   ☐ Explain how *p*-type semiconductors are formed
   ☐ Define *majority carrier* and *minority carrier*

Doping

The conductivity of silicon and germanium can be drastically increased by the controlled addition of impurities to the intrinsic (pure) semiconductor material. This process, called **doping,** increases the number of current carriers (electrons or holes), thus increasing the conductivity and decreasing the resistivity. The two categories of impurities are *n*-type and *p*-type.

N-Type Semiconductor

To increase the number of conduction-band electrons in intrinsic silicon, **pentavalent** impurity atoms are added. These are atoms with five valence electrons such as arsenic (As), phosphorus (P), bismuth (Bi), and antimony (Sb).

   As illustrated in Figure 1–15, each pentavalent atom (antimony, in this case) forms covalent bonds with four adjacent silicon atoms. Four of the antimony atom's valence electrons are used to form the covalent bonds with silicon atoms, leaving one extra electron. This extra electron becomes a conduction electron because it is not attached to any atom. Because the pentavalent atom gives up an electron, it is often called a *donor atom.* The number of conduction electrons can be carefully controlled by the number of impurity atoms added to the silicon. A conduction electron created by this doping process does not leave a hole in the valence band because it is in excess of the number required to fill the valence band.

Majority and Minority Carriers   Since most of the current carriers are electrons, silicon (or germanium) doped with pentavalent atoms is an *n*-type semiconductor material (the *n* stands for the negative charge on an electron). The electrons are called the **majority carriers** in *n*-type material. Although the majority of current carriers in *n*-type material are electrons, there are also a few holes that are created when electron-hole pairs are thermally generated. These holes are *not* produced by the addition of the pentavalent impurity atoms. Holes in an *n*-type material are called **minority carriers.**

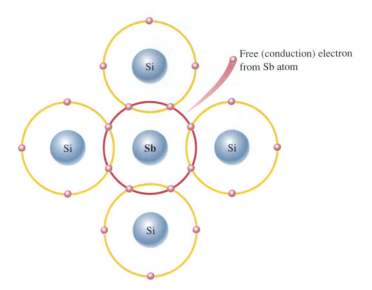

FIGURE 1–15

Pentavalent impurity atom in a silicon crystal structure. An antimony (Sb) impurity atom is shown in the center. The extra electron from the Sb atom becomes a free electron.

P-Type Semiconductor

To increase the number of holes in intrinsic silicon, **trivalent** impurity atoms are added. These are atoms with three valence electrons such as aluminum (Al), boron (B), indium (In), and gallium (Ga). As illustrated in Figure 1–16, each trivalent atom (boron, in this case) forms covalent bonds with four adjacent silicon atoms. All three of the boron atom's valence electrons are used in the covalent bonds; and, since four electrons are required, a hole results when each trivalent atom is added. Because the trivalent atom can take an electron, it is often referred to as an *acceptor atom*. The number of holes can be

FIGURE 1–16

Trivalent impurity atom in a silicon crystal structure. A boron (B) impurity atom is shown in the center.

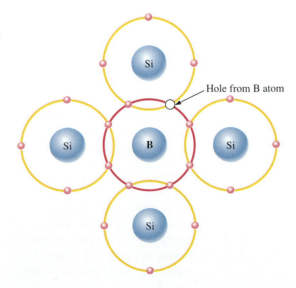

carefully controlled by the number of trivalent impurity atoms added to the silicon. A hole created by this doping process is *not* accompanied by a conduction (free) electron.

Majority and Minority Carriers   Since most of the current carriers are holes, silicon (or germanium) doped with trivalent atoms is called a *p*-type semiconductor material. Holes can be thought of as positive charges because the absence of an electron leaves a net positive charge on the atom. The holes are the majority carriers in *p*-type material. Although the majority of current carriers in *p*-type material are holes, there are also a few free electrons that are created when electron-hole pairs are thermally generated. These free electrons are *not* produced by the addition of the trivalent impurity atoms. Electrons in *p*-type material are the minority carriers.

SECTION 1–5 REVIEW

1. Define *doping*.
2. What is the difference between a pentavalent atom and a trivalent atom? What are other names for these atoms?
3. How is an *n*-type semiconductor formed?
4. How is a *p*-type semiconductor formed?
5. What is the majority carrier in an *n*-type semiconductor?
6. What is the majority carrier in a *p*-type semiconductor?
7. By what process are the majority carriers produced?
8. By what process are the minority carriers produced?
9. What is the difference between intrinsic and extrinsic semiconductors?

1–6 ■ THE *PN* JUNCTION

If you take a block of silicon and dope half of it with a trivalent impurity and the other half with a pentavalent impurity, a boundary called the pn junction is formed between the resulting p-type and n-type portions. The pn junction is the feature that allows diodes, transistors, and other devices to work. This section and the next one will provide a basis for the discussion of the diode in Section 1–9.

After completing this section, you should be able to

■ **Describe a *pn* junction and how it is formed**
  ☐ Discuss diffusion across a *pn* junction
  ☐ Explain the formation of the depletion region
  ☐ Define *barrier potential* and discuss its significance
  ☐ State the values of barrier potential in silicon and germanium

A *p*-type material consists of silicon atoms and trivalent impurity atoms such as boron. The boron atom adds a hole when it bonds with the silicon atoms. However, since the number of protons and the number of electrons are equal throughout the material, there is no net charge in the material and so it is neutral.

An *n*-type silicon material consists of silicon atoms and pentavalent impurity atoms such as antimony. As you have seen, an impurity atom releases an electron when it bonds with four silicon atoms. Since there is still an equal number of protons and electrons (including the free electrons) throughout the material, there is no net charge in the material and so it is neutral.

If a piece of intrinsic silicon is doped so that half is *n*-type and the other half is *p*-type, a ***pn* junction** forms between the two regions as indicated in Figure 1–17. The *p* region has many holes (majority carriers) from the impurity atoms and only a few thermally generated free electrons (minority carriers). The *n* region has many free electrons (majority carriers) from the impurity atoms and only a few thermally generated holes (minority carriers).

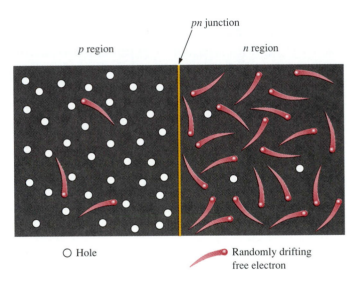

FIGURE 1–17
The basic pn structure at the instant of junction formation showing only the majority and minority carriers.

Formation of the Depletion Region

As you have seen, the free electrons in the *n* region are randomly drifting in all directions. At the instant of the *pn* junction formation, the free electrons near the junction in the *n* region begin to diffuse across the junction into the *p* region where they combine with holes near the junction, as shown in Figure 1–18(a).

Before the *pn* junction is formed, recall that there are as many electrons as protons in the *n*-type material making the material neutral in terms of net charge. The same is true for the *p*-type material.

When the *pn* junction is formed, the *n* region loses free electrons as they diffuse across the junction. This creates a layer of positive charges (pentavalent ions) near the junction. As the electrons move across the junction, the *p* region loses holes as the electrons and holes combine. This creates a layer of negative charges (trivalent ions) near the junction. These two layers of positive and negative charges form the **depletion region,** as

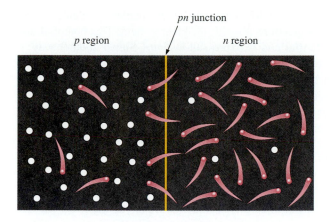

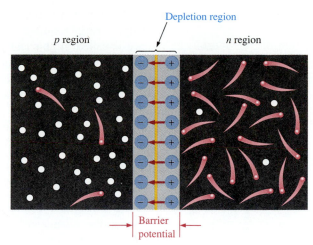

(a) At the instant of junction formation, free electrons in the *n* region near the *pn* junction begin to diffuse across the junction and fall into holes near the junction in the *p* region.

(b) For every electron that diffuses across the junction and combines with a hole, a positive charge is left in the *n* region and a negative charge is created in the *p* region, forming a barrier potential. This action continues until the voltage of the barrier repels further diffusion.

FIGURE 1–18

Formation of the depletion region.

shown in Figure 1–18(b). The term *depletion* refers to the fact that the region near the *pn* junction is depleted of *charge carriers* (electrons and holes) due to diffusion across the junction.

After the initial surge of free electrons across the *pn* junction, the depletion region has expanded to a point where equilibrium is established and there is no further diffusion of electrons across the junction. This occurs as follows. As electrons continue to diffuse across the junction, more and more positive and negative charges are created near the junction as the depletion region is formed. A point is reached where the total negative charge in the depletion region repels any further diffusion of electrons (negatively charged particles) into the *p* region (like charges repel) and the diffusion stops. In other words, the depletion region acts as a barrier to the further movement of electrons across the junction.

Keep in mind that the depletion region is formed very quickly and is very thin compared to the *n* region and *p* region. The width of the depletion region in Figure 1–18 is exaggerated for purposes of illustration.

Barrier Potential   Any time there is a positive charge and a negative charge near each other, there is a force acting on the charges as described by Coulomb's law. In the depletion region there are many positive charges and many negative charges on opposite sides of the *pn* junction. The forces between the opposite charges form a "field of forces" called an *electric field,* as illustrated in Figure 1–18(b) by the red arrows between the positive charges and the negative charges. This electric field is a barrier to the free electrons in the *n* region, and energy must be expended to move an electron through the electric field. That is, external energy must be applied to get the electrons to move across the barrier of the electric field in the depletion region.

The potential difference of the electric field across the depletion region is the amount of energy required to move electrons through the electric field. This potential difference is called the **barrier potential** and is expressed in *volts*. Stated another way: a certain amount of voltage equal to the barrier potential and with the proper polarity must be applied across a *pn* junction before electrons will begin to flow across the junction. You will learn more about this when we discuss *biasing* in Section 1–7.

The barrier potential of a *pn* junction depends on several factors, including the type of semiconductor material, the amount of doping, and the temperature. The typical barrier potential is approximately 0.7 V for silicon and 0.3 V for germanium at 25°C.

Energy Diagrams of the *PN* Junction and Depletion Region

The valence band and conduction band in an *n*-type material are at slightly lower energy levels than the valence and conduction bands in a *p*-type material. This is due to differences in the atomic characteristics of the pentavalent and the trivalent impurity atoms.

An energy diagram for a *pn* junction at the instant of formation is shown in Figure 1–19(a). As you can see, the valence and conduction bands in the *n* region are at lower energy levels than those in the *p* region, but there is a significant amount of overlapping.

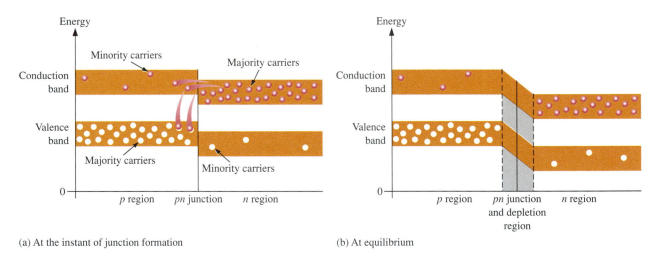

(a) At the instant of junction formation   (b) At equilibrium

FIGURE 1–19
Energy diagrams illustrating the formation of the pn junction and depletion region.

The free electrons in the *n* region that occupy the higher portion of the conduction band in terms of their energy can easily diffuse across the junction (they do not have to gain additional energy) and temporarily become free electrons in the lower energy portion of the *p*-region conduction band. After crossing the junction, the electrons quickly lose energy and fall into the holes in the *p*-region valence band as indicated in Figure 1–19(a).

As the diffusion continues, the depletion region begins to form and the energy level of the conduction band decreases. The decrease in the energy level of the conduction band is due to the loss of the high-energy electrons that have diffused across the junction to the *p* region. Soon, there are no electrons left in the *n*-region conduction band with enough energy to get across the junction to the *p*-region conduction band, as indicated by the alignment of the top of the *n*-region conduction band and the bottom of the *p*-region conduction band in Figure 1–19(b). At this point, the junction is at equilibrium; and the depletion region is complete because diffusion has ceased. There is an energy gradiant across the depletion region which acts as an energy "hill" that an *n*-region electron must climb to get to the *p* region.

Notice that as the energy level of the *n*-region conduction band has shifted downward, the energy level of the valence band has also shifted downward. It still takes the same amount of energy for a valence electron to become a free electron. In other words, the energy gap between the valence band and the conduction band remains the same.

SECTION 1–6 REVIEW	**1.** What is a *pn* junction?
	2. Explain what diffusion is.
	3. Describe the depletion region.
	4. Explain what the barrier potential is and how it is created.
	5. What is the typical value of the barrier potential for a silicon *pn* junction?
	6. What is the typical value of the barrier potential for a germanium *pn* junction?

1–7 ▪ BIASING THE *PN* JUNCTION

As you have learned, no electrons move through the pn junction at equilibrium; and since the flow of electrons is electrical current, there is no electrical current through the pn junction. In electronics, the term bias refers to the use of a dc voltage to establish certain operating conditions for an electronic device. In relation to a pn junction, there are two bias conditions: forward and reverse. Either of these bias conditions is established by connecting a sufficient dc voltage of the proper polarity across the pn junction.

After completing this section, you should be able to

▪ Discuss the bias of a *pn* junction
  ☐ Define *forward bias* and state the required conditions
  ☐ Define *reverse bias* and state the required conditions
  ☐ Discuss the effect of barrier potential on forward bias
  ☐ Explain how current is produced in forward bias
  ☐ Explain reverse current
  ☐ Describe reverse breakdown of a *pn* junction
  ☐ Explain forward bias and reverse bias in terms of energy diagrams

Forward Bias

To **bias** a *pn* junction, apply an external dc voltage across it. **Forward bias** is the condition that allows current through a *pn* junction. Figure 1–20 shows a dc voltage source connected by conductive material (contacts and wire) across a *pn* junction in the direction

FIGURE 1–20

A pn junction connected for forward bias.

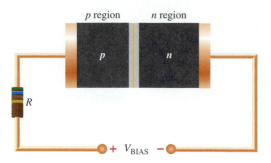

to produce forward bias. This external bias voltage is designated as V_{BIAS}. The resistor, R, limits the current to a value that will not damage the *pn* structure.

Notice that the negative side of V_{BIAS} is connected to the *n* region of the *pn* junction and the positive side is connected to the *p* region. This is one requirement for forward bias. A second requirement is that the bias voltage, V_{BIAS}, must be greater than the barrier potential.

A fundamental picture of what happens when a *pn* junction is forward-biased is shown in Figure 1–21. Because like charges repel, the negative side of the bias-voltage source "pushes" the free electrons, which are the majority carriers in the *n* region, toward the *pn* junction. This flow of free electrons is called *electron current.* The negative side of the source also provides a continuous flow of electrons through the external connection (conductor) and into the *n* region as shown.

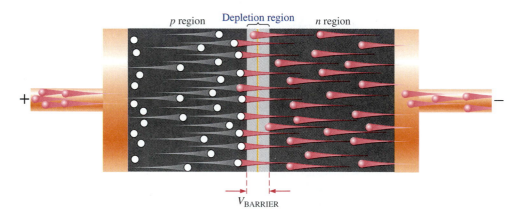

FIGURE 1–21

A forward-biased pn junction showing the flow of majority carriers and the voltage due to the barrier potential across the depletion region.

The bias-voltage source imparts sufficient energy to the free electrons for them to overcome the barrier potential of the depletion region and move on through into the *p* region. Once in the *p* region, these conduction electrons have lost enough energy to immediately combine with holes in the valence band.

Now, the electrons are in the valence band in the *p* region, simply because they have lost too much energy overcoming the barrier potential to remain in the conduction band. Since unlike charges attract, the positive side of the bias-voltage source attracts the valence electrons toward the left end of the *p* region. The holes in the *p* region provide the medium for these valence electrons to move through the *p* region. As was illustrated earlier in Figure 1–14, the electrons move from one hole to the next toward the left. The holes, which are the majority carriers in the *p* region, effectively (not actually) move to the right toward the junction, as you can see in Figure 1–21. This *effective* flow of holes is called the *hole current*. You can also view the hole current as the flow of valence electrons through the *p* region, with the holes providing the only means for these electrons to flow.

As the electrons flow out of the *p* region through the external connection (conductor) and to the positive side of the bias-voltage source, they leave holes behind in the *p* region; at the same time, these electrons become conduction electrons in the metal conductor. Recall that the conduction band in a conductor overlaps the valence band so that it takes much less energy for an electron to be a free electron in a conductor than in a semiconductor. So, there is a continuous availability of holes *effectively* moving toward the *pn* junction to combine with the continuous stream of electrons as they come across the junction into the *p* region.

The Effect of Forward Bias on the Depletion Region

As more electrons flow into the depletion region, the number of positive ions is reduced. As more holes effectively flow into the depletion region on the other side of the *pn* junction, the number of negative ions is reduced. This reduction in positive and negative ions during forward bias causes the depletion region to narrow, as indicated in Figure 1–22(a).

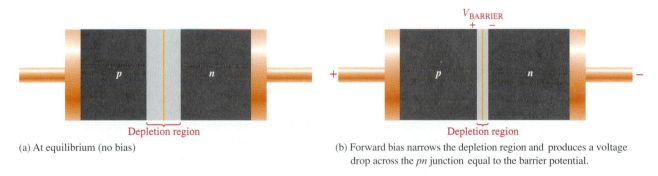

(a) At equilibrium (no bias)

(b) Forward bias narrows the depletion region and produces a voltage drop across the *pn* junction equal to the barrier potential.

FIGURE 1–22
The depletion region narrows and a voltage drop is produced across a pn junction when it is forward-biased.

Effect of the Barrier Potential During Forward Bias

Recall that the electric field between the positive and negative ions in the depletion region on either side of the junction creates an "energy hill" that prevents free electrons from diffusing across the junction at equilibrium (see Figure 1–19(b)). This is known as the *barrier potential*.

When forward bias is applied, the free electrons are provided with enough energy from the bias-voltage source to overcome the barrier potential and effectively "climb" the energy hill and cross the depletion region. The energy that the electrons require in order to pass through the depletion region is equal to the barrier potential. In other words, the electrons give up an amount of energy equivalent to the barrier potential when they cross the depletion region. This energy loss results in a voltage drop across the *pn* junction equal to the barrier potential (0.7 V for silicon or 0.3 V for germanium) as indicated in Figure 1–22(b). An additional small voltage drop occurs across the *p* and *n* regions due to the internal resistance of the material. For doped semiconductor material, this resistance, called the **dynamic resistance,** is very small and can usually be neglected. This is discussed in more detail in Section 1–8.

Reverse Bias

Reverse bias is the condition that prevents current through the *pn* junction. Figure 1–23 shows a dc voltage source connected across a *pn* junction in the direction to produce reverse bias. This external bias voltage is designated as V_{BIAS} just as it was for forward bias. Notice that the positive side of V_{BIAS} is connected to the *n* region of the *pn* junction and the negative side is connected to the *p* region. Also note that the depletion region is shown much wider than in forward bias or equilibrium.

FIGURE 1–23
A pn junction connected for reverse bias.

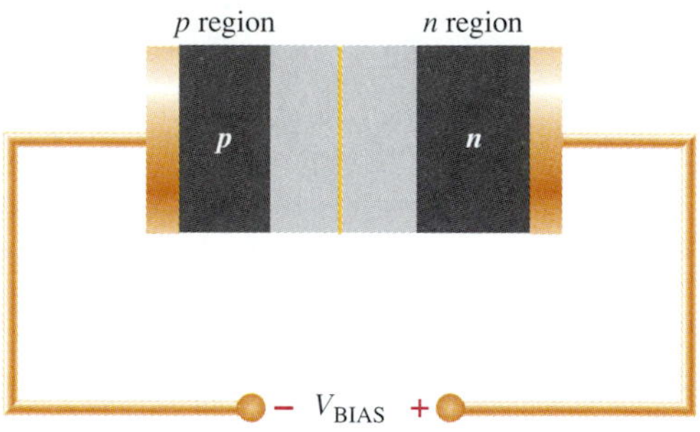

An illustration of what happens when a *pn* junction is reverse-biased is shown in Figure 1–24. Because unlike charges attract, the positive side of the bias-voltage source "pulls" the free electrons, which are the majority carriers in the *n* region, away from the *pn* junction.

In the *n* region, as the electrons flow toward the positive side of the voltage source, additional positive ions are created. This results in a widening of the depletion region and a depletion of majority carriers.

In the *p* region, electrons from the negative side of the voltage source enter as valence electrons and move from hole to hole toward the depletion region where they cre-

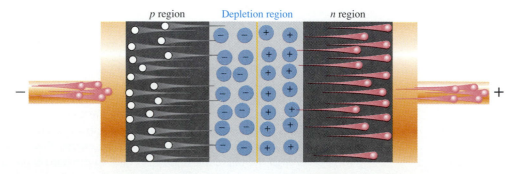

FIGURE 1–24
The pn junction during the short transition time immediately after reverse-bias voltage is applied.

ate additional negative ions. This results in a widening of the depletion region and a depletion of majority carriers. The flow of valence electrons can be viewed as holes being "pulled" toward the positive side.

The initial flow of charge carriers is transitional and lasts for only a very short time after the reverse-bias voltage is applied. As the depletion region widens, the availability of majority carriers decreases. As more of the *n* and *p* regions become depleted of majority carriers, the electric field between the positive and negative charges increases in strength until the potential across the depletion region equals the bias voltage, V_{BIAS}. At this point, the transition current essentially ceases except for a very small reverse current that can usually be neglected.

Reverse Current   The extremely small current that exists in reverse bias after the transition current dies out is caused by the minority carriers in the *n* and *p* regions that are produced by thermally generated electron-hole pairs. The small number of free minority electrons in the *p* region are "pushed" toward the *pn* junction by the negative bias voltage. When these electrons reach the wide depletion region, they "fall down the energy hill" and combine with the minority holes in the *n* region as valence electrons and flow toward the positive bias voltage, creating a small hole current.

The conduction band in the *p* region is at a higher energy level than the conduction band in the *n* region. Therefore, the minority electrons easily pass through the depletion region because they require no additional energy. Reverse current is illustrated in Figure 1–25.

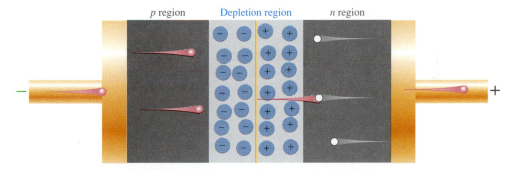

FIGURE 1–25
The extremely small reverse current in a reverse-biased pn junction is due to the minority carriers from thermally generated electron-hole pairs.

Reverse Breakdown   Normally, the reverse current is so small that it can be neglected. However, if the external reverse-bias voltage is increased to a value called the *breakdown voltage,* the reverse current will drastically increase.

This is what happens. The high reverse-bias voltage imparts energy to the free minority electrons so that as they speed through the *p* region, they collide with atoms with enough energy to knock valence electrons out of orbit and into the conduction band. The newly created conduction electrons are also high in energy and repeat the process. If one electron knocks only two others out of their valence orbit during its travel through the *p* region, the numbers quickly multiply. As these high-energy electrons go through the depletion region, they have enough energy to go through the *n* region as conduction electrons, rather than combining with holes.

The multiplication of conduction electrons just discussed is known as **avalanche** and results in a very high reverse current that can damage the *pn* structure because of excessive heat dissipation.

Energy Diagrams for Forward and Reverse Bias

Recall that at equilibrium (no bias) the top end of the conduction band in the *n* region and the conduction band in the *p* region do not overlap in terms of energy level. The equilibrium condition is shown again in Figure 1–26(a). None of the free electrons in the *n* region has sufficient energy to cross the *pn* junction into the *p* region.

When a *pn* junction is forward-biased, the *n*-region conduction band is raised to a higher energy level that overlaps with the *p*-region conduction band. Then large numbers of free electrons have enough energy to climb the "energy hill" and enter the *p* region where they combine with holes in the valence band. Forward bias is illustrated by the energy diagram in Figure 1–26(b).

When a *pn* junction is reverse-biased, the *n*-region conduction band remains at an energy level that prevents the free electrons from crossing into the *p* region. There are a few free minority electrons in the *p*-region conduction band that can easily flow down the "energy hill" into the *n* region where they combine with minority holes in the valence band. This reverse current is extremely small compared to the current in forward bias and can normally be considered negligible. Figure 1–26(c) illustrates reverse bias.

FIGURE 1–26

Energy diagrams for pn junction equilibrium, forward bias, and reverse bias.

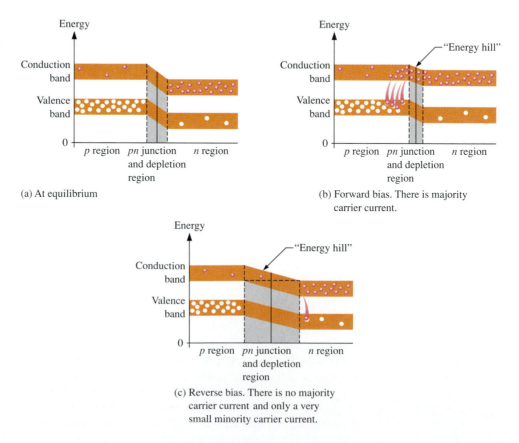

(a) At equilibrium

(b) Forward bias. There is majority carrier current.

(c) Reverse bias. There is no majority carrier current and only a very small minority carrier current.

SUMMARY OF *PN* JUNCTION BIAS

FORWARD BIAS: PERMITS CURRENT

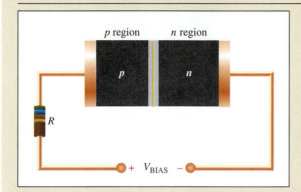

- Bias voltage connections: positive to *p* region; negative to *n* region.
- The bias voltage must be greater than the barrier potential.
- Barrier potential: 0.7 V for silicon; 0.3 V for germanium.
- Majority carriers provide the forward current.
- Majority carriers flow toward the *pn* junction.
- The depletion region narrows.

REVERSE BIAS: PREVENTS CURRENT

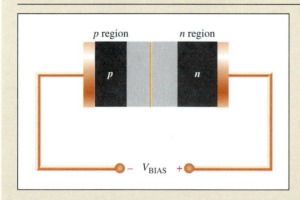

- Bias voltage connections: positive to *n* region; negative to *p* region.
- The bias voltage must be less than the breakdown voltage.
- Minority carriers provide the small reverse current.
- Majority carriers flow away from the *pn* junction during short transition time.
- There is no majority carrier current after transition time.
- The depletion region widens.

SECTION 1–7 REVIEW

1. Describe forward bias of a *pn* junction.
2. Explain how to forward-bias a *pn* junction.
3. Describe reverse bias of a *pn* junction.
4. Explain how to reverse-bias a *pn* junction.
5. Compare the depletion regions in forward bias and reverse bias.
6. Which bias condition produces majority carrier current?
7. How is reverse current in a *pn* junction produced?
8. When does reverse breakdown occur in a *pn* junction?
9. Define *avalanche* as applied to *pn* junctions.

1–8 ■ CURRENT-VOLTAGE CHARACTERISTIC OF A *PN* JUNCTION

As you have learned, forward bias produces current through a pn junction and reverse bias prevents current except for a negligible reverse current. Reverse bias prevents current as long as the reverse-bias voltage does not equal or exceed the breakdown voltage of the junction. In this section, we will examine more closely the relationship between the voltage and the current in a pn junction on a graphical basis.

After completing this section, you should be able to

■ **Analyze the current-voltage (*I-V*) characteristic curve of a *pn* junction**
   ☐ Explain the forward-bias portion of the curve
   ☐ Explain the reverse-bias portion of the curve
   ☐ Identify the barrier potential
   ☐ Identify the breakdown voltage
   ☐ Discuss temperature effects on a *pn* junction

I-V Characteristic for Forward Bias

As you have learned, when a forward-bias voltage is applied across a silicon *pn* junction, there is current through the junction. This current is called the *forward current* and is designated I_F. Figure 1–27 illustrates what happens as the forward-bias voltage is increased positively from 0 V. The resistor is used to limit the forward current to a value that will not overheat the *pn* junction and cause damage.

With 0 V across the *pn* junction, there is no forward current, as indicated in Figure 1–27(a). As you gradually increase the bias voltage, the forward current *and* the voltage across the *pn* junction gradually increase, as shown in part (b). A portion of the applied bias voltage is dropped across the limiting resistor. When the applied bias voltage is increased to a value where the voltage across the *pn* junction reaches approximately 0.7 V (barrier potential), the forward current begins to increase rapidly.

As you continue to increase the bias voltage, the current continues to increase very rapidly, but the voltage across the *pn* junction increases very gradually above 0.7 V, as illustrated in Figure 1–27(c). This small increase in the *pn* junction voltage above the barrier potential is due to the dynamic resistance of the semiconductor material.

Graphing the I-V Curve If the results of the type of measurements shown in Figure 1–27 are plotted on a graph, you get the *I-V* characteristic curve for a forward-biased *pn* junction, as shown in Figure 1–28(a) on page 30. The forward current (I_F) values increase upward along the vertical axis and the *pn* junction forward voltage (V_F) values increase to the right along the horizontal axis.

As you can see in Figure 1–28(a), the forward current increases very little until the forward voltage across the junction reaches approximately 0.7 V at the knee of the curve. After this point, the forward voltage remains at approximately 0.7 V, but I_F increases rapidly. As previously mentioned, there is a slight increase in V_F above 0.7 V as the current increases due mainly to dynamic resistance. *Normal operation for a forward-biased pn junction is above the knee of the curve.* The I_F scale is typically in mA, as indicated.

The three points *A, B,* and *C* shown on the curve in Figure 1–28(a), can be related to the measurements in Figure 1–27. Point *A* corresponds to Figure 1–27(a), which is a zero-bias condition. Point *B* corresponds to Figure 1–27(b) where the forward voltage is less than the barrier potential of 0.7 V. Point *C* corresponds to part (c) where the forward

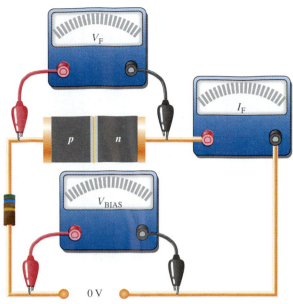

(a) No bias voltage. *PN* junction is at equilibrium.

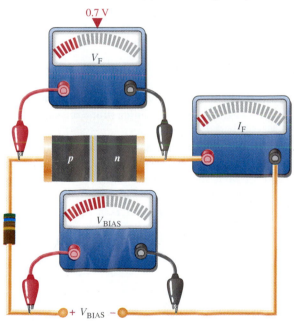

(b) Small forward-bias voltage ($V_F < 0.7$ V), very small forward current.

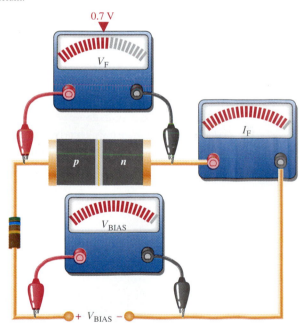

(c) Forward voltage reaches and remains at 0.7 V. Forward current continues to increase as the bias voltage is increased.

FIGURE 1–27

Current and voltage in a forward-biased pn junction.

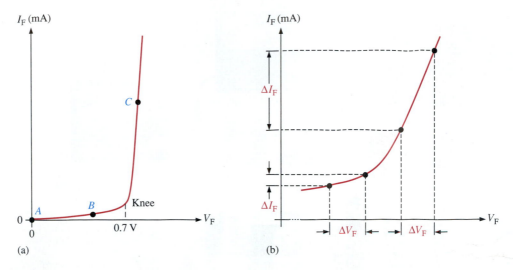

FIGURE 1–28

I-V characteristic curve for forward bias. Part (b) illustrates dynamic resistance.

voltage approximately equals the barrier potential and the external bias voltage and forward current have continued to increase.

Dynamic Resistance Unlike a linear resistance, the resistance of the forward-biased *pn* material is not constant over the entire curve. Because the resistance changes as you move along the *I-V* curve, it is called *dynamic* or *ac resistance*. Internal resistances of electronic devices are usually designated by lowercase italic *r* with a prime, instead of the standard *R*. The dynamic resistance of a diode is designated r'_d.

Below the knee of the curve the resistance is greatest because the current increases very little for a given change in voltage ($r'_d = \Delta V_F/\Delta I_F$). The resistance begins to decrease in the region of the knee of the curve and becomes smallest above the knee where there is a large change in current for a given change in voltage. This characteristic is illustrated in Figure 1–28(b) for equal changes in V_F (ΔV_F) on a magnified segment of the *I-V* curve below and above the knee.

I-V Characteristic for Reverse Bias

When a reverse-bias voltage is applied across a *pn* junction, there is only an extremely small reverse current (I_R) through the junction. Figure 1–29 illustrates what happens as the reverse-bias voltage is increased negatively from 0 V.

With 0 V across the *pn* junction, there is no reverse current. As you gradually increase the reverse-bias voltage, there is a very small reverse current and the voltage across the *pn* junction increases, as shown in Figure 1–29(a). When the applied bias voltage is increased to a value where the reverse voltage across the *pn* junction (V_R) reaches the breakdown value (V_{BR}), the reverse current begins to increase rapidly.

As you continue to increase the bias voltage, the current continues to increase very rapidly, but the voltage across the *pn* junction increases very little above V_{BR}, as illustrated in Figure 1–29(b). *Breakdown, with exceptions, is not a normal mode of operation for most pn junction devices.*

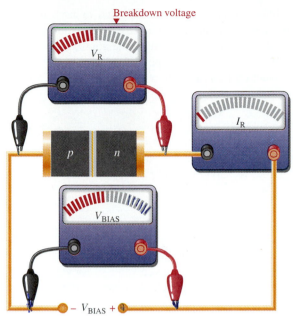

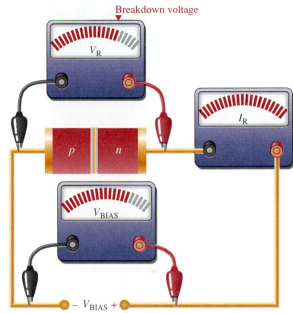

(a) A reverse-bias voltage less than breakdown value produces an extremely small reverse current.

(b) When the bias voltage exceeds the breakdown value, the reverse current rapidly increases and damage from excessive heat can destroy the *pn* material.

FIGURE 1–29
Current and voltage in a reverse-biased pn junction.

Graphing the I-V Curve If the results of the type of measurements shown in Figure 1–29, are plotted on a graph, you get the *I-V* characteristic curve for a reverse-biased *pn* junction. A typical curve is shown in Figure 1–30. The reverse current (I_R) values increase downward along the vertical axis and the *pn* junction reverse voltage (V_R) values increase to the left along the horizontal axis.

FIGURE 1–30
I-V characteristic curve for reverse bias.

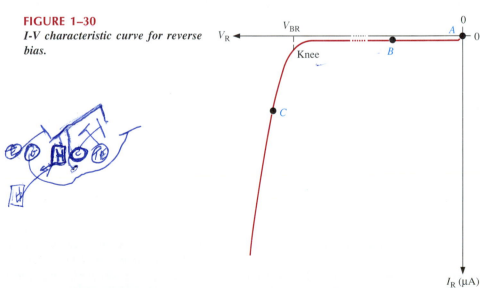

As you can see, there is very little reverse current (usually μA or nA) until the reverse voltage across the junction reaches approximately the breakdown value (V_{BR}) at the knee of the curve. After this point, the reverse voltage remains at approximately V_{BR}, but I_R increases very rapidly resulting in overheating and possible damage. The breakdown voltage for a typical silicon *pn* junction can vary, but a minimum value of 50 V is not unusual.

The Complete *I-V* Characteristic Curve

Combine the curves for both forward bias and reverse bias, and you have the complete *I-V* characteristic curve for a *pn* junction, as shown in Figure 1–31. Notice that the I_F scale is in mA compared to the I_R scale in μA.

Temperature Effects on the *I-V* Characteristic

For a forward-biased *pn* junction, as temperature is increased, the forward current increases for a given value of forward voltage. Also, for a given value of forward current, the forward voltage decreases. This is shown with the *I-V* characteristic curves in Figure 1–32. The red curve is at room temperature (25°C) and the blue curve is at an elevated temperature (25°C + Δ*T*). Notice that the barrier potential decreases as temperature increases.

For a reverse-biased *pn* junction, as temperature is increased, the reverse current increases. The difference in the two curves is exaggerated on the graph in Figure 1–32 for illustration. Keep in mind that the reverse current remains extremely small.

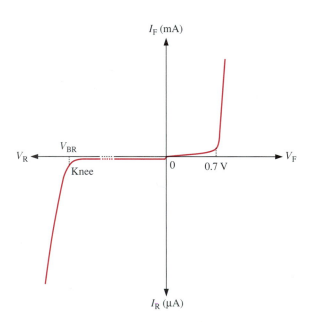

FIGURE 1–31
The complete I-V characteristic curve.

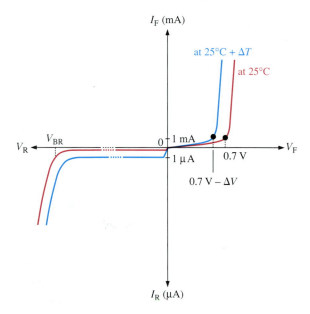

FIGURE 1–32
Temperature effect on the I-V characteristic of a pn junction. The 1 mA and 1 μA marks on the vertical axis are given as a basis for a relative comparison of the current scales.

1. Discuss the significance of the knee of the characteristic curve in forward bias.
2. On what part of the curve is a forward-biased *pn* junction normally operated?
3. Which is greater, the breakdown voltage or the barrier potential?
4. On what part of the curve is a reverse-biased *pn* junction normally operated?
5. What happens to the barrier potential when the temperature increases?

1–9 ■ THE DIODE

In this section, you will see that the diode is a pn junction device and learn its electrical symbol. You will also learn how the diode can be modeled for circuit applications using three levels of complexity.

After completing this section, you should be able to

■ **Discuss the operation of diodes and explain the three diode models**
  ☐ Explain the relation of the *pn* junction to a diode
  ☐ Recognize a diode symbol and indentify the diode terminals
  ☐ Recognize diodes in various physical configurations
  ☐ Explain the ideal diode model
  ☐ Explain the practical diode model
  ☐ Explain the complex diode model

Diode Structure and Symbol

The rectifier diode is a single *pn* junction device with conductive contacts and wire leads connected to each region, as shown in Figure 1–33(a). One half of the diode is an *n*-type semiconductor and the other half is a *p*-type semiconductor. You should recognize this as the *pn* junction device that was discussed in the preceding sections.

   The schematic symbol for a rectifier diode is shown in Figure 1–33(b). The *n* region is called the **cathode** and the *p* region is called the **anode.** The "arrow" in the symbol points in the direction opposite to electron flow.

FIGURE 1–33
Diode structure and schematic symbol.

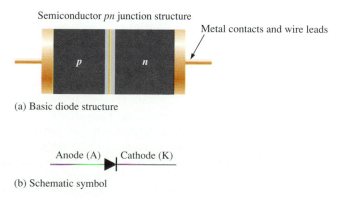

Semiconductor *pn* junction structure

Metal contacts and wire leads

(a) Basic diode structure

Anode (A)   Cathode (K)

(b) Schematic symbol

Forward and Reverse Bias of a Diode

The previous coverage of biasing with relation to a *pn* junction applies to the diode because the diode is a *pn* junction device.

Forward-Bias Connection A diode is forward-biased when a voltage source is connected as shown in Figure 1–34(a). The positive terminal of the source is connected to the anode through a current-limiting resistor. The negative terminal of the source is connected to the cathode. The forward current (I_F) is from cathode to anode as indicated. The forward voltage drop (V_F) due to the barrier potential is from positive at the anode to negative at the cathode.

FIGURE 1–34

Forward-bias and reverse-bias connections for a diode.

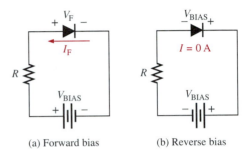

(a) Forward bias   (b) Reverse bias

Reverse-Bias Connection A diode is reverse-biased when a voltage source is connected as shown in Figure 1–34(b). The negative terminal is connected to the anode, and the positive terminal is connected to the cathode. A resistor is not necessary in reverse bias but it is shown for circuit consistency. The current is zero (neglecting the small reverse current). Notice that the entire bias voltage (V_{BIAS}) appears across the diode.

Typical Diodes

Several common physical configurations of diodes are illustrated in Figure 1–35. The anode and cathode are indicated on a diode in several ways, depending on the type of package. The cathode is usually marked by a band, a tab, or some other feature. On those packages where one lead is connected to the case, the case is the cathode. Always check the data sheet for the pin configuration, if there is uncertainty.

The Ideal Diode Model

The ideal model of a diode is a simple switch. When the diode is forward-biased, it acts like a closed (on) switch, as shown in Figure 1–36(a). When the diode is reverse-biased, it acts like an open (off) switch, as shown in part (b). The barrier potential, the forward dynamic resistance, and the reverse current are all neglected.

In Figure 1–36(c), the ideal **diode characteristic curve** graphically depicts the ideal diode operation. Since the barrier potential and the forward dynamic resistance are neglected, the diode is assumed to have a zero voltage across it when forward-biased, as indicated by the portion of the curve on the positive vertical axis.

$$V_F = 0 \text{ V}$$

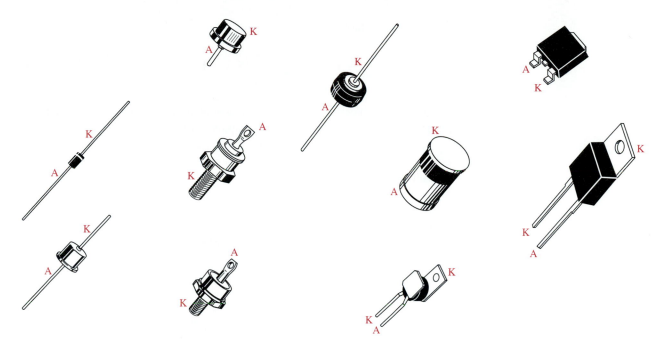

FIGURE 1–35
Typical diode packages and terminal identification.

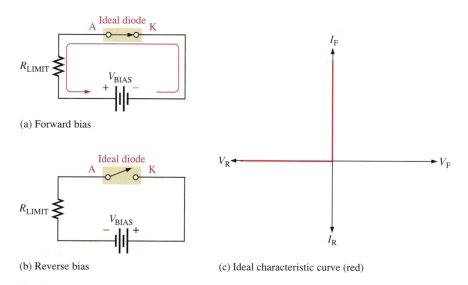

(a) Forward bias

(b) Reverse bias

(c) Ideal characteristic curve (red)

FIGURE 1–36
The ideal model of a diode.

The forward current is determined by the bias voltage and the limiting resistor.

$$I_F = \frac{V_{BIAS}}{R_{LIMIT}}$$

(1–2)

Since the reverse current is neglected, its value is assumed to be zero, as indicated in Figure 1–36(c) by the portion of the curve on the negative horizontal axis.

$$I_R = 0 \text{ A}$$

The reverse voltage equals the bias voltage:

$$V_R = V_{BIAS}$$

You may want to use the ideal model when you are troubleshooting or trying to figure out the operation of a circuit and are not concerned with more exact values of voltage or current.

The Practical Diode Model

The practical model, the one we will use most often, adds the barrier potential to the ideal switch model. When the diode is forward-biased, it acts as a closed switch in series with a small voltage (0.7 V for silicon) equal to the barrier potential with the positive side toward the anode, as indicated in Figure 1–37(a). The barrier potential voltage source rep-

FIGURE 1–37

The practical model of a diode.

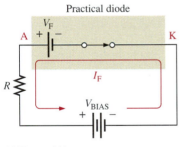

(a) Forward bias

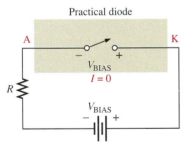

(b) Reverse bias

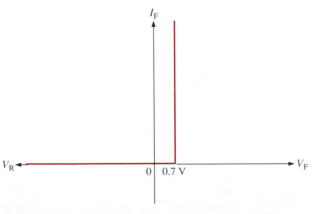

(c) Characteristic curve (silicon)

resents the fixed voltage drop (V_F) produced across the forward-biased *pn* junction of the diode and is not an active source of voltage.

When the diode is reverse-biased, it acts as an open switch just as in the ideal model, as shown in Figure 1–37(b). The barrier potential does not affect reverse bias, so it is not a factor.

The characteristic curve for the practical silicon diode model is shown in Figure 1–37(c). Since the barrier potential is included and the forward dynamic resistance is neglected, the diode is assumed to have a voltage across it when forward-biased, as indicated by the portion of the curve to the right of the origin.

$$V_F = 0.7 \text{ V} \qquad \text{(silicon)}$$
$$V_F = 0.3 \text{ V} \qquad \text{(germanium)}$$

The forward current is determined with the following formula:

$$I_F = \frac{V_{BIAS} - V_F}{R_{LIMIT}} \qquad \text{(1–3)}$$

Since the reverse current is neglected, the diode is assumed to have zero reverse current, as indicated by the portion of the curve on the negative horizontal axis.

$$I_R = 0 \text{ A}$$
$$V_R = V_{BIAS}$$

The Complex Diode Model

The complex model of a diode consists of the barrier potential, the small forward dynamic resistance (r_d'), and the large internal reverse resistance (r_R'). The reverse resistance is taken into account because it provides a path for the reverse current which is included in this diode model.

When the diode is forward-biased, it acts as a closed switch in series with the barrier potential voltage and the small forward dynamic resistance, as indicated in Figure 1–38(a). When the diode is reverse-biased, it acts as an open switch in parallel with the large internal reverse resistance (r_R'), as shown in Figure 1–38(b). The barrier potential does not affect reverse bias, so it is not a factor.

The characteristic curve for the complex silicon diode model is shown in Figure 1–38(c). Since the barrier potential and the forward dynamic resistance are included, the diode is assumed to have a voltage across it when forward-biased. This voltage (V_F) consists of the barrier potential voltage plus the small voltage drop across the dynamic resistance, as indicated by the portion of the curve to the right of the origin. The curve slopes because the voltage drop due to dynamic resistance increases as the current increases. For the complex model of a silicon diode, the following formulas apply:

$$V_F = 0.7 \text{ V} + I_F r_d' \qquad \text{(1–4)}$$

$$I_F = \frac{V_{BIAS} - 0.7 \text{ V}}{R_{LIMIT} - r_d'} \qquad \text{(1–5)}$$

The reverse current is taken into account with the parallel resistance and is indicated by the portion of the curve to the left of the origin. The breakdown portion of the

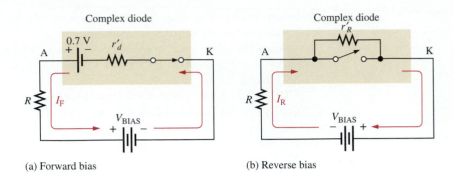

(a) Forward bias   (b) Reverse bias

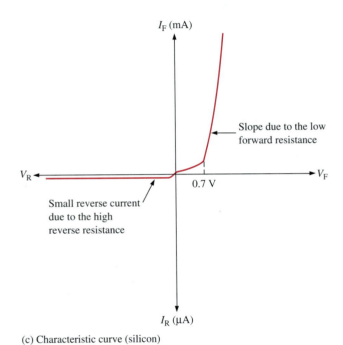

(c) Characteristic curve (silicon)

FIGURE 1–38

The complex model of a diode.

curve is not shown because breakdown is not a normal mode of operation for most diodes.

The following example illustrates the differences in the three diode models in the analysis of a simple circuit.

EXAMPLE 1–1

(a) Determine the forward voltage and forward current for the diode in Figure 1–39(a) for each of the diode models. Also find the voltage across the limiting resistor in each case. Assume $r'_d = 10 \ \Omega$ at the value of forward current.

(b) Determine the reverse voltage and reverse current for the diode in Figure 1–39(b) for each of the diode models. Also find the voltage across the limiting resistor in each case. Assume $I_R = 1\ \mu A$.

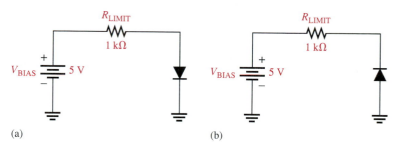

FIGURE 1–39

Solution

(a) Ideal model: $V_F = 0\ V$

$$I_F = \frac{V_{BIAS}}{R_{LIMIT}} = \frac{5\ V}{1\ k\Omega} = 5\ mA$$

$$V_{LIMIT} = I_F R_{LIMIT} = (5\ mA)(1\ k\Omega) = 5\ V$$

Practical model: $V_F = 0.7\ V$

$$I_F = \frac{V_{BIAS} - V_F}{R_{LIMIT}} = \frac{5\ V - 0.7\ V}{1\ k\Omega} = \frac{4.3\ V}{1\ k\Omega} = 4.7\ mA$$

$$V_{LIMIT} = I_F R_{LIMIT} = (4.7\ mA)(1\ k\Omega) = 4.7\ V$$

Complex model: $$I_F = \frac{V_{BIAS} - 0.7\ V}{R_{LIMIT} + r'_d} = \frac{5\ V - 0.7\ V}{1\ k\Omega + 10\ \Omega} = \frac{4.3\ V}{1010\ \Omega} = 4.26\ mA$$

$$V_F = 0.7\ V + I_F r'_d = 0.7\ V + (4.26\ mA)(10\ \Omega) = 743\ mV$$

$$V_{LIMIT} = I_F R_{LIMIT} = (4.26\ mA)(1\ k\Omega) = 4.26\ V$$

(b) Ideal model: $I_R = 0\ A$

$$V_R = V_{BIAS} = 5\ V$$

$$V_{LIMIT} = 0\ V$$

Practical model: $I_R = 0\ A$

$$V_R = V_{BIAS} = 5\ V$$

$$V_{LIMIT} = 0\ V$$

Complex model: $I_R = 1\ \mu A$

$$V_{LIMIT} = I_R R_{LIMIT} = (1\ \mu A)(1\ k\Omega) = 1\ mV$$

$$V_R = V_{BIAS} - V_{LIMIT} = 5\ V - 1\ mV = 4.999\ V$$

Related Exercise   Assume that the diode in Figure 1–39(a) fails open. What is the voltage across the diode and the voltage across the limiting resistor?

Testing a Diode

A multimeter can be used as a fast and simple way to check a diode. As you know, a good diode will show an extremely high resistance (or open) with reverse bias and a very low resistance with forward bias. A defective open diode will show an extremely high resistance (or open) for both forward and reverse bias. A defective shorted or resistive diode will show zero or a low resistance for both forward and reverse bias. An open diode is the most common type of failure.

The DMM Diode Test Position   Many digital multimeters (DMMs) have a diode test position which provides a convenient way to test a diode. A typical DMM, as shown in Figure 1–40, has a small diode symbol to mark the position of the function switch. When set to *diode test,* the meter provides an internal voltage sufficient to forward bias and reverse bias a diode. This internal voltage may vary among different makes of DMM, but 2.5 V to 3.5 V is a typical range of values. The meter provides a voltage reading or other indication to show the condition of the diode under test.

When the Diode Is Working   In Figure 1–40(a), the red (positive) lead of the meter is connected to the anode and the black (negative) lead is connected to the cathode to forward bias the diode. If the diode is good, you will get a reading of between 0.5 V and 0.9 V, with 0.7 V being typical for forward bias.

(a) Forward-bias test   (b) Reverse-bias test

FIGURE 1–40
Diode test on a properly functioning diode.

In Figure 1–40(b), the leads are switched to reverse bias the diode as shown. If the diode is working properly, you will get a voltage reading based on the meter's internal voltage source. The 2.6 V shown in the figure represents a typical value and indicates that the diode has an extremely high reverse resistance with essentially all of the internal voltage appearing across it.

When the Diode Is Defective   When a diode has failed open, you get an open circuit voltage reading (2.6 V is typical) for both the forward-bias and the reverse-bias condition, as illustrated in Figure 1–41(a) and (b). If a diode is shorted, the meter reads 0 V in both forward- and reverse-bias tests, as indicated in part (c). Sometimes, a failed diode may exhibit a small resistance for both bias conditions rather than a pure short. In this case, the meter will show a small voltage much less than the correct open voltage. For example, a resistive diode may result in a reading of 1.1 V in both directions rather than the correct readings of 0.7 V forward and 2.6 V reverse.

(a) Forward-bias test, open diode

(b) Reverse-bias test, open diode

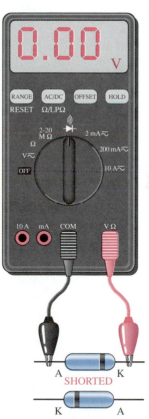

(c) Forward- and reverse-bias tests for a shorted diode give same 0 V reading. If the diode is resistive, the reading is less than 2.6 V.

FIGURE 1–41
Testing a defective diode.

Checking a Diode with the OHMs Function   DMMs that do not have a diode test position can be used to check a diode by setting the function switch on an OHMs range. For a forward-bias check of a good diode, you will get a resistance reading that can vary depending on the meter's internal battery. Many meters do not have sufficient voltage on the OHMs setting to fully forward-bias a diode and you may get a reading of from several hundred to several thousand ohms. For the reverse-bias check of a good diode, you will get an out-of-range indication on most DMMs because the reverse resistance is too high for the meter to measure.

Even though you may not get accurate forward- and reverse-resistance readings on a DMM, the relative readings indicate that a diode is functioning properly, and that is usually all you need to know. The out-of-range indication shows that the reverse resistance is extremely high, as you expect. The reading of a few hundred to a few thousand ohms for forward bias is relatively small compared to the reverse resistance, indicating that the diode is working properly. The actual resistance of a forward-biased diode is typically much less than 100 Ω.

SECTION 1–9 REVIEW	**1.** What are the two conditions under which the diode is operated?
	2. Under what condition is the diode never intentionally operated?
	3. What is the simplest way to visualize a diode?
	4. To more accurately represent a diode, what factors must be included?
	5. Which diode model will normally be used in this book?

■ **CHAPTER SUMMARY**

- According to the classical Bohr model, the atom is viewed as having a planetary-type structure with electrons orbiting at various distances around the central nucleus.

- The nucleus of an atom consists of protons and neutrons. The protons have a positive charge and the neutrons are uncharged. The number of protons and neutrons is the atomic weight of the atom.

- Electrons have a negative charge and orbit around the nucleus at distances that depend on their energy level. An atom has discrete bands of energy called *shells* in which the electrons orbit. Atomic structure allows a certain maximum number of electrons in each shell. These shells are designated *K, L, M,* and so on. In their natural state, all atoms are neutral because they have an equal number of protons and electrons.

- The outermost shell or band of an atom is called the *valence band,* and electrons that orbit in this band are called *valence electrons*. These electrons have the highest energy of all those in the atom. If a valence electron acquires enough energy from an outside source such as heat, it can jump out of the valence band and break away from its atom.

- Semiconductor atoms have four valence electrons. Silicon is the most widely used semiconductor material.

- Semiconductor atoms bond together to form a solid material called a *crystal*. The bonds that hold a crystal together are called *covalent bonds*. Within the crystal structure, the valence electrons that manage to escape from their parent atom are called *conduction electrons* or *free electrons*. They have more energy than the electrons in the valence band and are free to drift throughout the material. When an electron breaks away to become free, it leaves a hole in the valence band creating what is called an *electron-hole pair*. These electron-hole pairs are thermally produced because the electron has acquired enough energy from external heat to break away.

- A free electron will eventually lose energy and fall back into a hole. This is called *recombination*. But, electron-hole pairs are continuously being thermally generated so there are always free electrons in the material.
- When a voltage is applied across the semiconductor, the thermally produced free electrons move in a net direction and form the current. This is one type of current in an intrinsic (pure) semiconductor.
- Another type of current is the hole current. This occurs as valence electrons move from hole to hole creating, in effect, a movement of holes in the opposite direction.
- Materials that are conductors have a large number of free electrons and conduct current very well. Insulating materials have very few free electrons and do not conduct current at all under normal circumstances. Semiconductive materials fall in between conductors and insulators in their ability to conduct current.
- An *n*-type semiconductor material is created by adding impurity atoms that have five valence electrons. These impurities are pentavalent atoms. A *p*-type semiconductor is created by adding impurity atoms with only three valence electrons. These impurities are trivalent atoms.
- The process of adding pentavalent or trivalent impurities to a semiconductor is called *doping*.
- The majority carriers in an *n*-type semiconductor are free electrons acquired by the doping process, and the minority carriers are holes produced by thermally generated electron-hole pairs. The majority carriers in a *p*-type semiconductor are holes acquired by the doping process, and the minority carriers are free electrons produced by thermally generated electron-hole pairs.
- A *pn* junction is formed when part of a material is doped *n*-type and part of it is doped *p*-type. A depletion region forms starting at the junction that is devoid of any majority carriers. The depletion region is formed by ionization.
- Current flows through the *pn* junction only when it is forward-biased. Current does not flow when there is no bias or when there is reverse bias. There is a very small current in reverse bias due to the thermally generated minority carriers, but this can usually be neglected.
- Avalanche occurs in a reverse-biased *pn* junction if the bias voltage equals or exceeds the breakdown voltage.
- A diode conducts current when forward-biased and blocks current when reversed-biased.
- The forward-biased barrier potential is typically 0.7 V for a silicon diode and 0.3 V for a germanium diode. These values increase slightly with forward current.
- Reverse breakdown voltage for a diode is typically greater than 50 V.
- An ideal diode presents an open when reversed-biased and a short when forward-biased.

▪ GLOSSARY

Anode   The *p* region of a diode.

Atom   The smallest particle of an element that possesses the unique characteristics of that element.

Atomic number   The number of electrons in a neutral atom.

Atomic weight   Approximately the number of protons and neutrons in the nucleus of an atom.

Avalanche   The rapid buildup of conduction electrons due to excessive reverse-bias voltage.

Barrier potential   The amount of energy required to produce full conduction across the *pn* junction in forward bias.

Bias   The application of a dc voltage to a *pn* junction to make it either conduct or block current.

Carbon   A semiconductor material.

Cathode   The *n* region of a diode.

Conduction electron   A free electron.

Conductor   A material that conducts electrical current very well.

Core   The central part of an atom, includes the nucleus and all but the valence electrons.

Covalent   Related to the bonding of two or more atoms by the interaction of their valence electrons.

Crystal   The pattern or arrangement of atoms forming a solid material.

Current   The rate of flow of free electrons.

Depletion region   The area near a *pn* junction on both sides that has no majority carriers.

Diode characteristic curve   A graph showing the relationship of current versus voltage in a diode.

Doping   The process of imparting impurities to an intrinsic semiconductor material in order to control its conduction characteristics.

Dynamic resistance   The nonlinear internal resistance of a semiconductor material.

Electron   The basic particle of negative electrical charge.

Electron-hole pair   The conduction electron and the hole created when the electron leaves the valence band.

Forward bias   The condition in which a *pn* junction conducts current.

Free electron   An electron that has acquired enough energy to break away from the valence band of the parent atom, also called a *conduction electron*.

Germanium   A semiconductor material.

Hole   The absence of an electron in the valence band of an atom.

Insulator   A material that does not conduct current.

Intrinsic   The pure or natural state of a material.

Ionization   The removal or addition of an electron from or to a neutral atom so that the resulting atom (called an ion) has a net positive or negative charge.

Majority carrier   The most numerous charge carrier in a doped semiconductor material (either free electrons or holes).

Minority carrier   The least numerous charge carrier in a doped semiconductor material (either free electrons or holes).

Neutron   An uncharged particle found in the nucleus of an atom.

Nucleus   The central part of an atom containing protons and neutrons.

Orbit   The path an electron takes as it circles around the nucleus of an atom.

Pentavalent atom   An atom with five valence electrons.

***PN* junction**   The boundary between two different types of semiconductor materials.

Proton   The basic particle of positive charge.

Recombination   The process of a free (conduction band) electron falling into a hole in the valence band of an atom.

Reverse bias   The condition in which a *pn* junction prevents current.

Semiconductor   A material that lies between conductors and insulators in its conductive properties. Silicon, germanium, and carbon are examples.

Shell   An energy band in which electrons orbit the nucleus of an atom.

Silicon   A semiconductor material.

Trivalent atom   An atom with three valence electrons.

Valence   Related to the outer shell of an atom.

■ **FORMULAS**

(1–1) $N_e = 2n^2$ Maximum number of electrons in any shell

(1–2) $I_F = \dfrac{V_{BIAS}}{R_{LIMIT}}$ Forward current, ideal diode model

$$(1\text{--}3) \qquad I_F = \frac{V_{BIAS} - V_F}{R_{LIMIT}} \qquad\qquad \text{Forward current, practical diode model}$$

$$(1\text{--}4) \qquad V_F = 0.7\ V + I_F r_d' \qquad\qquad \text{Forward voltage, complex diode model}$$

$$(1\text{--}5) \qquad I_F = \frac{V_{BIAS} - 0.7\ V}{R_{LIMIT} - r_d'} \qquad\qquad \text{Forward current, complex diode model}$$

■ **SELF-TEST**

1. Every known element has
   (a) the same type of atoms   (b) the same number of atoms
   (c) a unique type of atom   (d) several different types of atoms

2. An atom consists of
   (a) one nucleus and only one electron   (b) one nucleus and one or more electrons
   (c) protons, electrons, and neutrons   (d) answers (b) and (c)

3. The nucleus of an atom is made up of
   (a) protons and neutrons   (b) electrons
   (c) electrons and protons   (d) electrons and neutrons

4. The atomic number of silicon is
   (a) 8   (b) 2   (c) 4   (d) 14

5. The atomic number of germanium is
   (a) 8   (b) 2   (c) 4   (d) 32

6. The valence shell in a silicon atom has the letter designation of
   (a) *A*   (b) *K*   (c) *L*   (d) *M*

7. Valence electrons are
   (a) in the closest orbit to the nucleus   (b) in the most distant orbit from the nucleus
   (c) in various orbits around the nucleus   (d) not associated with a particular atom

8. A positive ion is formed when
   (a) a valence electron breaks away from the atom
   (b) there are more holes than electrons in the outer orbit
   (c) two atoms bond together
   (d) an atom gains an extra valence electron

9. The most widely used semiconductor material in electronic devices is
   (a) germanium   (b) carbon   (c) copper   (d) silicon

10. The energy band in which free electrons exist is the
    (a) first band   (b) second band   (c) conduction band   (d) valence band

11. Electron-hole pairs are produced by
    (a) recombination   (b) thermal energy   (c) ionization   (d) doping

12. Recombination is when
    (a) an electron falls into a hole
    (b) a positive and a negative ion bond together
    (c) a valence electron becomes a conduction electron
    (d) a crystal is formed

13. In a semiconductor crystal, the atoms are held together by
    (a) the interaction of valence electrons   (b) forces of attraction
    (c) covalent bonds   (d) all of the above

14. Each atom in a silicon crystal has
    (a) four valence electrons
    (b) four conduction electrons
    (c) eight valence electrons, four of its own and four shared
    (d) no valence electrons because all are shared with other atoms

All of them

15. The current in a semiconductor is produced by
    (a) electrons only   (b) holes only   (c) negative ions   (d) both electrons and holes

16. In an intrinsic semiconductor,
    (a) there are no free electrons   (b) the free electrons are thermally produced
    (c) there are only holes   (d) there are as many electrons as there are holes
    (e) answers (b) and (d)

17. The difference between an insulator and a semiconductor is
    (a) a wider energy gap between the valence band and the conduction band
    (b) the number of free electrons
    (c) the atomic structure
    (d) all of the above

18. The process of adding an impurity to an intrinsic semiconductor is called
    (a) doping   (b) recombination   (c) atomic modification   (d) ionization

19. A trivalent impurity is added to silicon to create
    (a) germanium   (b) a *p*-type semiconductor
    (c) an *n*-type semiconductor   (d) a depletion region

20. The purpose of a pentavalent impurity is to
    (a) reduce the conductivity of silicon   (b) increase the number of holes
    (c) increase the number of free electrons   (d) create minority carriers

21. The majority carriers in an *n*-type semiconductor are
    (a) holes   (b) valence electrons   (c) conduction electrons   (d) protons

22. Holes in an *n*-type semiconductor are
    (a) minority carriers that are thermally produced
    (b) minority carriers that are produced by doping
    (c) majority carriers that are thermally produced
    (d) majority carriers that are produced by doping

23. A *pn* junction is formed by
    (a) the recombination of electrons and holes
    (b) ionization
    (c) the boundary of a *p*-type and an *n*-type material
    (d) the collision of a proton and a neutron

24. The depletion region is created by
    (a) ionization   (b) diffusion   (c) recombination   (d) all of the above

25. The depletion region consists of
    (a) nothing but minority carriers   (b) positive and negative ions
    (c) no majority carriers   (d) answers (b) and (c)

26. The term *bias* means
    (a) the ratio of majority carriers to minority carriers
    (b) the amount of current across the *pn* junction
    (c) a dc voltage that is applied to control the operation of a device
    (d) none of the above

27. To forward-bias a *pn* junction diode,
    (a) an external voltage is applied that is positive at the anode and negative at the cathode
    (b) an external voltage is applied that is negative at the anode and positive at the cathode
    (c) an external voltage is applied that is positive at the *p* region and negative at the *n* region
    (d) answers (a) and (c)

28. When a *pn* junction is forward-biased,
    (a) the only current is hole current
    (b) the only current is electron current
    (c) the only current is produced by majority carriers
    (d) the current is produced by both holes and electrons

All of them 29. Although current is blocked in reverse bias,
(a) there is some current due to majority carriers
(b) there is a very small current due to minority carriers
(c) there is an avalanche current

30. For a silicon diode, the value of the forward-bias voltage typically
(a) must be greater than 0.3 V
(b) must be greater than 0.7 V
(c) depends on the width of the depletion region
(d) depends on the concentration of majority carriers

31. When forward-biased, a diode
(a) blocks current (b) conducts current
(c) has a high resistance (d) drops a large voltage

32. When a voltmeter is placed across a forward-biased diode, it will read a voltage approximately equal to
(a) the bias battery voltage (b) 0 V
(c) the diode barrier potential (d) the total circuit voltage

33. A silicon diode is in series with a 1 kΩ resistor and a 5 V battery. If the anode is connected to the positive battery terminal, the cathode voltage with respect to the negative battery terminal is
(a) 0.7 V (b) 0.3 V (c) 5.7 V (d) 4.3 V

34. The positive lead of an ohmmeter is connected to the anode of a diode and the negative lead is connected to the cathode. The diode is
(a) reversed-biased (b) open (c) forward-biased
(d) faulty (e) answers (b) and (d)

■ BASIC PROBLEMS

SECTION 1–1 Atomic Structure

1. If the atomic number of an atom is 6, how many electrons does the atom have? How many protons?

2. What is the maximum number of electrons that can exist in the *M* shell of an atom?

SECTION 1–2 Semiconductors, Conductors, and Insulators

3. For each of the energy diagrams in Figure 1–42, determine the class of material based on relative comparisons.

4. A certain atom has four valence electrons. What type of atom is it?

FIGURE 1–42

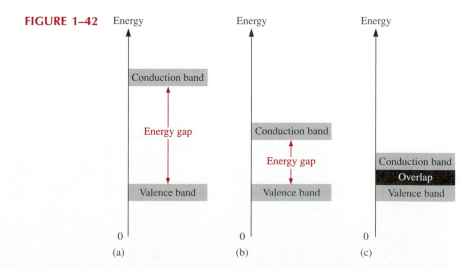

SECTION 1–3   Covalent Bonds

5. In a silicon crystal, how many covalent bonds does a single atom form?

SECTION 1–4   Conduction in Semiconductors

6. What happens when heat is added to silicon?
7. Name the two energy levels at which current is produced in silicon.

SECTION 1–5   *N*-type and *P*-type Semiconductors

8. Describe the process of doping and explain how it alters the atomic structure of silicon.
9. What is antimony? What is boron?

SECTION 1–6   The *PN* Junction

10. How is the electric field across the *pn* junction created?
11. Because of its barrier potential, can a *pn* junction device be used as a voltage source? Explain.

SECTION 1–7   Biasing the *PN* Junction

12. To forward-bias a *pn* junction, to which region must the positive terminal of a voltage source be connected?
13. Explain why a series resistor is necessary when a *pn* junction is forward-biased.

SECTION 1–8   Current-Voltage Characteristic of a *PN* Junction

14. Explain how to generate the forward-bias portion of the characteristic curve.
15. What would cause the barrier potential to decrease from 0.7 V to 0.6 V?

SECTION 1–9   The Diode

16. Determine whether each diode in Figure 1–43 is forward-biased or reverse-biased.
17. Determine the voltage across each diode in Figure 1–43, assuming that they are germanium diodes with a reverse resistance of 50 MΩ.

FIGURE 1–43

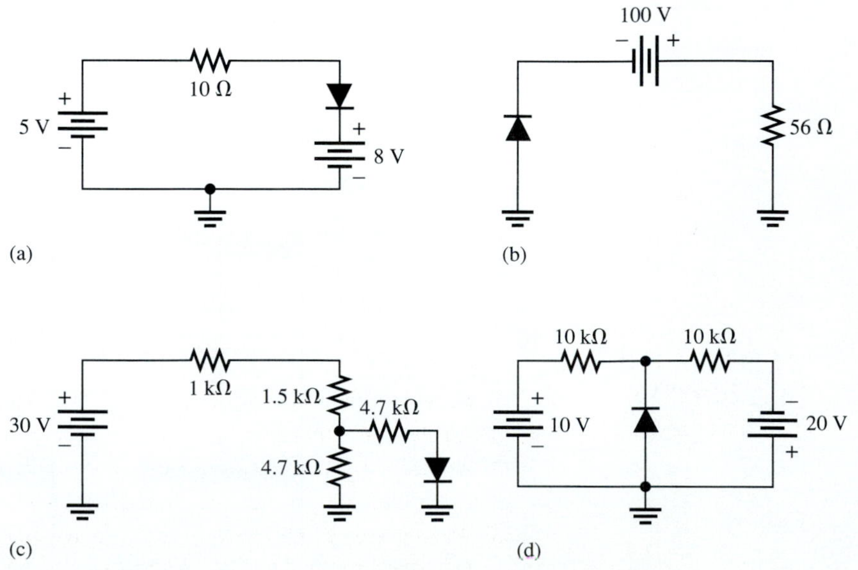

(a)   (b)   (c)   (d)

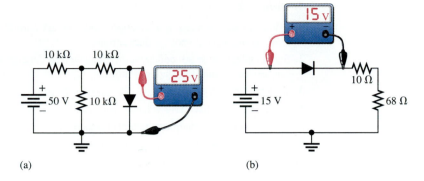

(a)

(b)

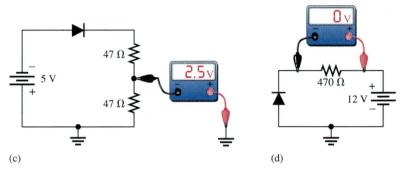

(c)

(d)

FIGURE 1–44

18. Consider the meter indications in each circuit of Figure 1–44, and determine whether the diode is functioning properly, or whether it is open or shorted.

19. Determine the voltage with respect to ground at each point in Figure 1–45. (The diodes are silicon.)

FIGURE 1–45

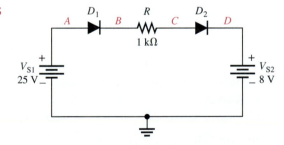

■ **ANSWERS TO SECTION REVIEWS**

Section 1–1

1. An atom is the smallest particle that retains the characteristics of its element.

2. An electron is the basic particle of negative electrical charge.

3. A valence electron is an electron in the outermost shell of an atom.

4. A free electron is one that has broken out of the valence band and entered the conduction band.

5. When a neutral atom loses an electron, the atom becomes a positive ion. When a neutral atom gains an electron, the atom becomes a negative ion.

Section 1–2

1. Conductors have many free electrons and conduct current well. Insulators have essentially no free electrons and do not conduct current.
2. Semiconductors do not conduct current as well as conductors do. In terms of conductivity, they are between conductors and insulators.
3. Conductors such as copper have one valence electron.
4. Semiconductors have four valence electrons.
5. Gold, silver, and copper are the best conductors.
6. Silicon is the most widely used semiconductor.
7. The valence electrons of a semiconductor are more tightly bound to the atom than those of conductors.

Section 1–3

1. Covalent bonds are formed by the sharing of valence electrons with neighboring atoms.
2. An intrinsic material is one that is in a pure state.
3. A crystal is a solid material formed by atoms bonding together in a fixed pattern.
4. There are eight shared valence electrons in each atom of a silicon crystal.

Section 1–4

1. Free electrons are in the conduction band.
2. Free (conduction) electrons are responsible for current in a material.
3. A hole is the absence of an electron in the valence band.
4. Hole current occurs at the valence level.

Section 1–5

1. Doping is the process of adding impurity atoms to a semiconductor in order to modify its conductive properties.
2. A pentavalent atom (donor) has five valence electrons and a trivalent atom (acceptor) has three valence electrons.
3. An *n*-type material is formed by the addition of pentavalent impurity atoms to the intrinsic semiconductor material.
4. A *p*-type material is formed by the addition of trivalent impurity atoms to the intrinsic semiconductor material.
5. The majority carrier in an *n*-type semiconductor is the free electron.
6. The majority carrier in a *p*-type semiconductor is the hole.
7. Majority carriers are produced by doping.
8. Minority carriers are thermally produced when electron-hole pairs are generated.
9. A pure semiconductor is intrinsic. A doped (impure) semiconductor is extrinsic.

Section 1–6

1. A *pn* junction is the boundary between *p*-type and *n*-type semiconductors.
2. Diffusion is the movement of the free electrons (majority carriers) in the *n*-region across the *pn* junction and into the *p* region.
3. The depletion region is the thin layers of positive and negative ions that exist on both sides of the *pn* junction.
4. The barrier potential is the potential difference of the electric field in the depletion region and is the amount of energy required to move electrons through the depletion region.
5. The barrier potential for a silicon *pn* junction is approximately 0.7 V.
6. The barrier potential for a germanium *pn* junction is approximately 0.3 V.

Section 1–7

1. When forward-biased, a *pn* junction conducts current. The free electrons in the *n* region move across the junction and combine with the holes in the *p* region.

2. To forward-bias a *pn* junction, the positive side of an external bias voltage is applied to the *p* region and the negative side to the *n* region.

3. When reverse-biased, a *pn* junction does not conduct current except for an extremely small reverse current.

4. To reverse-bias a *pn* junction, the positive side of an external bias voltage is applied to the *n* region and the negative side to the *p* region.

5. The depletion region for forward bias is much narrower than for reverse bias.

6. Majority carrier current is produced by forward bias.

7. Reverse current is produced by the minority carriers.

8. Reverse breakdown occurs when the reverse-bias voltage equals or exceeds the breakdown voltage of the *pn* junction.

9. Avalanche is the rapid multiplication of current carriers in reverse breakdown.

Section 1–8

1. The knee of the characteristic curve in forward bias is the point at which the barrier potential is overcome and the current increases drastically.

2. A forward-biased *pn* junction is normally operated above the knee of the curve.

3. Breakdown voltage is always much greater than the barrier potential.

4. A reverse-biased *pn* junction is operated below the breakdown point on the knee of the curve.

5. Barrier potential decreases as temperature increases.

Section 1–9

1. The diode is operated in forward bias and reverse bias.

2. The diode should never be operated in reverse breakdown.

3. The diode can be ideally viewed as a switch.

4. A diode includes barrier potential, dynamic resistance, and reverse resistance in the complex model.

5. The practical diode model (barrier potential) is generally used.

■ **ANSWER TO RELATED EXERCISE FOR EXAMPLE**

1–1 $V_D = 5$ V; $V_R = 0$ V

2

DIODE APPLICATIONS

■ CHAPTER OBJECTIVES

☐ Explain and analyze the operation of a half-wave rectifier

☐ Explain and analyze the operation of full-wave rectifiers

☐ Explain and analyze the operation and characteristics of power supply filters

☐ Explain and analyze the operation of diode limiting and clamping circuits

☐ Explain and analyze the operation of diode voltage multipliers

☐ Interpret and use a diode data sheet

☐ Troubleshoot diode circuits using accepted techniques

In Chapter 1, you learned that a semiconductor diode is a device with a single *pn* junction. The importance of the diode in electronic circuits cannot be overemphasized. Its ability to conduct current in one direction while blocking current in the other direction is essential to the operation of many types of circuits. One circuit in particular is the ac rectifier, which is covered in this chapter. Other important applications are circuits such as diode limiters, diode clampers, and diode voltage multipliers.

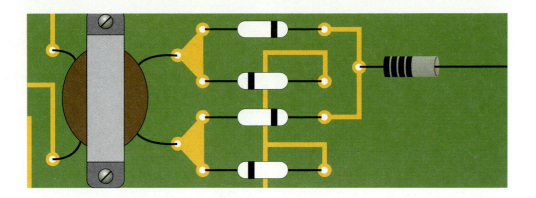

■ SYSTEM APPLICATION

As a technician for an electronics company, you are given the responsibility for the final design and testing of a power supply circuit board that your company plans to use in several of its products. Your assignment is to learn all you can about diode circuits by studying this chapter and then apply your knowledge to the system application in Section 2–8.

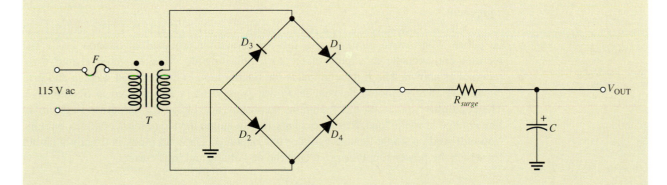

2–1 ■ HALF-WAVE RECTIFIERS

Because of their ability to conduct current in one direction and block current in the other direction, diodes are used in rectifier circuits that convert ac voltage into dc voltage. Rectifier circuits are found in all dc power supplies that operate from an ac voltage source. A power supply is an essential part of each electronic system from the simplest to the most complex. In this section, you will study the most basic type of rectifier circuit, the half-wave rectifier.

After completing this section, you should be able to

■ **Explain and analyze the operation of a half-wave rectifier**
  ☐ Describe a basic dc power supply and half-wave rectification
  ☐ Determine the average value of a half-wave rectified voltage
  ☐ Discuss the effect of barrier potential on a half-wave rectifier output
  ☐ Define peak inverse voltage (PIV)
  ☐ Describe the tranformer-coupled half-wave rectifier

The Basic DC Power Supply

The dc **power supply** converts the standard 110 V, 60 Hz ac available at wall outlets into a constant dc voltage. It is one of the most common electronic circuits that you will find. The dc voltage produced by a power supply is used to power all types of electronic circuits, such as television receivers, stereo systems, VCRs, CD players, and laboratory equipment.

A basic block diagram for a power supply is shown in Figure 2–1. The **rectifier** can be either a half-wave rectifier or a full-wave rectifier (covered in Section 2–2). The rectifier converts the ac input voltage to a pulsating dc voltage, which is half-wave rectified as shown. The **filter** eliminates the fluctuations in the rectified voltage and produces a relatively smooth dc voltage. Power supply filters are covered in Section 2–3. The **regulator** is a circuit that maintains a constant dc voltage for variations in the input line voltage or in the load. Regulators vary from a single device to more complex circuits. You will study a single-device regulator in Chapter 3. The load block is usually a circuit for which the power supply is producing the dc voltage and load current.

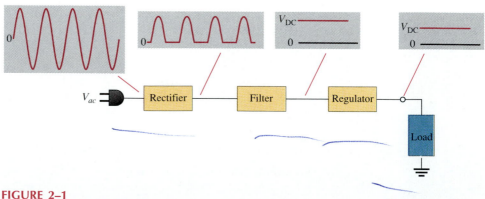

FIGURE 2–1
Block diagram of a dc power supply.

The Half-Wave Rectifier

Figure 2–2 illustrates the process called *half-wave rectification*. In part (a), a diode is connected to an ac source and a load resistor R_L, forming a **half-wave rectifier.** Let's examine what happens during one cycle of the input voltage using the ideal model for the diode. When the sinusoidal input voltage (V_{in}) goes positive, the diode is forward-biased and conducts current through the load resistor, as shown in part (b). The current produces an output voltage across the load R_L, which has the same shape as the positive half-cycle of the input voltage.

When the input voltage goes negative during the second half of its cycle, the diode is reverse-biased. There is no current, so the voltage across the load resistor is 0 V, as shown in Figure 2–2(c). The net result is that only the positive half-cycles of the ac input voltage appear across the load. Since the output does not change polarity, it is a *pulsating dc voltage* with a frequency of 60 Hz as shown in part (d).

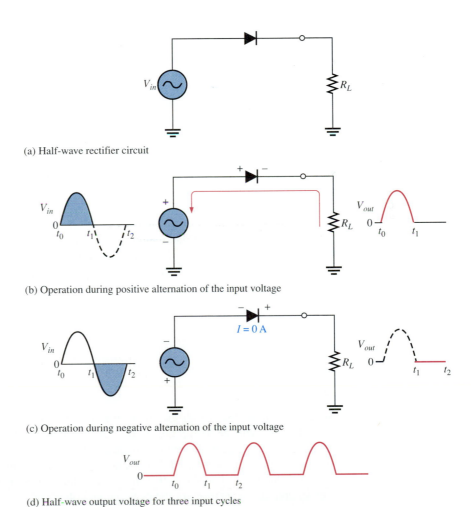

(a) Half-wave rectifier circuit

(b) Operation during positive alternation of the input voltage

(c) Operation during negative alternation of the input voltage

(d) Half-wave output voltage for three input cycles

FIGURE 2–2

Half-wave rectifier operation. The diode is considered to be ideal.

Average Value of the Half-Wave Output Voltage The average value of the half-wave rectified output voltage is the value you would measure on a dc voltmeter. It is determined by finding the area under the curve over a full cycle, as illustrated in Figure 2–3, and then dividing by the period, *T*. See Appendix B for detailed derivation.

$$V_{AVG} = \frac{V_p}{\pi}$$

(2–1)

V_p is the peak value of the voltage.

FIGURE 2–3

Average value of half-wave rectified signal.

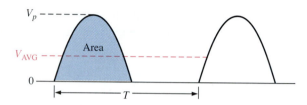

EXAMPLE 2–1   What is the average value of the half-wave rectified voltage in Figure 2–4?

FIGURE 2–4

Solution

$$V_{AVG} = \frac{V_p}{\pi} = \frac{100 \text{ V}}{\pi} = 31.8 \text{ V}$$

Related Exercise Determine the average value of the half-wave voltage if its peak amplitude is 12 V.

Effect of the Barrier Potential on the Half-Wave Rectifier Output

In the previous discussion, the diode was considered ideal. When the practical diode model is used and the barrier potential is taken into account, this is what happens. During the positive half-cycle, the input voltage must overcome the barrier potential before the diode becomes forward-biased. For a silicon diode this results in a half-wave output with a peak value that is 0.7 V less than the peak value of the input, as shown in Figure 2–5. For silicon, the expression for the peak output voltage is

$$V_{p(out)} = V_{p(in)} - 0.7 \text{ V}$$

(2–2)

In working with diode circuits, it is usually practical to neglect the effect of barrier potential when the peak value of the applied voltage is much greater than the barrier potential (at least 10 V, as a rule of thumb). We will always use the practical model of a silicon diode, taking the barrier potential into account unless stated otherwise.

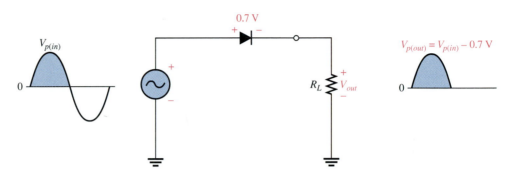

FIGURE 2–5
Effect of barrier potential on half-wave rectified output voltage (silicon diode shown).

EXAMPLE 2–2

Sketch the output voltages of each rectifier circuit for the indicated input voltages, as shown in Figure 2–6.

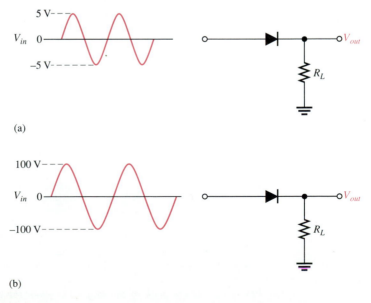

(a)

(b)

FIGURE 2–6

Solution   The peak output voltage for circuit (a) is

$$V_{p(out)} = V_{p(in)} - 0.7 \text{ V} = 5 \text{ V} - 0.7 \text{ V} = 4.30 \text{ V}$$

The peak output voltage for circuit (b) is

$$V_{p(out)} = V_{p(in)} - 0.7 \text{ V} = 100 \text{ V} - 0.7 \text{ V} = 99.3 \text{ V}$$

These outputs are shown in Figure 2–7. Note that the barrier potential could have been neglected in circuit (b) with very little error (0.7 percent); but, if it is neglected in circuit (a), a significant error results (14 percent).

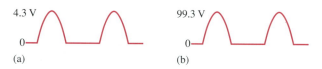

(a)   (b)

FIGURE 2–7
Output voltages for the circuits in Figure 2–6. They are not shown on the same scale.

Related Exercise   Determine the peak output voltages for the rectifiers in Figure 2–6 if the peak input in part (a) is 3 V and the peak input in part (b) is 210 V.

Peak Inverse Voltage (PIV)

The maximum value of reverse voltage, designated as *peak inverse voltage* (PIV), occurs at the peak of the negative alternation of the input cycle when the diode is reverse-biased. This condition is illustrated in Figure 2–8. The PIV equals the peak value of the input voltage, and the diode must be capable of withstanding this amount of repetitive reverse voltage.

$$PIV = V_{p(in)} \tag{2–3}$$

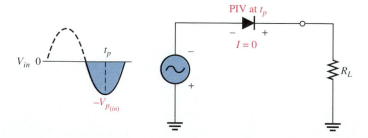

FIGURE 2–8
The PIV occurs at the peak of the half-cycle when the diode is reverse-biased. In this circuit, the PIV occurs at the peak of the negative half-cycle.

Half-Wave Rectifier with Transformer-Coupled Input Voltage

A transformer is often used to couple the ac input voltages from the source to the rectifier circuit, as shown in Figure 2–9. Transformer coupling provides two advantages. First, it allows the source voltage to be stepped up or stepped down as needed. Second, the ac power source is electrically isolated from the rectifier circuit, thus reducing the shock hazard.

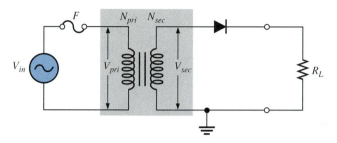

FIGURE 2–9
Half-wave rectifier with transformer-coupled input voltage.

From basic ac circuits recall that the secondary voltage of a transformer equals the turns ratio (N_{sec}/N_{pri}) times the primary voltage, as expressed in Equation (2–4).

$$V_{sec} = \left(\frac{N_{sec}}{N_{pri}}\right)V_{pri} \qquad \text{(2–4)}$$

If $N_{sec} > N_{pri}$, the secondary voltage is greater than the primary voltage. If $N_{sec} < N_{pri}$, the secondary voltage is less than the primary voltage. If $N_{sec} = N_{pri}$, then $V_{sec} = V_{pri}$.

EXAMPLE 2–3

Determine the peak value of the output voltage for Figure 2–10.

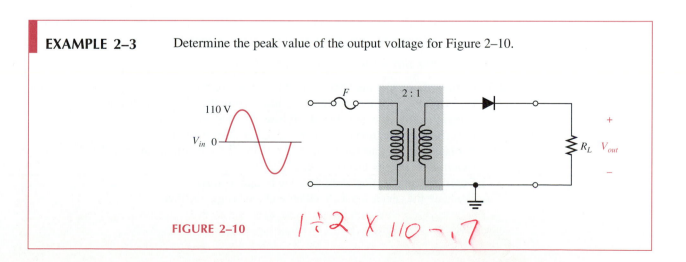

FIGURE 2–10

$$V_{p(pri)} = V_{p(in)} = 110 \text{ V}$$

The secondary peak voltage is

$$V_{p(sec)} = \left(\frac{N_{sec}}{N_{pri}}\right)V_{p(pri)} = 0.5(110 \text{ V}) = 55 \text{ V}$$

The peak rectified output voltage is

$$V_{p(out)} = V_{p(sec)} - 0.7 \text{ V} = 55 \text{ V} - 0.7 \text{ V} = 54.3 \text{ V}$$

Related Exercise
(a) Determine the peak value of the output voltage for Figure 2–10 if the turns ratio is 1:2 and $V_{p(in)} = 220$ V.
(b) What is the PIV across the diode?
(c) Describe the output voltage if the diode is turned around.

SECTION 2–1 REVIEW

1. At what point on the input cycle does the PIV occur? *Peak of neg Atturtleon*
2. For a half-wave rectifier, there is current through the load for approximately what percentage of the input cycle? *π 5%*
3. What is the average of a half-wave rectified voltage with a peak value of 10 V? *3.18*
4. What is the peak value of the output voltage of a half-wave rectifier with a peak sine wave input of 25 V? *24.3*
5. What PIV rating must a diode have to be used in a rectifier with a peak output voltage of 50 V? *50V*

2–2 ■ FULL-WAVE RECTIFIERS

Although half-wave rectifiers have some applications, the full-wave rectifier is the most commonly used type in dc power supplies. In this section, you will use what you learned about half-wave rectification and expand it to full-wave rectifiers. You will learn about two types of full-wave rectifiers: center-tapped and bridge.

After completing this section, you should be able to

■ **Explain and analyze the operation of full-wave rectifiers**
  ☐ Discuss how full-wave rectification differs from half-wave rectification
  ☐ Determine the average value of a full-wave rectified voltage
  ☐ Describe the operation of a full-wave center-tapped rectifier
  ☐ Explain how the transformer turns ratio affects the rectified output voltage
  ☐ Determine the peak inverse voltage (PIV)
  ☐ Describe the operation of a full-wave bridge rectifier
  ☐ Compare the center-tapped rectifier and the bridge rectifier

The difference between full-wave and half-wave rectification is that a **full-wave rectifier** allows unidirectional (one-way) current to the load during the entire 360° of the input cycle, and the half-wave rectifier allows this only during one half-cycle. The result of full-wave rectification is an output voltage with a frequency twice the input that pulsates every half-cycle of the input, as shown in Figure 2–11.

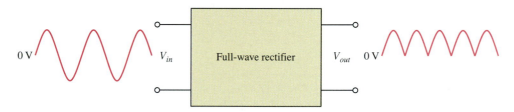

FIGURE 2–11
Full-wave rectification.

Since the number of positive alternations that make up the full-wave rectified voltage is twice that of the half-wave voltage for the same time interval, the average value for a full-wave rectified sinusoidal voltage is twice that of the half-wave, as shown in the following formula:

$$V_{\text{AVG}} = \frac{2V_p}{\pi}$$ (2–5)

EXAMPLE 2–4 Find the average value of the full-wave rectified voltage in Figure 2–12.

FIGURE 2–12

Solution
$$V_{\text{AVG}} = \frac{2V_p}{\pi} = \frac{2(15 \text{ V})}{\pi} = 9.55 \text{ V}$$

Related Exercise Find the average value of the full-wave rectified voltage if its peak is 155 V.

The Full-Wave Center-Tapped Rectifier

The full-wave **center-tapped rectifier** uses two diodes connected to the secondary of a center-tapped transformer, as shown in Figure 2–13. The input voltage is coupled through the transformer to the center-tapped secondary. Half of the total secondary voltage appears between the center tap and each end of the secondary winding as shown.

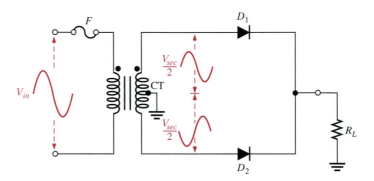

FIGURE 2–13
A full-wave center-tapped rectifier.

For a positive half-cycle of the input voltage, the polarities of the secondary voltages are as shown in Figure 2–14(a). This condition forward-biases the upper diode D_1 and reverse-biases the lower diode D_2. The current path is through D_1 and the load resistor R_L, as indicated. For a negative half-cycle of the input voltage, the voltage polarities on the secondary are as shown in Figure 2–14(b). This condition reverse-biases D_1 and forward-biases D_2. The current path is through D_2 and R_L, as indicated. Because the output current during both the positive and negative portions of the input cycle is in the same direction through the load, the output voltage developed across the load resistor is a full-wave rectified dc voltage.

Effect of the Turns Ratio on the Output Voltage   If the transformer's turns ratio is 1, the peak value of the rectified output voltage equals half the peak value of the primary input voltage less the barrier potential as illustrated in Figure 2–15. Incidentally, we will begin referring to the forward voltage due to the barrier potential as the **diode drop.** This is because half of the primary voltage appears across each half of the secondary winding $(V_{p(sec)} = V_{p(pri)})$.

In order to obtain an output voltage with a peak equal to the input peak (less the diode drop), a step-up transformer with a turns ratio of 2 must be used, as shown in Figure 2–16 on page 64. In this case, the total secondary voltage V_{sec} is twice the primary voltage $(2V_{pri})$, so the voltage across each half of the secondary is equal to V_{pri}.

In any case, the output voltage of a full-wave center-tapped rectifier is always one-half of the total secondary voltage less the diode drop, no matter what the turns ratio.

$$V_{out} = \frac{V_{sec}}{2} - 0.7 \text{ V}$$

(2–6)

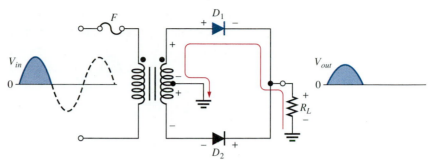

(a) During positive half-cycles, D_1 is forward-biased and D_2 is reverse-biased.

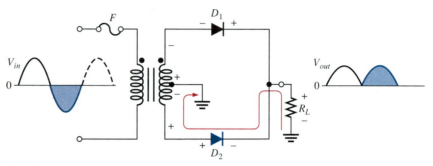

(b) During negative half-cycles, D_2 is forward-biased and D_1 is reverse-biased.

FIGURE 2–14

Basic operation of a full-wave center-tapped rectifier. Note that the current through the load resistor is in the same direction during the entire input cycle.

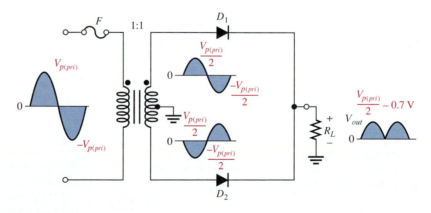

FIGURE 2–15

Full-wave center-tapped rectifier with a transformer turns ratio of 1. $V_{p(pri)}$ is the peak value of the primary voltage.

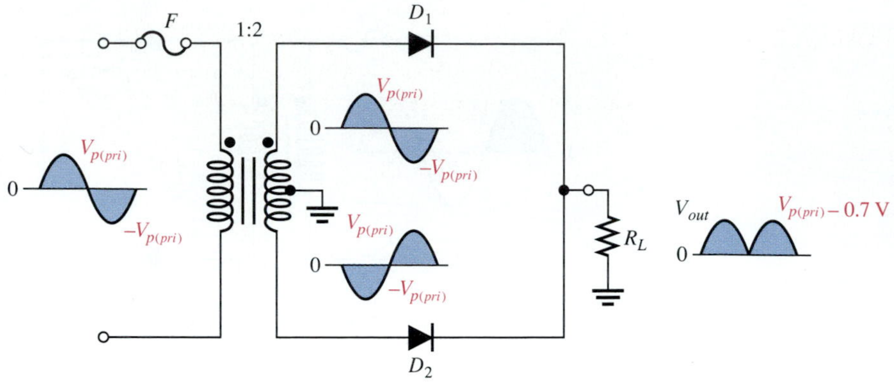

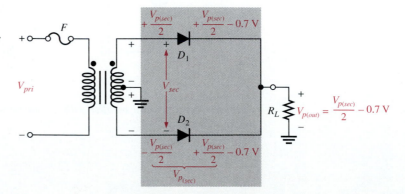

FIGURE 2–16

Full-wave center-tapped rectifier with a transformer turns ratio of 2.

Peak Inverse Voltage Each diode in the full-wave rectifier is alternately forward-biased and then reverse-biased. The maximum reverse voltage that each diode must withstand is the peak secondary voltage $V_{p(sec)}$. This is shown in Figure 2–17.

When the total secondary voltage has the polarity shown, the maximum anode voltage of D_1 is $+V_{p(sec)}/2$ and the maximum anode voltage of D_2 is $-V_{p(sec)}/2$. Since D_1 is forward-biased, its cathode is at the same voltage as its anode minus the diode drop; this is also the voltage on the cathode of D_2. Applying Kirchhoff's voltage law around the loop, the peak inverse voltage across D_2 is

$$\text{PIV} = \left(\frac{V_{p(sec)}}{2} - 0.7\text{ V}\right) - \left(\frac{-V_{p(sec)}}{2}\right) = \frac{V_{p(sec)}}{2} + \frac{V_{p(sec)}}{2} - 0.7\text{ V} = V_{p(sec)} - 0.7\text{ V}$$

Since

$$V_{p(out)} = \frac{V_{p(sec)}}{2} - 0.7\text{ V}$$

the peak inverse voltage across either diode in the full-wave center-tapped rectifier in terms of $V_{p(out)}$ is

$$\text{PIV} = 2V_{p(out)} + 0.7\text{ V} \tag{2–7}$$

FIGURE 2–17

Diode reverse voltage (D_2 shown reverse-biased). The PIV is twice the peak value of the output voltage plus a diode drop.

EXAMPLE 2–5 Show the voltage waveforms across each half of the secondary winding and across R_L when a 100 V peak sine wave is applied to the primary winding in Figure 2–18. Also, what PIV rating must the diodes have?

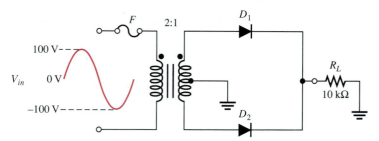

FIGURE 2–18

Solution The total peak secondary voltage is

$$V_{p(sec)} = \left(\frac{N_{sec}}{N_{pri}}\right)V_{p(pri)} = (0.5)100 \text{ V} = 50 \text{ V}$$

There is a 100 V peak across each half of the secondary. The output load voltage has a peak value of 100 V, less the 0.7 V drop across the diode. Each diode must have a minimum PIV rating of 50 V (neglecting diode drop). The waveforms are shown in Figure 2–19.

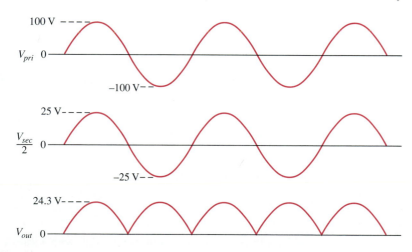

FIGURE 2–19

Related Exercise What diode PIV rating is required to handle a peak input of 160 V in Figure 2–18?

The Full-Wave Bridge Rectifier

The full-wave **bridge rectifier** uses four diodes, as shown in Figure 2–20. When the input cycle is positive as in part (a), diodes D_1 and D_2 are forward-biased and conduct current in the direction shown. A voltage is developed across R_L which looks like the positive half of the input cycle. During this time, diodes D_3 and D_4 are reverse-biased.

When the input cycle is negative as in Figure 2–20(b), diodes D_3 and D_4 are forward-biased and conduct current in the same direction through R_L as during the positive half-cycle. During the negative half-cycle, D_1 and D_2 are reverse-biased. A full-wave rectified output voltage appears across R_L as a result of this action.

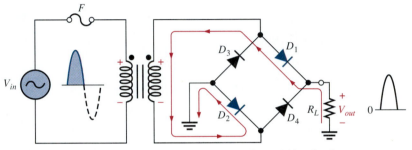

(a) During positive half-cycle of the input, D_1 and D_2 are forward- biased and conduct current. D_3 and D_4 are reverse-biased.

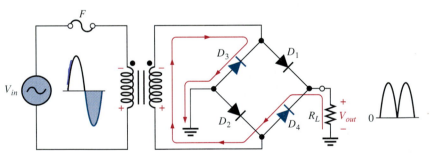

(b) During negative half-cycle of the input, D_3 and D_4 are forward-biased and conduct current. D_1 and D_2 are reverse-biased.

FIGURE 2–20
Operation of a full-wave bridge rectifier.

Bridge Output Voltage   A bridge rectifier with a transformer-coupled input is shown in Figure 2–21(a). During the positive half-cycle of the total secondary voltage, diodes D_1 and D_2 are forward-biased. Neglecting the diode drops, the secondary voltage V_{sec} appears across the load resistor. The same is true when D_3 and D_4 are forward-biased during the negative half-cycle.

$$V_{out} = V_{sec}$$

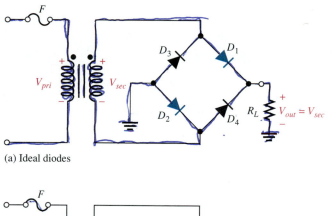

(a) Ideal diodes

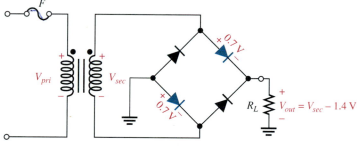

(b) Practical diodes (Diode drops included)

FIGURE 2–21
Bridge operation during positive half-cycle of the secondary voltage.

As you can see in Figure 2–21(b), two diodes are always in series with the load resistor during both the positive and negative half-cycles. If these diode drops are taken into account, the output voltage is

$$V_{out} = V_{sec} - 1.4 \text{ V} \tag{2–8}$$

Peak Inverse Voltage Let's assume that D_1 and D_2 are forward-biased and examine the reverse voltage across D_3 and D_4. Visualizing D_1 and D_2 as shorts (ideally), as in Figure 2–22(a), we can see that D_3 and D_4 have a peak inverse voltage equal to the peak secondary voltage. Since the output voltage is *ideally* equal to the secondary voltage,

$$\text{PIV} = V_{p(out)}$$

If the diode drops of the forward-biased diodes are included as shown in Figure 2–22(b), the peak inverse voltage across each reverse-biased diode in terms of $V_{p(out)}$ is

$$\text{PIV} = V_{p(out)} + 0.7 \text{ V} \tag{2–9}$$

The PIV rating of the bridge diodes is less than that required for the center-tapped configuration. If the diode drop is neglected, the bridge rectifier requires diodes with half the PIV rating of those in a center-tapped rectifier for the *same* output voltage.

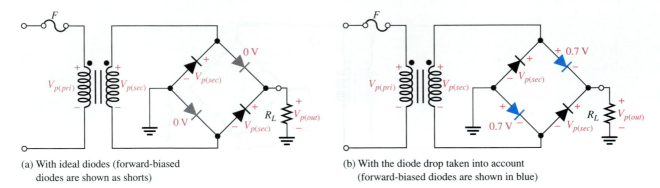

(a) With ideal diodes (forward-biased
diodes are shown as shorts)

(b) With the diode drop taken into account
(forward-biased diodes are shown in blue)

FIGURE 2–22

*Peak inverse voltage in a bridge rectifier during the positive half-cycle of the secondary
voltage.*

EXAMPLE 2–6   Determine the output voltage for the bridge rectifier in Figure 2–23. What PIV rating
is required for the silicon diodes? The transformer is specified to have a 12 V rms secondary voltage for the standard 110 V across the primary.

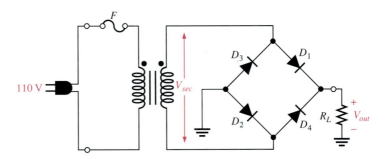

FIGURE 2–23

Solution   The peak output voltage is (taking into account the two diode drops)

$$V_{p(sec)} = 1.414V_{rms} = 1.414(12 \text{ V}) \cong 17 \text{ V}$$
$$V_{p(out)} = V_{p(sec)} - 1.4 \text{ V} = 17 \text{ V} - 1.4 \text{ V} = 15.6 \text{ V}$$

The PIV for each diode is

$$\text{PIV} = V_{p(out)} + 0.7 \text{ V} = 15.6 \text{ V} + 0.7 \text{ V} = 16.3 \text{ V}$$

Related Exercise   Determine the peak output voltage for the bridge rectifier in Figure 2–23 if the transformer produces an rms secondary voltage of 30 V. What is the
PIV rating for the diodes?

SUMMARY OF POWER SUPPLY RECTIFIERS

HALF-WAVE RECTIFIER

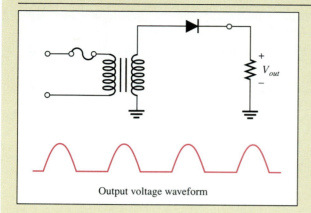

Output voltage waveform

- Peak value of output:

  $V_{p(out)} = V_{p(sec)} - 0.7 \text{ V}$

- Average value of output:

  $V_{AVG} = \dfrac{V_{p(out)}}{\pi}$

- Diode peak inverse voltage:

  $PIV = V_{p(sec)}$

FULL-WAVE CENTER-TAPPED RECTIFIER

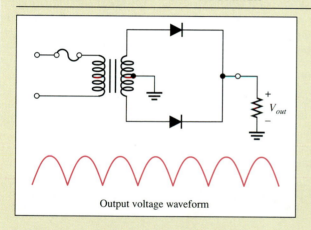

Output voltage waveform

- Peak value of output:

  $V_{p(out)} = \dfrac{V_{p(sec)}}{2} - 0.7 \text{ V}$

- Average value of output:

  $V_{AVG} = \dfrac{2V_{p(out)}}{\pi}$

- Diode peak inverse voltage:

  $PIV = 2V_{p(out)} + 0.7 \text{ V}$

FULL-WAVE BRIDGE RECTIFIER

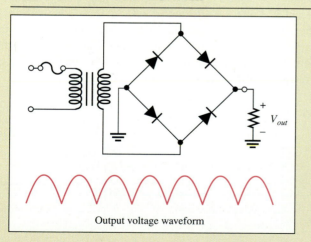

Output voltage waveform

- Peak value of output:

  $V_{p(out)} = V_{p(sec)} - 1.4 \text{ V}$

- Average value of output:

  $V_{AVG} = \dfrac{2V_{p(out)}}{\pi}$

- Diode peak inverse voltage:

  $PIV = V_{p(out)} + 0.7 \text{ V}$

1. How does a full-wave voltage differ from a half-wave voltage?

2. What is the average value of a full-wave rectified voltage with a peak value of 60 V?

3. Which type of full-wave rectifier has the greater output voltage for the same input voltage and transformer turns ratio?

4. For a peak output voltage of 45 V, in which type of rectifier would you use diodes with a PIV rating of 50 V?

5. What PIV rating is required for diodes used in the type of rectifier that was not selected in Question 4?

2–3 ■ POWER SUPPLY FILTERS

A power supply filter ideally eliminates the fluctuations in the output voltage of a half-wave or full-wave rectifier and produces a constant-level dc voltage. Filtering is necessary because electronic circuits require a constant source of dc voltage and current to provide power and biasing for proper operation. Filters are implemented with capacitors or combinations of capacitors and inductors, as you will see in this section.

After completing this section, you should be able to

■ **Explain and analyze the operation and characteristics of power supply filters**
  □ Explain the purpose of a filter
  □ Describe the capacitor filter
  □ Define *ripple voltage* and *ripple factor*
  □ Protect a capacitor filter from surge current
  □ Describe the *LC* filter
  □ Briefly discuss π-type and T-type filters

In most power supply applications, the standard 60 Hz ac power line voltage must be converted to a sufficiently constant dc voltage as discussed earlier. The 60 Hz pulsating dc output of a half-wave rectifier or the 120 Hz pulsating output of a full-wave rectifier must be filtered to reduce the large voltage variations. Figure 2–24 illustrates the filtering concept showing a nearly smooth dc output voltage from the filter. The small amount of fluctuation in the filter output voltage is called *ripple*.

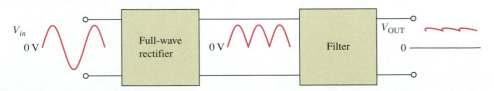

FIGURE 2–24
Power supply filtering.

Capacitor Filter

A half-wave rectifier with a capacitor filter is shown in Figure 2–25. R_L represents the equivalent resistance of a load. We will use the half-wave rectifier to illustrate the principle, and then expand the concept to full-wave rectification.

During the positive first quarter-cycle of the input, the diode is forward-biased and presents a low resistance path, allowing the capacitor to charge to within 0.7 V of the input peak, as illustrated in Figure 2–25(a). When the input begins to decrease below its peak, as shown in part (b), the capacitor retains its charge and the diode becomes reverse-biased since the cathode is more positive than the anode. During the remaining part of the cycle, the capacitor can discharge only through the load resistance at a rate determined by the $R_L C$ time constant, which is normally long compared to the period of the input. The

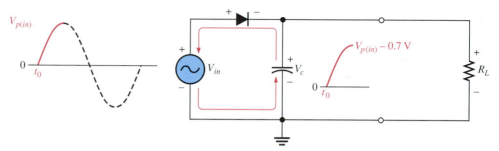

(a) Initial charging of capacitor (diode is forward-biased) happens only once when power is turned on.

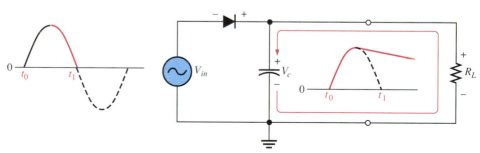

(b) Discharging through R_L after peak of positive alternation (diode is reverse-biased)

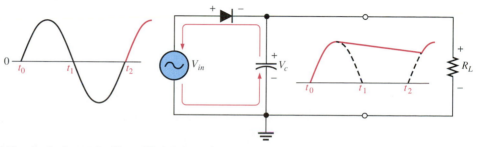

(c) Charging back to peak of input (diode is forward-biased)

FIGURE 2–25

Operation of a half-wave rectifier with a capacitor filter.

larger the time constant, the less the capacitor will discharge. During the first quarter of the next cycle, the diode will again become forward-biased when the input voltage exceeds the capacitor voltage by approximately 0.7 V. This is illustrated in part (c).

Ripple Voltage   As you have seen, the capacitor quickly charges at the beginning of a cycle and slowly discharges after the positive peak (when the diode is reverse-biased). The variation in the output voltage due to the charging and discharging is called the **ripple voltage.** Generally, ripple is undesirable; thus, the smaller the ripple, the better the filtering action, as illustrated in Figure 2–26.

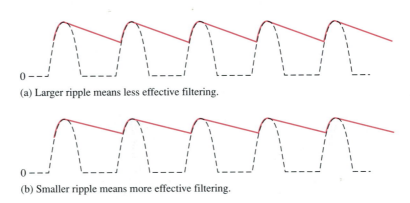

(a) Larger ripple means less effective filtering.

(b) Smaller ripple means more effective filtering.

FIGURE 2–26
Half-wave ripple voltage (solid waveform).

For a given input frequency, the output frequency of a full-wave rectifier is twice that of a half-wave rectifier, as illustrated in Figure 2–27. This makes a full-wave rectifier easier to filter. When filtered, the full-wave rectified voltage has a smaller ripple than does a half-wave voltage for the same load resistance and capacitor values. The capacitor discharges less during the shorter interval between full-wave pulses, as shown in Figure 2–28.

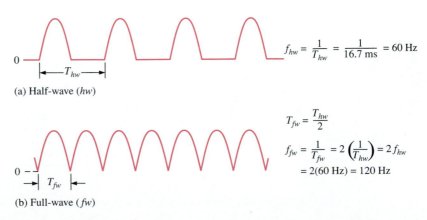

(a) Half-wave (*hw*)

$$f_{hw} = \frac{1}{T_{hw}} = \frac{1}{16.7 \text{ ms}} = 60 \text{ Hz}$$

$$T_{fw} = \frac{T_{hw}}{2}$$

$$f_{fw} = \frac{1}{T_{fw}} = 2\left(\frac{1}{T_{hw}}\right) = 2 f_{hw}$$

$$= 2(60 \text{ Hz}) = 120 \text{ Hz}$$

(b) Full-wave (*fw*)

FIGURE 2–27
Frequencies of half-wave and full-wave rectified voltages derived from a 60 Hz sine wave.

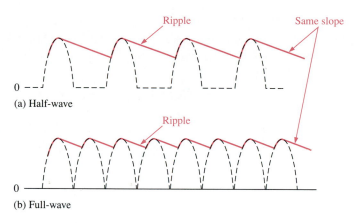

(a) Half-wave

(b) Full-wave

FIGURE 2–28

Comparison of ripple voltages for half-wave and full-wave signals with the same filter capacitor and load and derived from the same sine wave input.

Ripple Factor   The **ripple factor** is an indication of the effectiveness of the filter and is defined as

$$r = \frac{V_r}{V_{DC}} \tag{2–10}$$

where V_r is the peak-to-peak ripple voltage and V_{DC} is the dc (average) value of the filter's output voltage, as illustrated in Figure 2–29. The lower the ripple factor, the better the filter. *The ripple factor can be lowered by increasing the value of the filter capacitor or increasing the load resistance.*

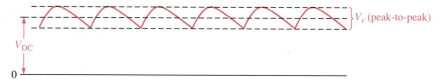

FIGURE 2–29

V_r and V_{DC} determine the ripple factor.

For a full-wave rectifier with a sufficiently high capacitance filter, if V_{DC} is within 10% of the peak rectified input voltage, then the expressions for the peak-to-peak ripple voltage, V_r, and V_{DC} are as follows. (Derivations are in Appendix B.)

$$V_r = \left(\frac{1}{fR_LC}\right)V_{p(in)} \tag{2–11}$$

$$V_{DC} = \left(1 - \frac{1}{2fR_LC}\right)V_{p(in)} \tag{2–12}$$

where $V_{p(in)}$ is the peak rectified full-wave voltage applied to the filter and f is 60 Hz for a half-wave rectifier or 120 Hz for a full-wave rectifier.

EXAMPLE 2–7   Determine the ripple factor for the filtered bridge rectifier in Figure 2–30.

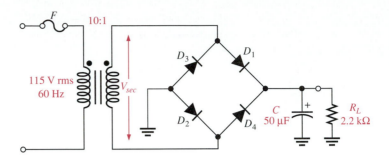

FIGURE 2–30

Solution   The peak primary voltage is

$$V_{p(pri)} = (1.414)115 \text{ V} = 163 \text{ V}$$

The peak secondary voltage is

$$V_{p(sec)} = \left(\frac{1}{10}\right)163 \text{ V} = 16.3 \text{ V}$$

The peak full-wave rectified voltage at the filter input is

$$V_{p(in)} = V_{p(sec)} - 1.4 \text{ V} = 16.3 \text{ V} - 1.4 \text{ V} = 14.9 \text{ V}$$

The filtered dc output voltage is determined as follows, where f is 120 Hz:

$$V_{DC} = \left(1 - \frac{1}{2fR_LC}\right)V_{p(in)} = \left(1 - \frac{1}{(240 \text{ Hz})(2.2 \text{ k}\Omega)(50 \text{ μF})}\right)14.9 \text{ V}$$
$$= (1 - 0.038)14.9 \text{ V} = 14.3 \text{ V}$$

The peak-to-peak ripple voltage is

$$V_r = \left(\frac{1}{fR_LC}\right)V_{p(in)} = \left(\frac{1}{(120 \text{ Hz})(2.2 \text{ k}\Omega)(50 \text{ μF})}\right)14.9 \text{ V} = 1.13 \text{ V}$$

The ripple factor is

$$r = \frac{V_r}{V_{DC}} = \frac{1.13 \text{ V}}{14.3 \text{ V}} = 0.079$$

The percent ripple is 7.9%.

Related Exercise   Determine the peak-to-peak ripple voltage if the filter capacitor in Figure 2–30 is increased to 100 μF and the load resistance changes to 12 kΩ.

Surge Current in the Capacitor Filter Before the switch in Figure 2–31(a) is closed, the filter capacitor is uncharged. At the instant the switch is closed, voltage is connected to the bridge and the capacitor appears as a short, as shown. This produces an initial surge of current I_{surge} through the two forward-biased diodes. The worst-case situation occurs when the switch is closed at a peak of the secondary voltage and a maximum surge current, $I_{surge(max)}$, is produced, as illustrated in the figure.

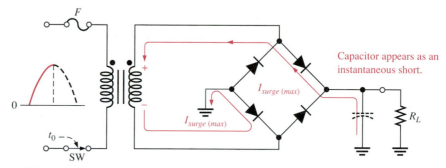

(a) Maximum surge current occurs when switch is closed at peak of input cycle.

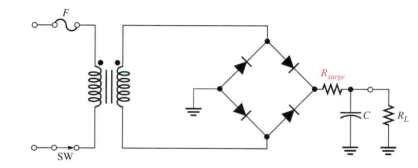

(b) A series resistor (R_{surge}) limits the surge current.

FIGURE 2–31

Surge current in a capacitor filter.

It is possible that the surge current could destroy the diodes, and for this reason a surge-limiting resistor is sometimes connected, as shown in Figure 2–31(b). The value of this resistor must be small compared to R_L. Also, the diodes must have a maximum forward surge current rating such that they can withstand the momentary surge of current. This rating is specified on diode data sheets as I_{FSM}. The minimum surge resistor value can be calculated as follows:

$$R_{surge} = \frac{V_{p(sec)} - 1.4\text{ V}}{I_{FSM}} \qquad (2\text{–}13)$$

The *LC* Filter

When an inductor is added to the filter, as in Figure 2–32, an additional reduction in the ripple voltage is achieved. The inductor has a high reactance at the 120 Hz ripple frequency, and the capacitive reactance is low compared to both X_L and R_L. The two reac-

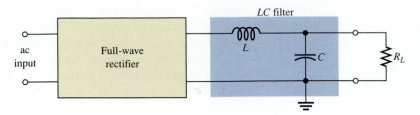

FIGURE 2–32
Rectifier with an LC filter.

tances form an ac voltage divider that tends to significantly reduce the ripple voltage from that of a straight capacitor filter, as shown in Figure 2–33.

The magnitude of the ripple voltage out of the filter is determined with the voltage divider equation.

$$V_{r(out)} = \left(\frac{X_C}{|X_L - X_C|} \right) V_{r(in)} \tag{2–14}$$

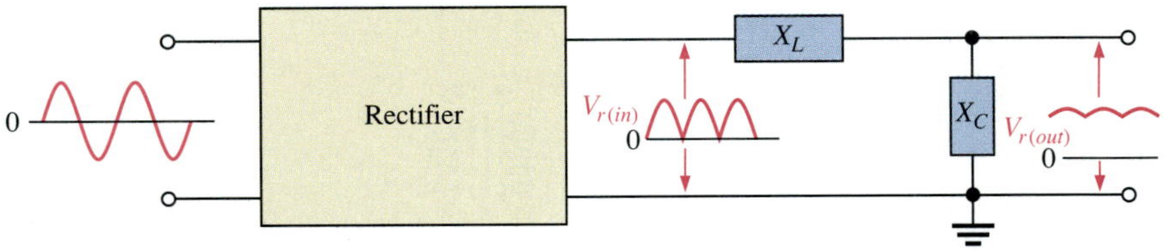

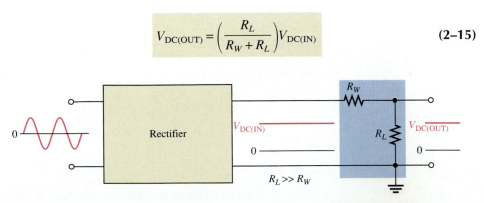

FIGURE 2–33
The LC filter as it looks to the ac component of the full-wave rectified voltage.

To the dc (average) value of the rectified input, the inductor presents a winding resistance (R_W) in series with the load resistance, as shown in Figure 2–34. This resistance produces an undesirable reduction of the dc value, and therefore R_W must be small compared to R_L. The dc output voltage is determined as follows:

$$V_{DC(OUT)} = \left(\frac{R_L}{R_W + R_L} \right) V_{DC(IN)} \tag{2–15}$$

FIGURE 2–34
The LC filter as it looks to the dc component of the full-wave rectified voltage.

EXAMPLE 2–8

A 120 Hz full-wave rectified voltage with a peak value of 162 V is applied to the *LC* filter in Figure 2–35. Determine the filter output in terms of its dc value and the rms ripple voltage.

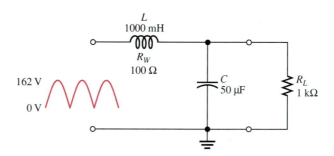

FIGURE 2–35

Solution First, determine the dc value of the full-wave rectified input using Equation (2–5).

$$V_{DC(IN)} = V_{AVG} = \frac{2V_p}{\pi} = \frac{2(162 \text{ V})}{\pi} = 103 \text{ V}$$

Next, find the amount of rms input ripple using a formula derived in Appendix B for the rms ripple of an unfiltered full-wave rectified signal.

$$V_{r(in)} = 0.308V_p = 0.308(162 \text{ V}) = 49.9 \text{ V}$$

Now that the input values are known, the output values can be calculated.

$$V_{DC(OUT)} = \left(\frac{R_L}{R_W + R_L}\right)V_{DC(IN)} = \left(\frac{1 \text{ k}\Omega}{1.1 \text{ k}\Omega}\right)103 \text{ V} = 93.6 \text{ V}$$

For the ripple voltage calculation, you need X_L and X_C.

$$X_L = 2\pi fL = 2\pi(120 \text{ Hz})(1000 \text{ mH}) = 754 \text{ }\Omega$$

$$X_C = \frac{1}{2\pi fC} = \frac{1}{2\pi(120 \text{ Hz})(50 \text{ }\mu\text{F})} = 26.5 \text{ }\Omega$$

$$V_{r(out)} = \left(\frac{X_C}{|X_L - X_C|}\right)V_{r(in)} = \left(\frac{26.5 \text{ }\Omega}{|754 \text{ }\Omega - 26.5 \text{ }\Omega|}\right)49.9 \text{ V} = 1.82 \text{ V rms}$$

Related Exercise A 120 Hz full-wave rectified voltage with a peak value of 50 V is applied to the *LC* filter in Figure 2–35 with $L = 300$ mH, $R_W = 50$ Ω, $C = 100$ μF, and $R_L = 10$ kΩ. Determine the filter output in terms of its dc value and the rms ripple voltage.

It should be noted that an *LC* rectifier filter produces an output with a dc value approximately equal to the average value of the rectified input. The capacitor filter, however, produces an output with a dc value approximately equal to the peak value of the input. Another point of comparison is that the amount of ripple voltage in the capacitor filter varies inversely with the load resistance. Ripple voltage in the *LC* filter is essentially independent of the load resistance and depends only on X_L and X_C, as long as X_C is sufficiently less than R_L.

π-Type and T-Type Filters

A one-section π-type filter is shown in Figure 2–36(a). It can be thought of as a capacitor filter followed by an *LC* filter. It is also known as a *capacitor-input filter*. The T-type filter in Figure 2–36(b) is basically an *LC* filter followed by an inductor. It is also known as an *inductor-input filter*.

FIGURE 2–36
π-type and T-type LC filters.

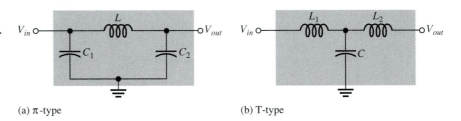

(a) π-type   (b) T-type

In the π-type capacitor-input filter, C_1 charges to the peak value of the input voltage and discharges slightly through the load during the remaining portion of the input cycle. The opposing action of the inductor tends to keep the discharge variation to a minimum, resulting in a relatively small amount of ripple voltage at the output. C_2 also helps to keep the output voltage constant.

In the T-type inductor-input filter, the voltage drop across the reactance reduces the output voltage well below the input peak. However, the smoothing action of L_1 and L_2 tends to provide an output with a smaller ripple than the π-type filter. In general, the T-type filter provides less output voltage for a given input voltage than does the π-type, but it provides improved ripple reduction.

SECTION 2–3 REVIEW

1. When a 60 Hz sinusoidal voltage is applied to the input of a half-wave rectifier, what is the output frequency?

2. When a 60 Hz sinusoidal voltage is applied to the input of a full-wave rectifier, what is the output frequency?

3. What causes the ripple voltage on the output of a capacitor filter?

4. If the load resistance connected to a power supply is decreased, what happens to the ripple voltage?

5. Define *ripple factor*.

6. Name one advantage of an *LC* filter over a capacitor filter. Name one disadvantage.

2–4 ▪ DIODE LIMITING AND CLAMPING CIRCUITS

Diode circuits, called limiters or clippers, are sometimes used to clip off portions of signal voltages above or below certain levels. Another type of diode circuit, called a clamper, is used to restore a dc level to an electrical signal. Both limiter and clamper diode circuits will be examined in this section.

After completing this section, you should be able to

▪ **Explain and analyze the operation of diode limiting and clamping circuits**
  - ☐ Explain the operation of diode limiters
  - ☐ Determine the output voltage of a biased limiter
  - ☐ Use voltage-divider bias to set the limiting level
  - ☐ Explain the operation of diode clampers

Diode Limiters

Figure 2–37(a) shows a diode **limiter** circuit (**clipper**) that limits or clips the positive part of the input voltage. As the input voltage goes positive, the diode becomes forward-biased. Since the cathode is at ground potential (0 V), the anode cannot exceed 0.7 V. So point A is limited to +0.7 V when the input voltage exceeds this value. When the input voltage goes back below 0.7 V, the diode is reverse-biased and appears as an open. The

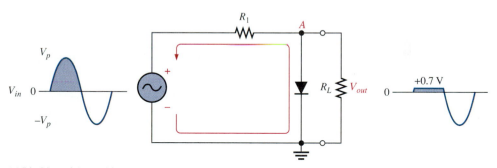

(a) Limiting of the positive alternation

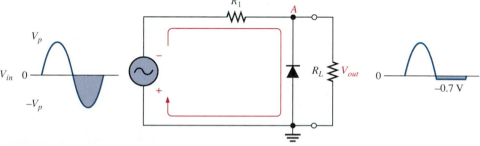

(b) Limiting of the negative alternation

FIGURE 2–37
Example of a diode limiter (clipper).

output voltage looks like the negative part of the input voltage, but with a magnitude determined by the voltage divider formed by R_1 and the load resistor, R_L, as follows:

$$V_{out} = \left(\frac{R_L}{R_1 + R_L} \right) V_{in}$$

If R_1 is small compared to R_L, then $V_{out} = V_{in}$.

Turn the diode around, as in Figure 2–37(b), and the negative part of the input voltage is clipped off. When the diode is forward-biased during the negative part of the input voltage, point A is held at -0.7 V by the diode drop. When the input voltage goes above -0.7 V, the diode is no longer forward-biased; and a voltage appears across R_L proportional to the input voltage.

EXAMPLE 2–9

What would you expect to see displayed on an oscilloscope connected across R_L in Figure 2–38?

FIGURE 2–38

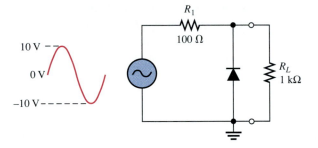

Solution The diode conducts when the input voltage goes below -0.7 V. So, a negative limiter with a peak output voltage can be determined by the following equation:

$$V_{p(out)} = \left(\frac{R_L}{R_1 + R_L} \right) V_{p(in)} = \left(\frac{1 \text{ k}\Omega}{1.1 \text{ k}\Omega} \right) 10 \text{ V} = 9.09 \text{ V}$$

The scope will display an output waveform as shown in Figure 2–39.

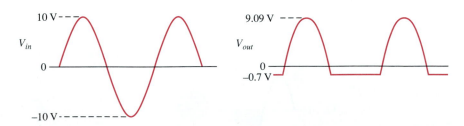

FIGURE 2–39
Waveforms for Figure 2–38.

Related Exercise Describe the output waveform for Figure 2–38 if the diode is germanium and R_L is changed to 680 Ω.

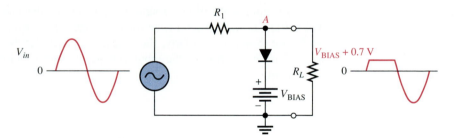

FIGURE 2–40
A positive limiter.

Biased Limiters   The level to which an ac voltage is limited can be adjusted by adding a bias voltage, V_{BIAS}, in series with the diode, as shown in Figure 2–40. The voltage at point A must equal $V_{BIAS} + 0.7$ V before the diode will conduct. Once the diode begins to conduct, the voltage at point A is limited to $V_{BIAS} + 0.7$ V so that all input voltage above this level is clipped off.

   If the bias voltage is varied up or down, the limiting level changes correspondingly, as shown in Figure 2–41.

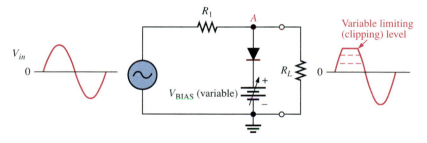

FIGURE 2–41
A positive limiter with variable bias.

   To limit a voltage to a specified negative level, the diode and bias voltage must be connected as in Figure 2–42. In this case, the voltage at point A must go below $-V_{BIAS} - 0.7$ V to forward-bias the diode and initiate limiting action as shown.

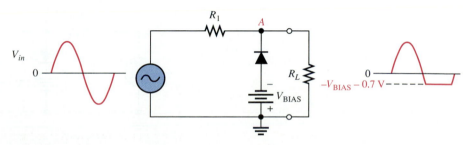

FIGURE 2–42
A negative limiter.

By turning the diode around, the positive limiter can be modified to limit the output voltage to the portion of the input voltage waveform above $V_{BIAS} - 0.7$ V, as shown in Figure 2–43(a). Similarly, the negative limiter can be modified to limit the output voltage to the portion of the input voltage waveform below $- V_{BIAS} + 0.7$ V, as shown in part (b).

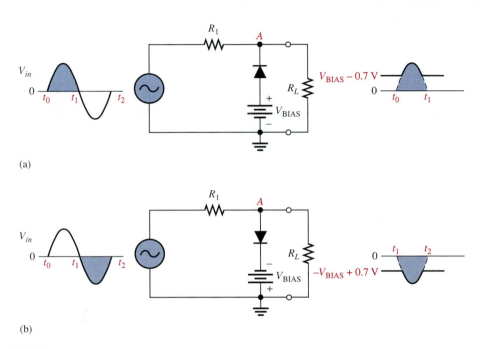

(a)

(b)

FIGURE 2–43

EXAMPLE 2–10   Figure 2–44 shows a circuit combining a positive limiter with a negative limiter. Determine the output voltage waveform.

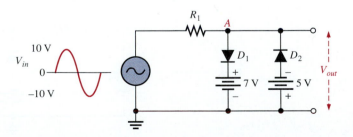

FIGURE 2–44

Solution   When the voltage at point A reaches +7.7 V, diode D_1 conducts and limits the waveform to +7.7 V. Diode D_2 does not conduct until the voltage reaches −5.7 V. Therefore, positive voltages above +7.7 V and negative voltages below −5.7 V are clipped off. The resulting output voltage waveform is shown in Figure 2–45.

FIGURE 2–45
Output voltage waveform for Figure 2–44.

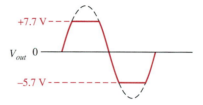

Related Exercise   Determine the output voltage waveform in Figure 2–44 if both dc sources are 10 V and the input voltage has a peak value of 20 V.

Voltage-Divider Bias   In practice, the bias voltage sources that have been used to illustrate the basic operation of diode limiters can be replaced by a resistive voltage divider that derives the desired bias voltage from the dc supply voltage as shown in Figure 2–46. The bias voltage is set by the resistor values according to the voltage-divider formula:

$$V_{\text{BIAS}} = \left(\frac{R_2}{R_1 + R_2} \right) V_{\text{SUPPLY}}$$

A positively biased limiter is shown in Figure 2–46(a), a negatively biased limiter is shown in part (b), and a variable positive bias circuit using a potentiometer voltage divider is shown in part (c). The bias resistors must be small compared to R_1.

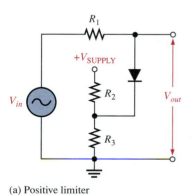

(a) Positive limiter

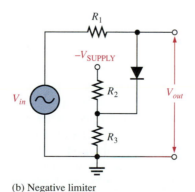

(b) Negative limiter

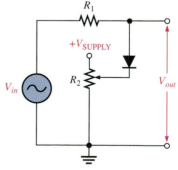

(c) Variable positive limiter

FIGURE 2–46
Diode limiters implemented with voltage-divider bias.

EXAMPLE 2–11   Describe the output voltage waveform for the diode limiter in Figure 2–47.

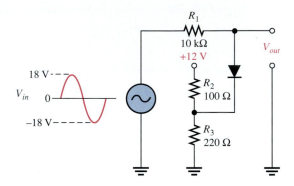

FIGURE 2–47

Solution   The circuit is a positive limiter and the bias voltage is determined using the voltage-divider formula.

$$V_{\text{BIAS}} = \left(\frac{R_3}{R_2 + R_3} \right) V_{\text{SUPPLY}} = \left(\frac{220 \ \Omega}{100 \ \Omega + 220 \ \Omega} \right) 12 \ \text{V} = 8.25 \ \text{V}$$

The output voltage waveform is shown in Figure 2–48. The positive part of the output voltage waveform is limited to $V_{\text{BIAS}} + 0.7$ V.

FIGURE 2–48

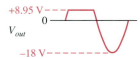

Related Exercise   Change the voltage divider in Figure 2–47 to limit the output voltage to +6.7 V.

Diode Clampers

A clamper adds a dc level to an ac voltage. **Clampers** are sometimes known as *dc restorers*. Figure 2–49 shows a diode clamper that inserts a positive dc level. The operation of this circuit can be seen by considering the first negative half-cycle of the input voltage. When the input voltage initially goes negative, the diode is forward-biased, allowing the capacitor to charge to near the peak of the input ($V_{p(in)} - 0.7$ V), as shown in Figure 2–49(a). Just after the negative peak, the diode is reverse-biased. This is because the cathode is held near $V_{p(in)} - 0.7$ V by the charge on the capacitor. The capacitor can only discharge through the high resistance of R_L. So, from the peak of one negative half-cycle to the next, the capacitor discharges very little. The amount that is discharged, of course, depends on the value of R_L. For good clamping action, the RC time constant should be at least ten times the period of the input frequency.

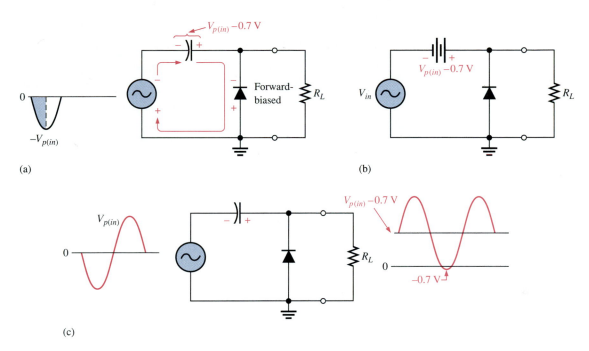

FIGURE 2–49
Positive clamper operation.

The net effect of the clamping action is that the capacitor retains a charge approximately equal to the peak value of the input less the diode drop. The capacitor voltage acts essentially as a battery in series with the input voltage, as shown in Figure 2–49(b). The dc voltage of the capacitor adds to the input voltage by superposition, as in Figure 2–49(c). If the diode is turned around, a negative dc voltage is added to the input voltage as shown in Figure 2–50.

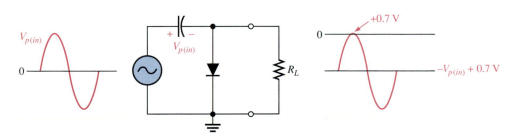

FIGURE 2–50
Negative clamper.

A Clamper Application   A clamping circuit is often used in television receivers as a dc restorer. The incoming composite video signal is normally processed through capacitively coupled amplifiers that eliminate the dc component, thus losing the black and white reference levels and the blanking level. Before being applied to the picture tube, these reference levels must be restored. Figure 2–51 illustrates this process in a general way.

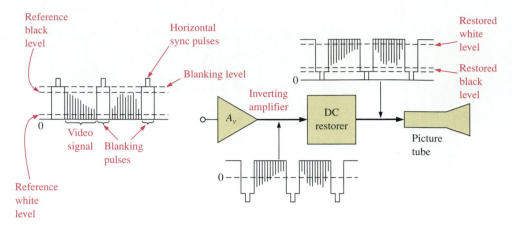

FIGURE 2–51
Clamper (dc restorer) in a TV receiver.

EXAMPLE 2–12

What is the output voltage that you would expect to observe across R_L in the clamper circuit of Figure 2–52? Assume that RC is large enough to prevent significant capacitor discharge.

FIGURE 2–52

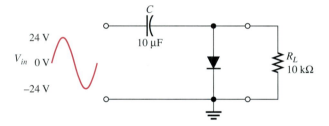

Solution Ideally, a negative dc value equal to the input peak less the diode drop is inserted by the clamping circuit.

$$V_{DC} \cong -(V_{p(in)} - 0.7 \text{ V}) = -(24 \text{ V} - 0.7 \text{ V}) = -23.3 \text{ V}$$

Actually, the capacitor will discharge slightly between peaks, and, as a result, the output voltage will have an average value of slightly less than that calculated above. The output waveform goes to approximately +0.7 V, as shown in Figure 2–53.

FIGURE 2–53
Output waveform for Figure 2–52.

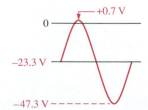

Related Exercise What is the output voltage that you would observe across R_L in Figure 2–52 for $C = 22$ μF and $R_L = 18$ kΩ?

1. Discuss how diode limiters and diode clampers differ in terms of their function.

2. What is the difference between a positive limiter and a negative limiter?

3. What is the maximum voltage across an unbiased positive silicon diode limiter during the positive alternation of the input voltage?

4. To limit the output voltage of a positive limiter to 5 V when a 10 V peak input is applied, what value must the bias voltage be?

5. What component in a clamper circuit effectively acts as a battery?

2–5 ■ VOLTAGE MULTIPLIERS

Voltage multipliers use clamping action to increase peak rectified voltages without the necessity of increasing the input transformer's voltage rating. Multiplication factors of two, three, and four are common. Voltage multipliers are used in high-voltage, low-current applications such as TV receivers.

After completing this section, you should be able to

■ **Explain and analyze the operation of diode voltage multipliers**
  ☐ Discuss voltage doublers
  ☐ Discuss voltage triplers
  ☐ Discuss voltage quadruplers

Voltage Doubler

Half-Wave Voltage Doubler A voltage doubler is a **voltage multiplier** with a multiplication factor of two. A half-wave voltage doubler is shown in Figure 2–54. During the positive half-cycle of the secondary voltage, diode D_1 is forward-biased and D_2 is reverse-biased. Capacitor C_1 is charged to the peak of the secondary voltage (V_p) less the diode drop with the polarity shown in part (a). During the negative half-cycle, diode D_2 is forward-biased and D_1 is reverse-biased, as shown in part (b). Since C_1 can't discharge,

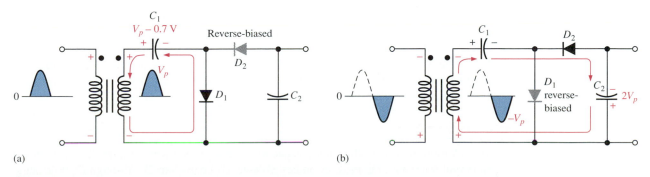

FIGURE 2–54
Half-wave voltage doubler operation. V_p is the peak secondary voltage.

the peak voltage on C_1 adds to the secondary voltage to charge C_2 to approximately $2V_p$. Applying Kirchhoff's law around the loop as shown in part (b),

$$V_{C1} - V_{C2} + V_p = 0$$
$$V_{C2} = V_p + V_{C1}$$

Neglecting the diode drop of D_2, $V_{C1} = V_p$. Therefore,

$$V_{C2} = V_p + V_p = 2V_p$$

Under a no-load condition, C_2 remains charged to approximately $2V_p$. If a load resistance is connected across the output, C_2 discharges slightly through the load on the next positive half-cycle and is again recharged to $2V_p$ on the following negative half-cycle. The resulting output is a half-wave, capacitor-filtered voltage. The peak inverse voltage across each diode is $2V_p$.

Full-Wave Voltage Doubler A full-wave doubler is shown in Figure 2–55. When the secondary voltage is positive, D_1 is forward-biased and C_1 charges to approximately V_p, as shown in part (a). During the negative half-cycle, D_2 is forward-biased and C_2 charges to approximately V_p, as shown in part (b). The output voltage, $2V_p$, is taken across the two capacitors in series.

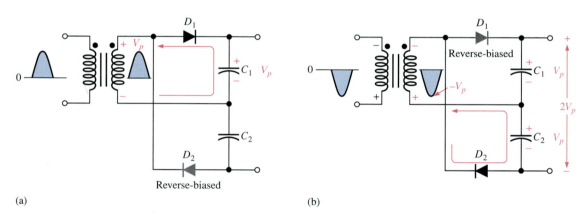

(a)   (b)

FIGURE 2–55
Full-wave voltage doubler operation.

Voltage Tripler

The addition of another diode-capacitor section to the half-wave voltage doubler creates a voltage tripler, as shown in Figure 2–56. The operation is as follows: On the positive half-cycle of the secondary voltage, C_1 charges to V_p through D_1. During the negative half-cycle, C_2 charges to $2V_p$ through D_2, as described for the doubler. During the next positive half-cycle, C_3 charges to $2V_p$ through D_3. The tripler output is taken across C_1 and C_3, as shown in the figure.

Voltage Quadrupler

The addition of still another diode-capacitor section, as shown in Figure 2–57, produces an output four times the peak secondary voltage. C_4 charges to $2V_p$ through D_4 on a negative half-cycle. The $4V_p$ output is taken across C_2 and C_4, as shown. In both the tripler and quadrupler circuits, the PIV of each diode is $2V_p$.

FIGURE 2–56
Voltage tripler

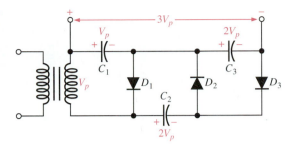

FIGURE 2–57
Voltage quadrupler.

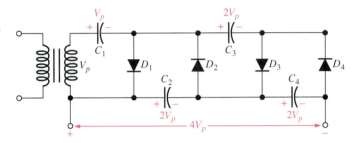

2–6 ■ THE DIODE DATA SHEET

A manufacturer's data sheet gives detailed information on a device so that it can be used properly in a given application. A typical data sheet provides maximum ratings, electrical characteristics, mechanical data, and graphs of various parameters. In this section, we use a specific example to illustrate a typical data sheet.

After completing this section, you should be able to

■ **Interpret and use a diode data sheet**
  ☐ Identify maximum voltage and current ratings
  ☐ Determine the electrical characteristics of a diode
  ☐ Analyze graphical data
  ☐ Select an appropriate diode for a given set of specifications

Table 2–1 shows the maximum ratings for a certain series of rectifier diodes (1N4001 through 1N4007). These are the absolute maximum values under which the diode can be operated without damage to the device. For greatest reliability and longer life, the diode should always be operated well under these maximums. Generally, the maximum ratings are specified at 25°C and must be adjusted downward for higher temperatures.

TABLE 2–1

Maximum ratings

Rating	Symbol	1N4001	1N4002	1N4003	1N4004	1N4005	1N4006	1N4007	Unit
Peak repetitive reverse voltage Working peak reverse voltage DC blocking voltage	V_{RRM} V_{RWM} V_R	50	100	200	400	600	800	1000	V
Nonrepetitive peak reverse voltage	V_{RSM}	60	120	240	480	720	1000	1200	V
RMS reverse voltage	$V_{R(rms)}$	35	70	140	280	420	560	700	V
Average rectified forward current (single-phase, resistive load, 60 Hz, $T_A = 75°C$)	I_O	←————————————1.0————————————→							A
Nonrepetitive peak surge current (surge applied at rated load conditions)	I_{FSM}	←——————30 (for 1 cycle)——————→							A
Operating and storage junction temperature range	T_J, T_{stg}	←——————−65 to +175——————→							°C

An explanation of the parameters from Table 2–1 is as follows:

V_{RRM} The maximum reverse peak voltage that can be applied repetitively across the diode. Notice that in this case, it is 50 V for the 1N4001 and 1 kV for the 1N4007. This is the same as PIV rating.

V_R The maximum reverse dc voltage that can be applied across the diode.

V_{RSM} The maximum reverse peak value of nonrepetitive voltage that can be applied across the diode.

I_O The maximum average value of a 120 Hz full-wave rectified forward current.

I_{FSM} The maximum peak value of nonrepetitive (one cycle) forward surge current. The graph in Figure 2–58 expands on this parameter to show values for more than one cycle at temperatures of 25°C and 175°C. The dashed lines represent values where typical failures occur. Notice what happens on the lower solid line when ten cycles of I_{FSM} are applied. The limit is 15 A rather than the one-cycle value of 30 A.

T_A Ambient temperature (temperature of surrounding air).

T_J The operating junction temperature.

T_{stg} The storage junction temperature.

Table 2–2 lists typical and maximum values of certain electrical characteristics. These items differ from the maximum ratings in that they are not selected by design but

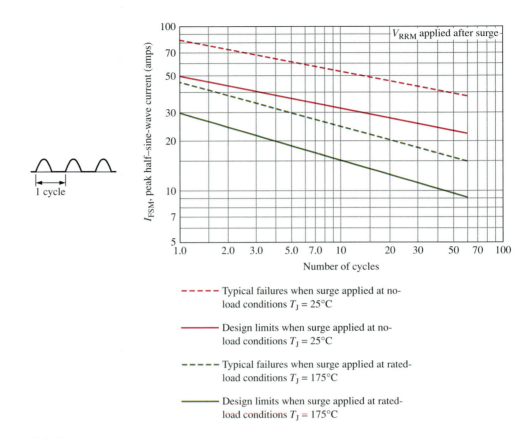

---- Typical failures when surge applied at no-
load conditions $T_J = 25°C$

—— Design limits when surge applied at no-
load conditions $T_J = 25°C$

---- Typical failures when surge applied at rated-
load conditions $T_J = 175°C$

—— Design limits when surge applied at rated-
load conditions $T_J = 175°C$

FIGURE 2–58
Nonrepetitive forward surge current capability.

TABLE 2–2
Electrical characteristics

Characteristic and Conditions	Symbol	Typical	Maximum	Unit
Maximum instantaneous forward voltage drop ($i_F = 1$ A, $T_J = 25°C$	v_F	0.93	1.1	V
Maximum full-cycle average forward voltage drop ($I_O = 1$ A, $T_L = 75°C$, 1 inch leads)	$V_{F(AVG)}$	—	0.8	V
Maximum reverse current (rated dc voltage) $T_J = 25°C$ $T_J = 100°C$	I_R	0.05 1.0	10.0 50.0	μA
Maximum full-cycle average reverse current ($I_O = 1$ A, $T_L = 75°C$, 1 inch leads)	$I_{R(AVG)}$	—	30.0	μA

are the result of operating the diode under specified conditions. A brief explanation of these parameters follows:

v_F The instantaneous voltage across the forward-biased diode when the forward current is 1 A at 25°C. Figure 2–59 shows how the forward voltages vary with forward current.

$V_{F(AVG)}$ The maximum forward voltage drop averaged over a full cycle.

I_R The maximum current when the diode is reverse-biased with a dc voltage.

$I_{R(AVG)}$ The maximum reverse current averaged over one cycle (when reverse-biased with an ac voltage).

T_L The lead temperature.

Figure 2–60 shows a selection of rectifier diodes arranged in order of increasing I_O, I_{FSM}, and V_{RRM} ratings.

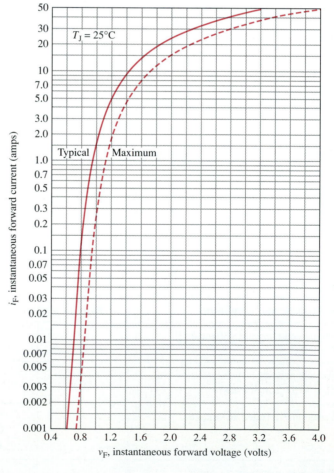

FIGURE 2–59
Forward voltage.

V_{RRM} (Volts)	I_O, Average Rectified Forward Current (Amperes)					
	1.0	**1.5**	**3.0**			**6.0**
	59-03 (DO-41) Plastic	59-04 Plastic	60-01 Metal	267-03 Plastic	267-02 Plastic	194-04 Plastic
50	1N4001	1N5391	1N4719	MR500	1N5400	MR750
100	1N4002	1N5392	1N4720	MR501	1N5401	MR751
200	1N4003	1N5393 MR5059	1N4721	MR502	1N5402	MR752
400	1N4004	1N5395 MR5060	1N4722	MR504	1N5404	MR754
600	1N4005	1N5397 MR5061	1N4723	MR506	1N5406	MR756
800	1N4006	1N5398	1N4724	MR508		MR758
1000	1N4007	1N5399	1N4725	MR510		MR760
I_{FSM} (Amps)	30	50	300	100	200	400
T_A @ Rated I_O (°C)	75	$T_L = 70$	75	95	$T_L = 105$	60
T_C @ Rated I_O (°C)						
T_J (Max) (°C)	175	175	175	175	175	175

V_{RRM} (Volts)	I_O, Average Rectified Forward Current (Amperes)										
	12	**20**	**24**	**25**	**30**		**40**	**50**	**25**	**35**	**40**
	245A-02 (DO-203AA) Metal		339-02 Plastic	193-04 Plastic	43-02 (DO-21) Metal		42A-01 (DO-203AB) Metal	43-04 Metal	309A-03	309A-02	
50	MR1120 1N1199,A,B	MR2000	MR2400	MR2500	1N3491	1N3659	1N1183A	MR5005	MDA2500	MDA3500	
100	MR1121 1N1200,A,B	MR2001	MR2401	MR2501	1N3492	1N3660	1N1184A	MR5010	MDA2501	MDA3501	
200	MR1122 1N1202,A,B	MR2002	MR2402	MR2502	1N3493	1N3661	1N1186A	MR5020	MDA2502	MDA3502	MDA4002
400	MR1124 1N1204,A,B	MR2004	MR2404	MR2504	1N3495	1N3663	1N1188A	MR5040	MDA2504	MDA3504	MDA4004
600	MR1126 1N1206,A,B	MR2006	MR2406	MR2506			1N1190A		MDA2506	MDA3506	MDA4006
800	MR1128	MR2008		MR2508					MDA2508	MDA3508	MDA4008
1000	MR1130	MR2010		MR2510					MDA2510	MDA3510	
I_{FSM} (Amps)	300	400	400	400	300	400	800	600	400	400	800
T_A @ Rated I_O (°C)											
T_C @ Rated I_O (°C)	150	150	125	150	130	100	150	150	55	55	35
T_J (Max) (°C)	190	175	175	175	175	175	190	195	175	175	175

FIGURE 2–60

A selection of rectifier diodes based on maximum ratings of I_O, I_{FSM}, and V_{RRM}.

1. List the three rating categories typically given on all diode data sheets.
2. Define each of the following parameters: V_F, I_R, I_O.
3. Define I_{FSM}, V_{RRM}, and V_{RSM}.
4. From Figure 2–60, select a diode to meet the following specifications: $I_O = 3$ A, $I_{FSM} = 300$ A, and $V_{RRM} = 100$ V.

2–7 ■ TROUBLESHOOTING

This section provides a general introduction to troubleshooting techniques that will apply throughout the book as well as specific troubleshooting examples of the power supply and other diode circuits you have studied in this chapter.

After completing this section, you should be able to

■ **Troubleshoot diode circuits using accepted techniques**
  ☐ Discuss the relationship between symptom and cause
  ☐ Discuss and apply the power check
  ☐ Discuss and apply the sensory check
  ☐ Discuss and apply the component replacement method
  ☐ Discuss and apply the signal tracing technique in its three variations.
  ☐ Explain fault analysis

You can approach the **troubleshooting** of a defective electronic circuit or group of circuits (system) in several ways. We will define a defective circuit or system as one with a known good input voltage but with no output voltage or an incorrect output voltage. For example, in Figure 2–61(a), a properly functioning dc power supply is represented by a single block with a known input voltage and a correct output voltage. A defective dc power supply is represented in part (b) as a block with an input voltage but no output voltage.

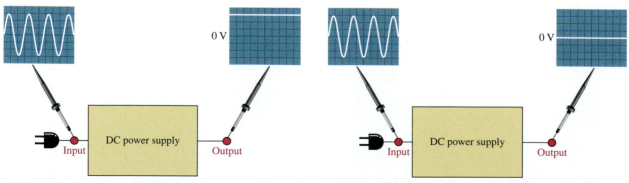

(a) The correct dc output voltage is measured. (b) 0 V dc is measured at the output. The input voltage is correct.

FIGURE 2–61
Block representations of functioning and nonfunctioning power supplies.

The Symptom and the Cause

The first step in troubleshooting a defective circuit or system is to identify the symptom. In the case of the power supply example, this is easy. The symptom is that there is no output voltage. In other situations, symptom identification may be a more crucial step.

The absence of an output voltage from the power supply does not tell much about what the specific cause may be. In some cases, however, a particular symptom may point to a given area where a fault is most likely to have occurred and give the technician an idea of where to begin to find the cause of the problem.

Troubleshooting Techniques

Several methods can be used to troubleshoot a defective circuit or system. The one to use depends on the type and complexity of the circuit or system, the nature of the problem, and the preference of the individual technician.

Power Check   The first thing you should do in checking out a defective circuit is to make sure the power cord is plugged in and the fuse is not burned out. Something this simple is often the cause of the problem.

Sensory Check   Beyond the power check, the simplest troubleshooting method relies on your senses of observation to check for obvious defects. For example, a burned resistor is often visible as are broken wires, poor solder connections, burned out fuses, etc. Also, when certain types of components fail, you may be able to detect a smell of smoke if you happen to be there when it fails or shortly afterward.

Since some failures are temperature dependent, you can sometimes use your sense of touch to detect an overheated component. This is an intermittent type of failure because the circuit works properly for a while and then fails when it heats up. As a rule, always perform a sensory check before proceeding with more sophisticated troubleshooting methods.

Component Replacement   This method relies largely on educated guesswork based on the symptom or symptoms. Based on your knowledge of the circuit operation, a certain symptom may indicate that a particular component in the circuit is defective. Using this method, you would change the suspected component and test the circuit for proper operation. If you were wrong, you then have to select the next most likely component.

This process can be very quick if you get it right the first time or it can be very time consuming and expensive if you don't. Obviously, component replacement is guaranteed to fix the problem because if you replace enough components, you will eventually find the defective one. This is not a recommended method unless the symptom/cause relationship is glaringly obvious.

Signal Tracing   This troubleshooting technique is very popular and effective. Basically, in signal tracing you look for a point in a circuit or system where you lose the signal. A point in a circuit where a known voltage can be measured is called a *test point* (TP).

Now, we will discuss three variations of the signal tracing technique and illustrate them in relation to finding a fault in the dc power supply.

Signal tracing method 1:   Start at the input of a circuit where there is a known input signal and work toward the output. Check the signal voltage at successive test points until you get an *incorrect* measurement. When you find no signal or an incorrect signal, you have isolated the problem to the portion of the circuit between the last test point and the

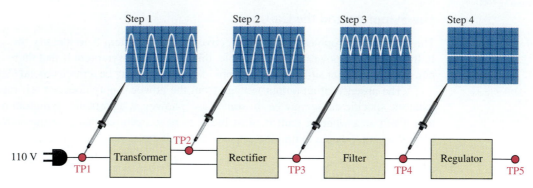

FIGURE 2–62
Example of signal tracing: input to output.

present test point. Of course, you must know how the signal is supposed to change as it goes through the circuit or have a schematic with correct voltages marked at the test points. This approach is illustrated in Figure 2–62 for the power supply which is shown as four functional blocks. The measurements indicate that the fault is in the filter circuit because the input to the filter (TP3) is the correct full-wave rectified voltage but there is no voltage on the output (TP4).

Signal tracing method 2: Start at the output of a circuit and work toward the input. Check for voltage at each test point until you get a *correct* measurement. At this point, you have isolated the problem to the portion of the circuit between the last test point and the current test point. This approach is illustrated in Figure 2–63. Again, the measurements indicate a faulty filter circuit because there is no output voltage from the filter (TP4) but the input is correct (TP3).

Signal tracing method 3: Start in the middle of the circuit. If the beginning test point has a correct signal, you know that the circuit is working properly from the input to that test point, and therefore the fault is somewhere between the test point and the output. So, you begin signal tracing from the test point toward the output. If the beginning test point has no signal or an incorrect signal, you know that the fault is between the input and the test point. So, you begin signal tracing from the test point toward the input. Figure 2–64

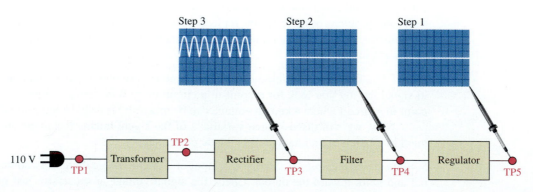

FIGURE 2–63
Example of signal tracing: output to input.

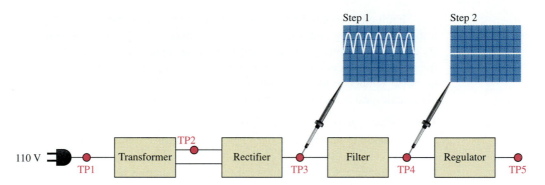

FIGURE 2–64
Example of signal tracing: midpoint.

illustrates this approach for the power supply. The correct voltage at TP3 indicates that the transformer and recifier are functioning properly. The measurement of 0 V at TP4 isolates the fault to the filter.

For this example, the midpoint method took two steps compared to four for method 1 and three for method 2. There is some advantage in method 3 especially for more complex systems because, statistically, you will find the problem more quickly.

Signal tracing has isolated the problem in our defective power supply to the filter circuit. The next step is to determine the defective component in the filter. The filter in this example consists of a capacitor and a surge resistor as shown in Figure 2–65.

FIGURE 2–65
Troubleshooting the filter circuit requires component replacement because of its simplicity.

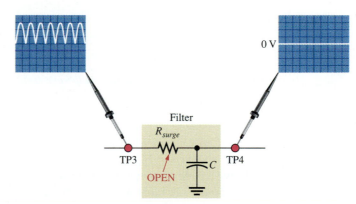

In this simple circuit the component replacement method is appropriate. Since we have found that the output voltage is zero for a correct input voltage, the surge resistor must be open. It is possible that the filter capacitor is shorted, but that is less likely to be the case.

Fault Analysis

Fault analysis is an area of troubleshooting that can best be described by this question: *If component X fails in circuit Y, what will be the symptom(s)?*

Fault analysis can be applied when you measure an incorrect voltage at a test point using signal tracing and isolate the fault to a specific circuit. You can perform fault analysis on a circuit using your knowledge of circuit operation and the effects of component failures on that operation to determine which component, if defective, would cause the incorrect voltage at the test point. We will discuss fault analysis by using the power supply circuits and common component failures as examples.

Effect of an Open Diode in a Half-Wave Rectifier A half-wave filtered rectifier with an open diode is shown in Figure 2–66. The resulting symptom is zero output voltage as indicated. This is obvious because the open diode breaks the current path from the transformer secondary winding to the load resistor and there is no load current.

Other faults that will cause the same symptom in this circuit are an open transformer winding, a shorted filter capacitor, an open fuse, or no input voltage.

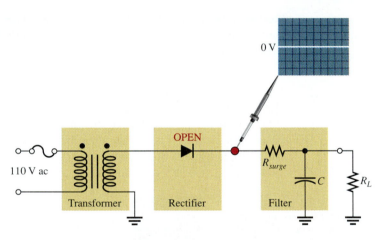

FIGURE 2–66
The effect of an open diode in a half-wave rectifier is an output of 0 V.

Effect of an Open Diode in a Full-Wave Rectifier A full-wave center-tapped filtered rectifier is shown in Figure 2–67. If either of the two diodes is open, the output voltage will have a larger than normal ripple voltage at 60 Hz rather than at 120 Hz, as indicated.

Another fault that will cause the same symptom is an open in one of the halves of the transformer secondary winding.

The reason for the increased ripple at 60 Hz rather than 120 Hz is as follows. If one of the diodes in Figure 2–67 is open, there is current through R_L only during one half-cycle of the input voltage. During the other half-cycle of the input, the open path caused by the open diode prevents current through R_L. The result is half-wave rectification, as shown in Figure 2–67, which produces the larger ripple voltage with a frequency of 60 Hz.

An open diode in a full-wave bridge rectifier will produce the same symptom as in the center-tapped circuit, as shown in Figure 2–68. The open diode prevents current through R_L during half of the input voltage cycle. The result is half-wave rectification which produces an increase in ripple voltage at 60 Hz.

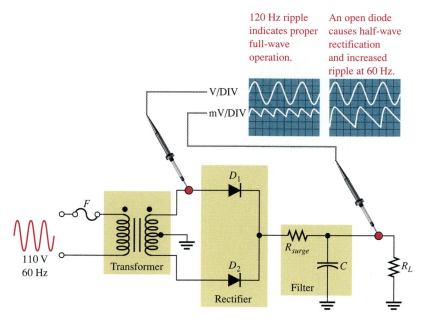

FIGURE 2–67

The effect of an open diode in a center-tapped rectifier is half-wave rectification and a larger ripple voltage at 60 Hz.

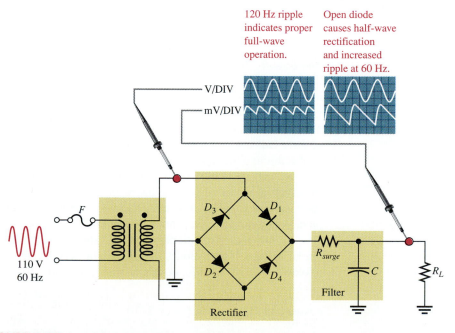

FIGURE 2–68

Effect of an open diode in a bridge rectifier.

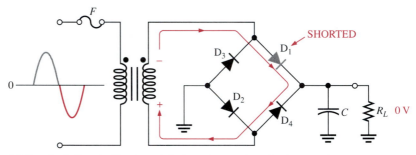

(a) Positive half-cycle: The shorted diode acts as a forward-biased diode,
so the load current is normal.

(b) Negative half-cycle: The shorted diode produces a short circuit across the source.
As a result the fuse should blow. If not properly fused, D_1, D_4, or the transformer
secondary will probably burn open.

FIGURE 2–69
Effect of a shorted diode in a bridge rectifier.

Effect of a Shorted Diode in a Full-Wave Rectifier

A shorted diode is one that has failed such that it has a very low resistance in both directions. If a diode suddenly shorts in a bridge rectifier, an excessively high current will flow during one half of the input voltage cycle, as illustrated in Figure 2–69 with diode D_1 shorted. This condition will most likely burn open one of the diodes if the circuit is not properly fused. During the positive half-cycle of the input voltage as shown in Figure 2–69(a), the current path to the load is through the shorted diode D_1 just the same as if it were forward-biased. During the negative half-cycle of the input, the current is shorted through D_1 and D_4 as shown in part (b). This is a very low resistance path because there is a forward-biased diode and a shorted diode in series. It is likely that the excessive current will burn either or both of the diodes open if the circuit is not properly fused.

If only one of the diodes, D_1 or D_4, should burn open, there is a half-wave rectified output voltage. If both diodes should burn open or if the excessive current causes the **fuse** to blow (which it should), the output voltage is zero.

Effects of a Faulty Filter Capacitor

Three types of defects of a filter capacitor are discussed and illustrated in Figure 2–70.

☐ *Open* If the filter capacitor for a full-wave rectifier opens, the output is a full-wave rectified voltage, as shown in the figure.

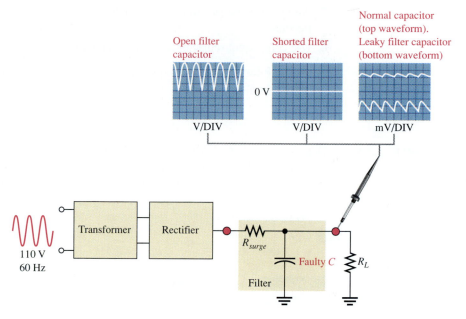

FIGURE 2–70
Effects of a faulty filter capacitor.

☐ *Shorted* If the filter capacitor shorts, the output is 0 V. A shorted capacitor may cause some or all of the diodes in the rectifier to burn open due to excessive current if there is no surge resistor and if the rectifier is not properly fused. If the rectifier is properly fused, the fuse should open and prevent damage to the circuit. In any event, the output is 0 V as shown in the figure.

☐ *Leaky* A leaky capacitor is equivalent to a capacitor with a parallel leakage resistance. The effect of the leakage resistance is to reduce the time constant and allow the capacitor to discharge more rapidly than normal. This results in an increase in the ripple voltage on the output, as shown in the figure.

Effects of a Faulty Transformer An open primary or secondary winding of a power supply transformer results in an output of 0 V, as mentioned before.

A partially shorted primary winding (which is less likely than an open) results in an increased rectifier output voltage because the turns ratio of the transformer is effectively increased. A partially shorted secondary winding results in a decreased rectifier output voltage because the turns ratio is effectively decreased.

The Complete Troubleshooting Process

Beginning with a malfunctioning or nonfunctioning circuit or system, a typical troubleshooting procedure may go as follows:

1. Identify the symptom(s).

2. Perform a power check.

3. Perform a sensory check.

4. Apply a signal tracing technique to isolate the fault to a single circuit.

5. Apply fault analysis to isolate the fault further to a single component or group of components.

6. Use component replacement to fix the problem.

The following example illustrates the steps and thought process for troubleshooting the power supply.

EXAMPLE 2–13   Determine the faulty component in the power supply represented by the block diagram in Figure 2–71 based on the indicated measurements in each step of the procedure. Assume that you find nothing during the power check and sensory check.

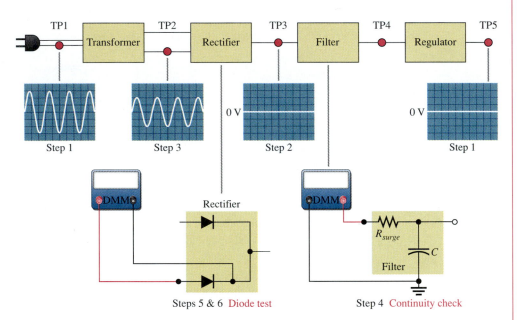

FIGURE 2–71

Solution   The step-by-step procedure in reference to Figure 2–71 is as follows:

Step 1:   Check the input and output voltages. The measurements indicate that the input voltage is correct but there is no output voltage.

Step 2:   Apply the midpoint method of signal tracing. There is no voltage at TP3. This indicates that the fault is between the input and TP3. It has to be the transformer or the rectifier, or does it? Since the voltage at TP3 is 0 V, there may be a short from the input of the filter to ground.

Step 3:   Check the voltage at TP2. If there is no voltage, then the transformer is defective. If the correct voltage appears at TP2, the problem has to be in the rectifier or a shorted filter input. The voltage at TP2 is correct. This narrows it down to a faulty rectifier or a shorted filter input.

Step 4:   With the power turned off, use a DMM to perform a continuity check for a short from the filter input to ground. Assume that the DMM indicates no short. The fault is now isolated to the rectifier circuit.

Step 5: Apply fault analysis. Determine the component failures in the rectifier that will produce an output of 0 V. If only one of the diodes in the rectifier is open, there should be a half-wave rectified output voltage, so this is not the problem. In order to have zero volts output, both of the diodes must be open.

Step 6: With the power turned off, use a DMM in the diode test mode to check each diode. Replace the defective diodes, turn on the power, and check for proper operation.

Related Exercise   If you had found a short in Step 4, what would have been the logical next step?

SECTION 2–7 REVIEW

1. What effect does an open diode have on the output voltage of a half-wave rectifier?
2. What effect does an open diode have on the output voltage of a full-wave rectifier?
3. If one of the diodes in a bridge rectifier shorts, what are some possible consequences?
4. What happens to the output voltage of a rectifier if the filter capacitor becomes very leaky?
5. The primary winding of the transformer in a power supply opens. What will you observe on the rectifier output?
6. The dc output voltage of a filtered rectifier is less than it should be. What may be the problem?

2–8 ■ SYSTEM APPLICATION

We will now put you in a hypothetical "real world" situation. Assume you are working for an electronics manufacturing company as an electronics technician. Your company designs, manufactures, and tests a wide variety of products. Your first on-the-job assignment is to develop and test a dc power supply that will be used in several different products such as an industrial counting system, a security alarm system, a voice intercom system, and a public address system. You will apply the knowledge you have gained to this point in completing your assignment.

The Power Supply Circuit

The dc power supply that will be produced by your company is shown in Figure 2–72. This is to be a universal design that can be used in several different systems with some modifications. In its basic form, this power supply is to be unregulated. Later modifications will include a regulator. The basic specifications are as follows:

1. Input voltage: 115 V rms at 60 Hz
2. Output voltage (unregulated): 12 V dc $\pm$ 10%
3. Maximum ripple factor: 3%
4. Maximum load current: 250 mA

FIGURE 2–72
Power supply preliminary schematic.

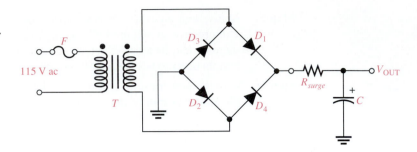

The Components

☐ *Transformer* Transformers are usually specified according to their rms output (secondary) voltage. Based on the power supply specifications, select a power transformer from the following list: 24 V, 18 V, 12 V, 12.6 V, 9 V, and 6.3 V.

☐ *Diodes* Select a diode type to be used in the rectifier bridge with minimum I_O, I_{FSM}, and V_{RRM} required to do the job. Refer to the diode selection chart in Figure 2–60.

☐ *Surge resistor* Determine a value of surge resistor based on the I_{FSM} of the selected diodes. Use the next higher standard value but keep it as small as possible to minimize voltage drop. Refer to table of standard values in Appendix A.

☐ *Primary fuse* Determine the smallest fuse rating that will handle the normal current but will blow if the load current exceeds its normal maximum or if the rectifier output is shorted. You can determine the turns ratio from the voltage ratings and use the basic formula learned in ac circuits to find the primary current. Select from the following standard fuse values: 250 mA, 500 mA, 1 A, 2 A, 3 A, 5 A, 7.5 A, and 10 A.

☐ *Filter capacitor* Select a minimum value of filter capacitor to meet the ripple factor requirements with a maximum load current (minimum load resistance). Choose from the following standard value electrolytics: 1 μF, 4.7 μF, 10 μF, 33 μF, 47 μF, 100 μF, 220 μF, 470 μF, 1000 μF, 1500 μF, 3300 μF, 4700 μF, 6800 μF, and 10,000 μF.

The Schematic

Complete the preliminary schematic in Figure 2–72 by adding component values and/or part numbers.

The Printed Circuit Board

☐ Check out the printed circuit board in Figure 2–73 to verify that it is correct according to the schematic. All black resistor bands indicate a value to be determined.

☐ Label a copy of the board with the component and input/output designations in agreement with the schematic.

A Test Procedure

Develop a step-by-step set of instructions on how to completely check the power supply for proper operation using the test points (circled numbers) indicated in the test bench setup of Figure 2–74. Specify voltage values and appropriate waveforms for all the measurements to be made. Provide a fault analysis for all possible component failures.

FIGURE 2–73
Power supply printed circuit board.

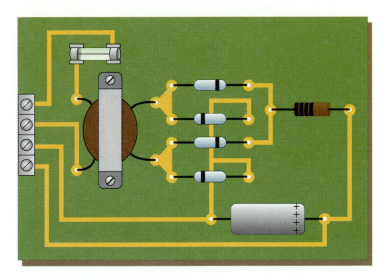

FIGURE 2–74
Power supply test bench.

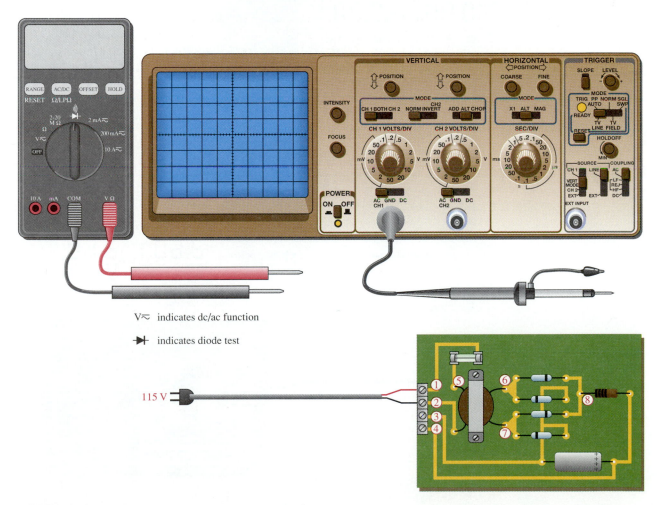

V⁓ indicates dc/ac function

⯈⊢ indicates diode test

115 V

Troubleshooting

A preliminary manufacturing run of the assembled power supply boards are coming off the assembly line. Three boards have been found to be defective. Troubleshooting techniques must be used to determine the problems.

The test bench setup is shown in Figure 2–74. Based on the sequence of measurements for each board indicated in Figure 2–75, determine the most likely fault(s) in each case.

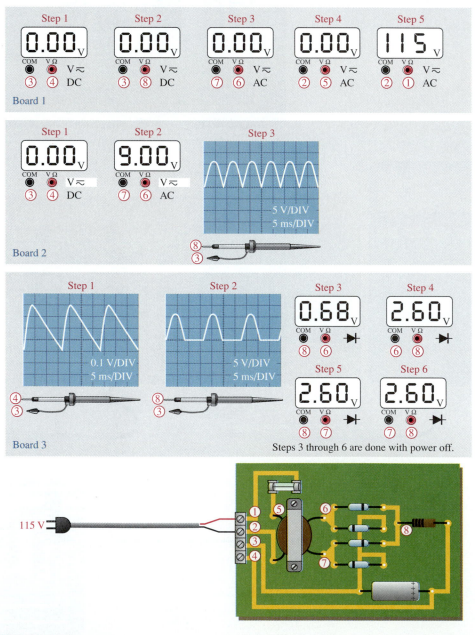

FIGURE 2–75

Signal tracing sequences for three faulty power supply boards.

The circled numbers indicate test point connections to the circuit board. The DMM function setting is indicated below the display, and the VOLT/DIV and SEC/DIV for the oscilloscope are shown on the screen in each case.

Final Report

Submit a final written report on the power supply circuit board using an organized format that includes the following:

1. A physical description of the power supply circuit.
2. A discussion of the operation of the power supply.
3. A list of the specifications.
4. A list of parts with part numbers if available.
5. A list of the types of problems on the three faulty circuit boards.
6. A complete description of how you determined the problem on each of the three faulty circuit boards.

■ **CHAPTER SUMMARY**

■ The single diode in a half-wave rectifier is forward-biased and conducts for 180° of the input cycle.
■ The output frequency of a half-wave rectifier equals the input frequency.
■ PIV (peak inverse voltage) is the maximum voltage appearing across the diode in reverse bias.
■ Each diode in a full-wave rectifier is forward-biased and conducts for 180° of the input cycle.
■ The output frequency of a full-wave rectifier is twice the input frequency.
■ The two basic types of full-wave rectifier are center-tapped and bridge.
■ The peak output voltage of a full-wave center-tapped rectifier is approximately one-half of the total peak secondary voltage less one diode drop.
■ The PIV for each diode in a full-wave center-tapped rectifier is twice the peak output voltage plus one diode drop.
■ The peak output voltage of a bridge rectifier equals the total peak secondary voltage less two diode drops.
■ The PIV for each diode in a bridge rectifier is approximately half that required for an equivalent center-tapped configuration and is equal to the peak output voltage plus one diode drop.
■ A capacitor filter provides a dc output approximately equal to the peak of its rectified input voltage.
■ Ripple voltage is caused by the charging and discharging of the filter capacitor.
■ The smaller the ripple voltage, the better the filter.
■ An *LC* filter provides improved ripple reduction over the capacitor filter.
■ An *LC* filter produces a dc output voltage approximately equal to the average value of its rectified input voltage.
■ Diode limiters cut off voltage above or below specified levels. Limiters are also called *clippers.*
■ Diode clampers add a dc level to an ac voltage.
■ A dc power supply typically consists of an input transformer, a diode rectifier, a filter, and a regulator.

▪ GLOSSARY

Bridge rectifier A type of full-wave rectifier consisting of four diodes arranged in a 4-cornered configuration.

Center-tapped rectifier A type of full-wave rectifier consisting of a center-tapped transformer and two diodes.

Clamper A circuit that adds a dc level to an ac voltage using a diode and a capacitor.

Clipper See *Limiter.*

Diode drop The voltage across the diode when it is forward-biased; approximately the same as the barrier potential and typically 0.7 V for silicon.

Filter A capacitor or combination of capacitor and inductor used to reduce the variation of the output voltage from a rectifier.

Full-wave rectifier A circuit that converts an ac sinusoidal input voltage into a pulsating dc voltage with two output pulses occurring for each input cycle.

Fuse A protective device that burns open when the current exceeds a rated limit.

Half-wave rectifier A circuit that converts an ac sinusoidal input voltage into a pulsating dc voltage with one output pulse occurring for each input cycle.

Limiter A diode circuit that clips off or removes part of a waveform above and/or below a specified level.

Power supply The circuit that converts ac line voltage to dc voltage and supplies constant power to operate a circuit or system.

Rectifier An electronic circuit that converts ac voltage into dc voltage.

Regulator An electronic device or circuit that maintains an essentially constant output voltage for a range of input voltage or load values; one part of a power supply.

Ripple factor A measure of effectiveness of a power supply filter in reducing the ripple voltage; ratio of the ripple voltage to the dc output voltage.

Ripple voltage The small variation in the dc output voltage of a filtered rectifier caused by the charging and discharging of the filter capacitor.

Troubleshooting The process and technique of identifying and locating faults in an electronic circuit or system.

Voltage multiplier A circuit using diodes and capacitors that increases the input voltage by two, three, or four times.

▪ FORMULAS

(2–1) $V_{AVG} = \dfrac{V_p}{\pi}$ Half-wave average value

(2–2) $V_{p(out)} = V_{p(in)} - 0.7\,V$ Peak half-wave rectifier output

(2–3) $PIV = V_{p(in)}$ Peak inverse voltage, half-wave rectifier

(2–4) $V_{sec} = \left(\dfrac{N_{sec}}{N_{pri}}\right)V_{pri}$ Transformer secondary voltage

(2–5) $V_{AVG} = \dfrac{2V_p}{\pi}$ Full-wave average value

(2–6) $V_{out} = \dfrac{V_{sec}}{2} - 0.7\,V$ Center-tapped full-wave output

(2–7) $PIV = 2V_{p(out)} + 0.7\,V$ Peak inverse voltage, center-tapped rectifier

(2–8) $V_{out} = V_{sec} - 1.4\,V$ Bridge full-wave output

(2–9) $PIV = V_{p(out)} + 0.7\,V$ Peak inverse voltage, bridge rectifier

$$(2\text{--}10) \qquad r = \frac{V_r}{V_{DC}} \qquad\qquad\qquad \text{Ripple factor}$$

$$(2\text{--}11) \qquad V_r = \left(\frac{1}{fR_LC}\right)V_{p(in)} \qquad\qquad \text{Peak-to-peak ripple voltage, capacitor filter}$$

$$(2\text{--}12) \qquad V_{DC} = \left(1 - \frac{1}{2fR_LC}\right)V_{p(in)} \qquad \text{DC output voltage, capacitor filter}$$

$$(2\text{--}13) \qquad R_{surge} = \frac{V_{p(sec)} - 1.4\ V}{I_{FSM}} \qquad\qquad \text{Surge resistance}$$

$$(2\text{--}14) \qquad V_{r(out)} = \left(\frac{X_C}{|X_L - X_C|}\right)V_{r(in)} \qquad \text{Ripple output, } LC \text{ filter}$$

$$(2\text{--}15) \qquad V_{DC(OUT)} = \left(\frac{R_L}{R_W + R_L}\right)V_{DC(IN)} \qquad \text{DC output, } LC \text{ filter}$$

▪ SELF-TEST

1. The average value of a half-wave rectified voltage with a peak value of 200 V is
   (a) 63.7 V (b) 127.3 V (c) 141 V (d) 0 V

2. When a 60 Hz sinusoidal voltage is applied to the input of a half-wave rectifier, the output frequency is
   (a) 120 Hz (b) 30 Hz (c) 60 Hz (d) 0 Hz

3. The peak value of the input to a half-wave rectifier is 10 V. The approximate peak value of the output is
   (a) 10 V (b) 3.18 V (c) 10.7 V (d) 9.3 V

4. For the circuit in Question 3, the diode must be able to withstand a reverse voltage of
   (a) 10 V (b) 5 V (c) 20 V (d) 3.18 V

5. The average value of a full-wave rectified voltage with a peak value of 75 V is
   (a) 53 V (b) 47.8 V (c) 37.5 V (d) 23.9 V

6. When a 60 Hz sinusoidal voltage is applied to the input of a full-wave rectifier, the output frequency is
   (a) 120 Hz (b) 60 Hz (c) 240 Hz (d) 0 Hz

7. The total secondary voltage in a center-tapped full-wave rectifier is 125 V rms. Neglecting the diode drop, the rms output voltage is
   (a) 125 V (b) 177 V (c) 100 V (d) 62.5 V

8. When the peak output voltage is 100 V, the PIV for each diode in a center-tapped full-wave rectifier is (neglecting the diode drop)
   (a) 100 V (b) 200 V (c) 141 V (d) 50 V

9. When the rms output voltage of a full-wave bridge rectifier is 20 V, the peak inverse voltage across the diodes is (neglecting the diode drop)
   (a) 20 V (b) 40 V (c) 28.3 V (d) 56.6 V

10. The ideal dc output voltage of a capacitor filter is equal to
    (a) the peak value of the rectified voltage
    (b) the average value of the rectified voltage
    (c) the rms value of the rectified voltage

11. A certain power-supply filter produces an output with a ripple of 100 mV peak-to-peak and a dc value of 20 V. The ripple factor is
    (a) 0.05 (b) 0.005 (c) 0.00005 (d) 0.02

12. A 60 V peak full-wave rectified voltage is applied to a capacitor filter. If $f = 120$ Hz, $R_L = 10$ kΩ and $C = 10$ μF, the ripple voltage is
    (a) 0.6 V (b) 6 mV (c) 5.0 V (d) 2.88 V

13. If the load resistance of a capacitor-filtered full-wave rectifier is reduced, the ripple voltage
   (a) increases   (b) decreases
   (c) is not affected   (d) has a different frequency

14. A 10 V peak-to-peak sinusoidal voltage is applied across a silicon diode and series resistor. The maximum voltage across the diode is
   (a) 9.3 V   (b) 5 V   (c) 0.7 V   (d) 10 V   (e) 4.3 V

15. If the input voltage to a voltage tripler has an rms value of 12 V, the dc output voltage is approximately
   (a) 36 V   (b) 50.9 V   (c) 33.9 V   (d) 32.4 V

16. If one of the diodes in a full-wave bridge rectifier opens, the output is
   (a) 0 V   (b) one-fourth the amplitude of the input voltage
   (c) a half-wave rectified voltage   (d) a 120 Hz voltage

17. If you are checking a 60 Hz full-wave bridge rectifier and observe that the output has a 60 Hz ripple,
   (a) the circuit is working properly   (b) there is an open diode
   (c) the transformer secondary is shorted   (d) the filter capacitor is leaky

---

▪ BASIC PROBLEMS

SECTION 2–1   Half-Wave Rectifiers

1. Draw the output voltage waveform for each circuit in Figure 2–76 and include the voltage values.

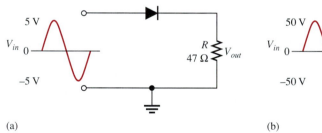

(a)   (b)

FIGURE 2–76

2. What is the peak forward current through each diode in Figure 2–76?

3. A power-supply transformer has a turns ratio of 5:1. What is the secondary voltage if the primary is connected to a 115 V rms source?

4. Determine the peak and average power delivered to R_L in Figure 2–77.

FIGURE 2–77

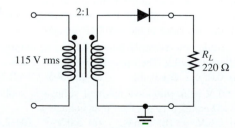

SECTION 2–2   Full-Wave Rectifiers

5. Find the average value of each voltage in Figure 2–78.

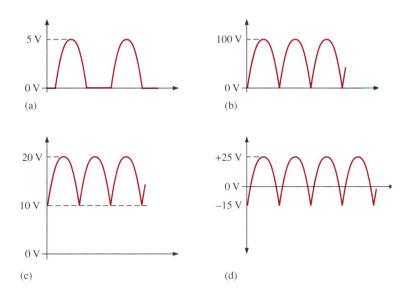

(a)   (b)

(c)   (d)

FIGURE 2–78

6. Consider the circuit in Figure 2–79.
   (a) What type of circuit is this?
   (b) What is the total peak secondary voltage?
   (c) Find the peak voltage across each half of the secondary.
   (d) Sketch the voltage waveform across R_L.
   (e) What is the peak current through each diode?
   (f) What is the PIV for each diode?

FIGURE 2–79

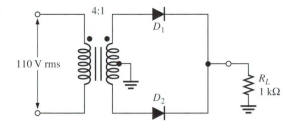

7. Calculate the peak voltage across each half of a center-tapped transformer used in a full-wave rectifier that has an average output voltage of 110 V.

8. Show how to connect the diodes in a center-tapped rectifier in order to produce a negative-going full-wave voltage across the load resistor.

9. What PIV rating is required for the diodes in a bridge rectifier that produces an average output voltage of 50 V?

10. The rms output voltage of a bridge rectifier is 20 V. What is the peak inverse voltage across the diodes?

FIGURE 2–80

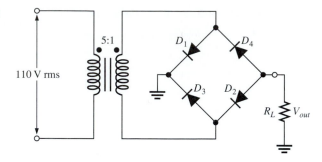

11. Sketch the output voltage of the bridge rectifier in Figure 2–80. Notice that all the diodes are reversed from previous circuits.

SECTION 2–3 Power Supply Filters

12. A certain rectifier filter produces a dc output voltage of 75 V with a peak-to-peak ripple voltage of 0.5 V. Calculate the ripple factor.

13. A certain full-wave rectifier has a peak output voltage of 30 V. A 50 μF capacitor filter is connected to the rectifier. Calculate the peak-to-peak ripple and the dc output voltage developed across a 600 Ω load resistance.

14. What is the percentage of ripple for the rectifier filter in Problem 13?

15. What value of filter capacitor is required to produce a 1% ripple factor for a full-wave rectifier having a load resistance of 1.5 kΩ? Assume the rectifier produces a peak output of 18 V.

16. A full-wave rectifier produces an 80 V peak rectified voltage from a 60 Hz ac source. If a 10 μF filter capacitor is used, determine the ripple factor for a load resistance of 10 kΩ.

17. Determine the peak-to-peak ripple and dc output voltages in Figure 2–81. The transformer has a 36 V secondary voltage rating, the line voltage has a frequency of 60 Hz, and the winding resistance of the coil is 100 Ω.

FIGURE 2–81

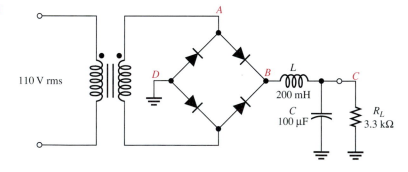

18. Refer to Figure 2–81 and sketch the following voltage waveforms in relationship to the input waveforms: V_{AB}, V_{AD}, V_{BD}, and V_{CD}. A double letter subscript indicates a voltage from one point to another.

SECTION 2–4 Diode Limiting and Clamping Circuits

19. Determine the output waveform for the circuit of Figure 2–82.

20. Determine the output voltage for the circuit in Figure 2–83(a) for each input voltage in (b), (c), and (d).

21. Determine the output voltage waveform for each circuit in Figure 2–84.

FIGURE 2–82

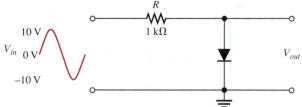

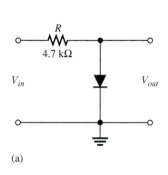

(a)

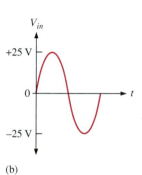

(b)

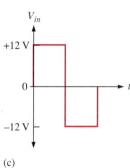

(c)

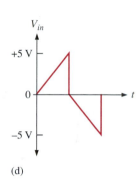

(d)

FIGURE 2–83

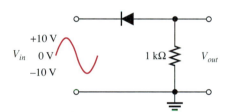

(a)

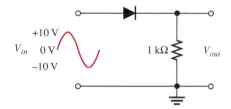

(b)

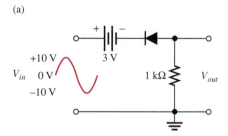

(c)

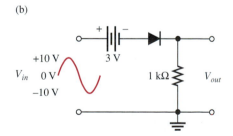

(d)

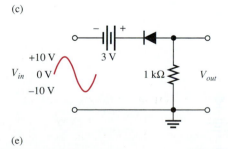

(e)

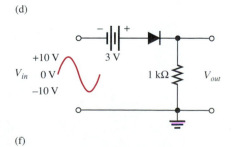

(f)

FIGURE 2–84

22. Determine the R_L voltage waveform for each circuit in Figure 2–85.

FIGURE 2–85

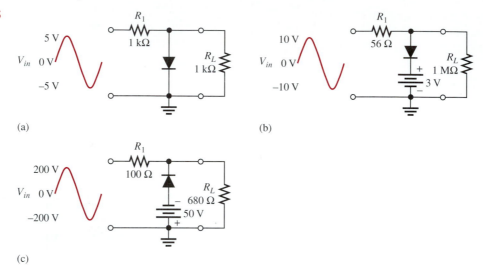

(a)

(b)

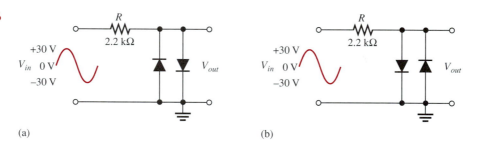

(c)

23. Sketch the output voltage waveform for each circuit in Figure 2–86.

FIGURE 2–86

(a)

(b)

24. Determine the output voltage waveform for each circuit in Figure 2–87.

FIGURE 2–87

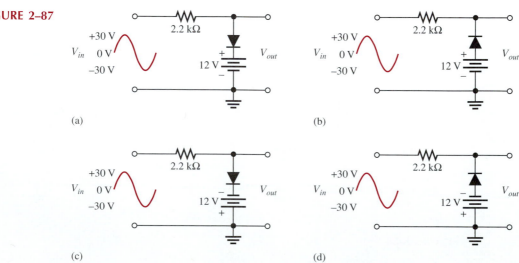

(a)

(b)

(c)

(d)

25. Describe the output waveform of each circuit in Figure 2–88. Assume the *RC* time constant is much greater than the period of the input.

26. Repeat Problem 25 with the diodes turned around.

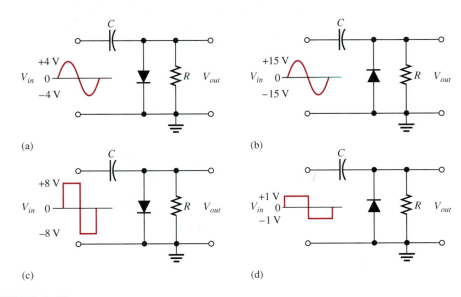

(a)

(b)

(c)

(d)

FIGURE 2–88

SECTION 2–5 Voltage Multipliers

27. A certain voltage doubler has 20 V rms on its input. What is the output voltage? Sketch the circuit, indicating the output terminals and PIV rating for the diode.

28. Repeat Problem 27 for a voltage tripler and quadrupler.

SECTION 2–6 The Diode Data Sheet

29. From the data sheet in Figure 2–60, determine how much peak inverse voltage that a 1N1183A diode can withstand.

30. Repeat Problem 29 for a 1N1188A.

31. If the peak output voltage of a full-wave bridge rectifier is 50 V, determine the minimum value of the surge limiting resistor required when 1N1183A diodes are used.

■ **TROUBLE-SHOOTING PROBLEMS**

SECTION 2–7   Troubleshooting

32. If one of the diodes in a bridge rectifier opens, what happens to the output?

33. From the meter readings in Figure 2–89, determine if the rectifier circuit is functioning properly. If it is not, determine the most likely failure(s).

34. Each part of Figure 2–90 shows oscilloscope displays of rectifier output voltages. In each case, determine whether or not the rectifier is functioning properly and if it is not, the most likely failure(s).

35. Based on the values given, would you expect the circuit in Figure 2–91 to fail? If so, why?

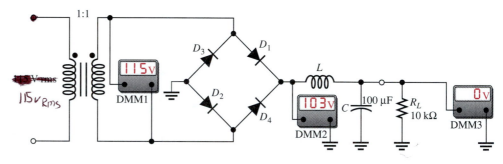
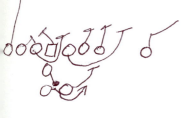
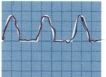
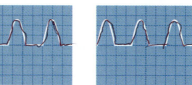

FIGURE 2–89

FIGURE 2–90

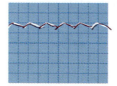
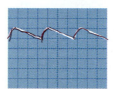

(a) Output of a half-wave unfiltered rectifier

(b) Output of a full-wave unfiltered rectifier

(c) Output of a full-wave filter

(d) Output of same full-wave filter as part (c)

FIGURE 2–91

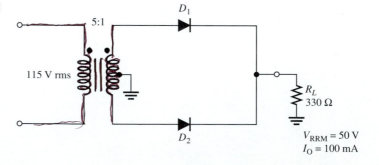

$V_{RRM} = 50$ V
$I_O = 100$ mA

SECTION 2–8   System Application

36. Determine the most likely failure in the circuit board of Figure 2–92 for each of the following symptoms. State the corrective action you would take in each case. The transformer has a rated output of 36 V.
   (a) No voltage measured from test point 1 to test point 2.
   (b) No voltage from test point 3 to test point 4, 110 V rms from test point 1 to test point 2.
   (c) 50 V rms from test point 3 to test point 4. Input is correct at 110 V rms.
   (d) 25 V rms from test point 3 to test point 4. Input is correct at 110 V rms.
   (e) A full-wave rectified voltage with a peak of approximately 50 V at test point 7 with respect to ground.
   (f) Excessive 120 Hz ripple voltage at test point 7.
   (g) The ripple voltage has a frequency of 60 Hz at test point 7.
   (h) No voltage at test point 7.

FIGURE 2–92

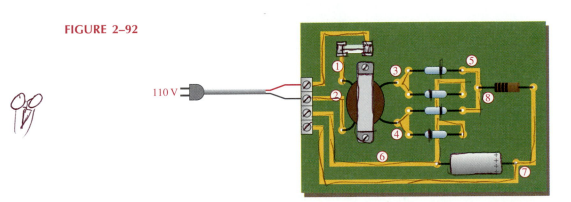

37. In testing the power supply board in Figure 2–92 with a 10 kΩ load resistor connected, you find the voltage at the positive side of the filter capacitor to have a 60 Hz ripple voltage. You replace all the diodes, plug in the board, and check the point again to verify proper operation and it still has the 60 Hz ripple voltage. What now?

38. If the top diode on the circuit board in Figure 2–92 is incorrectly installed backwards, what voltage would you measure at test point 8?

■ **ADVANCED PROBLEMS**

39. A full-wave rectifier with a capacitor filter provides a dc output voltage of 35 V to a 3.3 kΩ load. Determine the minimum value of filter capacitor if the maximum peak-to-peak ripple voltage is to be 0.5 V.

40. A certain unfiltered full-wave rectifier with 115 V, 60 Hz input produces an output with a peak of 15 V. When a capacitor filter and a 1 kΩ load are connected, the dc output voltage is 14 V. What is the peak-to-peak ripple voltage?

41. For a certain full-wave rectifier, the measured surge current in the capacitor filter is 50 A. The transformer is rated for a secondary voltage of 24 V with a 110 V, 60 Hz input. Determine the value of the surge resistor in this circuit.

42. Design a full-wave rectifier using an 18 V center-tapped transformer. The output ripple is not to exceed 5% of the output voltage with a load resistance of 680 Ω. Specify the I_O and PIV ratings of the diodes and select an appropriate diode from Figure 2–60.

43. Design a filtered power supply that can produce dc output voltages of +9 V ± 10% and −9 V ± 10% with a maximum load current of 100 mA. The voltages are to be switch selectable across one set of output terminals. The ripple voltage must not exceed 0.25 V rms.

44. Design a circuit to limit a 20 V rms sinusoidal voltage to a maximum positive amplitude of 18 V and a maximum negative amplitude of 10 V using a single 24 V dc voltage source.

45. Determine the voltage across each capacitor in the circuit of Figure 2–93.

FIGURE 2–93

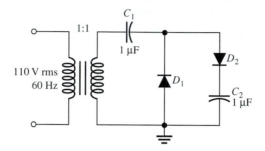

■ ANSWERS TO SECTION REVIEWS

Section 2–1

1. PIV across the diode occurs at the peak of the input.

2. There is current through the load for approximately half (50%) of the input cycle.

3. The average value is $10\ \text{V}/\pi = 3.18\ \text{V}$.

4. The peak output voltage is $25\ \text{V} - 0.7\ \text{V}$.

5. The PIV must be at least 50 V.

Section 2–2

1. A full-wave voltage occurs on each half of the input cycle and has a frequency of twice the input frequency. A half-wave voltage occurs once each input cycle and has a frequency equal to the input frequency.

2. The average value of $2(60\ \text{V})/\pi = 38.12\ \text{V}$.

3. The bridge rectifier has the greater output voltage.

4. The 50 V diodes must be used in the bridge rectifier.

5. In the center-tapped rectifier, diodes with a PIV rating of at least 90 V would be required.

Section 2–3

1. The output frequency is 60 Hz.

2. The output frequency is 120 Hz.

3. The ripple voltage is caused by the slight charging and discharging of the capacitor through the load resistor.

4. The ripple voltage amplitude increases when the load resistance decreases.

5. Ripple factor is the ratio of the ripple voltage to the average or dc voltage.

6. LC ripple voltage is independent of load resistance. LC output voltage equals the average value of the filter input rather than the peak value.

Section 2–4

1. Limiters clip off or remove portions of a waveform. Clampers insert a dc level.

2. A positive limiter clips off positive voltages. A negative limiter clips off negative voltages.

3. 0.7 V appears across the diode.

4. The bias voltage must be $5\ \text{V} - 0.7\ \text{V} = 4.3\ \text{V}$.

5. The capacitor acts as a battery.

Section 2–5

1. The peak voltage rating must be 100 V.

2. The PIV rating must be at least 310 V.

Section 2–6

1. The three rating categories on a diode data sheet are maximum ratings, electrical characteristics, and mechanical data.

2. V_F is forward voltage, I_R is reverse current, and I_O is peak average forward current.

3. I_{FSM} is maximum forward surge current, V_{RRM} is maximum reverse peak repetitive voltage, and V_{RSM} is maximum reverse peak nonrepetitive voltage.

4. The 1N4720 has an $I_O = 3.0$ A, $I_{FSM} = 300$ A, and $V_{RRM} = 100$ V.

Section 2–7

1. An open diode results in no output voltage.

2. An open diode produces a half-wave output voltage.

3. The shorted diode may burn open. Transformer will be damaged. Fuse will blow.

4. The amplitude of the ripple voltage increases with a leaky filter capacitor.

5. There will be no output voltage when the primary opens.

6. The problem may be a partially shorted secondary winding.

■ **ANSWERS TO RELATED EXERCISES FOR EXAMPLES**

2–1 3.82 V

2–2 (a) 2.3 V (b) 209.3 V

2–3 (a) 439.3 V (b) 440 V (c) negative half-cycles rather than positive half cycles

2–4 98.7 V

2–5 79.3 V including diode drop

2–6 41.0 V; 41.7 V

2–7 103 mV

2–8 $V_{DC} = 31.6$ V; $V_r = 0.96$ V

2–9 A positive peak of 8.71 V and clipped at −0.3 V

2–10 Limited at +10.7 V and −10.7 V

2–11 Change R_3 to 1 kΩ or R_2 to 2.2 kΩ.

2–12 Same voltage waveform as Figure 2–53

2–13 Check for a physical short in the circuit or check for shorted diodes in the rectifier.

3

SPECIAL-PURPOSE DIODES

■ CHAPTER OBJECTIVES

□ Describe the characteristics of a zener diode and analyze its operation

□ Explain how a zener is used in voltage regulation and limiting and analyze zener circuits

□ Describe the variable-capacitance characteristics of a varactor diode and analyze its operation in a typical circuit

□ Discuss the operation and characteristics of LEDs and photodiodes

□ Discuss the basic characteristics of the current regulator diode, the Schottky diode, the *pin* diode, the step-recovery diode, the tunnel diode, and the laser diode

□ Troubleshoot zener diode regulators

Chapter 2 was devoted to general-purpose and rectifier diodes, which are the most widely used types. In this chapter, we will cover several other types of diodes that are designed for specific applications, including the zener, varactor (variable-capacitance), light-emitting, photodiode, current regulator, Schottky, tunnel, *pin*, step-recovery, and laser diodes.

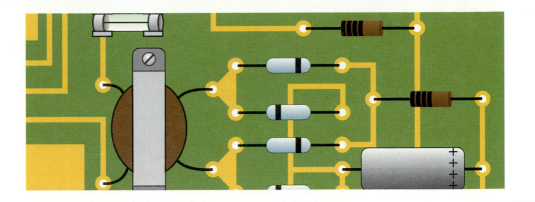

■ SYSTEM APPLICATION

You are given the assignment to analyze and test a new system that your company is manufacturing for counting and controlling items for packaging and shipment. The first system is to be installed in a sporting goods manufacturing plant to control the packaging of baseballs for shipment to consumers. The first step in your assignment is to learn all you can about various special-purpose diodes. You will then apply your knowledge to the system application in Section 3–7.

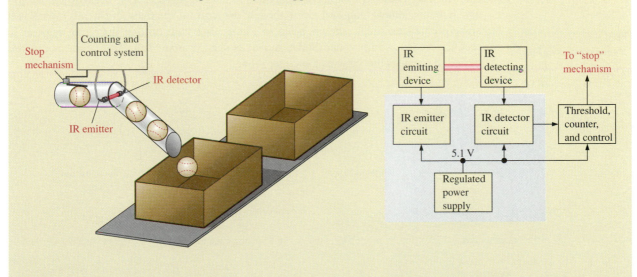

3–1 ■ ZENER DIODES

A major application for zener diodes is voltage regulation in dc power supplies. In this section, you will see how the zener maintains a nearly constant dc voltage under the proper operating conditions. You will learn the conditions and limitations for properly using the zener diode and the factors that affect its performance.

After completing this section, you should be able to

■ **Describe the characteristics of a zener diode and analyze its operation**
  □ Identify a zener diode by its symbol
  □ Discuss avalanche and zener breakdown
  □ Analyze the characteristic (*I-V*) curve of a zener
  □ Discuss the zener equivalent circuit
  □ Define temperature coefficient and apply it to zener analysis
  □ Discuss power dissipation in a zener and apply derating
  □ Interpret a zener data sheet

K

A

FIGURE 3–1
Zener diode symbol.

The symbol for a zener diode is shown in Figure 3–1. The **zener diode** is a silicon *pn* junction device that differs from rectifier diodes because it is designed for operation in the reverse breakdown region. The breakdown voltage of a zener diode is set by carefully controlling the doping level during manufacture. Recall, from the discussion of the diode characteristic curve in Chapter 1, that when a diode reaches reverse breakdown, its voltage remains almost constant even though the current changes drastically. This volt-ampere characteristic is shown again in Figure 3–2 with normal operating regions for rectifier diodes and for zener diodes shown as shaded areas. If a zener diode is forward-biased, it operates the same as a rectifier diode.

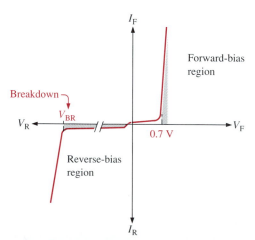

(a) The normal operating regions for a rectifier diode are shown as shaded areas.

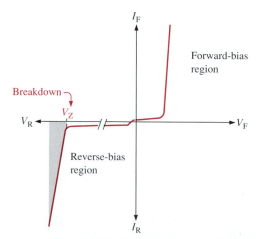

(b) The normal operating region for a zener diode is shaded.

FIGURE 3–2
General diode V-I characteristic.

Zener Breakdown

Zener diodes are designed to operate in reverse breakdown. Two types of reverse breakdown in a zener diode are avalanche and zener. The avalanche breakdown, discussed in Chapter 1, occurs in rectifier and zener diodes at a sufficiently high reverse voltage. **Zener breakdown** occurs in a zener diode at low reverse voltages. A zener diode is heavily doped to reduce the breakdown voltage. This causes a very thin depletion region. As a result, an intense electric field exists within the depletion region. Near the zener breakdown voltage (V_Z), the field is intense enough to pull electrons from their valence bands and create current.

Zener diodes with breakdown voltages of less than approximately 5 V operate predominately in zener breakdown. Those with breakdown voltages greater than approximately 5 V operate predominately in **avalanche breakdown.** Both types, however, are called *zener diodes*. Zeners are commercially available with breakdown voltages of 1.8 V to 200 V with specified tolerances from 1% to 20%.

Breakdown Characteristics

Figure 3–3 shows the reverse portion of a zener diode's characteristic curve. Notice that as the reverse voltage (V_R) is increased, the reverse current (I_R) remains extremely small up to the "knee" of the curve. The reverse current is also called the zener current, I_Z. At this point, the breakdown effect begins; the internal zener resistance, also called zener impedance (Z_Z) begins to decrease as the reverse current increases rapidly. From the bottom of the knee, the zener breakdown voltage (V_Z) remains essentially constant although it increases slightly as the zener current, I_Z, increases.

FIGURE 3–3

Reverse characteristic of a zener diode. V_Z is usually specified at the zener test current, I_{ZT}, and is designated V_{ZT}.

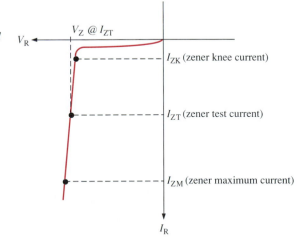

Zener Regulation   The ability to keep the voltage across its terminals essentially constant is the key feature of the zener diode. A zener diode operating in breakdown is a voltage regulator because it maintains a nearly constant voltage across its terminals over a specified range of reverse-current values.

A minimum value of reverse current, I_{ZK}, must be maintained in order to keep the diode in breakdown for voltage regulation. You can see on the curve that when the reverse current is reduced below the knee of the curve, the voltage changes drastically and regulation is lost. Also, there is a maximum current, I_{ZM}, above which the diode may be damaged due to excessive power dissipation. So, basically, the zener diode maintains a nearly constant voltage across its terminals for values of reverse current ranging from I_{ZK} to I_{ZM}. A nominal zener voltage, V_{ZT}, is usually specified on a data sheet at a value of reverse current called the *zener test current, I_{ZT}.*

Zener Equivalent Circuit

Figure 3–4(a) shows the ideal model of a zener diode in reverse breakdown. It has a constant voltage drop equal to the nominal zener voltage. This constant voltage drop is represented by a dc voltage source. The zener diode does not actually produce an emf voltage. The dc source simply indicates that the effect of reverse breakdown is a constant voltage across the zener terminals.

Figure 3–4(b) represents the practical model of a zener diode, where the zener impedance Z_Z is included. Since the voltage curve is not ideally vertical, a change in zener current produces a small change in zener voltage, as illustrated in Figure 3–4(c). The ratio of ΔV_Z to ΔI_Z is the impedance, as expressed in the following equation:

$$Z_Z = \frac{\Delta V_Z}{\Delta I_Z} \tag{3–1}$$

Normally, Z_Z is specified at I_{ZT}, the zener test current, and is designated Z_{ZT}. In most cases, you can assume that Z_Z is constant over the full linear range of zener current values and is purely resistive.

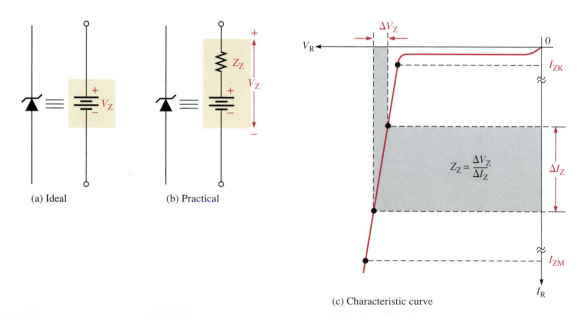

(a) Ideal   (b) Practical

(c) Characteristic curve

FIGURE 3–4

Zener diode equivalent circuit models and the characteristic curve illustrating Z_Z.

EXAMPLE 3–1

A zener diode exhibits a certain change in V_Z and a certain change in I_Z on a portion of the linear characteristic curve between I_{ZK} and I_{ZM} as illustrated in Figure 3–5. What is the zener impedance?

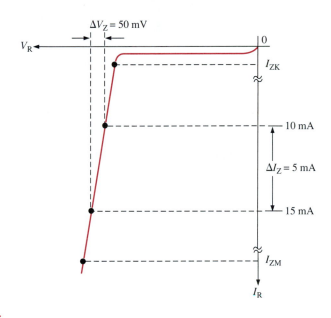

FIGURE 3–5

Solution

$$Z_Z = \frac{\Delta V_Z}{\Delta I_Z} = \frac{50 \text{ mV}}{5 \text{ mA}} = 10 \text{ } \Omega$$

Related Exercise Calculate the zener impedance if the change in zener voltage is 100 mV for a 20 mA change in zener current on the linear portion of the characteristic curve.

EXAMPLE 3–2

A certain zener diode has a Z_{ZT} of 5 Ω. The data sheet gives $V_{ZT} = 6.8$ V at $I_{ZT} = 20$ mA, $I_{ZK} = 1$ mA, and $I_{ZM} = 50$ mA. What is the voltage across the zener terminals when the current is 30 mA? When the current is 10 mA?

Solution Figure 3–6 represents the zener diode.

For $I_Z = 30$ mA: The 30 mA current is a 10 mA increase above $I_{ZT} = 20$ mA.

$$\Delta I_Z = I_Z - I_{ZT} = +10 \text{ mA}$$
$$\Delta V_Z = \Delta I_Z Z_{ZT} = (10 \text{ mA})(5 \text{ } \Omega) = +50 \text{ mV}$$

The change in voltage due to the increase in current above the I_{ZT} value causes the zener terminal voltage to increase. The zener voltage for $I_Z = 30$ mA is

$$V_Z = 6.8 \text{ V} + \Delta V_Z = 6.8 \text{ V} + 50 \text{ mV} = 6.85 \text{ V}$$

FIGURE 3–6

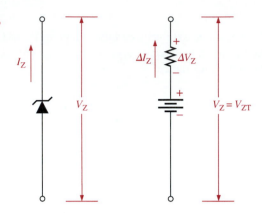

For $I_Z = 10$ mA: The 10 mA current is a 10 mA decrease below $I_{ZT} = 20$ mA.

$$\Delta I_Z = -10 \text{ mA}$$

$$\Delta V_Z = \Delta I_Z Z_{ZT} = (-10 \text{ mA})(5 \text{ }\Omega) = -50 \text{ mV}$$

The change in voltage due to the decrease in current below I_{ZT} causes the zener terminal voltage to decrease. The zener voltage for $I_Z = 10$ mA is

$$V_Z = 6.8 \text{ V} - \Delta V_Z = 6.8 \text{ V} - 50 \text{ mV} = 6.75 \text{ V}$$

Related Exercise Repeat the analysis for a zener with $V_{ZT} = 12$ V at $I_{ZT} = 50$ mA, $I_{ZK} = 0.5$ mA, $I_{ZM} = 100$ mA, and $Z_{ZT} = 20$ Ω for a current of 20 mA and for a current of 80 mA.

Temperature Coefficient

The temperature coefficient specifies the percent change in zener voltage for each C° change in temperature. For example, a 12 V zener diode with a positive temperature coefficient of 0.01%/C° will exhibit a 1.2 mV increase in V_Z when the junction temperature increases one Celsius degree. The formula for calculating the change in zener voltage for a given junction temperature change, for a specified temperature coefficient, is as follows:

$$\Delta V_Z = V_Z \times TC \times \Delta T \qquad (3\text{--}2)$$

where V_Z is the nominal zener voltage at 25°C, TC is the temperature coefficient, and ΔT is the change in temperature. A positive TC means that the zener voltage increases with an increase in temperature or decreases with a decrease in temperature. A negative TC means that the zener voltage decreases with an increase in temperature or increases with a decrease in temperature.

In some cases, the temperature coefficient is expressed in mV/C° rather than as %/C°. For these cases, ΔV_Z is calculated as

$$\Delta V_Z = TC \times \Delta T \qquad (3\text{--}3)$$

EXAMPLE 3–3   An 8.2 V zener diode (8.2 V at 25°C) has a positive temperature coefficient of 0.05%/C°. What is the zener voltage at 60°C?

Solution   The change in zener voltage is

$$\Delta V_Z = V_Z \times TC \times \Delta T = (8.2 \text{ V})(0.05\%/\text{C}°)(60°\text{C} - 25°\text{C})$$
$$= (8.2 \text{ V})(0.0005/\text{C}°)(35 \text{ C}°) = 144 \text{ mV}$$

The zener voltage at 60°C is

$$V_Z + \Delta V_Z = 8.2 \text{ V} + 144 \text{ mV} = 8.34 \text{ V}$$

Related Exercise   A 12 V zener has a positive temperature coefficient of 0.075%/C°. How much will the zener voltage change when the junction temperature decreases 50 C°?

Zener Power Dissipation and Derating

Zener diodes are specified to operate at a maximum power called the maximum dc power dissipation, $P_{\text{D(max)}}$. For example, the 1N746 zener is rated at a maximum P_D of 500 mW and the 1N3305A is rated at a maximum P_D of 50 W. The dc power dissipation is determined by the formula,

$$P_D = V_Z I_Z \tag{3–4}$$

Power Derating   The maximum power dissipation of a zener diode is typically specified for temperatures at or below a certain value (50°C, for example). Above the specified temperature, the maximum power dissipation is reduced according to a derating factor. The derating factor is expressed in mW/C°. The maximum derated power can be determined with the following formula:

$$P_{\text{D(derated)}} = P_{\text{D(max)}} - (\text{mW/C}°)\Delta T \tag{3–5}$$

EXAMPLE 3–4   A certain zener diode has a maximum power rating of 400 mW at 50°C and a derating factor of 3.2 mW/C°. Determine the maximum power the zener can dissipate at a temperature of 90°C.

Solution
$$P_{\text{D(derated)}} = P_{\text{D(max)}} - (\text{mW/C}°)\Delta T$$
$$= 400 \text{ mW} - (3.2 \text{ mW/C}°)(90°\text{C} - 50°\text{C})$$
$$= 400 \text{ mW} - 128 \text{ mW} = 272 \text{ mW}$$

Related Exercise   A certain 50 W zener diode must be derated with a derating factor of 0.5 W/C° above 75°C. Determine the maximum power it can dissipate at 160°C.

Zener Diode Data Sheet Information

The amount and type of information found on data sheets for zener diodes (or any category of electronic device) varies from one type of diode to the next. The data sheet for some zeners contains more information than for others. Figure 3–7 gives an example of the type of information that you have studied that can be found on a typical data sheet but does not represent the complete data sheet. This particular information is for a popular zener series, the 1N4728–1N4764.

Electrical Characteristics   The electrical characteristics are listed in a tabular form in Figure 3–7(a) with the zener type numbers in the first column. This feature is common to most device data sheets.

   Zener voltage   For each zener type number, the nominal zener voltage, V_Z, for a specified value of zener test current, I_{ZT}, is listed in the second column. The nominal value of V_Z can vary depending on the tolerance. For example, the 1N4738 has a nominal V_Z of 8.2 V. For 10% tolerance, this value can range from 7.38 V to 9.02 V.

   Zener test current   The value of zener current, I_{ZT}, in mA at which the nominal zener voltage is specified is listed in the third column of the table in Figure 3–7(a).

   Maximum zener impedance Z_{ZT} is the value of dynamic impedance in ohms measured at the test current. The values of Z_{ZT} for each zener type are listed in the fourth column. The term *dynamic* means that it is measured as an ac quantity; that is, the *change* in voltage for a specified *change* in current ($Z_{ZT} = \Delta V_Z / \Delta I_Z$). You cannot get Z_{ZT} using V_Z and I_{ZT}, which are dc values. The table also includes Z_{ZK}, which is the impedance measured at the zener knee current, I_{ZK}.

   Reverse leakage current   The values of leakage current are listed in the fifth column of the table. The leakage current is the current through the reverse-biased zener diode for values of reverse voltage less than the value at the knee of the characteristic curve. Notice that the values are extremely small as was the case for rectifier diodes.

   Maximum zener current   The maximum dc current, I_{ZM}, is not specified on this particular data sheet. However, it is worth mentioning because you will find it on some data sheets. The value of I_{ZM} is specified based on the power rating, the zener voltage at I_{ZM}, and the zener voltage tolerance. An approximate value for I_{ZM} can be calculated using the maximum power dissipation, $P_{D(max)}$ and V_Z at I_{ZT} as follows:

$$I_{ZM} = \frac{P_{D(max)}}{V_Z} \tag{3–6}$$

Graphical Data   Some data sheets provide various types of data in the form of graphs while others do not. Figure 3–7 includes graphs for data related to concepts covered in this section.

   Power derating   Figure 3–7(b) shows a power derating curve for this particular series of zener diodes. Notice that the zeners are rated for a maximum power dissipation of 1 W for temperatures of 50°C and below. Above 50°C the power rating decreases linearly as shown. For example at 140°C, the power rating is approximately 400 mW.

Maximum Ratings

Rating	Symbol	Value	Unit
DC power dissipation @ $T_A = 50°C$	P_D	1.0	Watt
Derate above 50°C		6.67	mW/C°
Operating and storage junction Temperature range	T_J, T_{stg}	−65 to +200	°C

Electrical Characteristics ($T_A = 25°C$ unless otherwise noted) $V_F = 1.2$ V max,
$I_F = 200$ mA for all types.

JEDEC Type No. (Note 1)	Nominal Zener Voltage V_Z @ I_{ZT} Volts	Test Current I_{ZT} mA	Maximum Zener Impedance			Leakage Current	
			Z_{ZT} @ I_{ZT} Ohms	Z_{ZK} @ I_{ZK} Ohms	I_{ZK} mA	I_R µA Max	V_R Volts
1N4728	3.3	76	10	400	1.0	100	1.0
1N4729	3.6	69	10	400	1.0	100	1.0
1N4730	3.9	64	9.0	400	1.0	50	1.0
1N4731	4.3	58	9.0	400	1.0	10	1.0
1N4732	4.7	53	8.0	500	1.0	10	1.0
1N4733	5.1	49	7.0	550	1.0	10	1.0
1N4734	5.6	45	5.0	600	1.0	10	2.0
1N4735	6.2	41	2.0	700	1.0	10	3.0
1N4736	6.8	37	3.5	700	1.0	10	4.0
1N4737	7.5	34	4.0	700	0.5	10	5.0
1N4738	8.2	31	4.5	700	0.5	10	6.0
1N4739	9.1	28	5.0	700	0.5	10	7.0
1N4740	10	25	7.0	700	0.25	10	7.6
1N4741	11	23	8.0	700	0.25	5.0	8.4
1N4742	12	21	9.0	700	0.25	5.0	9.1
1N4743	13	19	10	700	0.25	5.0	9.9
1N4744	15	17	14	700	0.25	5.0	11.4
1N4745	16	15.5	16	700	0.25	5.0	12.2
1N4746	18	14	20	750	0.25	5.0	13.7
1N4747	20	12.5	22	750	0.25	5.0	15.2
1N4748	22	11.5	23	750	0.25	5.0	16.7
1N4749	24	10.5	25	750	0.25	5.0	18.2
1N4750	27	9.5	35	750	0.25	5.0	20.6
1N4751	30	8.5	40	1000	0.25	5.0	22.8
1N4752	33	7.5	45	1000	0.25	5.0	25.1
1N4753	36	7.0	50	1000	0.25	5.0	27.4
1N4754	39	6.5	60	1000	0.25	5.0	29.7
1N4755	43	6.0	70	1500	0.25	5.0	32.7
1N4756	47	5.5	80	1500	0.25	5.0	35.8
1N4757	51	5.0	95	1500	0.25	5.0	38.8
1N4758	56	4.5	110	2000	0.25	5.0	42.6
1N4759	62	4.0	125	2000	0.25	5.0	47.1
1N4760	68	3.7	150	2000	0.25	5.0	51.7
1N4761	75	3.3	175	2000	0.25	5.0	56.0
1N4762	82	3.0	200	3000	0.25	5.0	62.2
1N4763	91	2.8	250	3000	0.25	5.0	69.2
1N4764	100	2.5	350	3000	0.25	5.0	76.0

NOTE 1 — Tolerance and Type Number Designation. The JEDEC type numbers listed have a standard tolerance on the nominal zener voltage of ±10%. A standard tolerance of ±5% on individual units is also available and is indicated by suffixing "A" to the standard type number. C for ±2.0%, D for ±1.0%.

(a) Electrical characteristics

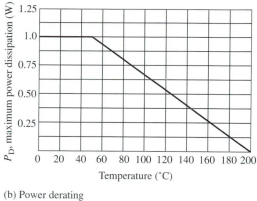

(b) Power derating

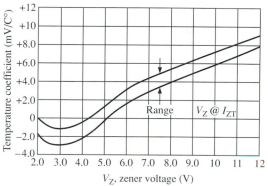

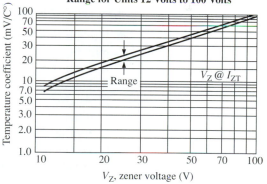

(c) Temperature coefficients

FIGURE 3–7

Partial data sheet for the 1N4728–1N4764 series 1 W zener diodes.

Temperature coefficients Figure 3–7(c) consists of two graphs showing the temperature coefficient in mW/C° versus zener voltage. The top graph is for zener voltages up to 12 V and the bottom graph is for zener voltages from 12 V to 100 V. The two curves on each graph define a range for the temperature coefficient. For example, a 6 V zener diode exhibits a temperature coefficient that can range from about 1.5 mV/C° to about 3 mV/C°.

SECTION 3–1 REVIEW	1. In what region of their characteristic curve are zener diodes operated? 2. At what value of zener current is the zener voltage normally specified? 3. How does the zener impedance affect the voltage across the terminals of the device? 4. For a certain zener diode, $V_Z = 10$ V at $I_{ZT} = 30$ mA. If $Z_Z = 8\ \Omega$, what is the terminal voltage at $I_Z = 50$ mA? 5. What does a positive temperature coefficient of 0.05%/C° mean? 6. Explain power derating.

3–2 ■ ZENER DIODE APPLICATIONS

The zener diode is often used as a voltage regulator in dc power supplies. In this section, the concept of voltage regulation is introduced and two types of regulation are examined—line regulation and load regulation. Also, you will see how zeners can be used as simple limiters or clippers.

After completing this section, you should be able to

■ **Explain how a zener is used in voltage regulation and limiting and analyze zener circuits**
  □ Discuss input or line regulation
  □ Discuss load regulation
  □ Analyze zener diode regulators under both varying input and varying load conditions
  □ Define *percent regulation* and explain its importance as a figure of merit in voltage regulation
  □ Analyze zener waveform-limiting circuits

Zener Regulation with a Varying Input Voltage

Zener diodes are widely used for voltage regulation. Figure 3–8 illustrates how a zener diode can be used to regulate a varying dc voltage. This is called *input* or **line regulation.** As the input voltage varies (within limits), the zener diode maintains a nearly constant output voltage across its terminals. However, as V_{IN} changes, I_Z will change proportionally so that the limitations on the input voltage variation are set by the minimum and maximum current values (I_{ZK} and I_{ZM}) with which the zener can operate. R is the series current-limiting resistor. The DMM displays indicate the relative values and trends.

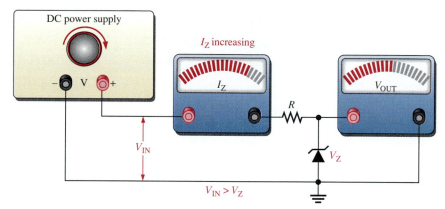

(a) As the input voltage increases, the output voltage remains constant ($I_{ZK} < I_Z < I_{ZM}$).

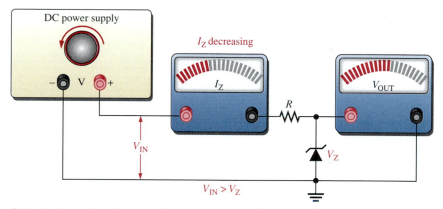

(b) As the input voltage decreases, the output voltage remains constant ($I_{ZK} < I_Z < I_{ZM}$).

FIGURE 3–8

Zener regulation of a varying input voltage.

For example, suppose that the 1N4740 10 V zener diode in Figure 3–9 can maintain regulation over a range of zener current values from $I_{ZK} = 0.25$ mA to $I_{ZM} = 100$ mA ($I_{ZM} = P_{D(max)}/V_Z = 1$ W/10 V = 100 mA). For the minimum current, the voltage across the 220 Ω resistor is

$$V_R = I_{ZK}R = (0.25 \text{ mA})(220 \text{ } \Omega) = 55 \text{ mV}$$

FIGURE 3–9

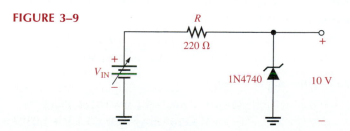

Since $V_R = V_{IN} - V_Z$,

$$V_{IN} = V_R + V_Z = 55 \text{ mV} + 10 \text{ V} = 10.055 \text{ V}$$

For the maximum current, the voltage across the 220 Ω resistor is

$$V_R = I_{ZM}R = (100 \text{ mA})(220 \text{ Ω}) = 22 \text{ V}$$

Therefore,

$$V_{IN} = 22 \text{ V} + 10 \text{ V} = 32 \text{ V}$$

This shows that this zener diode can regulate an input voltage from 10.055 V to 32 V and maintain an approximate 10 V output. (The output will vary slightly because of the zener impedance.)

EXAMPLE 3–5

Determine the minimum and the maximum input voltages that can be regulated by the zener diode in Figure 3–10.

FIGURE 3–10

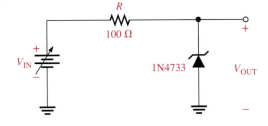

Solution From the data sheet in Figure 3–7, the following information for the 1N4733 is obtained: $V_Z = 5.1$ V at $I_{ZT} = 49$ mA, $I_{ZK} = 1$ mA, and $Z_Z = 7$ Ω at I_{ZT}. The equivalent circuit is shown in Figure 3–11. At $I_{ZK} = 1$ mA, the output voltage is

$$V_{OUT} = 5.1 \text{ V} - \Delta V_Z = 5.1 \text{ V} - (I_{ZT} - I_{ZK})Z_Z$$
$$= 5.1 \text{ V} - (48 \text{ mA})(7 \text{ Ω}) = 5.1 \text{ V} - 0.336 \text{ V} = 4.76 \text{ V}$$

Therefore,

$$V_{IN(min)} = I_{ZK}R + V_{OUT} = (1 \text{ mA})(100 \text{ Ω}) + 4.76 \text{ V} = 4.86 \text{ V}$$

To find the maximum input voltage, first calculate the maximum zener current. Assume the temperature is 50°C or below, so from the graph in Figure 3–7(b), the power dissipation is 1 W.

$$I_{ZM} = \frac{P_{D(max)}}{V_Z} = \frac{1 \text{ W}}{5.1 \text{ V}} = 196 \text{ mA}$$

At I_{ZM}, the output voltage is

$$V_{OUT} = 5.1 \text{ V} + \Delta V_Z = 5.1 \text{ V} + (I_{ZM} - I_{ZT})Z_Z$$
$$= 5.1 \text{ V} + (147 \text{ mA})(7 \text{ Ω}) = 5.1 \text{ V} + 1.03 \text{ V} = 6.13 \text{ V}$$

Therefore,

$$V_{\text{IN(max)}} = I_{\text{ZM}}R + V_{\text{OUT}} = (196 \text{ mA})(100 \ \Omega) + 6.13 \text{ V} = 25.7 \text{ V}$$

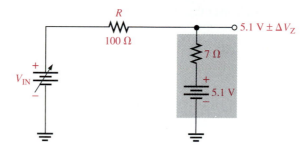

FIGURE 3–11
Equivalent of circuit in Figure 3–10.

Related Exercise   Determine the minimum and maximum input voltages that can be regulated by a 1N4736 zener diode used in Figure 3–10.

Zener Regulation with a Varying Load

Figure 3–12 shows a zener voltage regulator with a variable load resistor across the terminals. The zener diode maintains a nearly constant voltage across R_L as long as the zener current is greater than I_{ZK} and less than I_{ZM}. This is called **load regulation.**

FIGURE 3–12
Zener regulation with a variable load.

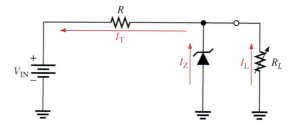

From No Load to Full Load

When the output terminals of the zener regulator are open ($R_L = \infty$), the load current is zero and *all* of the current is through the zener. When a load resistor (R_L) is connected, part of the total current is through the zener and part through R_L. As R_L is decreased, the load current, I_L increases and I_Z decreases. The zener diode continues to regulate until I_Z reaches its minimum value, I_{ZK}. At this point the load current is maximum. The total current through R remains essentially constant. The following example will illustrate this.

EXAMPLE 3–6   Determine the minimum and the maximum load currents for which the zener diode in Figure 3–13 will maintain regulation. What is the minimum R_L that can be used? $V_Z = 12$ V, $I_{ZK} = 1$ mA, and $I_{ZM} = 50$ mA. Assume $Z_Z = 0$ Ω and V_Z remains a constant 12 V over the range of current values, for simplicity.

FIGURE 3–13

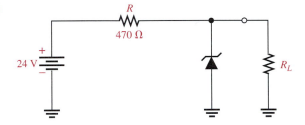

Solution   When $I_L = 0$ A ($R_L = \infty$), I_Z is maximum and equal to the total circuit current I_T.

$$I_{Z(max)} = I_T = \frac{V_{IN} - V_Z}{R} = \frac{24\ V - 12\ V}{470\ \Omega} = 25.5\ mA$$

Since $I_{Z(max)}$ is less than I_{ZM}, 0 A is an acceptable minimum value for I_L because the zener can handle all of the 25.5 mA.

$$I_{L(min)} = 0\ A$$

The maximum value of I_L occurs when I_Z is minimum ($I_Z = I_{ZK}$), so solve for $I_{L(max)}$ as follows:

$$I_{L(max)} = I_T - I_{ZK} = 25.5\ mA - 1\ mA = 24.5\ mA$$

The minimum value of R_L is

$$R_{L(min)} = \frac{V_Z}{I_{L(max)}} = \frac{12\ V}{24.5\ mA} = 490\ \Omega$$

Therefore, if R_L is less than 490 Ω, R_L will draw more current away from the zener and I_Z will be reduced below I_{ZK}. This will cause the zener to lose regulation.

Related Exercise   Find the minimum and maximum load currents for which the circuit in Figure 3–13 will maintain regulation. Determine the minimum R_L that can be used. $V_Z = 3.3$ V (constant), $I_{ZK} = 1$ mA, $I_{ZM} = 150$ mA. Assume $Z_Z = 0$ Ω for simplicity.

   In the last example, we made the assumption that Z_Z was zero and, therefore, the zener voltage remained constant over the range of currents. This assumption was made in order to more easily demonstrate the concept of how the regulator works with a varying load. Such a simplifying assumption is often acceptable and in some cases produces results that are accurate enough. In Example 3–7, we will take into account the zener impedance.

EXAMPLE 3–7   For the circuit in Figure 3–14:
(a) Determine V_{OUT} at I_{ZK} and at I_{ZM}.
(b) Calculate the value of R that should be used.
(c) Determine the minimum value of R_L that can be used.

FIGURE 3–14

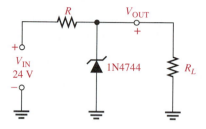

Solution   The 1N4744 zener used in the regulator circuit of Figure 3–14 is a 15 V diode. The data sheet in Figure 3–7(a) gives the following information: $V_Z = 15$ V @ I_{ZT}, $I_{ZK} = 0.25$ mA, $I_{ZT} = 17$ mA, and $Z_{ZT} = 14\ \Omega$.
(a) For I_{ZK}:

$$V_{OUT} = V_Z = 15\ \text{V} - \Delta I_Z Z_{ZT} = 15\ \text{V} - (I_{ZT} - I_{ZK})\,Z_{ZT}$$
$$= 15\ \text{V} - (16.75\ \text{mA})(14\ \Omega) = 15\ \text{V} - 0.235\ \text{V} = 14.76\ \text{V}$$

Calculate the zener maximum current. The power dissipation is 1 W.

$$I_{ZM} = \frac{P_{D(max)}}{V_Z} = \frac{1\ \text{W}}{15\ \text{V}} = 66.7\ \text{mA}$$

For I_{ZM}:

$$V_{OUT} = V_Z = 15\ \text{V} + \Delta I_Z Z_{ZT}$$
$$= 15\ \text{V} + (I_{ZM} - I_{ZT})Z_{ZT} = 15\ \text{V} + (49.7\ \text{mA})(14\ \Omega) = 15.7\ \text{V}$$

(b) The value of R is calculated for the maximum zener current that occurs when there is no load as shown in Figure 3–15(a).

$$R = \frac{V_{IN} - V_Z}{I_{ZM}} = \frac{24\ \text{V} - 15.7\ \text{V}}{66.7\ \text{mA}} = 124\ \Omega$$

$R = 130\ \Omega$ (nearest larger standard value).

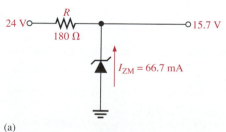

(a)

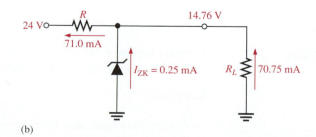

(b)

FIGURE 3–15

(c) For the minimum load resistance (maximum load current), the zener current is minimum ($I_{ZK} = 0.25$ mA) as shown in Figure 3–15(b).

$$I_T = \frac{V_{IN} - V_{OUT}}{R} = \frac{24\ V - 14.76\ V}{130\ \Omega} = 71.0\ \text{mA}$$

$$I_L = I_T - I_{ZK} = 71.0\ \text{mA} - 0.25\ \text{mA} = 70.75\ \text{mA}$$

$$R_{L(\text{min})} = \frac{V_{OUT}}{I_L} = \frac{14.76\ V}{70.75\ \text{mA}} = 209\ \Omega$$

Related Exercise   Repeat each part of the preceding analysis if the zener is changed to a 12 V device (1N4742).

Percent Regulation

The **percent regulation** is a figure of merit used to specify the performance of a voltage regulator. It can be in terms of input (line) regulation or load regulation. The *percent line regulation* specifies the change in the output voltage (ΔV_{OUT}) for a given change in input voltage (ΔV_{IN}).

$$\text{Percent line regulation} = \left(\frac{\Delta V_{OUT}}{\Delta V_{IN}} \right) 100\% \qquad (3\text{–}7)$$

Line regulation is usually expressed as the percent change in V_{OUT} for a one volt change in V_{IN} (%/V).

The *percent load regulation* specifies the change in the output voltage over a certain range of load current values, usually from minimum current (no load) to maximum current (full load). It can be calculated with the following formula:

$$\text{Percent load regulation} = \left(\frac{V_{NL} - V_{FL}}{V_{FL}} \right) 100\% \qquad (3\text{–}8)$$

where V_{NL} is the output voltage with no load, and V_{FL} is the output voltage with full (maximum current) load.

EXAMPLE 3–8   A certain regulator has a no-load output voltage of 6 V and has a full-load output of 5.82 V. What is the percent load regulation?

Solution

$$\text{Percent load regulation} = \left(\frac{V_{NL} - V_{FL}}{V_{FL}} \right) 100\% = \left(\frac{6\ V - 5.82\ V}{5.82\ V} \right) 100\% = 3.09\%$$

Related Exercise

(a) If the no-load output voltage of a regulator is 24.8 V and the full-load output is 23.9 V, what is the percent load regulation?

(b) If the output voltage changes 0.035 V for a 2 V change in the input voltage, what is the line regulation expressed as %/V?

Zener Limiting

In addition to voltage regulation applications, zener diodes can be used in ac applications to limit voltage swings to desired levels. Figure 3–16 shows three basic ways the limiting action of a zener diode can be used. Part (a) shows a zener used to limit the positive peak of a signal voltage to the selected zener voltage. During the negative alternation, the zener acts as a forward-biased diode and limits the negative voltage to −0.7 V. When the zener is turned around, as in part (b), the negative peak is limited by zener action and the positive voltage is limited to +0.7 V. Two back-to-back zeners limit both peaks to the zener voltage plus 0.7 V, as shown in part (c). During the positive alternation, D_2 is functioning as the zener limiter and D_1 is functioning as a forward-biased diode. During the negative alternation, the roles are reversed.

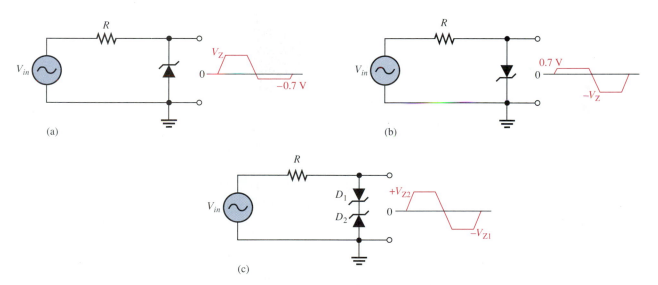

FIGURE 3–16

Basic zener limiting action with a sinusoidal input voltage.

EXAMPLE 3–9   Determine the output voltage for each zener limiting circuit in Figure 3–17.

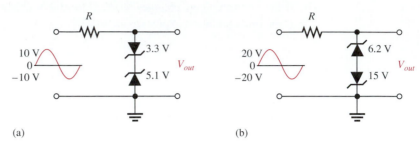

(a)   (b)

FIGURE 3–17

Solution   See Figure 3–18 for the resulting output voltages. Remember, when one zener is operating in breakdown, the other one is forward-biased with approximately 0.7 V across it.

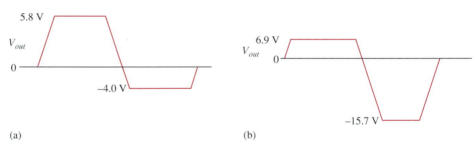

(a)   (b)

FIGURE 3–18

Related Exercise
(a) What is the output in Figure 3–17(a) if the input voltage is increased to a peak value of 20 V?
(b) What is the output in Figure 3–17(b) if the input voltage is decreased to a peak value of 5 V?

SECTION 3–2 REVIEW

1. Explain the difference between line regulation and load regulation.
2. In a zener diode regulator, what value of load resistance results in the maximum zener current?
3. Define the terms *no-load* and *full-load*.
4. A regulator has an output voltage of 12 V with no load and 11.9 V with full load. What is the percent load regulation?
5. How much voltage appears across a zener diode when it is forward-biased?

3–3 ■ VARACTOR DIODES

Varactor diodes are also known as voltage-variable-capacitance diodes because the junction capacitance varies with the amount of reverse-bias voltage. Varactors are specifically designed to take advantage of this variable-capacitance characteristic. The capacitance can be changed by changing the reverse voltage. These devices are commonly used in electronic tuning circuits used in communications systems.

After completing this section, you should be able to

■ **Describe the variable-capacitance characteristics of a varactor diode and analyze its operation in a typical circuit**
- ☐ Identify a varactor diode symbol
- ☐ Explain why a reverse-biased varactor exhibits capacitance
- ☐ Discuss how the capacitance varies with reverse-bias voltage
- ☐ Interpret a varactor data sheet
- ☐ Define *tuning ratio*
- ☐ Define *figure-of-merit*
- ☐ Discuss varactor temperature coefficients
- ☐ Analyze a varactor-tuned band-pass filter

A **varactor** is a *pn* junction diode that operates in reverse-bias and is doped to maximize the inherent capacitance of the depletion region. The depletion region, widened by the reverse bias, acts as a capacitor dielectric because of its nonconductive characteristic. The *p* and *n* regions are conductive and act as the capacitor plates, as illustrated in Figure 3–19.

FIGURE 3–19

The reverse-biased varactor diode acts as a variable capacitor.

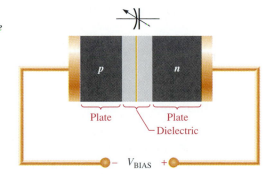

Basic Operation

As the reverse-bias voltage increases, the depletion region widens, effectively increasing the plate separation and the dielectric thickness and thus decreasing the capacitance. When the reverse-bias voltage decreases, the depletion region narrows, thus increasing the capacitance. This action is shown in Figure 3–20(a) and (b). A graph of diode capacitance (C_T) versus reverse voltage for an certain varactor is shown in Figure 3–20(c). For this particular device, C_T varies from 40 pF to slightly greater than 4 pF as V_R varies from 1 V to 40 V.

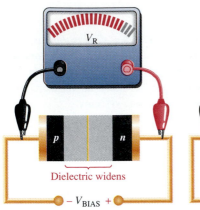

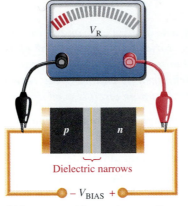

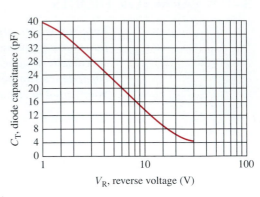

(a) Greater reverse bias, less capacitance   (b) Less reverse bias, greater capacitance   (c) Graph of diode capacitance versus reverse voltage

FIGURE 3–20
Varactor diode capacitance varies with reverse voltage.

Recall that capacitance is determined by the plate area (A), dielectric constant (ϵ), and dielectric thickness (d), as expressed in the following formula:

FIGURE 3–21
Varactor diode symbol.

$$C = \frac{A\epsilon}{d} \qquad (3\text{–}9)$$

In a varactor diode, the capacitance parameters are controlled by the method of doping near the *pn* junction and the size and geometry of the diode's construction. Nominal varactor capacitances are typically available from a few picofarads to several hundred picofarads. Figure 3–21(a) shows a common symbol for a varactor.

Varactor Data Sheet Information

A partial data sheet for a specific series of varactor diodes (1N5139–1N5148) is shown in Figure 3–22. The values of nominal diode capacitance, C_T, are measured at a reverse voltage of 4 V dc and range from 6.8 pF to 47 pF for this particular series.

Capacitance Tolerance Range   The minimum and maximum values of C_T are based on 10% tolerance. For example, this means that when reverse-biased at 4 V, the 1N5139 can exhibit a capacitance anywhere between 6.1 pF and 7.5 pF. This tolerance range should not be confused with the range of capacitance values that result from varying the reverse bias as determined by the tuning ratio, which we will discuss next.

Tuning Ratio   The varactor **tuning ratio** is also called the *capacitance ratio*. It is the ratio of the diode capacitance at a minimum reverse voltage to the diode capacitance at a maximum reverse voltage. For the varactor diodes represented in Figure 3–22, the tuning ratio is the ratio of C_T measured at a V_R of 4 V divided by C_T measured at a V_R of 60 V. The tuning ratio is designated as C_4/C_{60} in this case.

For the 1N5139, the typical tuning ratio is 2.9. This means that the capacitance value decreases by a factor of 2.9 as V_R is increased from 4 V to 60 V. The following calculation illustrates how to use the tuning ratio (*TR*) to find the capacitance range for

Maximum Ratings (T_C = 25°C unless otherwise noted)

Rating	Symbol	Value	Unit
Reverse voltage	V_R	60	Volts
Forward current	I_F	250	mA
RF power input*	P_{in}	5.0	Watts
Device dissipation @ T_A = 25°C Derate above 25°C	P_D	400 2.67	mW mW/C°
Device dissipation @ T_C = 25°C Derate above 25°C	P_C	2.0 13.3	Watts mW/C°
Junction temperature	T_J	+175	°C
Storage temperature range	T_{stg}	−65 to +200	°C

*The RF power input rating assumes that an adequate heatsink is provided.

Electrical Characteristics (T_A = 25°C unless otherwise noted)

Characteristic	Symbol	Min	Typ	Max	Unit
Reverse breakdown voltage (I_R = 10 μA dc)	$V_{(BR)R}$	60	70	–	V dc
Reverse voltage leakage current (V_R = 55 V dc, T_A = 25°C) (V_R = 55 V dc, T_A = 150°C)	I_R	– –	– –	0.02 20	μA dc
Series inductance (f = 250 MHz, $L \approx$ 1/16″)	L_S	–	5.0	–	nH
Case capacitance (f = 1.0 MHz, $L \approx$ 1/16″)	C_C	–	0.25	–	pF
Diode capacitance temperature coefficient (V_R = 4.0 V dc, f = 1.0 MHz)	TC_C	–	200	300	ppm/C°

Device	C_T, Diode Capacitance V_R = 4.0 V dc, f = 1.0 MHz pF			Q, Figure of Merit V_R = 4.0 V dc f = 50 MHz	TR, Tuning Ratio C_4/C_{60} f = 1.0 MHz	
	Min	Typ	Max	Min	Min	Typ
1N5139	6.1	6.8	7.5	350	2.7	2.9
1N5140	9.0	10	11	300	2.8	3.0
1N5141	10.8	12	13.2	300	2.8	3.0
1N5142	13.5	15	16.5	250	2.8	3.0
1N5143	16.2	18	19.8	250	2.8	3.0
1N5144	19.8	22	24.2	200	3.2	3.4
1N5145	24.3	27	29.7	200	3.2	3.4
1N5146	29.7	33	36.3	200	3.2	3.4
1N5147	36.1	39	42.9	200	3.2	3.4
1N5148	42.3	47	51.7	200	3.2	3.4

(a) Electrical characteristics

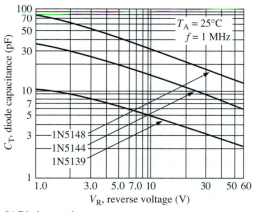

(b) Diode capacitance

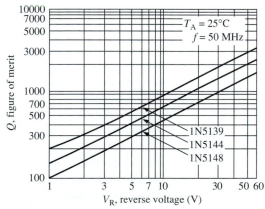

(c) Figure of merit

FIGURE 3–22
Partial data sheet for the 1N5139–1N5148 varactor diodes.

the 1N5139. From the data table in Figure 3–22(a), $C_4 = 6.8$ pF, and the typical $TR = C_4/C_{60} = 2.9$. Therefore,

$$C_{60} = \frac{C_4}{TR} = \frac{6.8 \text{ pF}}{2.9} = 2.3 \text{ pF}$$

The diode capacitance varies from 6.8 pF to 2.3 pF when V_R is increased from 4 V to 60 V.

The capacitance range can also be determined from the graph in Figure 3–22(b), which shows how the varactor capacitance varies for reverse voltages from 1 V to 60 V. On the graph, you can see that the capacitance for the 1N5139 is approximately 10.5 pF at $V_R = 1$ V and approximately 2.3 pF at $V_R = 60$ V.

The 1N51xx series of varactors are abrupt junction devices. The doping in the n and p regions is made uniform so that at the pn junction there is a relatively abrupt change from n to p instead of the more gradual change found in the rectifier diodes. The abruptness of the pn junction determines the tuning ratio. Other types of varactors such as the MV1401 are hyper-abrupt devices in which the doping pattern results in an even more abrupt junction. Many hyper-abrupt varactors exhibit tuning ratios from 10 to 15.

Figure of Merit The **figure of merit** or quality factor (Q) of a reactive component is the ratio of energy stored by the capacitor (or inductor) and returned to the energy dissipated in the resistance. The 1N5139 has a minimum Q of 350, which indicates that the energy stored and returned by the diode capacitance is 350 times greater than the energy lost in the resistance of the device. High values of Q are desirable. Figure 3–22(c) is a graph showing how the figure of merit increases with increasing reverse voltage for three varactors in the series.

Temperature Coefficients The diode capacitance has a positive temperature coefficient so C_T increases a small amount as the temperature increases. The figure of merit has a negative temperature coefficient, so Q decreases as the temperature increases.

An Application

A major application of varactors is in tuning circuits. For example, electronic tuners in TV and other commercial receivers utilize varactors. When used in a resonant circuit, the varactor acts as a variable capacitor, thus allowing the resonant frequency to be adjusted by a variable voltage level, as illustrated in Figure 3–23 where the varactor diode provides the total variable capacitance in the parallel resonant band-pass filter.

The varactor diode and the inductor form a parallel resonant circuit from the output to ac ground. Capacitors C_1, C_2, C_3 and C_4 are coupling capacitors to prevent the dc bias circuit from being loaded by the filter circuit. These capacitors have no effect on the filter's frequency response because their reactances are negligible at the resonant frequencies. C_1 prevents a dc path from the potentiometer wiper back to the ac source through the inductor and R_1. C_2 prevents a dc path from the cathode to the anode of the varactor through the inductor. C_3 prevents a dc path from the wiper to a load on the output through the inductor. C_4 prevents a dc path from the wiper to ground.

Resistors R_2, R_3, R_5, and potentiometer R_4 form a variable dc voltage divider for biasing the varactor. The reverse-bias voltage across the varactor can be varied with the potentiometer.

Recall that the parallel resonant frequency is

$$f_r \cong \frac{1}{2\pi\sqrt{LC}} \tag{3–10}$$

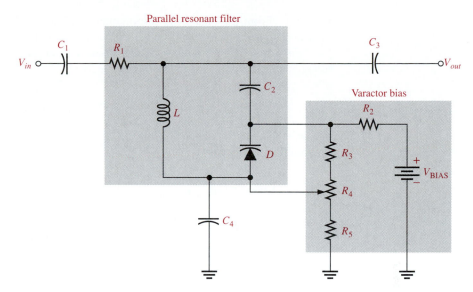

FIGURE 3–23

A resonant band-pass filter using a varactor diode for adjusting the resonant frequency over a specified range.

EXAMPLE 3–10

For the varactor-tuned band-pass filter in Figure 3–24, determine the range of resonant frequencies over which it can be adjusted. The values of the bias resistors are selected to prevent significant ac loading on the filter.

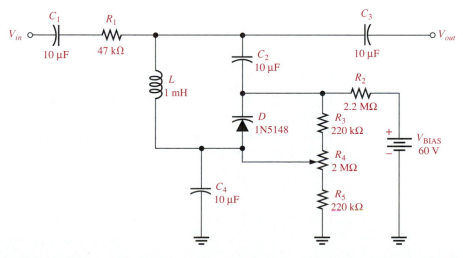

FIGURE 3–24

Solution   From the data sheet information in Figure 3–22(a), the 1N5148 varactor has a nominal capacitance of 47 pF at a reverse bias of 4 V.

First, determine the range of reverse-bias voltages for the filter circuit. The dc voltage at the cathode (V_K) of the varactor is fixed at

$$V_K = \left(\frac{R_3 + R_4 + R_5}{R_2 + R_3 + R_4 + R_5} \right) V_{\text{BIAS}} = \left(\frac{2.44 \text{ M}\Omega}{4.64 \text{ M}\Omega} \right) 60 \text{ V} = 31.6 \text{ V}$$

The dc voltage at the anode (V_A) of the varactor can be varied from a minimum to a maximum with the potentiometer R_4.

$$V_{A(\text{min})} = \left(\frac{R_5}{R_2 + R_3 + R_4 + R_5} \right) V_{\text{BIAS}} = \left(\frac{220 \text{ k}\Omega}{4.64 \text{ M}\Omega} \right) 60 \text{ V} = 2.85 \text{ V}$$

$$V_{A(\text{max})} = \left(\frac{R_4 + R_5}{R_2 + R_3 + R_4 + R_5} \right) V_{\text{BIAS}} = \left(\frac{2.22 \text{ M}\Omega}{4.64 \text{ M}\Omega} \right) 60 \text{ V} = 28.7 \text{ V}$$

The minimum and maximum values for the reverse voltage, V_R, are determined as follows:

$$V_{R(\text{min})} = V_K - V_{A(\text{max})} = 31.6 \text{ V} - 28.7 \text{ V} = 2.9 \text{ V}$$
$$V_{R(\text{max})} = V_K - V_{A(\text{min})} = 31.6 \text{ V} - 2.85 \text{ V} = 29 \text{ V}$$

Although it is difficult to get exact figures from the graph in Figure 3–22(b), the approximate capacitance values of the varactor at 2.9 V and 29 V are $C_{2.9} \cong 55$ pF and $C_{29} \cong 17$ pF. The minimum resonant frequency for the filter is

$$f_{r(min)} \cong \frac{1}{2\pi\sqrt{LC}} = \frac{1}{2\pi\sqrt{(1 \text{ mH})(55 \text{ pF})}} = 679 \text{ kHz}$$

The maximum resonant frequency for the filter is

$$f_{r(max)} \cong \frac{1}{2\pi\sqrt{LC}} = \frac{1}{2\pi\sqrt{(1 \text{ mH})(17 \text{ pF})}} = 1.22 \text{ MHz}$$

Related Exercise   If the bias voltage source in Figure 3–24 is reduced to 30 V, determine the range of the reverse voltage across the varactor.

SECTION 3–3 REVIEW

1. What is the key feature of a varactor diode?
2. Under what bias condition is a varactor operated?
3. What part of the varactor produces the capacitance?
4. Based on the graph in Figure 3–22(b), what happens to the diode capacitance when the reverse voltage is increased?
5. Define *tuning ratio*.

3–4 ■ OPTICAL DIODES

In this section, two types of optoelectronic devices—the light-emitting diode (LED) and the photodiode—are introduced. As the name implies, the LED is a light emitter. The photodiode, on the other hand, is a light detector. We will examine the characteristics of both devices, and you will see an example of their use in a system application in the last section of the chapter.

After completing this section, you should be able to

■ **Discuss the operation and characteristics of LEDs and photodiodes**
  □ Identify LED and photodiode symbols
  □ Explain basically how an LED emits light
  □ Analyze the spectral output curves and radiation patterns of LEDs
  □ Interpret an LED data sheet
  □ Define *radiant intensity* and *irradiance*
  □ Use an LED seven-segment display
  □ Explain how a photodiode detects light
  □ Analyze the response curve of a photodiode
  □ Interpret a photodiode data sheet
  □ Discuss photodiode sensitivity

The Light-Emitting Diode (LED)

FIGURE 3–25
Symbol for an LED.

The symbol for an LED is shown in Figure 3–25. The basic operation of the **light-emitting diode (LED)** is as follows. When the device is forward-biased, electrons cross the *pn* junction from the *n*-type material and recombine with holes in the *p*-type material. Recall from Chapter 1 that these free electrons are in the conduction band and at a higher energy level than the holes in the valence band. When recombination takes place, the recombining electrons release energy in the form of heat and light. A large exposed surface area on one layer of the semiconductor material permits the **photons** to be emitted as visible light. Figure 3–26 illustrates this process, called **electroluminescence.** Various impurities are added during the doping process to establish the wavelength of the emitted light. The wavelength determines the color of the light and if it is visible or invisible (infrared).

Semiconductor Materials LEDs are made of gallium arsenide (GaAs), gallium arsenide phosphide (GaAsP), or gallium phosphide (GaP). Silicon and germanium are not used because they are essentially heat-producing materials and are very poor at producing light. GaAs LEDs emit **infrared (IR)** radiation, which is nonvisible, GaAsP produces either red or yellow visible light, and GaP emits red or green visible light. Red is the most common.

LED Biasing The forward voltage across an LED is considerably greater than for a silicon diode. Typically the maximum V_F for LEDs is between 1.2 V and 3.2 V, depending on the device. Reverse breakdown for an LED is much less than for a silicon rectifier diode (3 V to 10 V is typical).

The LED emits light in response to a sufficient forward current, as shown in Figure 3–27(a). The amount of power output translated into light is directly proportional to the forward current, as indicated in Figure 3–27(b). The greater I_F is, the greater the light output.

Light energy

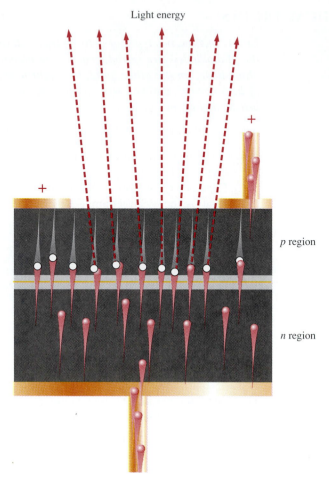

p region

n region

FIGURE 3–26
Electroluminescence in a forward-biased LED.

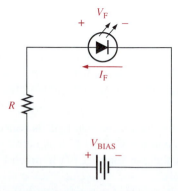

(a) Forward-biased operation

(b) General light output versus forward current

FIGURE 3–27
Basic operation of an LED.

Light Emission The **wavelength** of light determines whether it is visible or infrared. An LED emits light over a specified range of wavelengths as indicated by the **spectral** output curves in Figure 3–28. The curves in part (a) represent the light output versus wavelength for typical visible LEDs and the curve in part (b) is for a typical infrared LED. The wavelength (λ) is expressed in nanometers (nm). The normalized output of the visible red LED peaks at 660 nm, the yellow at 590 nm, and green at 540 nm. The output for the infrared LED peaks at 940 nm.

The graph in Figure 3–29 is the **radiation** pattern for a typical LED. It shows how directional the emitted light is. The radiation pattern depends on the type of lens structure of the LED. The narrower the radiation pattern, the more the light is concentrated in a particular direction. Also, colored lenses are used to enhance the color.

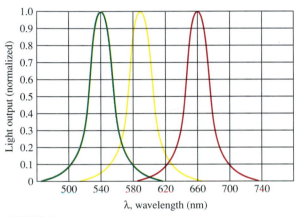

(a) Visible light

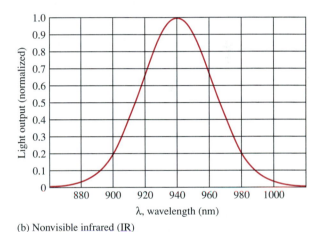

(b) Nonvisible infrared (IR)

FIGURE 3–28
Examples of typical spectral output curves for LEDs.

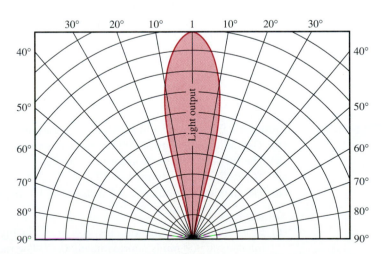

FIGURE 3–29
General radiation pattern of a typical LED.

Typical LEDs are shown in Figure 3–30. Photodiodes, to be studied next, generally have the same appearance.

LED Data Sheet Information

A partial data sheet for an MLED81 infrared light-emitting diode is shown in Figure 3–31. Notice that the maximum reverse voltage is only 5 V, the maximum continuous forward current is 100 mA, and the forward voltage drop is 1.35 V for $I_F = 100$ mA.

From the graph in part (c), you can see that the peak power output for this device occurs at a wavelength of 940 nm and its radiation pattern is shown in part (d). At 30° on either side of the maximum orientation, the output power drops to approximately 60% of maximum.

Radiant Intensity and Irradiance In Figure 3–31(a), the axial **radiant intensity, I_e** (symbol not to be confused with current), is the output power per steradian and is specified as 15 mW/sr. The steradian (sr) is the unit of solid angular measurement. **Irradiance, H,** is the power per unit area at a given distance from the LED source expressed in mW/cm^2 and can be calculated using radiant intensity and the distance in centimeters (cm) using the following formula:

$$H = \frac{I_e}{d^2}$$

(3–11)

Irradiance is important because the response of a detector (photodiode) used in conjunction with the LED depends on the irradiance of the light it receives. We will discuss this further in relation to photodiodes.

EXAMPLE 3–11

From the LED data sheet in Figure 3–31 determine the following:
(a) The radiant intensity at 900 nm if the maximum output is 15 mW/sr.
(b) The forward voltage drop for $I_F = 20$ mA.
(c) The radiant intensity for $I_F = 30$ mA.
(d) The maximum irradiance at a distance of 10 cm from the LED source.

Solution
(a) From the relative spectral emission graph in Figure 3–31(c), the relative radiant intensity at 900 nm is approximately 0.75. The radiant intensity is, therefore,

$$I_e = 0.75(15\text{mW/sr}) = 11.3 \text{ mW/sr}$$

Maximum Ratings

Rating	Symbol	Value	Unit
Reverse voltage	V_R	5	Volts
Forward current — continuous	I_F	100	mA
Forward current — peak pulse	I_F	1	A
Total power dissipation @ $T_A = 25°C$ Derate above 25°C	P_D	100 2.2	mW mW/C°
Ambient operating temperature range	T_A	−30 to +70	°C
Storage temperature	T_{stg}	−30 to +80	°C
Lead soldering temperature, 5 seconds max, 1/16 inch from case	—	260	°C

Electrical Characteristics ($T_A = 25°C$ unless otherwise noted)

Characteristic	Symbol	Min	Typ	Max	Unit
Reverse leakage current ($V_R = 3$ V)	I_R	—	10	—	nA
Reverse leakage current ($V_R = 5$ V)	I_R	—	1	10	μA
Forward voltage ($I_F = 100$ mA)	V_F	—	1.35	1.7	V
Temperature coefficient of forward voltage	ΔV_F	—	− 1.6	—	mV/K
Capacitance ($f = 1$ MHz)	C	—	25	—	pF

Optical Characteristics ($T_A = 25°C$ unless otherwise noted)

Characteristic	Symbol	Min	Typ	Max	Unit
Peak wavelength ($I_F = 100$ mA)	λp	—	940	—	nm
Spectral half-power bandwidth	$\Delta \lambda$	—	50	—	nm
Total power output ($I_F = 100$ mA)	ϕe	—	16	—	mW
Temperature coefficient of total power output	$\Delta \phi e$	—	− 0.25	—	%/K
Axial radiant intensity ($I_F = 100$ mA)	I_e	10	15	—	mW/sr
Temperature coefficient of axial radiant intensity	ΔI_e	—	− 0.25	—	%/K
Power half-angle	ϕ	—	±30	—	°

MLED 81

(a) Ratings and characteristics

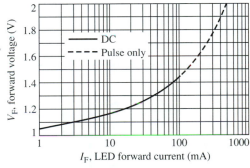

(b) LED forward voltage versus forward current

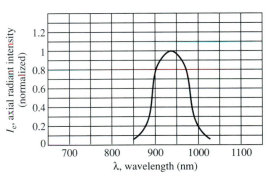

(c) Relative spectral emission

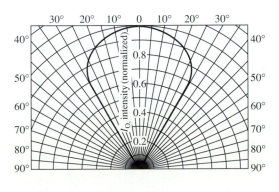

(d) Spatial radiation pattern

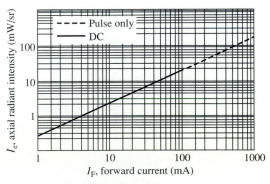

(e) Intensity versus forward current

FIGURE 3–31

Partial data sheet for an MLED81 IR light-emitting diode.

(b) From the graph in part (b), $V_F = 1.23$ V for $I_F = 20$ mA.

(c) From the graph in part (e), $I_e = 5$ mW/sr for $I_F = 30$ mA.

(d) $H = \dfrac{I_e}{d^2} = \dfrac{15 \text{ mW/sr}}{(10 \text{ cm})^2} = 0.15 \text{ mW/cm}^2$

Related Exercise If $I_e = 12$ mW/sr, at a wavelength of 940 nm, determine the radiant intensity at 1000 nm.

Applications LEDs are used for indicator lamps and readout displays on a wide variety of instruments, ranging from consumer appliances to scientific apparatus. A common type of display device using LEDs is the seven-segment display. Combinations of the segments form the ten decimal digits as illustrated in Figure 3–32. Each segment in the display is an LED. By forward-biasing selected combinations of segments, any decimal digit and a decimal point can be formed. Two types of LED circuit arrangements are the common anode and common cathode as shown.

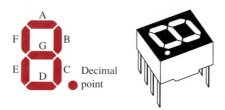

(a) LED segment arrangement and typical device

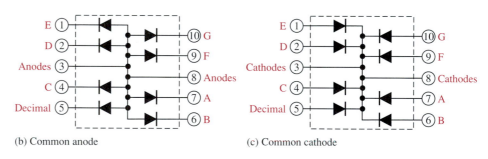

(b) Common anode

(c) Common cathode

FIGURE 3–32
The 7-segment LED display.

IR-emitting diodes are employed in optical coupling applications, often in conjunction with fiber optics. Areas of application include industrial processing and control, position encoders, bar graph readers, and optical switching.

The Photodiode

The **photodiode** is a *pn* junction device that operates in reverse bias, as shown in Figure 3–33(a), where I_λ is the reverse current. The photodiode has a small transparent window that allows light to strike the *pn* junction. An alternate photodiode symbol is shown in Figure 3–33(b).

FIGURE 3–33
Photodiode.

I_λ

V_R

(a) Reverse-bias operation   (b) Alternate symbol

Recall that when reverse-biased, a rectifier diode has a very small reverse leakage current. The same is true for the photodiode. The reverse-biased current is produced by thermally generated electron-hole pairs in the depletion region, which are swept across the junction by the electric field created by the reverse voltage. In a rectifier diode, the reverse leakage current increases with temperature due to an increase in the number of electron-hole pairs.

A photodiode differs from a rectifier diode in that when its *pn* junction is exposed to light, the reverse current increases with the light intensity. When there is no incident light, the reverse current I_λ is almost negligible and is called the **dark current.** An increase in the amount of light intensity, expressed as irradiance (mW/cm^2), produces an increase in the reverse current, as shown by the graph in Figure 3–34(a).

From the graph in Figure 3–34(b), the light current for this particular device is approximately 0.9 μA at a reverse-bias voltage of 10 V. Therefore, the resistance of the device with an irradiance of 0.5 mW/cm^2 is

$$R_R = \frac{V_R}{I_\lambda} = \frac{10 \text{ V}}{0.9 \text{ μA}} = 11 \text{ MΩ}$$

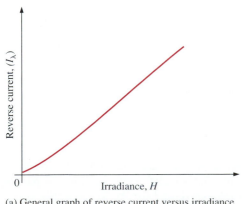

(a) General graph of reverse current versus irradiance

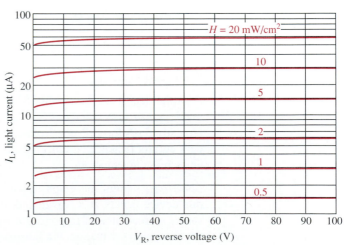

(b) Example of a graph of light current versus reverse voltage for several values of irradiance

FIGURE 3–34
Typical photodiode characteristics.

At 20 mW/cm^2, the current is approximately 55 μA at $V_R = 10$ V. The resistance under this condition is

$$R_R = \frac{V_R}{I_\lambda} = \frac{10 \text{ V}}{55 \text{ μA}} = 182 \text{ k}\Omega$$

These calculations show that the photodiode can be used as a variable-resistance device controlled by light intensity.

Figure 3–35 illustrates that the photodiode allows essentially no reverse current (except for a very small dark current) when there is no incident light. When a light beam strikes the photodiode, it conducts an amount of reverse current that is proportional to the light intensity (irradiance).

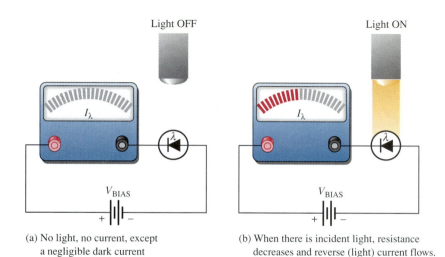

(a) No light, no current, except a negligible dark current

(b) When there is incident light, resistance decreases and reverse (light) current flows.

FIGURE 3–35
Operation of a photodiode.

Photodiode Data Sheet Information

A partial data sheet for an MRD821 photodiode is shown in Figure 3–36. Notice that the maximum reverse voltage is 35 V and the dark current (reverse current with no light) is typically 3 nA for a reverse voltage of 10 V. As the graphs in parts (b) and (c) show, the dark current (leakage current) increases with an increase in reverse voltage and also with an increase in temperature.

Sensitivity   From the graph in part (d), you can see that the maximum sensitivity for this device occurs at a wavelength of 940 nm. The angular response graph in part (e) shows a broad area of response measured as relative sensitivity. At 50° on either side of the maximum orientation, the sensitivity drops to approximately 80% of maximum.

In Figure 3–36(a), the typical sensitivity is specified as 50 μA/mW/cm^2 for a wavelength of 940 nm and a reverse voltage of 20 V. This means, for example, that if the irradiance is 1 mW/cm^2, there are 50 μA of reverse (light) current and if the irradiance is 0.5 mW/cm^2, there are 25 μA of reverse current.

Maximum Ratings

Rating	Symbol	Value	Unit
Reverse voltage	V_R	35	Volts
Forward current — continuous	I_F	100	mA
Total power dissipation @ $T_A = 25°C$ Derate above 25°C	P_D	150 3.3	mW mW/C°
Ambient operating temperature range	T_A	−30 to +70	°C
Storage temperature	T_{stg}	−40 to +80	°C
Lead soldering temperature, 5 seconds max, 1/16 inch from case	—	260	°C

Electrical Characteristics ($T_A = 25°C$ unless otherwise noted)

Characteristic	Symbol	Min	Typ	Max	Unit
Dark current ($V_R = 10$ V)	I_D	—	3	30	nA
Capacitance ($f = 1$ MHz, $V = 0$)	C_J	—	175	—	pF

Optical Characteristics ($T_A = 25°C$ unless otherwise noted)

Characteristic	Symbol	Min	Typ	Max	Unit
Wavelength of maximum sensitivity	λmax	—	940	—	nm
Spectral range	Δλ	—	170	—	nm
Sensitivity ($\lambda = 940$ nm, $V_R = 20$ V)	S	—	50	—	μA/mW/cm²
Temperature coefficient of sensitivity	ΔS	—	0.18	—	%/K
Acceptance half-angle	φ	—	±70	—	°
Short circuit current (Ev = 1000 lux)	I_S	—	50	—	μA
Open circuit voltage (Ev = 1000 lux)	V_L	—	0.3	—	V

(a) Ratings and characteristics

MRD 821

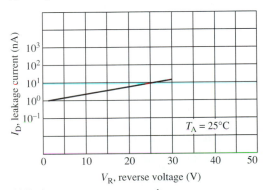

(b) Dark current versus reverse voltage

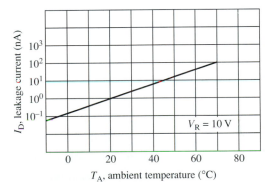

(c) Dark current versus temperature

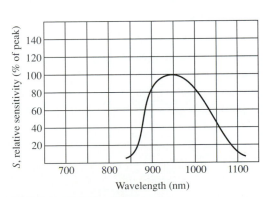

(d) Relative spectral sensitivity

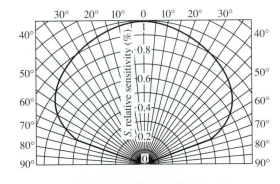

(e) Angular response

FIGURE 3–36

Partial data sheet for the MRD821 photodiode.

EXAMPLE 3–12

An MRD821 photodiode is exposed to a 1000 nm infrared light with an irradiance (H) of 2.5 mW/cm². The angle at which the light strikes the photodiode is 35°. Determine the response of the photodiode in terms of the reverse current (I_λ) through the device.

Solution From the photodiode data sheet in Figure 3–36, the sensitivity of the photodiode is 50 μA/mW/cm² at 940 nm. The light on the photodiode is at a wavelength of 1000 nm. From the data sheet graph in part (d), the sensitivity (S) at 1000 nm is approximately 83% of the sensitivity at 940 nm.

$$S_{1000} = 0.83S_{940} = 0.83(50 \text{ μA/mW/cm}^2) = 41.5 \text{ μA/mW/cm}^2$$

Also, the angle at which the light strikes the photodiode reduces the sensitivity further. From the graph in Figure 3–36(e), at an angle of 35° from the maximum orientation (0°), the relative sensitivity is approximately 90%.

$$S = 0.9(41.5 \text{ μA/mW/cm}^2) = 37.4 \text{ μA/mW/cm}^2$$

For an irradiance, H, of 2.5 mW/cm², the reverse (light) current is

$$I_\lambda = S \times H = (37.4 \text{ μA/mW/cm}^2)(2.5 \text{ mW/cm}^2) = 93.5 \text{ μA}$$

Related Exercise Determine the MRD821 response (reverse current) to an irradiance of 1 mW/cm² for a wavelength of 900 nm at an angle of 60° from maximum orientation.

SECTION 3–4 REVIEW

1. Name two types of LEDs in terms of their light-emission spectrum.
2. Which has the greater wavelength, visible light or infrared?
3. In what bias condition is an LED normally operated?
4. What happens to the light emission of an LED as the forward current increases?
5. The forward voltage drop of an LED is 0.7 V (true or false).
6. In what bias condition is a photodiode normally operated?
7. When the intensity of the incident light (irradiance) on a photodiode increases, what happens to its internal reverse resistance?
8. What is *dark current?*

3–5 ■ OTHER TYPES OF DIODES

In this section, we introduce several types of diodes that you are less likely to encounter as a technician but are nevertheless important. Among these are the current regulator, the Schottky diode, the tunnel diode, the pin diode, the step-recovery diode, and the laser diode.

After completing this section, you should be able to

■ **Discuss the basic characteristics of the current regulator diode, the Schottky diode, the *pin* diode, the step-recovery diode, the tunnel diode, and the laser diode**
  ☐ Identify the various diode symbols
  ☐ Discuss how the current regulator diode maintains a constant forward current
  ☐ Describe the characteristics of the Schottky diode
  ☐ Describe the characteristics of the *pin* diode
  ☐ Describe the characteristics of the step-recovery diode
  ☐ Describe the characteristics of the tunnel diode and explain its negative resistance
  ☐ Describe the laser diode and how it differs from an LED

Current Regulator Diode

The current regulator diode is often referred to as a constant-current diode. Rather than maintaining a constant voltage, as the zener diode does, this diode maintains a constant current. The symbol is shown in Figure 3–37.

The current regulator diode operates in forward bias and the forward current becomes a specified constant value at forward voltages ranging from about 1.5 V to about 6 V, depending on the diode type. The constant forward current is called the *regulator current* and is designated I_P. For example, the 1N5283–1N5314 series have nominal regulator currents ranging from 220 μA to 4.7 mA. These diodes may be used in parallel to obtain higher currents.

A typical characteristic curve is shown in Figure 3–38. First, notice that this diode does not have a sharply defined reverse breakdown, so the reverse current begins to increase for V_R values of less than 1 V. This device should not be operated in reverse bias.

In forward bias, the diode regulation begins at the limiting voltage, V_L, and extends up to the POV (peak operating voltage). Notice that between V_K and POV, the current is essentially constant. V_T is the test voltage at which I_P and the diode impedance, Z_T, are specified on a data sheet. The impedance Z_T has very high values ranging from 235 kΩ to 25 MΩ for the diode series mentioned before.

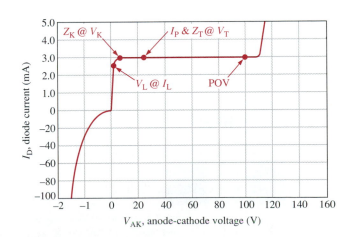

FIGURE 3–37
Symbol for a current regulator diode.

FIGURE 3–38
Typical characteristic curve for a current regulator diode.

The Schottky Diode

FIGURE 3–39
Schottky diode symbol.

Schottky diodes are used primarily in high-frequency and fast-switching applications. They are also known as *hot-carrier diodes*. A Schottky diode symbol is shown in Figure 3–39. A Schottky diode is formed by joining a doped semiconductor region (usually *n*-type) with a metal such as gold, silver, or platinum. So, rather than a *pn* junction, there is a metal-to-semiconductor junction, as shown in Figure 3–40. The forward voltage drop is typically around 0.3 V.

FIGURE 3–40

Basic internal construction of Schottky diode.

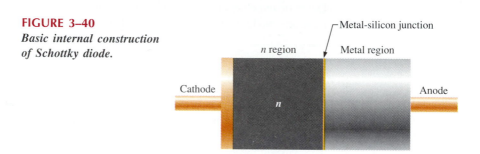

The Schottky diode operates only with majority carriers. There are no minority carriers and thus no reverse leakage current as in other types of diodes. The metal region is heavily occupied with conduction-band electrons, and the *n*-type semiconductor region is lightly doped. When forward-biased, the higher energy electrons in the *n* region are injected into the metal region where they give up their excess energy very rapidly. Since there are no minority carriers, as in a conventional rectifier diode, there is a very rapid response to a change in bias. The Schottky is a very fast-switching diode, and most of its applications make use of this property. It can be used in high-frequency applications and in many digital circuits to decrease switching times.

The *PIN* Diode

The *pin* diode consists of heavily doped *p* and *n* regions separated by an intrinsic (*i*) region, as shown in Figure 3–41(a). When reverse-biased, the *pin* diode acts like a nearly constant capacitance. When forward-biased, it acts like a current-controlled variable resistance. This is shown in Figure 3–41(b) and (c). The low forward resistance of the intrinsic region decreases with increasing current.

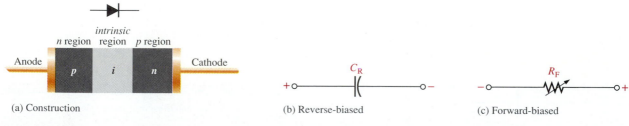

(a) Construction   (b) Reverse-biased   (c) Forward-biased

FIGURE 3–41
PIN diode.

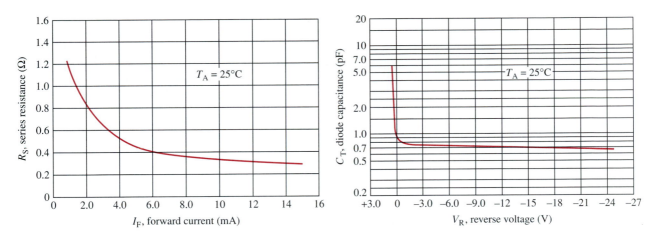

FIGURE 3–42
PIN diode characteristics.

The forward series resistance characteristic and the reverse capacitance characteristic are shown graphically in Figure 3–42 for a typical *pin* diode.

The *pin* diode is used as a dc-controlled microwave switch operated by rapid changes in bias or as a modulating device that takes advantage of the variable forward-resistance characteristic. Since no rectification occurs at the *pn* junction, a high-frequency signal can be modulated (varied) by a lower-frequency bias variation. A *pin* diode can also be used in attenuator applications because its resistance can be controlled by the amount of current. Certain types of *pin* diodes are used as photodetectors in fiber-optic systems.

The Step-Recovery Diode

The step-recovery diode employs graded doping where the doping level of the semiconductor materials is reduced as the *pn* junction is approached. This produces an abrupt turn-off time by allowing a very fast release of stored charge when switching from forward to reverse bias. It also allows a rapid re-establishment of forward current when switching from reverse to forward bias. This diode is used in very high frequency (VHF) and fast-switching applications.

The Tunnel Diode

The **tunnel diode** exhibits a special characteristic known as *negative resistance*. This feature makes it useful in oscillator and microwave amplifier applications. Three alternate symbols are shown in Figure 3–43. Tunnel diodes are constructed with germanium or gallium arsenide by doping the *p* and *n* regions much more heavily than in a conventional rectifier diode. This heavy doping results in a very narrow depletion region. The heavy doping allows conduction for all reverse voltages so that there is no breakdown effect as with the conventional rectifier diode. This is shown in Figure 3–44.

Also, the extremely narrow depletion region permits electrons to "tunnel" through the *pn* junction at very low forward-bias voltages, and the diode acts as a conductor. This is shown in Figure 3–44 between points *A* and *B*. At point *B*, the forward voltage begins

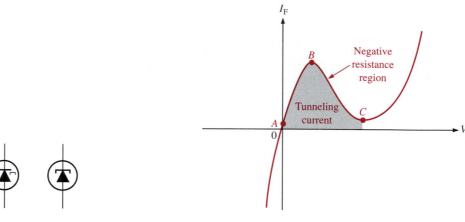

FIGURE 3–43
Tunnel diode symbols.

FIGURE 3–44
Tunnel diode characteristic curve.

to develop a barrier, and the current begins to decrease as the forward voltage continues to increase. This is the *negative-resistance region*.

$$R_F = \frac{\Delta V_F}{\Delta I_F}$$

This effect is opposite to that described in Ohm's law, where an increase in voltage results in an increase in current. At point C, the diode begins to act as a conventional forward-biased diode.

An Application   A parallel resonant circuit can be represented by a capacitance, inductance, and resistance in parallel, as in Figure 3–45(a). R_P is the parallel equivalent of the series winding resistance of the coil. When the tank circuit is "shocked" into oscillation by an application of voltage as in Figure 3–45(b), a damped sinusoidal output results. The damping is due to the resistance of the tank, which prevents a sustained oscillation because energy is lost when there is current through the resistance.

If a tunnel diode is placed in series with the tank circuit and biased at the center of the negative-resistance portion of its characteristic curve, as shown in Figure 3–46, a sustained oscillation (constant sinusoidal voltage) will result on the output. This is because

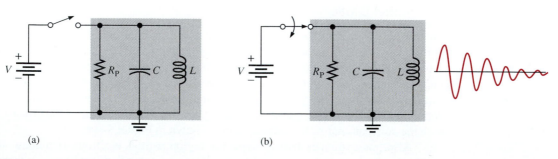

(a)   (b)

FIGURE 3–45
Parallel resonant circuit.

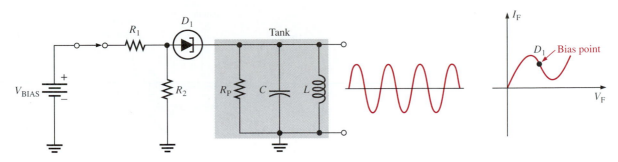

FIGURE 3–46
Basic tunnel diode oscillator.

the negative-resistance characteristic of the tunnel diode counteracts the positive-resistance characteristic of the tank resistance.

The Laser Diode

The term **laser** stands for *l*ight *a*mplification by *s*timulated *e*mission of *r*adiation. Laser light is **monochromatic,** which means that it consists of a single color and not a mixture of colors. Laser light is also called **coherent light,** a single wavelength, as compared to incoherent light, which consists of a wide band of wavelengths. The laser diode normally emits coherent light, whereas the LED emits incoherent light. The symbols are the same as shown in Figure 3–47(a).

The basic construction of a laser diode is shown in Figure 3–47(b). A *pn* junction is formed by two layers of doped gallium arsenide. The length of the *pn* junction bears a precise relationship with the wavelength of the light to be emitted. There is a highly reflective surface at one end of the junction and a partially reflective surface at the other end produced by "polishing" the ends. External leads provide the anode and cathode connections.

The basic operation is as follows. The *pn* junction is forward-biased by an external voltage source. As electrons move through the junction, recombination occurs just as in an ordinary diode. As electrons fall into holes to recombine, photons are released. A released photon can strike an atom, causing another photon to be released. As the forward current is increased, more electrons enter the depletion region and cause more photons to be emitted. Eventually some of the photons that are randomly drifting within the depletion region strike the reflected surfaces perpendicularly. These reflected photons move along the depletion region, striking atoms and releasing additional photons due to the avalanche effect. This back-and-forth movement of photons increases as the generation of photons "snowballs" until a very intense beam of laser light is formed by the photons that pass through the partially reflective end of the *pn* junction.

Each photon produced in this process is identical to the other photons in energy level, phase relationship, and frequency. So a single wavelength of intense light emerges from the laser diode, as indicated in Figure 3–47(c). Laser diodes have a threshold level of current above which the laser action occurs and below which the diode behaves essentially as an LED, emitting incoherent light.

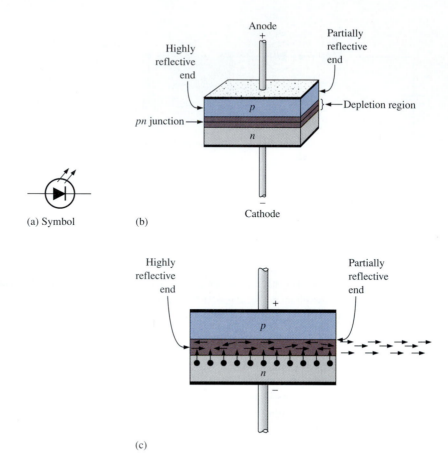

FIGURE 3–47
Basic laser diode construction and operation.

Application   Laser diodes and photodiodes are used in the pick-up system of compact disk (CD) players. Audio information (sound) is digitally recorded in stereo on the surface of a compact disk in the form of microscopic "pits" and "flats." A lens arrangement focuses the laser beam from the diode onto the CD surface. As the CD rotates, the lens and beam follow the track under control of a servomotor. The laser light, which is altered by the pits and flats along the recorded track, is reflected back from the track through a lens and optical system to infrared photodiodes. The signal from the photodiodes is then used to reproduce the digitally recorded sound.

**SECTION 3–5
REVIEW**

1. Between what two voltages does a current regulator diode operate?
2. What are the primary application areas for Schottky diodes?
3. What is a hot-carrier diode?
4. What is the key characteristic of a tunnel diode?
5. What is one application for a tunnel diode?
6. Name the three regions of a *pin* diode.
7. What does *laser* mean?
8. What is the difference between incoherent and coherent light and which is produced by a laser diode?

3–6 ■ TROUBLESHOOTING

In this section, you will see how a faulty zener diode can affect the output of a regulated dc power supply. Like other diodes, the zener can fail open, it can exhibit degraded performance in which its internal resistance increases significantly, or it can short out.

After completing this section, you should be able to

■ **Troubleshoot zener diode regulators**
  □ Recognize the effects of an open zener
  □ Recognize the effects of a zener with excessive impedance

A Zener-Regulated DC Power Supply

Figure 3–48 shows a filtered dc power supply that produces a constant 24 V before it is regulated down to 15 V by the zener regulator. The 1N4744 zener diode is the same as the one in Example 3–7. A no-load check of the regulated output voltage shows 15.5 V as indicated in part (a). The voltage expected at maximum zener current (I_{ZM}) for this particular diode is 15.7 V. In part (b), a potentiometer is connected to provide a variable load resistance. It is adjusted to a minimum value for a full-load test as determined by the following calculations. The full-load test is at minimum zener current (I_{ZK}). The meter reading of 14.8 V indicates approximately the expected output voltage of 14.76 V.

$$I_T = \frac{24 \text{ V} - 14.76 \text{ V}}{180 \ \Omega} = 51.3 \text{ mA}$$

$$I_L = 51.3 \text{ mA} - 0.25 \text{ mA} = 51.1 \text{ mA}$$

$$R_{L(\text{min})} = \frac{14.76 \text{ V}}{51.1 \text{ mA}} = 289 \ \Omega$$

Case 1: Zener Diode Open   If the zener diode fails open, the power supply test gives the approximate results indicated in Figure 3–49. In the no-load check shown in part (a), the output voltage is 24 V because there is no voltage dropped between the filtered output of the power supply and the output terminal. This definitely indicates an open between

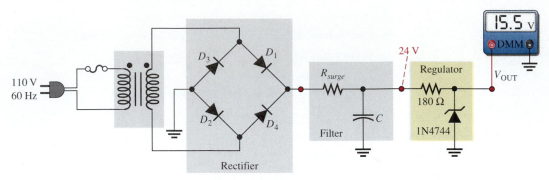

(a) Correct output voltage with no load

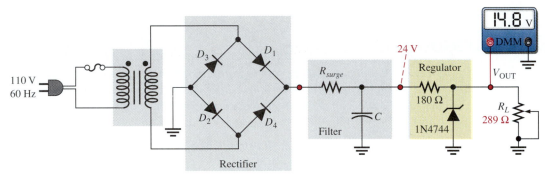

(b) Correct output voltage with full load

FIGURE 3–48
Zener-regulated power supply test.

FIGURE 3–49
Indications of an open zener.

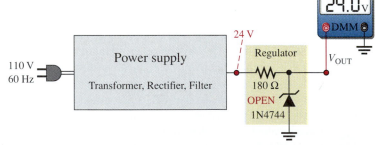

(a) Open zener diode with no load

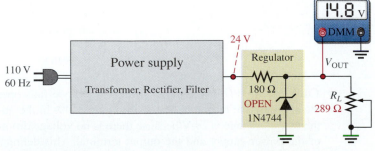

(b) Open zener diode cannot be detected by full-load measurement in this case.

the output terminal and ground. In the full-load check, the voltage of 14.8 V results from the voltage-divider action of the 180 Ω series resistor and the 289 Ω load. In this case, the result is too close to the normal reading to be a reliable fault indication but the no-load check will verify the problem. Also, if R_L is varied, V_{OUT} will vary if the zener diode is open.

Case 2: Excessive Zener Impedance As indicated in Figure 3–50, a no-load check that results in an output voltage greater than the maximum zener voltage but less than the power supply output voltage indicates that the zener has failed such that its internal impedance is more than it should be. The 20 V output in this case is 4.5 V higher than the expected value of 15.5 V. That additional voltage is caused by the drop across the excessive internal impedance of the zener.

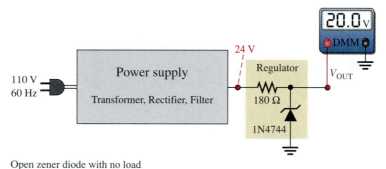

Open zener diode with no load

FIGURE 3–50
Indication of excessive zener impedance.

1. In a filtered power supply with a zener regulator, what are the symptoms of an open zener diode?
2. If a zener regulator fails so that the zener impedance is greater than the specified value, is the output voltage more or less than it should be?
3. If you measure 0 V at the output of a zener-regulated power supply, what is the most likely fault(s)?
4. The zener diode regulator in a power supply is open. What will you observe on the output with a voltmeter if the load resistance is varied within its specified range?

3–7 ■ SYSTEM APPLICATION

You have been assigned to modify the power supply circuit board from the system application of Chapter 2. You are to incorporate voltage regulation and light emission and detection circuits to be used in a new system your company is developing. This system will be used in a sporting goods manufacturing plant for counting and controlling the number of baseballs going into various sizes of boxes for shipment. You will apply the knowledge you have gained to this chapter in completing your assignment.

The Counting and Control System

This particular system is used to count baseballs as they are fed down a chute into a box for shipping. It can also be applied to inventory and shipping control for many other types of products. The portion of the system for which you are responsible consists of the regulated power supply, an infrared emitter circuit and an infrared detector circuit which are all on the same board.

The complete system also includes a threshold circuit that senses the output of the infrared detector and provides a pulse output to a digital counter. The output of the counter goes to a display and a control mechanism for stopping the baseballs when a box is full. The system concept and block diagram are shown in Figure 3–51.

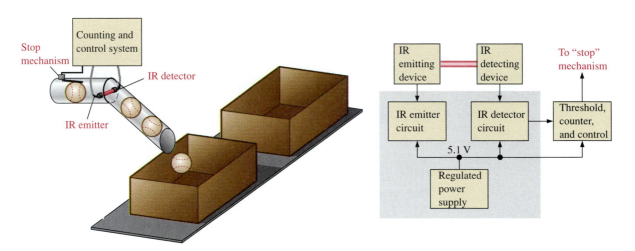

FIGURE 3–51
Basic system concept and block diagram of the counting and control system.

The Power Supply Circuit

The dc power supply circuit is the same as the one developed in the system application of Chapter 2 except that it is modified by the addition of a zener voltage regulator as shown in the schematic of Figure 3–52. Also, the new circuit board will include the IR emitter and IR detector circuits. The basic power supply specifications are

1. Input voltage: 115 V rms at 60 Hz
2. Unregulated output voltage: 12 V dc ± 10%
3. Regulated output voltage: 5.1 V ± 10%
4. Maximum ripple factor: 3%
5. Maximum load current: 100 mA

The Power Supply Components

☐ *The unregulated portion of the power supply* This portion of the circuit is the same as the one developed in Chapter 2.

☐ *The regulator* Select a zener diode to be used in the regulator. Refer to the data sheet in Figure 3–7(a).

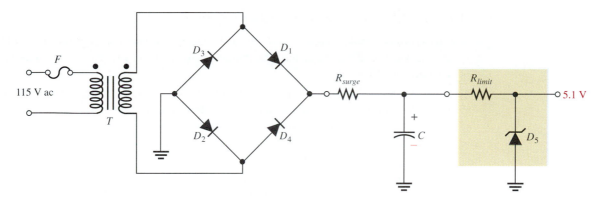

FIGURE 3–52
Regulated power supply preliminary schematic.

☐ *The limiting resistor*   Determine a value of limiting resistor to be used in the regulator.

☐ *The fuse*   Determine a rating for the fuse to be used in the power supply.

The IR Emitter and IR Detector Circuits

The MLED81 light-emitting diode is used as the IR emitter, and the MRD821 photodiode is used as the IR detector. These devices are located on either side of the tube through which the baseballs are routed. The diameter of the tube is 1.5 cm greater than the diameter of a baseball. The LED emits a constant beam of infrared light directly toward the photodiode; this beam of light is interrupted as a baseball passes in the tube.

The IR detector senses the interruption in the LED emission and produces a minimum positive-going output transition of 3 V for the threshold circuit that generates a pulse to advance the digital counter. The counter is advanced by one count for each baseball that passes the IR beam. When a preset number of baseballs has been packed into a box, the control circuit produces a signal to activate the stop mechanism. The system is reset for the next box.

Both the LED and photodiode are connected to their series limiting resistors and the voltage source on the circuit board with a four-wire cable. Figure 3–53 shows the basic IR emitter and IR detector circuits.

FIGURE 3–53
Basic schematics of the IR emitter and IR detector circuits.

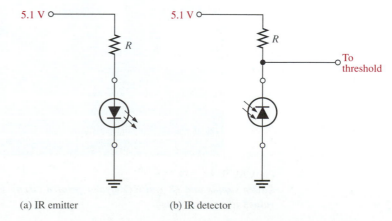

(a) IR emitter   (b) IR detector

The IR Emitter and IR Detector Components

☐ *The distance between emitter and detector* Determine the distance from the LED to the photodiode (diameter of the tube) based on the diameter of a baseball (7.3 cm) plus a clearance of 1.5 cm.

☐ *The IR emitter series-limiting resistor* Determine a value for the current-limiting series resistor to achieve the maximum possible irradiance (H) at the detector. Assume the emitter and detector are aligned for maximum angular response. Refer to the data sheet in Figure 3–31.

☐ *The IR detector resistor* Determine a value of resistor in the IR detector to produce minimum voltage transitions of 3 V on the threshold output when the photodiode turns on and off. Refer to the data sheet in Figure 3–36.

The Schematic

Produce a complete schematic that includes the regulated power supply, the IR emitter, and the IR detector.

The Printed Circuit Board

☐ Check out the printed circuit board in Figure 3–54 to verify that it is correct according to the schematic.

☐ Label a copy of the board with the component and input/output designations in agreement with the schematic.

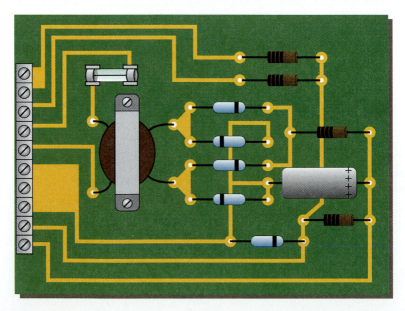

FIGURE 3–54
Power supply and IR emitter/detector printed circuit board. All black resistor bands indicate values to be determined.

A Test Procedure

☐ Develop a step-by-step set of instructions on how to completely check the power supply and IR emitter/detector for proper operation using the test points (circled numbers) indicated in the test bench setup of Figure 3–55. There is a test fixture for simulating the operation of the LED and photodiode by momentarily inserting an obstruction for the IR beam in a tube-type mounting device.

☐ Specify voltage values and appropriate waveforms for all the measurements to be made.

☐ Provide a fault analysis for all possible component failures.

☐ Expand the test procedure you developed for the unregulated portion of the power supply circuit in Chapter 2 to include the additional circuits.

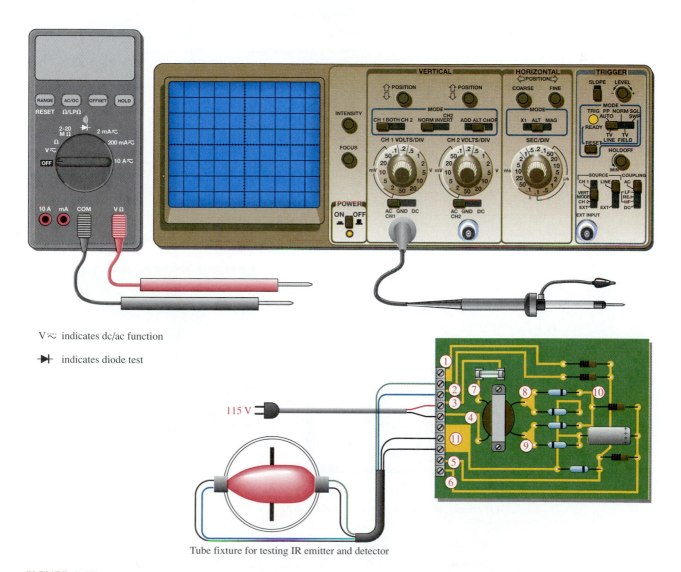

V∿ indicates dc/ac function

⊣⊢ indicates diode test

115 V

Tube fixture for testing IR emitter and detector

FIGURE 3–55

Power supply and IR emitter/detector test bench.

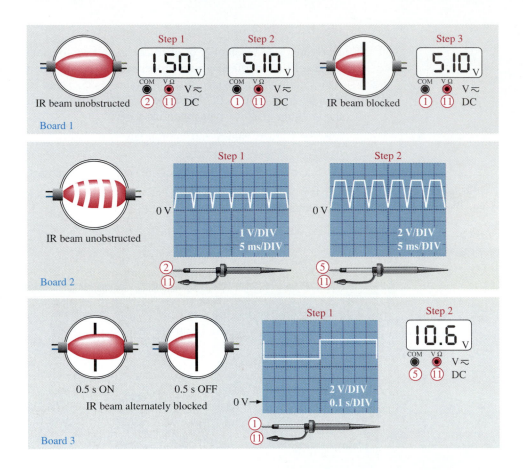

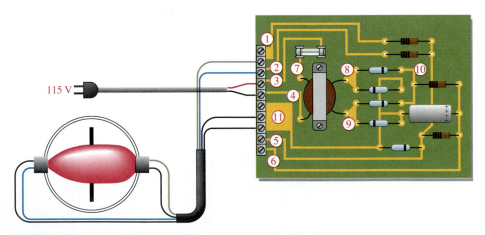

FIGURE 3–56

Test results of three prototype circuit boards.

Troubleshooting

Several prototype boards have been assembled and are ready for testing. The test bench setup is shown in Figure 3–55. Based on the sequence of measurements for each board indicated in Figure 3–56, determine the most likely fault in each case.

The circled numbers indicate test point connections to the circuit board. The DMM function setting is indicated below the display and the VOLT/DIV and SEC/DIV settings for the oscilloscope are shown on the screen in each case.

Final Report

Submit a final written report on the power supply and IR emitter/detector circuit board using an organized format that includes the following:

1. A physical description of the circuits.
2. A discussion of the operation of each circuit.
3. A list of the specifications.
4. A list of parts with part numbers if available.
5. A list of the types of problems on the three circuit boards.
6. A complete description of how you determined the problem on each of the three circuit boards.

■ CHAPTER SUMMARY

- The zener diode operates in reverse breakdown.
- There are two breakdown mechanisms in a zener diode: avalanche breakdown and zener breakdown.
- When $V_Z < 5$ V, zener breakdown is predominant.
- When $V_Z > 5$ V, avalanche breakdown is predominant.
- A zener diode maintains a nearly constant voltage across its terminals over a specified range of zener currents.
- Zener diodes are used as voltage regulators and limiters.
- Zener diodes are available in many voltage ratings ranging from 1.8 V to 200 V.
- Regulation of output voltage over a range of input voltages is called *input* or *line regulation*.
- Regulation of output voltage over a range of load currents is called *load regulation*.
- The smaller the percent regulation, the better.
- A varactor diode acts as a variable capacitor under reverse-bias conditions.
- The capacitance of a varactor varies inversely with reverse-bias voltage.
- The current regulator diode keeps its forward current at a constant specified value.
- The Schottky diode has a metal-to-semiconductor junction. It is used in fast-switching applications.
- The tunnel diode is used in oscillator circuits.
- An LED emits light when forward-biased.
- LEDs are available for either infrared or visible light.
- The photodiode exhibits an increase in reverse current with light intensity.
- The *pin* diode has a *p* region, an *n* region, and an intrinsic (*i*) region and displays a variable resistance characteristic when forward-biased and a constant capacitance when reverse-biased.

■ A laser diode is similar to an LED except that it emits coherent (single wavelength) light when the forward current exceeds a threshold value.

■ A summary of special-purpose diode symbols is given in Figure 3–57.

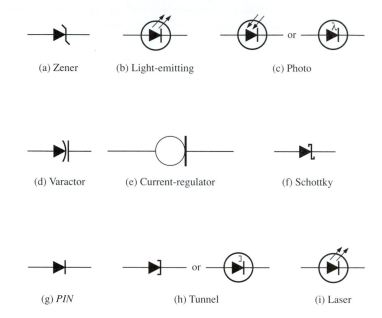

(a) Zener   (b) Light-emitting   (c) Photo

(d) Varactor   (e) Current-regulator   (f) Schottky

(g) *PIN*   (h) Tunnel   (i) Laser

FIGURE 3–57
Diode symbols.

■ **GLOSSARY**

Avalanche breakdown   The higher voltage breakdown in a zener diode.

Coherent light   Light having only one wavelength.

Dark current   The amount of thermally generated reverse current in a photodiode in the absence of light.

Electroluminescence   The process of releasing light energy by the recombination of electrons in a semiconductor.

Figure of merit   The ratio of energy stored and returned by a reactive component to the energy dissipated; also called *quality factor, Q*.

Infrared (IR)   Light that has a range of wavelengths greater than visible light.

Irradiance (H)   The power per unit area at a specified distance from the LED; the light intensity.

Laser   *L*ight *a*mplification by *s*timulated *e*mission of *r*adiation.

Light-emitting diode (LED)   A type of diode that emits light when forward current flows.

Line regulation   The percent change in output voltage for a given change in line (input) voltage.

Load regulation   The percent change in output voltage for a given change in load current.

Monochromatic   Related to light of a single frequency; one color.

Percent regulation   A figure of merit used to specify the performance of a voltage regulator.

Photodiode   A diode in which the reverse current varies directly with the amount of light.

Photon   A particle of light energy.

Radiant intensity (I_e)   The output power of an LED per steradian in units of mW/sr.

Radiation The process of emitting electromagnetic or light energy.

Schottky diode A diode using only majority carriers and intended for high-frequency operation.

Spectral Pertaining to a range of frequencies.

Tuning ratio The ratio of varactor capacitances at minimum and at maximum reverse voltages.

Tunnel diode A diode exhibiting a negative resistance characteristic.

Varactor A variable capacitance diode.

Wavelength The distance in space occupied by one cycle of an electromagnetic or light wave.

Zener breakdown The lower voltage breakdown in a zener diode.

Zener diode A diode designed for limiting the voltage across its terminals in reverse bias.

■ FORMULAS

(3–1)	$$Z_Z = \dfrac{\Delta V_Z}{\Delta I_Z}$$		Zener impedance
(3–2)	$$\Delta V_Z = V_Z \times TC \times \Delta T$$		V_Z temperature change when TC is %/C°
(3–3)	$$\Delta V_Z = TC \times \Delta T$$		V_Z temperature change when TC is mV/C°
(3–4)	$$P_D = V_Z I_Z$$		Zener power dissipation
(3–5)	$$P_{D(derated)} = P_{D(max)} - (mW/C°)\Delta T$$		Derated power dissipation
(3–6)	$$I_{ZM} = \dfrac{P_{D(max)}}{V_Z}$$		Maximum zener current
(3–7)	$$\text{Percent line regulation} = \left(\dfrac{\Delta V_{OUT}}{\Delta V_{IN}}\right)100\%$$		Line regulation
(3–8)	$$\text{Percent load regulation} = \left(\dfrac{V_{NL} - V_{FL}}{V_{FL}}\right)100\%$$		Load regulation
(3–9)	$$C = \dfrac{A\epsilon}{d}$$		Capacitance
(3–10)	$$f_r \cong \dfrac{1}{2\pi\sqrt{LC}}$$		Resonant frequency
(3–11)	$$H = \dfrac{I_e}{d^2}$$		Irradiance

■ SELF-TEST

1. The cathode of a zener diode in a voltage regulator is normally
   (a) more positive than the anode   (b) more negative than the anode
   (c) at + 0.7 V   (d) grounded

2. If a certain zener diode has a zener voltage of 3.6 V, it operates in
   (a) regulated breakdown   (b) zener breakdown
   (c) forward conduction   (d) avalanche breakdown

3. For a certain 12 V zener diode, a 10 mA change in zener current produces a 0.1 V change in zener voltage. The zener impedance for this current range is
   (a) 1 Ω   (b) 100 Ω   (c) 10 Ω   (d) 0.1 Ω

4. The data sheet for a particular zener gives V_Z = 10 V at I_{ZT} = 500 mA. Z_Z for these conditions is
   (a) 50 Ω   (b) 20 Ω   (c) 10 Ω   (d) unknown

5. Line regulation is determined by
   (a) load current
   (b) zener current and load current
   (c) changes in load resistance and output voltage
   (d) changes in output voltage and input voltage

6. Load regulation is determined by
   (a) changes in load current and input voltage
   (b) changes in load current and output voltage
   (c) changes in load resistance and input voltage
   (d) changes in zener current and load current

7. A no-load condition means that
   (a) the load has infinite resistance   (b) the load has zero resistance
   (c) the output terminals are open   (d) answers (a) and (c)

8. A varactor diode exhibits
   (a) a variable capacitance that depends on reverse voltage
   (b) a variable resistance that depends on reverse voltage
   (c) a variable capacitance that depends on forward current
   (d) a constant capacitance over a range of reverse voltages

9. An LED
   (a) emits light when reverse-biased   (b) senses light when reverse-biased
   (c) emits light when forward-biased   (d) acts as a variable resistance

10. Compared to a visible red LED, an infrared LED
    (a) produces light with shorter wavelengths   (b) produces light of all wavelengths
    (c) produces only one color of light   (d) produces light with longer wavelengths

11. The internal resistance of a photodiode
    (a) increases with light intensity when reverse-biased
    (b) decreases with light intensity when reverse-biased
    (c) increases with light intensity when forward-biased
    (d) decreases with light intensity when forward-biased

12. A diode that has a negative resistance characteristic is the
    (a) Schottky diode   (b) tunnel diode   (c) laser diode   (d) hot-carrier diode

13. An infrared LED is optically coupled to a photodiode. When the LED is turned off, the reading on an ammeter in series with the reverse-biased photodiode will
    (a) not change   (b) decrease   (c) increase   (d) fluctuate

14. In order for a system to function properly, the various types of circuits that make up the system must be
    (a) properly biased   (b) properly connected   (c) properly interfaced
    (d) all of the above   (e) answers (a) and (b)

■ BASIC PROBLEMS

SECTION 3–1   Zener Diodes

1. A certain zener diode has a $V_Z = 7.5$ V and an $Z_Z = 5$ Ω at a certain current. Sketch the equivalent circuit.

2. From the characteristic curve in Figure 3–58, what is the approximate minimum zener current (I_{ZK}) and the approximate zener voltage at I_{ZK}?

3. When the reverse current in a particular zener diode increases from 20 mA to 30 mA, the zener voltage changes from 5.6 V to 5.65 V. What is the impedance of this device?

4. A zener has a impedance of 15 Ω. What is its terminal voltage at 50 mA if $V_{ZT} = 4.7$ V at $I_{ZT} = 25$ mA?

5. A certain zener diode has the following specifications: $V_Z = 6.8$ V at 25°C and $TC = +0.04\%$/C°. Determine the zener voltage at 70°C.

FIGURE 3–58

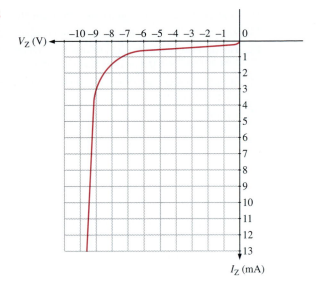

SECTION 3–2 Zener Diode Applications

6. Determine the minimum input voltage required for regulation to be established in Figure 3–59. Assume an ideal zener diode with $I_{ZK} = 1.5$ mA and $V_Z = 14$ V.

7. Repeat Problem 6 with $Z_Z = 20$ Ω and $V_{ZT} = 14$ V at 30 mA.

8. To what value must R be adjusted in Figure 3–60 to make $I_Z = 40$ mA? Assume $V_Z = 12$ V at 30 mA and $Z_Z = 30$ Ω.

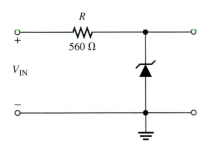

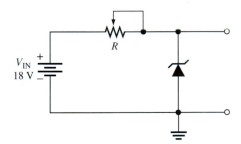

FIGURE 3–59   **FIGURE 3–60**

9. A 20 V peak sinusoidal voltage is applied to the circuit in Figure 3–60 in place of the dc source. Sketch the output waveform. Use the parameter values established in Problem 8.

10. A loaded zener regulator is shown in Figure 3–61. $V_Z = 5.1$ V at 35 mA, $I_{ZK} = 1$ mA, $Z_Z = 12$ Ω, and $I_{ZM} = 70$ mA. Determine the minimum and maximum permissible load currents.

FIGURE 3–61

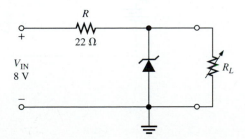

11. Find the percent load regulation in Problem 10.

12. Analyze the circuit in Figure 3–61 for an input voltage from 6 V to 12 V with no load.

13. The no-load output voltage of a certain zener regulator is 8.23 V, and the full-load output is 7.98 V. Calculate the percent load regulation.

14. In a certain zener regulator, the output voltage changes 0.2 V when the input voltage goes from 5 V to 10 V. What is the percent input regulation?

15. The output voltage of a zener regulator is 3.6 V at no load and 3.4 V at full load. Determine the percent load regulation.

SECTION 3–3 Varactor Diodes

16. Figure 3–62 is a curve of reverse voltage versus capacitance for a certain varactor. Determine the change in capacitance if V_R varies from 5 V to 20 V.

17. Refer to Figure 3–62 and determine the value of V_R that produces 25 pF.

18. What capacitance value is required for each of the varactors in Figure 3–63 to produce a resonant frequency of 1 MHz?

19. At what value must the voltage V_R be set in Problem 18 if the varactors have the characteristic curve in Figure 3–62?

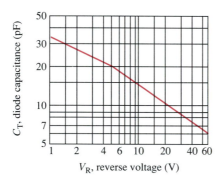

FIGURE 3–62

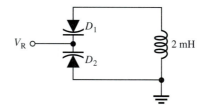

FIGURE 3–63

SECTION 3–4 Optical Diodes

20. The LED in Figure 3–64(a) has a light-producing characteristic as shown in 3–64(b). Neglecting the forward voltage drop of the LED, determine the amount of radiant (light) power produced in mW.

FIGURE 3–64

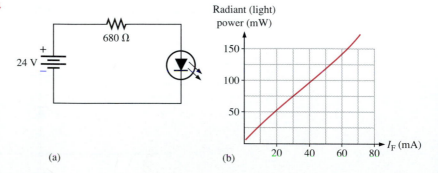

(a)

(b)

21. Determine how to connect the seven-segment display in Figure 3–65 to display "5." The maximum continuous forward current for each LED is 30 mA and a +5 V dc source is to be used.

FIGURE 3–65

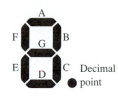

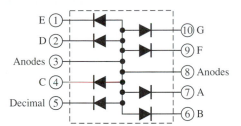

22. For a certain photodiode at a given irradiance, the reverse resistance is 200 kΩ and the reverse voltage is 10 V. What is the current through the device?

23. What is the resistance of each photodiode in Figure 3–66?

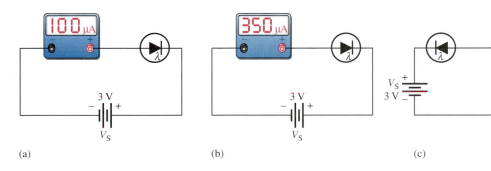

(a)   (b)   (c)

FIGURE 3–66

24. When the switch in Figure 3–67 is closed, will the microammeter reading increase or decrease? Assume D_1 and D_2 are optically coupled.

FIGURE 3–67

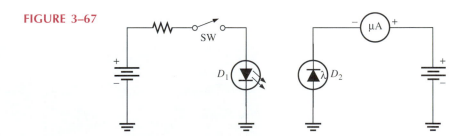

SECTION 3–5   Other Types of Diodes

25. The *I-V* characteristic of a certain tunnel diode shows that the current changes from 0.25 mA to 0.15 mA when the voltage changes from 125 mV to 200 mV. What is the resistance?

26. In what type of circuit are tunnel diodes commonly used?

27. What purpose do the reflective surfaces in the laser diode serve? Why is one end only partially reflective?

▪ **TROUBLE-SHOOTING PROBLEMS**

SECTION 3–6   Troubleshooting

28. For each set of measured voltages at the points (1, 2, and 3) indicated in Figure 3–68, determine if they are correct and if not, identify the most likely fault(s). State what you would do to correct the problem once it is isolated. The zener is rated at 12 V.

   (a) $V_1 = 110$ V rms, $V_2 = 30$ V dc, $V_3 = 12$ V dc

   (b) $V_1 = 100$ V rms, $V_2 = 30$ V dc, $V_3 = 30$ V dc

   (c) $V_1 = 0$ V, $V_2 = 0$ V, $V_3 = 0$ V

   (d) $V_1 = 110$ V rms, $V_2 = 30$ V peak full-wave 120 Hz, $V_3 = 12$ V 120 Hz pulsating voltage

   (e) $V_1 = 110$ V rms, $V_2 = 9$ V, $V_3 = 0$ V

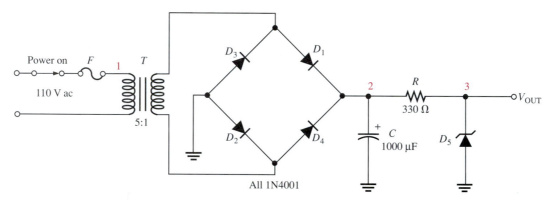

FIGURE 3–68

29. What is the output voltage in Figure 3–68 for each of the following faults?

   (a) D_5 open   (b) R open   (c) C leaky   (d) C open   (e) D_3 open

   (f) D_2 open   (g) T open   (h) F open

SECTION 3–7   System Application

30. The counting and control system has been installed at a customer's location. The system is performing erratically, so you decide to first check the power supply and IR emitter/detector board. Based on Figure 3–69, determine the problem.

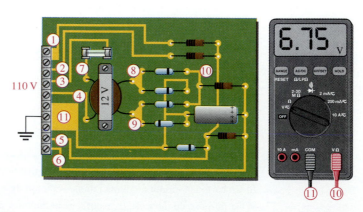

FIGURE 3–69

31. Another problem has developed with the counter and control system. This time, the system completely quits working and again you decide to first check the power supply board. Based on Figure 3–70, determine the problem.

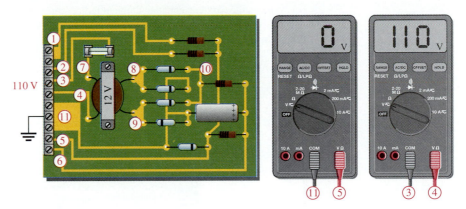

FIGURE 3–70

32. List the possible reasons for the LED in Figure 3–55 not emitting infrared light when the power supply is plugged in.

33. List the possible reasons for the photodiode in Figure 3–55 not responding to the infrared light from the LED. List the steps in sequence that you would take to isolate the problem.

■ **DATA SHEET PROBLEMS**

34. Refer to the zener diode data sheet in Figure 3–7.
   (a) What is the maximum dc power dissipation at 25°C for a 1N4738?
   (b) Determine the maximum power dissipation at 70°C and at 100°C for a 1N4751.
   (c) What is the minimum current required by the 1N4738 for regulation?
   (d) What is the maximum current for the 1N4750 at 25°C?
   (e) The current through a 1N4740 changes from 25 mA to 0.25 mA. How much does the zener impedance change?
   (f) What is the maximum zener voltage of a 1N4736 at 50°C?
   (g) What is the minimum zener voltage for a 1N4747 at 75°C?

35. Refer to the varactor diode data sheet in Figure 3–22.
   (a) What is the maximum reverse voltage for the 1N5139?
   (b) Determine the maximum power dissipation for a 1N5141 at an ambient temperature of 60°C.
   (c) Determine the maximum power dissipation for a 1N5148 at a case temperature of 80°C.
   (d) What is the capacitance of a 1N5148 at a reverse voltage of 20 V?
   (e) If figure-of-merit were the only criteria, which varactor diode would you select?
   (f) What is the typical capacitance at $V_R = 60$ V for a 1N5142?

36. Refer to the LED data sheet in Figure 3–31.
   (a) Can 9 V be applied in reverse across an MLED81?
   (b) Determine the minimum value of series resistor for the MLED81 when a voltage of 5.1 V is used to forward-bias the diode.
   (c) Assume the forward current is 50 mA and the forward voltage drop is 1.5 V at an ambient temperature of 45°C. Is the maximum power rating exceeded?
   (d) Determine the axial radiant intensity for a forward current of 30 mA.
   (e) What is the radiant intensity at an angle of 20° from the axis if the forward current is 100 mA?

37. Refer to the photodiode data sheet in Figure 3–36.

    (a) An MRD821 is connected in series with a 10 kΩ resistor and a reverse-bias voltage source. There is no incident light on the diode. What is the voltage drop across the resistor?

    (b) At what wavelength will the reverse current be the greatest for a given irradiance?

    (c) What is the dark current at an ambient temperature of 60°C?

    (d) At what wavelength is sensitivity of the MRD821 at a maximum?

    (e) If the maximum sensitivity is 50 µA/mW/cm², what is the sensitivity at 900 nm?

    (f) An infrared light with a wavelength of 900 nm strikes an MRD821 with an irradiance of 3 mW/cm² and an angle of 40° from the maximum axis. Determine the reverse current.

■ **ADVANCED PROBLEMS**

38. Develop the schematic for the circuit board in Figure 3–71 and determine what type of circuit it is.

FIGURE 3–71

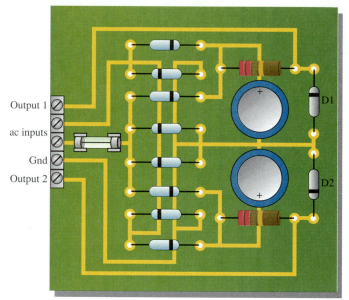

Rectifier diodes: 1N4001
Zener diodes: D1-1N4736, D2-1N4749
Filter capacitors: 100 µF

39. If a 110 V rms, 60 Hz input voltage is connected to the ac inputs, determine the output voltages on the circuit board in Figure 3–71.

40. If each output of the board in Figure 3–71 is loaded with 1 kΩ, what fuse rating should be used?

41. Design a zener voltage regulator to meet the following specifications: The input voltage is 24 V dc, the load current is 35 mA, and the load voltage is 8.2 V.

42. The varactor-tuned band-pass filter in Figure 3–24 is to be redesigned to produce a bandwidth of from 350 kHz to 850 kHz within a 10% tolerance. Using the basic circuit in Figure 3–72, determine all components necessary to meet the specification. Use the nearest standard values.

43. Design a seven-segment LED display circuit in which any of the ten digits can be displayed using a set of switches. Each LED segment is to have a current of 20 mA ± 10% from a 12 V source and the circuit must be designed with a minimum number of switches.

44. If you used a common anode seven-segment display in Problem 43, redesign it for a common-cathode display or vice versa.

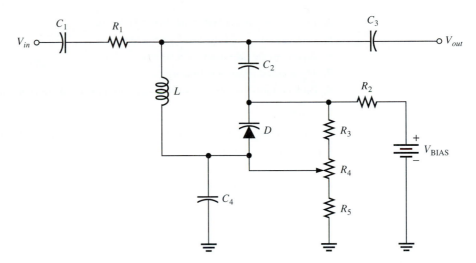

FIGURE 3–72

■ **ANSWERS TO SECTION REVIEWS**

Section 3–1

1. Zener diodes are operated in the reverse-bias region.
2. The test current, I_{ZT}
3. The zener impedance causes the voltage to vary slightly with current.
4. $V_Z = 10 \text{ V} + (20 \text{ mA})(8 \ \Omega) = 10.16 \text{ V}$.
5. The zener voltage increases (or decreases) 0.05% for each centrigrade degree increase (or decrease).
6. Power derating is the reduction in the power rating of a device as a result of an increase in temperature.

Section 3–2

1. Line regulation is a change in output voltage for a given change in input voltage. Load regulation is a change in output voltage for a given change in load current.
2. A infinite resistance (open)
3. With no load, there is no current to a load. With full load, there is maximum current to the load.
4. Percent load regulation = ((12 V − 11.9 V)/11.9 V) 100% = 0.84%
5. Approximately 0.7 V, just like a rectifier diode

Section 3–3

1. A varactor exhibits variable capacitance.
2. A varactor is operated in reverse bias.
3. The depletion region
4. Capacitance decreases with more reverse bias.
5. The tuning ratio is the ratio of a varactor's capacitance at a specified minimum voltage to the capacitance at a specified maximum voltage.

Section 3–4

1. Infrared and visible light
2. Infrared has the greater wavelengths.
3. An LED operates in forward bias.

4. Light emission increases with forward current.

5. False, V_F of an LED is usually greater than 1.2 V.

6. A photodiode operates in reverse bias.

7. The internal resistance decreases.

8. Dark current is the reverse photodiode current when there is no light.

Section 3–5

1. A current regulator operates between V_L (limiting voltage) and POV (peak operating voltage).

2. High-frequency and fast-switching circuits

3. *Hot carrier* is another name for Schottky diodes.

4. Tunnel diodes have negative resistance.

5. Oscillators

6. *p* region, *n* region, and intrinsic (*i*) region

7. *l*ight *a*mplification by *s*timulated *e*mission of *r*adiation

8. Coherent light has only a single wavelength, but incoherent light has a wide band of wavelengths. A laser diode produces coherent light.

Section 3–6

1. The output voltage is too high equal to the rectifier output.

2. More

3. Series limiting resistor open, fuse blown

4. The output voltage changes as the load resistance changes.

▪ **ANSWERS TO RELATED EXERCISES FOR EXAMPLES**

3–1 5 Ω

3–2 $V_Z = 11.4$ V at 20 mA; $V_Z = 12.6$ V at 80 mA

3–3 0.45 V to 11.55 V

3–4 7.5 W

3–5 $V_{IN(min)} = 6.77$ V; $V_{IN(max)} = 21.9$ V

3–6 $I_{L(min)} = 0$ A; $I_{L(max)} = 43$ mA; $R_{L(min)} = 76.7$ Ω

3–7 **(a)** 11.8 V at I_{ZK}; 12.6 V at I_{ZM} **(b)** 144 Ω **(c)** 140 Ω

3–8 **(a)** 3.77% **(b)** 1.75%/V

3–9 **(a)** A waveform identical to Figure 3–18(a)

(b) A sine wave with a peak value of 5 V

3–10 $V_{R(min)} = 1.43$ V; $V_{R(max)} = 14.4$ V

3–11 2.4 mW/sr

3–12 29.8 μA

4

BIPOLAR
JUNCTION
TRANSISTORS

■ **CHAPTER OBJECTIVES**

☐ Describe the basic structure of the bipolar
   junction transistor
☐ Explain how a transistor is biased and discuss
   the transistor currents and voltages
☐ Discuss transistor parameters and characteristics
   and use these to analyze a transistor circuit
☐ Discuss how a transistor is used as a voltage
   amplifier
☐ Discuss how a transistor is used as an electronic
   switch
☐ Identify various types of transistor package
   configurations
☐ Troubleshoot various faults in transistor circuits

The transistor was invented by a team of three men
at Bell Laboratories in 1947. Although this first
transistor was not a bipolar junction device, it was the
beginning of a technological revolution that is still
continuing. All of the complex electronic devices and
systems today are an outgrowth of early developments
in semiconductor transistors.

Two basic types of transistors are the bipolar
junction transistor (BJT), which we will begin to
study in this chapter, and the field-effect transistor
(FET), which we will cover in later chapters. The BJT
is used in two broad areas—as a linear amplifier to
boost or amplify an electrical signal and as an
electronic switch. Both of these applications are
introduced in this chapter.

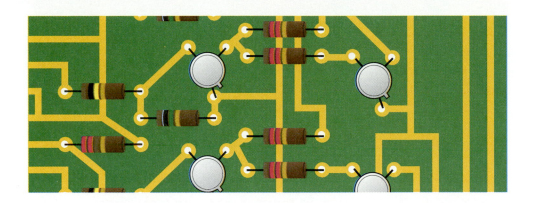

■ SYSTEM APPLICATION

Another system in the diversified product line of your company is a security alarm system for protecting homes and places of business against illegal entry. You are given the responsibility for final development and for testing each system before it goes to the customer's location. The first step in your assignment is to learn all you can about transistor operation. You will then apply your knowledge to the system application in Section 4–8.

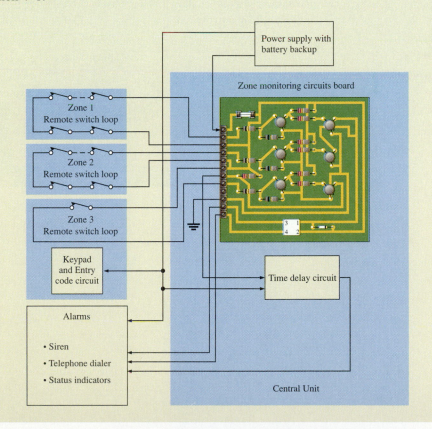

4–1 ▪ TRANSISTOR CONSTRUCTION

The basic structure of the bipolar junction transistor (BJT) determines its operating characteristics. In this section, you will see how semiconductor materials are used to form a transistor, and you will learn the standard transistor symbols.

After completing this section, you should be able to

▪ **Describe the basic structure of the bipolar junction transistor**
   ☐ Explain the difference between the structure of an *npn* and a *pnp* transistor
   ☐ Identify the symbols for *npn* and *pnp* transistors
   ☐ Name the three regions of a BJT and their labels

The **bipolar junction transistor (BJT)** is constructed with three doped semiconductor regions separated by two *pn* junctions, as shown in the epitaxial planar structure in Figure 4–1(a). The three regions are called **emitter, base,** and **collector.** Physical representations of the two types of bipolar transistors are shown in Figure 4–1(b) and (c). One type consists of two *n* regions separated by a *p* region (*npn*), and the other consists of two *p* regions separated by an *n* region (*pnp*).

The *pn* junction joining the base region and the emitter region is called the *base-emitter junction*. The junction joining the base region and the collector region is called the *base-collector junction,* as indicated in Figure 4–1(b). A wire lead connects to each of the three regions, as shown. These leads are labeled E, B, and C for emitter, base, and collector, respectively.

FIGURE 4–1

Basic bipolar transistor construction.

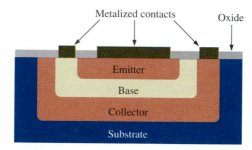

(a) Basic epitaxial planar structure

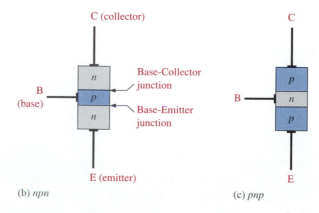

(b) *npn*   (c) *pnp*

The base region is lightly doped and very thin compared to the heavily doped emitter and the moderately doped collector regions. (The reason for this is discussed in the next section.) Figure 4–2 shows the schematic symbols for the *npn* and *pnp* bipolar transistors. The term **bipolar** refers to the use of both holes and electrons as carriers in the transistor structure.

FIGURE 4–2

Standard bipolar junction transistor (BJT) symbols.

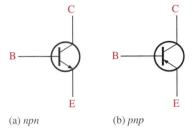

(a) *npn*   (b) *pnp*

SECTION 4–1 REVIEW

1. Name the two types of BJTs according to their construction.
2. The BJT is a three-terminal device. Name the three terminals.
3. What separates the three regions in a BJT?

4–2 ■ BASIC TRANSISTOR OPERATION

In order for the transistor to operate properly as an amplifier, the two pn junctions must be correctly biased with external dc voltages. In this section, we use the npn transistor for illustration. The operation of the pnp is the same as for the npn except that the roles of the electrons and holes, the bias voltage polarities, and the current directions are all reversed.

After completing this section, you should be able to

■ **Explain how a transistor is biased and discuss the transistor currents and their relationships**
   □ Describe forward-reverse bias
   □ Show how to connect a transistor to the bias-voltage sources
   □ Describe the basic internal operation of a transistor
   □ State the formula relating the collector, emitter, and base currents in a transistor

Figure 4–3 shows the proper **bias** arrangement for both *npn* and *pnp* **transistors** for active operation as an **amplifier.** Notice that in both cases the base-emitter (BE) junction is forward-biased and the base-collector (BC) junction is reverse-biased.

To illustrate transistor action let's examine what happens inside the *npn* transistor when the junctions are forward-reverse biased. The forward bias from base to emitter narrows the BE depletion region, and the reverse bias from base to collector widens the BC depletion region, as depicted in Figure 4–4. The heavily doped *n*-type emitter region is teeming with conduction-band (free) electrons that easily diffuse through the forward-

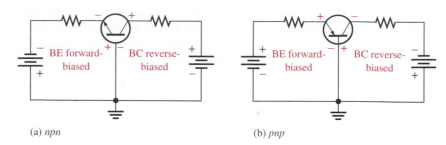

(a) *npn* (b) *pnp*

FIGURE 4–3

Forward-reverse bias of a bipolar transistor.

biased BE junction into the *p*-type base region where they become minority carriers, just as in a forward-biased diode. The base region is lightly doped and very thin so that it has a very limited number of holes. Thus, only a small percentage of all the electrons flowing through the BE junction can combine with the available holes in the base. These relatively few recombined electrons flow out of the base lead as valence electrons, forming the small base electron current, as shown in Figure 4–4.

Most of the electrons flowing from the emitter into the thin, lightly doped base region do not recombine but diffuse into the BC depletion region. Once in this region they are pulled through the reverse-biased BC junction by the electric field set up by the force of attraction between the positive and negative ions. Actually, you can think of the electrons as being pulled across the reverse-biased BC junction by the attraction of the collector supply voltage. The electrons now move through the collector region, out

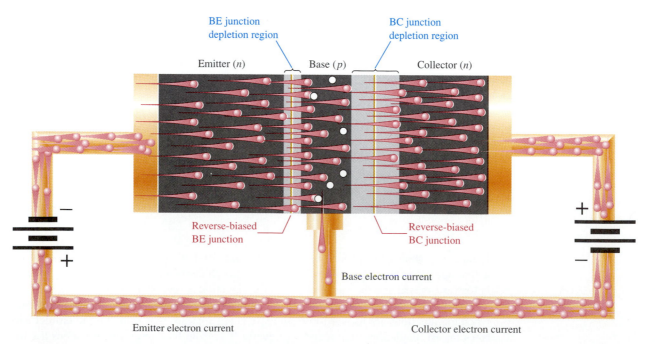

FIGURE 4–4

Illustration of BJT action.

through the collector lead, and into the positive terminal of the collector voltage source. This forms the collector electron current, as shown in Figure 4–4.

Transistor Currents

The directions of the currents in an *npn* transistor are as shown in Figure 4–5(a), and those for a *pnp* are shown in Figure 4–5(b). The currents are indicated on the corresponding schematic symbols in parts (c) and (d) of the figure. These diagrams show that the emitter current (I_E) is the sum of the collector current (I_C) and the base current (I_B), expressed as follows:

$$I_E = I_C + I_B \tag{4–1}$$

As mentioned before, I_B is very small compared to I_E or I_C. The uppercase subscripts indicate dc values.

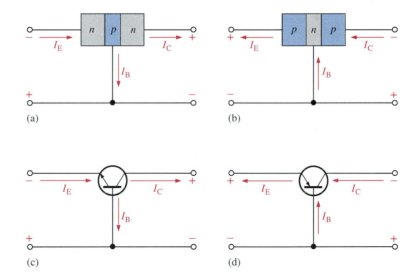

FIGURE 4–5
Transistor currents.

SECTION 4–2 REVIEW

1. What are the bias conditions of the base-emitter and base-collector junctions for a transistor to operate as an amplifier?

2. Which is the largest of the three transistor currents?

3. Is the base current smaller or larger than the emitter current?

4. Is the base region much thinner or much wider than the collector and emitter regions?

5. If the collector current is 1 mA and the base current is 10 μA, what is the emitter current?

4–3 ■ TRANSISTOR CHARACTERISTICS AND PARAMETERS

In this section, you will first learn how to set up a dc circuit to properly bias a transistor. Two important parameters, β_{DC} (dc current gain) and α_{DC} are introduced and used to analyze a transistor circuit. Also, transistor characteristic curves are covered, and you will learn how a transistor's operation can be determined from these curves. Finally, maximum ratings of a transistor are discussed.

After completing this section, you should be able to

■ **Discuss transistor parameters and characteristics and use these to analyze a transistor circuit**
  □ Define dc *beta* (β_{DC})
  □ Define dc *alpha* (α_{DC})
  □ State the mathematical relationship between β_{DC} and α_{DC}
  □ Identify all currents and voltages in a transistor circuit
  □ Analyze a basic transistor dc circuit
  □ Interpret collector characteristic curves and use a dc load line
  □ Describe how β_{DC} varies with temperature and collector current
  □ Discuss and apply maximum transistor ratings
  □ Derate a transistor for power dissipation
  □ Interpret a transistor data sheet

When a transistor is connected to bias voltages, as shown in Figure 4–6 for both *npn* and *pnp* types, V_{BB} forward-biases the base-emitter junction, and V_{CC} reverse-biases the base-collector junction. Although in this chapter we are using battery symbols to represent the bias voltages, in practice the voltages are usually derived from a dc power supply. For example, V_{CC} is usually taken directly from the power supply output and V_{BB} (which is smaller) can be produced with a voltage divider. Bias arrangements are examined thoroughly in Chapter 5.

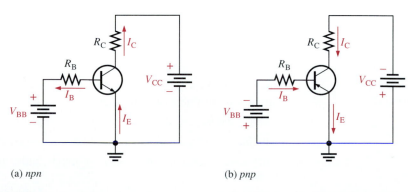

(a) *npn*   (b) *pnp*

FIGURE 4–6
Transistor dc bias circuits.

DC Beta (β_{DC}) and DC Alpha (α_{DC})

The ratio of the collector current I_C to the base current I_B is the dc **beta** (β_{DC}), which is the dc current **gain** of a transistor.

$$\beta_{DC} = \frac{I_C}{I_B} \tag{4–2}$$

Typical values of β_{DC} range from less than 20 to 200 or higher. β_{DC} is usually designated as h_{FE} on transistor data sheets.

$$h_{FE} = \beta_{DC}$$

The ratio of the collector current I_C to the emitter current I_E is the dc **alpha** (α_{DC}).

$$\alpha_{DC} = \frac{I_C}{I_E} \tag{4–3}$$

Typically, values of α_{DC} range from 0.95 to 0.99 or greater, but α_{DC} is always less than 1. The reason is that I_C is always slightly less than I_E by the amount of I_B. For example, if $I_E = 100$ mA and $I_B = 1$ mA, then $I_C = 99$ mA and $\alpha_{DC} = 0.99$.

Relationship of β_{DC} and α_{DC}

Let's start with the current formula $I_E = I_C + I_B$ and divide each current by I_C:

$$\frac{I_E}{I_C} = \frac{I_C}{I_C} + \frac{I_B}{I_C} = 1 + \frac{I_B}{I_C}$$

Since $\beta_{DC} = I_C/I_B$ and $\alpha_{DC} = I_C/I_E$, we can substitute the reciprocals into the equation:

$$\frac{1}{\alpha_{DC}} = 1 + \frac{1}{\beta_{DC}}$$

By rearranging and solving for β_{DC}, we get

$$\frac{1}{\alpha_{DC}} = \frac{\beta_{DC} + 1}{\beta_{DC}}$$

$$\beta_{DC} = \alpha_{DC}(\beta_{DC} + 1)$$

$$\beta_{DC} = \alpha_{DC}\beta_{DC} + \alpha_{DC}$$

$$\beta_{DC} - \alpha_{DC}\beta_{DC} = \alpha_{DC}$$

$$\beta_{DC}(1 - \alpha_{DC}) = \alpha_{DC}$$

$$\beta_{DC} = \frac{\alpha_{DC}}{1 - \alpha_{DC}} \tag{4–4}$$

Equation (4–4) shows that the closer α_{DC} is to 1, the higher the value of β_{DC}.

EXAMPLE 4–1   Determine β_{DC}, I_E, and α_{DC} for a transistor where $I_B = 50 \ \mu A$ and $I_C = 3.65 \ mA$.

Solution

$$\beta_{DC} = \frac{I_C}{I_B} = \frac{3.65 \ mA}{50 \ \mu A} = 73$$

$$I_E = I_C + I_B = 3.65 \ mA + 50 \ \mu A = 3.70 \ mA$$

$$\alpha_{DC} = \frac{I_C}{I_E} = \frac{3.65 \ mA}{3.70 \ mA} = 0.986$$

Related Exercise   A certain transistor has a β_{DC} of 200. When the base current is 50 μA, determine the collector current. What is α_{DC}?

Current and Voltage Analysis

Consider the circuit configuration in Figure 4–7. Three transistor currents and three dc voltages can be identified.

I_B: base current (dc)

I_E: emitter current (dc)

I_C: collector current (dc)

V_{BE}: dc voltage at base with respect to emitter

V_{CB}: dc voltage at collector with respect to base

V_{CE}: dc voltage at collector with respect to emitter

FIGURE 4–7
Transistor currents and voltages.

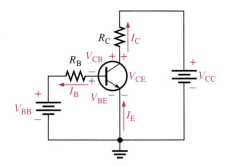

V_{BB} forward-biases the base-emitter junction and V_{CC} reverse-biases the base-collector junction. When the base-emitter junction is forward-biased, it is like a forward-biased diode and has a forward voltage drop of

$$V_{BE} \cong 0.7 \ V \qquad\qquad (4\text{–}5)$$

Since the emitter is at ground (0 V), by Kirchhoff's voltage law, the voltage across R_B is

$$V_{R_B} = V_{BB} - V_{BE}$$

Also, by Ohm's law,

$$V_{R_B} = I_B R_B$$

Substituting for V_{R_B} yields

$$I_B R_B = V_{BB} - V_{BE}$$

Solving for I_B, you get

$$I_B = \frac{V_{BB} - V_{BE}}{R_B} \qquad (4\text{–}6)$$

The drop across R_C is

$$V_{R_C} = I_C R_C$$

The voltage at the collector with respect to the emitter, which is grounded, is

$$V_{CE} = V_{CC} - I_C R_C \qquad (4\text{–}7)$$

where $I_C = \beta_{DC} I_B$. The voltage across the reverse-biased collector-base junction is

$$V_{CB} = V_{CE} - V_{BE} \qquad (4\text{–}8)$$

EXAMPLE 4–2

Determine I_B, I_C, I_E, V_{BE}, V_{CE}, and V_{CB} in the circuit of Figure 4–8. The transistor has a $\beta_{DC} = 150$.

FIGURE 4–8

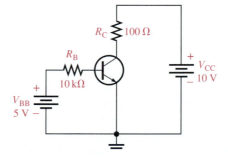

Solution From Equation (4–5) $V_{BE} = 0.7$ V. Calculate the base, collector, and emitter currents as follows:

$$I_B = \frac{V_{BB} - V_{BE}}{R_B} = \frac{5\text{ V} - 0.7\text{ V}}{10\text{ k}\Omega} = 430\ \mu\text{A}$$

$$I_C = \beta_{DC} I_B = (150)(430\ \mu\text{A}) = 64.5\text{ mA}$$

$$I_E = I_C + I_B = 64.5\text{ mA} + 430\ \mu\text{A} = 64.9\text{ mA}$$

Solve for V_{CE} and V_{CB}:

$$V_{CE} = V_{CC} - I_C R_C = 10 \text{ V} - (64.5 \text{ mA})(100 \text{ }\Omega)$$
$$= 10 \text{ V} - 6.45 \text{ V} = 3.55 \text{ V}$$

$$V_{CB} = V_{CE} - V_{BE} = 3.55 \text{ V} - 0.7 \text{ V} = 2.85 \text{ V}$$

Since the collector is at a higher voltage than the base, the collector-base junction is reverse-biased.

Related Exercise   Determine I_B, I_C, I_E, V_{CE}, and V_{CB} in Figure 4–8 for the following values: $R_B = 22 \text{ k}\Omega$, $R_C = 220 \text{ }\Omega$, $V_{BB} = 6 \text{ V}$, $V_{CC} = 9 \text{ V}$, and $\beta_{DC} = 90$.

Collector Characteristic Curves

Using a circuit like that shown in Figure 4–9(a), you can generate a set of collector characteristic curves that show how the collector current, I_C, varies with the collector-to-emitter voltage, V_{CE}, for specified values of base current, I_B. Notice in the circuit diagram that both V_{BB} and V_{CC} are variable sources of voltage.

Assume that V_{BB} is set to produce a certain value of I_B and V_{CC} is zero. For this condition, both the base-emitter junction and the base-collector junction are forward-biased because the base is at approximately 0.7 V while the emitter and the collector are at 0 V. The base current is through the base-emitter junction because of the low impedance path to ground and, therefore, I_C is zero. When both junctions are forward-biased, the transistor is in the **saturation** region of its operation.

As V_{CC} is increased, V_{CE} increases gradually as the collector current increases. This is indicated by the portion of the characteristic curve between points A and B in Figure 4–9(b). I_C increases as V_{CC} is increased because V_{CE} remains less than 0.7 V due to the forward-biased base-collector junction.

When V_{CE} exceeds 0.7 V, the base-collector junction becomes reverse-biased and the transistor goes into the *active* or **linear** region of its operation. Once the base-collector junction is reverse-biased, I_C levels off and remains essentially constant for a given value of I_B as V_{CE} continues to increase. Actually, I_C increases very slightly as V_{CE} increases due to widening of the base-collector depletion region. This results in fewer holes for recombination in the base region which effectively causes a slight increase in β_{DC}. This is shown by the portion of the characteristic curve between points B and C in Figure 4–9(b). For this portion of the characteristic curve, the value of I_C is determined only by the relationship expressed as $I_C = \beta_{DC} I_B$.

When V_{CE} reaches a sufficiently high voltage, the reverse-biased base-collector junction goes into breakdown; and the collector current increases rapidly as indicated by the part of the curve to the right of point C in Figure 4–9(b). A transistor should never be operated in this breakdown region.

A family of collector characteristic curves is produced when I_C versus V_{CE} is plotted for several values of I_B, as illustrated in Figure 4–9(c). When $I_B = 0$, the transistor is in the **cutoff** region although there is a very small collector leakage current as indicated. The amount of collector leakage current for $I_B = 0$ is exaggerated on the graph for purposes of illustration.

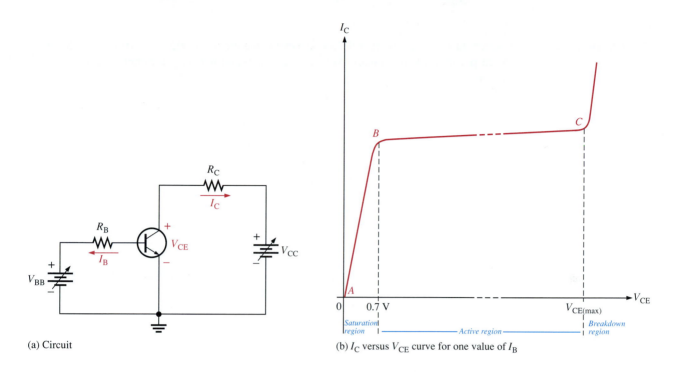

(a) Circuit

(b) I_C versus V_{CE} curve for one value of I_B

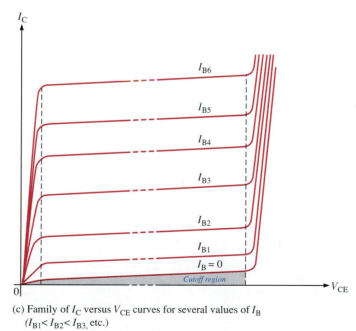

(c) Family of I_C versus V_{CE} curves for several values of I_B
($I_{B1} < I_{B2} < I_{B3}$, etc.)

FIGURE 4–9

Collector characteristic curves.

EXAMPLE 4–3

Sketch an ideal family of collector curves for the circuit in Figure 4–10 for $I_B = 5$ μA to 25 μA in 5 μA increments. Assume $\beta_{DC} = 100$ and that V_{CE} does not exceed breakdown.

FIGURE 4–10

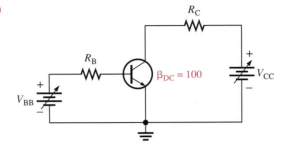

Solution Using the relationship $I_C = \beta_{DC} I_B$, values of I_C are calculated and tabulated in Table 4–1. The resulting curves are plotted in Figure 4–11. These are ideal curves because the slight increase in I_C for a given value of I_B as V_{CE} increases in the active region is neglected.

TABLE 4–1

I_B	I_C
5 μA	0.5 mA
10 μA	1.0 mA
15 μA	1.5 mA
20 μA	2.0 mA
25 μA	2.5 mA

FIGURE 4–11

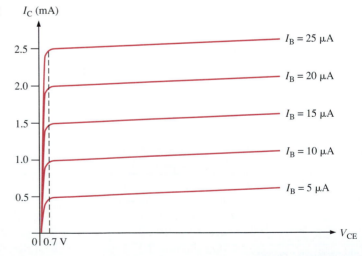

Related Exercise Where would the curve for $I_B = 0$ appear on the graph in Figure 4–11, neglecting collector leakage current?

Cutoff

As previously mentioned, when $I_B = 0$, the transistor is in the cutoff region of its operation. This is shown in Figure 4–12 with the base lead open, resulting in a base current of zero. Under this condition, there is a very small amount of collector leakage current, I_{CEO}, due mainly to thermally produced carriers. Because I_{CEO} is extremely small, it will usually be neglected in circuit analysis so that $V_{CE} = V_{CC}$. In cutoff, both the base-emitter and the base-collector junctions are reverse-biased.

FIGURE 4–12

Collector leakage current (I_{CEO}) in cutoff.

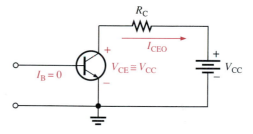

Saturation

When the base current in Figure 4–13(a) is increased, the collector current also increases $(I_C = \beta_{DC}I_B)$ and V_{CE} decreases as a result of more drop across the collector resistor $(V_{CE} = V_{CC} - I_C R_C)$. This is illustrated in Figure 4–13. When V_{CE} reaches its saturation value, $V_{CE(sat)}$, the base-collector junction becomes forward-biased and I_C can increase no further even with a continued increase in I_B. At the point of saturation, the relation $I_C = \beta_{DC}I_B$ is no longer valid. $V_{CE(sat)}$ for a transistor occurs somewhere below the knee of the collector curves, and it is usually only a few tenths of a volt for silicon transistors.

FIGURE 4–13

As I_B increases due to increasing V_{BB}, I_C also increases and V_{CE} decreases due to the increased voltage drop across R_C.

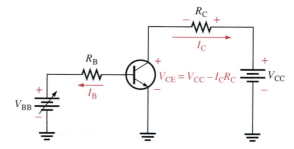

DC Load Line

Cutoff and saturation can be illustrated in relation to the collector characteristic curves by the use of a load line. Figure 4–14 shows a dc load line drawn on a family of curves connecting the cutoff point and the saturation point. The bottom of the load line is at ideal cutoff where $I_C = 0$ and $V_{CE} = V_{CC}$. The top of the load line is at saturation where $I_C = I_{C(sat)}$ and $V_{CE} = V_{CE(sat)}$. In between cutoff and saturation along the load line is the active region of the transistor's operation. Load line operation is discussed more in a later chapter.

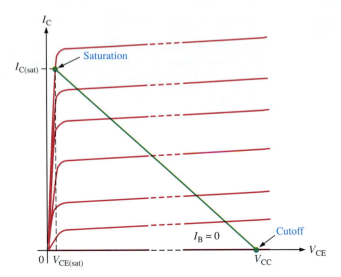

FIGURE 4–14

DC load line on a family of collector characteristic curves illustrating the cutoff and saturation conditions.

EXAMPLE 4–4

Determine whether or not the transistor in Figure 4–15 is in saturation. Assume $V_{CE(sat)} = 0.2$ V.

FIGURE 4–15

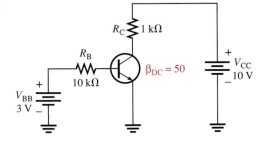

Solution   First, determine $I_{C(sat)}$.

$$I_{C(sat)} = \frac{V_{CC} - V_{CE(sat)}}{R_C} = \frac{10\text{ V} - 0.2\text{ V}}{1\text{ k}\Omega} = \frac{9.8\text{ V}}{1\text{ k}\Omega} = 9.8\text{ mA}$$

Now, see if I_B is large enough to produce $I_{C(sat)}$.

$$I_B = \frac{V_{BB} - V_{BE}}{R_B} = \frac{3\text{ V} - 0.7\text{ V}}{10\text{ k}\Omega} = \frac{2.3\text{ V}}{10\text{ k}\Omega} = 0.23\text{ mA}$$

$$I_C = \beta_{DC}I_B = (50)(0.23\text{ mA}) = 11.5\text{ mA}$$

This shows that with the specified β_{DC}, this base current is capable of producing an I_C greater than $I_{C(sat)}$. Therefore, the transistor is saturated, and the collector current value of 11.5 mA is never reached. If you further increase I_B, the collector current remains at its saturation value.

Related Exercise Determine whether or not the transistor in Figure 4–15 is saturated for the following values: $\beta_{DC} = 125$, $V_{BB} = 1.5$ V, $R_B = 6.8$ kΩ, $R_C = 180$ Ω, and $V_{CC} = 12$ V.

More About β_{DC}

The β_{DC} or h_{FE} is a very important bipolar junction transistor parameter that we need to examine further. β_{DC} is not truly constant but varies with both collector current and with temperature. Keeping the junction temperature constant and increasing I_C causes β_{DC} to increase to a maximum. A further increase in I_C beyond this maximum point causes β_{DC} to decrease. If I_C is held constant and the temperature is varied, β_{DC} changes directly with the temperature. If the temperature goes up, β_{DC} goes up and vice versa. Figure 4–16 shows the variation of β_{DC} with I_C and junction temperature (T_J) for a typical transistor.

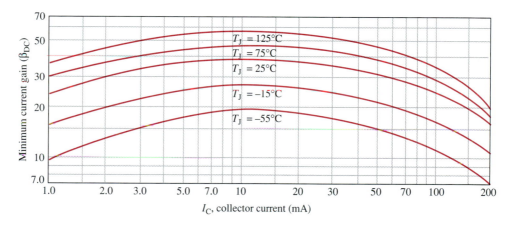

FIGURE 4–16
Variation of β_{DC} with I_C for several temperatures.

A transistor data sheet usually specifies β_{DC} (h_{FE}) at specific I_C values. Even at fixed values of I_C and temperature, β_{DC} varies from device to device for a given transistor due to inconsistencies in the manufacturing process that are unavoidable. The β_{DC} specified at a certain value of I_C is usually the minimum value, $\beta_{DC(min)}$, although the maximum and typical values are also sometimes specified.

Maximum Transistor Ratings

The transistor, like any other electronic device, has limitations on its operation. These limitations are stated in the form of maximum ratings and are normally specified on the manufacturer's data sheet. Typically, maximum ratings are given for collector-to-base voltage, collector-to-emitter voltage, emitter-to-base voltage, collector current, and power dissipation.

The product of V_{CE} and I_C must not exceed the maximum power dissipation. Both V_{CE} and I_C cannot be maximum at the same time. If V_{CE} is maximum, I_C can be calculated as

$$I_C = \frac{P_{D(max)}}{V_{CE}} \tag{4–9}$$

If I_C is maximum, V_{CE} can be calculated by rearranging Equation (4–9) as follows:

$$V_{CE} = \frac{P_{D(max)}}{I_C} \tag{4–10}$$

For any given transistor, a maximum power dissipation curve can be plotted on the collector characteristic curves, as shown in Figure 4–17(a). These values are tabulated in Figure 4–17(b). Assume $P_{D(max)}$ is 500 mW, $V_{CE(max)}$ is 20 V, and $I_{C(max)}$ is 50 mA. The curve shows that this particular transistor cannot be operated in the shaded portion of the graph. $I_{C(max)}$ is the limiting rating between points A and B, $P_{D(max)}$ is the limiting rating between points B and C, and $V_{CE(max)}$ is the limiting rating between points C and D.

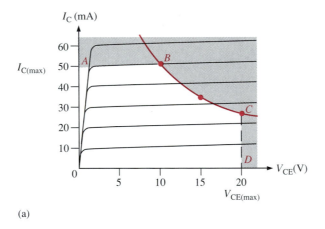

$P_{D\,(max)}$	V_{CE}	I_C
500 mW	5 V	100 mA
500 mW	10 V	50 mA
500 mW	15 V	33 mA
500 mW	20 V	25 mA

(a)   (b)

FIGURE 4–17
Maximum power dissipation curve.

EXAMPLE 4–5

A certain transistor is to be operated with $V_{CE} = 6$ V. If its maximum power rating is 250 mW, what is the most collector current that it can handle?

Solution

$$I_C = \frac{P_{D(max)}}{V_{CE}} = \frac{250\ \text{mW}}{6\ \text{V}} = 41.7\ \text{mA}$$

Remember that this is not necessarily the maximum I_C. The transistor can handle more collector current if V_{CE} is reduced, as long as $P_{D(max)}$ is not exceeded.

Related Exercise If $P_{D(max)} = 1$ W, how much voltage is allowed from collector to emitter if the transistor is operating with $I_C = 100$ mA?

EXAMPLE 4–6

The transistor in Figure 4–18 has the following maximum ratings: $P_{D(max)} = 800$ mW, $V_{CE(max)} = 15$ V, and $I_{C(max)} = 100$ mA. Determine the maximum value to which V_{CC} can be adjusted without exceeding a rating. Which rating would be exceeded first?

FIGURE 4–18

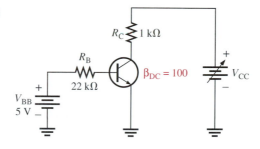

Solution   First, find I_B so that you can determine I_C.

$$I_B = \frac{V_{BB} - V_{BE}}{R_B} = \frac{5\text{ V} - 0.7\text{ V}}{22\text{ k}\Omega} = 195\ \mu\text{A}$$

$$I_C = \beta_{DC}I_B = (100)(195\ \mu\text{A}) = 19.5\text{ mA}$$

I_C is much less than $I_{C(max)}$ and will not change with V_{CC}. It is determined only by I_B and β_{DC}.

The voltage drop across R_C is

$$V_{RC} = I_C R_C = (19.5\text{ mA})(1\text{ k}\Omega) = 19.5\text{ V}$$

Now you can determine the value of V_{CC} when $V_{CE} = V_{CE(max)} = 15$ V.

$$V_{RC} = V_{CC} - V_{CE}$$

So,

$$V_{CC(max)} = V_{CE(max)} + V_{RC} = 15\text{ V} + 19.5\text{ V} = 34.5\text{ V}$$

V_{CC} can be increased to 34.5 V, under the existing conditions, before $V_{CE(max)}$ is exceeded. However, at this point it is not known whether or not $P_{D(max)}$ has been exceeded.

$$P_D = V_{CE(max)}I_C = (15\text{ V})(19.5\text{ mA}) = 293\text{ mW}$$

Since $P_{D(max)}$ is 800 mW, it is *not* exceeded when $V_{CC} = 34.5$ V. So, $V_{CE(max)} = 15$ V is the limiting rating in this case. If the base current is removed causing the transistor to turn off, $V_{CE(max)}$ will be exceeded because the entire supply voltage, V_{CC}, will be dropped across the transistor.

Related Exercise   The transistor in Figure 4–18 has the following maximum ratings: $P_{D(max)} = 500$ mW, $V_{CE(max)} = 25$ V, and $I_{C(max)} = 200$ mA. Determine the maximum value to which V_{CC} can be adjusted without exceeding a rating. Which rating would be exceeded first?

Derating $P_{D(max)}$

$P_{D(max)}$ is usually specified at 25°C. For higher temperatures, $P_{D(max)}$ is less. Data sheets often give derating factors for determining $P_{D(max)}$ at any temperature above 25°C. For example, a derating factor of 2 mW/C° indicates that the maximum power dissipation is reduced 2 mW for each Centigrade degree increase in temperature.

EXAMPLE 4–7

A certain transistor has a $P_{D(max)}$ of 1 W at 25°C. The derating factor is 5 mW/C°. What is the $P_{D(max)}$ at a temperature of 70°C?

Solution   The change (reduction) in $P_{D(max)}$ is

$$\Delta P_{D(max)} = (5 \text{ mW/C°})(70°C - 25°C) = (5 \text{ mW/C°})(45 \text{ C°}) = 225 \text{ mW}$$

Therefore, the $P_{D(max)}$ at 70°C is

$$1 \text{ W} - 225 \text{ mW} = 775 \text{ mW}$$

Related Exercise   A transistor has a $P_{D(max)} = 5$ W at 25°C. The derating factor is 10 mW/C°. What is the $P_{D(max)}$ at 70°C?

Transistor Data Sheet

A partial data sheet for the 2N3903 and 2N3904 *npn* transistors is shown in Figure 4–19. Notice that the maximum collector-emitter voltage (V_{CEO}) is 40 V. The "O" in the subscript indicates that the voltage is measured from collector (C) to emitter (E) with the base open (O). In the text, we use $V_{CE(max)}$ for purposes of clarity. Also notice that the maximum collector current is 200 mA.

The β_{DC} (h_{FE}) is specified for several values of I_C and, as you can see, h_{FE} varies with I_C as we previously discussed.

The collector-emitter saturation voltage, $V_{CE(sat)}$ is 0.2 V maximum for $I_{C(sat)} = 10$ mA and increases with the current.

SECTION 4–3 REVIEW

1. Define β_{DC} and α_{DC}. What is h_{FE}?
2. If the dc current gain of a transistor is 100, determine β_{DC} and α_{DC}.
3. What two variables are plotted on a collector characteristic curve?
4. What bias conditions must exist for a transistor to operate as an amplifier?
5. Does β_{DC} increase or decrease with temperature?
6. For a given type of transistor, can β_{DC} be considered to be a constant?

Maximum Ratings

Rating	Symbol	Value	Unit
Collector-Emitter voltage	V_{CEO}	40	V dc
Collector-Base voltage	V_{CBO}	60	V dc
Emitter-Base voltage	V_{EBO}	6.0	V dc
Collector current — continuous	I_C	200	mA dc
Total device dissipation @ $T_A = 25°C$ Derate above 25°C	P_D	625 5.0	mW mW/C°
Total device dissipation @ $T_C = 25°C$ Derate above 25°C	P_D	1.5 12	Watts mW/C°
Operating and storage junction Temperature range	T_J, T_{stg}	−55 to +150	°C

Thermal Characteristics

Characteristic	Symbol	Max	Unit
Thermal resistance, junction to case	$R_{\theta JC}$	83.3	C°/W
Thermal resistance, junction to ambient	$R_{\theta JA}$	200	C°/W

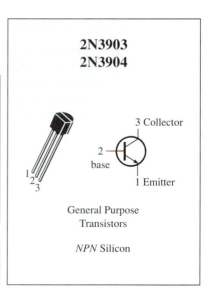

2N3903
2N3904

3 Collector

2
base

1 Emitter

General Purpose
Transistors

NPN Silicon

Electrical Characteristics ($T_A = 25°C$ unless otherwise noted.)

Characteristic		Symbol	Min	Max	Unit
OFF Characteristics					
Collector-Emitter breakdown voltage ($I_C = 1.0$ mA dc, $I_B = 0$)		$V_{(BR)CEO}$	40	–	V dc
Collector-Base breakdown voltage ($I_C = 10$ µA dc, $I_E = 0$)		$V_{(BR)CBO}$	60	–	V dc
Emitter-Base breakdown voltage ($I_E = 10$ µA dc, $I_C = 0$)		$V_{(BR)EBO}$	6.0	–	V dc
Base cutoff current ($V_{CE} = 30$ V dc, $V_{EB} = 3.0$ V dc)		I_{BL}	–	50	nA dc
Collector cutoff current ($V_{CE} = 30$ V dc, $V_{EB} = 3.0$ V dc)		I_{CEX}	–	50	nA dc
ON Characteristics					
DC current gain		h_{FE}			–
($I_C = 0.1$ mA dc, $V_{CE} = 1.0$ V dc)	2N3903 2N3904		20 40	– –	
($I_C = 1.0$ mA dc, $V_{CE} = 1.0$ V dc)	2N3903 2N3904		35 70	– –	
($I_C = 10$ mA dc, $V_{CE} = 1.0$ V dc)	2N3903 2N3904		50 100	150 300	
($I_C = 50$ mA dc, $V_{CE} = 1.0$ V dc)	2N3903 2N3904		30 60	– –	
($I_C = 100$ mA dc, $V_{CE} = 1.0$ V dc)	2N3903 2N3904		15 30	– –	

Collector-Emitter saturation voltage
    ($I_C = 10$ mA dc, $I_B = 1.0$ mA dc)
    ($I_C = 50$ mA dc, $I_B = 5.0$ mA dc)
		$V_{CE(sat)}$	–
			–
				0.2
				0.3
					V dc

Base-Emitter saturation voltage
    ($I_C = 10$ mA dc, $I_B = 1.0$ mA dc)
    ($I_C = 50$ mA dc, $I_B = 5.0$ mA dc)
		$V_{BE(sat)}$	0.65
			–
				0.85
				0.95
					V dc

FIGURE 4–19
Partial transistor data sheet.

4–4 ■ THE TRANSISTOR AS AN AMPLIFIER

Amplification is the process of linearly increasing the amplitude of an electrical signal and is one of the major properties of a transistor. As you learned, the transistor exhibits current gain (called β). When a transistor is biased in the active (or linear) region, as previously described, the BE junction has a low resistance due to forward bias and the BC junction has a high resistance due to reverse bias.

After completing this section, you should be able to

■ **Discuss how a transistor is used as a voltage amplifier**
  □ Describe amplification
  □ Develop the ac equivalent circuit for a basic transistor amplifier
  □ Determine the voltage gain of a basic transistor amplifier

DC and AC Quantities

Before introducing the concept of transistor **amplification,** the designations that we will use for the circuit quantities of current, voltage, and resistance must be explained because amplifier circuits have both dc and ac quantities.

In this text, italic capital letters are used for both dc and ac currents (I) and voltages (V). This rule applies to rms, average, peak, and peak-to-peak ac values. AC current and voltage values are always rms unless stated otherwise. Although some texts use lowercase i and v for ac current and voltage, we reserve the use of lowercase i and v only for instantaneous values, as you learned in your dc/ac circuits course. In this text, the distinction between a dc current or voltage and an ac current or voltage is in the subscript.

DC quantities always carry an uppercase roman (nonitalic) subscript. For example, I_B, I_C, and I_E are the dc transistor currents. V_{BE}, V_{CB}, and V_{CE} are the dc voltages from one transistor terminal to another. Single subscripted voltages such as V_B, V_C, and V_E are dc voltages from the transistor terminals to ground.

AC and all time-varying quantities always carry a lowercase italic subscript. For example, I_b, I_c, and I_e are the ac transistor currents. V_{be}, V_{cb}, and V_{ce} are the ac voltages from one transistor terminal to another. Single subscripted voltages such as V_b, V_c, and V_e are ac voltages from the transistor terminals to ground.

The rule is different for *internal* transistor resistances. As you will see later, transistors have internal ac resistances that are designated by lowercase r' with an appropriate subscript. For example, the internal ac emitter resistance is designated as r'_e.

Circuit resistances external to the transistor itself use the standard italic capital R with a subscript that identifies the resistance as dc or ac (when applicable), just as for current and voltage. For example R_E is an external dc emitter resistance and R_e is an external ac emitter resistance.

Transistor Amplification

As you have learned, a transistor amplifies current because the collector current is equal to the base current multiplied by the current gain, β. The base current in a transistor is very small compared to the collector and emitter currents. Because of this, the collector current is approximately equal to the emitter current.

With this in mind, let's look at the circuit in Figure 4–20(a). An ac voltage, V_{in} is superimposed on the dc bias voltage V_{BB} by connecting them in series with the base resistor, R_B, as shown. The dc bias voltage V_{CC} is connected to the collector through the collector resistor, R_C.

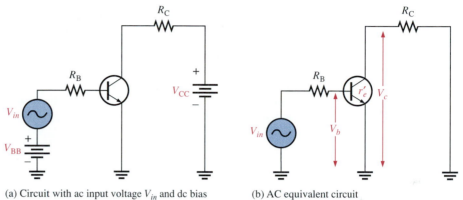

(a) Circuit with ac input voltage V_{in} and dc bias   (b) AC equivalent circuit
   voltage superimposed

FIGURE 4–20
Basic transistor amplifier circuit.

The ac input voltage produces an ac base current, which results in a much larger ac collector current. The ac collector current produces an ac voltage across R_C, thus producing an apparent amplified reproduction of the ac input voltage in the active region of operation.

The AC Equivalent Circuit   The dc bias sources ideally appear as shorts to the ac voltage. Therefore, the ac equivalent circuit can be represented as shown in Figure 4–20(b) with V_{CC} and V_{BB} replaced by shorts.

The forward-biased base-emitter junction presents a very low resistance to the ac signal. This internal ac emitter resistance is designated r'_e. In Figure 4–20(b), the ac emitter current is

$$I_e = \frac{V_b}{r'_e}$$

The ac collector voltage, V_c, equals the ac voltage drop across R_C:

$$V_c = I_c R_C$$

Since $I_c \cong I_e$, the ac collector voltage is

$$V_c \cong I_e R_C$$

V_b can be considered the transistor ac input voltage where $V_b = V_{in} - I_b R_B$. V_c can be considered the transistor ac output voltage. The ratio of V_c to V_b is the ac voltage gain, A_v, of the transistor circuit:

$$A_v = \frac{V_c}{V_b}$$

Substituting $I_e R_C$ for V_c and $I_e r'_e$ for V_b, we get

$$A_v = \frac{V_c}{V_b} \cong \frac{I_e R_C}{I_e r'_e}$$

The I_e terms cancel; therefore,

$$A_v \cong \frac{R_C}{r'_e} \qquad \qquad \textbf{(4–11)}$$

Equation (4–11) shows that the transistor in Figure 4–20 provides amplification or voltage gain dependent on the value of R_C and r'_e.

Since R_C is always considerably larger in value than r'_e, the output voltage is always greater than the input voltage. Various types of amplifiers are covered in detail in later chapters.

EXAMPLE 4–8   Determine the voltage gain and the ac output voltage in Figure 4–21 if $r'_e = 50\ \Omega$.

FIGURE 4–21

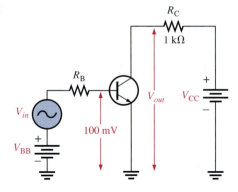

Solution   The voltage gain is

$$A_v \cong \frac{R_C}{r'_e} = \frac{1\ \text{k}\Omega}{50\ \Omega} = 20$$

Therefore, the ac output voltage is

$$V_{out} = A_v V_b = (20)(100\ \text{mV}) = 2\ \text{V rms}$$

Related Exercise   What value of R_C in Figure 4–21 will it take to have a voltage gain of 50?

4–5 ▪ THE TRANSISTOR AS A SWITCH

In the previous section, we discussed the transistor as a linear amplifier. The second major application area is switching applications. When used as an electronic switch, a transistor is normally operated alternately in cutoff and saturation.

After completing this section, you should be able to

▪ **Discuss how a transistor is used as an electronic switch**
  ☐ Define *cutoff* and *saturation*
  ☐ Describe the conditions that produce cutoff
  ☐ Describe the conditions that produce saturation
  ☐ Analyze a transistor switching circuit for cutoff and saturation
  ☐ Discuss a basic application of a transistor switching circuit

Figure 4–22 illustrates the basic operation of the transistor as a switching device. In part (a), the transistor is in the cutoff region because the base-emitter junction is not forward-biased. In this condition, there is, ideally, an *open* between collector and emitter as indicated by the switch equivalent. In part (b), the transistor is in the saturation region because the base-emitter junction and the base-collector junction are forward-biased and the base current is made large enough to cause the collector current to reach its saturation value. In this condition, there is, ideally, a *short* between collector and emitter as indi-

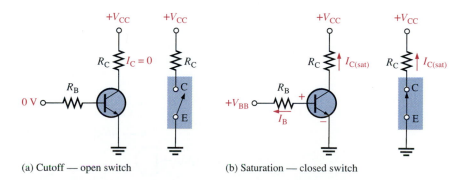

(a) Cutoff — open switch   (b) Saturation — closed switch

FIGURE 4–22
Ideal switching action of a transistor.

cated by the switch equivalent. Actually, a voltage drop of up to a few tenths of a volt normally occurs, which is the saturation voltage, $V_{CE(sat)}$.

Conditions in Cutoff

As mentioned before, a transistor is in the cutoff region when the base-emitter junction is *not* forward-biased. Neglecting leakage current, all of the currents are zero, and V_{CE} is equal to V_{CC}.

$$V_{CE(cutoff)} = V_{CC} \qquad \text{(4–12)}$$

Conditions in Saturation

As you have learned, when the base-emitter junction is forward-biased and there is enough base current to produce a maximum collector current, the transistor is saturated. The formula for collector saturation current is

$$I_{C(sat)} = \frac{V_{CC} - V_{CE(sat)}}{R_C} \qquad \text{(4–13)}$$

Since $V_{CE(sat)}$ is very small compared to V_{CC}, it can usually be neglected.
The minimum value of base current needed to produce saturation is

$$I_{B(min)} = \frac{I_{C(sat)}}{\beta_{DC}} \qquad \text{(4–14)}$$

I_B should be significantly greater than $I_{B(min)}$ to keep the transistor well into saturation.

EXAMPLE 4–9

(a) For the transistor circuit in Figure 4–23, what is V_{CE} when $V_{IN} = 0$ V?

(b) What minimum value of I_B is required to saturate this transistor if β_{DC} is 200? Neglect $V_{CE(sat)}$.

(c) Calculate the maximum value of R_B when $V_{IN} = 5$ V.

FIGURE 4–23

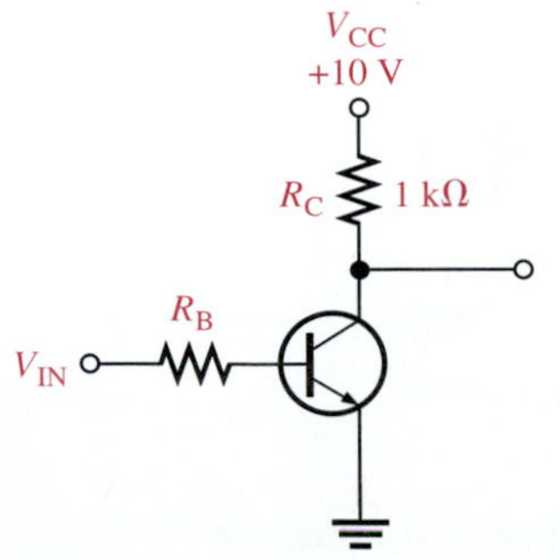

Solution

(a) When $V_{IN} = 0$ V, the transistor is in cutoff (acts like an open switch) and $V_{CE} = V_{CC} = 10$ V.

(b) Since $V_{CE(sat)}$ is neglected (0 V),

$$I_{C(sat)} = \frac{V_{CC}}{R_C} = \frac{10 \text{ V}}{1 \text{ k}\Omega} = 10 \text{ mA}$$

$$I_B = \frac{I_{C(sat)}}{\beta_{DC}} = \frac{10 \text{ mA}}{200} = 50 \text{ }\mu\text{A}$$

This is the value of I_B necessary to drive the transistor to the point of saturation. Any further increase in I_B will drive the transistor deeper into saturation but will not increase I_C.

(c) When the transistor is on, $V_{BE} = 0.7$ V. The voltage across R_B is

$$V_{R_B} = V_{IN} - V_{BE} = 5 \text{ V} - 0.7 \text{ V} = 4.3 \text{ V}$$

Calculate the maximum value of R_B needed to allow a minimum I_B of 50 μA by Ohm's law as follows:

$$R_B = \frac{V_{R_B}}{I_B} = \frac{4.3 \text{ V}}{50 \text{ }\mu\text{A}} = 86 \text{ k}\Omega$$

Related Exercise Determine the minimum value of I_B required to saturate the transistor in Figure 4–23 if β_{DC} is 125 and $V_{CE(sat)}$ is 0.2 V.

A Simple Application of a Transistor Switch

The transistor in Figure 4–24 is used as a switch to turn the LED on and off. For example, a square wave input voltage with a period of 2 s is applied to the input as indicated. When the square wave is at 0 V, the transistor is in cutoff and, since there is no collector current, the LED does not emit light. When the square wave goes to its high level, the transistor saturates. This forward-biases the LED, and the resulting collector current through the LED causes it to emit light. So, we have a blinking LED that is on for 1 s and off for 1 s.

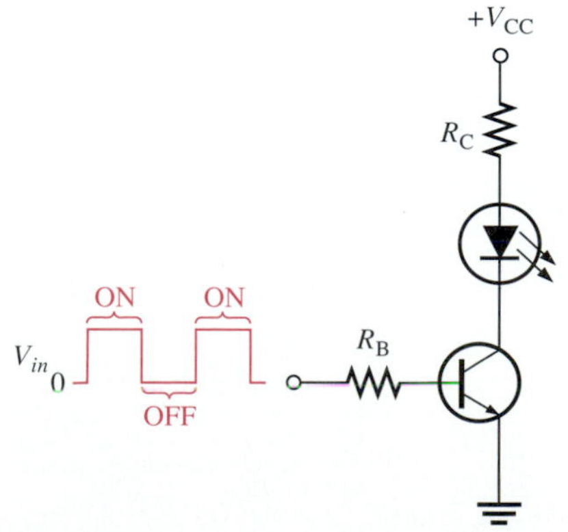

FIGURE 4–24
A transistor used to switch an LED on and off.

EXAMPLE 4–10

The LED in Figure 4–24 requires 30 mA to emit a sufficient level of light. Therefore, the collector current should be approximately 30 mA. For the following circuit values, determine the amplitude of the square wave input voltage necessary to make sure that the transistor saturates. Use double the minimum value of base current as a safety margin to ensure saturation. $V_{CC} = 9$ V, $V_{CE(sat)} = 0.3$ V, $R_C = 270$ Ω, $R_B = 3.3$ kΩ, and $\beta_{DC} = 50$.

Solution

$$I_{C(sat)} = \frac{V_{CC} - V_{CE(sat)}}{R_C} = \frac{9 \text{ V} - 0.3 \text{ V}}{270 \ \Omega} = 32.2 \text{ mA}$$

$$I_{B(min)} = \frac{I_{C(sat)}}{\beta_{DC}} = \frac{32.2 \text{ mA}}{50} = 644 \ \mu\text{A}$$

To ensure saturation, use twice the value of $I_{B(min)}$, which gives 1.29 mA. Then

$$I_B = \frac{V_{R_B}}{R_B} = \frac{V_{in} - V_{BE}}{R_B} = \frac{V_{in} - 0.7 \text{ V}}{3.3 \text{ k}\Omega}$$

Solve for the voltage amplitude of the square wave input V_{in}:

$$V_{in} - 0.7 \text{ V} = 2I_{B(min)}R_B = (1.29 \text{ mA})(3.3 \text{ k}\Omega)$$

$$V_{in} = (1.29 \text{ mA})(3.3 \text{ k}\Omega) + 0.7 \text{ V} = 4.96 \text{ V}$$

Related Exercise If you change the LED in Figure 4–24 to one that requires 50 mA for a specified light emission and you can't increase the input amplitude above 5 V or V_{CC} above 9 V, how would you modify the circuit? Specify the component(s) to be changed and the value(s).

SECTION 4–5 REVIEW

1. When a transistor is used as a switch, in what two states is it operated?
2. When is the collector current maximum?
3. When is the collector current approximately zero?
4. Under what condition is $V_{CE} = V_{CC}$?
5. When is V_{CE} minimum?

SUMMARY OF BIPOLAR JUNCTION TRANSISTORS

SYMBOLS

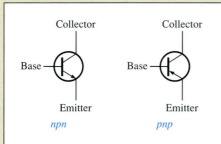

npn

pnp

CURRENTS AND VOLTAGES

$I_E = I_C + I_B$

AMPLIFICATION

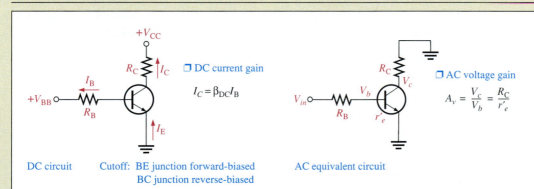

❏ DC current gain

$$I_C = \beta_{DC} I_B$$

❏ AC voltage gain

$$A_v = \frac{V_c}{V_b} = \frac{R_C}{r'_e}$$

DC circuit   Cutoff: BE junction forward-biased
              BC junction reverse-biased

AC equivalent circuit

SUMMARY OF BIPOLAR JUNCTION TRANSISTORS, *continued*

CUTOFF AND SATURATION

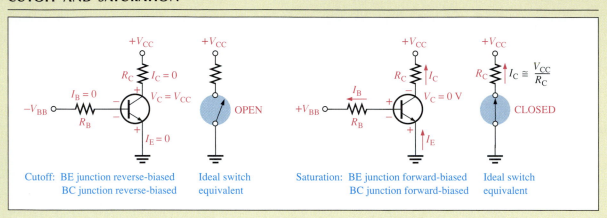

Cutoff: BE junction reverse-biased   Ideal switch   Saturation: BE junction forward-biased   Ideal switch
        BC junction reverse-biased   equivalent   BC junction forward-biased   equivalent

4–6 ∎ TRANSISTOR PACKAGES AND TERMINAL IDENTIFICATION

Transistors are available in a wide range of package types for various applications. Those with mounting studs or heat sinks are usually power transistors. Low and medium power transistors are usually found in smaller metal or plastic cases. Still another package classification is for high-frequency devices. You should be familiar with common transistor packages and be able to identify the emitter, base, and collector terminals. This section is about transistor packages and terminal identification.

After completing this section, you should be able to

∎ **Identify various types of transistor package configurations**
  ☐ List three broad categories of transistors
  ☐ Recognize various types of cases and identify the pin configurations

Transistor Categories

Manufacturers generally classify their bipolar junction transistors into three broad categories: general-purpose/small-signal devices, power devices, and RF (radio frequency/ microwave) devices. Although each of these categories, to a large degree, has its own unique package types, you will find certain types of packages used in more than one device category. While keeping in mind there is some overlap, we will look at transistor packages for each of the three categories, so that you will be able to recognize a transistor when you see one on a circuit board and have a good idea of what general category it is in.

General-Purpose/Small-Signal Transistors General-purpose/small-signal transistors are generally used for low or medium power amplifiers or switching circuits. The packages are either plastic or metal cases. Certain types of packages contain multiple transistors. Figure 4–25 illustrates common plastic cases, Figure 4–26 shows packages called *metal cans,* and Figure 4–27 shows multiple-transistor packages. Some of the multiple-transistor packages such as the dual-in-line (DIP) and the small-outline (SO) are the same as those used for many integrated circuits. Typical pin connections are shown so you can identify the emitter, base, and collector.

(a) TO-92 or TO-226AA   (b) TO-92 or TO-226AE   (c) SOT-23 or TO-236AB

FIGURE 4–25
Plastic cases for general-purpose/small-signal transistors. Both old and new JEDEC TO numbers are given. Pin configurations may vary. Always check the data sheet. (Copyright of Motorola, Inc. Used by permission.)

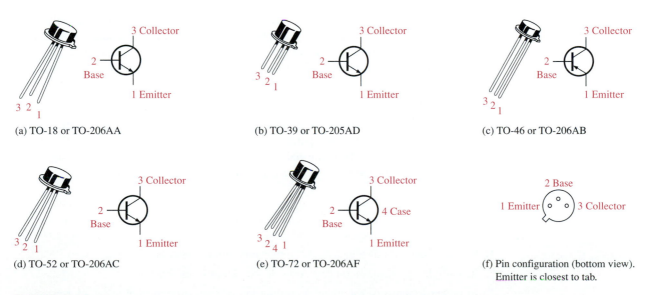

(a) TO-18 or TO-206AA   (b) TO-39 or TO-205AD   (c) TO-46 or TO-206AB

(d) TO-52 or TO-206AC   (e) TO-72 or TO-206AF   (f) Pin configuration (bottom view). Emitter is closest to tab.

FIGURE 4–26
Metal cases for general-purpose/small-signal transistors (copyright of Motorola, Inc. Used by permission).

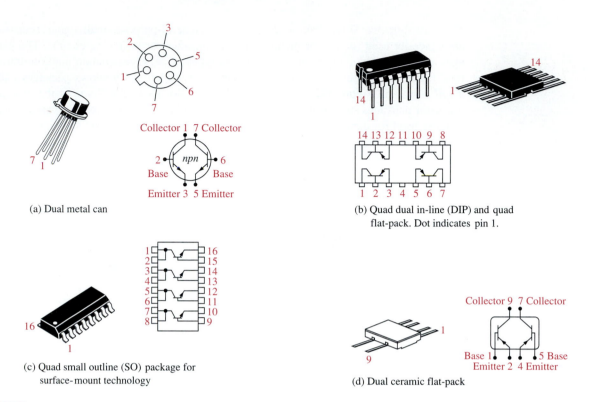

(a) Dual metal can

Collector 1 7 Collector

2 Base — npn — 6 Base

Emitter 3 5 Emitter

(b) Quad dual in-line (DIP) and quad
flat-pack. Dot indicates pin 1.

(c) Quad small outline (SO) package for
surface-mount technology

(d) Dual ceramic flat-pack

Collector 9 7 Collector

Base 1 — 5 Base
Emitter 2 4 Emitter

FIGURE 4–27

Typical multiple-transistor packages (copyright of Motorola, Inc. Used by permission).

Power Transistors Power transistors are used to handle large currents (typically more than 1 A) and/or large voltages. For example, the final audio stage in a stereo system uses a power transistor amplifier to drive the speakers. Figure 4–28 shows some common package configurations. In most applications, the metal tab or the metal case is common to the collector and is thermally connected to a heat sink for heat dissipation. Notice in part (g) how the small transistor chip is mounted inside the much larger package.

RF Transistors RF transistors are designed to operate at extremely high frequencies and are commonly used for various purposes in communications systems and other high-frequency applications. Their unusual shapes and lead configurations are designed to optimize certain high-frequency parameters. Figure 4–29 shows some examples.

SECTION 4–6 REVIEW

1. List the three broad categories of bipolar junction transistors.
2. In a single-transistor metal case, how do you identify the leads?
3. In power transistors, the metal mounting tab or case is connected to which transistor region?

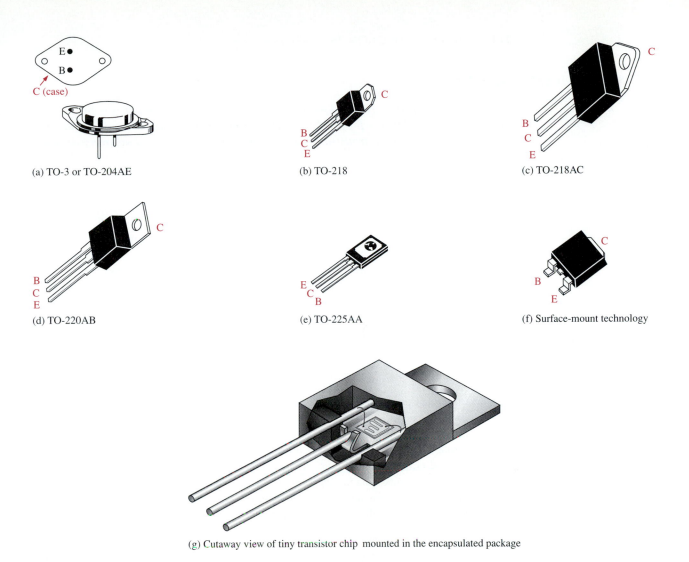

(a) TO-3 or TO-204AE

(b) TO-218

(c) TO-218AC

(d) TO-220AB

(e) TO-225AA

(f) Surface-mount technology

(g) Cutaway view of tiny transistor chip mounted in the encapsulated package

FIGURE 4–28
Typical power transistors (copyright of Motorola, Inc. Used by permission).

FIGURE 4–29
Examples of RF transistors (copyright of Motorola, Inc. Used by permission).

(a)

(b)

(c)

(d)

4–7 ■ TROUBLESHOOTING

As you already know, a critical skill in electronics is the ability to identify a circuit malfunction and to isolate the failure to a single component if possible. In this section, the basics of troubleshooting transistor bias circuits and testing individual transistors are covered.

After completing this section, you should be able to

■ **Troubleshoot various faults in transistor circuits**
  □ Define *floating point*
  □ Use voltage measurements to identify a fault in a transistor circuit
  □ Use a DMM to test a transistor
  □ Explain how a transistor can be viewed in terms of a diode equivalent
  □ Discuss in-circuit and out-of-circuit testing
  □ Discuss point-of-measurement in troubleshooting
  □ Discuss leakage and gain measurements

Troubleshooting a Biased Transistor

Several faults can occur in a simple transistor bias circuit. Possible faults are open bias resistors, open or resistive connections, shorted connections, and opens or shorts internal to the transistor itself. Figure 4–30 is a basic transistor bias circuit with all voltages referenced to ground. The two bias voltages are $V_{BB} = 3$ V and $V_{CC} = 9$ V. The correct voltages at the base and collector are shown. Analytically, these voltages are determined as follows. A $\beta_{DC} = 200$ is taken as midway between the minimum and maximum values of h_{FE} given on the data sheet for the 2N3904 in Figure 4–19. A different h_{FE} (β_{DC}), of course, will produce different results for the given circuit.

$$V_B = V_{BE} = 0.7 \text{ V}$$

$$I_B = \frac{3 \text{ V} - 0.7 \text{ V}}{56 \text{ k}\Omega} = \frac{2.3 \text{ V}}{56 \text{ k}\Omega} = 41.1 \text{ μA}$$

$$I_C = \beta_{DC}I_B = 200(41.1 \text{ μA}) = 8.2 \text{ mA}$$

$$V_C = 9 \text{ V} - I_C R_C = 9 \text{ V} - (8.2 \text{ mA})(560 \text{ }\Omega) = 4.4 \text{ V}$$

Several faults that can occur in the circuit and the accompanying symptoms are illustrated in Figure 4–31. Symptoms are shown in terms of measured voltages that are incorrect. The term **floating point** refers to a point in the circuit that is not electrically

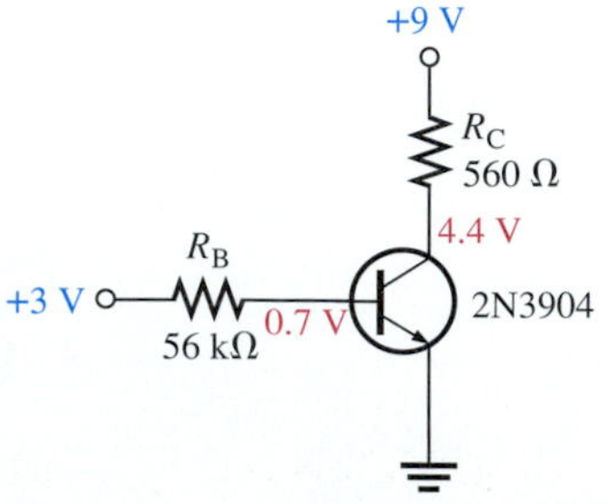

FIGURE 4–30

A basic transistor bias circuit.

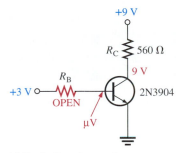

(a) *Fault:* Open base resistor.
   Symptoms: A few μV at base due
   to floating point. 9 V at collector
   because transistor is in cutoff.

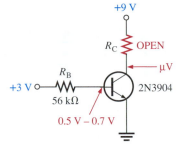

(b) *Fault:* Open collector resistor.
   Symptoms: A few μV at collector
   due to floating point. 0.5 V – 0.7 V
   at base due to forward voltage drop
   across the base-emitter junction.

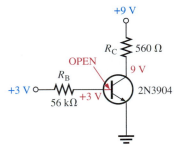

(c) *Fault:* Base internally open.
   Symptoms: 3 V at base lead.
   9 V at collector because transistor
   is in cutoff.

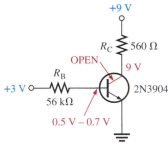

(d) *Fault:* Collector internally open.
   Symptoms: 0.5 V – 0.7 V at base
   lead due to forward voltage drop
   across base-emitter junction. 9 V at
   collector because the open prevents
   collector current.

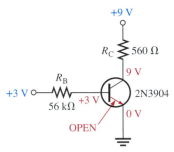

(e) *Fault:* Emitter internally open.
   Symptoms: 3 V at base lead. 9 V
   at collector because there is no
   collector current. 0 V at the emitter
   as normal.

(f) *Fault:* Open ground connection.
   Symptoms: 3 V at base lead. 9 V
   at collector because there is no
   collector current. 2.5 V or more at the
   emitter due to the forward voltage
   drop across the base-emitter junction.
   The measuring voltmeter provides a
   forward current path through its
   internal resistance.

FIGURE 4–31
Typical faults and symptoms in the basic transistor bias circuit.

connected to ground or a "solid" voltage. Normally, very small and sometimes fluctuating
voltages in the μV range are observed at floating points. The faults in Figure 4–31 are
typical but do not represent all possible faults that may occur.

Testing a Transistor with a DMM

A digital multimeter can be used as a fast and simple way to check a transistor for open or
shorted junctions. For this test, you can view the transistor as two diodes connected as
shown in Figure 4–32 for both *npn* and *pnp* transistors. The base-collector junction is one
diode and the base-emitter junction is the other.

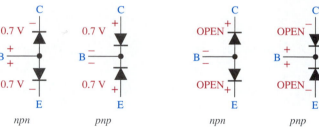

(a) Both junctions should read 0.7 V ± 0.2 V when forward-biased.

(b) Both junctions should ideally read OPEN when reverse-biased.

FIGURE 4–32

A transistor viewed as two diodes.

Recall that a good diode will show an extremely high resistance (or open) with reverse bias and a very low resistance with forward bias. A defective open diode will show an extremely high resistance (or open) for both forward and reverse bias. A defective shorted or resistive diode will show zero or a very low resistance for both forward and reverse bias. An open diode is the most common type of failure. Since the transistor *pn* junctions are, in effect diodes, the same basic characteristics apply.

The DMM Diode Test Position   Many digital multimeters (DMMs) have a *diode test* position that provides a convenient way to test a transistor. A typical DMM, as shown in Figure 4–33, has a small diode symbol to mark the position of the function switch. When set to diode test, the meter provides an internal voltage sufficient to forward-bias and reverse-bias a transistor junction. This internal voltage may vary among different makes of DMM, but 2.5 V to 3.5 V is a typical range of values. The meter provides a voltage reading to indicate the condition of the transistor junction under test.

When the Transistor Is Not Defective   In Figure 4–33(a), the red (positive) lead of the meter is connected to the base of an *npn* transistor and the black (negative) lead is connected to the emitter to forward-bias the base-emitter junction. If the junction is good, you will get a reading of between 0.5 V and 0.9 V, with 0.7 V being typical for forward bias.

In Figure 4–33(b), the leads are switched to reverse-bias the base-emitter junction, as shown. If the transistor is working properly, you will get a voltage reading based on the meter's internal voltage source. The 2.6 V shown in the figure represents a typical value and indicates that the junction has an extremely high reverse resistance with essentially all of the internal voltage appearing across it.

The process just described is repeated for the base-collector junction as shown in Figure 4–33(c) and (d). For a *pnp* transistor the polarity of the meter leads are reversed for each test.

When the Transistor Is Defective   When a transistor has failed with an open junction or internal connection, you get an open circuit voltage reading (2.6 V is typical) for both the forward-bias and the reverse-bias conditions for that junction, as illustrated in Figure 4–34(a) and (b). If a junction is shorted, the meter reads 0 V in both forward and reverse, as indicated in part (c). Sometimes, a failed junction may exhibit a small resistance for both bias conditions rather than a pure short. In this case, the meter will show a small

FIGURE 4–33
DMM test of a properly functioning npn transistor.
Leads are reversed for a pnp test.

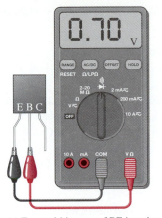

(a) Forward-bias test of BE junction

(b) Reverse-bias test of BE junction

(c) Forward-bias test of BC junction

(d) Reverse-bias test of BC junction

(a) Forward-bias test, open BC junction

(b) Reverse-bias test, open BC junction

(c) Forward and reverse tests for a shorted junction give the same 0 V reading. If the junction is resistive, the reading is less than 2.6 V.

FIGURE 4–34
Testing a defective npn transistor. Leads are reversed for a pnp.

voltage much less than the correct open voltage. For example, a resistive junction may result in a reading of 1.1 V in both directions rather than the correct readings of 0.7 V forward and 2.6 V reverse.

Some DMMs provide a test socket on their front panel for testing a transistor for the h_{FE} (β_{DC}) value. If the transistor is inserted improperly in the socket or if it is not functioning properly due to a faulty junction or internal connection, a typical meter will flash a 1 or display a 0. If a value of β_{DC} within the normal range for the specific transistor is displayed, the device is functioning properly. The normal range of β_{DC} can be determined from the data sheet.

Checking a Transistor with the OHMs Function DMMs that do not have a diode test position or an h_{FE} socket can be used to test a transistor for open or shorted junctions by setting the function switch to an OHMs range. For the forward-bias check of a good transistor junction, you will get a resistance reading that can vary depending on the meter's internal battery. Many DMMs do not have sufficient voltage on the OHMs range to fully forward-bias a junction, and you may get a reading of from several hundred to several thousand ohms.

For the reverse-bias check of a good transistor, you will get an out-of-range indication on most DMMs because the reverse resistance is too high to measure. An out-of-range indication may be a flashing 1 or a display of dashes, depending on the particular DMM.

Even though you may not get accurate forward and reverse resistance readings on a DMM, the relative readings are sufficient to indicate a properly functioning transistor junction. The out-of-range indication shows that the reverse resistance is very high, as you expect. The reading of a few hundred to a few thousand ohms for forward bias indicates that the forward resistance is small compared to the reverse resistance, as you expect.

Transistor Testers

An individual transistor can be tested either in-circuit or out-of-circuit with a transistor tester. For example, let's say that an amplifier on a particular printed circuit (pc) board has malfunctioned. Good troubleshooting practice dictates that you do not unsolder a component from a circuit board unless you are reasonably sure that it is bad or you simply cannot isolate the problem down to a single component. When components are removed, there is a risk of damage to the pc board contacts and traces.

The first step is to do an in-circuit check of the transistor using a transistor tester similar to the one shown in Figure 4–35. The three clip-leads are connected to the transistor terminals and the tester gives a positive indication if the transistor is good.

Case 1 If the transistor tests defective, it should be carefully removed and replaced with a known good one. An out-of-circuit check of the replacement device is usually a good idea, just to make sure it is OK. The transistor is plugged into the socket on the transistor tester for out-of-circuit tests.

Case 2 If the transistor tests good in-circuit but the circuit is not working properly, examine the circuit board for a poor connection at the collector pad or for a break in the connecting trace. A poor solder joint often results in an open or a highly resistive contact. The physical point at which you actually measure the voltage is very important in this

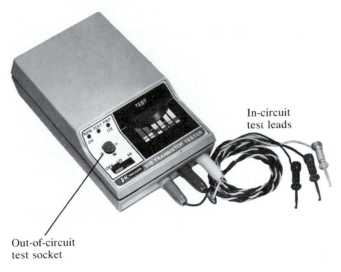

In-circuit
test leads

Out-of-circuit
test socket

FIGURE 4–35
Transistor tester (courtesy of B & K Precision).

FIGURE 4–36
The indication of an open, when it is in the external circuit, depends on where you measure.

Meter reads dc
supply voltage.

A few μV indicates
a floating point.

E B C

$--V_{CC}$

OPEN connection
at pad

case. For example, if you measure on the collector lead when there is an external open at the collector pad, you will get a floating point. If you measure on the connecting trace or on the R_C lead, you will read V_{CC}. This situation is illustrated in Figure 4–36.

Importance of Point-of-Measurement in Troubleshooting   In case 2, if you had taken the initial measurement on the transistor lead itself and the open were *internal* to the transistor as shown in Figure 4–37, you would have measured V_{CC}. This would have indicated a defective transistor even before the tester was used. This simple concept emphasizes the importance of point-of-measurement in certain troubleshooting situations.

FIGURE 4–37

Illustration of an internal open. Compare with Figure 4–36.

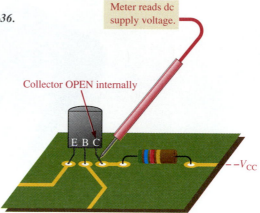

EXAMPLE 4–11   What fault do the measurements in Figure 4–38 indicate?

FIGURE 4–38

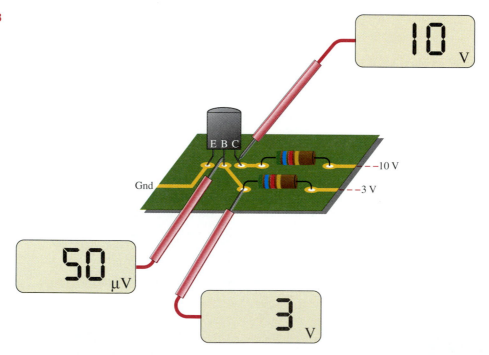

Solution   The transistor is in cutoff, as indicated by the 10 V measurement on the collector lead. The base bias voltage of 3 V appears on the pc board contact but not on the transistor lead as indicated by the floating point measurement. This shows that there is an open external to the transistor between the two measured base points. Check the solder joint at the base contact on the pc board. If the open were internal, there would be 3 V on the base lead.

Related Exercise   If the meter in Figure 4–38 that now reads 3 V indicates a floating point when touching the circuit board pad, what is the most likely fault?

Leakage Measurement

Very small leakage currents exist in all transistors and in most cases are small enough to neglect (usually nA). When a transistor is connected as shown in Figure 4–39(a) with the base open ($I_B = 0$), it is in cutoff. Ideally $I_C = 0$; but actually there is a small current from collector to emitter, as mentioned earlier, called I_{CEO} (collector-to-emitter current with base open). This leakage current is usually in the nA range for silicon. A faulty transistor will often have excessive leakage current and can be checked in a transistor tester, which connects an ammeter as shown in part (a). Another leakage current in transistors is the reverse collector-to-base current, I_{CBO}. This is measured with the emitter open, as shown in Figure 4–39(b). If it is excessive, a shorted collector-base junction is likely.

FIGURE 4–39
Leakage current test circuits.

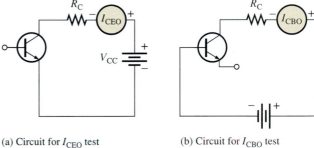

(a) Circuit for I_{CEO} test   (b) Circuit for I_{CBO} test

Gain Measurement

In addition to leakage tests, the typical transistor tester also checks the β_{DC}. A known value of I_B is applied and the resulting I_C is measured. The reading will indicate the value of the I_C/I_B ratio, although in some units only a relative indication is given. Most testers provide for an in-circuit β_{DC} check, so that a suspected device does not have to be removed from the circuit for testing.

Curve Tracers

The *curve tracer* is an oscilloscope type of instrument that can display transistor characteristics such as a family of collector curves. In addition to the measurement and display of various transistor characteristics, diode curves can also be displayed.

SECTION 4–7 REVIEW

1. If a transistor on a circuit board is suspected of being faulty, what should you do?
2. In a transistor bias circuit, such as the one in Figure 4–30, what happens if R_B opens?
3. In a circuit such as the one in Figure 4–30, what are the base and collector voltages if there is an external open between the emitter and ground?

4–8 ■ SYSTEM APPLICATION

You have been assigned to work on an electronic security alarm system that your company plans to market commercially. This system has several parts, but you will concentrate on the transistor circuits that detect a break (open) in the loops containing remote sensors for windows and doors. In this particular application, the transistors are used as switching devices. You will apply the knowledge you have gained in this chapter in completing your assignment.

The Security Alarm System

The block diagram for a three-zone security alarm system is shown in Figure 4–40. A detail of the circuit board containing the zone-monitoring circuits is included because this is the part of the system that is the focus of this assignment.

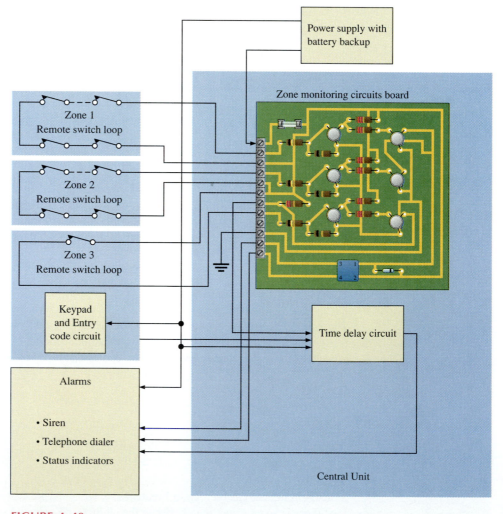

FIGURE 4–40
Basic block diagram of the security alarm system.

Two of the zones contain conductive loops with several magnetic switches that protect the windows and doors in that particular zone. The third zone protects the main entry door.

When an intrusion occurs, a magnetic switch at that point of entry breaks contact and opens the zone loop. This causes the input to the monitoring circuit for that zone to go to a 0 V level activating the circuit which, in turn, energizes the relay. The closure of the relay contacts sets off the audible alarm (siren) and/or initiates an automatic telephone dialing sequence. The monitoring circuits for either zone 1 or zone 2 can energize the common relay.

For zone 3, which is the main entry, the output of the monitoring circuit goes to the time-delay circuit to allow time for keying in an entry code that will disarm the system. If, after a preset time interval, no code or an incorrect code has been entered, the time-delay circuit will set off the alarm and/or the telephone dialing sequence.

There are many features in a typical security system, and systems can range from very simple to very complex. For this assignment, you will concentrate on the detection circuit board.

The Zone-Monitoring Circuits

There are three identical transistor monitoring circuits, one for each zone. The magnetic switches in the zone loop are normally closed and open only when there is an intrusion (a window or door opened). Because the magnetic switches are in series, the entire loop is open when one switch is open.

The circuit for one zone consists of two transistors and is shown in Figure 4–41. Transistor Q_1 detects when one of the remote magnetic switches is open, and transistor Q_2 drives the relay that activates the alarm. The transistors operate only in cutoff or saturation.

The zone loop is normally closed, keeping the input to R_1 at 12 V and transistor Q_1 in saturation. Since Q_1 is operating in saturation, its collector voltage (with respect to ground) is no more than 0.2 V. This keeps transistor Q_2 in cutoff; and since there is no Q_2 collector current through the relay coil, the system is unactivated.

When a magnetic switch in the zone loop opens, the 12 volts at the input is removed and transistor Q_1 switches to cutoff because R_2 pulls the base to ground. When Q_1 goes into cutoff, Q_2 is biased into saturation by the 12 V and the base current through R_3 and R_4. The resulting Q_2 collector current through the relay coil energizes the normally open relay and causes the contact to close; this activates the alarm. Diode D_1 across the relay coil suppresses the induced transient voltage when the relay is deactivated.

The Components

☐ *The transistors* The transistors are 2N3947s.

☐ *The resistors* Determine the minimum power rating for each of the resistors in the circuit and specify a standard rating.

☐ *The relay* Based on the system requirement that the relay must provide up to 5 A to the load (alarm circuits) and on the power supply voltage, select one of the following relays:

Relay A: coil voltage, 12 V; coil resistance, 730 Ω; contact current, 1 A
Relay B: coil voltage, 12 V; coil resistance, 95 Ω; contact current, 30 A
Relay C: coil voltage, 24 V; coil resistance, 660 Ω; contact current, 30 A

☐ *The diode* Select a diode with a PIV rating of at least 100 V.

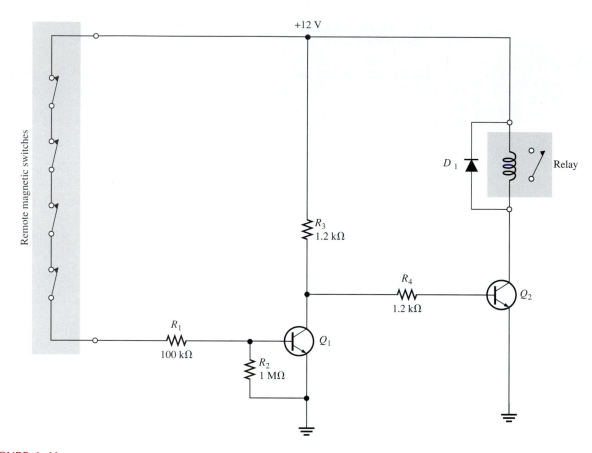

FIGURE 4–41

Zone-monitoring circuit. All three are identical with the exception that the zone 3 circuit does not drive the relay and has a different value of Q_2 collector resistor.

Analysis of the Circuit

Verify that the resistors in Figure 4–41 have values that produce adequate currents for the transistors to operate in cutoff and saturation and to drive the selected relay. Refer to the transistor data sheet in Figure 4–42.

The Printed Circuit Board

☐ Using the complete zone-monitoring circuit schematic in Figure 4–43(b) on page 226, check out the printed circuit board in Figure 4–43(a) to verify that it is correct.

☐ Label a copy of the board with the component and input/output designations in agreement with the schematic.

A Test Procedure

☐ Develop a step-by-step set of instructions on how to completely check the zone-monitoring circuit board for proper operation using the test points (circled numbers) and connector terminals indicated in the test bench setup of Figure 4–44 on page 227. The zone loop for each circuit is simulated with a single SPST switch.

Maximum Ratings

Rating	Symbol	Value	Unit
Collector-Emitter voltage	V_{CEO}	40	V dc
Collector-Base voltage	V_{CBO}	60	V dc
Emitter-Base voltage	V_{EBO}	6.0	V dc
Collector current — continuous	I_C	200	mA dc
Total device dissipation @ $T_A = 25°C$ Derate above 25°C	P_D	0.36 2.06	Watts mW/C°
Total device dissipation @ $T_C = 25°C$ Derate above 25°C	P_D	1.2 6.9	Watts mW/C°
Operating and storage junction Temperature range	T_J, T_{stg}	−65 to +200	°C

Thermal Characteristics

Characteristic	Symbol	Max	Unit
Thermal resistance, junction to case	$R_{\theta JC}$	0.15	C°/mW
Thermal resistance, junction to ambient	$R_{\theta JA}$	0.49	C°/mW

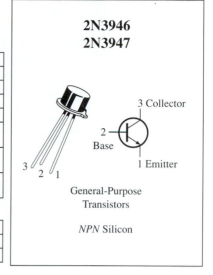

2N3946
2N3947

3 Collector

2 Base

1 Emitter

3 2 1

General-Purpose
Transistors

NPN Silicon

Electrical Characteristics ($T_A = 25°C$ unless otherwise noted.)

Characteristic	Symbol	Min	Max	Unit
OFF Characteristics				
Collector-Emitter breakdown voltage ($I_C = 10$ mA dc)	$V_{(BR)CEO}$	40	–	V dc
Collector-Base breakdown voltage ($I_C = 10$ μA dc, $I_E = 0$)	$V_{(BR)CBO}$	60	–	V dc
Emitter-Base breakdown voltage ($I_E = 10$ μA dc, $I_C = 0$)	$V_{(BR)EBO}$	6.0	–	V dc
Collector cutoff current ($V_{CE} = 40$ V dc, $V_{OB} = 3.0$ V dc) ($V_{CE} = 40$ V dc, $V_{OB} = 3.0$ V dc, $T_A = 150°C$)	I_{CEX}	– –	0.010 15	μA dc
Base cutoff current ($V_{CE} = 40$ V dc, $V_{OB} = 3.0$ V dc)	I_{BL}	–	.025	μA dc
ON Characteristics				
DC current gain ($I_C = 0.1$ mA dc, $V_{CE} = 1.0$ V dc) 2N3946 2N3947	h_{FE}	 30 60	– – –	–
($I_C = 1.0$ mA dc, $V_{CE} = 1.0$ V dc) 2N3946 2N3947		45 90	– –	
($I_C = 10$ mA dc, $V_{CE} = 1.0$ V dc) 2N3946 2N3947		50 100	150 300	
($I_C = 50$ mA dc, $V_{CE} = 1.0$ V dc) 2N3946 2N3947		20 40	– –	
Collector-Emitter saturation voltage ($I_C = 10$ mA dc, $I_B = 1.0$ mA dc) ($I_C = 50$ mA dc, $I_B = 5.0$ mA dc)	$V_{CE(sat)}$	– –	0.2 0.3	V dc
Base-Emitter saturation voltage ($I_C = 10$ mA dc, $I_B = 1.0$ mA dc) ($I_C = 50$ mA dc, $I_B = 5.0$ mA dc)	$V_{BE(sat)}$	0.6 –	0.9 1.0	V dc
Small-Signal Characteristics				
Current gain — Bandwidth product ($I_C = 10$ mA dc, $V_{CE} = 20$ V dc, $f = 100$ MHz) 2N3946 2N3947	f_T	250 300	– –	MHz
Output capacitance ($V_{CB} = 10$ V dc, $I_E = 0, f = 100$ kHz)	C_{obo}	–	4.0	pF

FIGURE 4–42

Partial data sheet for the 2N3947 npn transistor.

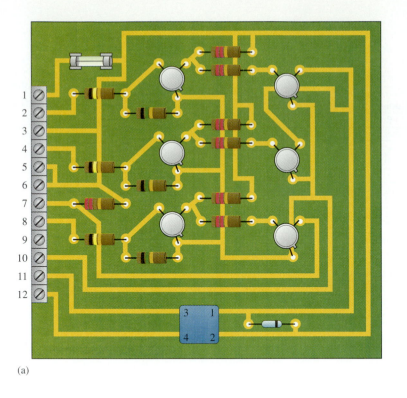

(a)

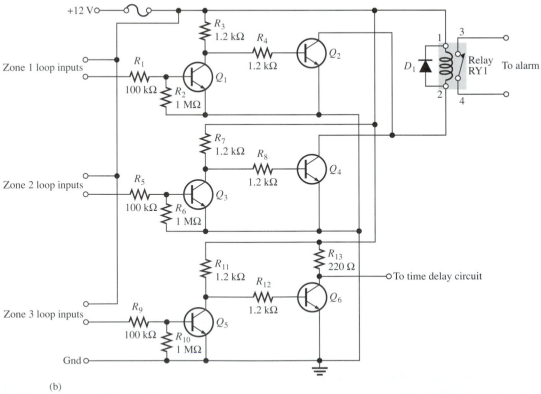

(b)

FIGURE 4–43

Zone-monitoring circuit schematic and printed circuit board.

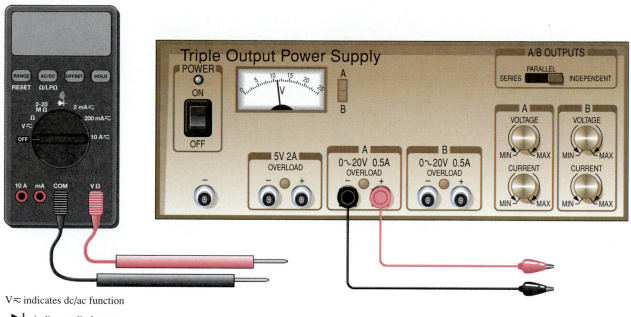

V≈ indicates dc/ac function

▶▎ indicates diode test

SPST switch

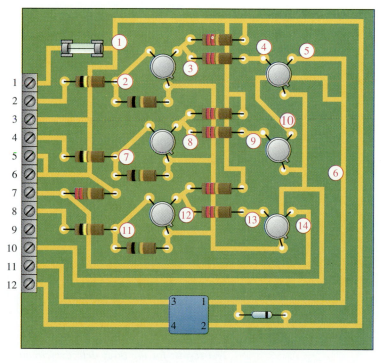

FIGURE 4–44
Zone-monitoring circuit board test bench.

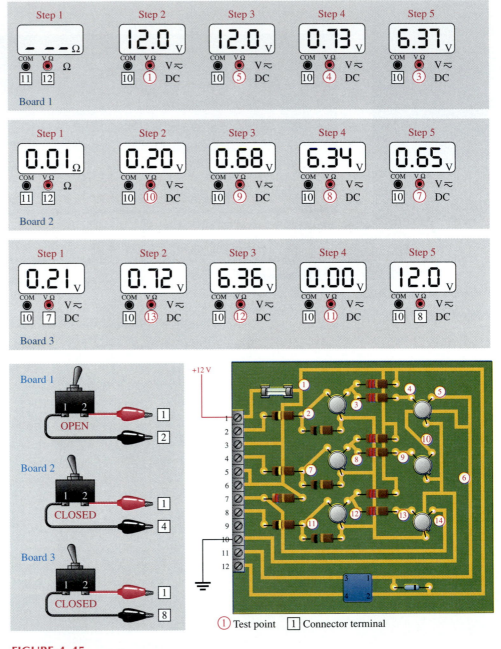

FIGURE 4–45

Test results for the three defective circuit boards.

☐ Specify voltage values and/or resistance for all the measurements to be made. Provide a fault analysis for all possible component failures.

Troubleshooting

Three boards in the initial manufacturing run have been found to be defective. Troubleshooting must be used to determine the problems.

The test results are shown in Figure 4–45. Based on the sequence of measurements and the loop switch configuration for each defective board, determine the most likely fault(s) in each case.

The red circled numbers indicate test points on the circuit board and the black numbers in a square are for connector terminals on the circuit board. The DMM function setting is indicated below the display.

Final Report

Submit a final written report on the zone-monitoring circuit board using an organized format that includes the following:

1. A physical description of the circuits.
2. A discussion of the operation of each circuit.
3. A list of the specifications.
4. A list of parts with part numbers if available.
5. A list of the types of problems on the three defective circuit boards.
6. A complete description of how you determined the problem on each of the three defective boards.

■ **CHAPTER SUMMARY**

- The bipolar junction transistor (BJT) is constructed with three regions: base, collector, and emitter.
- The BJT has two *pn* junctions, the base-emitter junction and the base-collector junction.
- Current in a BJT consists of both free electrons and holes, thus the term *bipolar*.
- The base region is very thin and lightly doped compared to the collector and emitter regions.
- The two types of bipolar junction transistor are the *npn* and the *pnp*.
- To operate as an amplifier, the base-emitter junction must be forward-biased and the base-collector junction must be reverse-biased. This is called *forward-reverse bias*.
- The three currents in the transistor are the base current (I_B), emitter current (I_E), and collector current (I_C).
- I_B is very small compared to I_C and I_E.
- The dc current gain of a transistor is the ratio of I_C to I_B and is designated β_{DC}. Values typically range from less than 20 to several hundred.
- β_{DC} is usually referred to as h_{FE} on transistor data sheets.
- The ratio of I_C to I_E is called α_{DC}. Values typically range from 0.95 to 0.99.
- When a transistor is forward-reverse biased, the voltage gain depends on the internal emitter resistance and the external collector resistance.
- A transistor can be operated as an electronic switch in cutoff and saturation.

- In cutoff, both *pn* junctions are reverse-biased and there is essentially no collector current. The transistor ideally behaves like an open switch between collector and emitter.
- In saturation, both *pn* junctions are forward-biased and the collector current is maximum. The transistor ideally behaves like a closed switch between collector and emitter.
- There is a variation in β_{DC} over temperature and also from one transistor to another of the same type.
- There are many types of transistor packages using plastic, metal, or ceramic.
- It is best to check a transistor in-circuit before removing it.
- Common faults are open junctions, low β_{DC}, excessive leakage currents, and external opens and shorts on the circuit board.

■ GLOSSARY

Alpha (α)   The ratio of collector current to emitter current in a bipolar junction transistor.

Amplification   The process of increasing the power, voltage, or current by electronic means.

Amplifier   An electronic circuit having the capability to amplify power, voltage, or current.

Base   One of the semiconductor regions in a bipolar junction transistor. The base is very thin and lightly doped compared to the other regions.

Beta (β)   The ratio of collector current to base current in a bipolar junction transistor; current gain from base to collector.

Bias   The necessary application of a dc voltage to a transistor or other device to produce a desired mode of operation.

Bipolar   Characterized by both free electrons and holes as current carriers.

Bipolar junction transistor (BJT)   A transistor constructed with three doped semiconductor regions separated by two *pn* junctions.

Collector   The largest of the three semiconductor regions of a BJT.

Cutoff   The nonconducting state of a transistor.

Emitter   The most heavily doped of the three semiconductor regions of a BJT.

Floating point   A point in the circuit that is not electrically connected to ground or a "solid" voltage.

Gain   The amount by which an electrical signal is increased or amplified.

Linear   Characterized by a straight-line relationship.

Saturation   The state of a BJT in which the collector current has reached a maximum and is independent of the base current.

Transistor   A semiconductor device used for amplification and switching applications.

■ FORMULAS

(4–1) $I_E = I_C + I_B$ Transistor currents

(4–2) $\beta_{DC} = \dfrac{I_C}{I_B}$ DC current gain

(4–3) $\alpha_{DC} = \dfrac{I_C}{I_E}$ DC alpha

(4–4) $\beta_{DC} = \dfrac{\alpha_{DC}}{1 - \alpha_{DC}}$ α_{DC}-to-β_{DC} conversion

(4–5) $V_{BE} \cong 0.7 \text{ V}$ Base-to-emitter voltage (silicon)

(4–6) $I_B = \dfrac{V_{BB} - V_{BE}}{R_B}$ Base current

(4–7) $V_{CE} = V_{CC} - I_C R_C$ Collector-to-emitter voltage (common-emitter)

(4–8) $V_{CB} = V_{CE} - V_{BE}$ Collector-to-base voltage

(4–9) $I_C = \dfrac{P_{D(max)}}{V_{CE}}$ Maximum I_C for given V_{CE}

(4–10) $V_{CE} = \dfrac{P_{D(max)}}{I_C}$ Maximum V_{CE} for given I_C

(4–11) $A_v \cong \dfrac{R_C}{r'_e}$ Approximate ac voltage gain

(4–12) $V_{CE(cutoff)} = V_{CC}$ Cutoff condition

(4–13) $I_{C(sat)} = \dfrac{V_{CC} - V_{CE(sat)}}{R_C}$ Collector saturation current

(4–14) $I_{B(min)} = \dfrac{I_{C(sat)}}{\beta_{DC}}$ Minimum base current for saturation

---

■ **SELF-TEST**

1. The three terminals of a bipolar junction transistor are called
      (a) *p, n, p*   (b) *n, p, n*
      (c) input, output, ground   (d) base, emitter, collector

2. In a *pnp* transistor, the *p*-regions are
      (a) base and emitter   (b) base and collector   (c) emitter and collector

3. For operation as an amplifier, the base of an *npn* transistor must be
      (a) positive with respect to the emitter   (b) negative with respect to the emitter
      (c) positive with respect to the collector   (d) 0 V

4. The emitter current is always
      (a) greater than the base current   (b) less than the collector current
      (c) greater than the collector current   (d) answers (a) and (c)

5. The β_{DC} of a transistor is its
      (a) current gain   (b) voltage gain   (c) power gain   (d) internal resistance

6. If I_C is 50 times larger than I_B, then β_{DC} is
      (a) 0.02   (b) 100   (c) 50   (d) 500

7. If β_{DC} is 100, the value of α_{DC} is
      (a) 99   (b) 0.99   (c) 101   (d) 0.01

8. The approximate voltage across the forward-biased base-emitter junction of a silicon BJT is
      (a) 0 V   (b) 0.7 V   (c) 0.3 V   (d) V_{BB}

9. The bias condition for a transistor to be used as a linear amplifier is called
      (a) forward-reverse   (b) forward-forward   (c) reverse-reverse   (d) collector bias

10. If the output of a transistor amplifier is 5 V rms and the input is 100 mV rms, the voltage gain is
       (a) 5   (b) 500   (c) 50   (d) 100

11. When operated in cutoff and saturation, the transistor acts like
       (a) a linear amplifier   (b) a switch   (c) a variable capacitor   (d) a variable resistor

12. In cutoff, V_{CE} is
       (a) 0 V   (b) minimum   (c) maximum
       (d) equal to V_{CC} (e) answers (a) and (b)   (f) answers (c) and (d)

13. In saturation, V_{CE} is
       (a) 0.7 V   (b) equal to V_{CC} (c) minimum   (d) maximum

14. To saturate a BJT,

    (a) $I_B = I_{C(sat)}$ **(b)** $I_B > I_{C(sat)}/\beta_{DC}$

    (c) V_{CC} must be at least 10 V   **(d)** the emitter must be grounded

15. Once in saturation, a further increase in base current will

    (a) cause the collector current to increase   **(b)** not affect the collector current

    (c) cause the collector current to decrease   **(d)** turn the transistor off

16. If the base-emitter junction is open, the collector voltage is

    (a) V_{CC} **(b)** 0 V   **(c)** floating   **(d)** 0.2 V

■ BASIC PROBLEMS

SECTION 4–1   Transistor Construction

1. What are the majority carriers in the base region of an *npn* transistor called?

2. Explain the purpose of a thin, lightly doped base region.

SECTION 4–2   Basic Transistor Operation

3. Why is the base current in a transistor so much less than the collector current?

4. In a certain transistor circuit, the base current is 2 percent of the 30 mA emitter current. Determine the collector current.

5. For normal operation of a *pnp* transistor, the base must be (+ or −) with respect to the emitter, and (+ or −) with respect to the collector.

6. What is the value of I_C for $I_E = 5.34$ mA and $I_B = 475$ μA?

SECTION 4–3   Transistor Characteristics and Parameters

7. What is the α_{DC} when $I_C = 8.23$ mA and $I_E = 8.69$ mA?

8. A certain transistor has an $I_C = 25$ mA and an $I_B = 200$ μA. Determine the β_{DC}.

9. What is the β_{DC} of a transistor if α_{DC} is 0.96?

10. What is the α_{DC} if β_{DC} is 30?

11. A certain transistor exhibits an α_{DC} of 0.96. Determine I_C when $I_E = 9.35$ mA.

12. A base current of 50 μA is applied to the transistor in Figure 4–46, and a voltage of 5 V is dropped across R_C. Determine the β_{DC} of the transistor.

13. Calculate α_{DC} for the transistor in Problem 12.

14. Determine each current in Figure 4–47. What is the β_{DC}?

15. Find V_{CE}, V_{BE}, and V_{CB} in both circuits of Figure 4–48.

16. Determine whether or not the transistors in Figure 4–48 are saturated.

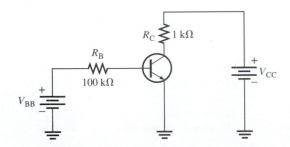

FIGURE 4–46

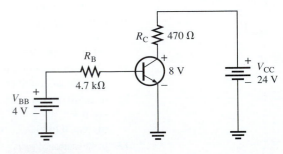

FIGURE 4–47

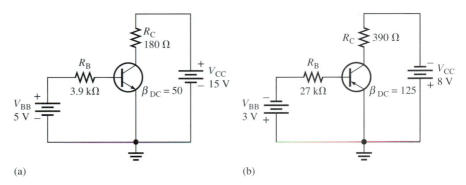

FIGURE 4–48

17. Find I_B, I_E, and I_C in Figure 4–49. $\alpha_{DC} = 0.98$.

18. Determine the terminal voltages of each transistor with respect to ground for each circuit in Figure 4–50. Also determine V_{CE}, V_{BE}, and V_{BC}.

19. If the β_{DC} in Figure 4–50(a) changes from 100 to 150 due to a temperature increase, what is the change in collector current?

20. A certain transistor is to be operated at a collector current of 50 mA. How high can V_{CE} go without exceeding a $P_{D(max)}$ of 1.2 W?

21. The power dissipation derating factor for a certain transistor is 1 mW/C°. The $P_{D(max)}$ is 0.5 W at 25°C. What is $P_{D(max)}$ at 100°C?

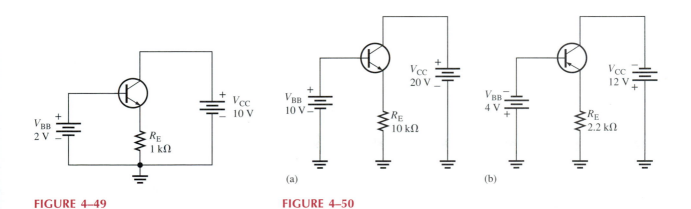

FIGURE 4–49

FIGURE 4–50

SECTION 4–4   The Transistor as an Amplifier

22. A transistor amplifier has a voltage gain of 50. What is the output voltage when the input voltage is 100 mV?

23. To achieve an output of 10 V with an input of 300 mV, what voltage gain is required?

24. A 50 mV signal is applied to the base of a properly biased transistor with $r'_e = 10\ \Omega$ and $R_C = 560\ \Omega$. Determine the signal voltage at the collector.

SECTION 4–5 The Transistor as a Switch

25. Determine $I_{C(sat)}$ for the transistor in Figure 4–51. What is the value of I_B necessary to produce saturation? What minimum value of V_{IN} is necessary for saturation? Assume $V_{CE(sat)} = 0$ V.

26. The transistor in Figure 4–52 has a β_{DC} of 50. Determine the value of R_B required to ensure saturation when V_{IN} is 5 V. What must V_{IN} be to cut off the transistor? Assume $V_{CE(sat)} = 0$ V.

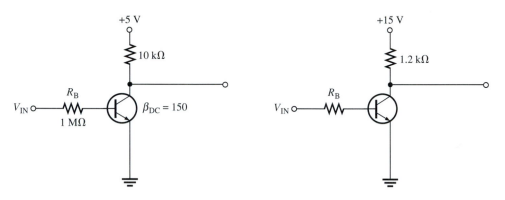

FIGURE 4–51   **FIGURE 4–52**

SECTION 4–6 Transistor Packages and Terminal Identification

27. Identify the leads on the transistors in Figure 4–53. Bottom views are shown.

28. What is the most probable category of each transistor in Figure 4–54?

FIGURE 4–53

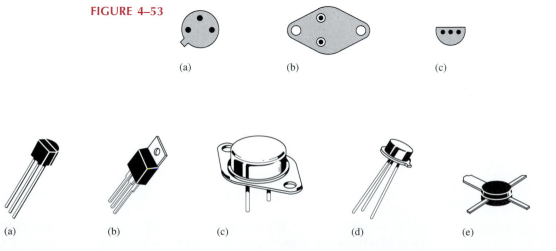

FIGURE 4–54

■ TROUBLE-SHOOTING PROBLEMS

SECTION 4–7 Troubleshooting

29. In an out-of-circuit test of a good *npn* transistor, what should an analog ohmmeter indicate when its positive probe is touching the emitter and the negative probe is touching the base? When its positive probe is touching the base and the negative probe is touching the collector?

30. What is the most likely problem, if any, in each circuit of Figure 4–55? Assume a β_{DC} of 75.

31. What is the value of the β_{DC} of each transistor in Figure 4–56?

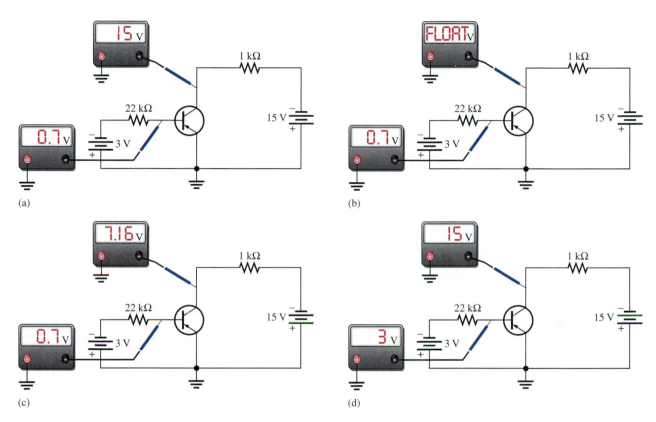

(a)

(b)

(c)

(d)

FIGURE 4–55

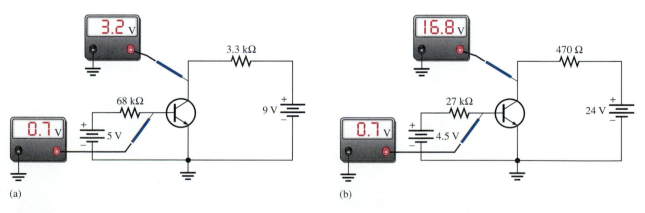

(a)

(b)

FIGURE 4–56

SECTION 4–8 System Application

32. This problem relates to the circuit board and schematic in Figure 4–43. A remote switch loop is connected between pins 2 and 3. When the remote switches are closed, the relay (RY1) contacts between pin 11 and pin 12 are normally open. When a remote switch is opened, the relay contacts do not close. Determine the possible causes of this malfunction.

33. This problem relates to Figure 4–43. The relay contact remains closed between pins 11 and 12 of the circuit board no matter what any of the inputs are. This means that the relay is energized continuously. What are the possible faults?

34. This problem relates to Figure 4–43. Pin 7 stays at approximately 0.1 V, regardless of the input at pin 8. What do you think is wrong? What would you check first?

■ DATA SHEET PROBLEMS

35. Refer to the partial transistor data sheet in Figure 4–19.
(a) What is the maximum collector-to-emitter voltage for a 1N3903?
(b) How much dc collector current can the 2N3904 handle?
(c) How much power can a 2N3903 dissipate if the surrounding air is at a temperature of 25°C?
(d) How much power can a 2N3904 dissipate if the case is at a temperature of 25°C?
(e) What is the minimum h_{FE} of a 2N3903 if the collector current is 1 mA?

36. Refer to the transistor data sheet in Figure 4–19. A 2N3904 is operating in an environment where the ambient temperature is 65°C. What is the most power that it can dissipate?

37. Refer to the transistor data sheet in Figure 4–19. A 2N3903 is operating with a case temperature of 45°C. What is the most power that it can dissipate?

38. Refer to the transistor data sheet in Figure 4–19. Determine if any rating is exceeded in each circuit of Figure 4–57 based on minimum specified values.

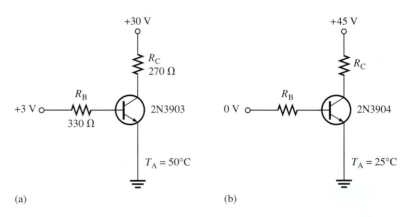

(a)   (b)

FIGURE 4–57

FIGURE 4–58

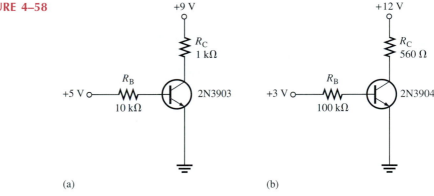

(a)　　　　　　　　　　　　　　(b)

FIGURE 4–59

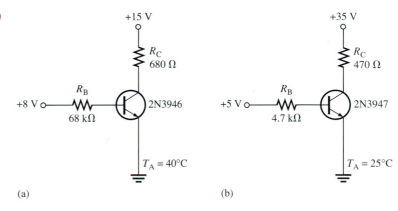

(a)　　　　　　　　　　　　　　(b)

39. Refer to the transistor data sheet in Figure 4–19. Determine whether or not the transistor is saturated in each circuit of Figure 4–58 based on the maximum specified value of h_{FE}.

40. Refer to the partial transistor data sheet in Figure 4–42. Determine the minimum and maximum base currents required to produce a collector current of 10 mA in a 2N3946. Assume that the transistor is not in saturation and $V_{CE} = 1$ V.

41. For each of the circuits in Figure 4–59, determine if there is a problem based on the data sheet information in Figure 4–42. Use the maximum specified h_{FE}.

■ **ADVANCED PROBLEMS**

42. Derive a formula for α_{DC} in terms of β_{DC}.

43. A certain 2N3904 dc bias circuit with the following values is in saturation. $I_B = 500$ μA, $V_{CC} = 10$ V, and $R_C = 180$ Ω, $h_{FE} = 150$. If you increase V_{CC} to 15 V, does the transistor come out of saturation? If so, what is the collector-to-emitter voltage and the collector current?

44. Design a dc bias circuit for a 2N3904 operating from a collector supply voltage of 9 V and a base-bias voltage of 3 V that will supply 150 mA to a resistive load which acts as the collector resistor. The circuit must not be in saturation. Assume the minimum specified β_{DC} from the data sheet.

45. Modify the design in Problem 44 to use a single 9 V dc source rather than two different sources. Other requirements remain the same.

46. Design a dc bias circuit for an amplifier in which the voltage gain is to be a minimum of 50 and the output signal voltage is to be "riding" on a dc level of 5 V. The maximum input signal voltage at the base is 10 mV rms, $V_{CC} = 12$ V, and $V_{BB} = 4$ V. Assume $r'_e = 8$ Ω.

■ **ANSWERS TO SECTION REVIEWS**

Section 4–1

1. The two types of BJTs are *npn* and *pnp*.
2. The terminals of a BJT are base, collector, and emitter.
3. The three regions of a BJT are separated by two *pn* junctions.

Section 4–2

1. To operate as an amplifier, the base-emitter is forward-biased and the base-collector is reverse-biased.
2. The emitter current is the largest.
3. The base current is much smaller than the emitter current.
4. The base region is very narrow compared to the other two regions.
5. $I_E = 1\,\text{mA} + 10\,\mu\text{A} = 1.01\,\text{mA}$

Section 4–3

1. $\beta_{DC} = I_C/I_B$; $\alpha_{DC} = I_C/I_E$; h_{FE} is β_{DC}.
2. $\beta_{DC} = 100$; $\alpha_{DC} = 100/(100 + 1) = 0.99$
3. I_C is plotted versus V_{CE}.
4. Forward-reverse bias is required for amplifier operation.
5. β_{DC} increases with temperature.
6. β_{DC} generally varies some from one device to the next for a given type.

Section 4–4

1. Amplification is the process where a smaller signal is used to produce a larger identical signal.
2. Voltage gain is the ratio of output voltage to input voltage.
3. R_C and r'_e determine the voltage gain.
4. $A_v = 5\,\text{V}/250\,\text{mV} = 20$
5. $A_v = 1200\,\Omega/20\,\Omega = 60$

Section 4–5

1. A transistor switch operates in cutoff and saturation.
2. The collector current is maximum in saturation.
3. The collector current is approximately zero in cutoff.
4. $V_{CE} = V_{CC}$ in cutoff.
5. V_{CE} is minimum in saturation.

Section 4–6

1. Three categories of BJT are small signal/general purpose, power, and RF.
2. Going clockwise from tab: emitter, base, and collector (bottom view).
3. The metal mounting tab or case in power transistors is the collector.

Section 4–7

1. First, test it in-circuit.
2. If R_B opens, the transistor is in cutoff.
3. The base voltage is +3 V and the collector voltage is +9 V.

■ **ANSWERS TO RELATED EXERCISES FOR EXAMPLES**

4–1 10 mA; 0.995

4–2 $I_B = 241$ μA; $I_C = 21.7$ mA; $I_E = 21.94$ mA; $V_{CE} = 4.23$ V; $V_{CB} = 3.53$ V

4–3 Along the horizontal axis

4–4 Not saturated

4–5 10 V

4–6 $V_{CC(max)} = 44.5$ V; $V_{CE(max)}$ is exceeded first.

4–7 4.55 W

4–8 2.5 kΩ

4–9 78.4 μA

4–10 Reduce R_C to 160 Ω and R_B to 2.2 kΩ.

4–11 R_B open

5

TRANSISTOR BIAS CIRCUITS

■ **CHAPTER OBJECTIVES**

□ Discuss the concept of dc bias in a linear amplifier
□ Analyze a base bias circuit
□ Analyze an emitter bias circuit
□ Analyze a voltage-divider bias circuit
□ Analyze a collector-feedback bias circuit
□ Troubleshoot various faults in transistor bias circuits

As you learned in the last chapter, a transistor must be properly biased in order to operate as an amplifier. DC biasing is used to establish a steady level of transistor current and voltage called the *dc operating point* or *quiescent point (Q-point)*. In this chapter, several types of bias circuits are studied. This material lays the groundwork for the study of amplifiers, oscillators, and other circuits that cannot operate without proper biasing.

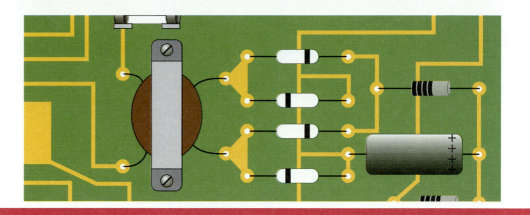

■ SYSTEM APPLICATION

The system application in this chapter focuses on a system for controlling temperature in an industrial chemical process. You will be dealing with a circuit that converts a temperature measurement to a proportional voltage that is used to adjust the temperature of a liquid chemical. You will also be working on the power supply for the system. The first step in your assignment is to learn all you can about transistor operation. You will then apply your knowledge to the system application in Section 5–7.

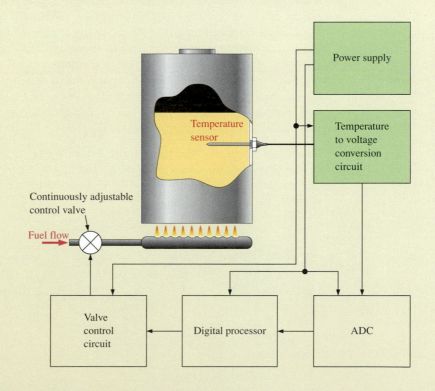

5–1 ■ THE DC OPERATING POINT

A transistor must be dc-biased in order to operate as an amplifier. A dc operating point must be set so that signal variations at the input terminal are amplified and accurately reproduced at the output terminal. As you learned in Chapter 4, when you bias a transistor, you establish a certain current and voltage condition. This means, for example, that at the dc operating point, I_C and V_{CE} have specified values. The dc operating point is often referred to as the Q-point (quiescent point).

After completing this section, you should be able to

■ **Discuss the concept of dc bias in a linear amplifier**
  ☐ Describe how to generate collector characteristic curves for a biased transistor
  ☐ Draw a dc load line for a given biased transistor circuit
  ☐ Explain Q-point
  ☐ Explain the conditions for linear operation
  ☐ Explain the conditions for saturation and cutoff
  ☐ Discuss the reasons for output waveform distortion

DC Bias

Bias provides for proper operation of an amplifier. If an amplifier is not biased correctly, it can go into saturation or cutoff when an input signal is applied. Figure 5–1 shows the effects of proper and improper biasing of an inverting amplifier. In part (a), the output signal is an amplified replica of the input signal except that it is inverted or 180° out of phase. The output signal swings equally above and below the dc bias level of the output. Improper biasing can cause distortion in the output signal, as illustrated in parts (b) and (c). Part (b) illustrates limiting of the positive portion of the output voltage as a result of a dc operating point (**Q-point**) being too close to cutoff. Part (c) shows limiting of the negative portion of the output voltage as a result of a dc operating point being too close to saturation.

Graphical Analysis The transistor in Figure 5–2(a) is biased with variables V_{CC} and V_{BB} to obtain certain values of I_B, I_C, I_E, and V_{CE}. The collector characteristic curves for this particular transistor are shown in Figure 5–2(b); we will use these to graphically illustrate the effects of dc bias.

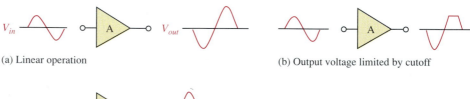

(a) Linear operation

(b) Output voltage limited by cutoff

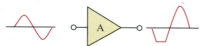

(c) Output voltage limited by saturation

FIGURE 5–1

Examples of linear and nonlinear operation of an inverting amplifier (the triangle symbol).

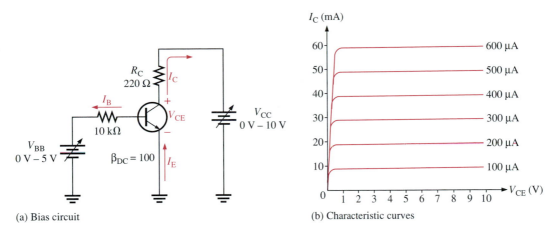

(a) Bias circuit   (b) Characteristic curves

FIGURE 5–2

A transistor circuit with variable bias voltages.

In Figure 5–3, we assign three values to I_B and observe what happens to I_C and V_{CE}. First, V_{BB} is adjusted to produce an I_B of 200 µA, as shown in Figure 5–3(a). Since $I_C = \beta_{DC}I_B$, the collector current is 20 mA, as indicated, and

$$V_{CE} = V_{CC} - I_C R_C = 10\ V - (20\ mA)(220\ \Omega) = 10\ V - 4.4\ V = 5.6\ V$$

This Q-point is shown on the graph of Figure 5–3(a) as Q_1.

Next, as shown in Figure 5–3(b), V_{BB} is increased to produce an I_B of 300 µA and an I_C of 30 mA.

$$V_{CE} = 10\ V - (30\ mA)(220\ \Omega) = 10\ V - 6.6\ V = 3.4\ V$$

The Q-point for this condition is indicated by Q_2 on the graph.

Finally, as in Figure 5–3(c), V_{BB} is increased to give an I_B of 400 µA and an I_C of 40 mA.

$$V_{CE} = 10\ V - (40\ mA)(220\ \Omega) = 10\ V - 8.8\ V = 1.2\ V$$

Q_3 is the corresponding Q-point on the graph.

DC Load Line Notice that when I_B increases, I_C increases and V_{CE} decreases. When I_B decreases, I_C decreases and V_{CE} increases. So, as V_{BB} is adjusted up or down, the dc operating point of the transistor moves along a sloping straight line, called the **dc load line,** connecting each Q-point. At any point along the line, values of I_B, I_C, and V_{CE} can be picked off the graph, as shown in Figure 5–4.

The dc load line intersects the V_{CE} axis at 10 V, the point where $V_{CE} = V_{CC}$. This is the transistor cutoff point because I_B and I_C are zero (ideally). Actually, there is a small leakage current, I_{CBO}, at cutoff as indicated, and therefore V_{CE} is slightly less than 10 V but normally this can be neglected.

The dc load line intersects the I_C axis at 45.5 mA ideally. This is the transistor saturation point because I_C is maximum (ideally 50 mA) at the point where $V_{CE} = 0\ V$ and $I_C = V_{CC}/R_C$. Actually, there is a small voltage ($V_{CE(sat)}$) across the transistor, and $I_{C(sat)}$ is

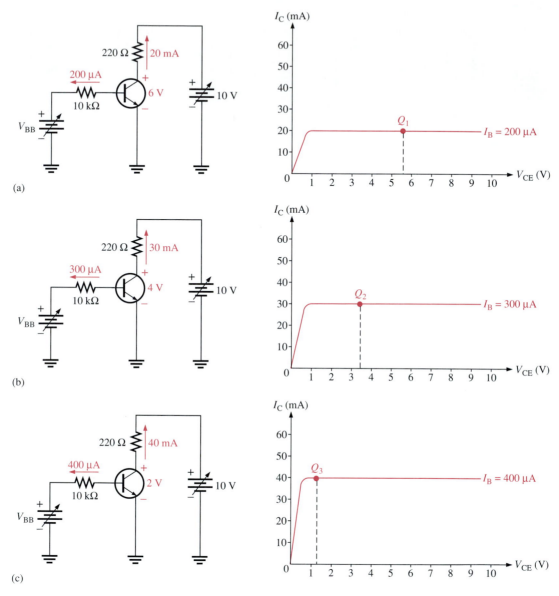

FIGURE 5–3

Illustration of Q-point adjustments.

slightly less than 50 mA, as indicated in Figure 5–4. Note that Kirchhoff's voltage law applied around the collector loop gives

$$V_{CC} - I_C R_C - V_{CE} = 0$$

This results in a straight line equation for the load line of the form $y = mx + b$ as follows:

$$I_C = -\left(\frac{1}{R_E}\right)V_{CE} + \frac{V_{CC}}{R_C}$$

where $-1/R_C$ is the slope and V_{CC}/R_C is the y-axis intercept point.

FIGURE 5–4
The dc load line.

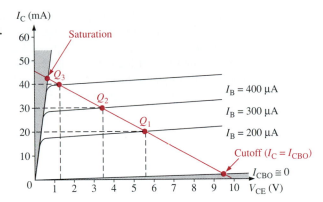

Linear Operation   The region along the load line including all points between saturation and cutoff is generally known as the *linear region* of the transistor's operation. As long as the transistor is operated in this region, the output voltage is ideally a linear reproduction of the input.

Figure 5–5 shows an example of the linear operation of a transistor. Assume a sinusoidal voltage, V_{in}, is superimposed on V_{BB}, causing the base current to vary sinusoidally 100 μA above and below its Q-point value of 300 μA. This, in turn, causes the collector current to vary 10 mA above and below its Q-point value of 30 mA. As a result of the variation in collector current, the collector-to-emitter voltage varies 2.2 V above and below its Q-point value of 3.4 V. Point *A* on the load line in Figure 5–5 corresponds to the positive peak of the sinusoidal input voltage. Point *B* corresponds to the negative peak, and point *Q* corresponds to the zero value of the sine wave, as indicated. V_{CEQ} and I_{CQ} are dc Q-point values with no input sinusoidal voltage applied.

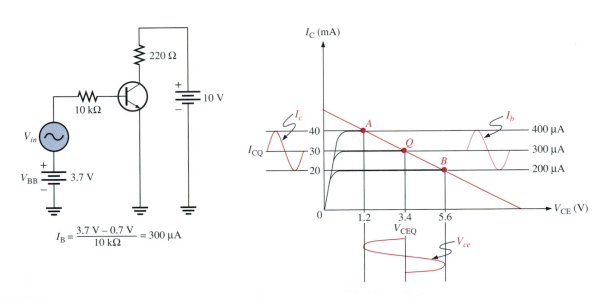

FIGURE 5–5
Variations in collector current and collector-to-emitter voltage as a result of a variation in base current. Notice that ac quantities are indicated by lowercase italic subscripts.

Waveform Distortion As previously mentioned, under certain input signal conditions the location of the Q-point on the load line can cause one peak of V_{ce} to be limited or clipped, as shown in parts (a) and (b) of Figure 5–6. In each case the input signal is too large for the Q-point location and is driving the transistor into cutoff or saturation during a portion of the input cycle. When both peaks are limited as in Figure 5–6(c), the transistor is being driven into both saturation and cutoff by an excessively large input signal.

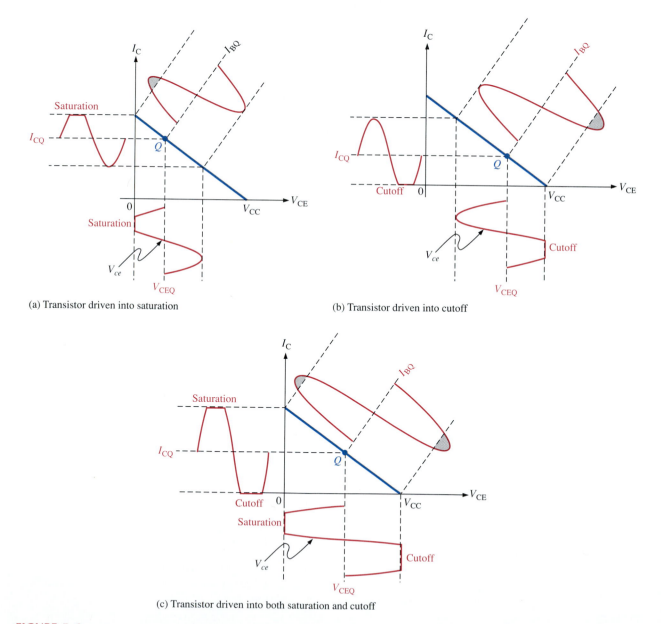

(a) Transistor driven into saturation

(b) Transistor driven into cutoff

(c) Transistor driven into both saturation and cutoff

FIGURE 5–6

Graphical illustration of saturation and cutoff.

When only the positive peak is limited, the transistor is being driven into cutoff but not saturation. When only the negative peak is limited, the transistor is being driven into saturation but not cutoff. A good rule-of-thumb for a distortion-free output is to limit the maximum peak V_{ce} swing to $0.95V_{CC}$ and the minimum peak to $0.05V_{CC}$ with the Q-point centered on the load line.

EXAMPLE 5–1

Determine the Q-point in Figure 5–7, and find the maximum peak value of base current for linear operation ($\beta_{DC} = 200$).

FIGURE 5–7

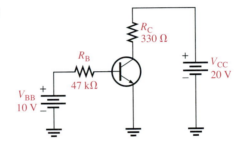

Solution   The Q-point is defined by I_C and V_{CE}. Find these values by using formulas you learned in Chapter 4.

$$I_B = \frac{V_{BB} - V_{BE}}{R_B} = \frac{10 \text{ V} - 0.7 \text{ V}}{47 \text{ k}\Omega} = 198 \text{ }\mu\text{A}$$

$$I_C = \beta_{DC}I_B = (200)(198 \text{ }\mu\text{A}) = 39.6 \text{ mA}$$

$$V_{CE} = V_{CC} - I_C R_C = 20 \text{ V} - 13.07 \text{ V} = 6.93 \text{ V}$$

The Q-point is at $I_C = 39.6$ mA and at $V_{CE} = 6.93$ V. Since $I_{C(cutoff)} = 0$, you need to know $I_{C(sat)}$ to determine how much variation in collector current can occur and still maintain linear operation of the transistor.

$$I_{C(sat)} = \frac{V_{CC}}{R_C} = \frac{20 \text{ V}}{330 \text{ }\Omega} = 60.6 \text{ mA}$$

The dc load line is graphically illustrated in Figure 5–8, showing that before saturation is reached, I_C can increase an amount ideally equal to

$$I_{C(sat)} - I_{CQ} = 60.6 \text{ mA} - 39.6 \text{ mA} = 21 \text{ mA}$$

However, I_C can decrease by 39.6 mA before cutoff is reached. Therefore, the limiting excursion is 21 mA because the Q-point is closer to saturation than to cutoff. The 21 mA is the maximum peak variation of the collector current. Actually, it would be slightly less in practice because $V_{CE(sat)}$ is not quite zero. The maximum peak variation of the base current is determined as follows:

$$I_{b(peak)} = \frac{I_{c(peak)}}{\beta_{DC}} = \frac{21 \text{ mA}}{200} = 105 \text{ }\mu\text{A}$$

FIGURE 5–8

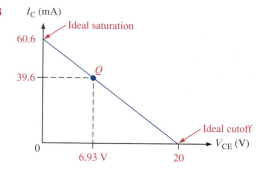

Related Exercise Find the Q-point in Figure 5–7, and determine the maximum peak value of base current for linear operation for the following circuit values: $\beta_{DC} = 100$, $R_C = 1 \text{ k}\Omega$, and $V_{CC} = 24$ V.

SECTION 5–1 REVIEW

1. What are the upper and lower limits on a dc load line in terms of V_{CE} and I_C?
2. Define *Q-point*.
3. At what point on the load line does saturation begin? At what point does cutoff occur?
4. For maximum V_{ce}, where should the Q-point be placed?

5–2 ■ BASE BIAS

In this section and several following sections, you will learn various methods for dc biasing a transistor circuit without using a separate base bias source. You will see the advantages and disadvantages of each method.

After completing this section, you should be able to

■ **Analyze a base bias circuit**
  □ Recognize a base-biased transistor circuit
  □ Describe how β_{DC} affects the Q-point
  □ Describe how V_{BE} and I_{CBO} affect the Q-point
  □ Discuss the stability of a base bias circuit

Up to this point, we used a separate dc source, V_{BB}, to bias the base-emitter junction strictly for convenience in illustrating transistor operation since it could be varied independently of V_{CC}. A more practical method is to use V_{CC} as the single bias source, as shown in Figure 5–9(a). To simplify the schematics, the battery symbol can be omitted

FIGURE 5–9
Base bias.

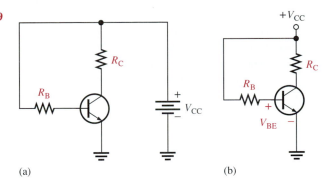

(a)   (b)

and replaced by a line termination circle with a voltage indicator, as shown in Figure 5–9(b).

The analysis of the circuit in Figure 5–9 for the linear region is as follows. The voltage drop across R_B is $V_{CC} - V_{BE}$. Therefore,

$$I_B = \frac{V_{CC} - V_{BE}}{R_B} \qquad (5-1)$$

Kirchhoff's voltage law applied around the collector circuit in Figure 5–9(a) gives the following equation:

$$V_{CC} - I_C R_C - V_{CE} = 0$$

Solving for V_{CE}, we get

$$V_{CE} = V_{CC} - I_C R_C \qquad (5-2)$$

We can neglect the leakage current, I_{CBO}. We already know that $I_C = \beta_{DC} I_B$; therefore,

$$I_C = \beta_{DC}\left(\frac{V_{CC} - V_{BE}}{R_B}\right) \qquad (5-3)$$

Effect of β_{DC} on the Q-Point

Notice that Equation (5–3) shows that I_C is dependent on β_{DC}. The disadvantage of this is that a variation in β_{DC} causes both I_C and V_{CE} to change, thus changing the Q-point of the transistor and making the base bias circuit extremely beta-dependent.

Recall that β_{DC} varies with temperature and collector current. In addition, there is a large spread of values from one device to another of the same type due to manufacturing variations. Therefore, a circuit using base bias may suddenly produce a distorted output if, due to a fault, the transistor is replaced with one having a different β_{DC} or if a temperature change causes a sufficient shift in the value of β_{DC}.

EXAMPLE 5–2

The base bias circuit in Figure 5–10 is subjected to an increase in temperature from 25°C to 75°C. If $\beta_{DC} = 100$ at 25°C and 150 at 75°C, determine the percent change in Q-point values (I_C and V_{CE}) over the temperature range. Neglect any change in V_{BE} and the effects of any leakage current.

FIGURE 5–10

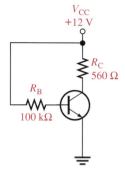

Solution At 25°C, I_C and V_{CE} are determined as follows:

$$I_C = \beta_{DC}\left(\frac{V_{CC} - V_{BE}}{R_B}\right) = 100\left(\frac{12\ \text{V} - 0.7\ \text{V}}{100\ \text{k}\Omega}\right) = 11.3\ \text{mA}$$

$$V_{CE} = V_{CC} - I_C R_C = 12\ \text{V} - (11.3\ \text{mA})(560\ \Omega) = 5.67\ \text{V}$$

At 75°C, I_C and V_{CE} are determined as follows:

$$I_C = \beta_{DC}\left(\frac{V_{CC} - V_{BE}}{R_B}\right) = 150\left(\frac{12\ \text{V} - 0.7\ \text{V}}{100\ \text{k}\Omega}\right) = 17.0\ \text{mA}$$

$$V_{CE} = V_{CC} - I_C R_C = 12\ \text{V} - (17.0\ \text{mA})(560\ \Omega) = 2.48\ \text{V}$$

The percent change in I_C is

$$\%\ \Delta I_C = \frac{I_{C(75°)} - I_{C(25°)}}{I_{C(25°)}} \times 100\%$$

$$= \frac{17.0\ \text{mA} - 11.3\ \text{mA}}{11.3\ \text{mA}} \times 100\% \cong 50\% \quad \text{(an increase)}$$

Notice the I_C changes by the same percentage as β_{DC}. The percent change in V_{CE} is

$$\%\ \Delta V_{CE} = \frac{V_{CE(75°)} - V_{CE(25°)}}{V_{CE(25°)}} \times 100\%$$

$$= \frac{2.48\ \text{V} - 5.67\ \text{V}}{5.67\ \text{V}} \times 100\% = -56.3\ \% \quad \text{(a decrease)}$$

As you can see, the Q-point is extremely dependent on β_{DC} in this circuit and therefore makes the base bias arrangement very unstable. Consequently, base bias is not normally used if linear operation is required. It can be used for switching operation.

Related Exercise If $\beta_{DC} = 50$ at 0°C and 125 at 100°C for the circuit in Figure 5–10, determine the percent change in the Q-point values over the temperature range.

Other Factors Influencing Bias Stability

In addition to being affected by a change in β_{DC}, the bias point can be affected by changes in V_{BE} and I_{CBO}. The base-to-emitter voltage, V_{BE}, decreases with an increase in temperature. As you can see from the equation for I_B, a decrease in V_{BE} increases I_B.

$$I_B = \frac{V_{CC} - V_{BE}}{R_B}$$

The effect of a change in V_{BE} is negligible if $V_{CC} \gg V_{BE}$ (V_{CC} at least 10 times greater than V_{BE}).

The reverse leakage current, I_{CBO}, has the effect of decreasing the net base current and thus increasing the base voltage because it creates a voltage drop across R_B with a polarity that adds to the base bias voltage, as shown in Figure 5–11. In modern transistors, I_{CBO} is usually less than 100 nA, and its effect on the bias is negligible if $V_{BB} \gg I_{CBO}R_B$.

FIGURE 5–11
Effect of I_{CBO}.

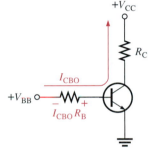

1. What is an advantage of base bias over using two separate dc supplies?
2. What is the main disadvantage of the base bias method?
3. Explain why the base bias Q-point changes with temperature.

5–3 ■ EMITTER BIAS

Although this method of biasing requires two separate dc voltage sources, one positive and the other negative, it does have an important advantage as you will learn.

After completing this section, you should be able to

■ **Analyze an emitter bias circuit**
   □ Discuss the effect of β_{DC} and V_{CE} on the Q-point
   □ Explain how to minimize or essentially eliminate the effects of β_{DC} and V_{BE} on the stability of the Q-point
   □ Discuss emitter bias for a *pnp* transistor

An emitter bias circuit uses both a positive and a negative supply voltage, as shown in Figure 5–12. In this circuit, the V_{EE} supply voltage forward-biases the base-emitter junction. Kirchhoff's voltage law applied around the base-emitter circuit in Figure 5–12(a), which has been redrawn in Figure 5–12(b) for analysis, gives the following equation:

$$V_{EE} - I_B R_B - V_{BE} - I_E R_E = 0$$

Solving for V_{EE}, we get

$$V_{EE} = I_B R_B + I_E R_E + V_{BE}$$

Since $I_C \cong I_E$ and $I_C = \beta_{DC} I_B$,

$$I_B \cong \frac{I_E}{\beta_{DC}}$$

Substituting for I_B, we get

$$\left(\frac{I_E}{\beta_{DC}}\right) R_B + I_E R_E + V_{BE} = V_{EE}$$

Factoring out I_E yields

$$I_E \left(\frac{R_B}{\beta_{DC}} + R_E\right) + V_{BE} = V_{EE}$$

Transposing V_{BE} and then solving for I_E, we get

$$I_E = \frac{V_{EE} - V_{BE}}{R_E + R_B/\beta_{DC}} \tag{5–4}$$

Since $I_C \cong I_E$,

$$I_C \cong \frac{V_{EE} - V_{BE}}{R_E + R_B/\beta_{DC}} \tag{5–5}$$

The emitter voltage with respect to ground is

$$V_E = -V_{EE} + I_E R_E \tag{5–6}$$

The base voltage with respect to ground is

$$V_B = V_E + V_{BE} \tag{5–7}$$

The collector voltage with respect to ground is

$$V_C = V_{CC} - I_C R_C \tag{5–8}$$

Subtracting V_E from V_C and using the approximation $I_C \cong I_E$, we get

$$V_{CE} = V_{CC} - I_C R_C - (-V_{EE} + I_E R_E)$$

$$V_{CE} \cong V_{CC} + V_{EE} - I_C(R_C + R_E) \tag{5–9}$$

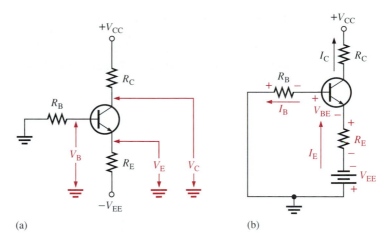

FIGURE 5–12

An npn transistor with emitter bias. Polarities are reversed for a pnp. Single subscripts indicate voltages with respect to ground.

EXAMPLE 5–3

Find I_E, I_C, and V_{CE} in Figure 5–13 for $\beta_{DC} = 100$ and $V_{BE} = 0.7$ V. Draw the dc load line.

FIGURE 5–13

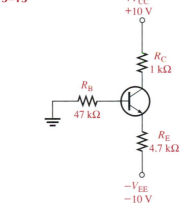

Solution

$$I_E = \frac{V_{EE} - V_{BE}}{R_E + R_B/\beta_{DC}} = \frac{10\ V - 0.7\ V}{4.7\ k\Omega + 47\ k\Omega/100} = \frac{9.3\ V}{5.17\ k\Omega} = 1.80\ mA$$

$$I_C \cong I_E = 1.80\ mA$$

$$V_{CE} \cong V_{CC} + V_{EE} - I_C(R_C + R_E) = 10\ V + 10\ V - 1.80\ mA\ (5.7\ k\Omega) = 9.74\ V$$

I_C and V_{CE} are the Q-point values for the circuit in Figure 5–13. The dc load line is graphically illustrated in Figure 5–14. The approximate collector saturation current is determined as follows:

$$I_{C(sat)} = \frac{V_{CC} - (-V_{EE})}{R_C + R_E} = \frac{10\ V - (-10\ V)}{5.7\ k\Omega} = \frac{20\ V}{5.7\ k\Omega} = 3.51\ mA$$

The collector-to-emitter voltage at cutoff is

$$V_{CE(cutoff)} = V_{CC} - (-V_{EE}) = 10 \text{ V} - (-10 \text{ V}) = 20 \text{ V}$$

The dc load line, illustrated in Figure 5–14, shows that I_C can increase an amount ideally equal to

$$\Delta I_{C(max)} = I_{C(sat)} - I_{CQ} = 3.51 \text{ mA} - 1.80 \text{ mA} = 1.71 \text{ mA}$$

before saturation is reached. I_C can decrease by 1.80 mA before cutoff is reached. As you can see, this circuit is biased slightly closer to saturation than to cutoff.

FIGURE 5–14

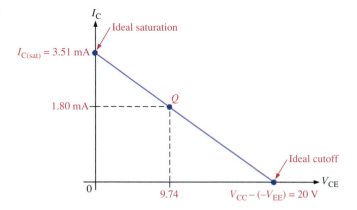

Related Exercise Find I_E, I_C, and V_{CE} for the circuit in Figure 5–13 with the following component values: $R_B = 100 \text{ k}\Omega$, $R_C = 680 \text{ }\Omega$, $R_E = 3.3 \text{ k}\Omega$, $V_{CC} = +15 \text{ V}$, and $-V_{EE} = -15 \text{ V}$. Assume $\beta_{DC} = 150$.

Stability of Emitter Bias

The formula for I_E shows that the emitter bias circuit is dependent on V_{BE} and β_{DC}, both of which change with temperature and current.

$$I_E = \frac{V_{EE} - V_{BE}}{R_E + R_B/\beta_{DC}}$$

If $R_E \gg R_B/\beta_{DC}$, the equation becomes

$$I_E \cong \frac{V_{EE} - V_{BE}}{R_E}$$

This condition makes I_E independent of β_{DC}.

A further modification can be made if $V_{EE} \gg V_{BE}$.

$$I_E \cong \frac{V_{EE}}{R_E}$$

This condition makes I_E independent of V_{BE}.

If I_E is independent of β_{DC} and V_{BE}, then the Q-point is not affected appreciably by variations in these parameters. Thus, emitter bias can provide a stable Q-point if properly designed.

EXAMPLE 5–4

Determine how much the Q-point in Figure 5–15 will change over a temperature range where β_{DC} increases from 85 to 100 and V_{BE} decreases from 0.7 V to 0.6 V.

FIGURE 5–15

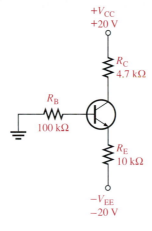

Solution For $\beta_{DC} = 85$ and $V_{BE} = 0.7$ V,

$$I_C \cong I_E = \frac{V_{EE} - V_{BE}}{R_E + R_B/\beta_{DC}} = \frac{20\ V - 0.7\ V}{10\ k\Omega + 100\ k\Omega/85} = 1.73\ mA$$

$$V_C = V_{CC} - I_C R_C = 20\ V - (1.73\ mA)(4.7\ k\Omega) = 11.9\ V$$

$$V_E = -V_{EE} + I_E R_E = -20\ V + (1.73\ mA)(10\ k\Omega) = -2.7\ V$$

Therefore,

$$V_{CE} = V_C - V_E = 11.9\ V - (-2.7\ V) = 14.6\ V$$

For $\beta_{DC} = 100$ and $V_{BE} = 0.6$ V,

$$I_C \cong I_E = \frac{V_{EE} - V_{BE}}{R_E + R_B/\beta_{DC}} = \frac{20\ V - 0.6\ V}{10\ k\Omega + 100\ k\Omega/100} = 1.85\ mA$$

$$V_C = V_{CC} - I_C R_C = 20\ V - (1.85\ mA)(4.7\ k\Omega) = 11.3\ V$$

$$V_E = -V_{EE} + I_E R_E = -20\ V + (1.85\ mA)(10\ k\Omega) = -1.5\ V$$

Therefore,

$$V_{CE} = V_C - V_E = 11.3\ V - (-1.5\ V) = 12.8\ V$$

The percent change in I_C as β_{DC} changes from 85 to 100 is

$$\frac{1.85\ mA - 1.73\ mA}{1.73\ mA} \times 100\% = 6.94\%$$

The percent change in V_{CE} is

$$\frac{12.8\ V - 14.6\ V}{12.8\ V} \times 100\% = 14.0\%$$

Related Exercise Determine how much the Q-point in Figure 5–15 changes over a temperature range where β_{DC} increases from 65 to 75 and V_{BE} decreases from 0.75 V to 0.59 V. The supply voltages are ±10 V.

Emitter-Biased *PNP*

Figure 5–16 shows a *pnp* transistor with emitter bias. The basic difference is that the polarities of the supply voltages are reversed from those of the *npn*. The operation and analysis are basically the same, as Example 5–5 illustrates.

FIGURE 5–16

Emitter-biased pnp transistor.

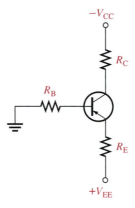

EXAMPLE 5–5

Determine V_C, V_E, and V_{CE} in the circuit of Figure 5–17. Assume $\beta_{DC} = 100$ and $V_{BE} = 0.7$ V. Notice how the transistor is oriented in this diagram with the positive emitter supply at the top and the negative collector supply at the bottom.

FIGURE 5–17

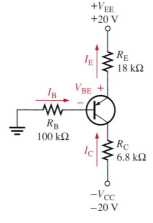

Solution

$$I_C \cong I_E = \frac{V_{EE} - V_{BE}}{R_E + R_B/\beta_{DC}} = \frac{20\text{ V} - 0.7\text{ V}}{18\text{ k}\Omega + 100\text{ k}\Omega/100} = 1.02\text{ mA}$$

$$V_C = -V_{CC} + I_C R_C = -20\text{ V} + (1.02\text{ mA})(6.8\text{ k}\Omega) = 13.1\text{ V}$$

$$V_E = V_{EE} - I_E R_E = 20\text{ V} - (1.02\text{ mA})(18\text{ k}\Omega) = 1.64\text{ V}$$

Therefore,

$$V_{CE} = V_C - V_E = -13.1\text{ V} - 1.64\text{ V} = -14.7\text{ V}$$

Related Exercise Does V_C become more or less negative if R_C is increased? Will an increase in R_C change the collector current significantly? Calculate the collector voltage if R_C is 8.2 kΩ.

SECTION 5–3 REVIEW

1. Why is emitter bias more stable than base bias?

2. For an emitter-biased *npn* transistor, what is the approximate relationship of V_B and V_E? For a *pnp* transistor?

3. What is the main disadvantage of emitter bias?

4. An emitter-biased *pnp* transistor has dc supply voltages of ±15 V. If R_E is 10 kΩ, what is the emitter current?

5–4 ■ VOLTAGE-DIVIDER BIAS

Next, you will study a method of biasing a transistor for linear operation using a resistive voltage-divider. This is the most widely used biasing method, for reasons that you will discover in this section. This method is actually a form of emitter bias using a single supply voltage.

After completing this section, you should be able to

■ **Analyze a voltage-divider bias circuit**
   □ Discuss the effect of the input resistance on the bias circuit
   □ Discuss the stability of voltage-divider bias
   □ Explain how to minimize or essentially eliminate the effects of β_{DC} and V_{BE} on the stability of the Q-point
   □ Discuss voltage-divider bias for a *pnp* transistor

A dc bias voltage at the base of the transistor is developed by a resistive voltage-divider consisting of R_1 and R_2 as shown in Figure 5–18. At point A, there are two current paths to ground: one through R_2 and the other through the base-emitter junction of the transistor.

FIGURE 5–18
Voltage-divider bias.

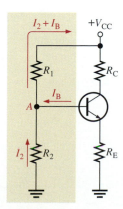

If the base current is much smaller than the current through R_2, the bias circuit can be viewed as a simplified voltage-divider consisting of R_1 and R_2, as indicated in Figure 5–19(a). If I_B is *not* small enough to neglect compared to I_2, then the dc input resistance, $R_{IN(base)}$, from the base of the transistor to ground must be considered. $R_{IN(base)}$ appears in parallel with R_2, as shown in Figure 5–19(b).

FIGURE 5–19
Simplified voltage-divider.

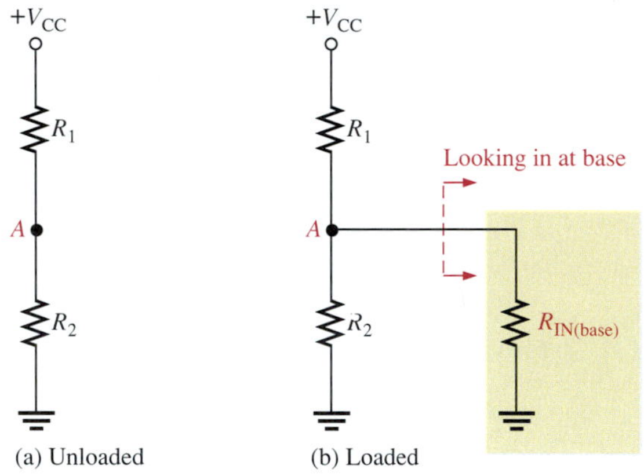

(a) Unloaded   (b) Loaded

Input Resistance at the Base

To develop a formula for the dc input resistance at the base of a transistor, we will use the diagram in Figure 5–20. V_{IN} is applied between base and ground, and I_{IN} is the current out of the base as shown. By Ohm's law,

$$R_{IN(base)} = \frac{V_{IN}}{I_{IN}}$$

Kirchhoff's voltage law applied around the base-emitter circuit yields

$$V_{IN} = V_{BE} + I_E R_E$$

With the assumption that $V_{BE} \ll I_E R_E$, the equation reduces to

$$V_{IN} \cong I_E R_E$$

Now, since $I_E \cong I_C = \beta_{DC} I_B$,

$$V_{IN} \cong \beta_{DC} I_B R_E$$

The input current is the base current:

$$I_{IN} = I_B$$

By substitution,

$$R_{IN(base)} = \frac{V_{IN}}{I_{IN}} \cong \frac{\beta_{DC} I_B R_E}{I_B}$$

So,

$$R_{IN(base)} \cong \beta_{DC} R_E \qquad \qquad (5\text{–}10)$$

FIGURE 5–20
DC input resistance is V_{IN}/I_{IN}.

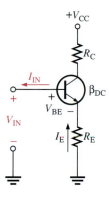

EXAMPLE 5–6

Determine the dc input resistance at the base of the transistor in Figure 5–21. $\beta_{DC} = 125$.

FIGURE 5–21

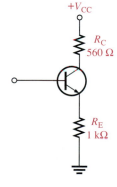

Solution $\quad\quad\quad\quad R_{IN(base)} \cong \beta_{DC}R_E = (125)(1 \text{ k}\Omega) = 125 \text{ k}\Omega$

Related Exercise What is $R_{IN(base)}$ if $\beta_{DC} = 60$ and $R_E = 910 \ \Omega$ in Figure 5–21?

Analysis of a Voltage-Divider Bias Circuit

A voltage-divider biased *npn* transistor is shown in Figure 5–22(a). Let's begin the analysis by determining the voltage at the base using the voltage-divider formula, which is developed as follows.

$$R_{IN(base)} \cong \beta_{DC}R_E$$

The total resistance from base to ground is

$$R_2 \parallel \beta_{DC}R_E$$

FIGURE 5–22

An npn transistor with voltage-divider bias.

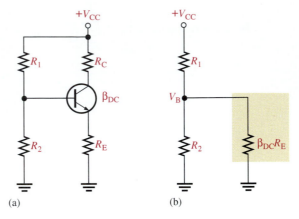

(a)   (b)

A voltage-divider is formed by R_1 and the resistance from base to ground in parallel with R_2 as shown in Figure 5–22(b). Applying the voltage-divider formula yields

$$V_B = \left(\frac{R_2 \parallel \beta_{DC}R_E}{R_1 + (R_2 \parallel \beta_{DC}R_E)} \right) V_{CC} \qquad (5\text{–}11)$$

If $\beta_{DC}R_E \gg R_2$, then the formula simplifies to

$$V_B \cong \left(\frac{R_2}{R_1 + R_2} \right) V_{CC} \qquad (5\text{–}12)$$

Once you know the base voltage, you can determine the emitter voltage, which equals V_B less the value of the base-emitter drop (V_{BE}).

$$V_E = V_B - V_{BE}$$

The emitter current can be found by using Ohm's law.

$$I_E = \frac{V_E}{R_E}$$

Once you know I_E, you can find all the other circuit values.

$$I_C \cong I_E$$

$$I_C \cong \frac{V_B - V_{BE}}{R_E} \qquad (5\text{–}13)$$

$$V_C = V_{CC} - I_C R_C$$
$$V_{CE} = V_C - V_E$$

Also, you can express V_{CE} in terms of I_C by using Kirchhoff's voltage law as follows:

$$V_{CC} - I_C R_C - I_E R_E - V_{CE} = 0$$

Since $I_C \cong I_E$,

$$V_{CE} \cong V_{CC} - I_C R_C - I_C R_E$$

$$V_{CE} \cong V_{CC} - I_C(R_C + R_E) \qquad (5\text{–}14)$$

EXAMPLE 5–7   Determine V_{CE} and I_C in Figure 5–23, where $\beta_{DC} = 100$.

FIGURE 5–23

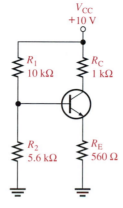

Solution   First, determine the dc input resistance at the base to see if it can be neglected.

$$R_{IN(base)} = \beta_{DC} R_E = (100)(560\ \Omega) = 56\ k\Omega$$

A common rule-of-thumb is that if two resistors are in parallel and one is at least ten times the other, the total resistance is taken to be approximately equal to the smallest value although, in some cases, this may result in unacceptable inaccuracy.

In this case, $R_{IN(base)} = 10R_2$, so neglect $R_{IN(base)}$. In the related exercise, you will rework this example taking $R_{IN(base)}$ into account and compare the difference. Proceed with the analysis by determining the base voltage:

$$V_B \cong \left(\frac{R_2}{R_1 + R_2} \right) V_{CC} = \left(\frac{5.6\ k\Omega}{15.6\ k\Omega} \right) 10\ V = 3.59\ V$$

So,

$$V_E = V_B - V_{BE} = 3.59\ V - 0.7\ V = 2.89\ V$$

and

$$I_E = \frac{V_E}{R_E} = \frac{2.89\ V}{560\ \Omega} = 5.16\ mA$$

Therefore,

$$I_C \cong 5.16\ mA$$

and

$$V_{CE} \cong V_{CC} - I_C(R_C + R_E) = 10 \text{ V} - 5.16 \text{ mA}(1.56 \text{ k}\Omega) = 1.95 \text{ V}$$

Since $V_{CE} > 0$ V (or a few tenths of a volt), you know that the transistor is *not* in saturation.

Related Exercise Rework this example taking into account $R_{IN(base)}$ and compare the results.

Stability of Voltage-Divider Bias

Another way to analyze a voltage-divider biased transistor circuit is to apply Thevenin's theorem. We will use this method to evaluate the stability of the circuit. First, let's get an equivalent base-emitter circuit for Figure 5–22 using Thevenin's theorem. Looking out from the base terminal, the bias circuit can be redrawn as shown in Figure 5–24(a). Applying Thevenin's theorem to the circuit left of point A, we get the following results:

$$R_{TH} = \frac{R_1 R_2}{R_1 + R_2}$$

and

$$V_{TH} = \left(\frac{R_2}{R_1 + R_2}\right)V_{CC}$$

The Thevenin equivalent is shown in Figure 5–24(b). Applying Kirchhoff's voltage law around the equivalent base-emitter loop gives

$$V_{TH} = I_B R_{TH} + V_{BE} + I_E R_E$$

Substituting I_E/β_{DC} for I_B, we get

$$V_{TH} = I_E(R_E + R_{TH}/\beta_{DC}) + V_{BE}$$

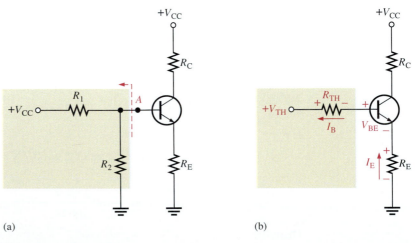

(a)   (b)

FIGURE 5–24
Thevenizing the bias circuit.

or

$$I_E = \frac{V_{TH} - V_{BE}}{R_E + R_{TH}/\beta_{DC}}$$

If $R_E \gg R_{TH}/\beta_{DC}$, then

$$I_E \cong \frac{V_{TH} - V_{BE}}{R_E}$$

   This last equation shows that I_E is essentially independent of β_{DC} (notice that β_{DC} does not appear in the equation) under the condition indicated. This can be achieved in practice by selecting a value for R_E that is at least ten times the resistance of the parallel combination of the voltage-divider resistors (R_{TH}) divided by the minimum β_{DC}. The voltage-divider bias is popular because reasonably good stability is achieved with a single supply voltage.

Voltage-Divider Biased *PNP*

As you know, a *pnp* transistor requires bias polarities opposite to the *npn*. This can be accomplished with a negative collector supply voltage, as in Figure 5–25(a), or with a positive emitter supply voltage, as in Figure 5–25(b). In a schematic, the *pnp* is often drawn upside down so that the supply voltage line can be drawn across the top of the schematic and ground at the bottom, as in Figure 5–26. The analysis procedure is basically the same as for an *npn* circuit, as demonstrated in the following steps with reference to Figure 5–26. The base voltage is determined by using the voltage-divider formula.

$$V_B = \left(\frac{R_1}{R_1 + R_2 \,\|\, \beta_{DC}R_E} \right) V_{EE} \qquad (5\text{–}15)$$

and

$$V_E = V_B + V_{BE}$$

FIGURE 5–25
Voltage-divider biased pnp transistor.

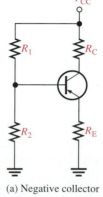

(a) Negative collector
supply voltage

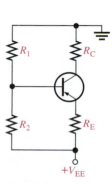

(b) Positive emitter
supply voltage

FIGURE 5–26

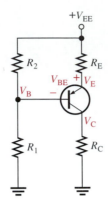

By Ohm's law,

$$I_E = \frac{V_{EE} - V_E}{R_E}$$

and

$$V_C = I_C R_C$$
$$V_{EC} = V_E - V_C$$

EXAMPLE 5–8

Find I_C and V_{EC} in Figure 5–27.

FIGURE 5–27

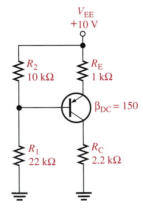

Solution   First, check to see if $R_{IN(base)}$ can be neglected.

$$R_{IN(base)} = \beta_{DC}R_E = (150)(1 \text{ k}\Omega) = 150 \text{ k}\Omega$$

Since 150 kΩ is more than ten times R_2, the condition $\beta_{DC}R_E \gg R_2$ is met and $R_{IN(base)}$ can be neglected. Now, calculate V_B.

$$V_B \cong \left(\frac{R_1}{R_1 + R_2} \right) V_{EE} = \left(\frac{22 \text{ k}\Omega}{32 \text{ k}\Omega} \right) 10 \text{ V} = 6.88 \text{ V}$$

Then

$$V_E = V_B - V_{BE} = 6.88 \text{ V} + 0.7 \text{ V} = 7.58 \text{ V}$$

and

$$I_E = \frac{V_{EE} - V_E}{R_E} = \frac{10 \text{ V} - 7.58 \text{ V}}{1 \text{ k}\Omega} = 2.42 \text{ mA}$$

From I_E, you can determine I_C and V_{CE} as follows:

$$I_C \cong I_E = 2.42 \text{ mA}$$

and

$$V_C = I_C R_C = (2.42 \text{ mA})(2.2 \text{ k}\Omega) = 5.32 \text{ V}$$

Therefore,

$$V_{EC} = V_E - V_C = 7.58 \text{ V} - 5.32 \text{ V} = 2.26 \text{ V}$$

Related Exercise Determine I_C and V_{EC} in Figure 5–27 with $R_{IN(base)}$ taken into account.

EXAMPLE 5–9

Find I_C and V_{CE} in Figure 5–25(a) for $R_1 = 68 \text{ k}\Omega$, $R_2 = 47 \text{ k}\Omega$, $R_C = 1.8 \text{ k}\Omega$, $R_E = 2.2 \text{ k}\Omega$, $-V_{CC} = -6 \text{ V}$, and $\beta_{DC} = 75$.

Solution $\qquad R_{IN(base)} = \beta_{DC} R_E = 75(2.2 \text{ k}\Omega) = 165 \text{ k}\Omega$

Since $R_{IN(base)}$ is not ten times greater than R_2, it must be taken into account. Now, determine the base voltage:

$$V_B = \left(\frac{R_2 \| R_{IN(base)}}{R_1 + R_2 \| R_{IN(base)}} \right)(-V_{CC}) = \left(\frac{47 \text{ k}\Omega \| 165 \text{ k}\Omega}{68 \text{ k}\Omega + 47 \text{ k}\Omega \| 165 \text{ k}\Omega} \right)(-6 \text{ V})$$

$$= \left(\frac{36.6 \text{ k}\Omega}{68 \text{ k}\Omega + 36.6 \text{ k}\Omega} \right)(-6 \text{ V}) = -2.1 \text{ V}$$

Next, calculate the emitter voltage and current:

$$V_E = V_B + V_{BE} = -2.1 \text{ V} + 0.7 \text{ V} = -1.4 \text{ V}$$

$$I_E = \frac{V_E}{R_E} = \frac{-1.4 \text{ V}}{2.2 \text{ k}\Omega} = -636 \text{ }\mu\text{A}$$

From I_E, you can determine I_C and V_{CE} as follows:

$$I_C \cong I_E = -636 \text{ }\mu\text{A}$$
$$V_C = -V_{CC} - I_C R_C = -6 \text{ V} - (-636 \text{ }\mu\text{A})(1.8 \text{ k}\Omega) = -4.86 \text{ V}$$
$$V_{CE} = V_C - V_E = -4.86 \text{ V} - (-1.4 \text{ V}) = -3.46 \text{ V}$$

Related Exercise What value of β_{DC} is required in this example in order to neglect $R_{IN(base)}$ in keeping with the basic ten-times rule?

1. If the voltage at the base of a transistor is 5 V and the base current is 5 μA, what is the dc input resistance at the base?

2. If a transistor has a dc beta of 190 and its emitter resistor is 1 kΩ, what is the dc input resistance at the base?

3. What bias voltage is developed at the base of a transistor if both resistors in the voltage divider are equal and $V_{CC} = +10$ V? Assume the input resistance at the base is large enough to neglect.

4. What are two advantages of voltage-divider bias?

5–5 ■ COLLECTOR-FEEDBACK BIAS

Another type of bias arrangement is the collector-feedback circuit. This type of circuit is a negative feedback connection that provides a relatively stable Q-point by reducing the effect of variations in β_{DC}. It is also a simple circuit in terms of the components required.

After completing this section, you should be able to

■ **Analyze a collector-feedback bias circuit**
   □ Discuss the effect of negative feedback
   □ Explain how the collector feedback maintains a relatively stable Q-point over temperature

In Figure 5–28, the base resistor R_B is connected to the collector rather than to V_{CC}, as in the base bias arrangement discussed earlier. The collector voltage provides the bias for the base-emitter junction. The negative **feedback** creates an "offsetting" effect that tends to keep the Q-point stable. If I_C tries to increase, it drops more voltage across R_C, thereby causing V_C to decrease. When V_C decreases, there is a decrease in voltage across R_B, which decreases I_B. The decrease in I_B produces less I_C which, in turn, drops less voltage across R_C and thus offsets the decrease in V_C.

FIGURE 5–28
Collector-feedback bias.

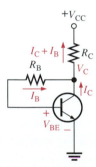

Analysis of Collector Feedback

By Ohm's law, we can express the base current as

$$I_B = \frac{V_C - V_{BE}}{R_B} \qquad (5\text{–}16)$$

Let's assume that $I_C \gg I_B$. The collector voltage is

$$V_C \cong V_{CC} - I_C R_C$$

Also,

$$I_B = \frac{I_C}{\beta_{DC}}$$

Substituting for I_B and V_C in Equation (5–16), we get

$$\frac{I_C}{\beta_{DC}} = \frac{V_{CC} - I_C R_C - V_{BE}}{R_B}$$

We can rearrange the terms so that

$$\frac{I_C R_B}{\beta_{DC}} + I_C R_C = V_{CC} - V_{BE}$$

Then we can solve for I_C as follows:

$$I_C(R_C + R_B/\beta_{DC}) = V_{CC} - V_{BE}$$

$$I_C = \frac{V_{CC} - V_{BE}}{R_C + R_B/\beta_{DC}} \qquad (5\text{–}17)$$

Since the emitter is ground, $V_{CE} = V_C$.

$$V_{CE} = V_{CC} - I_C R_C \qquad (5\text{–}18)$$

Stability Over Temperature

Equation (5–17) shows that the collector current is dependent to some extent on β_{DC} and V_{BE}. This dependency, of course, can be minimized by making $R_C \gg R_B/\beta_{DC}$ and $V_{CC} \gg V_{BE}$. An important feature of collector-feedback bias is that it essentially eliminates the β_{DC} and V_{BE} dependency even if the stated conditions are not met.

As you have learned, β_{DC} varies directly with temperature, and V_{BE} varies inversely with temperature. Refer to Figure 5–29. The circuit has initial values of I_B, I_C, and V_{CE} as indicated in part (a). In part (b), as the temperature goes up, β_{DC} goes up and V_{BE} goes down. The increase in β_{DC} acts to increase I_C. The decrease in V_{BE} acts to increase I_B which, in turn also acts to increase I_C. As I_C tries to increase, the voltage drop across R_C also tries to increase. This tends to reduce the collector voltage and therefore the voltage

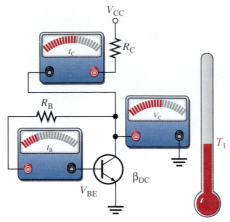

(a) Stabilized at initial temperature, T_1

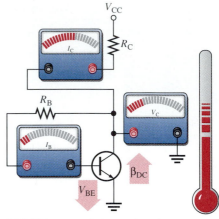

(b) Initial response to temperature rise

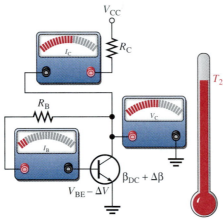

(c) Stabilized at higher temperature, T_2

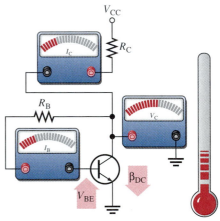

(d) Initial response to temperature drop

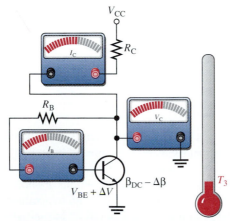

(e) Stabilized at lower temperature, T_3

FIGURE 5–29

Collector-feedback stabilization of Q-point values over temperature.

across R_B, thus reducing I_B and offsetting the attempted increase in I_C and the attempted decrease in V_C. The result is that the collector-feedback circuit maintains a stable Q-point, as indicated in Figure 5–29(c).

   The reverse action occurs when the temperature decreases, as illustrated in Figure 5–29(d) and (e).

EXAMPLE 5–10   Calculate the Q-point values (I_C and V_{CE}) for the circuit in Figure 5–30.

FIGURE 5–30

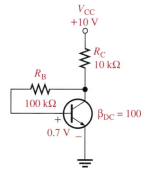

Solution   Use Equation (5–17):

$$I_C = \frac{V_{CC} - V_{BE}}{R_C + R_B/\beta_{DC}} = \frac{10\text{ V} - 0.7\text{ V}}{10\text{ k}\Omega + 100\text{ k}\Omega/100} = 845\text{ }\mu\text{A}$$

The collector-to-emitter voltage is

$$I_{CE} = V_{CC} - I_C R_C = 10\text{ V} - (845\text{ }\mu\text{A})(10\text{ k}\Omega) = 1.55\text{ V}$$

Related Exercise   Calculate the Q-point values in Figure 5–30 for $\beta_{DC} = 250$.

**SECTION 5–5
REVIEW**

1. Explain how an increase in β_{DC} causes a reduction in base current in a collector-feedback circuit.

2. In a certain collector-feedback circuit, $R_B = 47$ kΩ, $R_C = 2.2$ kΩ, and $V_{CC} = 15$ V. If $I_C = 5$ mA, what is I_B?

SUMMARY OF TRANSISTOR BIAS CIRCUITS

npn transistors are shown. Supply voltage polarities are reversed for *pnp* transistors.

BASE BIAS

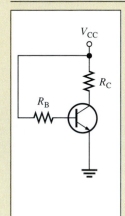

- ■ Q-point values ($I_C \cong I_E$)
- ■ Collector current:

$$I_C = \beta_{DC}\left(\frac{V_{CC} - V_{BE}}{R_B}\right)$$

- ■ Collector-to-emitter voltage:

$$V_{CE} = V_{CC} - I_C R_C$$

EMITTER BIAS

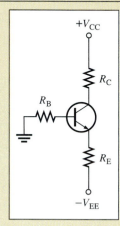

- ■ Q-point values ($I_C \cong I_E$)
- ■ Collector current:

$$I_C \cong \frac{V_{EE} - V_{BE}}{R_E + R_B/\beta_{DC}}$$

- ■ Collector-to-emitter voltage:

$$V_{CE} \cong V_{CC} + V_{EE} - I_C(R_C + R_E)$$

VOLTAGE-DIVIDER BIAS

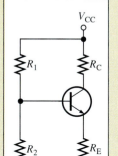

- ■ Q-point values ($I_C \cong I_E$)
- ■ Collector current:

$$I_C \cong \frac{\left(\dfrac{R_2}{R_1 + R_2}\right)V_{CC} - V_{BE}}{R_E}$$

- ■ Collector-to-emitter voltage:

$$V_{CE} \cong V_{CC} - I_C(R_C + R_E)$$

COLLECTOR-FEEDBACK BIAS

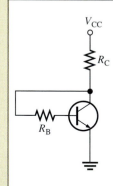

- ■ Q-point values ($I_C \cong I_E$)
- ■ Collector current:

$$I_C = \frac{V_{CC} - V_{BE}}{R_C + R_B/\beta_{DC}}$$

- ■ Collector-to-emitter voltage:

$$V_{CE} = V_{CC} - I_C R_C$$

5–6 ■ TROUBLESHOOTING

In a biased transistor circuit, the transistor can fail or sometimes a resistor in the bias circuit can fail. We will examine several possibilities in this section using the voltage-divider bias arrangement for illustration. Many circuit failures result from open resistors, internally open transistor leads and junctions, or shorted junctions. Often, these failures can produce an apparent cutoff or saturation condition when voltage is measured at the collector.

After completing this section, you should be able to

■ **Troubleshoot various faults in transistor bias circuits**
  ☐ Use voltage measurements to identify a fault in a transistor bias circuit
  ☐ Analyze a transistor bias circuit for several common faults

---

Troubleshooting a Voltage-Divider Biased Transistor

An example of a transistor with voltage-divider bias is shown in Figure 5–31. For the specific component values shown, you should get the voltage readings approximately as indicated when the circuit is operating properly.

FIGURE 5–31

A voltage-divider biased transistor with correct voltages.

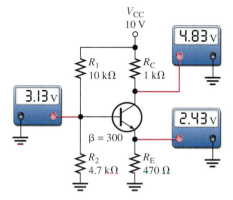

For this type of bias circuit, a particular group of faults will cause the transistor collector to be at V_{CC} when measured with respect to ground. Five faults are shown in Figure 5–32(a). The collector voltage is equal to 10 V with respect to ground for each of the faults as indicated. Also, for each of the faults, the base voltage and the emitter voltage with respect to ground are displayed in part (b).

Fault 1: Resistor R_1 Open This fault removes the bias voltage from the base, thus connecting the base to ground through R_2 and forcing the transistor into cutoff because $V_B = 0$ V and $I_B = 0$ A. The transistor is nonconducting so there is no I_C and, therefore, no voltage drop across R_C. This makes the collector voltage equal to V_{CC} (10 V). Since there is no base current or collector current, there is also no emitter current and $V_E = 0$ V.

Fault 2: Resistor R_E Open This fault prevents base current, emitter current, and collector current except for a very small I_{CBO} that can be neglected. Since $I_C = 0$ A, there is no voltage drop across R_C and, therefore, $V_C = V_{CC} = 10$ V. The voltage divider produces a voltage at the base with respect to ground as follows:

$$V_B = \left(\frac{R_2}{R_1 + R_2}\right)V_{CC} = \left(\frac{4.7 \text{ k}\Omega}{14.7 \text{ k}\Omega}\right)10 \text{ V} = 3.20 \text{ V}$$

When a voltmeter is connected to the emitter, it provides a current path through its high internal impedance, resulting in a forward-biased emitter-base junction. Therefore, the emitter voltage is $V_E = V_B - V_{BE}$. The amount of the forward voltage drop across the

FIGURE 5–32
Faults for which $V_C = V_{CC}$.

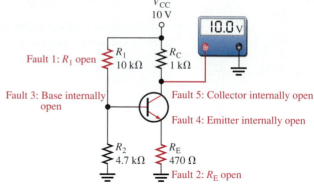

(a) For all indicated faults, $V_C = V_{CC}$

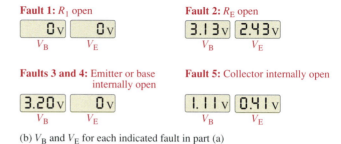

(b) V_B and V_E for each indicated fault in part (a)

BE junction depends on the current. $V_{BE} = 0.7$ V is assumed for purposes of illustration, but it may be much less. The result is an emitter voltage as follows:

$$V_E = V_B - V_{BE} = 3.2 \text{ V} - 0.7 \text{ V} = 2.5 \text{ V}$$

Fault 3: Base Lead Internally Open An internal transistor fault is more likely to happen than an open resistor. Again, the transistor is nonconducting so $I_C = 0$ A and $V_C = V_{CC} = 10$ V. Just as for the case of the open R_E, the voltage divider produces 3.2 V at the external base connection. The voltage at the external emitter connection is 0 V because there is no emitter current through R_E and, thus, no voltage drop.

Fault 4: BE Junction or Emitter Connection Internally Open Again, the transistor is nonconducting, so $I_C = 0$ A and $V_C = V_{CC} = 10$ V. Just as for the case of the open R_E and the internally open base, the voltage divider produces 3.2 V at the base. The voltage at the external emitter lead is 0 V because that point is open and connected to ground through R_E. Notice that Faults 3 and 4 produce identical symptoms.

Fault 5: BC Junction or Collector Connection Internally Open Since there is an internal open in the transistor collector, there is no I_C and, therefore, $V_C = V_{CC} = 10$ V. In this situation, the voltage divider is loaded by R_E through the forward-biased BE junction,

FIGURE 5–33

Equivalent bias circuit for an internally open collector.

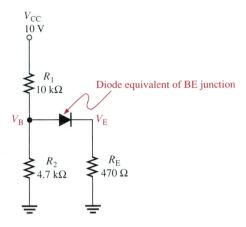

as shown by the approximate equivalent circuit in Figure 5–33. The base voltage and emitter voltage are determined as follows:

$$V_B \cong \left(\frac{R_2 \parallel R_E}{R_1 + R_2 \parallel R_E} \right) V_{CC} + 0.7 \text{ V}$$

$$= \left(\frac{427 \ \Omega}{10.427 \ \text{k}\Omega} \right) 10 \text{ V} + 0.7 \text{ V} = 0.41 \text{ V} + 0.7 \text{ V} = 1.11 \text{ V}$$

$$V_E = V_B - V_{BE} = 1.11 \text{ V} - 0.7 \text{ V} = 0.41 \text{ V}$$

There are two possible additional faults for which the transistor is conducting or appears to be conducting, based on the collector voltage measurement. These are indicated in Figure 5–34.

Fault 6: Resistor R_C Open For this fault, which is illustrated in Figure 5–34(a), the collector voltage may lead you to think that the transistor is in saturation, but actually it is nonconducting. Obviously, if R_C is open, there can be no collector current. In this situa-

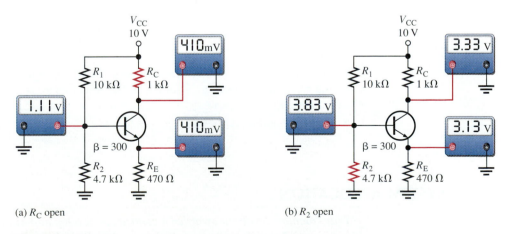

(a) R_C open

(b) R_2 open

FIGURE 5–34

Faults for which the transistor is conducting or appears to be conducting.

tion, the equivalent bias circuit is the same as for Fault 4, as illustrated in Figure 5–32. Therefore, $V_B = 1.11$ V and since the BE junction is forward-biased, $V_E = V_B - V_{BE} = 1.11$ V $- 0.7$ V $= 0.41$ V. When a voltmeter is connected to the collector to measure V_C, a current path is provided through the internal impedance of the meter and the BC junction is forward-biased by V_B. Therefore, $V_C = V_B - V_{BC} = 1.11$ V $- 0.7$ V $= 0.41$ V. Again the forward drops across the internal transistor junctions depend on the current. We are using 0.7 V for illustration but the forward drops may be much less.

Fault 7: Resistor R_2 Open When R_2 opens as shown in Figure 5–34(b), the normal base voltage and base current increases because the voltage divider is now formed by R_1 and $\beta_{DC}R_{IN(base)}$. In this case, the base voltage is determined by the emitter voltage ($V_B = V_E + V_{BE}$).

First, verify whether the transistor is in saturation or not. The collector saturation current and the base current *required* to produce saturation are determined as follows (assuming $V_{CE(sat)} = 0.2$ V):

$$I_{C(sat)} = \frac{V_{CC} - V_{CE(sat)}}{R_C + R_E} = \frac{9.8 \text{ V}}{1.47 \text{ k}\Omega} = 6.67 \text{ mA}$$

$$I_{B(sat)} = \frac{I_{C(sat)}}{\beta_{DC}} = \frac{6.67 \text{ mA}}{300} = 22.2 \text{ μA}$$

Assuming the transistor is saturated, the maximum base current is

$$R_{IN(base)} = \beta_{DC}R_E = 300(470 \text{ }\Omega) = 141 \text{ k}\Omega$$

$$I_{B(max)} \cong \frac{V_{CC}}{R_1 + R_{IN(base)}} = \frac{10 \text{ V}}{151 \text{ k}\Omega} = 66.2 \text{ μA}$$

Since this amount of base current is more than enough to produce saturation, the transistor is definitely saturated. Therefore, V_E and V_B are as follows:

$$V_E \cong I_{C(sat)} R_E = (6.67 \text{ mA})(470 \text{ }\Omega) = 3.13 \text{ V}$$

$$V_B = V_E + V_{BE} = 3.13 \text{ V} + 0.7 \text{ V} = 3.83 \text{ V}$$

SECTION 5–6 REVIEW

1. How do you determine when a transistor is saturated? When a transistor is in cutoff?

2. In a voltage-divider biased *npn* transistor circuit, you measure V_{CC} at the collector and an emitter voltage 0.7 V less than the base voltage. Is the transistor functioning in cutoff, or is R_E open?

3. What symptoms does an open R_C produce?

5–7 ■ SYSTEM APPLICATION

Your company has been awarded a contract to develop and build an industrial tempera-ture control system to be installed in a chemical plant. Several departments will be working on various portions of the system, and you have been assigned responsibility for the temperature-to-voltage conversion circuit and the power supply. You will have to coordinate with the other departments to determine the interface requirements between your part of the system and theirs. You will apply the knowledge you have gained in this chapter in completing your assignment.

The Industrial Temperature Control System

This system is to be used for maintaining a preset sequence of specified temperatures in a liquid chemical substance during the mixing process in a large vat. The system is an example of a process-control system that is based on the principle of closed-loop feedback as shown in block diagram form in Figure 5–35.

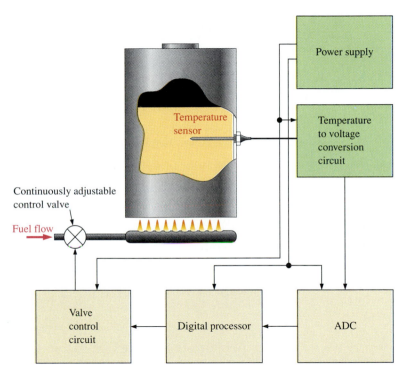

FIGURE 5–35
Basic system concept and block diagram of the industrial temperature control system.

The temperature of the chemical is sensed by a temperature-sensitive resistor called a **thermistor.** The resistance varies inversely with temperature; that is, it has a negative temperature coefficient. Essentially, the thermistor converts temperature to resistance.

The thermistor is a part of the temperature-to-voltage conversion circuit, for which you are responsible. The temperature of the chemical mixture determines the resistance of the thermistor. The temperature-to-voltage conversion circuit produces an output voltage that is proportional to the temperature. This output voltage goes to an analog-to-digital converter (ADC), which converts the voltage to digital form. The digital processor adds the appropriate scaling and linearization to the digitized voltage and sends it to the valve control circuit that is programmed to sequence through a series of set-point temperatures during the mixing process. The valve control circuit compares the set-point temperature to the actual temperature as measured by the temperature-to-voltage conversion circuit and makes the proper adjustments to the burner.

If the temperature of the chemical is below its set-point value, the valve control circuit adjusts the valve to allow increased fuel flow to the burner to raise the temperature of the chemical. When the temperature of the chemical reaches the set-point value, the valve control circuit reduces the fuel flow and continuously adjusts it to maintain the constant set-point temperature. A similar process occurs when the temperature of the chemical is above the set-point value.

In addition to the temperature-to-voltage conversion circuit, you are also responsible for the dc power supply that will provide dc voltages and currents to each part of the system. Although you do not need to know how the other blocks work, you must know their voltage and current requirements and the load impedance that the ADC presents to the temperature-to-voltage conversion circuit.

The System Requirements

☐ There are three set-point temperatures in the mixing process.

☐ Each set-point temperature is maintained for a preprogrammed time interval determined by the valve control circuit.

☐ The set-point temperatures are 46°C, 50°C, and 54°C.

☐ The dc voltages and currents for each circuit are as follows:

Temperature-to-voltage conversion circuit: 9.1 V regulated, 10 mA

Analog-to-digital converter (ADC): 5.1 V regulated, 50 mA

Digital processor: 5.1 V regulated, 25 mA

Valve control circuit: 9.1 V regulated, 40 mA

☐ The input resistance of the ADC is 100 kΩ.

The Temperature-to-Voltage Conversion Circuit

The schematic for the temperature-to-voltage conversion circuit is shown in Figure 5–36. As you can see, it is simply a voltage-divider biased transistor with the thermistor used as one of the voltage-divider resistors.

Basic Operation When the temperature increases, the resistance of the thermistor decreases and reduces the base bias voltage which, in turn increases the collector (output) voltage proportionally. When the temperature decreases, the resistance of the thermistor increases and increases the base bias voltage, which then causes a decrease in the output voltage.

The output voltage of the temperature-to-voltage conversion circuit tracks the changing temperature of the chemical and, therefore, must operate in its linear or active region and not in saturation or cutoff. The output must interface with the ADC. Notice the schematic symbol for the thermistor in Figure 5–36.

FIGURE 5–36

Schematic of the temperature-to-voltage conversion circuit.

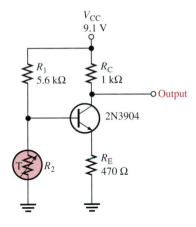

The Thermistor Characteristics

☐ The response characteristic of the thermistor is nonlinear. This means that the resistance does not vary in a straight-line relationship with temperature.

☐ The resistance versus temperature characteristic of the thermistor for the range of 45°C to 55°C is given in Figure 5–37.

Analysis of the Temperature-to-Voltage Conversion Circuit

A 100 kΩ load resistance is connected to simulate the ADC input.

☐ Determine the output voltage at each specified set-point temperature.

☐ Determine if the transistor is operating in its linear (active) region over the prescribed temperature range. That is, verify that the transistor does not go into saturation or cutoff. If it does, suggest a modification to the circuit to correct the design flaw.

FIGURE 5–37

Thermistor resistance versus temperature.

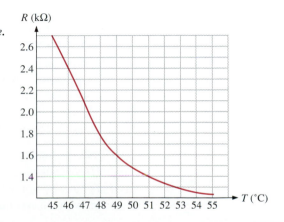

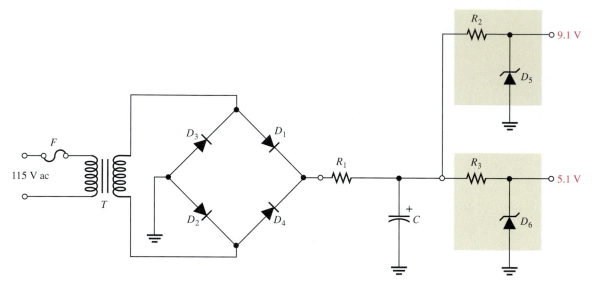

FIGURE 5–38
Power supply schematic.

The Power Supply Circuit

The dc power supply circuit is the same as the one developed in previous system applications except that it needs to be modified for different voltage and current requirements, as shown in the schematic of Figure 5–38. The basic power supply specifications are

1. Input voltage: 115 V rms at 60 Hz
2. Regulated output voltages: 5.1 V dc $\pm$ 5%; 9.1 V dc $\pm$ 5%
3. Maximum ripple factor: 3%
4. Maximum load currents: 75 mA @ 5.1 V; 50 mA @ 9.1 V

Component Values and the Schematic

☐ Determine the values of all the unspecified components in the power supply. You may wish to refer to previous system application assignments.

☐ The decision has been made to combine the power supply and the temperature-to-voltage conversion circuit on one printed circuit board. Produce a complete schematic that includes both circuits.

The Printed Circuit Board

☐ Check out the printed circuit board in Figure 5–39 to verify that it is correct according to the schematic.

☐ Label a copy of the board with the component and input/output designations in agreement with the schematic.

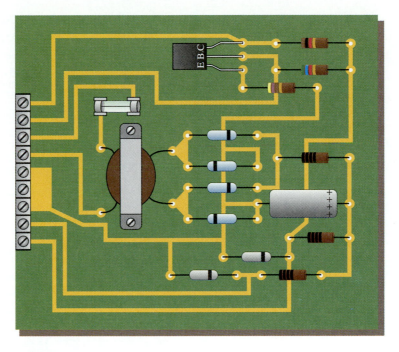

FIGURE 5–39
Power supply and temperature detector printed circuit board. All black resistor bands indicate values to be determined.

Test Procedure

☐ Develop a step-by-step set of instructions on how to completely check the power supply and temperature-to-voltage conversion circuit for proper operation using the test points (circled numbers) indicated in the test bench setup of Figure 5–40. This particular thermistor is a tubular shaft with a threaded mount for inserting in a vessel. The oven-type device is a controlled-temperature chamber.

☐ Specify voltage values for all the measurements to be made. Provide a fault analysis for all possible component failures. Utilize and modify the test procedure you developed for the power supply circuit in Chapters 2 and 3 to include the changes.

Troubleshooting

Three boards in the initial manufacturing run have been found to be defective. The test results are shown in Figure 5–41. Based on the sequence of measurements for each defective board, determine the most likely fault(s) in each case.

The red circled numbers indicate test points on the circuit board. The DMM function setting is indicated below the display.

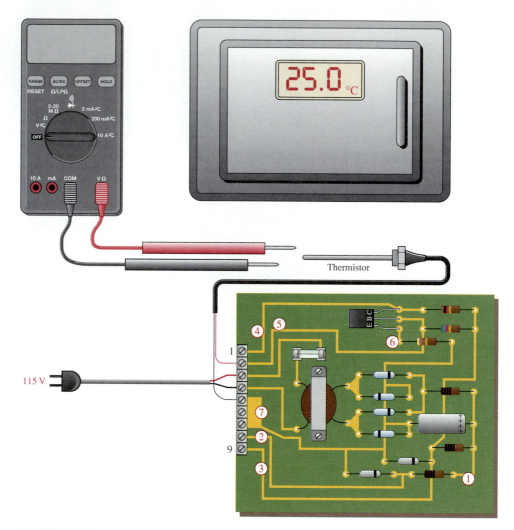

FIGURE 5–40

Test bench setup for the circuit board.

Final Report

Submit a final written report on the power supply and temperature-to-voltage conversion circuit board using an organized format that includes the following:

1. A physical description of the circuits.

2. A discussion of the operation of each circuit.

3. A list of the specifications.

4. A list of parts with part numbers if available.

5. A list of types of problems on the three circuit boards.

6. A complete description of how you determined the problem on each of the three circuit boards.

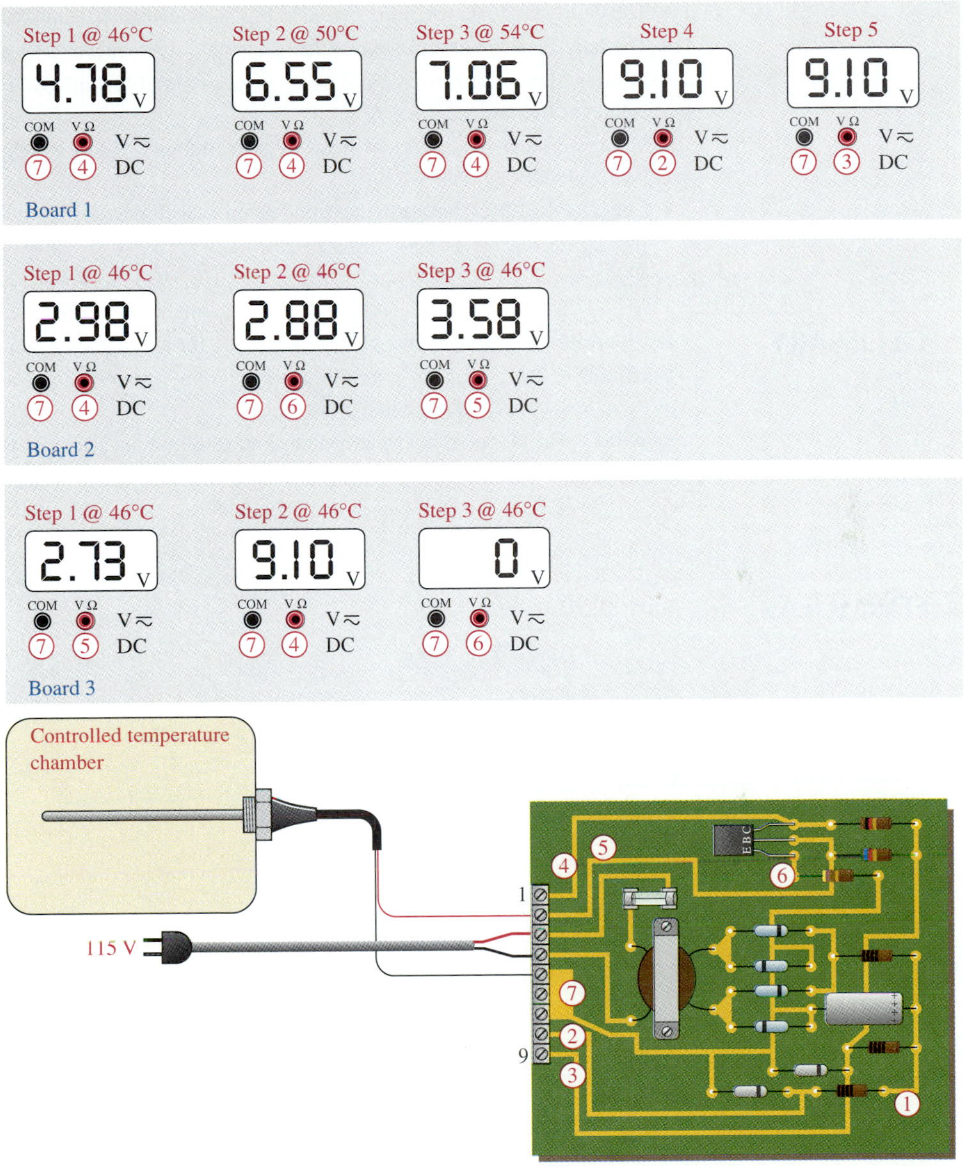

FIGURE 5–41

Test results for the three faulty circuit boards.

■ **CHAPTER SUMMARY**

- The purpose of biasing a circuit is to establish a proper stable dc operating point (Q-point).
- The Q-point of a circuit is defined by specific values for I_C and V_{CE}. These values are called the coordinates of the Q-point.
- A dc load line passes through the Q-point on a transistor's collector curves intersecting the vertical axis at approximately $I_{C(sat)}$ and the horizontal axis at $V_{CE(off)}$.
- The linear (active) operating region of a transistor lies along the load line below saturation and above cutoff.

- The dc input resistance at the base of a bipolar transistor is approximately $\beta_{DC}R_E$.
- The base bias circuit arrangement has poor stability because its Q-point varies widely with β_{DC}.
- Emitter bias generally provides good Q-point stability but requires both positive and negative supply voltages.
- Voltage-divider bias provides good Q-point stability with a single-polarity supply voltage. It is the most common bias circuit.
- Collector-feedback bias provides good stability using negative feedback from collector to base.

■ GLOSSARY

DC load line   A straight line plot of I_C and V_{CE} for a transistor circuit.

Feedback   The process of returning a portion of a circuit's output back to the input in such a way as to oppose or aid a change in the output.

Q-point   The dc operating (bias) point of an amplifier specified by voltage and current values.

Thermistor   A temperature-sensitive resistor with a negative temperature coefficient.

■ FORMULAS

Base Bias

$$(5\text{–}1) \qquad I_B = \frac{V_{CC} - V_{BE}}{R_B}$$

$$(5\text{–}2) \qquad V_{CE} = V_{CC} - I_C R_C$$

$$(5\text{–}3) \qquad I_C = \beta_{DC}\left(\frac{V_{CC} - V_{BE}}{R_B}\right)$$

Emitter Bias

$$(5\text{–}4) \qquad I_E = \frac{V_{EE} - V_{BE}}{R_E + R_B/\beta_{DC}}$$

$$(5\text{–}5) \qquad I_C \cong \frac{V_{EE} - V_{BE}}{R_E + R_B/\beta_{DC}}$$

$$(5\text{–}6) \qquad V_E = -V_{EE} + I_E R_E$$

$$(5\text{–}7) \qquad V_B = V_E + V_{BE}$$

$$(5\text{–}8) \qquad V_C = V_{CC} - I_C R_C$$

$$(5\text{–}9) \qquad V_{CE} \cong V_{CC} + V_{EE} - I_C(R_C + R_E)$$

Voltage-Divider Bias

$$(5\text{–}10) \qquad R_{IN(base)} \cong \beta_{DC}R_E$$

$$(5\text{–}11) \qquad V_B = \left(\frac{R_2 \,\|\, \beta_{DC}R_E}{R_1 + (R_2 \,\|\, \beta_{DC}R_E)}\right)V_{CC} \qquad npn$$

$$(5\text{–}12) \qquad V_B \cong \left(\frac{R_2}{R_1 + R_2}\right)V_{CC} \qquad \beta_{DC}R_E \gg R_2$$

$$(5\text{–}13) \qquad I_C \cong \frac{V_B - V_{BE}}{R_E}$$

$$(5\text{–}14) \qquad V_{CE} \cong V_{CC} - I_C(R_C + R_E)$$

$$(5\text{–}15) \qquad V_B = \left(\frac{R_1}{R_1 + R_2 \,\|\, \beta_{DC}R_E}\right)V_{EE} \qquad pnp$$

Collector-Feedback Bias

$$(5\text{–}16) \qquad I_B = \frac{V_C - V_{BE}}{R_B}$$

$$(5\text{–}17) \qquad I_C = \frac{V_{CC} - V_{BE}}{R_C + R_B/\beta_{DC}}$$

$$(5\text{–}18) \qquad V_{CE} = V_{CC} - I_C R_C$$

■ **SELF-TEST**

1. The maximum value of collector current in a biased transistor is
   (a) $\beta_{DC}I_B$ (b) $I_{C(sat)}$ (c) greater than I_E (d) $I_E - I_B$

2. Ideally, a dc load line is a straight line drawn on the collector characteristic curves between
   (a) the Q-point and cutoff   (b) the Q-point and saturation
   (c) $V_{CE(cutoff)}$ and $I_{C(sat)}$ (d) $I_B = 0$ and $I_B = I_C/\beta_{DC}$

3. If a sinusoidal voltage is applied to the base of a biased *npn* transistor and the resulting sinusoidal collector voltage is clipped near zero volts, the transistor is
   (a) being driven into saturation   (b) being driven into cutoff
   (c) operating nonlinearly   (d) answers (a) and (c)   (e) answers (b) and (c)

4. The dc beta (h_{FE}) for a given type of transistor
   (a) varies with temperature   (b) is a fixed constant
   (c) varies from device to device   (d) answers (a) and (c)

5. The disadvantage of base bias is that
   (a) it is very complex   (b) it produces low gain
   (c) it is too beta dependent   (d) it produces high leakage current

6. Emitter bias is
   (a) essentially independent of β_{DC} (b) very dependent on β_{DC}
   (c) provides a stable bias point   (d) answers (a) and (c)

7. In an emitter bias circuit, $R_E = 2.7 \text{ k}\Omega$ and $V_{EE} = 15 \text{ V}$. The emitter current
   (a) is 5.3 mA   (b) is 2.7 mA   (c) is 180 mA   (d) cannot be determined

8. The input resistance at the base of a biased transistor depends mainly on
   (a) β_{DC} (b) R_B (c) R_E (d) β_{DC} and R_E

9. In a voltage-divider biased transistor circuit such as in Figure 5–22, $R_{IN(base)}$ can generally be neglected in calculations when
   (a) $R_{IN(base)} > R_2$ (b) $R_2 > 10R_{IN(base)}$ (c) $R_{IN(base)} > 10R_2$ (d) $R_1 << R_2$

10. In a certain voltage-divider biased *npn* transistor, V_B is 2.95 V. The dc emitter voltage is approximately
    (a) 2.25 V   (b) 2.95 V   (c) 3.65 V   (d) 0.7 V

11. Voltage-divider bias
    (a) cannot be independent of β_{DC}
    (b) can be essentially independent of β_{DC}
    (c) is not widely used
    (d) requires fewer components than all the other methods

12. Collector-feedback bias is
    (a) based on the principle of positive feedback   (b) based on beta multiplication
    (c) based on the principle of negative feedback   (d) not very stable

13. In a voltage-divider biased *npn* transistor, if the upper voltage-divider resistor (the one connected to V_{CC}) opens,
    (a) the transistor goes into cutoff   (b) the transistor goes into saturation
    (c) the transistor burns out   (d) the supply voltage is too high

14. In a voltage-divider biased *npn* transistor, if the lower voltage-divider resistor (the one connected to ground) opens,
    (a) the transistor is not affected
    (b) the transistor may be driven into cutoff
    (c) the transistor may be driven into saturation
    (d) the collector current will decrease

15. In a voltage-divider biased *pnp* transistor, there is no base current, but the base voltage is approximately correct. The most likely problem(s) is
    (a) a bias resistor is open   (b) the collector resistor is open
    (c) the base-emitter junction is open   (d) the emitter resistor is open
    (e) answers (a) and (c)   (f) answers (c) and (d)

■ **BASIC PROBLEMS**

SECTION 5–1   The DC Operating Point

1. The output (collector voltage) of a biased transistor amplifier is shown in Figure 5–42. Is the transistor biased too close to cutoff or too close to saturation?

FIGURE 5–42

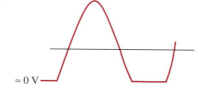

2. What is the Q-point for a biased transistor as in Figure 5–2 with $I_B = 150$ μA, $\beta_{DC} = 75$, $V_{CC} = 18$ V, and $R_C = 1$ kΩ?

3. What is the saturation value of collector current in Problem 2?

4. What is the cutoff value of V_{CE} in Problem 2?

5. Determine the intercept points of the dc load line on the vertical and horizontal axes of the collector-characteristic curves for the circuit in Figure 5–43.

FIGURE 5–43

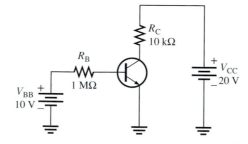

6. Assume that you wish to bias the transistor in Figure 5–43 with $I_B = 20$ μA. To what voltage must you change the V_{BB} supply? What are I_C and V_{CE} at the Q-point, given that $\beta_{DC} = 50$?

7. Design a biased-transistor circuit using $V_{BB} = V_{CC} = 10$ V, for a Q-point of $I_C = 5$ mA and $V_{CE} = 4$ V. Assume $\beta_{DC} = 100$. The design involves finding R_B, R_C, and the *minimum* power rating of the transistor. (The actual power rating should be greater.) Sketch the circuit.

8. Determine whether the transistor in Figure 5–44 is biased in cutoff, saturation, or the linear region. Keep in mind that $I_C = \beta_{DC}I_B$ is valid only in the linear region.

FIGURE 5–44

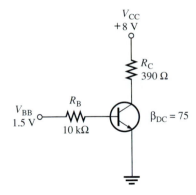

SECTION 5–2   Base Bias

9. Determine I_B, I_C, and V_{CE} for a base-biased transistor circuit with the following values: $\beta_{DC} = 90$, $V_{CC} = 12$ V, $R_B = 22$ kΩ, and $R_C = 100$ Ω.

10. If β_{DC} in Problem 9 doubles over temperature, what are the Q-point values?

11. You have two base bias circuits connected for testing. They are identical except that one is biased with a separate V_{BB} source and the other is biased with the base resistor connected to V_{CC}. Ammeters are connected to measure collector current in each circuit. You vary the V_{CC} supply voltage and observe that the collector current varies in one circuit, but not in the other. In which circuit does the collector current change? Explain your observation.

12. The data sheet for a particular transistor specifies a minimum β_{DC} of 50 and a maximum β_{DC} of 125. What range of Q-point values can be expected if an attempt is made to mass-produce the circuit in Figure 5–45? Is this range acceptable if the Q-point must remain in the transistor's linear region?

FIGURE 5–45

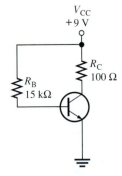

13. The base bias circuit in Figure 5–45 is subjected to a temperature variation from 0°C to 70°C. The β_{DC} decreases by 50 percent at 0°C and increases by 75 percent at 70°C from its nominal value of 110 at 25°C. What are the changes in I_C and V_{CE} over the temperature range of 0°C to 70°C?

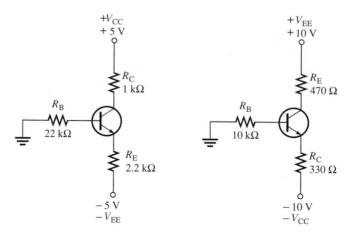

FIGURE 5–46   **FIGURE 5–47**

SECTION 5–3   Emitter Bias

14. Analyze the circuit in Figure 5–46 to determine the correct voltage at the transistor terminals with respect to ground. Assume $\beta_{DC} = 100$.

15. To what value can R_E in Figure 5–46 be reduced without the transistor going into saturation?

16. Taking V_{BE} into account in Figure 5–46, how much will I_E change with a temperature increase from 25°C to 100°C? The V_{BE} is 0.7 V at 25°C and decreases 2.5 mV per Celsius degree. Neglect β_{DC}.

17. When can the effect of a change in β_{DC} be neglected in the emitter bias circuit?

18. Determine I_C and V_{CE} in the *pnp* emitter bias circuit of Figure 5–47. Assume $\beta_{DC} = 100$.

SECTION 5–4   Voltage-Divider Bias

19. What is the minimum value of β_{DC} in Figure 5–48 that makes $R_{IN(base)} \geq 10R_2$?

20. The bias resistor R_2 in Figure 5–48 is replaced by a 15 kΩ potentiometer. What minimum resistance setting causes saturation?

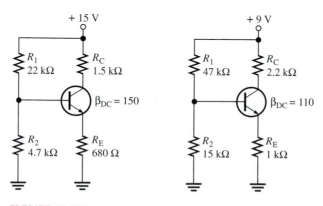

FIGURE 5–48   **FIGURE 5–49**

21. If the potentiometer described in Problem 20 is set at 2 kΩ, what are the values for I_C and V_{CE}?

22. Determine all transistor terminal voltages with respect to ground in Figure 5–49. Do not neglect the input resistance at the base or V_{BE}.

23. Show the connections required to replace the transistor in Figure 5–49 with a *pnp* device.

24. (a) Determine V_B in Figure 5–50.
(b) If R_E is doubled, what is the value of V_B?

25. (a) Find the Q-point values for Figure 5–50.
(b) Find the minimum power rating of the transistor in Figure 5–50.

FIGURE 5–50

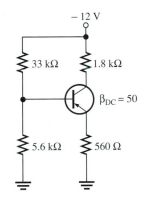

SECTION 5–5   Collector-Feedback Bias

26. Determine V_B, V_C, and I_C in Figure 5–51.

FIGURE 5–51

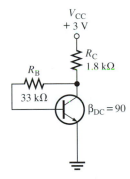

27. What value of R_C can be used to decrease I_C in Problem 26 by 25 percent?

28. What is the minimum power rating for the transistor in Problem 27?

29. A collector-feedback circuit uses an *npn* transistor with $V_{CC} = 12$ V, $R_C = 1.2$ kΩ, and $R_B = 47$ kΩ. Determine the collector voltage and the collector current if $\beta_{DC} = 200$.

■ TROUBLE-SHOOTING PROBLEMS

SECTION 5–6   Troubleshooting

30. Assume the emitter becomes shorted to ground in Figure 5–52 by a solder splash or stray wire clipping. What do the meters read? When you correct the problem, what do the meters read?

31. Determine the most probable failures, if any, in each circuit of Figure 5–53, based on the indicated measurements.

FIGURE 5–52

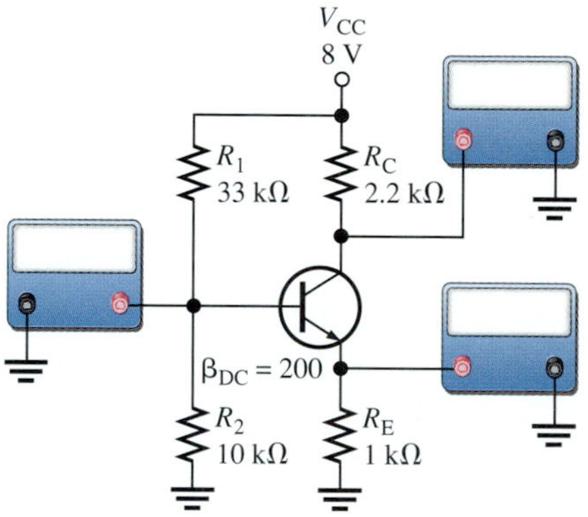

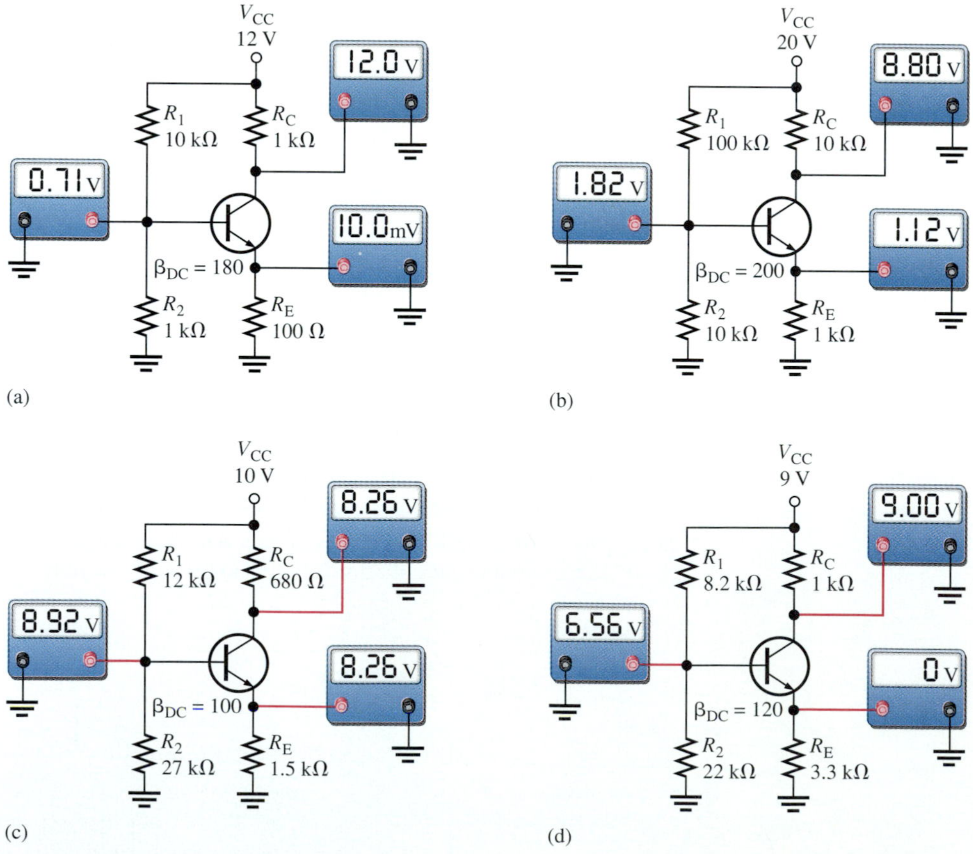

(a)

(b)

(c)

(d)

FIGURE 5–53

32. Determine if the DMM readings 2 through 4 in the breadboard circuit of Figure 5–54 are correct. If they are not, isolate the problem(s). The transistor is a *pnp* device with a specified dc beta range of 35 to 100.

33. Determine each meter reading in Figure 5–54 for each of the following faults:
   (a) the 680 Ω resistor open
   (b) the 5.6 kΩ resistor open
   (c) the 10 kΩ resistor open
   (d) the 1 kΩ resistor open
   (e) a short from emitter to ground
   (f) an open base-emitter junction

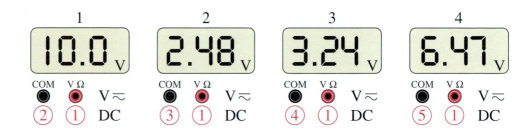

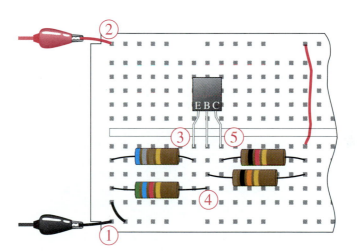

FIGURE 5–54

SECTION 5–7   System Application

34. Determine V_B, V_E, and V_C in the temperature-to-voltage conversion circuit in Figure 5–36 if R_1 fails open.

35. What faults will cause the transistor in the temperature-to-voltage conversion circuit to go into cutoff?

36. Assume that a 5.1 V zener diode is inadvertently installed in the place of the 9.1 V zener on the circuit board in Figure 5–39. What voltages will you measure at the base, emitter, and collector of the temperature-to-voltage conversion circuit when the temperature is 45°C?

37. Explain how you would identify an open collector-base junction in the transistor in Figure 5–36.

■ **DATA SHEET PROBLEMS**

38. Analyze the temperature-to-voltage conversion circuit in Figure 5–55(a) at the temperature extremes for both minimum and maximum specified data sheet values of h_{FE}. Refer to the partial data sheet in Figure 5–56.

39. Verify that no maximum ratings are exceeded in the temperature-to-voltage conversion circuit. Refer to the partial data sheet in Figure 5–56.

40. Refer to the partial data sheet in Figure 5–57 on page 292.
  (a) What is the maximum collector current for a 2N2222A?
  (b) What is the maximum reverse base-emitter voltage for a 2N2118?

41. Determine the maximum power dissipation for a 2N2222 at 100°C.

42. When you increase the collector current in a 2N2219 from 1 mA to 500 mA, how much does the minimum β_{DC} (h_{FE}) change?

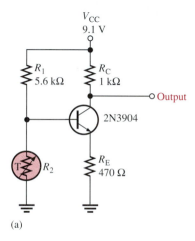

(a)

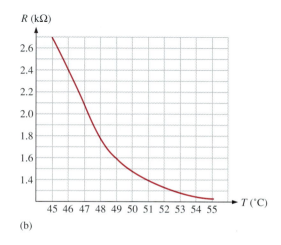

(b)

FIGURE 5–55

Maximum Ratings

Rating	Symbol	Value	Unit
Collector-Emitter voltage	V_{CEO}	40	V dc
Collector-Base voltage	V_{CBO}	60	V dc
Emitter-Base voltage	V_{EBO}	6.0	V dc
Collector current — continuous	I_C	200	mA dc
Total device dissipation @ $T_A = 25°C$ Derate above 25°C	P_D	625 5.0	mW mW/C°
Total device dissipation @ $T_C = 25°C$ Derate above 25°C	P_D	1.5 12	Watts mW/C°
Operating and storage junction Temperature range	T_J, T_{stg}	−55 to +150	°C

Thermal Characteristics

Characteristic	Symbol	Max	Unit
Thermal resistance, junction to case	$R_{\theta JC}$	83.3	C°/W
Thermal resistance, junction to ambient	$R_{\theta JA}$	200	C°/W

Electrical Characteristics ($T_A = 25°C$ unless otherwise noted.)

Characteristic	Symbol	Min	Max	Unit

Off Characteristics

Characteristic		Symbol	Min	Max	Unit
Collector-Emitter breakdown voltage $I_C = 1.0$ mA dc, $I_B = 0$)		$V_{(BR)CEO}$	40	—	V dc
Collector-Base breakdown voltage $I_C = 10$ µA dc, $I_E = 0$)		$V_{(BR)CBO}$	60	—	V dc
Emitter-Base breakdown voltage $I_E = 10$ µA dc, $I_C = 0$)		$V_{(BR)EBO}$	6.0	—	V dc
Base cutoff current ($V_{CE} = 30$ V dc, $V_{EB} = 3.0$ V dc)		I_{BL}	—	50	nA dc
Collector cutoff current ($V_{CE} = 30$ V dc, $V_{EB} = 3.0$ V dc)		I_{CEX}	—	50	nA dc

On Characteristics

Characteristic		Symbol	Min	Max	Unit
DC current gain		h_{FE}			—
($I_C = 0.1$ mA dc, $V_{CE} = 1.0$ V dc)	2N3903 2N3904		20 40	— —	
($I_C = 1.0$ mA dc, $V_{CE} = 1.0$ V dc)	2N3903 2N3904		35 70	— —	
($I_C = 10$ mA dc, $V_{CE} = 1.0$ V dc)	2N3903 2N3904		50 100	150 300	
($I_C = 50$ mA dc, $V_{CE} = 1.0$ V dc)	2N3903 2N3904		30 60	— —	
($I_C = 100$ mA dc, $V_{CE} = 1.0$ V dc)	2N3903 2N3904		15 30	— —	
Collector-Emitter saturation voltage ($I_C = 10$ mA dc, $I_B = 1.0$ mA dc) ($I_C = 50$ mA dc, $I_B = 5.0$ mA dc)		$V_{CE(sat)}$	— —	0.2 0.3	V dc
Base-Emitter saturation voltage ($I_C = 10$ mA dc, $I_B = 1.0$ mA dc) ($I_C = 50$ mA dc, $I_B = 5.0$ mA dc)		$V_{BE(sat)}$	0.65 —	0.85 0.95	V dc

2N3903
2N3904

3 Collector

2
Base

1 Emitter

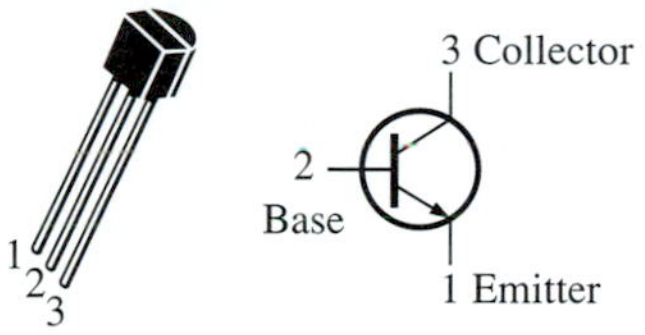

General Purpose Transistors

NPN Silicon

FIGURE 5–56

Maximum Ratings

Rating	Symbol	2N2218 2N2219 2N2221 2N2222	2N2218A 2N2219A 2N2221A 2N2222A	2N5581 2N5582	Unit
Collector-Emitter voltage	V_{CEO}	30	40	40	V dc
Collector-Base voltage	V_{CBO}	60	75	75	V dc
Emitter-Base voltage	V_{EBO}	5.0	6.0	6.0	V dc
Collector current — continuous	I_C	800	800	800	mA dc
		2N2218,A 2N2219,A	2N2221,A 2N2222,A	2N5581 2N5582	
Total device dissipation @ T_A = 25°C Derate above 25°C	P_D	0.8 4.57	0.5 2.28	0.6 3.33	Watt mW/C°
Total device dissipation @ T_C = 25°C Derate above 25°C	P_D	3.0 17.1	1.2 6.85	2.0 11.43	Watt mW/C°
Operating and storage junction Temperature range	T_J, T_{stg}	−65 to +200			°C

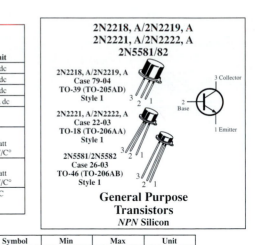

2N2218, A/2N2219, A
2N2221, A/2N2222, A
2N5581/82

2N2218, A/2N2219, A
Case 79-04
TO-39 (TO-205AD)
Style 1

2N2221, A/2N2222, A
Case 22-03
TO-18 (TO-206AA)
Style 1

2N5581/2N5582
Case 26-03
TO-46 (TO-206AB)
Style 1

General Purpose Transistors
NPN Silicon

Electrical Characteristics (T_A = 25°C unless otherwise noted.)

Characteristic		Symbol	Min	Max	Unit
Off Characteristics					
Collector-Emitter breakdown voltage (I_C = 10 mA dc, I_B = 0)	Non-A Suffix A-Suffix, 2N5581, 2N5582	$V_{(BR)CEO}$	30 40	— —	V dc
Collector-Base breakdown voltage (I_C = 10 μA dc, I_E = 0)	Non-A Suffix A-Suffix, 2N5581, 2N5582	$V_{(BR)CBO}$	60 75	— —	V dc
Emitter-Base breakdown voltage (I_E = 10 μA dc, I_C = 0)	Non-A Suffix A-Suffix, 2N5581, 2N5582	$V_{(BR)EBO}$	5.0 6.0	— —	V dc
Collector cutoff current (V_{CE} = 60 V dc, $V_{EB(off)}$ = 3.0 V dc)	A-Suffix, 2N5581, 2N5582	I_{CEX}	—	10	nA dc
Collector cutoff current (V_{CB} = 50 V dc, I_E = 0) (V_{CB} = 60 V dc, I_E = 0) (V_{CB} = 50 V dc, I_E = 0, T_A = 150°C) (V_{CB} = 60 V dc, I_E = 0, T_A = 150°C)	Non-A Suffix A-Suffix, 2N5581, 2N5582 Non-A Suffix A-Suffix, 2N5581, 2N5582	I_{CBO}	— — — —	0.01 0.01 10 10	μA dc
Emitter cutoff current (V_{EB} = 3.0 V dc, I_C = 0)	A-Suffix, 2N5581, 2N5582	I_{EBO}	—	10	nA dc
Base cutoff current (V_{CE} = 60 V dc, $V_{EB(off)}$ = 3.0 V dc)	A-Suffix	I_{BL}	—	20	nA dc
On Characteristics					
DC current gain (I_C = 0.1 mA dc, V_{CE} = 10 V dc)	2N2218,A, 2N2221,A, 2N5581(1) 2N2219,A, 2N2222,A, 2N5582(1)	h_{FE}	20 35	— —	—
(I_C = 1.0 mA dc, V_{CE} = 10 V dc)	2N2218,A, 2N2221,A, 2N5581 2N2219,A, 2N2222,A, 2N5582		25 50	— —	
(I_C = 10 mA dc, V_{CE} = 10 V dc)	2N2218,A, 2N2221,A, 2N5581(1) 2N2219,A, 2N2222,A, 2N5582(1)		35 75	— —	
(I_C = 10 mA dc, V_{CE} = 10 V dc, T_A = −55°C)	2N2218,A, 2N2221,A, 2N5581 2N2219,A, 2N2222,A, 2N5582		15 35	— —	
(I_C = 150 mA dc, V_{CE} = 10 V dc)	2N2218,A, 2N2221,A, 2N5581 2N2219,A, 2N2222,A, 2N5582		40 100	120 300	
(I_C = 150 mA dc, V_{CE} = 1.0 V dc)	2N2218,A, 2N2221,A, 2N5581 2N2219,A, 2N2222,A, 2N5582		20 50	— —	
(I_C = 500 mA dc, V_{CE} = 10 V dc)	2N2218, 2N2221 2N2219, 2N2222 2N2218A, 2N2221A, 2N5581 2N2219A, 2N2222A, 2N5582		20 30 25 40	— — — —	
Collector-Emitter saturation voltage (I_C = 150 mA dc, I_B = 15 mA dc)	Non-A Suffix A-Suffix, 2N5581, 2N5582	$V_{CE(sat)}$	— —	0.4 0.3	V dc
(I_C = 500 mA dc, I_B = 50 mA dc)	Non-A Suffix A-Suffix, 2N5581, 2N5582		— —	1.6 1.0	
Base-Emitter saturation voltage (I_C = 150 mA dc, I_B = 15 mA dc)	Non-A Suffix A-Suffix, 2N5581, 2N5582	$V_{BE(sat)}$	0.6 0.6	1.3 1.2	V dc
(I_C = 500 mA dc, I_B = 50 mA dc)	Non-A Suffix A-Suffix, 2N5581, 2N5582		— —	2.6 2.0	

FIGURE 5–57

■ **ADVANCED PROBLEMS**

43. Design a circuit using base bias that operates from a 15 V dc voltage and draws a maximum current from the dc source ($I_{CC(max)}$) of 10 mA. The Q-point values are to be I_C = 5 mA and V_{CE} = 5 V. The transistor is a 2N3903. Assume a midpoint value for β_{DC}.

44. Design a circuit using emitter bias that operates from dc voltages of +12 V and −12 V. The maximum I_{CC} is to be 20 mA and the Q-point is at 10 mA and 4 V. The transistor is a 2N3904.

45. Design a circuit using voltage-divider bias for the following specifications: V_{CC} = 9 V, $I_{CC(max)}$ = 5 mA, I_C = 1.5 mA, and V_{CE} = 3 V. The transistor is a 2N3904.

46. Design a collector-feedback circuit using a 2N2222 with V_{CC} = 5 V, I_C = 10 mA, and V_{CE} = 1.5 V.

47. Can you replace the 2N3904 in Figure 5–55 with a 2N2222A and maintain the same range of output voltage over a temperature range from 45°C to 55°C?

48. Refer to the data sheet graph in Figure 5–58 and the partial data sheet in Figure 5–57. Determine the minimum dc current gain for a 2N2222 at −55°C, 25°C, and 175°C for V_{CE} = 1 V.

49. A design change is required in the ADC block of the industrial temperature control system shown in Figure 5–35. The new design will have an ADC input resistance of 10 kΩ. Determine the effect this change has on the temperature-to-voltage conversion circuit.

50. Investigate the feasibility of redesigning the temperature-to-voltage conversion circuit in Figure 5–36 to operate from a dc supply voltage of 5.1 V and produce the same range of output voltages over the required thermistor temperature range from 46°C to 54°C.

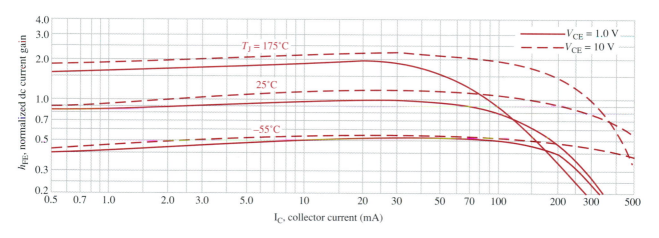

FIGURE 5–58

■ **ANSWERS TO SECTION REVIEWS**

Section 5–1

1. The upper load line limit is $I_{C(sat)}$ and $V_{CE(sat)}$. The lower limit is I_C = 0 and $V_{CE(cutoff)}$.

2. The Q-point is the dc point at which a transistor is biased specified by V_{CE} and I_C.

3. Saturation begins at the intersection of the load line and the vertical portion of the collector curve. Cutoff occurs at the intersection of the load line and the I_B = 0 curve.

4. The Q-point must be centered on the load line for maximum V_{CC}.

Section 5–2

1. Base bias does not require two supply voltages.
2. Base bias is beta-dependent.
3. The Q-point changes due to changes in β_{DC} and V_{CE} over temperature.

Section 5–3

1. Emitter bias is much less dependent on the value of beta than is base bias.
2. $V_E = V_B - 0.7$ V; $V_E = V_B + 0.7$ V
3. Emitter bias requires two separate supply voltages.
4. $I_E = 14.3$ V/10 kΩ = 1.43 mA

Section 5–4

1. $R_{IN(base)} = V_{IN}/I_{IN} = 5$ V/5 μA = 1 MΩ
2. $R_{IN(base)} = \beta_{DC}R_E = 190(1$ k$\Omega) = 190$ kΩ
3. $V_B = 5$ V
4. Voltage-divider bias is stable and requires only one supply voltage.

Section 5–5

1. I_C increases with β_{DC}, causing a reduction in V_C and, therefore, less voltage across R_B, thus less I_B.
2. $I_B = (V_C - V_{BE})/R_B = (4$ V $- 0.7$ V)/47 kΩ = 70.2 μA

Section 5–6

1. A transistor is saturated when $V_{CE} = 0$ V. A transistor is in cutoff when $V_{CE} = V_{CC}$.
2. R_E is open because the BE junction of the transistor is still forward-biased.
3. If R_C is open, V_C is about 0.7 V less than V_B.

■ **ANSWERS TO RELATED EXERCISES FOR EXAMPLES**

5–1 $I_{CQ} = 19.8$ mA; $V_{CEQ} = 4.2$ V; $I_{b(peak)} = 42$ μA

5–2 % $\Delta I_C = 150\%$; % $\Delta V_{CE} = 53.7\%$

5–3 $I_E = 3.61$ mA; $I_C \cong 3.61$ mA; $V_{CE} = 15.6$ V

5–4 % $\Delta I_C = 2.40\%$; % $\Delta V_{CE} = -3.87\%$

5–5 Less negative; No; -11.3 V

5–6 54.6 kΩ

5–7 $V_{CE} = 2.56$ V, $I_C = 4.77$ mA

5–8 $I_C \cong 2.29$ mA, $V_{EC} = 2.67$ V

5–9 214

5–10 $I_C = 0.894$ mA, $V_{CE} = 1.06$ V

6

SMALL-SIGNAL BIPOLAR AMPLIFIERS

■ CHAPTER OBJECTIVES

☐ Understand the concept of small-signal amplifiers

☐ Identify and apply internal transistor parameters

☐ Understand and analyze the operation of common-emitter amplifiers

☐ Understand and analyze the operation of common-collector amplifiers

☐ Understand and analyze the operation of common-base amplifiers

☐ Discuss multistage amplifiers and analyze their operation

☐ Troubleshoot amplifier circuits

The things you learned about biasing a transistor in Chapter 5 are carried forward into this chapter where bipolar-junction transistor circuits are used as small-signal amplifiers. The term *small-signal* refers to the use of signals that take up a relatively small percentage of an amplifier's operational range, that is, signals that use only a small portion of the load line. Additionally, you will learn how to reduce an amplifier to an equivalent dc and ac circuit for easier analysis, and you will learn about multistage amplifiers.

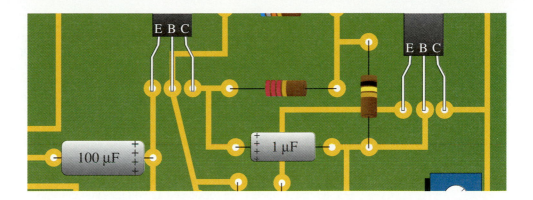

■ SYSTEM APPLICATION

The system application in this chapter involves a preamplifier circuit for a public address and paging system. The complete system includes the preamplifier, a power amplifier, and a dc power supply. You will focus on the preamplifier in this chapter and then on the power amplifier in the next chapter. The first step in your assignment is to learn all you can about amplifier operation. You will then apply your knowledge to the system application in Section 6–8.

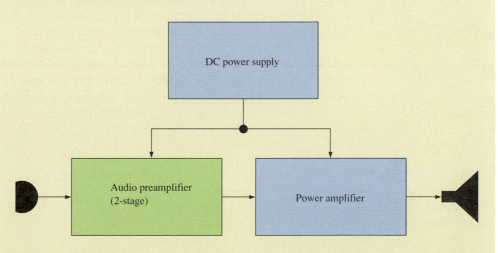

6–1 ▪ SMALL-SIGNAL AMPLIFIER OPERATION

The biasing of a transistor is purely a dc operation. The purpose of biasing is to establish a Q-point about which variations in current and voltage can occur in response to an ac input signal. In applications where small signal voltages must be amplified, such as from an antenna, a microphone, etc., variations about the Q-point are relatively small. Amplifiers designed to handle these small ac signals are called small-signal amplifiers.

After completing this section, you should be able to

▪ **Understand the concept of small-signal amplifiers**
  ☐ Interpret labels used for dc and ac voltages and currents
  ☐ Discuss the general operation of a small-signal amplifier
  ☐ Analyze ac load line operation
  ☐ Describe phase inversion

AC Quantities

In the previous chapters, dc quantities were identified by nonitalic uppercase (capital) subscripts such as I_C, I_E, V_C, and V_{CE}. Lowercase italic subscripts are used to indicate ac quantities of rms, peak, and peak-to-peak currents and voltages, for example, I_c, I_e, I_b, V_c, and V_{ce} (rms values are assumed unless otherwise stated). Instantaneous quantities are represented by both lowercase letters and subscripts such as i_c, i_e, i_b, and v_{ce}. Figure 6–1 illustrates these quantities for a specific voltage waveform.

In addition to currents and voltages, resistances often have different values when a circuit is analyzed from an ac viewpoint as opposed to a dc viewpoint. Lowercase subscripts are used to identify ac resistance values. For example, R_c is the ac collector resistance, and R_C is the dc collector resistance. You will see the need for this distinction later.

FIGURE 6–1

V_{ce} can represent rms, average, peak, or peak-to-peak, but rms is always assumed unless stated otherwise. v_{ce} can be any instantaneous value on the curve.

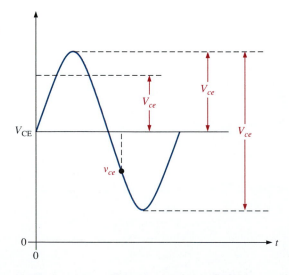

Resistance values *internal* to the transistor use a lowercase r'. An example is the internal ac emitter resistance, r'_e.

The Small-Signal Amplifier

A voltage-divider biased transistor with a sinusoidal ac source capacitively coupled to the base and a load capacitively coupled to the collector is shown in Figure 6–2. The coupling capacitors block dc and thus prevent the source resistance, R_s, and the load resistance R_L from changing the dc bias voltages at the base and collector. The sinusoidal source voltage causes the base voltage to vary sinusoidally above and below its dc bias level. The resulting variation in base current produces a larger variation in collector current because of the current gain of the transistor.

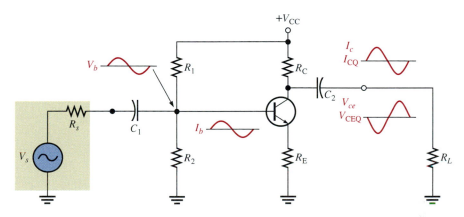

FIGURE 6–2
An amplifier with voltage-divider bias driven by an ac voltage source with an internal resistance, R_s.

As the sinusoidal collector current increases, the collector voltage decreases. The collector current varies above and below its Q-point value in phase with the base current. The sinusoidal collector-to-emitter voltage varies above and below its Q-point value 180° out of phase with the base voltage, as illustrated in Figure 6–2. A transistor always produces a phase inversion between the base voltage and the collector voltage.

A Graphical Picture The operation just described can be illustrated graphically on the collector-characteristic curves, as shown in Figure 6–3. The sinusoidal voltage at the base produces a base current that varies above and below the Q-point on the ac load line, as shown by the arrows. Lines projected from the peaks of the base current, across to the I_C axis, and down to the V_{CE} axis, indicate the peak-to-peak variations of the collector current and collector-to-emitter voltage, as shown. The ac load line differs from the dc load line because the effective ac collector resistance is R_L in parallel with R_C and is less than the dc collector resistance R_C alone without R_L in parallel. This difference between the dc and the ac load lines is covered in Chapter 7 in relation to power amplifiers.

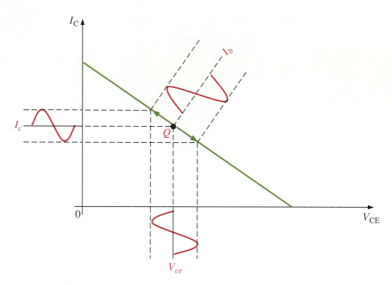

FIGURE 6–3

Graphical operation of the amplifier showing the base current, collector current, and collector-to-emitter voltage. I_b and I_c are on different scales.

EXAMPLE 6–1

The ac load line operation of a certain amplifier extends 10 μA above and below the Q-point base current value of 50 μA, as shown in Figure 6–4. Determine the resulting peak-to-peak values of collector current and collector-to-emitter voltage from the graph.

FIGURE 6–4

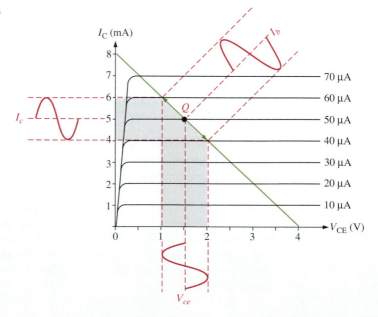

Solution   Projections on the graph of Figure 6–4 show the collector current varying from 6 mA to 4 mA for a peak-to-peak value of 2 mA and the collector-to-emitter voltage varying from 1 V to 2 V for a peak-to-peak value of 1 V.

Related Exercise   What are the Q-point values of I_C and V_{CE} in Figure 6–4?

**SECTION 6–1
REVIEW**

1. When I_b is at its positive peak, I_c is at its _____ peak, and V_{ce} is at its _____ peak.
2. What is the difference between V_{CE} and V_{ce}?
3. What is the difference between R_e and r'_e?

6–2 ■ TRANSISTOR AC EQUIVALENT CIRCUITS

To visualize the operation of a transistor in an amplifier circuit, it is often useful to represent the device by an equivalent circuit. An equivalent circuit uses various internal transistor parameters to represent the transistor's operation. Two types of equivalent circuit representations are described in this section. One is based on hybrid or h parameters and the other is based on resistance or r parameters.

After completing this section, you should be able to

■ **Identify and apply internal transistor parameters**
   ☐ Define the *h* parameters
   ☐ Represent a transistor by a hybrid equivalent circuit
   ☐ Convert from *h* parameters to *r* parameters
   ☐ Represent a transistor by an *r* parameter equivalent circuit
   ☐ Distinguish between the dc beta and the ac beta

h Parameters

Because they are typically specified on a manufacturer's data sheet, (as illustrated in Figure 6–5), *h* (hybrid) parameters (h_i, h_r, h_f, and h_o) are important. These parameters are specified by manufacturers because they are relatively easy to measure.

The four basic ac *h* parameters and their descriptions are given in Table 6–1. Each of the four *h* parameters carries a second subscript letter to designate the common-emitter (*e*), common-base (*b*), or common-collector (*c*) configuration, as listed in Table 6–2.

The three amplifier circuit configurations are described as follows:

☐ The circuit is a **common-emitter** type when the emitter is connected to ac ground, the input signal is applied to the base, and the output signal is on the collector. (The term *common* refers to the ground point or some other common reference point.)

☐ The circuit is a **common-collector** type when the collector is connected to ac ground, the input signal is applied to the base, and the output signal is on the emitter.

☐ The circuit is a **common-base** type when the base is connected to ac ground, the input signal is applied to the emitter, and the output signal is on the collector.

Small Signal Characteristics ($T_A = 25°C$ unless otherwise noted.)

Characteristic		Symbol	Min	Max	Unit
Output capacitance ($V_{CB} = 5.0$ V dc, $I_E = 0$, $f = 1.0$ MHz)		C_{obo}	—	4.0	pF
Input capacitance ($V_{BE} = 0.5$ V dc, $I_C = 0$, $f = 1.0$ MHz)		C_{ibo}	—	8.0	pF
Input impedance ($I_C = 1.0$ mA dc, $V_{CE} = 10$ V dc, $f = 1.0$ kHz)	2N3903 2N3904	h_{ie}	1.0 1.0	8.0 10	k ohms
Voltage feedback ratio ($I_C = 1.0$ mA dc, $V_{CE} = 10$ V dc, $f = 1.0$ kHz)	2N3903 2N3904	h_{re}	0.1 0.5	5.0 8.0	$\times 10^{-4}$
Small-signal current gain ($I_C = 1.0$ mA dc, $V_{CE} = 10$ V dc, $f = 1.0$ kHz)	2N3903 2N3904	h_{fe}	50 100	200 400	—
Output admittance ($I_C = 1.0$ mA dc, $V_{CE} = 10$ V dc, $f = 1.0$ kHz)		h_{oe}	1.0	40	μmhos or μS
Noise figure ($I_C = 100$ μA dc, $V_{CE} = 5.0$ V dc, $R_S = 1.0$ k ohms, $f = 1.0$ kHz)	2N3903 2N3904	NF	— —	6.0 5.0	dB

FIGURE 6–5

Part of the data sheet for the 2N3903 and 2N3904 transistors.

TABLE 6–1

Basic ac h parameters

h Parameter	Description	Condition
h_i	Input impedance (resistance)	Output shorted
h_r	Voltage feedback ratio	Input open
h_f	Forward current gain	Output shorted
h_o	Output admittance (conductance)	Input open

TABLE 6–2

Subscripts of h parameters for the three circuit configurations

Configuration	h Parameters
Common-Emitter	h_{ie}, h_{re}, h_{fe}, h_{oe}
Common-Base	h_{ib}, h_{rb}, h_{fb}, h_{ob}
Common-Collector	h_{ic}, h_{rc}, h_{fc}, h_{oc}

The characteristics of each of the three bipolar transistor amplifier configurations are examined later in this chapter.

Definitions of h Parameters   Each of the h parameters is derived from ac measurements taken from operating transistor characteristic curves. h_i is the ac resistance looking in at the input terminal of the transistor with the output shorted, as indicated in Figure 6–6(a), for a common-emitter connection. h_{ie} is the ratio of input (V_b) voltage to input current (I_b), expressed as follows:

$$h_{ie} = \frac{V_b}{I_b} \tag{6–1}$$

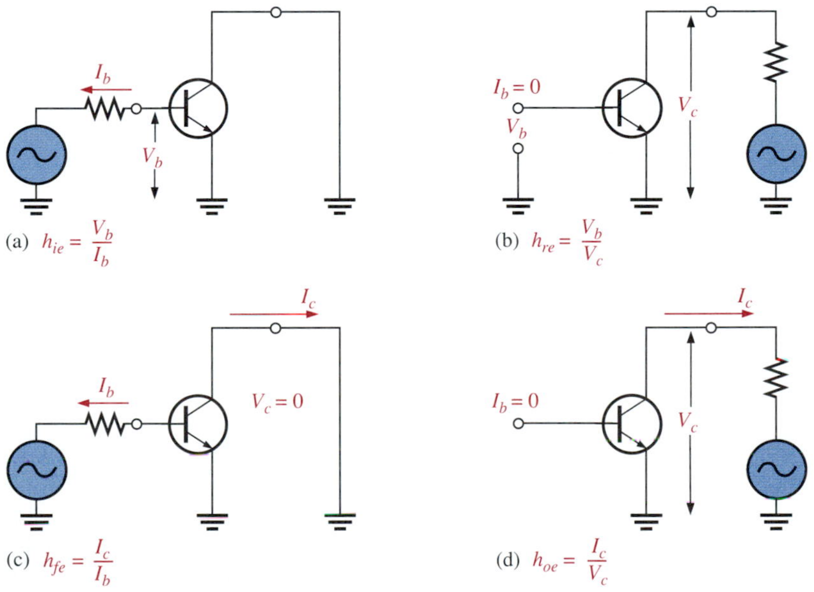

(a) $h_{ie} = \dfrac{V_b}{I_b}$ (b) $h_{re} = \dfrac{V_b}{V_c}$

(c) $h_{fe} = \dfrac{I_c}{I_b}$ (d) $h_{oe} = \dfrac{I_c}{V_c}$

FIGURE 6–6

AC equivalent circuits for defining h parameters using the common-emitter configuration.

h_r is a measure of the amount of output voltage that is reflected (fed back) to the input with the input open. The common-emitter measurement circuit is shown in Figure 6–6(b). h_{re} is the ratio of input voltage (V_b) to output voltage (V_c).

$$h_{re} = \frac{V_b}{V_c} \tag{6–2}$$

h_f is the forward current gain measured with the output (collector) shorted, as shown in Figure 6–6(c). For the common-emitter configuration, h_{fe} is expressed as

$$h_{fe} = \frac{I_c}{I_b} \tag{6–3}$$

Finally, h_o is the conductance looking in at the output with the input open, as shown in Figure 6–6(d). For the common-emitter configuration, h_o is expressed as

$$h_{oe} = \frac{I_c}{V_c} \tag{6–4}$$

The units of h_{oe} are siemens (S). Table 6–3 summarizes the h-parameter formulas for each amplifier configuration.

TABLE 6–3

h-parameter ratios for the three amplifier configurations

Common-Emitter	Common-Base	Common-Collector
$h_{ie} = V_b/I_b$	$h_{ib} = V_e/I_b$	$h_{ic} = V_b/I_b$
$h_{re} = V_b/V_c$	$h_{rb} = V_e/V_c$	$h_{rc} = V_b/V_e$
$h_{fe} = I_c/I_b$	$h_{fb} = I_c/I_e$	$h_{fc} = I_e/I_b$
$h_{oe} = I_c/V_c$	$h_{ob} = I_c/V_c$	$h_{oc} = I_e/V_e$

Hybrid Equivalent Circuits

The general form of the h-parameter equivalent circuit is shown in Figure 6–7. The input resistance, h_i, appears in series at the input. The reverse voltage ratio, h_r, is multiplied by the output voltage ($h_r V_{out}$) to produce an equivalent voltage source in series with the input. The forward current gain, h_f, is multiplied by the input current ($h_f I_{in}$) and appears as an equivalent current source in the output. The output conductance, h_o, appears across the output terminals. Specifically, there are three hybrid equivalent circuits—one for the common-emitter configuration, one for the common-base configuration, and one for the common-collector configuration, as shown in Figure 6–8.

r Parameters

The resistance, r, parameters are perhaps easier to work with than the h parameters. The five r parameters are given in Table 6–4. As previously mentioned, the italic lowercase letter r with a prime denotes resistance values internal to the transistor.

FIGURE 6–7

Generalized h parameter equivalent circuit for a bipolar junction transistor.

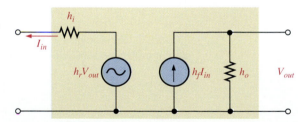

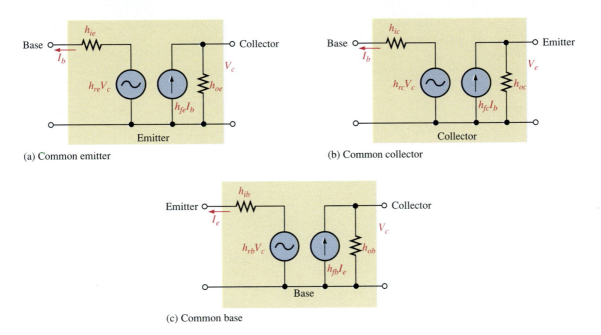

(a) Common emitter

(b) Common collector

(c) Common base

FIGURE 6–8

h parameter equivalent circuits for the three amplifier configurations.

TABLE 6–4	*r* Parameter	Description
r parameters	α_{ac}	ac alpha (I_c/I_e)
	β_{ac}	ac beta (I_c/I_b)
	r'_e	ac emitter resistance
	r'_b	ac base resistance
	r'_c	ac collector resistance

Relationships of *h* Parameters and *r* Parameters

The ac current ratios, α_{ac} and β_{ac}, convert directly from *h* parameters as follows:

$$\alpha_{ac} = h_{fb} \qquad\qquad \textbf{(6–5)}$$

$$\beta_{ac} = h_{fe} \qquad\qquad \textbf{(6–6)}$$

Recall that we used α_{DC} and β_{DC} in previous chapters. These are dc parameters and they sometimes have values different from those of the ac parameters. The difference between β_{DC} and β_{ac} will be discussed later.

Because data sheets often provide only common-emitter h parameters, the following formulas show how to convert them to r parameters.

$$r_e' = \frac{h_{re}}{h_{oe}}$$

(6–7)

$$r_c' = \frac{h_{re} + 1}{h_{oe}}$$

(6–8)

$$r_b' = h_{ie} - \frac{h_{re}}{h_{oe}}(1 + h_{fe})$$

(6–9)

We will use r parameters throughout the text.

r-Parameter Equivalent Circuits

An r-parameter equivalent circuit is shown in Figure 6–9(a). For most general analysis work, Figure 6–9(a) can be simplified as follows: The effect of the ac base resistance r_b' is usually small enough to neglect, so that it can be replaced by a short. The ac collector resistance is usually several hundred kilohms and can be replaced by an open. The resulting simplified r-parameter equivalent circuit is shown in Figure 6–9(b).

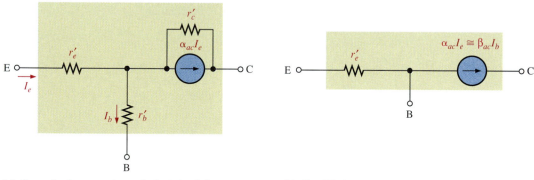

(a) Generalized r-parameter equivalent circuit for a bipolar junction transistor

(b) Simplified r-parameter equivalent circuit for a bipolar junction transistor

FIGURE 6–9

r-parameter equivalent circuits.

The interpretation of this equivalent circuit in terms of a transistor's ac operation is as follows: A resistance r_e' appears between the emitter and base terminals. This is the resistance "seen" looking into the emitter of a forward-biased transistor. The collector effectively acts as a current source of $\alpha_{ac}I_e$ or, equivalently, $\beta_{ac}I_b$. These factors are shown with a transistor symbol in Figure 6–10.

FIGURE 6–10

Relation of transistor symbol to r-parameter equivalent.

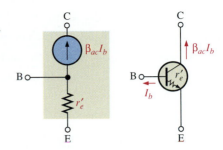

Determining r'_e by a Formula

For amplifier analysis, r'_e is the most important of the r parameters. Instead of using h parameters to get r'_e, you can use the simple formula of Equation (6–10) to calculate an approximate value.

$$r'_e \cong \frac{25 \text{ mV}}{I_E} \qquad (6\text{–}10)$$

Although the formula is simple, its derivation is not and is therefore reserved for Appendix B.

EXAMPLE 6–2   Determine the r'_e of a transistor that is operating with a dc emitter current of 2 mA.

Solution $$r'_e \cong \frac{25 \text{ mV}}{I_E} = \frac{25 \text{ mV}}{2 \text{ mA}} = 12.5 \ \Omega$$

Related Exercise   What is I_E if $r'_e = 8 \ \Omega$?

Comparison of the AC Beta (β_{ac}) to the DC Beta (β_{DC})

For a typical transistor, a graph of I_C versus I_B is nonlinear, as shown in Figure 6–11(a). If you pick a Q-point on the curve and cause the base current to vary an amount ΔI_B, then the collector current will vary an amount ΔI_C as shown in part (b). At different points on the nonlinear curve, the ratio $\Delta I_C/\Delta I_B$ will be different, and it may also differ from the I_C/I_B ratio at the Q-point. Since $\beta_{DC} = I_C/I_B$ and $\beta_{ac} = \Delta I_C/\Delta I_B$, the values of these two quantities can differ. Remember that $\beta_{DC} = h_{FE}$ and $\beta_{ac} = h_{fe}$.

FIGURE 6–11

I_C-versus-I_B curve illustrates the difference between $\beta_{DC} = I_C/I_B$ and $\beta_{ac} = \Delta I_C/\Delta I_B$.

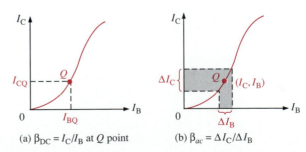

(a) $\beta_{DC} = I_C/I_B$ at Q point   (b) $\beta_{ac} = \Delta I_C/\Delta I_B$

SECTION 6–2
REVIEW

1. Define each of the parameters: h_{ie}, h_{re}, h_{fe}, and h_{oe}.
2. Which h parameter is equivalent to β_{ac}?
3. If $I_E = 15$ mA, what is the approximate value of r_e'?

6–3 ■ COMMON-EMITTER AMPLIFIERS

Now that you have an idea of how a transistor can be modeled in an ac circuit, a complete amplifier circuit will be examined. The common-emitter (CE) configuration is covered in this section. CE amplifiers exhibit high voltage and current gains. The common-collector and common-base configurations are covered in the following sections.

After completing this section, you should be able to

■ **Understand and analyze the operation of common-emitter amplifiers**
- ☐ Represent a CE amplifier by its dc equivalent circuit
- ☐ Analyze the dc operation of a CE amplifier
- ☐ Represent a CE amplifier by its ac equivalent circuit
- ☐ Analyze the ac operation of a CE amplifier
- ☐ Determine the input resistance
- ☐ Determine the output resistance
- ☐ Determine the voltage gain
- ☐ Explain the effects of an emitter-bypass capacitor
- ☐ Describe swamping and discuss its purpose and effects
- ☐ Describe the effect of a load resistor on the voltage gain
- ☐ Discuss phase inversion in a CE amplifier
- ☐ Determine current gain
- ☐ Determine power gain

Figure 6–12 shows a common-emitter amplifier with voltage-divider bias and coupling capacitors, C_1 and C_3, on the input and output and a bypass capacitor C_2 from emitter to ground. The circuit has a combination of dc and ac operation, both of which must be considered.

FIGURE 6–12
A common-emitter amplifier.

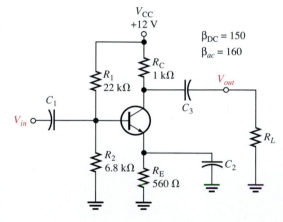

DC Analysis

To analyze the amplifier in Figure 6–12, the dc bias values must first be determined. To do this, a dc equivalent circuit is developed by replacing the coupling and bypass capacitors with opens (remember, a capacitor appears open to dc), as shown in Figure 6–13. Recall from Chapter 5 that the dc input resistance at the base is determined as follows:

$$R_{\text{IN(base)}} = \beta_{\text{DC}}R_{\text{E}} = (150)(560 \ \Omega) = 84 \ \text{k}\Omega$$

Since $R_{\text{IN(base)}}$ is more than ten times R_2, it will be neglected when calculating the dc base voltage.

$$V_{\text{B}} \cong \left(\frac{R_2}{R_1 + R_2} \right)V_{\text{CC}} = \left(\frac{6.8 \ \text{k}\Omega}{28.8 \ \text{k}\Omega} \right)12 \ \text{V} = 2.83 \ \text{V}$$

and

$$V_{\text{E}} = V_{\text{B}} - V_{\text{BE}} = 2.83 \ \text{V} - 0.7 \ \text{V} = 2.13 \ \text{V}$$

Therefore,

$$I_{\text{E}} = \frac{V_{\text{E}}}{R_{\text{E}}} = \frac{2.13 \ \text{V}}{560 \ \Omega} = 3.80 \ \text{mA}$$

Since $I_{\text{C}} \cong I_{\text{E}}$, then

$$V_{\text{C}} = V_{\text{CC}} - I_{\text{C}}R_{\text{C}} = 12 \ \text{V} - 3.80 \ \text{V} = 8.20 \ \text{V}$$

Finally,

$$V_{\text{CE}} = V_{\text{C}} - V_{\text{E}} = 8.20 \ \text{V} - 2.13 \ \text{V} = 6.07 \ \text{V}$$

FIGURE 6–13

DC equivalent circuit for the amplifier in Figure 6–12.

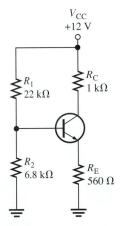

The AC Equivalent Circuit

To analyze the ac signal operation of an amplifier, an ac equivalent circuit is developed as follows: The capacitors C_1, C_2, and C_3 are replaced by effective shorts because we assume that $X_C \cong 0 \ \Omega$ at the signal frequency.

AC Ground   The dc source is replaced by a ground. We assume that the voltage source has an internal resistance of approximately 0 Ω, so that no ac voltage is developed across the source terminals. Therefore, the V_{CC} terminal is at a zero-volt ac potential and is called *ac ground.*

The ac equivalent circuit for the common-emitter amplifier in Figure 6–12 is shown in Figure 6–14(a). Notice that both R_C and R_1 have one end connected to ac ground because, in the actual circuit, they are connected to V_{CC} which is, in effect, ac ground.

In ac analysis, the ac ground and the actual ground are treated as the same point electrically. The amplifier in Figure 6–12 is a common-emitter type because the bypass capacitor C_2 keeps the emitter at ac ground (ground is the common point in the circuit).

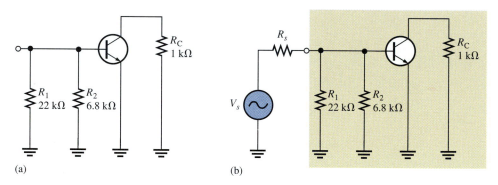

(a)   (b)

FIGURE 6–14

AC equivalent circuit for the amplifier in Figure 6–12.

Signal (AC) Voltage at the Base   An ac voltage source is shown connected to the input in Figure 6–14(b). If the internal resistance of the ac source is 0 Ω, then all of the source voltage appears at the base terminal. If, however, the ac source has a nonzero internal resistance, then three factors must be taken into account in determining the actual signal voltage at the base. These are the source resistance, the bias resistance, and the input resistance at the base. This is illustrated in Figure 6–15(a) and is simplified by combining

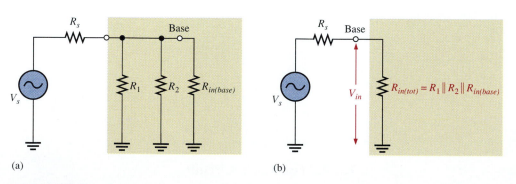

(a)   (b)

FIGURE 6–15

AC equivalent base circuit.

R_1, R_2, and $R_{in(base)}$ in parallel to get the total input resistance, $R_{in(tot)}$, as shown in Figure 6–15(b). As you can see, the source voltage V_s is divided down by R_s (source resistance) and R_{in}, so that the signal voltage at the base of the transistor is found by the voltage-divider formula as follows:

$$V_b = \left(\frac{R_{in}}{R_s + R_{in}} \right) V_s$$

If $R_s \ll R_{in(tot)}$, then $V_b \cong V_s$. V_b is the input voltage, V_{in}, to the amplifier.

Input Resistance To develop an expression for the input resistance as seen by an ac source looking in at the base, we will use the simplified r-parameter model of the transistor. Figure 6–16 shows the transistor connected with the external resistor R_C. The input resistance looking in at the base is

$$R_{in(base)} = \frac{V_{in}}{I_{in}} = \frac{V_b}{I_b}$$

The base voltage is

$$V_b = I_e r'_e$$

and since $I_e \cong I_c$,

$$I_b \cong \frac{I_e}{\beta_{ac}}$$

Substituting for V_b and I_b, we get

$$R_{in(base)} = \frac{V_b}{I_b} = \frac{I_e r'_e}{I_e / \beta_{ac}}$$

Cancelling I_e, we get

$$R_{in(base)} = \beta_{ac} r'_e \qquad \qquad \textbf{(6–11)}$$

FIGURE 6–16

r-parameter transistor model (inside shaded block) connected to external circuit.

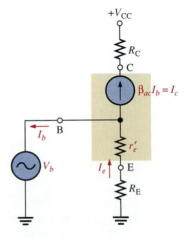

The total input resistance seen by the source is the parallel combination of R_1, R_2, and $R_{in(base)}$.

$$R_{in(tot)} = R_1 \parallel R_2 \parallel R_{in(base)} \qquad (6\text{--}12)$$

Output Resistance The output resistance of the common-emitter amplifier looking in at the collector is approximately equal to the collector resistor.

$$R_{out} \cong R_C \qquad (6\text{--}13)$$

Actually, $R_{out} = R_C \parallel r'_c$, but since the internal ac collector resistance of the transistor, r'_c, is typically much larger than R_C, the approximation is usually valid. For example, the partial data sheet in Figure 6–5 gives minimum values for the 2N3904 of $h_{re} = 0.5 \times 10^{-4}$ and $h_{oe} = 1.0\ \mu S$. (Siemens and mhos are the same unit; mhos is the older designation but Siemens is the current standard.) From these values, r'_c can be calculated using Equation (6–8).

$$r'_c = \frac{h_{re} + 1}{h_{oe}} = \frac{0.5 \times 10^{-4} + 1}{1.0\ \mu S} \cong 1\ M\Omega$$

EXAMPLE 6–3

Determine the signal voltage at the base in Figure 6–17. This circuit is the ac equivalent of the amplifier in Figure 6–12 with a 10 mV rms, 300 Ω signal source. I_E was previously found to be 3.80 mA.

FIGURE 6–17

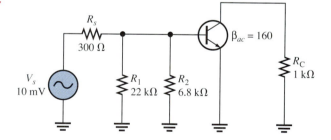

Solution First, determine the ac emitter resistance:

$$r'_e = \frac{25\ mV}{I_E} = \frac{25\ mV}{3.80\ mA} = 6.58\ \Omega$$

Then,

$$R_{in(base)} = \beta_{ac}\, r'_e = 160(6.58\ \Omega) = 1.05\ k\Omega$$

Next, determine the total input resistance viewed from the source:

$$R_{in(tot)} = R_1 \parallel R_2 \parallel R_{in(base)} = \cfrac{1}{\dfrac{1}{22\ k\Omega} + \dfrac{1}{6.8\ k\Omega} + \dfrac{1}{1.05\ k\Omega}} = 873\ \Omega$$

The source voltage is divided down by R_s and $R_{in(tot)}$, so the signal voltage at the base is the voltage across $R_{in(tot)}$.

$$V_b = \left(\frac{R_{in(tot)}}{R_s + R_{in(tot)}} \right) V_s = \left(\frac{873\ \Omega}{1173\ \Omega} \right) 10\ \text{mV} = 7.44\ \text{mV}$$

As you can see, there is attenuation (reduction) of the source voltage due to the source resistance and amplifier's input resistance acting as a voltage divider.

Related Exercise Determine the signal voltage at the base of Figure 6–17 if the source resistance is 75 Ω and another transistor with an ac beta of 200 is used.

Voltage Gain of the CE Amplifier

The ac voltage gain expression is developed using the equivalent circuit in Figure 6–18. The gain is the ratio of ac output voltage (V_c) to ac input voltage at the base (V_b).

$$A_v = \frac{V_{out}}{V_{in}} = \frac{V_c}{V_b}$$

Notice in the figure that $V_c = \alpha_{ac} I_e R_C \cong I_e R_C$ and $V_b = I_e r'_e$. Therefore,

$$A_v = \frac{I_e R_C}{I_e r'_e}$$

The I_e terms cancel, so

$$A_v = \frac{R_C}{r'_e} \qquad\qquad \textbf{(6–14)}$$

Equation (6–14) is the voltage gain from base to collector. To get the overall gain of the amplifier from the source voltage to collector, the attenuation of the input circuit must be included. **Attenuation** is defined as a gain of less than 1 and results in a reduction of the signal voltage. The attenuation from source to base multiplied by the gain from base to collector is the overall amplifier gain. Suppose the source produces 10 mV and the

FIGURE 6–18
Equivalent circuit for obtaining ac voltage gain.

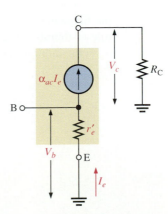

source resistance and input resistance is such that the base voltage is 5 mV. The attenuation is therefore 5 mV/10 mV = 0.5. Now assume the amplifier has a voltage gain from base to collector of 20. The output voltage is 5 mV × 20 = 100 mV. Therefore, the overall gain is 100 mV/10 mV = 10 and is equal to the attenuation times the gain (0.5 × 20 = 10). Overall gain is illustrated in Figure 6–19.

The expression for the attenuation in the base circuit where R_s and $R_{in(tot)}$ act as a voltage divider is

$$\text{Attenuation} = \frac{V_b}{V_s} = \frac{R_{in(tot)}}{R_s + R_{in(tot)}}$$

The overall gain, A_v', is the product of the attenuation and the gain from base to collector A_v.

$$A_v' = \left(\frac{V_b}{V_s}\right) A_v \qquad\qquad (6\text{–}15)$$

FIGURE 6–19

Base circuit attenuation and overall gain.

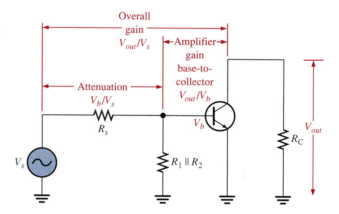

Effect of the Emitter-Bypass Capacitor on Voltage Gain   The emitter-bypass capacitor provides an effective short to the ac signal around the emitter resistor, thus keeping the emitter at ac ground, as you have seen. With the bypass capacitor, the gain of a given amplifier is maximum and equal to R_C/r_e'.

The value of the bypass capacitor must be large enough so that its reactance over the frequency range of the amplifier is very small (ideally 0 Ω) compared to R_E. A good rule-of-thumb is that X_C of the bypass capacitor should be at least 10 times smaller than R_E at the minimum frequency for which the amplifier must operate.

$$10X_C \leq R_E$$

EXAMPLE 6–4

Select a minimum value for the emitter-bypass capacitor in Figure 6–20 if the amplifier must operate over a frequency range from 2 kHz to 10 kHz.

FIGURE 6–20

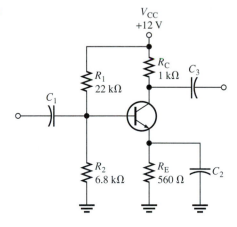

Solution   Since $R_E = 560\ \Omega$, X_C of the bypass capacitor should be no greater than

$$10X_C = R_E$$

$$X_C = \frac{R_E}{10} = \frac{560\ \Omega}{10} = 56\ \Omega$$

The capacitance value is determined at the minimum frequency of 2 kHz as follows:

$$C_2 = \frac{1}{2\pi f X_C} = \frac{1}{2\pi(2\ \text{kHz})(56\ \Omega)} = 1.42\ \mu\text{F}$$

This is the minimum value for the bypass capacitor for this circuit. You can use a larger value, although cost and physical size usually impose limitations.

Related Exercise   If the minimum frequency is reduced to 1 kHz, what value of bypass capacitor must you use?

Voltage Gain Without the Bypass Capacitor   To see how the bypass capacitor affects ac voltage gain, let's remove it from the circuit and compare voltage gains.

Without the bypass capacitor, the emitter is no longer at ac ground. Instead, R_E is seen by the ac signal between the emitter and ground and effectively adds to r'_e in the voltage gain formula.

$$A_v = \frac{R_C}{r'_e + R_E} \qquad (6\text{–}16)$$

The effect of R_E is to decrease the ac voltage gain.

EXAMPLE 6–5

Calculate the base-to-collector voltage gain of the amplifier in Figure 6–12 without and with an emitter bypass capacitor if there is no load resistor.

Solution   From Example 6–3, $r_e' = 6.58 \, \Omega$ for this particular amplifier. Without C_2, the gain is

$$A_v = \frac{R_C}{r_e' + R_E} = \frac{1 \text{ k}\Omega}{566.58 \, \Omega} = 1.76$$

With C_2, the gain is

$$A_v = \frac{R_C}{r_e'} = \frac{1 \text{ k}\Omega}{6.58 \, \Omega} = 152$$

What a difference the bypass capacitor makes!

Related Exercise   Determine the base-to-collector voltage gain in Figure 6–12 with R_E bypassed, for the following circuit values: $R_C = 1.8 \text{ k}\Omega$, $R_E = 1 \text{ k}\Omega$, $R_1 = 33 \text{ k}\Omega$, and $R_2 = 6.8 \text{ k}\Omega$.

Effect of a Load on Voltage Gain   When a **load,** R_L, is connected to the output through the coupling capacitor C_3, as shown in Figure 6–21(a), the collector resistance at the signal frequency is effectively R_C in parallel with R_L. Remember, the upper end of R_C is effectively at ac ground. The ac equivalent circuit is shown in Figure 6–21(b). The total ac collector resistance is

$$R_c = \frac{R_C R_L}{R_C + R_L}$$

Replacing R_C with R_c in the voltage gain expression gives

$$A_v = \frac{R_c}{r_e'} \qquad\qquad (6\text{–}17)$$

FIGURE 6–21

A common-emitter amplifier with an ac (capacitively) coupled load.

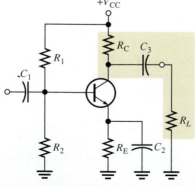

(a) Complete amplifier

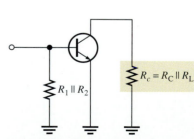

(b) AC equivalent ($X_{C2} = 0$)

When $R_c < R_C$, the voltage gain is reduced. If $R_L \gg R_C$, then $R_c \cong R_C$ and the load has very little effect on the gain.

EXAMPLE 6–6

Calculate the base-to-collector voltage gain of the amplifier in Figure 6–12 when a load resistance of 5 kΩ is connected to the output. The emitter is effectively bypassed and $r'_e = 6.58$ Ω.

Solution   The ac collector resistance is

$$R_c = \frac{R_C R_L}{R_C + R_L} = \frac{(1 \text{ k}\Omega)(5 \text{ k}\Omega)}{6 \text{ k}\Omega} = 833 \text{ }\Omega$$

Therefore,

$$A_v = \frac{R_c}{r'_e} = \frac{833 \text{ }\Omega}{6.58 \text{ }\Omega} = 127$$

The unloaded gain was found to be 152 in Example 6–5.

Related Exercise   Determine the base-to-collector voltage gain in Figure 6–12 when a 10 kΩ load resistance is connected from collector to ground. Change the resistance values as follows: $R_C = 1.8$ kΩ, $R_E = 1$ kΩ, $R_1 = 33$ kΩ, and $R_2 = 6.8$ kΩ. The emitter resistor is effectively bypassed and $r'_e = 10$ Ω.

Stability of the Voltage Gain

Although bypassing R_E does produce the maximum voltage gain, there is a **stability** problem because the ac voltage gain is dependent on r'_e ($A_v = R_C/r'_e$) and r'_e depends on I_E and varies considerably with temperature. This causes the gain to be unstable over temperature because when r'_e increases, the gain decreases and vice versa.

With no bypass capacitor, the gain is decreased because R_E is now in the ac circuit ($A_v = R_C/(r'_e + R_E)$). However, with R_E, the gain is much less dependent on r'_e; if $R_E \gg r'_e$, the gain is essentially independent of r'_e because

$$A_v \cong \frac{R_C}{R_E}$$

Swamping r'_e to Stabilize the Voltage Gain   *Swamping* is a method used to minimize the effect of r'_e without reducing the voltage gain to its minimum value. This method "swamps" out the effect of r'_e on the voltage gain. Swamping is, in effect, a compromise between having a bypass capacitor across R_E and having no bypass capacitor at all.

In a swamped amplifier, R_E is partially bypassed so that a reasonable gain can be achieved, and the effect of r'_e on the gain is greatly reduced or eliminated. The total external emitter resistance, R_E, is formed with two separate emitter resistors, R_{E1} and R_{E2}, as indicated in Figure 6–22. One of the resistors, R_{E2}, is bypassed and the other is not.

Both resistors ($R_{E1} + R_{E2}$) affect the dc bias while only R_{E1} affects the ac voltage gain.

$$A_v = \frac{R_C}{r'_e + R_{E1}}$$

If R_{E1} is several times larger than r'_e, then the effect of r'_e is minimized and the approximate voltage gain is

$$A_v \cong \frac{R_C}{R_{E1}}$$

(6–18)

FIGURE 6–22
A swamped amplifier uses a partially bypassed emitter resistance to achieve gain stability.

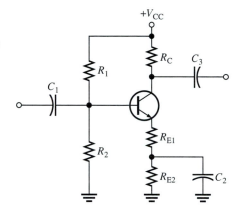

EXAMPLE 6–7

Determine the voltage gain of the swamped amplifier in Figure 6–23. Assume that the bypass capacitor has a negligible reactance for the frequency at which the amplifier is operated. Assume $r'_e = 20\ \Omega$.

FIGURE 6–23

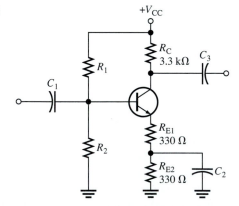

Solution R_{E2} is bypassed by C_2. R_{E1} is more than ten times r'_e so the approximate voltage gain is

$$A_v = \frac{R_C}{R_{E1}} = \frac{3.3\ \text{k}\Omega}{330\ \Omega} = 10$$

Related Exercise   What would be the voltage gain without C_2? What would be the voltage gain with C_2 bypassing both R_{E1} and R_{E2}?

The Effect of Swamping on the Amplifier's Input Resistance The ac input resistance looking in at the base of a common-emitter amplifier with R_E completely bypassed is $R_{in} = \beta_{ac}r'_e$. When the emitter resistance is partially bypassed, the portion of the resistance that is unbypassed is seen by the ac signal and contributes to the input resistance by appearing in series with r'_e. The formula is

$$R_{in(base)} = \beta_{ac}(r'_e + R_{E1}) \qquad (6\text{--}19)$$

Phase Inversion in a Common-Emitter Amplifier

The output voltage at the collector of a common-emitter amplifier is 180° out of phase with the input voltage at the base. The phase inversion is sometimes indicated by a negative sign in front of voltage gain, $-A_v$. The next example pulls together the concepts covered so far as they relate to the common-emitter amplifier.

EXAMPLE 6–8

For the amplifier in Figure 6–24, determine the total collector voltage (dc and ac).

FIGURE 6–24

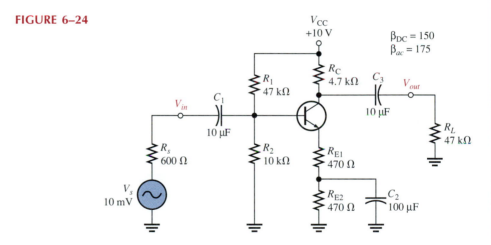

Solution Two sets of calculations are necessary to determine the total collector voltage.

Step 1: Determine the dc bias values. Refer to the dc equivalent circuit in Figure 6–25.

$$R_{IN(base)} = \beta_{DC}(R_{E1} + R_{E2}) = 150(940\ \Omega) = 141\ k\Omega$$

Since $R_{IN(base)}$ is more than ten times larger than R_2, it can be neglected in the dc base voltage calculation.

$$V_B = \left(\frac{R_2}{R_1 + R_2}\right)V_{CC} = \left(\frac{10\ k\Omega}{47\ k\Omega + 10\ k\Omega}\right)10\ V = 1.75\ V$$

$$V_E = V_B - 0.7\ V = 1.75\ V - 0.7\ V = 1.05\ V$$

$$I_E = \frac{V_E}{R_{E1} + R_{E2}} = \frac{1.05}{940\ \Omega} = 1.12\ mA$$

$$V_C = V_{CC} - I_C R_C = 10\ V - (1.12\ mA)(4.7\ k\Omega) = 4.74\ V$$

FIGURE 6–25

DC equivalent for the circuit in Figure 6–24.

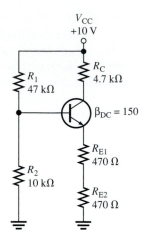

Step 2: The ac analysis is based on the ac equivalent circuit in Figure 6–26. The first thing to do in the ac analysis is calculate r'_e.

$$r'_e \cong \frac{25 \text{ mV}}{I_E} = \frac{25 \text{ mV}}{1.12 \text{ mA}} = 22 \text{ }\Omega$$

Next, determine the attenuation in the base circuit. Looking from the 600 Ω source, the total R_{in} is

$$R_{in(tot)} = R_1 \parallel R_2 \parallel R_{in(base)}$$
$$R_{in(base)} = \beta_{ac}(r'_e + R_{E1}) = 175(492 \text{ }\Omega) = 86.1 \text{ k}\Omega$$

Therefore,

$$R_{in(tot)} = 47 \text{ k}\Omega \parallel 10 \text{ k}\Omega \parallel 86.1 \text{ k}\Omega = 7.53 \text{ k}\Omega$$

The attenuation from source to base is

$$\text{Attenuation} = \frac{V_b}{V_s} = \frac{R_{in(tot)}}{R_s + R_{in(tot)}} = \frac{7.53 \text{ k}\Omega}{600 \text{ }\Omega + 7.53 \text{ k}\Omega} = 0.93$$

Before A_v can be determined, you must know the ac collector resistance R_c.

$$R_c = \frac{R_C R_L}{R_C + R_L} = \frac{(4.7 \text{ k}\Omega)(47 \text{ k}\Omega)}{4.7 \text{ k}\Omega + 47 \text{ k}\Omega} = 4.27 \text{ k}\Omega$$

The voltage gain from base to collector is

$$A_v \cong \frac{R_c}{R_{E1}} = \frac{4.27 \text{ k}\Omega}{470 \text{ }\Omega} = 9.09$$

The overall voltage gain is the attenuation times the amplifier voltage gain:

$$A'_v = \left(\frac{V_b}{V_s}\right) A_v = (0.93)(9.09) = 8.45$$

The source produces 10 mV rms, so the rms voltage at the collector is

$$V_c = A'_v V_{in} = (8.45)(10 \text{ mV}) = 84.5 \text{ mV}$$

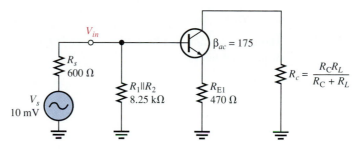

FIGURE 6–26

AC equivalent for the circuit in Figure 6–24.

The total collector voltage is the signal of 84.5 mV rms riding on a dc level of 4.74 V, as shown in Figure 6–27(a), where approximate peak values are shown. The coupling capacitor C_3 keeps the dc level from getting to the output. So, V_{out} is equal to the ac portion of the collector voltage, as indicated in Figure 6–27(b). The source voltage is shown to emphasize the phase inversion.

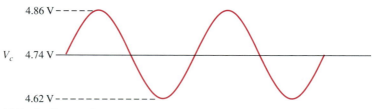

(a) Total collector voltage

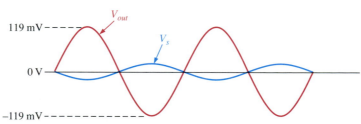

(b) Source and output ac voltages

FIGURE 6–27

Voltages for Figure 6–24.

Related Exercise What is A_v in Figure 6–24 with R_L removed?

Current Gain

The current gain from base to collector is I_c/I_b or β_{ac}. However, the overall current gain of the amplifier is

$$A_i = \frac{I_c}{I_s} \qquad \text{(6–20)}$$

I_s is the total signal current from the source, part of which is base current and part of which is the signal current in the bias network ($R_1 \parallel R_2$), as shown in Figure 6–28. The total current signal from the source is

$$I_s = \frac{V_s}{R_{in(tot)} + R_s}$$

Power Gain

The power gain is the product of the overall voltage gain and the current gain.

$$A_p = A_v' A_i \qquad \text{(6–21)}$$

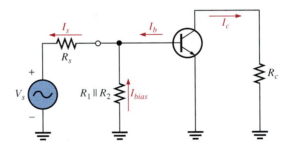

FIGURE 6–28

Total input signal current (directions shown are for the positive half-cycle of V_s).

SUMMARY OF THE COMMON-EMITTER AMPLIFIER

CIRCUIT WITH VOLTAGE-DIVIDER BIAS

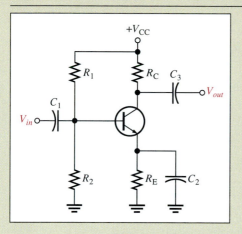

- Input is at the base. Output is at the collector.
- There is a phase inversion from input to output.
- C_1 and C_3 are coupling capacitors for the input and output signals.
- C_2 is the emitter-bypass capacitor.
- All capacitors must have a negligible reactance at the frequency of operation.
- Emitter is at ac ground due to the bypass capacitor.

EQUIVALENT CIRCUITS AND FORMULAS

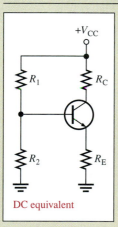

DC equivalent

- DC formulas:

$$V_B = \left(\frac{R_2 \| \beta_{DC}R_E}{R_1 + R_2 \| \beta_{DC}R_E} \right) V_{CC}$$

$$V_E = V_B - V_{BE}$$

$$I_E = \frac{V_E}{R_E}$$

$$V_C = V_{CC} - I_C R_C$$

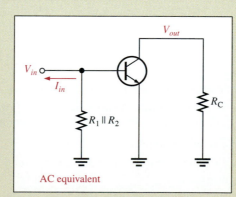

AC equivalent

- AC formulas:

$$r'_e = \frac{25\ \text{mV}}{I_E}$$

$$R_{in(base)} = \beta_{ac} r'_e$$

$$R_{out} \cong R_C$$

$$A_v = \frac{R_C}{r'_e}$$

$$A'_v = \left(\frac{V_b}{V_s} \right) A_v$$

$$A_i = \frac{I_C}{I_{in}}$$

$$A_p = A'_v A_i$$

SUMMARY OF THE COMMON-EMITTER AMPLIFIER, *continued*

SWAMPED AMPLIFIER WITH RESISTIVE LOAD

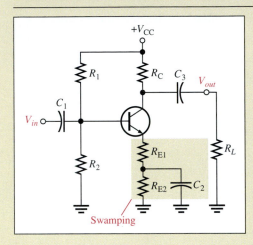

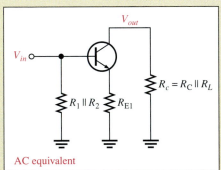

AC equivalent

- ■ AC formulas:

$$A_v \cong \frac{R_C \parallel R_L}{R_{E1}}$$

$$R_{in(base)} = \beta_{ac} = \beta_{ac}(r_e' + R_{E1})$$

- ■ Swamping stabilizes gain by minimizing the effect of r_e'.
- ■ Swamping reduces voltage gain from its unswamped value.
- ■ Swamping increases input resistance.
- ■ Load resistance reduces voltage gain.

SECTION 6–3 REVIEW

1. In the dc equivalent circuit of an amplifier, how are the capacitors treated?
2. When the emitter resistor is bypassed with a capacitor, how is the gain of the amplifier affected?
3. Explain swamping.
4. List the elements included in the total input resistance of a common-emitter amplifier.
5. What elements determine the overall voltage gain of a common-emitter amplifier?
6. When a load resistor is capacitively coupled to the collector of a CE amplifier, is the voltage gain increased or decreased?
7. What is the phase relationship of the input and output voltages of a CE amplifier?

6–4 ■ COMMON-COLLECTOR AMPLIFIERS

The common-collector (CC) amplifier is usually referred to as an emitter-follower (EF). The input is applied to the base through a coupling capacitor, and the output is at the emitter. The voltage gain of a CC amplifier is approximately 1, and its main advantages are its high input resistance and current gain, as you will see.

After completing this section, you should be able to

■ **Understand and analyze the operation of common-collector amplifiers**
  ☐ Represent a CC amplifier by its dc and ac equivalent circuits
  ☐ Analyze the dc and ac operation of a CC amplifier
  ☐ Determine the voltage gain
  ☐ Determine the input resistance
  ☐ Determine the output resistance
  ☐ Determine the current gain
  ☐ Determine the power gain
  ☐ Discuss the darlington pair and describe its main advantage

An emitter-follower circuit with voltage-divider bias is shown in Figure 6–29. Notice that the input is at the base, the output is at the emitter, and the collector is at ac ground.

FIGURE 6–29
Emitter-follower with voltage-divider bias.

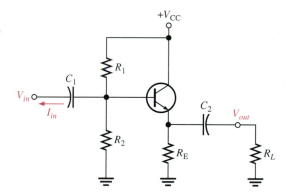

Voltage Gain

As in all amplifiers, the voltage gain is $A_v = V_{out}/V_{in}$. The capacitive reactances are assumed to be negligible. For the emitter-follower, as shown in the ac equivalent model in Figure 6–30,

$$V_{out} = I_e R_e \quad \text{and} \quad V_{in} = I_e(r'_e + R_e)$$

Therefore, the voltage gain is

$$A_v = \frac{I_e R_e}{I_e(r'_e + R_e)}$$

The currents cancel, and the base-to-emitter voltage gain expression simplifies to

$$A_v = \frac{R_e}{r'_e + R_e} \qquad \text{(6–22)}$$

FIGURE 6–30

Emitter-follower model for voltage gain derivation.

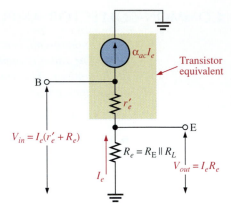

where R_e is the parallel combination of R_E and R_L. If there is no load, then $R_e = R_E$. Notice that the gain is always slightly less than 1. If $R_e \gg r'_e$, then a good approximation is

$$A_v \cong 1$$

Since the output voltage is at the emitter, it is in phase with the base voltage, so there is no inversion from input to output. Because there is no inversion and because the voltage gain is approximately 1, the output voltage closely *follows* the input voltage; thus the term **emitter-follower.**

Input Resistance

The emitter-follower is characterized by a high input resistance; this is what makes it a useful circuit. Because of the high input resistance, it can be used as a buffer to minimize loading effects when a circuit is driving a low resistance load. The derivation of the input resistance looking in at the base of the common-collector amplifier is similar to that for the common-emitter amplifier. In a common-collector circuit, however, the emitter resistor is *never* bypassed because the output is taken across R_e.

$$R_{in(base)} = \frac{V_{in}}{I_{in}} = \frac{V_b}{I_b} = \frac{I_e(r'_e + R_e)}{I_b}$$

Since $I_e \cong I_c = \beta_{ac}I_b$,

$$R_{in(base)} \cong \frac{\beta_{ac}I_b(r'_e + R_e)}{I_b}$$

The I_b terms cancel; therefore,

$$R_{in(base)} \cong \beta_{ac}(r'_e + R_e) \qquad (6\text{–}23)$$

If $R_e \gg r'_e$, then the input resistance at the base is simplified to

$$R_{in(base)} \cong \beta_{ac}R_e$$

The bias resistors in Figure 6–29 appear in parallel with $R_{in(base)}$ looking from the input source, and just as in the common-emitter circuit, the total input resistance is

$$R_{in(tot)} = R_1 \parallel R_2 \parallel R_{in(base)} \qquad \textbf{(6–24)}$$

Output Resistance

With the load removed, the output resistance looking into the emitter of the emitter-follower is approximated as follows:

$$R_{out} \cong \left(\frac{R_s}{\beta_{ac}} \right) \parallel R_E \qquad \textbf{(6–25)}$$

R_s is the resistance of the input source. The derivation of this expression is relatively involved and several simplifying assumptions have been made, as shown in Appendix B. The output resistance is very low, making the emitter-follower useful for driving low resistance loads.

Current Gain

The overall gain for the emitter-follower in Figure 6–29 is I_e/I_{in}. You can calculate I_{in} as $V_{in}/R_{in(tot)}$. If the parallel combination of the voltage-divider bias resistors R_1 and R_2 is much greater than $R_{in(base)}$, then most of the input current goes into the base; thus, the current gain of the amplifier approaches the current gain of the transistor, β_{ac}, which is I_c/I_b. This is because very little signal current is in the bias resistors. Stated concisely, if

$$R_1 \parallel R_2 >> \beta_{ac}R_e$$

then

$$A_i \cong \beta_{ac}$$

Otherwise,

$$A_i = \frac{I_e}{I_{in}} \qquad \textbf{(6–26)}$$

β_{ac} is the maximum achievable current gain in both common-collector and common-emitter amplifiers.

Power Gain

The common-collector power gain is the product of the voltage gain and the current gain. For the emitter-follower, the power gain is approximately equal to the current gain because the voltage gain is approximately one.

$$A_p = A_v A_i$$

Since $A_v \cong 1$,

$$A_p \cong A_i \qquad \textbf{(6–27)}$$

EXAMPLE 6–9

Determine the total input resistance of the emitter-follower in Figure 6–31. Also find the voltage gain, current gain, and power gain. Assume $\beta_{ac} = 175$ and that the capacitive reactances are negligible at the frequency of operation.

FIGURE 6–31

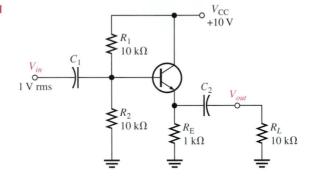

Solution The ac emitter resistance external to the transistor is

$$R_e = R_E \parallel R_L = 1 \text{ k}\Omega \parallel 10 \text{ k}\Omega = 909 \text{ }\Omega$$

The approximate resistance looking in at the base is

$$R_{in(base)} \cong \beta_{ac}R_e = (175)(909 \text{ }\Omega) = 159 \text{ k}\Omega$$

The total input resistance is

$$R_{in(tot)} = R_1 \parallel R_2 \parallel R_{in(base)} = 10 \text{ k}\Omega \parallel 10 \text{ k}\Omega \parallel 159 \text{ k}\Omega = 4.85 \text{ k}\Omega$$

The voltage gain is $A_v \cong 1$. By using r'_e, you can determine a more precise value of A_v if necessary.

$$I_E = \frac{V_E}{R_E} = \frac{4.3 \text{ V}}{1 \text{ k}\Omega} = 4.3 \text{ mA}$$

and

$$r'_e \cong \frac{25 \text{ mV}}{I_E} = \frac{25 \text{ mV}}{4.3 \text{ mA}} = 5.8 \text{ }\Omega$$

So,

$$A_v = \frac{R_e}{r'_e + R_e} = \frac{909 \text{ }\Omega}{914.8 \text{ }\Omega} = 0.994$$

The small difference in A_v as a result of considering r'_e is insignificant in most cases.

The current gain is $A_i = I_e/I_{in}$. The calculations are as follows:

$$I_e = \frac{V_e}{R_e} = \frac{A_vV_b}{R_e} \cong \frac{1 \text{ V}}{909 \text{ }\Omega} = 1.10 \text{ mA}$$

$$I_{in} = \frac{V_{in}}{R_{in}} = \frac{1 \text{ V}}{4.85 \text{ k}\Omega} = 206 \text{ }\mu\text{A}$$

$$A_i = \frac{I_e}{I_{in}} = \frac{1.10 \text{ mA}}{206 \text{ }\mu\text{A}} = 5.34$$

The power gain is

$$A_p \cong A_i = 5.34$$

Related Exercise   Find the total input resistance, voltage gain, current gain, and power gain for the circuit in Figure 6–31 for the following circuit values: $V_{CC} = 12$ V, $R_1 = R_2 = 22$ kΩ, $R_E = 1.8$ kΩ, $R_L = 10$ kΩ, and $\beta_{ac} = 100$.

The Darlington Pair

As you have seen, β_{ac} is a major factor in determining the input resistance of an amplifier. The β_{ac} of the transistor limits the maximum achievable input resistance you can get from a given emitter-follower circuit.

One way to boost input resistance is to use a **darlington pair,** as shown in Figure 6–32. The collectors of two transistors are connected, and the emitter of the first drives the base of the second. This configuration achieves β_{ac} multiplication as shown in the following steps. The emitter current of the first transistor is

$$I_{e1} \cong \beta_{ac1} I_{b1}$$

This emitter current becomes the base current for the second transistor, producing a second emitter current of

$$I_{e2} \cong \beta_{ac2} I_{e1} = \beta_{ac1}\beta_{ac2} I_{b1}$$

Therefore, the effective current gain of the darlington pair is

$$\beta_{ac} = \beta_{ac1}\beta_{ac2}$$

Neglecting r'_e by assuming that it is much smaller than R_E, the input resistance is

$$R_{in} = \beta_{ac1}\beta_{ac2}R_E \qquad\qquad (6\text{–}28)$$

An Application   The emitter-follower can be used as an interface between a circuit with a high output resistance and a low-resistance load. In such an application, the emitter-follower is called a *buffer*.

For example, suppose a common-emitter amplifier with a 1 kΩ collector resistance (output resistance) must drive a low-resistance load such as an 8 Ω low-power speaker. If

FIGURE 6–32
A darlington pair multiplies β_{ac}.

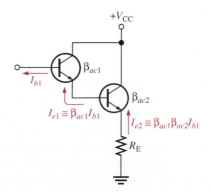

the speaker is capacitively coupled to the output of the amplifier, the 8 Ω load appears—to the ac signal—in parallel with the 1 kΩ collector resistor. This results in an ac collector resistance of

$$R_c = R_C \| R_L = 1 \text{ k}\Omega \| 8 \Omega = 7.94 \Omega$$

Obviously, this is not acceptable because most of the voltage gain is lost ($A_v = R_c/r'_e$). For example, if $r'_e = 5 \Omega$, the voltage gain is reduced from

$$A_v = \frac{R_C}{r'_e} = \frac{1 \text{ k}\Omega}{5 \Omega} = 200$$

with no load to

$$A_v = \frac{R_c}{r'_e} = \frac{7.94 \Omega}{5 \Omega} = 1.59$$

with an 8 Ω speaker load.

An emitter-follower using a darlington pair can be used to interface the amplifier and the speaker, as shown in Figure 6–33.

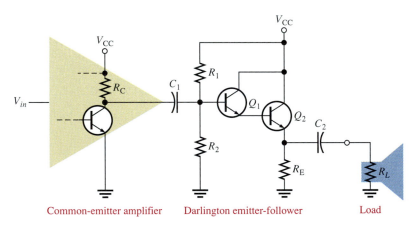

Common-emitter amplifier   Darlington emitter-follower   Load

FIGURE 6–33

A darlington emitter-follower used as a buffer between a common-emitter amplifier and a low-resistance load.

EXAMPLE 6–10

In Figure 6–33 for the common-emitter amplifier, $V_{CC} = 12$ V, $R_C = 1$ kΩ and $r'_e = 5$ Ω. For the darlington emitter-follower, $R_1 = 10$ kΩ, $R_2 = 22$ kΩ, $R_E = 100$ Ω, $V_{CC} = 12$ V, and $\beta_{DC} = \beta_{ac} = 200$ for each transistor.

(a) Determine the voltage gain of the common-emitter amplifier.

(b) Determine the voltage gain of the darlington emitter-follower.

(c) Determine the overall gain and compare to the gain of the common-emitter amplifier driving the speaker directly without the darlington emitter-follower.

Solution

(a) To determine A_v for the common-emitter amplifier, first, find r'_e for the emitter-follower transistor.

$$V_B = \left(\frac{R_2 \parallel \beta_{DC}^2 R_E}{R_1 + R_2 \parallel \beta_{DC}^2 R_E}\right)V_{CC} = \left(\frac{20 \text{ k}\Omega}{30 \text{ k}\Omega}\right)12 \text{ V} = 8.0 \text{ V}$$

$$I_E = \frac{V_E}{R_E} = \frac{V_B - 2V_{BE}}{R_E} = \frac{8.0 \text{ V} - 1.4 \text{ V}}{22 \text{ }\Omega} = \frac{6.6 \text{ V}}{22 \text{ }\Omega} = 300 \text{ mA}$$

$$r'_e = \frac{25 \text{ mV}}{I_E} = \frac{25 \text{ mV}}{300 \text{ mA}} = 83 \text{ m}\Omega$$

Note that R_E must dissipate a power of

$$P_{R_E} = I_E^2 R_E = (300 \text{ mA})^2\, 22 \text{ }\Omega = 1.98 \text{ W}$$

and transistor Q_2 must dissipate

$$P_{Q2} = (V_{CC} - V_E)I_E = (5.4 \text{ V})(300 \text{ mA}) = 1.62 \text{ W}$$

Next, the ac emitter resistance of the darlington emitter-follower is

$$R_e = R_E \parallel R_L = 22 \text{ }\Omega \parallel 8 \text{ }\Omega = 5.87 \text{ }\Omega$$

The total input resistance of the darlington emitter-follower is

$$R_{in(tot)} = R_1 \parallel R_2 \parallel \beta_{ac}^2(r'_e + R_e)$$
$$= 10 \text{ k}\Omega \parallel 22 \text{ k}\Omega \parallel 100^2(83 \text{ m}\Omega + 5.87 \text{ }\Omega) = 6.16 \text{ k}\Omega$$

The effective ac collector resistance of the common-emitter amplifier is

$$R_c = R_C \parallel R_{in(tot)} = 1 \text{ k}\Omega \parallel 6.16 \text{ k}\Omega = 860 \text{ }\Omega$$

The voltage gain of the common-emitter amplifier is

$$A_v = \frac{R_c}{r'_e} = \frac{860 \text{ }\Omega}{5 \text{ }\Omega} = 172$$

(b) The effective ac emitter resistance was found in part (a) to be 5.87 Ω. The voltage gain for the darlington emitter-follower is

$$A_v = \frac{R_e}{r'_e + R_e} = \frac{5.87 \text{ }\Omega}{83 \text{ m}\Omega + 5.87 \text{ }\Omega} = 0.99$$

(c) The overall gain is

$$A'_v = A_{v(EF)}A_{v(CE)} = (0.99)(172) = 170$$

If the common-emitter amplifier drives the speaker directly, the gain is 1.59 (previously calculated).

Related Exercise Using the same circuit values, determine the voltage gain of the common-emitter amplifier in Figure 6–33 if a single transistor is used in the emitter-follower in place of the darlington pair. Assume $\beta_{DC} = \beta_{ac} = 100$. Explain the difference in the voltage gain without the darlington pair.

SUMMARY OF THE COMMON-COLLECTOR AMPLIFIER

CIRCUIT WITH VOLTAGE-DIVIDER BIAS

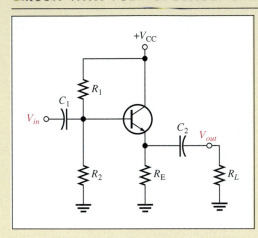

- Input is at the base. Output is at the emitter.
- There is no phase inversion from input to output.
- Input resistance is high. Output resistance is low.
- Maximum voltage gain is 1.
- Collector is at ac ground.
- Capacitors must have a negligible reactance at the frequency of operation.

EQUIVALENT CIRCUITS AND FORMULAS

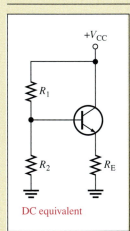

DC equivalent

- DC formulas:

$$V_B = \left(\frac{R_2 \parallel \beta_{DC} R_E}{R_1 + R_2 \parallel \beta_{DC} R_E} \right) V_{CC}$$

$$V_E = V_B - V_{BE}$$

$$I_E = \frac{V_E}{R_E}$$

$$V_C = V_{CC}$$

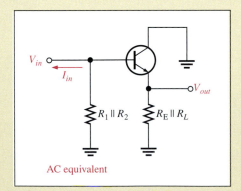

AC equivalent

- AC formulas:

$$r'_e = \frac{25 \text{ mV}}{I_E}$$

$$R_{in(base)} = \beta_{ac}(r'_e + R_e)$$

$$R_{out} = \left(\frac{R_s}{\beta_{ac}} \right) \parallel R_E$$

$$A_v = \frac{R_e}{r'_e + R_e}$$

$$A_i = \frac{I_e}{I_{in}}$$

$$A_p = A_i$$

SECTION 6–4 REVIEW

1. What is a common-collector amplifier called?
2. What is the ideal maximum voltage gain of a common-collector amplifier?
3. What characteristic of the common-collector amplifier makes it a useful circuit?

6–5 ▪ COMMON-BASE AMPLIFIERS

The common-base (CB) amplifier provides high voltage gain with a maximum current gain of 1. Since it has a low input resistance, the CB amplifier is the most appropriate type for certain applications where sources tend to have very low-resistance outputs.

After completing this section, you should be able to

▪ **Understand and analyze the operation of common-base amplifiers**
  ☐ Represent a CB amplifier by its dc and ac equivalent circuits
  ☐ Analyze the dc and ac operation of a CB amplifier
  ☐ Determine the voltage gain
  ☐ Determine the input resistance
  ☐ Determine the output resistance
  ☐ Determine the current gain
  ☐ Determine the power gain

A typical common-base amplifier is shown in Figure 6–34. The base is the common terminal and is at ac ground because of capacitor C_2. The input signal is applied at the emitter. The output is capacitively coupled from the collector to a load resistor.

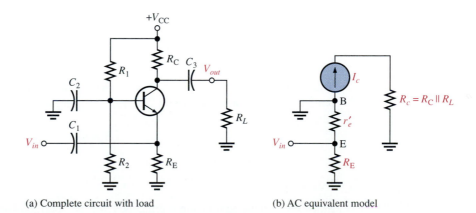

(a) Complete circuit with load   (b) AC equivalent model

FIGURE 6–34

Common-base amplifier with voltage-divider bias.

Voltage Gain

The voltage gain from emitter to collector is developed as follows ($V_{in} = V_e$, $V_{out} = V_c$).

$$A_v = \frac{V_{out}}{V_{in}} = \frac{V_c}{V_e} = \frac{I_c R_c}{I_e(r_e' \parallel R_E)} \cong \frac{I_e R_c}{I_e(r_e' \parallel R_E)}$$

If $R_E \gg r_e'$, then

$$A_v \cong \frac{R_c}{r_e'} \qquad (6\text{–}29)$$

where $R_c = R_C \parallel R_L$. Notice that the gain expression is the same as for the common-emitter amplifier. However, there is no phase inversion from emitter to collector.

Input Resistance

The resistance looking in at the emitter is

$$R_{in(emitter)} = \frac{V_{in}}{I_{in}} = \frac{V_e}{I_e} = \frac{I_e(r_e' \parallel R_E)}{I_e}$$

If $R_E \gg r_e'$, then

$$R_{in(emitter)} \cong r_e' \qquad (6\text{–}30)$$

R_E is typically much greater than r_e', so the assumption that $r_e' \parallel R_E \cong r_e'$ is usually valid.

Output Resistance

Looking into the collector, the ac collector resistance, r_c', appears in parallel with R_C. As you have previously seen in connection with the CE amplifier, r_c' is typically much larger than R_C, so good approximation for the output resistance is

$$R_{out} \cong R_C \qquad (6\text{–}31)$$

Current Gain

The current gain is the output current divided by the input current. I_c is the ac output current, and I_e is the ac input current. Since $I_c \cong I_e$, the current gain is approximately 1.

$$A_i \cong 1 \qquad (6\text{–}32)$$

Power Gain

Since the current gain is approximately 1 for the common-base amplifier and $A_p = A_v A_i$, the power gain is approximately equal to the voltage gain.

$$A_p \cong A_v \qquad (6\text{–}33)$$

EXAMPLE 6–11

Find the input resistance, voltage gain, current gain, and power gain for the amplifier in Figure 6–35. $\beta_{DC} = 250$.

FIGURE 6–35

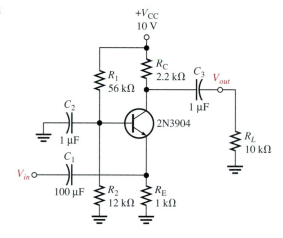

Solution First, find I_E so that you can determine r'_e. Then $R_{in} \cong r'_e$. Since $\beta_{DC}R_E \gg R_2$, then

$$V_B \cong \left(\frac{R_2}{R_1 + R_2} \right) V_{CC} = \left(\frac{12 \text{ k}\Omega}{68 \text{ k}\Omega} \right) 10 \text{ V} = 1.76 \text{ V}$$

$$V_E = V_B - 0.7 \text{ V} = 1.76 \text{ V} - 0.7 \text{ V} = 1.06 \text{ V}$$

$$I_E = \frac{V_E}{R_E} = \frac{1.06 \text{ V}}{1 \text{ k}\Omega} = 1.06 \text{ mA}$$

Therefore,

$$R_{in} \cong r'_e = \frac{25 \text{ mV}}{1.06 \text{ mA}} = 23.6 \text{ }\Omega$$

The voltage gain is determined as follows:

$$R_c = R_C \| R_L = 2.2 \text{ k}\Omega \| 10 \text{ k}\Omega = 1.8 \text{ k}\Omega$$

$$A_v = \frac{R_c}{r'_e} = \frac{1.8 \text{ k}\Omega}{23.6 \text{ }\Omega} = 76.3$$

Also, $A_i \cong 1$ and $A_p \cong A_v = 76.3$.

Related Exercise Find A_v in Figure 6–35 if $\beta_{DC} = 50$.

SUMMARY OF THE COMMON-BASE AMPLIFIER

CIRCUIT WITH VOLTAGE-DIVIDER BIAS

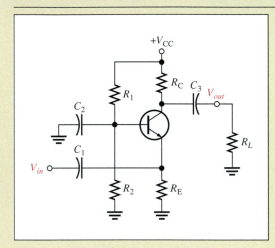

■ Input is at the emitter. Output is at the collector.

■ There is no phase inversion from input to output.

■ Input resistance is low. Output resistance is high.

■ Maximum current gain is 1.

■ Base is at ac ground.

EQUIVALENT CIRCUITS AND FORMULAS

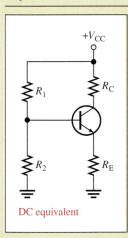

DC equivalent

■ DC formulas:

$$V_B = \left(\frac{R_2 \parallel \beta_{DC}R_E}{R_1 + R_2 \parallel \beta_{DC}R_E} \right) V_{CC}$$

$$V_E = V_B - V_{BE}$$

$$I_E = \frac{V_E}{R_E}$$

$$V_C = V_{CC} - I_C R_C$$

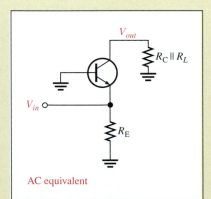

AC equivalent

■ AC formulas:

$$r'_e = \frac{25 \text{ mV}}{I_E}$$

$$R_{in(emitter)} \cong r'_e$$

$$R_{out} \cong R_C$$

$$A_v \cong \frac{R_c}{r'_e}$$

$$A_i \cong 1$$

$$A_p \cong A_v$$

SECTION 6–5 REVIEW

1. Can the same voltage gain be achieved with a common-base as with a common-emitter amplifier?

2. Does the common-base amplifier have a low or a high input resistance?

3. What is the maximum current gain in a common-base amplifier?

6–6 ▪ MULTISTAGE AMPLIFIERS

Several amplifiers can be connected in a cascaded arrangement with the output of one amplifier driving the input of the next. Each amplifier in the cascaded arrangement is known as a stage. The basic purpose of a multistage arrangement is to increase the overall voltage gain.

After completing this section, you should be able to

▪ **Discuss multistage amplifiers and analyze their operation**
  ☐ Determine multistage voltage gain
  ☐ Express the voltage gain in decibels (dB)
  ☐ Determine the loading effects in a multistage amplifier
  ☐ Analyze each stage to determine the overall voltage gain
  ☐ Discuss capacitive coupling in multistage amplifiers
  ☐ Describe a basic direct-coupled multistage amplifier
  ☐ Describe a basic transformer-coupled multistage amplifier

Multistage Voltage Gain

The overall voltage gain, A'_v, of **cascaded** amplifiers, as shown in Figure 6–36, is the product of the individual gains.

$$A'_v = A_{v1}A_{v2}A_{v3} \cdots A_{vn} \qquad \text{(6–34)}$$

where n is the number of **stages.**

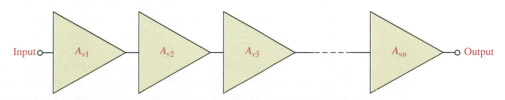

FIGURE 6–36
Cascaded amplifiers. Each triangular symbol represents a separate amplifier.

Voltage Gain Expressed in Decibels

Amplifier voltage gain is often expressed in **decibels** (dB) as follows:

$$A_{v(dB)} = 20 \log A_v \qquad (6\text{–}35)$$

This is particularly useful in **multistage** systems because the overall voltage gain in dB is the *sum* of the individual gains in dB.

$$A'_{v(dB)} = A_{v1(dB)} + A_{v2(dB)} + \cdots + A_{vn(dB)} \qquad (6\text{–}36)$$

EXAMPLE 6–12

A certain cascaded amplifier arrangement has the following voltage gains: $A_{v1} = 10$, $A_{V2} = 15$, and $A_{v3} = 20$. What is the overall voltage gain? Also express each gain in decibels (dB) and determine the total voltage gain in dB.

Solution

$$A'_v = A_{v1}A_{v2}A_{v3} = (10)(15)(20) = 3000$$
$$A_{v1(dB)} = 20 \log 10 = 20.0 \text{ dB}$$
$$A_{v2(dB)} = 20 \log 15 = 23.5 \text{ dB}$$
$$A_{v3(dB)} = 20 \log 20 = 26.0 \text{ dB}$$
$$A'_{v(dB)} = 20 \text{ dB} + 23.5 \text{ dB} + 26.0 \text{ dB} = 69.5 \text{ dB}$$

Related Exercise   In a certain multistage amplifier, the individual stages have the following voltage gains: $A_{v1} = 25$, $A_{v2} = 5$, and $A_{v3} = 12$. What is the overall gain? Express each gain in dB and determine the total voltage gain in dB.

Multistage Amplifier Analysis

For purposes of illustration, we will use the two-stage capacitively coupled amplifier in Figure 6–37. Notice that both stages are identical common-emitter amplifiers with the output of the first stage capacitively coupled to the input of the second stage. Capacitive coupling prevents the dc bias of one stage from affecting that of the other but allows the ac signal to pass without attenuation because $X_C \cong 0 \ \Omega$ at the frequency of operation. Notice, also, that the transistors are labeled Q_1 and Q_2.

Loading Effects   In determining the voltage gain of the first stage, you must consider the loading effect of the second stage. Because the coupling capacitor C_3 effectively appears as a short at the signal frequency, the total input resistance of the second stage presents an ac load to the first stage. Looking from the collector of Q_1, the two biasing resistors in the second stage, R_5 and R_6, appear in parallel with the input resistance at the base of Q_2. In other words, the signal at the collector of Q_1 "sees" R_3 and R_5, R_6 and $R_{in(base2)}$ of the second stage all in parallel to ac ground. Thus, the effective ac collector resistance of Q_1 is the total of all these resistances in parallel, as Figure 6–38 illustrates. The voltage gain of the first stage is reduced by the loading of the second stage, because the effective ac collector resistance of the first stage is less than the actual value of its collector resistor, R_3. Remember that $A_v = R_c/r'_e$.

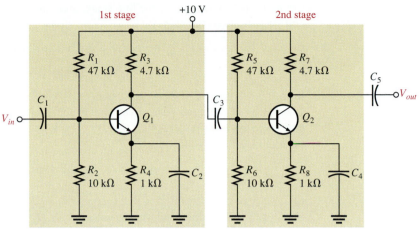

$$\beta_{DC} = \beta_{ac} = 150, \text{ for } Q_1 \text{ and } Q_2$$

FIGURE 6–37
A two-stage common-emitter amplifier.

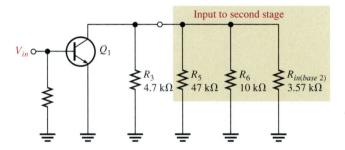

FIGURE 6–38
AC equivalent of first stage in Figure 6–37, showing loading from second stage.

Voltage Gain of the First Stage   The ac collector resistance of the first stage is

$$R_{c1} = R_3 \parallel R_5 \parallel R_6 \parallel R_{in(base2)}$$

Remember that lowercase italic subscripts denote ac quantities such as for R_c.

You can verify that $I_E = 1.05$ mA, $r'_e = 23.8\ \Omega$, and $R_{in(base2)} = 3.57$ kΩ. The effective ac collector resistance of the first stage is as follows:

$$R_{c1} = 4.7 \text{ k}\Omega \parallel 47 \text{ k}\Omega \parallel 10 \text{ k}\Omega \parallel 3.57 \text{ k}\Omega = 1.63 \text{ k}\Omega$$

Therefore, the base-to-collector voltage gain of the first stage is

$$A_{v1} = \frac{R_{c1}}{r'_e} = \frac{1.63 \text{ k}\Omega}{23.8\ \Omega} = 68.5$$

Voltage Gain of the Second Stage   The second stage has no load resistor, so the ac collector resistance is R_7, and the gain is

$$A_{v2} = \frac{R_7}{r'_e} = \frac{4.7 \text{ k}\Omega}{23.8\ \Omega} = 197$$

Compare this to the gain of the first stage, and notice how much the loading from the second stage reduced the gain.

Overall Voltage Gain   The overall amplifier gain with no load on the output is

$$A'_v = A_{v1}A_{v2} = (68.5)(197) \cong 13,495$$

If an input signal of 100 μV, for example, is applied to the first stage and if there is no attenuation in the input base circuit due to the source resistance, an output from the second stage of (100 μV)(13,495) ≅ 1.35 V will result. The overall voltage gain can be expressed in dB as follows:

$$A'_{v(dB)} = 20 \log (13,495) = 82.6 \text{ dB}$$

DC Voltages in the Capacitively Coupled Multistage Amplifier   Since both stages in Figure 6–37 are identical, the dc voltages for Q_1 and Q_2 are the same. Since $\beta_{DC}R_4 \gg R_2$ and $\beta_{DC}R_8 \gg R_6$, the dc base voltage for Q_1 and Q_2 is

$$V_B \cong \left(\frac{R_2}{R_1 + R_2} \right) 10 \text{ V} = \left(\frac{10 \text{ k}\Omega}{57 \text{ k}\Omega} \right) 10 \text{ V} = 1.75 \text{ V}$$

The dc emitter and collector voltages are as follows:

$$V_E = V_B - 0.7 \text{ V} = 1.05 \text{ V}$$

$$I_E = \frac{V_E}{R_4} = \frac{1.05 \text{ V}}{1 \text{ k}\Omega} = 1.05 \text{ mA}$$

$$I_C \cong I_E = 1.05 \text{ mA}$$

$$V_C = V_{CC} - I_C R_3 = 10 \text{ V} - (1.05 \text{ mA})(4.7 \text{ k}\Omega) = 5.07 \text{ V}$$

Direct-Coupled Multistage Amplifiers

A basic two-stage, direct-coupled amplifier is shown in Figure 6–39. Notice that there are no coupling or bypass capacitors in this circuit. The dc collector voltage of the first stage provides the base-bias voltage for the second stage. Because of the direct coupling, this type of amplifier has a better low-frequency response than the capacitively coupled type

FIGURE 6–39

A basic two-stage direct-coupled amplifier.

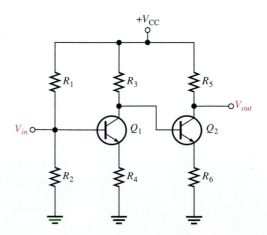

in which the reactance of coupling and bypass capacitors at very low frequencies may become excessive. The increased reactance at lower frequencies produces signal loss and gain reduction.

Direct-coupled amplifiers, on the other hand, can be used to amplify low frequencies all the way down to dc (0 Hz) without loss of voltage gain because there are no capacitive reactances in the circuit. The disadvantage of direct-coupled amplifiers is that small changes in the dc bias voltages from temperature effects or power-supply variation are amplified by the succeeding stages, which can result in a significant drift in the dc levels throughout the circuit.

Transformer-Coupled Multistage Amplifiers

A basic transformer-coupled two-stage amplifier is shown in Figure 6–40. Transformer coupling is often used in high-frequency amplifiers such as those in the RF (radio frequency) and IF (intermediate frequency) sections of radio and TV receivers. At lower frequency ranges such as **audio,** the size of transformers is usually prohibitive. Capacitors are usually connected across the primary windings of the transformers to obtain resonance and increased selectivity for the band of frequencies to be amplified.

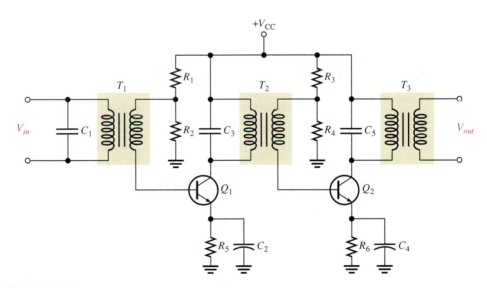

FIGURE 6–40
A basic two-stage transformer-coupled amplifier.

SECTION 6–6 REVIEW

1. What does the term *stage* mean?
2. How is the overall voltage gain of a multistage amplifier determined?
3. Express a voltage gain of 500 in dB.
4. Discuss a disadvantage of a capacitively coupled amplifier.

6–7 ■ TROUBLESHOOTING

In working with any circuit, you must first know how it is supposed to work before you can troubleshoot it for a failure. The two-stage capacitively coupled amplifier discussed in Section 6–6 is used to illustrate a typical troubleshooting procedure.

After completing this section, you should be able to

■ **Troubleshoot amplifier circuits**
   ☐ Discuss the complete troubleshooting process
   ☐ Apply the troubleshooting process to a two-stage amplifier
   ☐ Use the signal-tracing method
   ☐ Apply fault analysis

At this point, you should review the general troubleshooting techniques discussed in Chapter 2. The two-stage common-emitter amplifier that was discussed in the last section is used to illustrate basic multistage amplifier troubleshooting.

When you are faced with having to troubleshoot a circuit, the first thing you must have is a schematic with the proper dc and signal voltages labeled. You must know what the correct voltages in the circuit should be before you can identify an incorrect voltage. Schematics of some circuits are available with voltages indicated at certain points. If this is not the case, you must use your knowledge of the circuit operation to determine the correct voltages. Figure 6–41 is the schematic for the two-stage amplifier that was analyzed in the previous section. The correct voltages are indicated at each point.

The Complete Troubleshooting Process

Beginning with a malfunctioning or nonfunctioning circuit or system, a typical troubleshooting procedure may go as follows:

1. Identify the symptom(s).

2. Perform a power check.

3. Perform a sensory check.

4. Apply a signal-tracing technique to isolate the fault to a single circuit.

5. Apply fault analysis to isolate the fault further to a single component or group of components.

6. Use replacement or repair to fix the problem.

Applying the Troubleshooting Process to a Two-Stage Amplifier   To determine the faulty component in a multistage amplifer, use the general six-step troubleshooting procedure which is illustrated as follows.

Step 1: *Check the input and output voltages.* Assume the measurements indicate that the input signal voltage is correct. However, there is no output signal voltage or the output signal voltage is much less than it should be, as shown by the diagram in Figure 6–42.

Step 2: *Perform a power check.* Assume the dc supply voltage is correct as indicated in Figure 6–42.

Step 3: *Perform a sensory check.* Assume there are no detectable signs of a fault.

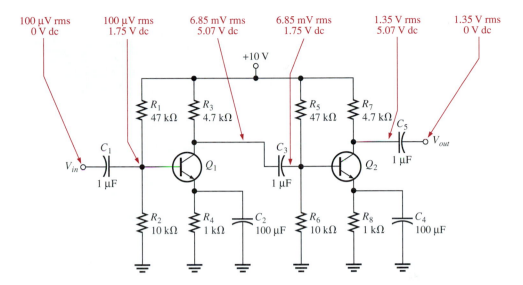

FIGURE 6–41
Two-stage common-emitter amplifier with correct voltages indicated.

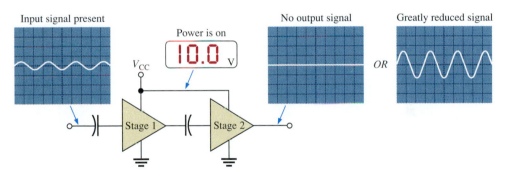

FIGURE 6–42
Initial check of a faulty two-stage amplifier.

Step 4: *Apply the midpoint method of signal tracing.* Check the voltages at the output of the first stage. No signal voltage or a reduced signal voltage indicates that the problem is in the first stage. An incorrect dc voltage also indicates a first stage problem. If the signal voltage and the dc voltage are correct at the output of the first stage, the problem is in the second stage. After this check, you have narrowed the problem to one of the two stages. This step is illustrated in Figure 6–43.

Step 5: *Apply fault analysis.* Focus on the faulty stage and determine the component failure that can produce the incorrect output.

Symptom: DC voltages incorrect.
Faults: A failure of any resistor or the transistor will produce an incorrect dc bias voltage. A leaky bypass or coupling capacitor will also affect the dc bias voltages. Further measurements in the stage are necessary to isolate the faulty component.

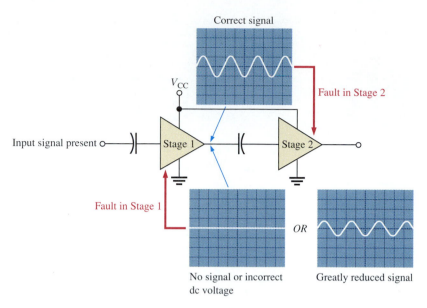

FIGURE 6–43

Midpoint signal tracing isolates the faulty stage.

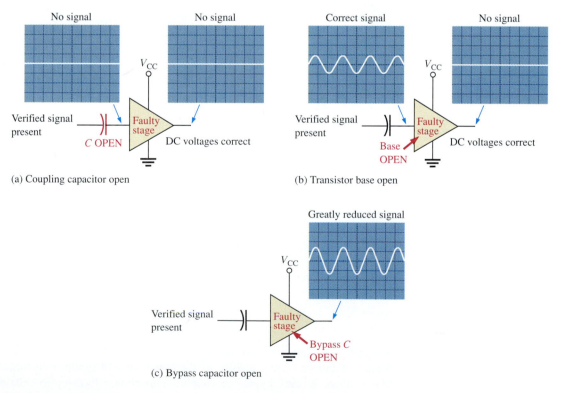

FIGURE 6–44

Troubleshooting a faulty stage.

Incorrect ac voltages and the most likely fault(s) are illustrated in Figure 6–44 as follows:

(a) *Symptom 1:* Signal voltage at output missing; dc voltage correct.
 Symptom 2: Signal voltage at base missing; dc voltage correct.
 Fault: Input coupling capacitor open. This prevents the signal from getting to the base.

(b) *Symptom:* Correct signal at base but no output signal.
 Fault: Transistor base open.

(c) *Symptom:* Signal voltage at output greatly reduced; dc voltage correct.
 Fault: Bypass capacitor open.

Step 6: *Replace or repair.* With the power turned off, replace the defective component or repair the defective connection. Turn on the power, and check for proper operation.

EXAMPLE 6–13

The two-stage amplifier in Figure 6–41 has malfunctioned. Specify the step-by-step troubleshooting procedure for an assumed fault.

Solution The troubleshooting procedure for a certain fault scenario is as follows:

Step 1: There is a verified input signal voltage, but no output signal voltage is measured.

Step 2: There is power to the circuit as indicated by a correct V_{CC} measurement.

Step 3: Assume there are no visual or other indications of a problem such as a charred resistor, solder splash, wire clipping, broken connection, or extremely hot component.

Step 4: The signal voltage and the dc voltage at the collector of Q_1 are correct. This means that the problem is in the second stage or the coupling capacitor C_3 between the stages.

Step 5: The correct signal voltage and dc bias voltage are measured at the base of Q_2. This eliminates the possibility of a fault in C_3 or the second stage bias circuit.

 The collector of Q_2 is at 10 V and there is no signal voltage. This measurement, made directly on the transistor collector, indicates that either the collector is shorted to V_{CC} or the transistor is internally open.

 A visual inspection reveals no physical shorts between V_{CC} and the collector connection. It is unlikely that the collector resistor R_7 is shorted but to verify, turn off the power and use an ohmmeter to check.

 The possibility of a short is eliminated by the ohmmeter check. The other possible faults are (a) transistor Q_2 internally open or (b) emitter resistor or connection open. Use a transistor tester and/or ohmmeter to check each of these possible faults with power off.

Step 6: Replace the faulty component or repair open connection and retest the circuit for proper operation.

Related Exercise Determine the possible fault(s) if, in Step 5, you find no signal voltage at the base of Q_2 but the dc voltage is correct.

1. If C_4 in Figure 6–41 were open, how would the output signal be affected? How would the dc level at the collector of Q_2 be affected?

2. If R_5 in Figure 6–41 were open, how would the output signal be affected?

3. If the coupling capacitor C_3 in Figure 6–41 shorted out, would any of the dc voltages in the amplifier be changed? If so, which ones?

6–8 ■ SYSTEM APPLICATION

Your company develops, manufactures, and markets public address and voice paging systems for use in service garages, supermarkets, department stores, and the like. You have been assigned responsibility for the most basic product in this line, which is a simple preamplifier with a microphone input followed by a power amplifier and a speaker. You will apply the knowledge you have gained in this chapter in completing your assignment.

The Public Address System

This system includes a magnetic microphone, a two-stage preamplifier, a power amplifier, and a speaker. This design is the most basic public address system that your company plans to produce and market and is intended for limited application in small businesses for paging employees or making announcements.

A block diagram of the system is shown in Figure 6–45. The microphone input goes to the preamplifier, which has an adjustable volume control. The output of the pre-amplifier goes to a power amplifier that drives a horn speaker. A dc power supply provides the required dc voltage for the circuits and is included as part of the system.

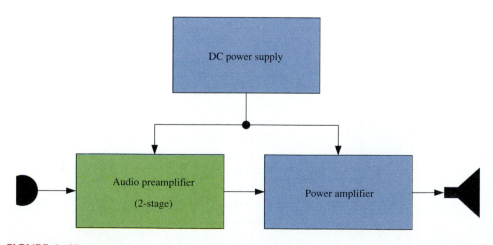

FIGURE 6–45
Basic public address/paging system block diagram.

The System Requirements

☐ The magnetic microphone has an output impedance of 30 Ω and produces an average output voltage of 2 mV rms in response to a typical speaking voice. The frequency response is from 100 Hz to 8,000 Hz.

☐ The speaker has an impedance of 8 Ω and can handle up to 15 W. The frequency response is from 100 Hz to 7500 Hz.

☐ The frequency range of the amplifier is to be at least 300 Hz to 5000 Hz.

Basic Operation The schematic of the two-stage preamplifier is shown in Figure 6–46. The first stage is a common-base amplifier, and the second stage is a common-emitter amplifier with a swamping resistor (R_8) and an emitter resistor (R_9) with variable bypass for volume control.

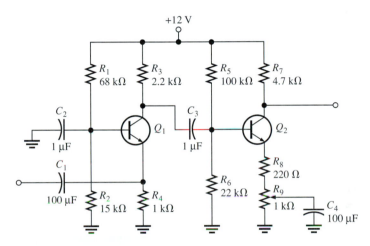

FIGURE 6–46
Schematic of the preamplifier circuit. Both transistors are 2N3904.

This particular design uses a common-base amplifier as the input stage. The second stage common-emitter amplifier has a potentiometer in the emitter to provide an adjustable voltage gain for volume control.

The output of the amplifier will connect to a power amplifier to be added later in the development process (Chapter 7). For now, the assignment is to check out the preamplifier which is in a preproduction phase.

Analysis of the Preamplifier Circuit

☐ Determine the input resistance of the first stage.

☐ For each milliwatt of power produced by the microphone, how much is transferred to the amplifier?

☐ Determine the dc voltages at the base, collector, and emitter of Q_1 and Q_2.

☐ Determine the overall voltage gain of the amplifier.

☐ Calculate the current that is drawn from the dc power supply when there is no input signal to the amplifier.

☐ Specify the minimum standard power rating for each resistor.

☐ *Optional analysis:* Determine the lowest frequency that the circuit can amplify without the gain decreasing more than 3 dB. The coupling capacitors and the input resistances of the amplifier stages determine the low-frequency response.

The Power Supply Circuit

The dc power supply circuit is the same as the one developed in the system application for Chapter 2 except that it must produce a regulated voltage of approximately +12 V. Specify the changes necessary to adapt the power supply to this system application.

The Printed Circuit Board

☐ Check out the printed circuit board in Figure 6–47 to verify that it is correct according to the schematic.

☐ Label a copy of the board with component and input/output designations in agreement with the schematic.

Test Procedure

☐ Develop a step-by-step set of instructions on how to check the preamplifier circuit board for proper operation at a test frequency of 5 kHz using the test points (circled numbers) indicated in the test bench setup of Figure 6–48. A function generator is the

FIGURE 6–47
Preamplifier circuit board.

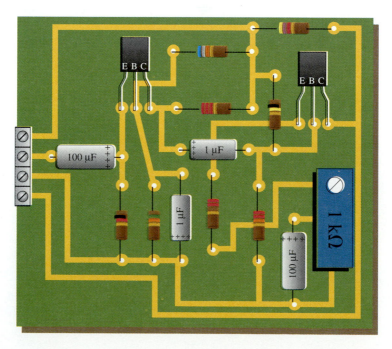

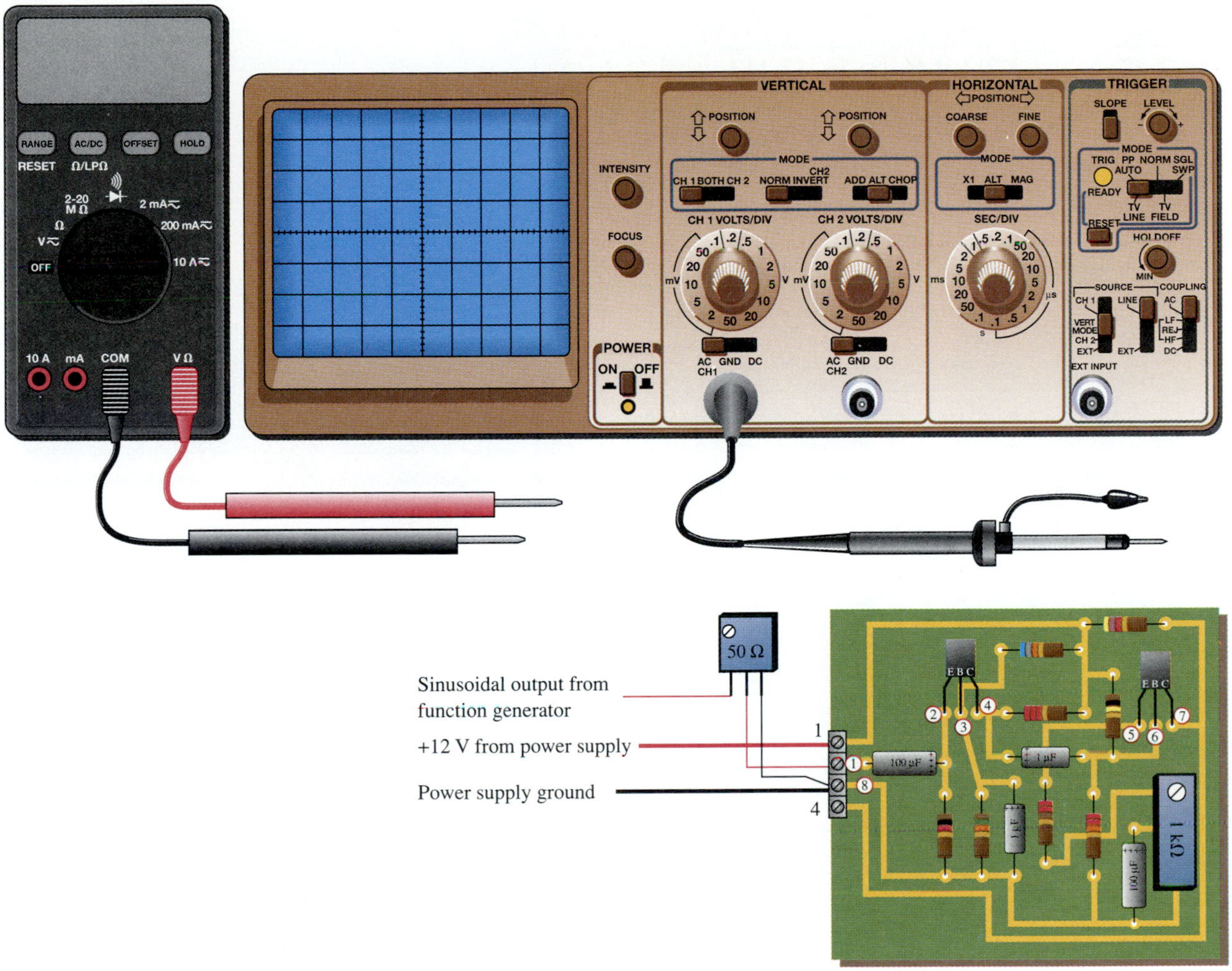

FIGURE 6–48
Test bench setup for the preamplifier board.

signal source with a potentiometer used as an attenuator, if necessary, to reduce the signal to a value small enough to simulate the microphone input.

☐ Specify voltage values for all the measurements to be made. Provide a fault analysis for all possible component failures.

Troubleshooting

Several prototype boards have been assembled and are ready for testing. Based on the sequence of test bench measurements for each board indicated in Figure 6–49, determine the most likely fault in each case. The circled numbers indicate test point connections to the circuit board.

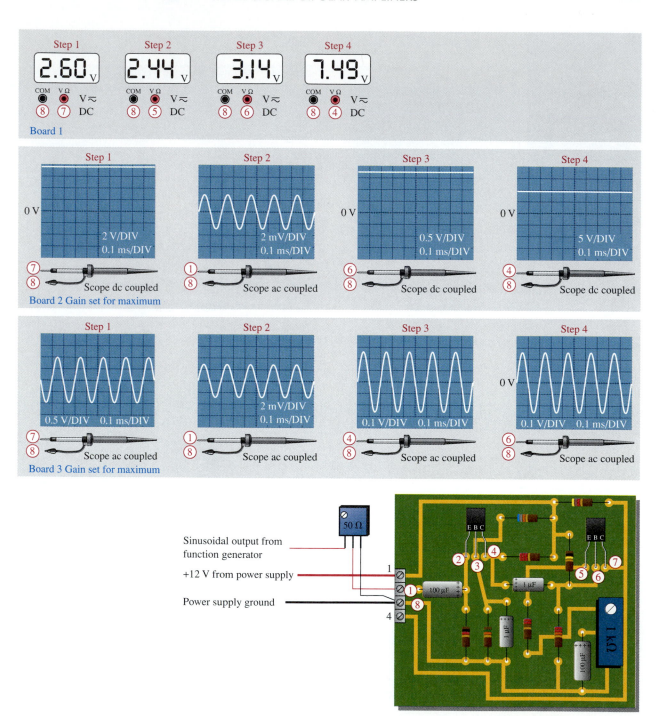

FIGURE 6–49

Test results for three faulty circuit boards.

Final Report

Submit a final written report on the audio preamplifier circuit board using an organized format that includes the following:

1. A physical description of the circuits.

2. A discussion of the operation of the circuit.

3. A list of the specifications.

4. A list of parts with part numbers if available.

5. A list of the types of problems on the three faulty circuit boards.

6. A complete description of how you determined the problem on each of the faulty circuit boards.

■ **CHAPTER SUMMARY**

■ A small-signal amplifier uses only a small portion of its load line under signal conditions.

■ h parameters are important to technicians and technologists because manufacturers' data sheets specify transistors using h parameters.

■ r parameters are easily identifiable and applicable with a transistor's circuit operation.

■ A common-emitter amplifier has good voltage, current, and power gains, but a relatively low input resistance.

■ A common-collector amplifier has high input resistance and good current gain, but its voltage gain is approximately one.

■ The common-base amplifier has a good voltage gain, but it has a very low input resistance and its current gain is approximately one.

■ A darlington pair provides beta multiplication for increased input resistance.

■ The total gain of a multistage amplifier is the product of the individual gains (sum of dB gains).

■ Single-stage amplifiers can be connected in sequence with various coupling methods to form multistage amplifiers.

■ Common-emitter, common-collector, and common-base amplifier configurations are summarized in Table 6–5.

TABLE 6–5

Relative comparison of amplifier configurations. The current gains and the input and output resistances are the approximate maximum achievable values, with the bias resistors neglected.

	CE	CC	CB
Voltage gain, A_v	High R_C/r'_e	Low $\cong 1$	High R_C/r'_e
Current gain, $A_{i(max)}$	High β_{ac}	High β_{ac}	Low $\cong 1$
Power gain, A_p	Very high $A_i A_v$	High $\cong A_i$	High $\cong A_v$
Input resistance, $R_{in(max)}$	Low $\beta_{ac} r'_e$	High $\beta_{ac} R_E$	Very low r'_e
Output resistance, R_{out}	High R_C	Very low $(R_s/\beta_{ac}) \parallel R_E$	High R_C

■ GLOSSARY

Attenuation   The reduction in the level of power, current, or voltage.

Audio   Related to the range of sound waves that can be heard by the human ear.

Cascade   An arrangement of circuits in which the output of one circuit becomes the input to the next.

Common-base (CB)   A BJT amplifier configuration in which the base is the common terminal to an ac signal or ground.

Common-collector (CC)   A BJT amplifier configuration in which the collector is the common terminal to an ac signal or ground.

Common-emitter (CE)   A BJT amplifier configuration in which the emitter is the common terminal to an ac signal or ground.

Darlington pair   A configuration of two transistors in which the collectors are connected and the emitter of the first drives the base of the second to achieve beta multiplication.

Decibel (dB)   The unit of the logarithmic expression of a ratio, such as power or voltage.

Emitter-follower   A popular term for a common-collector amplifier.

Load   The amount of current drawn from the output of a circuit through a load resistance.

Multistage   Characterized by having more than one stage; a cascaded arrangement of two or more amplifiers.

Stability   A measure of how well an amplifier maintains its design values (Q-point, gain, etc.) over changes in temperature.

Stage   One of the amplifier circuits in a multistage configuration.

■ FORMULAS

h Parameters

(6–1) $h_{ie} = \dfrac{V_b}{I_b}$ Input resistance, CE

(6–2) $h_{re} = \dfrac{V_b}{V_c}$ Reverse voltage ratio, CE

(6–3) $h_{fe} = \dfrac{I_c}{I_b}$ Forward current gain, CE

(6–4) $h_{oe} = \dfrac{I_c}{V_c}$ Output admittance, CE

Conversions from *h* Parameters to *r* Parameters

(6–5) $\alpha_{ac} = h_{fb}$ AC alpha

(6–6) $\beta_{ac} = h_{fe}$ AC beta

(6–7) $r'_e = \dfrac{h_{re}}{h_{oe}}$ Internal ac emitter resistance

(6–8) $r'_c = \dfrac{h_{re} + 1}{h_{oe}}$ Internal ac collector resistance

(6–9) $r'_b = h_{ie} - \dfrac{h_{re}}{h_{oe}}(1 + h_{fe})$ Internal ac base resistance

(6–10) $r'_e \cong \dfrac{25\ \text{mV}}{I_E}$ Internal ac emitter resistance

Common-Emitter

(6–11)	$R_{in(base)} = \beta_{ac} r'_e$	Input resistance at base
(6–12)	$R_{in(tot)} = R_1 \| R_2 \| R_{in(base)}$	Total amplifier input resistance, voltage-divider bias
(6–13)	$R_{out} \cong R_{\mathrm{C}}$	Output resistance
(6–14)	$A_v = \dfrac{R_{\mathrm{C}}}{r'_e}$	Voltage gain, base-to-collector, unloaded
(6–15)	$A'_v = \left(\dfrac{V_b}{V_s} \right) A_v$	Overall voltage gain
(6–16)	$A_v = \dfrac{R_{\mathrm{C}}}{r'_e + R_{\mathrm{E}}}$	Voltage gain without bypass capacitor
(6–17)	$A_v = \dfrac{R_c}{r'_e}$	Voltage gain, base-to-collector, loaded, bypassed R_{E}
(6–18)	$A_v \cong \dfrac{R_{\mathrm{C}}}{R_{\mathrm{E1}}}$	Voltage gain, swamped amplifier
(6–19)	$R_{in(base)} = \beta_{ac} (r'_e + R_{\mathrm{E1}})$	Input resistance at base, swamped amplifier
(6–20)	$A_i = \dfrac{I_c}{I_s}$	Current gain
(6–21)	$A_p = A'_v A_i$	Power gain

Common-Collector

(6–22)	$A_v = \dfrac{R_e}{r'_e + R_e}$	Voltage gain, base-to-emitter
(6–23)	$R_{in(base)} \cong \beta_{ac}(r'_e + R_e)$	Input resistance at base, loaded
(6–24)	$R_{in(tot)} = R_1 \| R_2 \| R_{in(base)}$	Total amplifier input resistance, voltage-divider bias
(6–25)	$R_{out} \cong \left(\dfrac{R_s}{\beta_{ac}} \right) \| R_{\mathrm{E}}$	Output resistance
(6–26)	$A_i = \dfrac{I_e}{I_{in}}$	Current gain
(6–27)	$A_p \cong A_i$	Power gain
(6–28)	$R_{in} = \beta_{ac1} \beta_{ac2} R_{\mathrm{E}}$	Input resistance, darlington pair

Common-Base

(6–29)	$A_v \cong \dfrac{R_c}{r'_e}$	Voltage gain, emitter-to-collector
(6–30)	$R_{in(emitter)} \cong r'_e$	Input resistance at emitter
(6–31)	$R_{out} \cong R_{\mathrm{C}}$	Output resistance
(6–32)	$A_i \cong 1$	Current gain
(6–33)	$A_p \cong A_v$	Power gain

Multistage Amplifier

(6–34)	$A'_v = A_{v1} A_{v2} A_{v3} \cdots A_{vn}$	Overall voltage gain
(6–35)	$A_{v(\mathbf{dB})} = 20 \log A_v$	Voltage gain expressed in dB
(6–36)	$A'_{v(\mathbf{dB})} = A_{v1(\mathbf{dB})} + A_{v2(\mathbf{dB})} + \cdots + A_{vn(\mathbf{dB})}$	Overall voltage gain in dB

▪ **SELF-TEST**

1. A small-signal amplifier
   (a) uses only a small portion of its load line
   (b) always has an output signal in the mV range
   (c) goes into saturation once on each input cycle
   (d) is always a common-emitter amplifier

2. The parameter h_{fe} corresponds to
   (a) β_{DC} (b) β_{ac} (c) r'_e (d) r'_c

3. If the dc emitter current in a certain transistor amplifier is 3 mA, the approximate value of r'_e is
   (a) 3 kΩ   (b) 3 Ω   (c) 8.33 Ω   (d) 0.33 kΩ

4. A certain common-emitter amplifier has a voltage gain of 100. If the emitter bypass capacitor is removed,
   (a) the circuit will become unstable   (b) the voltage gain will decrease
   (c) the voltage gain will increase   (d) the Q-point will shift

5. For a common-collector amplifier, $R_E = 100$ Ω, $r'_e = 10$ Ω, and $\beta_{ac} = 150$. The ac input resistance at the base is
   (a) 1500 Ω   (b) 15 kΩ   (c) 110 Ω   (d) 16.5 kΩ

6. If a 10 mV signal is applied to the base of the common-collector circuit is Question 5, the output signal is approximately
   (a) 100 mV   (b) 150 mV   (c) 1.5 V   (d) 10 mV

7. For a common-emitter amplifier, $R_C = 1$ kΩ, $R_E = 390$ Ω, $r'_e = 15$ Ω, and $\beta_{ac} = 75$. Assuming that R_E is completely bypassed at the operating frequency, the voltage gain is
   (a) 66.7   (b) 2.56   (c) 2.47   (d) 75

8. In the circuit of Question 7, if the frequency is reduced to the point where $X_{C(bypass)} = R_E$, the voltage gain
   (a) remains the same   (b) is less   (c) is greater

9. In a certain common-collector circuit, the current gain is 50. The power gain is approximately
   (a) $50A_v$ (b) 50   (c) 1   (d) answers (a) and (b)

10. In a darlington pair configuration, each transistor has an ac beta of 125. If R_E is 560 Ω, the input resistance is
    (a) 560 Ω   (b) 70 kΩ   (c) 8.75 MΩ   (d) 140 kΩ

11. The input resistance of a common-base amplifier is
    (a) very low   (b) very high   (c) the same as a CE   (d) the same as a CC

12. In a common-emitter amplifier with voltage-divider bias, $R_{in(base)} = 68$ kΩ, $R_1 = 33$ kΩ, and $R_2 = 15$ kΩ. The total input resistance is
    (a) 68 kΩ   (b) 8.95 kΩ   (c) 22.2 kΩ   (d) 12.3 kΩ

13. A CE amplifier is driving a 10 kΩ load. If $R_C = 2.2$ kΩ and $r'_e = 10$ Ω, the voltage gain is approximately
    (a) 220   (b) 1000   (c) 10   (d) 180

14. Each stage of a four-stage amplifier has a voltage gain of 15. The overall voltage gain is
    (a) 60   (b) 15   (c) 50,625   (d) 3078

15. The overall gain found in Question 14 can be expressed in decibels as
    (a) 94.1 dB   (b) 47.0 dB   (c) 35.6 dB   (d) 69.8 dB

◼ **BASIC PROBLEMS**

SECTION 6–1 Small-Signal Amplifier Operation

1. What is the lowest value of dc collector current to which a transistor having the characteristic curves in Figure 6–4 can be biased and still retain linear operation with a peak-to-peak base current swing of 20 μA?

2. What is the highest value of I_C under the conditions described in Problem 1?

SECTION 6–2 Transistor AC Equivalent Circuits

3. What are the hybrid parameters that can be measured with each of the test circuits in Figure 6–50, and what is the value of each?

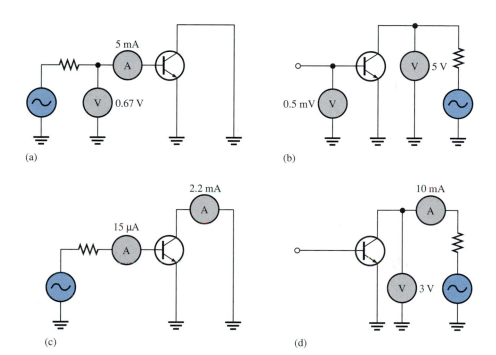

(a)   (b)

(c)   (d)

FIGURE 6–50

4. Use the *h*-parameter values obtained in Problem 3 to determine the *r* parameters β_{ac}, α_{ac}, r'_e, r'_b, and r'_c.

5. A certain transistor has a dc beta (h_{FE}) of 130. If the dc base current is 10 μA, determine r'_e. $\alpha_{DC} = 0.99$.

6. At the dc bias point of a certain transistor circuit, $I_B = 15$ μA and $I_C = 2$ mA. Also, a variation in I_B of 3 μA about the Q-point produces a variation in I_C of 0.35 mA about the Q-point. Determine β_{DC} and β_{ac}.

SECTION 6–3 Common-Emitter Amplifiers

7. Draw the dc equivalent circuit and the ac equivalent circuit for the unloaded amplifier in Figure 6–51.

8. Determine the following values for the amplifier in Figure 6–51.
   (a) $R_{in(base)}$ (b) R_{in} (c) A_v

9. Connect a bypass capacitor across R_E in Figure 6–51, and repeat Problem 8.

10. Connect a 10 kΩ load resistor to the output in Figure 6–51, and repeat Problem 9.

11. Determine the following dc values for the amplifier in Figure 6–52.
    (a) V_B (b) V_E (c) I_E (d) I_C (e) V_C (f) V_{CE}

12. Determine the following ac values for the amplifier in Figure 6–52.
    (a) $R_{in(base)}$ (b) R_{in} (c) A_v (d) A_i (e) A_p

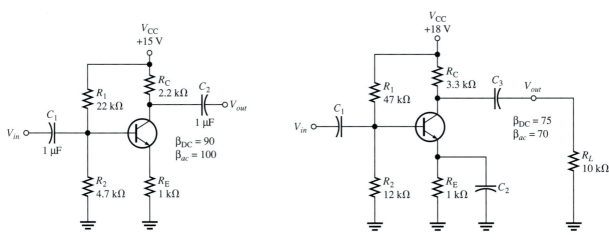

FIGURE 6–51   **FIGURE 6–52**

13. Assume that a 600 Ω, 12 μV rms voltage source is driving the amplifier in Figure 6–52. Determine the overall voltage gain by taking into account the attenuation in the base circuit, and find the *total* output voltage (ac and dc). What is the phase relationship of the collector signal voltage to the base signal voltage?

14. The amplifier in Figure 6–53 has a variable gain control, using a 100 Ω potentiometer for R_E with the wiper ac-grounded. As the potentiometer is adjusted, more or less of R_E is bypassed to ground, thus varying the gain. The total R_E remains constant to dc, keeping the bias fixed. Determine the maximum and minimum gains for this unloaded amplifier.

FIGURE 6–53

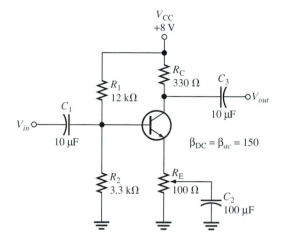

15. If a load resistance of 600 Ω is placed on the output of the amplifier in Figure 6–53, what are the maximum and minimum gains?

16. Find the overall maximum voltage gain for the amplifier in Figure 6–53 with a 1 kΩ load if it is being driven by a 300 Ω source.

17. Modify the schematic to show how you would "swamp out" the temperature effects of r'_e in Figure 6–52 by making R_e at least ten times larger than r'_e. Keep the same total R_E. How does this affect the voltage gain?

SECTION 6–4   Common-Collector Amplifiers

18. Determine the *exact* voltage gain for the unloaded emitter-follower in Figure 6–54.

FIGURE 6–54

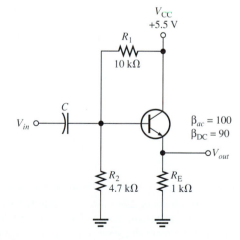

19. What is the total input resistance in Figure 6–54? What is the dc output voltage?

20. A load resistance is capacitively coupled to the emitter in Figure 6–54. In terms of signal operation, the load appears in parallel with R_E and reduces the effective emitter resistance. How does this affect the voltage gain?

21. In Problem 20, what value of R_L will cause the voltage gain to drop to 0.9?

22. For the circuit in Figure 6–55, determine the following:
    (a) Q_1 and Q_2 dc terminal voltages   (b) overall β_{ac}
    (c) r_e' for each transistor   (d) total input resistance
23. Find the overall current gain A_i in Figure 6–55.

FIGURE 6–55

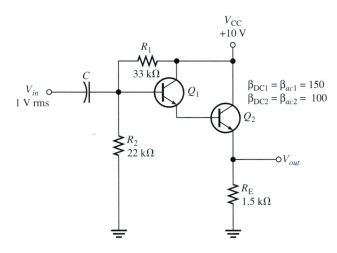

SECTION 6–5   Common-Base Amplifiers

24. What is the main disadvantage of the common-base amplifier compared to the common-emitter and the emitter-follower amplifiers?

FIGURE 6–56

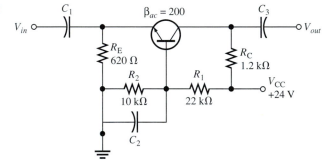

25. Find $R_{in(emitter)}$, A_v, A_i, and A_p for the unloaded amplifier in Figure 6–56.
26. Match the following generalized characteristics with the appropriate amplifier configuration.
    (a) Unity current gain, good voltage gain, very low input resistance
    (b) Good current gain, good voltage gain, low input resistance
    (c) Good current gain, unity voltage gain, high input resistance

SECTION 6–6   Multistage Amplifiers

27. Each of two cascaded amplifier stages has an $A_v = 20$. What is the overall gain?
28. Each of three cascaded amplifier stages has a dB voltage gain of 10 dB. What is the overall voltage gain in dB? What is the actual overall voltage gain?

29. For the two-stage, capacitively coupled amplifier in Figure 6–57, find the following values:
    (a) voltage gain of each stage   (b) overall voltage gain
    (c) Express the gains found in (a) and (b) in dB.

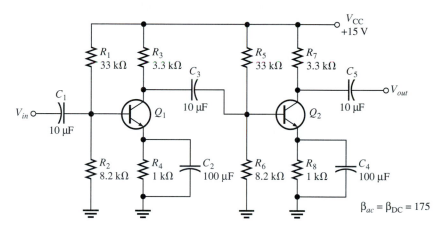

FIGURE 6–57

30. If the multistage amplifier in Figure 6–57 is driven by a 75 Ω, 50 μV source and the second stage is loaded with an $R_L = 18$ kΩ, determine
    (a) voltage gain of each stage   (b) overall voltage gain
    (c) Express in dB the gains found in (a) and (b).

31. Figure 6–58 shows a direct-coupled (that is, with no coupling capacitors between stages) two-stage amplifier. The dc bias of the first stage sets the dc bias of the second. Determine all dc voltages for both stages and the overall ac voltage gain.

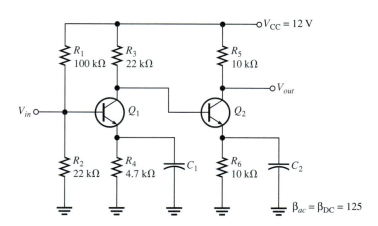

FIGURE 6–58

32. Express the following voltage gains in dB:
    (a) 12   (b) 50   (c) 100   (d) 2500

33. Express the following voltage gains in dB as standard voltage gains:
    (a) 3 dB   (b) 6 dB   (c) 10 dB   (d) 20 dB   (e) 40 dB

■ **TROUBLE-SHOOTING PROBLEMS**

SECTION 6–7 Troubleshooting

34. Assume that the coupling capacitor C_3 is shorted in Figure 6–37. What dc voltage will appear at the collector of Q_1?

35. Assume that R_5 opens in Figure 6–37. Will Q_2 be in cutoff or in conduction? What dc voltage will you observe at the Q_2 collector?

36. Refer to Figure 6–57 and determine the general effect of each of the following failures:
    (a) C_2 open   **(b)** C_3 open
    (c) C_4 open   **(d)** C_2 shorted
    (e) base-collector junction of Q_1 open   **(f)** base-emitter junction of Q_2 open

37. Assume that you must troubleshoot the amplifier in Figure 6–57. Set up a table of test point values, input, output, and all transistor terminals that include both dc and rms values that you expect to observe when a 300 Ω test signal source with a 25 µV rms output is used.

SECTION 6–8 System Application

38. Refer to the public address/paging system block diagram in Figure 6–45. You are asked to repair a system that is not working. After a preliminary check, you find that there is no output from the power amplifier or from the preamplifier. Based on this check and assuming that only one of the blocks is faulty, which block can you eliminate as the faulty one? What would you check next?

39. Determine the dc and ac voltages at the output (collector of Q_2) for each of the following faults in the amplifier of Figure 6–59. Assume a 2 mV rms input signal and $\beta_{ac} = 200$.
    (a) Open C_2 **(b)** Open C_1 **(c)** Open C_3
    (d) Open C_4 **(e)** Q_1 collector internally open   **(f)** Q_2 emitter shorted to ground

40. Suppose a 220 kΩ resistor is incorrectly installed in the R_6 position of the amplifier in Figure 6–59. What effect does this have on the circuit and what is the output voltage (dc and ac) if the input voltage is 3 mV rms?

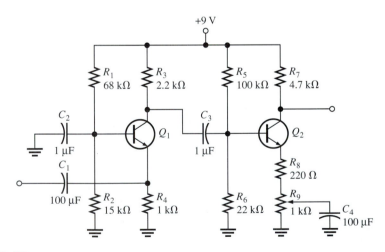

FIGURE 6–59

41. The connection from R_1 to the supply voltage in Figure 6–59 has opened.
    (a) What happens to Q_1?
    (b) What is the voltage at the Q_1 collector?
    (c) What is the voltage at the Q_2 collector?

■ **DATA SHEET PROBLEMS**

42. Refer to the partial data sheet in Figure 6–5. Determine the minimum value for each of the following r parameters for a 2N3904.
   (a) β_{ac} **(b)** r_e' **(c)** r_c'

43. Repeat Problem 42 for maximum values.

44. Should you use a 2N3903 or a 2N3904 transistor in a certain application if the criteria is maximum current gain?

■ **ADVANCED PROBLEMS**

45. In an amplifier such as the one in Figure 6–59, explain the general effect that a leaky coupling capacitor would have on circuit performance.

46. Draw the dc and ac equivalent circuits for the amplifier in Figure 6–59.

47. Redesign the 2-stage amplifier in Figure 6–59 using *pnp* transistors (2N3906). Maintain the same voltage gain.

48. Design a single-stage common-emitter amplifier with a voltage gain of 40 dB that operates from a dc supply voltage of +12 V. Use a 2N2222 transistor, voltage-divider bias, and a 330 Ω swamping resistor. The maximum input signal is 25 mV rms.

49. Design an emitter-follower with a minimum input resistance of 50 kΩ using a 2N3904.

50. Repeat Problem 49 using a 2N3906.

51. Design a single-stage common-base amplifier for a voltage gain of 75. Use a 2N3904 with emitter bias. The dc supply voltages are to be ±6 V.

52. Refer to the amplifier in Figure 6–59 and determine the minimum value of coupling capacitors necessary for the amplifier to produce the same output voltage at 100 Hz that it does at 5000 Hz.

53. Prove that for any unloaded common-emitter amplifier with a collector resistor R_C and R_E bypassed, the voltage gain is $A_v \cong 40V_{R_\text{C}}$.

■ **ANSWERS TO SECTION REVIEWS**

Section 6–1

1. Positive, negative

2. V_{CE} is a dc quantity and V_{ce} is an ac quantity.

3. R_e is the external emitter ac resistance, r_e' is the internal emitter ac resistance.

Section 6–2

1. h_{ie}—input resistance common-emitter configuration, h_{re}—voltage feedback ratio common-emitter configuration, h_{fe}—forward current gain common-emitter configuration, h_{oe}—output conductance common-emitter configuration

2. h_{fe} is equivalent to β_{ac}.

3. $r_e' = 25$ mV/15 mA = 1.67 Ω

Section 6–3

1. The capacitors are treated as opens.

2. The gain increases with a bypass capacitor.

3. Swamping eliminates the effects of r_e' by partially bypassing R_E.

4. Total input resistance includes the bias resistors, r_e', and any unbypassed R_E.

5. The gain is determined by R_c, r_e', and any unbypassed R_E.

6. The voltage gain decreases with a load.

7. The input and output are 180° out of phase.

Section 6–4

1. A common-collector amplifier is an emitter-follower.

2. The maximum voltage gain of a common-collector amplifier is one.

3. A common-collector amplifier has a high input resistance.

Section 6–5

1. Yes

2. The common-base amplifier has a low input resistance.

3. Maximum current gain is 1 in a CB amplifier.

Section 6–6

1. A stage is one amplifier in a cascaded arrangement.

2. The overall voltage gain is the product of the individual gains.

3. $20 \log(500) = 54.0$ dB

4. At lower frequencies, X_C becomes large enough to affect the gain.

Section 6–7

1. If C_4 opens, the gain drops. The dc level would not be affected.

2. Q_2 would be biased in cutoff.

3. The collector voltage of Q_1 and the base, emitter, and collector voltages of Q_2 would change.

■ **ANSWERS TO RELATED EXERCISES FOR EXAMPLES**

6–1 $I_C = 5$ mA; $V_{CE} = 1.5$ V

6–2 3.13 mA

6–3 9.3 mV

6–4 $C_2 = 28.4$ μF

6–5 97.3

6–6 153

6–7 4.85; 165

6–8 9.56

6–9 $R_{in(tot)} = 10.4$ kΩ; $A_v = 0.995$; $A_i = 6.7$; $A_p \cong 6.7$

6–10 71

6–11 55.9

6–12 $A'_v = 1500$; $A_{v1(dB)} = 27.96$ dB; $A_{v2(dB)} = 13.98$ dB; $A_{v3(dB)} = 21.58$ dB; $A'_{v(dB)} = 63.52$ dB

6–13 C_3 open

7

POWER AMPLIFIERS

■ CHAPTER OBJECTIVES

☐ Explain and analyze the operation of large-signal class A amplifiers

☐ Explain and analyze the operation of class B and class AB amplifiers

☐ Discuss and analyze the operation of class C amplifiers

☐ Troubleshoot power amplifiers

Power amplifiers are large-signal amplifiers. This generally means that a much larger portion of the load line is used during signal operation than in a small-signal amplifier. In this chapter, we will cover four classes of large-signal amplifiers: class A, class B, class AB, and class C. These amplifier classifications are based on the percentage of the input cycle for which the amplifier operates in its linear region. Each class has a unique circuit configuration because of the way it must be operated.

The emphasis in large-signal amplifiers is on power amplification. Power amplifiers are normally used as the final stage of a communications receiver or transmitter to provide signal power to speakers or to a transmitting antenna.

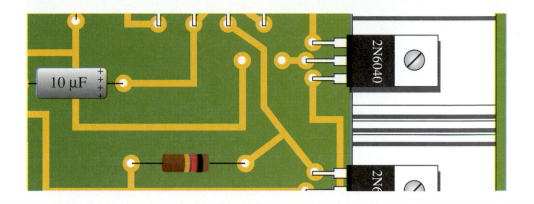

■ SYSTEM APPLICATION

The system application in this chapter continues with the public address and paging system started in the last chapter. Recall that the complete system includes the preamplifier, a power amplifier, and a dc power supply. You will focus on the power amplifier in this chapter and complete the total system. The first step in your assignment is to learn all you can about power amplifier operation. You will then apply your knowledge to the system application in Section 7–5.

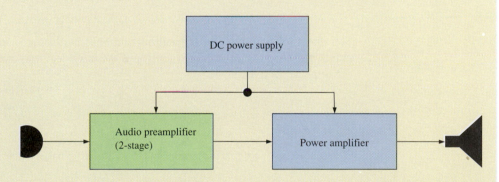

7–1 ■ CLASS A AMPLIFIERS

When a common-emitter, common-collector, or common-base amplifier is biased so that it operates in the linear region for the full 360° of the input cycle, it is a class A amplifier. In this mode of operation, the amplifier does not go into cutoff or saturation; therefore, the output voltage waveform has the same shape as the input waveform. A class A amplifier can be either inverting or noninverting. All of the small-signal amplifiers that you studied in Chapter 6 were class A. In this section, you will learn about large-signal class A amplifiers.

After completing this section, you should be able to

■ **Explain and analyze the operation of large-signal class A amplifiers**
  ☐ Explain class A operation
  ☐ Discuss how the location of the Q-point on the ac load line affects the operation of the amplifier
  ☐ Analyze ac load line operation
  ☐ Explain how to center the Q-point on the load line
  ☐ Determine large-signal voltage gain
  ☐ Discuss distortion
  ☐ Calculate power gain and quiescent power
  ☐ Calculate ac output power for several Q-point conditions
  ☐ Determine the efficiency of a class A amplifier
  ☐ Determine the maximum load power

The operation of a class A **large-signal** amplifier is illustrated in Figure 7–1, where the output waveform is an amplified replica of the input and may be either in phase or 180° out of phase with the input.

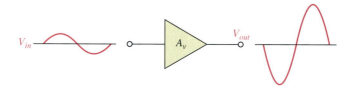

FIGURE 7–1

Basic class A amplifier operation (inverting).

Centered Q-Point for Maximum Output Signal

When the Q-point is at the center of the ac load line (midway between saturation and cutoff), a maximum class A signal can be obtained. This is graphically illustrated in the ac load line in Figure 7–2(a). Ideally, the collector current can vary from its Q-point value, I_{CQ}, up to its saturation value $I_{c(sat)}$, and down to its cutoff value of zero. This operation is indicated in Figure 7–2(b).

As you can see in Figure 7–2(b), the peak value of the collector current equals I_{CQ}, and the peak value of the collector-to-emitter voltage equals V_{CEQ}. This is the largest signal achievable from a class A amplifier. When the input signal is too large, the amplifier is driven into cutoff and saturation, as illustrated in Figure 7–3.

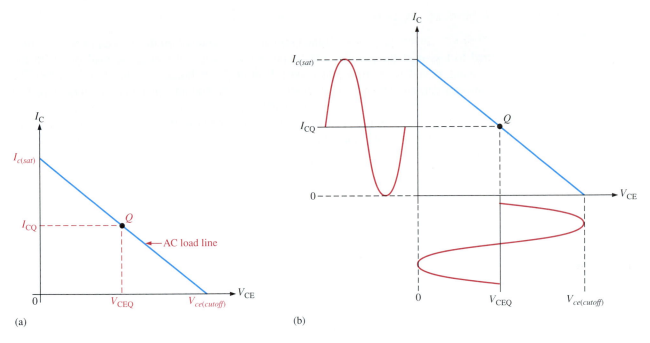

FIGURE 7–2

For maximum class A operation, the Q-point is centered on the ac load line.

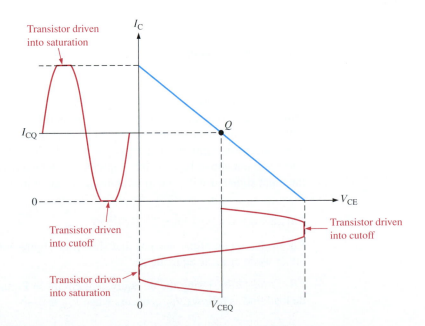

FIGURE 7–3

Waveforms are clipped at cutoff and saturation because the amplifier is overdriven (too large an input signal).

A Noncentered Q-Point Limits the Output

Q-Point Closer to Cutoff If the Q-point is not centered on the ac load line, V_{ce} is limited to less than the possible maximum. Figure 7–4(a) shows an ac load line with the Q-point moved away from center toward cutoff. V_{ce} is limited by cutoff in this case. The collector current can swing only down to near zero and an equal amount above I_{CQ}. The collector-to-emitter voltage can swing only up to its cutoff value and an equal amount below V_{CEQ}. If the amplifier is driven any further than this by an increase in the input signal, it will go into cutoff, as shown in Figure 7–4(b), and the waveforms will be clipped off on one peak.

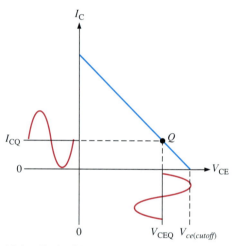

(a) Amplitude of V_{ce} and I_c limited by cutoff

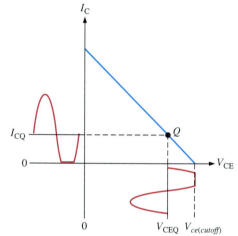

(b) Transistor driven into cutoff by a further increase in input amplitude

FIGURE 7–4

Q-point closer to cutoff.

Q-Point Closer to Saturation Figure 7–5(a) shows an ac load line with the Q-point moved away from center toward saturation. In this case, V_{ce} is limited by saturation. The collector current can swing only up to near saturation and an equal amount below I_{CQ}. The collector-to-emitter voltage can swing only down to near its saturation value and an equal amount above V_{CEQ}. If the amplifier is driven any further than this by an increase in the input signal, it will go into saturation, as shown in Figure 7–5(b).

Large-Signal Load Line Operation

Recall that an amplifier such as that shown in Figure 7–6 can be represented in terms of either its dc or its ac equivalent.

DC Load Line Using the dc equivalent circuit in Figure 7–7(a), you can determine the dc load line as follows: $I_{C(sat)}$ occurs when $V_{CE} \cong 0$, so

$$I_{C(sat)} \cong \frac{V_{CC}}{R_C + R_E}$$

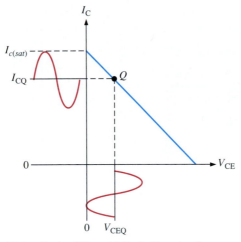

(a) Amplitude of V_{ce} and I_c limited by saturation

(b) Transistor driven into saturation by a further increase in input amplitude

FIGURE 7–5
Q-point closer to saturation.

FIGURE 7–6
Common-emitter amplifier with signal source.

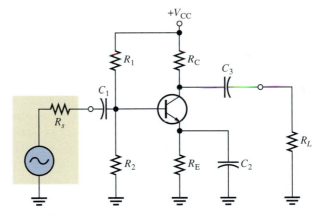

FIGURE 7–7
DC equivalent circuit and dc load line for the amplifier in Figure 7–6.

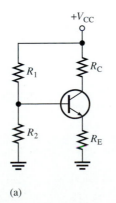

(a)

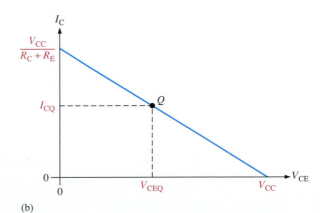

(b)

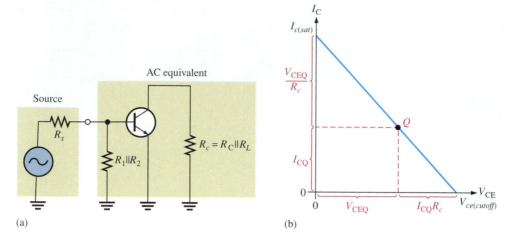

FIGURE 7–8

AC equivalent circuit and ac load line for the amplifier in Figure 7–6.

$V_{\text{CE(cutoff)}}$ occurs when $I_C \cong 0$, so

$$V_{\text{CE(cutoff)}} \cong V_{\text{CC}}$$

The dc load line is shown in Figure 7–7(b).

AC Load Line   From the ac viewpoint, the circuit in Figure 7–6 looks different than it does from the dc viewpoint. The collector resistance is different because R_L is in parallel with R_C due to the coupling capacitor C_3, and the emitter resistance is zero due to the bypass capacitor C_2; therefore, the ac load line is different from the dc load line. How much collector current can there be under ac conditions before saturation occurs? To answer this question, refer to the ac equivalent circuit and ac load line in Figure 7–8. Remember that a lowercase italic subscript indicates an ac quantity and an uppercase nonitalic subscript indicates a dc quantity. For example, R_c is the ac collector resistance and R_C is the dc collector resistance. I_{CQ} and V_{CEQ} are the dc Q-point coordinates. Going from the Q-point to the saturation point, the collector-to-emitter voltage swings from V_{CEQ} to near 0; that is, $\Delta V_{\text{CE}} = V_{\text{CEQ}}$. The change in collector current going from the Q-point to saturation is therefore

$$\Delta I_C = \frac{V_{\text{CEQ}}}{R_c}$$

where $R_c = R_C \| R_L$ is the ac collector resistance. The ac collector current at saturation is

$$I_{c(sat)} = I_{CQ} + \Delta I_C$$

Thus,

$$I_{c(sat)} = I_{CQ} + \frac{V_{CEQ}}{R_c} \qquad \text{(7–1)}$$

Going from the Q-point to the cutoff point, the collector current swings from I_{CQ} to near 0; that is, $\Delta I_C = I_{CQ}$. The change in collector-to-emitter voltage going from the Q-point to cutoff is therefore

$$\Delta V_{CE} = (\Delta I_C)R_c = I_{CQ}R_c$$

The cutoff value of ac collector-to-emitter voltage is

$$V_{ce(cutoff)} = V_{CEQ} + I_{CQ}R_c \qquad \text{(7–2)}$$

These results are shown on the ac load line of Figure 7–9. The corresponding dc load line is shown for comparison.

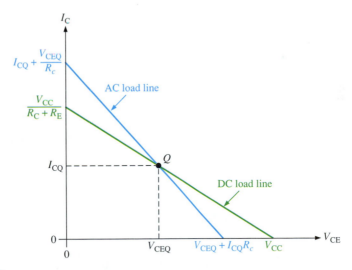

FIGURE 7–9
DC and ac load lines.

EXAMPLE 7–1   Determine the collector current and the collector-to-emitter voltage at the points of saturation and cutoff in Figure 7–10 with an ac signal input. Assume $X_{C1} = X_{C2} = X_{C3} \cong 0\ \Omega$.

FIGURE 7–10

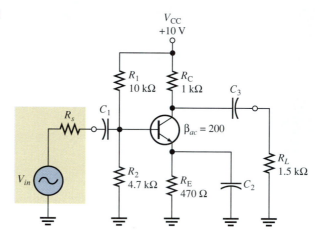

Solution   The Q-point values for this amplifier are determined as follows (neglecting $R_{\text{IN(base)}}$):

$$V_{\text{BQ}} \cong \left(\frac{R_2}{R_1 + R_2}\right)V_{\text{CC}} = \left(\frac{4.7\ \text{k}\Omega}{14.7\ \text{k}\Omega}\right)10\ \text{V} = 3.2\ \text{V}$$

Then,

$$I_{\text{EQ}} = \frac{V_{\text{BQ}} - 0.7\ \text{V}}{R_{\text{E}}} = \frac{3.2\ \text{V} - 0.7\ \text{V}}{470\ \Omega} = 5.3\ \text{mA}$$

Therefore, since $I_{\text{C}} \cong I_{\text{E}}$,

$$I_{\text{CQ}} \cong 5.3\ \text{mA}$$

The collector voltage is

$$V_{\text{CQ}} \cong V_{\text{CC}} - I_{\text{CQ}}R_{\text{C}} = 10\ \text{V} - (5.3\ \text{mA})(1\ \text{k}\Omega) = 4.7\ \text{V}$$

Therefore, the collector-to-emitter voltage is

$$V_{\text{CEQ}} = V_{\text{CQ}} - I_{\text{EQ}}R_{\text{E}} = 4.7\ \text{V} - 2.5\ \text{V} = 2.2\ \text{V}$$

The point of saturation under ac conditions is determined as follows:

$$V_{ce(sat)} \cong 0\ \text{V}$$

$$I_{c(sat)} = I_{\text{CQ}} + \frac{V_{\text{CEQ}}}{R_c}$$

The ac collector resistance is

$$R_c = R_{\text{C}} \parallel R_L = \frac{R_{\text{C}}R_L}{R_{\text{C}} + R_L} = \frac{(1\ \text{k}\Omega)(1.5\ \text{k}\Omega)}{2.5\ \text{k}\Omega} = 600\ \Omega$$

Thus,

$$I_{c(sat)} = 5.3 \text{ mA} + \frac{2.2 \text{ V}}{600 \text{ }\Omega} = 5.3 \text{ mA} + 3.67 \text{ mA} = 8.97 \text{ mA}$$

The point of cutoff under ac conditions is determined as follows:

$$I_{c(cutoff)} = 0 \text{ A}$$
$$V_{ce(cutoff)} = V_{CEQ} + I_{CQ}R_c = 2.2 \text{ V} + (5.3 \text{ mA})(600 \text{ }\Omega) = 5.38 \text{ V}$$

These results show that the collector current can swing up to almost 8.97 mA or the collector-to-emitter voltage can swing up to almost 5.38 V without the peaks being clipped due to saturation or cutoff.

Related Exercise Determine I_c and V_{ce} at the points of saturation and cutoff in Figure 7–10 for the following circuit values: $V_{CC} = 15$ V and $\beta_{ac} = 150$.

As described earlier, a centered Q-point allows a maximum unclipped output swing. A closer look at Example 7–1 shows that the Q-point is not centered, and therefore, the unclipped output swing is somewhat less than it could be if the Q-point were centered on the ac load line. This is examined further in Example 7–2.

EXAMPLE 7–2

Figure 7–11 shows the ac load line for the circuit in Example 7–1. Notice that the Q-point is *not* centered. The maximum output swings are

$$\Delta I_C = I_{c(sat)} - I_{CQ} = 8.97 \text{ mA} - 5.3 \text{ mA} = 3.67 \text{ mA}$$

and

$$\Delta V_{CE} = V_{CEQ} = 2.2 \text{ V}$$

Determine the maximum output swings for collector current and collector-to-emitter voltage when the Q-point is centered, assuming the same load line is maintained.

FIGURE 7–11

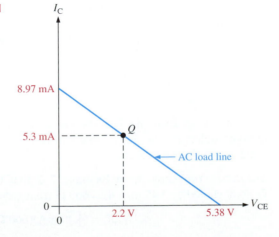

Solution The maximum output swings for a centered Q-point are

$$\Delta I_C = I_{CQ} = \frac{I_{c(sat)}}{2} = \frac{8.97 \text{ mA}}{2} = 4.49 \text{ mA}$$

$$\Delta V_{CE} = V_{CEQ} = \frac{V_{ce(cutoff)}}{2} = \frac{5.38 \text{ V}}{2} = 2.69 \text{ V}$$

Related Exercise Explain what happens to the output voltage if the Q-point is shifted to $V_{CE} = 3$ V.

Centering the Q-Point on the AC Load Line

To have a centered Q-point on the ac load line in Figure 7–9, V_{CEQ} must be midway between zero and the lower (cutoff) end of the ac load line, expressed as

$$V_{CEQ} = \frac{V_{CEQ} + I_{CQ}R_c}{2}$$

I_{CQ} must be midway between zero and the upper saturation end of the ac load line, expressed as

$$I_{CQ} = \frac{I_{CQ} + V_{CEQ}/R_c}{2}$$

The condition for a centered Q-point is developed as follows using the preceding equation for V_{CEQ}:

$$2V_{CEQ} = V_{CEQ} + I_{CQ}R_c$$
$$2V_{CEQ} - V_{CEQ} = I_{CQ}R_c$$

$$\boxed{V_{CEQ} = I_{CQ}R_c} \qquad\qquad (7\text{–}3)$$

The Q-point can be moved to an approximate center position on the load line by changing I_{CQ} until both sides of Equation (7–3) are approximately equal.

To move the Q-point toward cutoff on the ac load line without affecting the load line itself, I_{CQ} should be decreased by increasing R_E. To move the Q-point toward saturation on the load line, I_{CQ} should be increased by decreasing R_E.

EXAMPLE 7–3 As you saw in Example 7–2, the circuit in Figure 7–10 does *not* have a centered Q-point. Select a value for R_E that will produce a Q-point that is approximately centered on the ac load line.

Solution You know from Example 7–2 that the collector current for a centered Q-point should be 4.49 mA. $R_c = 600$ Ω from Example 7–1.

$$I_{CQ}R_c = (4.49 \text{ mA})(600 \text{ Ω}) = 2.69 \text{ V}$$

Determine R_E by the following steps:

$$V_{CEQ} = V_{CC} - I_{CQ}(R_C + R_E)$$
$$= V_{CC} - I_{CQ}R_C - I_{CQ}R_E$$

Transpose terms:

$$V_{CEQ} - V_{CC} + I_{CQ}R_C = -I_{CQ}R_E$$

Then,

$$I_{CQ}R_E = V_{CC} - V_{CEQ} - I_{CQ}R_C$$
$$R_E = \frac{V_{CC} - V_{CEQ} - I_{CQ}R_C}{I_{CQ}} = \frac{10 \text{ V} - 2.69 \text{ V} - (4.49 \text{ mA})(1 \text{ k}\Omega)}{4.49 \text{ mA}} = 628 \ \Omega$$

Choose the nearest standard value:

$$R_E = 620 \ \Omega$$

This will produce an approximately centered Q-point.

Related Exercise If R_C in Figure 7–10 is increased to 1.2 kΩ, what should you do to R_E to keep a centered Q-point?

Voltage Gain

The voltage gain of a class A large-signal amplifier is determined in the same way as for a small-signal amplifier with the exception that the formula $r_e' \cong 25 \text{ mV}/I_E$ is not valid for the large-signal amplifier. This is because the signal swings over a large portion of the transconductance curve. Since $r_e' = \Delta V_{BE}/\Delta I_C$, the value is different for large-signal operation than it is for small-signal conditions because of the nonlinearity of the curve.

The large-signal ac emitter resistance, r_e', can be determined graphically from the transconductance curve, as shown in Figure 7–12, using the relationship in Equation (7–4):

$$r_e' = \frac{\Delta V_{BE}}{\Delta I_C} \tag{7–4}$$

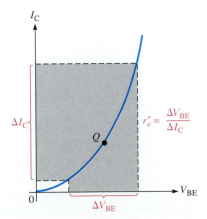

FIGURE 7–12

Determination of r_e' from the transconductance curve.

The voltage-gain formula for a common-emitter, large-signal amplifier with R_E completely bypassed is

$$A_v = \frac{R_c}{r'_e}$$

(7–5)

EXAMPLE 7–4

Find the large-signal voltage gain of the amplifier in Figure 7–13. Assume that r'_e has been found to be 8 Ω from graphical data.

FIGURE 7–13

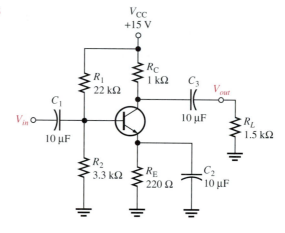

Solution

$$R_c = \frac{(R_C)(R_L)}{R_C + R_L} = \frac{(1\ \text{k}\Omega)(1.5\ \text{k}\Omega)}{2.5\ \text{k}\Omega} = 600\ \Omega$$

$$A_v = \frac{R_c}{r'_e} = \frac{600\ \Omega}{8\ \Omega} = 75$$

Related Exercise Find the large-signal voltage gain of the amplifier in Figure 7–13 if the supply voltage is changed to 9 V, R_L is changed to 1 kΩ, and r'_e is 10 Ω.

Distortion

When the collector current swings over a large portion of the transconductance curve, distortion can occur on the negative half-cycle because of the greater nonlinearity on the lower end of the curve, as shown in Figure 7–14. Distortion can be sufficiently reduced by keeping the collector current on the more linear part of the curve (at higher values of I_{CQ} and V_{BEQ}). Increasing the base bias voltage will result in more collector current and an increase in V_{BE} due to more voltage drop across r'_e.

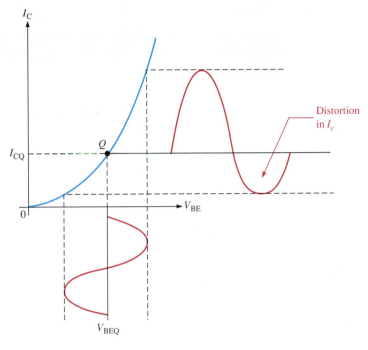

FIGURE 7–14
Example of distortion.

Power Gain

The main purpose of a large-signal amplifier is to achieve power gain. If we assume that the large-signal current gain, A_i, is approximately equal to β_{DC}, then the power gain, A_p, for a common-emitter amplifier is

$$A_p = A_i A_v = \beta_{DC} A_v$$

$$A_p = \beta_{DC}\left(\frac{R_c}{r'_e}\right) \qquad (7\text{--}6)$$

Quiescent Power

The power dissipation of a transistor with no signal input is the product of its Q-point current and voltage.

$$P_{DQ} = I_{CQ} V_{CEQ} \qquad (7\text{--}7)$$

The quiescent power is the maximum power that the class A transistor must handle; therefore, its power rating should exceed this value.

Output Power

In general, for any Q-point location on the ac load line, the output power of a common-emitter amplifier is the product of the rms collector current and the rms collector-to-emitter voltage.

$$P_{out} = V_{ce}I_c$$

Let's now consider the output power for three cases of Q-point location.

Q-Point Closer to Saturation When the Q-point is closer to saturation, the maximum collector-to-emitter voltage swing is V_{CEQ}, and the maximum collector current swing is V_{CEQ}/R_c, as shown in Figure 7–15(a). The ac output power is, therefore,

$$P_{out} = \left(\frac{0.707V_{CEQ}}{R_c}\right)0.707V_{CEQ}$$

$$P_{out} = \frac{0.5V_{CEQ}^2}{R_c} \qquad\qquad (7\text{–}8)$$

where $R_c = R_C \parallel R_L$.

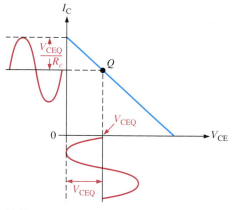

(a) Limited by saturation

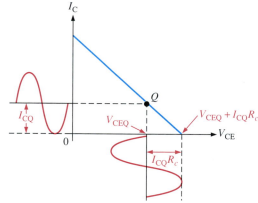

(b) Limited by cutoff

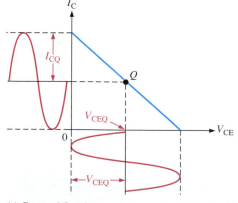

(c) Centered Q-point

FIGURE 7–15
AC load line operation showing limitations of output voltage swings.

Q-Point Closer to Cutoff   When the Q-point is closer to cutoff, the maximum collector current swing is I_{CQ}, and the collector-to-emitter voltage swing is $I_{CQ}R_c$, as shown in Figure 7–15(b). The ac output power is, therefore,

$$P_{out} = (0.707I_{CQ})(0.707I_{CQ}R_c)$$

$$P_{out} = 0.5I_{CQ}^2R_c \qquad\qquad (7\text{--}9)$$

Q-Point Centered   When the Q-point is centered, the maximum collector current swing is I_{CQ}, and the maximum collector-to-emitter voltage swing is V_{CEQ}, as shown in Figure 7–15(c). The ac output power is, therefore,

$$P_{out} = (0.707V_{CEQ})(0.707I_{CQ})$$

$$P_{out} = 0.5V_{CEQ}I_{CQ} \qquad\qquad (7\text{--}10)$$

This is the maximum ac output power from a class A amplifier under signal conditions. Notice that it is one-half the quiescent power dissipation.

Efficiency

Efficiency (η) of an amplifier is the ratio of ac output power to dc input power. The dc input power is the dc supply voltage times the current drawn from the supply.

$$P_{DC} = V_{CC}I_{CC}$$

The average supply current I_{CC} equals I_{CQ}, and the supply voltage V_{CC} is twice V_{CEQ} when the Q-point is centered. The maximum efficiency is therefore

$$\eta_{max} = \frac{P_{out}}{P_{DC}} = \frac{0.5V_{CEQ}I_{CQ}}{V_{CC}I_{CC}} = \frac{0.5V_{CEQ}I_{CQ}}{2V_{CEQ}I_{CQ}} = \frac{0.5}{2}$$

$$\eta_{max} = 0.25 \qquad\qquad (7\text{--}11)$$

Thus, 0.25 or 25 percent is the highest possible efficiency available from a class A amplifier and is approached only when the Q-point is at the center of the ac load line.

EXAMPLE 7–5   Determine the following values for the amplifier in Figure 7–16 when operated with the maximum possible output signal:
(a) Minimum transistor power rating
(b) AC output power
(c) Efficiency

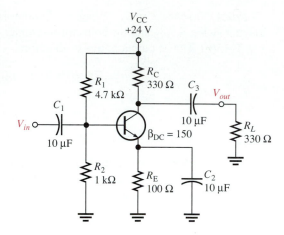

Solution First determine the dc values. Neglect $R_{\text{IN(base)}}$.

$$V_{\text{B}} \cong \left(\frac{R_2}{R_1 + R_2} \right) V_{\text{CC}} = \left(\frac{1 \text{ k}\Omega}{5.7 \text{ k}\Omega} \right) 24 \text{ V} = 4.2 \text{ V}$$

Then

$$V_{\text{E}} = V_{\text{B}} - V_{\text{BE}} = 4.2 \text{ V} - 0.7 \text{ V} = 3.5 \text{ V}$$

$$I_{\text{E}} = \frac{V_{\text{E}}}{R_{\text{E}}} = \frac{3.5 \text{ V}}{100 \text{ }\Omega} = 35 \text{ mA}$$

Therefore, the collector current at the Q-point is

$$I_{\text{CQ}} \cong 35 \text{ mA}$$

and

$$V_{\text{C}} = V_{\text{CC}} - I_{\text{CQ}}R_{\text{C}} = 24 \text{ V} - (35 \text{ mA})(330 \text{ }\Omega) = 12.5 \text{ V}$$
$$V_{\text{CEQ}} = V_{\text{C}} - V_{\text{E}} = 12.5 \text{ V} - 3.5 \text{ V} = 9.0 \text{ V}$$

(a) The transistor power rating must be greater than

$$P_{\text{D}} = V_{\text{CEQ}}I_{\text{CQ}} = (9.0 \text{ V})(35 \text{ mA}) = 315 \text{ mW}$$

(b) To make a calculation of ac output power under a *maximum* signal condition, you must know the location of the Q-point relative to the center. This will tell you whether I_{CQ} or V_{CEQ} is the limiting factor if the Q-point is not centered. The ac load line values are as follows:

$$R_c = R_{\text{C}} \| R_L = 330 \text{ }\Omega \| 330 \text{ }\Omega = 165 \text{ }\Omega$$

$$I_{c(sat)} = I_{\text{CQ}} + \frac{V_{\text{CEQ}}}{R_c} = 35 \text{ mA} + \frac{9.0 \text{ V}}{165 \text{ }\Omega} = 89.5 \text{ mA}$$

and

$$V_{ce(cutoff)} = V_{\text{CEQ}} + I_{\text{CQ}}R_c = 9.0 \text{ V} + (35 \text{ mA})(165 \text{ }\Omega) = 14.8 \text{ V}$$

A *centered* Q-point is at

$$I_{\text{CQ}} = \frac{89.5 \text{ mA}}{2} = 44.8 \text{ mA}$$

and

$$V_{CEQ} = \frac{14.8\ V}{2} = 7.37\ V$$

These Q-point values are shown on the ac load line in Figure 7–17. The *actual* Q-point for this amplifier is closer to cutoff, as shown in the figure. Therefore, the maximum collector current swing is I_{CQ}, and the ac output power is

$$P_{out} = 0.5I_{CQ}^2 R_c = 0.5(35\ mA)^2(165\ \Omega) = 101\ mW$$

FIGURE 7–17

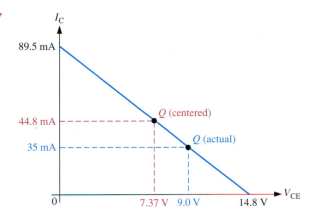

(c) The efficiency is

$$\eta = \frac{P_{out}}{P_{DC}} = \frac{P_{out}}{V_{CC}I_{CC}} = \frac{P_{out}}{V_{CC}I_{CQ}} = \frac{101\ mW}{(24\ V)(35\ mA)} = 0.12$$

This efficiency of 12 percent is considerably less than the maximum possible efficiency (25 percent) because the actual Q-point is not centered.

Related Exercise Suggest ways to increase the efficiency of the amplifier in Figure 7–16.

Maximum Load Power

The maximum power to the load in a class A amplifier occurs when the Q-point is centered, as you can see in Figure 7–15(c). The maximum peak load voltage equals V_{CEQ}, assuming a negligible voltage drop across the output coupling capacitor.

$$P_L = \frac{V_L^2}{R_L} = \frac{(0.707V_{CEQ})^2}{R_L}$$

$$P_L = \frac{0.5V_{CEQ}^2}{R_L} \tag{7–12}$$

Sometimes the load power is defined as the output power.

EXAMPLE 7–6

Determine the maximum load power for the amplifier in Example 7–5.

Solution   When the Q-point is centered, $V_{CEQ} = 7.37$ V.

$$P_{L(max)} = \frac{0.5V_{CEQ}^2}{R_L} = \frac{0.5(7.37 \text{ V})^2}{330 \text{ }\Omega} = 82.3 \text{ mW}$$

Related Exercise   If the Q-point in the amplifier is shifted to $V_{CEQ} = 5$ V, what is the load power?

SECTION 7–1 REVIEW

1. Why does the ac load line differ from the dc load line?
2. What is the optimum Q-point location for class A amplifiers?
3. What is the maximum efficiency of a class A amplifier?
4. How do you approach maximum efficiency in a class A amplifier?

7–2 ■ CLASS B AND CLASS AB PUSH-PULL AMPLIFIERS

When an amplifier is biased at cutoff such that it operates in the linear region for 180° of the input cycle and is in cutoff for 180°, it is a class B amplifier. Class AB amplifiers are biased to conduct for slightly more than 180°. The primary advantage of a class B or class AB amplifier over a class A amplifier is that they are more efficient; you can get more output power for a given amount of input power. A disadvantage of class B or class AB is that it is more difficult to implement the circuit in order to get a linear reproduction of the input waveform. As you will see in this section, the term push-pull refers to a common type of class B or class AB amplifier circuit in which the input wave shape is reproduced at the output.

After completing this section, you should be able to

■ **Explain and analyze the operation of class B and class AB amplifiers**
  ☐ Explain class B operation
  ☐ Discuss Q-point location for class B
  ☐ Describe class B push-pull operation
  ☐ Explain crossover distortion and its cause
  ☐ Explain class AB operation
  ☐ Analyze class AB push-pull amplifiers
  ☐ Calculate maximum output power
  ☐ Calculate dc input power
  ☐ Determine class B maximum efficiency
  ☐ Find the input resistance
  ☐ Analyze a darlington push-pull amplifier
  ☐ Discuss methods of driving a push-pull amplifier

FIGURE 7–18

Basic class B amplifier operation (noninverting).

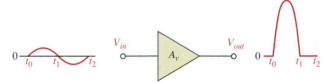

Class B Operation

The class B operation is illustrated in Figure 7–18, where the output waveform is shown relative to the input.

The Q-Point Is at Cutoff The class B amplifier is biased at the cutoff point so that $I_{CQ} = 0$ and $V_{CEQ} = V_{CE(cutoff)}$. It is brought out of cutoff and operates in its linear region when the input signal drives it into conduction. This is illustrated in Figure 7–19 with an emitter-follower circuit. Obviously, the output is not a replica of the input. Therefore, a two-transistor configuration, known as a **push-pull** amplifier, is necessary to get a sufficiently good reproduction of the input waveform.

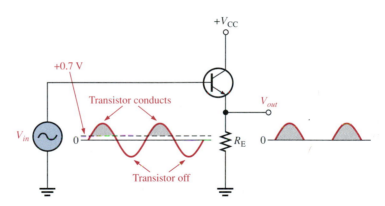

FIGURE 7–19

Common-collector class B amplifier.

Push-Pull Class B Operation Figure 7–20 shows one type of push-pull class B amplifier using two emitter-followers. This is called a complementary amplifier because one emitter-follower uses an *npn* transistor and the other a matched *pnp;* the transistors conduct on *opposite* alternations of the input cycle. A matched **complementary pair** of transistors have identical characteristics, except one is an *npn* and the other a *pnp*. The 2N3904 and 2N3906 are examples. Notice that there is no dc base bias voltage ($V_B = 0$); thus, only the signal voltage drives the transistors into conduction. Q_1 conducts during the positive half of the input cycle, and Q_2 conducts during the negative half.

Crossover Distortion When the dc base voltage is zero, both transistors are off and the input signal voltage must exceed V_{BE} before a transistor conducts. Because of this, there is a time interval between the positive and negative alternations of the input when neither transistor is conducting, as shown in Figure 7–21. The resulting distortion in the output waveform is quite common and is called **crossover distortion.**

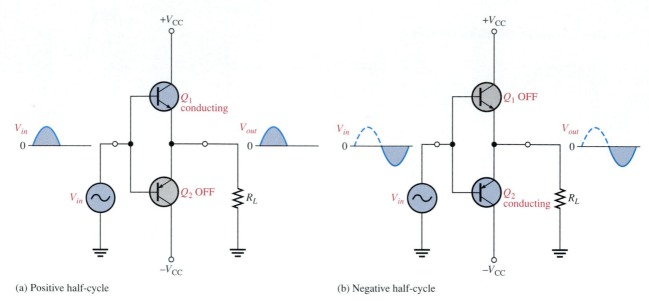

(a) Positive half-cycle   (b) Negative half-cycle

FIGURE 7–20
Class B push-pull ac operation.

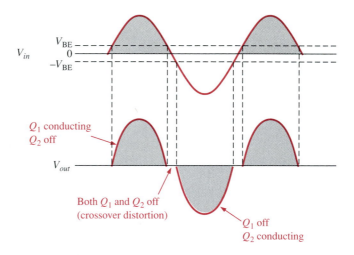

FIGURE 7–21
Illustration of crossover distortion in a class B push-pull amplifier. The transistors conduct only during portions of the input indicated by the shaded areas.

Class AB Operation

To eliminate crossover distortion, both transistors in the push-pull arrangement must be biased slightly above cutoff when there is no signal. This variation of the class B push-pull amplifier is designated as class AB. Class AB biasing can be done with a voltage-divider arrangement, as shown in Figure 7–22(a). It is, however, difficult to maintain a stable bias point with this circuit due to changes in V_{BE} over temperature changes. (The requirement for dual-polarity power supplies is eliminated when R_L is capacitively cou-

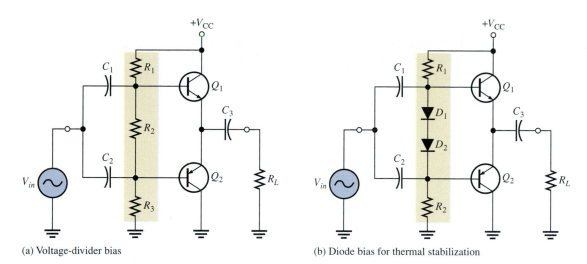

(a) Voltage-divider bias   (b) Diode bias for thermal stabilization

FIGURE 7–22

Biasing the push-pull amplifier for class AB operation to eliminate crossover distortion.

pled.) A more suitable arrangement is shown in Figure 7–22(b). When the diode characteristics of D_1 and D_2 are closely matched to the transconductance characteristics of the
transistors, a stable bias over temperature can be maintained. This stabilization can also
be accomplished by using the base-emitter junctions of two additional matched transistors instead of D_1 and D_2. Although technically incorrect, class AB amplifiers are often
referred to as class B in common practice.

The dc equivalent circuit of the push-pull amplifier is shown in Figure 7–23. Resistors R_1 and R_2 are of equal value; therefore the voltage at point A between the two diodes
is $V_{CC}/2$. Assuming that both diodes and both transistors are identical, the drop across D_1
equals the V_{BE} of Q_1, and the drop across D_2 equals the V_{BE} of Q_2. As a result, the voltage
at the emitters is also $V_{CC}/2$, and therefore, $V_{CEQ1} = V_{CEQ2} = V_{CC}/2$, as indicated. Because
both transistors are biased near cutoff, $I_{CQ} \cong 0$.

FIGURE 7–23

DC equivalent of push-pull amplifier.

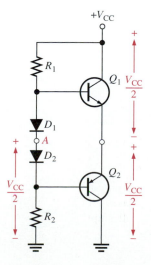

EXAMPLE 7–7

Determine the dc voltages at the bases and emitters of the matched complementary transistors Q_1 and Q_2 in Figure 7–24. Also determine V_{CEQ} for each transistor. Assume $V_{D1} = V_{D2} = V_{BE} = 0.7$ V.

FIGURE 7–24

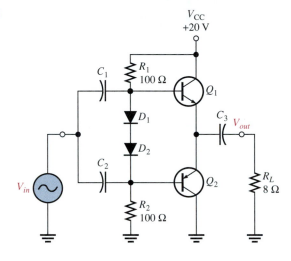

Solution   The equivalent for the bias circuit is shown in Figure 7–25(a).

$$I_T = \frac{V_{CC} - V_{D1} - V_{D2}}{R_1 + R_2} = \frac{20 \text{ V} - 1.4 \text{ V}}{200 \text{ } \Omega} = 93 \text{ mA}$$

$$V_{B1} = V_{CC} - I_T R_1 = 20 \text{ V} - (93 \text{ mA})(100 \text{ } \Omega) = 10.7 \text{ V}$$

and

$$V_{B2} = V_{B1} - V_{D1} - V_{D2} = 10.7 \text{ V} - 1.4 \text{ V} = 9.3 \text{ V}$$

$$V_{E1} = V_{E2} = 10.7 \text{ V} - 0.7 \text{ V} = 10 \text{ V}$$

Therefore,

$$V_{CEQ1} = V_{CEQ2} = \frac{V_{CC}}{2} = \frac{20 \text{ V}}{2} = 10 \text{ V}$$

These values are shown in Figure 7–25(b).

FIGURE 7–25

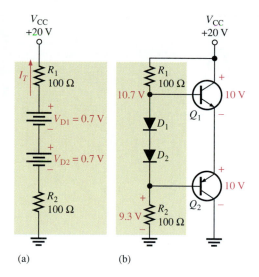

(a)

(b)

Related Exercise If R_1 and R_2 are changed to 220 Ω, what are the base and emitter voltages in Figure 7–24?

AC Operation Under maximum conditions, transistors Q_1 and Q_2 in a class AB amplifier are alternately driven from near cutoff to near saturation. During the positive alternation of the input signal, the Q_1 emitter is driven from its Q-point value of $V_{CC}/2$ to near V_{CC}, producing a positive peak voltage approximately equal to V_{CEQ}. At the same time, the Q_1 current swings from its Q-point value near zero to near-saturation value, as shown in Figure 7–26(a).

During the negative alternation of the input signal, the Q_2 emitter is driven from its Q-point value of $V_{CC}/2$ to near zero, producing a negative peak voltage approximately equal to V_{CEQ}. Also, the Q_2 current swings from near zero to near-saturation value, as shown in Figure 7–26(b).

In terms of the ac load line operation, the V_{ce} of both transistors swings from $V_{CC}/2$ to near zero, and the current swings from zero to $I_{c(sat)}$, as shown in Figure 7–26(c). Because the peak voltage across each transistor is V_{CEQ}, the ac collector saturation current is

$$I_{c(sat)} = \frac{V_{CEQ}}{R_L}$$

(7–13)

Since $I_e \cong I_c$ and the output current is the emitter current, the peak output current is also V_{CEQ}/R_L.

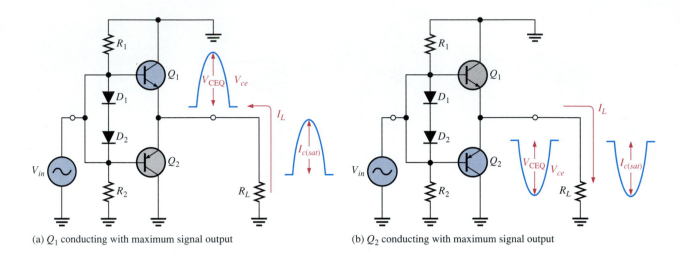

(a) Q_1 conducting with maximum signal output   (b) Q_2 conducting with maximum signal output

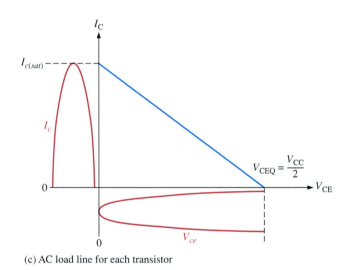

(c) AC load line for each transistor

FIGURE 7–26

Ideal ac push-pull operation for maximum signal operation.

EXAMPLE 7–8 Determine the maximum ideal peak values for the output voltage and current in Figure 7–27.

FIGURE 7–27

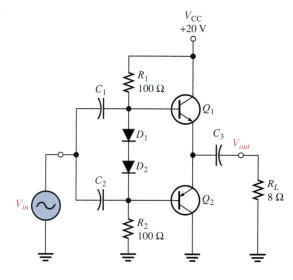

Solution The maximum peak output voltage is

$$V_{out(peak)} \cong V_{CEQ} = \frac{V_{CC}}{2} = \frac{20 \text{ V}}{2} = 10 \text{ V}$$

The maximum peak output current is

$$I_{out(peak)} \cong I_{c(sat)} = \frac{V_{CEQ}}{R_L} = \frac{10 \text{ V}}{8 \text{ }\Omega} = 1.25 \text{ A}$$

Related Exercise Find the maximum peak values for the output voltage and current in Figure 7–27 if V_{CC} is lowered to 15 V and the load resistance is changed to 16 Ω.

Maximum Output Power It has been shown that the maximum peak output current is approximately $I_{c(sat)}$, and the maximum peak output voltage is approximately V_{CEQ}. The maximum average output power is, therefore,

$$P_{out} = V_{out(rms)}I_{out(rms)}$$

Since

$$V_{out(rms)} = 0.707V_{out(peak)} = 0.707V_{CEQ}$$

and

$$I_{out(rms)} = 0.707I_{out(peak)} = 0.707I_{c(sat)}$$

then

$$P_{out} = 0.5V_{CEQ}I_{c(sat)}$$

Substituting $V_{CC}/2$ for V_{CEQ}, you get

$$P_{out} = 0.25 V_{CC} I_{c(sat)} \qquad \text{(7–14)}$$

DC Input Power   The dc input power comes from the V_{CC} supply and is

$$P_{DC} = V_{CC} I_{CC}$$

Since each transistor draws current for a half-cycle, the current is a half-wave signal with an average of

$$I_{CC} = \frac{I_{c(sat)}}{\pi} \qquad \text{(7–15)}$$

So,

$$P_{DC} = \frac{V_{CC} I_{c(sat)}}{\pi} \qquad \text{(7–16)}$$

Efficiency   An advantage of push-pull class B and class AB amplifiers over class A is a much higher efficiency. This advantage usually overrides the difficulty of biasing the class AB push-pull amplifier to eliminate crossover distortion. Recall that efficiency is defined as the ratio of ac output power to dc input power.

$$\text{Efficiency} = \frac{P_{out}}{P_{DC}}$$

The maximum efficiency for a class B amplifier (class AB is slightly less) is designated η_{max} and is developed as follows, starting with Equation (7–14).

$$P_{out} = 0.25 V_{CC} I_{c(sat)}$$

$$\eta_{max} = \frac{P_{out}}{P_{DC}} = \frac{0.25 V_{CC} I_{c(sat)}}{V_{CC} I_{c(sat)}/\pi} = 0.25\pi$$

$$\eta_{max} = 0.79 \qquad \text{(7–17)}$$

or, as a percentage

$$\eta_{max} = 79\%$$

Recall that the maximum efficiency for class A is 0.25 (25 percent).

Input Resistance   The complementary push-pull configuration used in class B/class AB amplifiers is, in effect, two emitter-followers. The input resistance is, therefore, the same as developed in Chapter 6 for the emitter-follower:

$$R_{in} = \beta_{ac}(r'_e + R_E)$$

Since $R_E = R_L$, the formula is

$$R_{in} = \beta_{ac}(r'_e + R_L) \qquad \text{(7–18)}$$

EXAMPLE 7–9   Find the maximum ac output power and the dc input power of the amplifier in Figure 7–28. Also, determine the input resistance assuming $\beta_{ac} = 50$ and $r'_e = 6\ \Omega$.

FIGURE 7–28

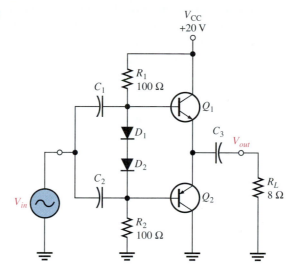

Solution $I_{c(sat)}$ for this same circuit was found to be 1.25 A in Example 7–8.

$$P_{out} = 0.25 V_{CC} I_{c(sat)} = 0.25(20\text{ V})(1.25\text{ A}) = 6.25\text{ W}$$

$$P_{DC} = \frac{V_{CC} I_{c(sat)}}{\pi} = \frac{(20\text{ V})(1.25\text{ A})}{\pi} = 7.96\text{ W}$$

$$R_{in} = \beta_{ac}(r'_e + R_L) = 50(6\ \Omega + 8\ \Omega) = 700\ \Omega$$

Related Exercise   Determine the maximum ac output power and the dc input power in Figure 7–28 for $V_{CC} = 15$ V and $R_L = 16\ \Omega$.

Darlington Class AB Amplifier

In many applications where the push-pull configuration is used, the load resistance is relatively small. For example, an 8 Ω speaker is a common load for a class AB push-pull amplifier.

As a result of low-resistance loading, push-pull amplifiers generally present a quite low input resistance to the preceding amplifier that drives it. Depending on the output resistance of the preceding amplifier, the low push-pull input resistance can load it severely and significantly reduce the voltage gain. As an example, if the complementary transistors in a push-pull amplifier exhibit an ac beta of 50 and the load resistance is 8 Ω, the input resistance (assuming $r'_e = 5\ \Omega$) is

$$R_{in} = \beta_{ac}(r'_e + R_L) = 50(5\ \Omega + 8\ \Omega) = 650\ \Omega$$

If the output resistance of the driving amplifier is, for example, 1 kΩ, the input resistance of the push-pull amplifier reduces the effective collector resistance of the driving amplifier (assuming a common-emitter) to $R_{out} \parallel R_{in} = 1 \text{ k}\Omega \parallel 650 \ \Omega = 394 \ \Omega$. This drastically reduces the voltage gain of the driving amplifier.

In certain applications with low-resistance loads, a push-pull amplifier using darlington transistors can be used to increase the input resistance presented to the driving amplifier and avoid severely reducing the voltage gain. The ac beta of a darlington pair is generally between a few hundred and several thousand.

In the previous case, for example, if $\beta_{ac} = 50$ for each transistor in a darlington pair, the overall ac beta is $\beta_{ac} = (50)(50) = 2500$. The input resistance is greatly increased, as the following calculation shows.

$$R_{in} = \beta_{ac}(r'_e + R_L) = 2500(5 \ \Omega + 8 \ \Omega) = 32.5 \text{ k}\Omega$$

A darlington class AB push-pull amplifier is shown in Figure 7–29. Four diodes are required in the bias circuit to match the four base-emitter junctions of the two darlington pairs.

FIGURE 7–29

A darlington class AB push-pull amplifier.

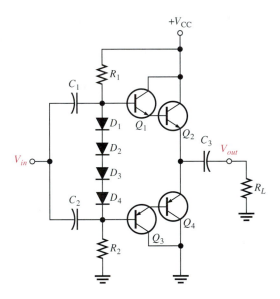

Push-Pull Amplifier with Class A Driver

Up to this point, the input signal to class B and class AB push-pull amplifiers has been capacitively coupled from the input source to each of the two transistors. Another method is to use a class A stage as the driver, as shown in Figure 7–30. As you can see, the driver is a voltage-divider-biased amplifier with a swamped emitter (no emitter-bypass capacitor).

The dc bias voltages for the push-pull amplifier are established by the dc collector current of the driver transistor Q_3. When an input signal is coupled to the base of Q_3, the resulting ac collector current of Q_3 produces signal voltages at the bases of Q_1 and Q_2, as shown in Figure 7–30. The ac voltage dropped across the dynamic resistances of the bias diodes is negligible, so the Q_1 and Q_2 base signals are essentially equal. The reason for this is that the dynamic resistances of the diodes are very small compared to R_1 and, for all practical purposes, the diodes can be considered shorts to the ac signal. This driver circuit has proved to be an effective way to interface the push-pull amplifier with a preamplifier circuit that precedes it.

FIGURE 7–30

A push-pull amplifier with a class A driver.

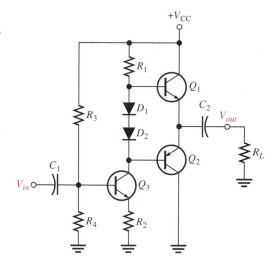

SECTION 7–2 REVIEW

1. Where is the Q-point for a class B amplifier?
2. What causes crossover distortion?
3. What is the maximum efficiency of a push-pull class B amplifier?
4. Explain the purpose of the push-pull configuration for class B.
5. How does a class AB differ from a class B amplifier?

7–3 ■ CLASS C AMPLIFIERS

Class C amplifiers are biased so that conduction occurs for much less than 180°. Class C amplifiers are more efficient than either class A or push-pull class B and class AB, which means that more output power can be obtained from class C operation. Because the output waveform is severely distorted, class C amplifiers are normally limited to applications as tuned amplifiers at radio frequencies (rf) as you will see in this section.

After completing this section, you should be able to

■ **Discuss and analyze the operation of class C amplifiers**
  □ Explain class C operation
  □ Discuss class C power dissipation
  □ Describe tuned operation
  □ Calculate maximum output power
  □ Determine efficiency
  □ Explain clamper bias in a class C amplifier

Basic Class C Operation

The basic concept of class C operation is illustrated in Figure 7–31. A common-emitter class C amplifier with a resistive load is shown in Figure 7–32(a). It is biased below cutoff with the $-V_{BB}$ supply. The ac source voltage has a peak value that is slightly greater

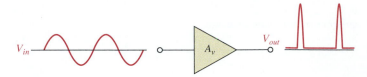

FIGURE 7–31

Basic class C amplifier operation (noninverting).

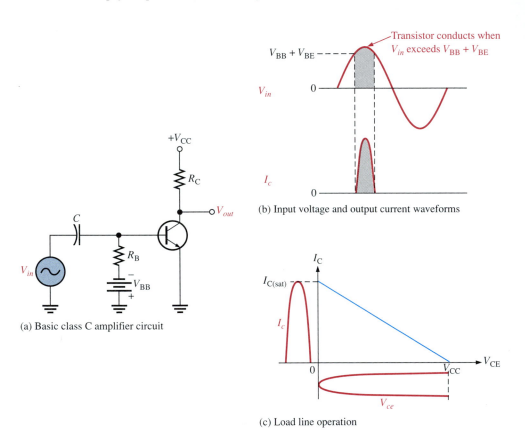

(a) Basic class C amplifier circuit

(b) Input voltage and output current waveforms

(c) Load line operation

FIGURE 7–32

Basic class C operation.

than $V_{BB} + V_{BE}$ so that the base voltage exceeds the barrier potential of the base-emitter junction for a short time near the positive peak of each cycle, as illustrated in Figure 7–32(b). During this short interval, the transistor is turned on. When the entire ac load line is used, as shown in Figure 7–30(c), the ideal maximum collector current is approximately $I_{C(sat)}$, and the ideal minimum collector voltage is approximately $V_{CE(sat)}$.

Power Dissipation

The power dissipation of the transistor in a class C amplifier is low because it is on for only a small percentage of the input cycle. Figure 7–33(a) shows the collector current pulses. The time between the pulses is the period (T) of the ac input voltage. To avoid complex mathematics, we will use ideal pulse approximations for the collector current and the collector voltage during the *on* time of the transistor, as shown in Figure 7–33(b). Using this simplification, the maximum current amplitude is $I_{C(sat)}$ and the minimum voltage amplitude is $V_{CE(sat)}$ during the time the transistor is on, if the output swings over the entire load line. The power dissipation during the *on* time is, therefore,

$$P_{D(on)} = V_{CE(sat)}I_{C(sat)}$$

The transistor is on for a short time, t_{ON}, and off for the rest of the input cycle. Therefore, assuming the entire load line is used, the power dissipation averaged over the entire cycle is

$$P_{D(avg)} = \left(\frac{t_{ON}}{T}\right)P_{D(on)}$$

$$P_{D(avg)} = \left(\frac{t_{ON}}{T}\right)V_{CE(sat)}I_{C(sat)} \tag{7–19}$$

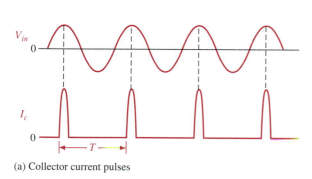

(a) Collector current pulses

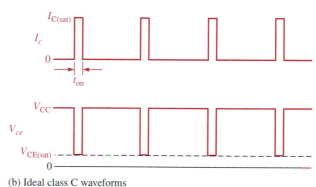

(b) Ideal class C waveforms

FIGURE 7–33
Class C waveforms.

EXAMPLE 7–10   A class C amplifier is driven by a 200 kHz signal. The transistor is on for 1 μs, and the amplifier is operating over 100 percent of its load line. If $I_{C(sat)} = 100$ mA and $V_{CE(sat)} = 0.2$ V, what is the average power dissipation?

Solution   The period is

$$T = \frac{1}{200 \text{ kHz}} = 5 \text{ μs}$$

Therefore,

$$P_{D(avg)} = \left(\frac{t_{ON}}{T}\right)V_{CE(sat)}I_{C(sat)} = (0.2)(0.2 \text{ V})(100 \text{ mA}) = 4 \text{ mW}$$

Related Exercise   If the frequency is reduced from 200 kHz to 150 kHz, what is the average power dissipation?

Tuned Operation

Because the collector voltage (output) is not a replica of the input, the resistively loaded class C amplifier is of no value in linear applications. It is therefore necessary to use a class C amplifier with a parallel resonant circuit (tank), as shown in Figure 7–34(a). The resonant frequency of the tank circuit is determined by the formula $f_r = 1/(2\pi\sqrt{LC})$. The short pulse of collector current on each cycle of the input initiates and sustains the oscillation of the tank circuit so that an output sinusoidal voltage is produced, as illustrated in Figure 7–34(b).

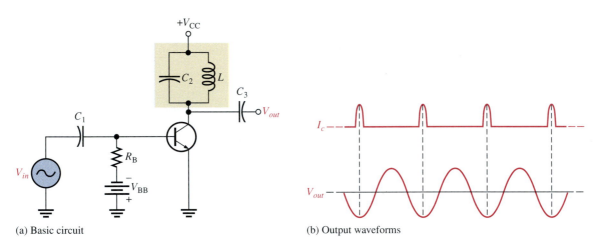

(a) Basic circuit

(b) Output waveforms

FIGURE 7–34

Tuned class C amplifier.

The current pulse charges the capacitor to approximately $+V_{CC}$, as shown in Figure 7–35(a). After the pulse, the capacitor quickly discharges, thus charging the inductor. Then, after the capacitor completely discharges, the inductor's magnetic field collapses and then quickly recharges C to near V_{CC} in a direction opposite to the previous charge. This completes one half-cycle of the oscillation, as shown in parts (b) and (c) of Figure 7–35. Next, the capacitor discharges again, quickly increasing the inductor's magnetic field. The inductor then quickly recharges the capacitor back to a positive peak less than the previous one, due to energy loss in the winding resistance. This completes the second half-cycle, as shown in parts (d) and (e) of Figure 7–35. The peak-to-peak output voltage is therefore approximately equal to $2V_{CC}$.

The amplitude of each successive cycle of the oscillation will be less than that of the previous cycle because of energy loss in the resistance of the tank circuit, as shown in Figure 7–36(a) on page 398, and the oscillation will eventually die out. However, the regular recurrences of the collector current pulse re-energizes the resonant circuit and sustains the oscillations at a constant amplitude.

When the tank circuit is tuned to the frequency of the input signal (fundamental), re-energizing occurs on each cycle of the tank voltage V_r, as shown in Figure 7–36(b). When the tank circuit is tuned to the second harmonic of the input signal, re-energizing occurs on alternate cycles as shown in Figure 7–36(c). In this case, a class C amplifier operates as a frequency multiplier ($\times 2$). By tuning the resonant tank circuit to higher harmonics, further frequency multiplication factors are achieved.

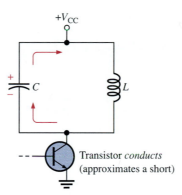

(a) *C* charges to +V_{CC} at the input peak
when transistor is conducting.

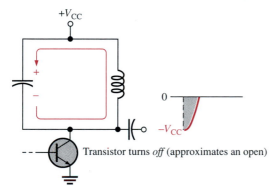

(b) *C* discharges to 0 volts.

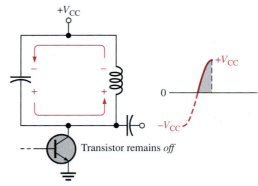

(c) *L* recharges *C* in opposite direction.

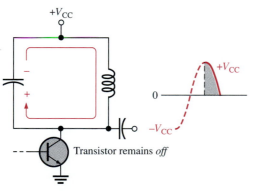

(d) *C* discharges to 0 volts.

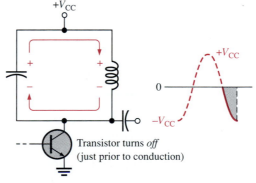

(e) *L* recharges *C*.

FIGURE 7–35
Resonant circuit action.

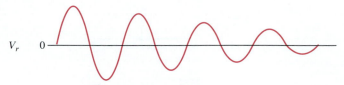

(a) Oscillation gradually dies out due to energy loss.
   The rate of decay depends on the efficiency of the tank circuit.

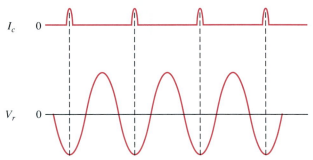

(b) Oscillation at the fundamental frequency is sustained by short
   pulses of collector current.

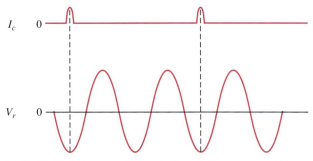

(c) Oscillation at the second harmonic frequency

FIGURE 7–36

Tank circuit oscillations. V_r is the voltage across the tank circuit.

Maximum Output Power

Since the voltage developed across the tank circuit has a peak-to-peak value of approximately $2V_{CC}$, the maximum output power can be expressed as

$$P_{out} = \frac{V_{rms}^2}{R_c} = \frac{(0.707V_{CC})^2}{R_c}$$

$$P_{out} = \frac{0.5V_{CC}^2}{R_c} \qquad (7\text{–}20)$$

R_c is the equivalent parallel resistance of the collector tank circuit and represents the parallel combination of the coil resistance and the load resistance. It usually has a low value. The total power that must be supplied to the amplifier is

$$P_T = P_{out} + P_{D(avg)}$$

Therefore, the efficiency is

$$\eta = \frac{P_{out}}{P_{out} + P_{D(avg)}} \qquad (7\text{–}21)$$

When $P_{out} \gg P_{D(avg)}$, the class C efficiency closely approaches 1 (100 percent).

EXAMPLE 7–11

Suppose the class C amplifier described in Example 7–10 has a V_{CC} equal to 24 V, and the R_C is 100 Ω. Determine the efficiency.

Solution   From Example 7–10, $P_{D(avg)} = 4$ mW.

$$P_{out} = \frac{0.5V_{CC}^2}{R_c} = \frac{0.5(24\text{ V})^2}{100\text{ Ω}} = 2.88\text{ W}$$

Therefore,

$$\eta = \frac{P_{out}}{P_{out} + P_{D(avg)}} = \frac{2.88\text{ W}}{2.88\text{ W} + 4\text{ mW}} = 0.999$$

or

$$\eta \times 100\% = 99.9\%$$

Related Exercise   What happens to the efficiency of the amplifier if R_c is increased?

Clamper Bias for a Class C Amplifier

Figure 7–37 shows a class C amplifier with a base bias clamping circuit. The base-emitter junction functions as a diode. When the input signal goes positive, capacitor C_1 is charged to the peak value with the polarity shown in Figure 7–38(a). This action produces

FIGURE 7–37
Tuned class C amplifier with clamper bias.

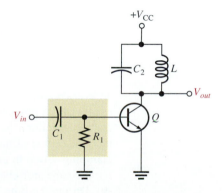

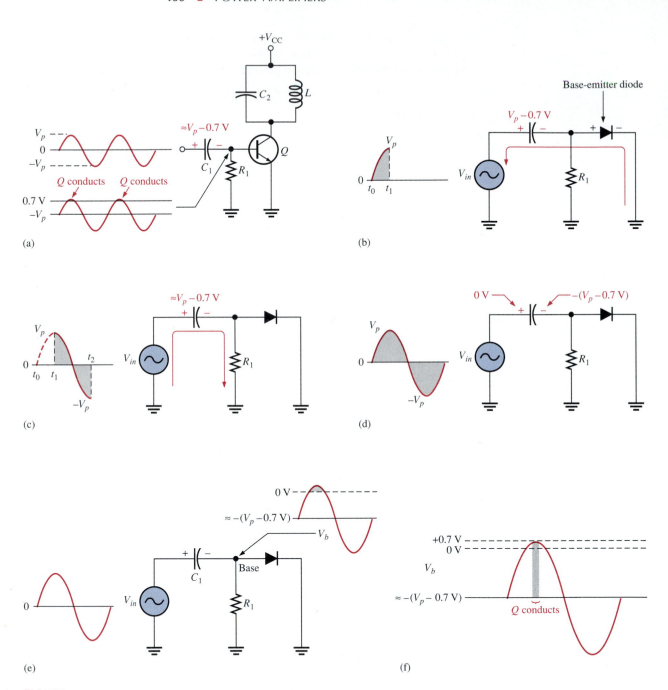

FIGURE 7–38

Clamper bias action.

an average voltage at the base of approximately $-V_p$. This places the transistor in cutoff except at the positive peaks, when the transistor conducts for a short interval. For good clamping action, the R_1C_1 time constant of the clamping circuit must be much greater than the period of the input signal. Parts (b) through (f) of Figure 7–38 illustrate the bias clamping action in more detail. During the time up to the positive peak of the input (t_0 to t_1), the capacitor charges to $V_p - 0.7$ V through the base-emitter diode, as shown in part (b). During the time from t_1 to t_2, as shown in part (c), the capacitor discharges very little because of the large RC time constant. The capacitor, therefore, maintains an average charge slightly less than $V_p - 0.7$ V.

Since the dc value of the input signal is zero (positive side of C_1), the dc voltage at the base (negative side of C_1) is slightly more positive than $-(V_p - 0.7$ V), as indicated in Figure 7–38(d). As shown in Figure 7–38(e), the capacitor couples the ac input signal through to the base so that the voltage at the transistor's base is the ac signal riding on a dc level slightly more positive than $-(V_p - 0.7$ V). Near the positive peaks of the input voltage, the base voltage goes slightly above 0.7 V and causes the transistor to conduct for a short time, as shown in Figure 7–38(f).

<table>
<tr><td>SECTION 7–3
REVIEW</td><td>1. At what point is a class C amplifier normally biased?

2. What is the purpose of the tuned circuit in a class C amplifier?

3. A certain class C amplifier has a power dissipation of 100 mW and an output power of 1 W. What is its percent efficiency?</td></tr>
</table>

7–4 ■ TROUBLESHOOTING

In this section, an example of isolating a component failure in a circuit is presented. We will use a class A amplifier with the output voltage monitored by an oscilloscope. Several incorrect output waveforms will be examined and the most likely faults will be discussed.

After completing this section, you should be able to

■ **Troubleshoot power amplifiers**
   □ Analyze various faults in a class A amplifier

As shown in Figure 7–39, the class A amplifier should have a normal sine wave output when a sinusoidal input signal is applied. Now we will consider several incorrect output waveforms and the most likely causes in each case. In Figure 7–40(a), the scope displays a dc level equal to the dc supply voltage, indicating that the transistor is in cutoff. The

FIGURE 7–39

Class A amplifier with proper output voltage swing.

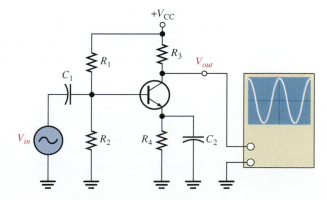

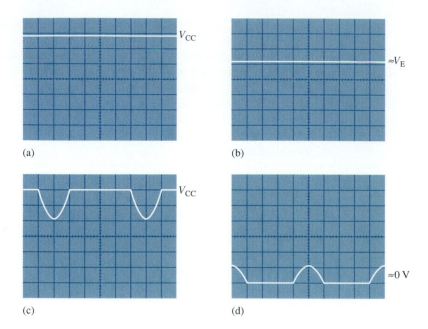

(a)

(b)

(c)

(d)

FIGURE 7–40

Oscilloscope displays of output voltage for the amplifier in Figure 7–39, illustrating several types of failures.

two most likely causes of this condition are (1) the transistor has an open *pn* junction, or (2) R_4 is open, preventing collector and emitter current.

In Figure 7–40(b), the scope displays a dc level at the collector approximately equal to the dc emitter voltage. The two probable causes of this indication are (1) the transistor is shorted from collector to emitter, or (2) R_2 is open, causing the transistor to be biased in saturation. In the second case, a sufficiently large input signal can bring the transistor out of saturation on its negative peaks, resulting in short pulses on the output.

In Figure 7–40(c), the scope displays an output waveform that indicates the transistor is cut off except during a small portion of the input cycle. Possible causes of this indication are (1) the Q-point has shifted down due to a drastic out-of-tolerance change in a resistor value, or (2) R_1 is open, biasing the transistor in cutoff. The display shows that the input signal is sufficient to bring it out of cutoff for a small portion of the cycle.

In Figure 7–40(d), the scope displays an output waveform that indicates the transistor is saturated except during a small portion of the input cycle. Again, it is possible that a resistance change has caused a drastic shift in the Q-point up toward saturation, or R_2 is open, causing the transistor to be biased in saturation, and the input signal is bringing it out of saturation for a small portion of the cycle.

SECTION 7–4 REVIEW

1. What would you check for if you noticed clipping at both peaks of the output waveform?

2. A significant loss of gain in the amplifier of Figure 7–39 would most likely be caused by what type of failure?

7–5 ■ SYSTEM APPLICATION

In the system application in Chapter 6, you were assigned responsibility for a public address system and you completed the audio preamplifier circuit. In this phase of the project, you will complete work on the power amplifier circuit and then put the complete system, including the power supply, together. You will apply the knowledge you have gained in this chapter in completing your assignment.

The Public Address System

Recall from the last chapter that this system includes a magnetic microphone, a two-stage preamplifier, a power amplifier, and a horn speaker as shown in the block diagram of Figure 7–41. The focus of this assignment is the power amplifier circuit board.

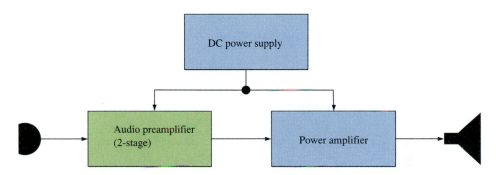

FIGURE 7–41
Basic public address/paging system block diagram.

Basic Operation   The schematic of the push-pull power amplifier is shown in Figure 7–42. The circuit is a class AB amplifier implemented with darlington power transistors. The base-emitter junctions of two additional darlington transistors of the same type are used in the bias circuit to assure matched thermal characteristics.

Rather than applying the signal to each base using two input coupling capacitors, this design capacitively couples the output of the preamplifier to the midpoint between the bias diodes. Since the diodes are forward-biased, the signal voltage is equally developed at the bases of Q_1 and Q_3. Darlington transistors are used in this application because of their high betas to prevent excessive loading of the preamplifier and a resulting loss of voltage gain.

Each darlington pair is packaged in a single case which looks like one transistor. The darlington transistors are mounted to a heat sink to prevent overheating and reduction of maximum power dissipation. Heat sinks come in many forms. In this application, a metal bracket with cooling fins is attached to the printed circuit board. The transistors are mounted with their collectors thermally connected but electrically isolated from the heat sink using a mica insulator.

FIGURE 7–42
Schematic of the power amplifier circuit.

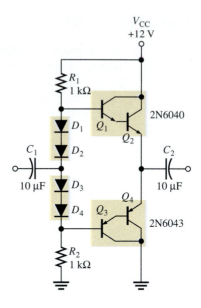

Analysis of the Power Amplifier Circuit

Refer to the partial data sheet in Figure 7–43.

☐ Determine the minimum input resistance of the push-pull amplifier.

☐ Determine the overall maximum voltage gain of the preamplifier and the power amplifier when connected together.

☐ Verify that the power rating of each transistor in the push-pull amplifier is not exceeded under maximum signal conditions.

The Printed Circuit Board

☐ Check out the printed circuit board in Figure 7–44 on page 406 to verify that it is correct according to the schematic. One interconnection is on the back side of the board between the two blank pads. Each blank pad provides a feedthrough.

☐ Label a copy of the board with component and input/output designations in agreement with the schematic.

Test Procedure

☐ Develop a step-by-step set of instructions on how to check the power amplifier circuit board for proper operation at a test frequency of 5 kHz using the test points (circled numbers) indicated in the test bench setup of Figure 7–45 on page 406. A function generator is the signal source.

☐ Specify voltage values for all the measurements to be made. Provide a fault analysis for all possible component failures.

☐ Combine the power amplifier test procedure with the test procedure developed for the preamplifier in Chapter 6 and the test precedure for the power supply board to form a complete test procedure for the system which is shown in its physical form in Figure 7–46 on page 406.

☐ Make sure that the interconnecting wiring in Figure 7–46 is correct and produce a complete schematic for the system.

<table>
<tr><td colspan="2">

**Plastic Medium-Power
Complementary Silicon Transistors**

. . . designed for general-purpose amplifier and low-speed switching applications.

• High DC current gain —
h_{FE} = 2500 (Typ) @ I_C = 4.0 A dc

• Collector-Emitter sustaining voltage — @ 100 mA dc
$V_{CEO(sus)}$ = 60 V dc (Min) — 2N6040, 2N6043
= 80 V dc (Min) — 2N6041, 2N6044
= 100 V dc (Min) — 2N6042, 2N6045

• Low collector-emitter saturation voltage
$V_{CE(sat)}$ = 2.0 V dc (Max) @ I_C = 4.0 A dc — 2N6040,41,2N6043,44
= 2.0 V dc (Max) @ I_C = 3.0 A dc — 2N6042, 2N6045

• Monolithic construction with built-in base-emitter shunt resistors

</td><td>

**Darlington
8 ampere**

**Complementary Silicon
Power Transistors**

**60-80-100 VOLTS
75 WATTS**

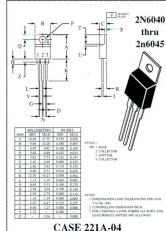

**2N6040
thru
2n6045**

**CASE 221A-04
TO-220AB**

</td></tr>
</table>

Maximum Ratings

Rating	Symbol	2N6040 2N6043 MJE6040 MJE6043	2N6041 2N6044 MJE6041 MJE6044	2N6042 2N6045 MJE6045	Unit
Collector-Emitter voltage	V_{CEO}	60	80	100	V dc
Collector-Base voltage	V_{CB}	60	80	100	V dc
Emitter-Base voltage	V_{EB}	← 5.0 →			V dc
Collector Current—Continuous Peak	I_C	← 8.0 → 16			A dc
Base current	I_B	← 120 →			mA dc
Total power dissipation @ T_C = 25°C Derate above 25°C	P_D	← 75 → 0.60			Watts W/C°
Total power dissipation @ T_A = 25°C Derate above 25°C	P_D	← 2.2 → 0.0175			Watts W/C°
Operating and storage junction, Temperature range	T_J, T_{stg}	← −65 to +150 →			°C

Electrical Characteristics (T_C = 25°C unless otherwise noted)

Characteristic	Symbol	Min	Max	Unit
DC current gain (I_C = 4.0 A dc, V_{CE} = 4.0 V dc) 2N6040,41,2N6043,44,MJE6040,41, MJE6043,44 (I_C = 3.0 A dc, V_{CE} = 4.0 V dc) 2N6042, 2N6045, MJE6045 (I_C = 8.0 A dc, V_{CE} = 4.0 V dc) All Types	h_{FE}	1000 1000 100	20,000 20,000 –	–
Collector-Emitter saturation voltage (I_C = 4.0 A dc, I_B = 16 mA dc) 2N6040,41,2N6043,44,MJE6040,41,MJE6043,44 (I_C = 3.0 A dc, I_B = 12 mA dc) 2N6042,2N6045,MJE6045 (I_C = 8.0 A dc, I_B = 80 mA dc) All Types	$V_{CE(sat)}$	– – –	2.0 2.0 4.0	V dc
Base-Emitter saturation voltage (I_C = 8.0 A dc, I_B = 80 mA dc)	$V_{CE(sat)}$	–	4.5	V dc
Base-Emitter on voltage (I_C = 4.0 A dc, V_{CE} = 4.0 V dc)	$V_{BE(on)}$	–	2.8	V dc

Dynamic Characteristics

	Symbol	Min	Max	Unit
Small-signal current gain (I_C = 3.0 A dc, V_{CE} = 4.0 V dc, f = 1.0 MHz)	$|h_{fe}|$	4.0	–	
Output capacitance (V_{CB} = 10 V dc, I_E = 0, f = 1.0 MHz) 2N6040/2N6042, MJE6040 2N6043/2N6045, MJE6043/MJE6045	C_{ob}	– –	300 200	pF
Small-signal current gain (I_C = 3.0 A dc, V_{CE} = 4.0 V dc, f = 1.0 kHz)	h_{fe}	300	–	–

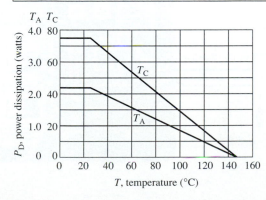

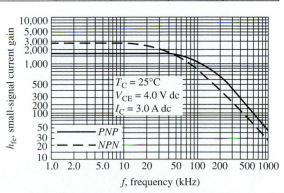

FIGURE 7–43

Partial data sheet for complementary darlington transistors 2N6040 (npn) and 2N6043 (pnp).

FIGURE 7–44

*Power amplifier circuit board and transistor
pin configuration.*

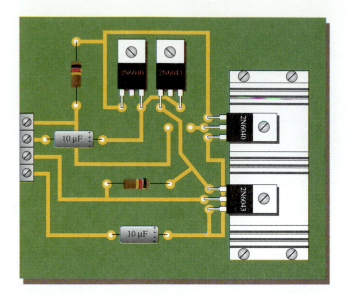

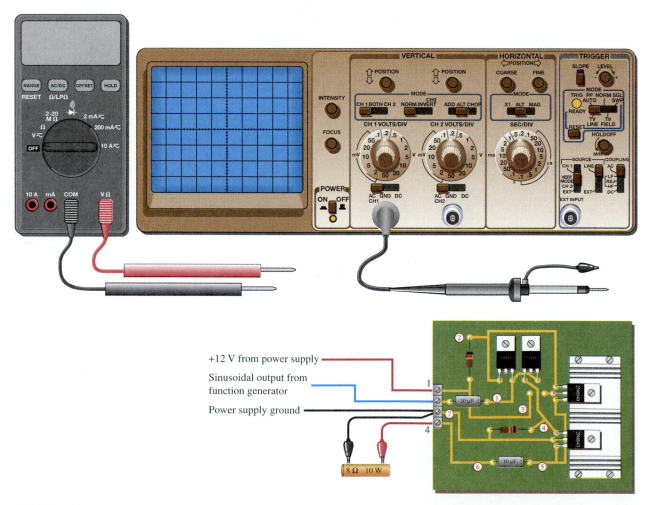

+12 V from power supply

Sinusoidal output from
function generator

Power supply ground

FIGURE 7–45
Test bench setup for the power amplifier board.

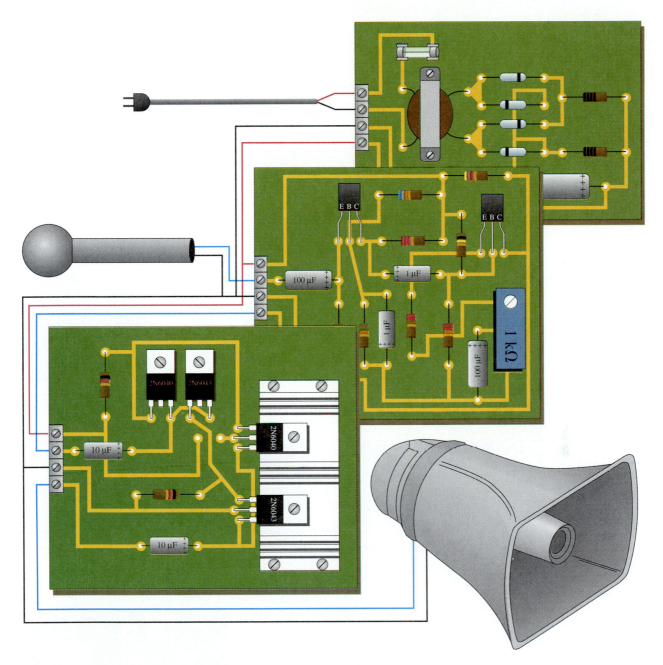

FIGURE 7–46
The complete public address/paging system before packaging.

Troubleshooting

Problems have developed in two prototype boards. Based on the sequence of test bench measurements for each board indicated in Figure 7–47, determine the most likely fault in each case. The circled numbers indicate test point connections to the circuit board.

Final Report

Submit a final written report on the power amplifier circuit board using an organized format that includes the following:

1. A physical description of the circuits.
2. A discussion of the operation of the circuits.

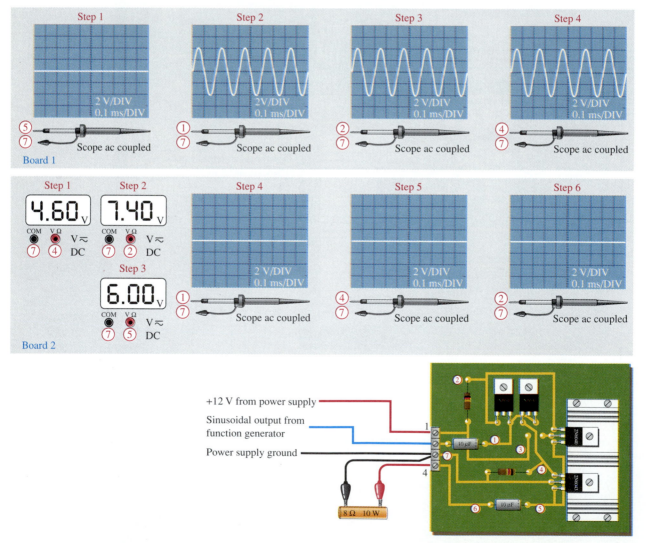

FIGURE 7–47

Test results for two faulty circuit boards.

3. A list of the specifications.

4. A list of parts with part numbers if available.

5. A list of the types of problems on the two faulty circuit boards.

6. A complete description of how you determined the problem on each of the faulty circuit boards.

■ **CHAPTER SUMMARY**

- A class A amplifier operates entirely in the linear region of the transistor's characteristic curves. The transistor conducts during the full 360° of the input cycle.
- The Q-point must be centered on the load line for maximum class A output signal swing.
- The maximum efficiency of a class A amplifier is 25 percent.
- A class B amplifier operates in the linear region for half of the input cycle (180°), and it is in cutoff for the other half.
- The Q-point is at cutoff for class B operation.
- Class B amplifiers are normally operated in a push-pull configuration in order to produce an output that is a replica of the input.
- The maximum efficiency of a class B amplifier is 79 percent.
- A class AB amplifier is biased slightly above cutoff and operates in the linear region for slightly more than 180° of the input cycle.
- Class AB eliminates crossover distortion found in pure class B.
- A class C amplifier operates in the linear region for only a small part of the input cycle.
- The class C amplifier is biased below cutoff.
- Class C amplifiers are normally operated as tuned amplifiers to produce a sinusoidal output.
- The maximum efficiency of a class C amplifier is higher than that of either class A or class B amplifiers. Under conditions of low power dissipation and high output power, the efficiency can approach 100 percent.

■ **GLOSSARY**

Complementary pair Two transistors, one *npn* and one *pnp,* having matched characteristics.

Crossover distortion Distortion in the output of a class B push-pull amplifier at the point where each transistor changes from the cutoff state to the *on* state.

Large-signal A signal that operates an amplifier over a significant portion of its load line.

Push-Pull A type of class B amplifier with two transistors in which one transistor conducts for one half-cycle and the other conducts for the other half-cycle.

■ **FORMULAS**

Large-Signal Class A Amplifiers

(7–1) $I_{c(sat)} = \dfrac{I_{CQ} + V_{CEQ}}{R_c}$ Saturation current

(7–2) $V_{ce(cutoff)} = V_{CEQ} + I_{CQ}R_c$ Cutoff voltage

(7–3) $V_{CEQ} = I_{CQ}R_c$ Centered Q-point

(7–4) $r_e' = \dfrac{\Delta V_{BE}}{\Delta I_C}$ AC emitter resistance

(7–5) $A_v = \dfrac{R_c}{r_e'}$ Voltage gain

(7–6)	$A_p = \beta_{DC}\left(\dfrac{R_c}{r'_e}\right)$	Power gain
(7–7)	$P_{DQ} = I_{CQ}V_{CEQ}$	Quiescent power
(7–8)	$P_{out} = \dfrac{0.5V^2_{CEQ}}{R_c}$	Output power (Q-point closer to saturation)
(7–9)	$P_{out} = 0.5I^2_{CQ}R_c$	Output power (Q-point closer to cutoff)
(7–10)	$P_{out} = 0.5V_{CEQ}I_{CQ}$	Output power (Q-point centered)
(7–11)	$\eta_{max} = 0.25$	Maximum efficiency
(7–12)	$P_L = \dfrac{0.5V^2_{CEQ}}{R_L}$	Load power

Class B and Class AB Push-Pull Amplifiers

(7–13)	$I_{c(sat)} = \dfrac{V_{CEQ}}{R_L}$	Saturation current
(7–14)	$P_{out} = 0.25V_{CC}I_{c(sat)}$	Output power
(7–15)	$I_{CC} = \dfrac{I_{c(sat)}}{\pi}$	DC supply current
(7–16)	$P_{DC} = \dfrac{V_{CC}I_{c(sat)}}{\pi}$	DC input power
(7–17)	$\eta_{max} = 0.79$	Maximum efficiency
(7–18)	$R_{in} = \beta_{ac}(r'_e + R_L)$	Input resistance

Class C Amplifiers

(7–19)	$P_{D(avg)} = \left(\dfrac{t_{ON}}{T}\right)V_{CE(sat)}I_{C(sat)}$	Average power
(7–20)	$P_{out} = \dfrac{0.5V^2_{CC}}{R_c}$	Output power
(7–21)	$\eta = \dfrac{P_{out}}{P_{out} + P_{D(avg)}}$	Efficiency

---

■ **SELF-TEST**

1. When the Q-point of an inverting class A amplifier is closer to saturation than cutoff and the input sine wave is gradually increased, clipping on the output will first appear on
   (a) the positive peaks **(b)** the negative peaks **(c)** both peaks simultaneously

2. The saturation value of ac collector current for an amplifier with an ac collector resistance of 3 kΩ and Q-point values of $I_{CQ} = 2$ mA and $V_{CEQ} = 3$ V is
   (a) 2 mA **(b)** 0 mA **(c)** 1 mA **(d)** 3 mA

3. The cutoff value of the ac collector-to-emitter voltage for the amplifier in Question 2 is
   (a) 9 V **(b)** 3 V **(c)** 0 V **(d)** 6 V

4. The maximum peak-to-peak collector voltage for the amplifier in Question 2 is
   (a) 3 V **(b)** 9 V **(c)** 6 V **(d)** 18 V

5. If $r'_e = 18$ Ω and $R_c = 500$ Ω in a class A amplifier, the large-signal voltage gain is
   (a) 500 **(b)** 27.8 **(c)** 43.9 **(d)** 180

6. A certain class A amplifier has a current gain of 75 and a voltage gain of 50. The power gain is
   (a) 75 **(b)** 125 **(c)** 3750 **(d)** 50

7. A class A amplifier is biased with a centered Q-point at $V_{CEQ} = 5$ V and $I_{CEQ} = 10$ mA. The maximum output power is
  (a) 25 mW   (b) 50 mW   (c) 10 mW   (d) 37.5 mW

8. The transistors in a class B amplifier are biased
  (a) into cutoff   (b) in saturation   (c) at midpoint   (d) right at cutoff

9. The emitters of a certain push-pull class B amplifier have a Q-point value of 10 V. If R_e is 50 Ω, the value of $I_{c(sat)}$ is
  (a) 5 mA   (b) 0.2 A   (c) 2 mA   (d) 20 mA

10. The output power of the amplifier in Question 9 under maximum signal conditions is
  (a) 0.5 W   (b) 0.1 W   (c) 1 W   (d) 5 W

11. The power dissipation of a class C amplifier is normally
  (a) very low   (b) very high
  (c) the same as a class B   (d) the same as a class A

12. The efficiency of a class C amplifier is
  (a) less than class A   (b) less than class B
  (c) less than class AB   (d) greater than class A, class B, or class AB

■ **BASIC PROBLEMS**

SECTION 7–1   Class A Amplifiers

1. Determine the approximate values for $I_{C(sat)}$ and $V_{CE(cutoff)}$ in Figure 7–48. Assume $V_{CE(sat)} \cong$ 0 V.

FIGURE 7–48

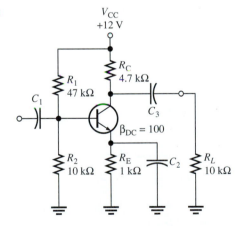

2. Sketch the ac equivalent circuit for the amplifier in Figure 7–48. Calculate the saturation value of the collector current under signal conditions.

3. Find the cutoff value of the ac collector-to-emitter voltage in Figure 7–48.

4. In Figure 7–49 assume β_{DC} for each transistor increases by 50 percent with a certain temperature rise. How is the maximum output voltage affected in each circuit?

5. What is the maximum peak value of collector current that can be realized in each circuit of Figure 7–50? What is the maximum peak value of output voltage?

6. Find the large-signal voltage gain for each circuit in Figure 7–50.

7. Determine the maximum rms value of input voltage that can be applied to each amplifier in Figure 7–50 without producing a clipped output voltage.

8. Determine the minimum power rating for each of the transistors in Figure 7–51.

9. Find the maximum output signal power and efficiency for each amplifier in Figure 7–51 with a 10 kΩ load resistor.

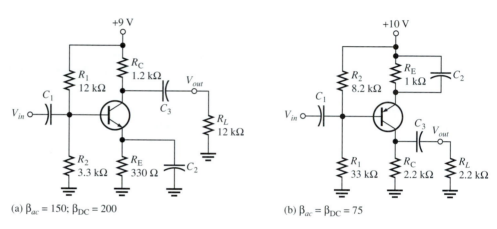

(a) $\beta_{ac} = 150$; $\beta_{DC} = 200$ (b) $\beta_{ac} = \beta_{DC} = 75$

FIGURE 7–49

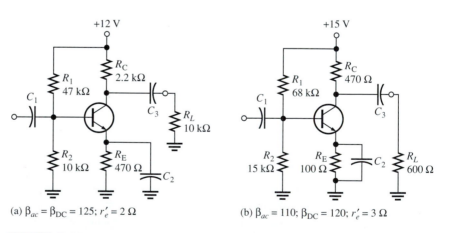

(a) $\beta_{ac} = \beta_{DC} = 125$; $r'_e = 2\ \Omega$ (b) $\beta_{ac} = 110$; $\beta_{DC} = 120$; $r'_e = 3\ \Omega$

FIGURE 7–50

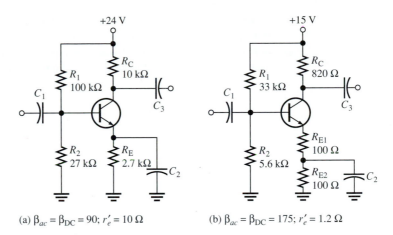

(a) $\beta_{ac} = \beta_{DC} = 90$; $r'_e = 10\ \Omega$ (b) $\beta_{ac} = \beta_{DC} = 175$; $r'_e = 1.2\ \Omega$

FIGURE 7–51

SECTION 7–2   Class B and Class AB Push-Pull Amplifiers

10. What dc voltage would you expect to measure with each dc meter in Figure 7–52?

11. Determine the dc voltages at the bases and emitters of Q_1 and Q_2 in Figure 7–53. Also determine V_{CEQ} for each transistor.

12. Determine the maximum peak output voltage and peak load current for the circuit in Figure 7–53.

13. Find the maximum signal power achievable with the class B push-pull amplifier in Figure 7–53. Find the dc input power.

14. The efficiency of a certain class B push-pull amplifier is 0.71, and the dc input power is 16.3 W. What is the ac output power?

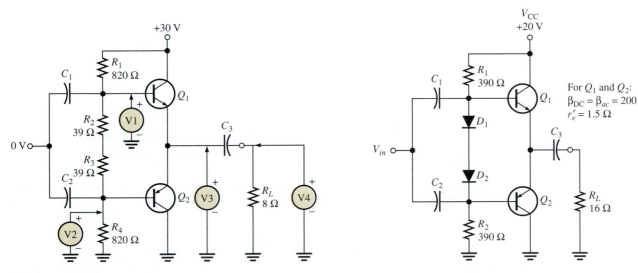

FIGURE 7–52   **FIGURE 7–53**

15. A certain class B push-pull amplifier has an R_e of 8 Ω and a V_{CEQ} of 12 V. Determine I_{CC}, P_{DC}, and P_{out}. Assuming this amplifier is operating under maximum output conditions, what is V_{CC}?

16. In Figure 7–54, what rms input voltage will produce a maximum output voltage swing? Use an ideal voltage gain of one for the emitter-followers and assume the maximum peak output voltage equals $V_{CC}/2$.

FIGURE 7–54

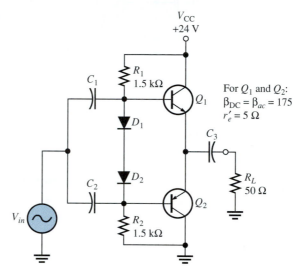

For Q_1 and Q_2:
$\beta_{DC} = \beta_{ac} = 175$
$r'_e = 5\ \Omega$

SECTION 7–3 Class C Amplifiers

17. A certain class C amplifier transistor is on for 10 percent of the input cycle. If $V_{CE(sat)} = 0.18$ V and $I_{C(sat)} = 25$ mA, what is the average power dissipation for maximum output?

18. What is the resonant frequency of a tank circuit with $L = 10$ mH and $C = 0.001\ \mu$F?

19. What is the maximum peak-to-peak output voltage of a tuned class C amplifier with $V_{CC} = 12$ V?

20. Determine the efficiency of the class C amplifier described in Problem 17 if $V_{CC} = 15$ V and the equivalent parallel resistance in the collector tank circuit is 50 Ω.

■ **TROUBLE-SHOOTING PROBLEMS**

SECTION 7–4 Troubleshooting

21. Refer to Figure 7–54. What would you expect to observe across R_L if C_1 opened?

22. Your oscilloscope displays a half-wave output when connected across R_L in Figure 7–54. What is the probable cause?

23. Determine the possible fault or faults, if any, for each circuit in Figure 7–55 based on the indicated dc voltage measurements.

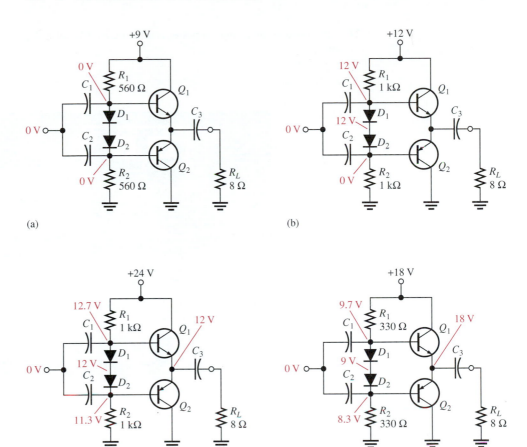

FIGURE 7–55

SECTION 7–5 System Application

24. Assume that the public address/paging system represented by the block diagram in Figure 7–41 has quit working. You find there is no signal output from the power amplifier or the preamplifier, but you have verified that the microphone is working. Which two blocks are the most likely to be the problem? How would you narrow the choice down to one block?

25. Describe the output that would be observed in the push-pull amplifier of Figure 7–42 with a 3 V rms sinusoidal input voltage if the base-emitter junction of the 2N6043 opened.

26. Describe the output that would be observed in Figure 7–42 if the base-emitter junction of the 2N6040 opened for the same input as in Problem 25.

27. After visually inspecting the power amplifier circuit board in Figure 7–56, describe any problems.

FIGURE 7–56

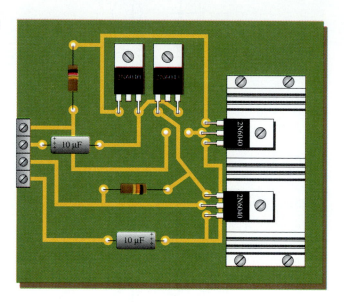

■ **DATA SHEET PROBLEMS**

28. Referring to the data sheet in Figure 7–43, determine the following:
    (a) minimum β_{DC} for the 2N6040 and the conditions
    (b) maximum collector-to-emitter voltage for the 2N6041
    (c) maximum power dissipation for the 2N6043 at a case temperature of 25°C
    (d) maximum continuous collector current for the 2N6040

29. Determine the maximum power dissipation for a 2N6040 at a case temperature of 65°C.

30. Determine the maximum power dissipation for a 2N6043 at an ambient temperature of 80°C.

31. Describe what happens to the small-signal current gain as the frequency increases.

32. Determine the approximate h_{fe} for the 2N6040 at a frequency of 2 kHz. At 100 kHz.

■ **ADVANCED PROBLEMS**

33. Explain why the specified maximum power dissipation of a power transistor at an ambient temperature of 25°C is much less than maximum power dissipation at a case temperature of 25°C.

34. Draw the dc and the ac load lines for the amplifier in Figure 7–57.

35. Design a swamped class A power amplifier that will operate from a dc supply of +15 V with an approximate voltage gain of 50. The quiescent collector current should be approximately 500 mA, and the total dc current from the supply should not exceed 750 mA. The output power must be at least 1 W.

36. The public address/paging system in Figure 7–46 is to be converted to a portable unit that is independent of 115 V ac power. Determine the modifications necessary for the system to operate for 8 hours on a continuous basis.

FIGURE 7–57

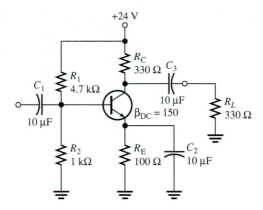

■ **ANSWERS TO SECTION REVIEWS**

Section 7–1

1. The load lines differ because the ac load resistance has a value different from that of the dc load resistance.
2. Optimum class A Q-point is centered on the load line.
3. Maximum efficiency of class A is 25%.
4. Center the Q-point for maximum efficiency.

Section 7–2

1. The class B Q-point is at cutoff.
2. The barrier potential of the base-emitter junction causes crossover distortion.
3. Maximum efficiency of class B is 79%.
4. Push-pull reproduces both positive and negative alternations of the input signal with greater efficiency.

Section 7–3

1. Class C is biased well into cutoff.
2. The purpose of the tuned circuit is to produce a sinusoidal voltage output.
3. $\eta = [1 \text{ W}/(1 \text{ W} + 0.1 \text{ W})]100 = 90.9\%$

Section 7–4

1. Excess input signal voltage
2. Open bypass capacitor, C_2

■ **ANSWERS TO RELATED EXERCISES FOR EXAMPLES**

7–1 $V_{ce(sat)} = 0 \text{ V}; I_{c(sat)} = 12.4 \text{ mA}; V_{ce(cutoff)} = 7.41 \text{ V}; I_{c(cutoff)} = 0 \text{ A}$

7–2 The output voltage will clip at 2.38 V.

7–3 Decrease R_E to 240 Ω.

7–4 $A_v = 50$

7–5 Shift the Q-point toward the center of the load line by increasing I_{CQ} and decreasing V_{CEQ}.

7–6 37.9 mW

7–7 The base and emitter voltages will remain the same.

7–8 7.5 V; 469 mA

7–9 $P_{out} = 1.76 \text{ W}; P_{DC} = 2.24 \text{ W}$

7–10 3 mW

7–11 The efficiency decreases.

8

FIELD-EFFECT TRANSISTORS AND BIASING

■ CHAPTER OBJECTIVES

☐ Explain the operation of JFETs
☐ Define, discuss, and apply important JFET parameters
☐ Discuss and analyze JFET bias circuits
☐ Explain the operation of MOSFETs
☐ Define, discuss, and apply important MOSFET parameters
☐ Discuss and analyze MOSFET bias circuits
☐ Troubleshoot FET circuits

Bipolar junction transistors (BJTs) were covered in previous chapters. Now we will discuss the second major type of transistor, the **field effect transistor (FET).** FETs are unipolar devices because, unlike bipolar transistors that use both electron and hole current, they operate only with one type of charge carrier. The two main types of FETs are the junction field-effect transistor (JFET) and the metal oxide semiconductor field-effect transistor (MOSFET). You will learn about both types in this chapter.

Recall that the bipolar junction transistor is a current-controlled device; that is, the base current controls the amount of collector current. The FET is different. It is a voltage-controlled device, where the voltage between two of the terminals (gate and source) controls the current through the device. As you will learn, a major feature of FETs is their very high input resistance.

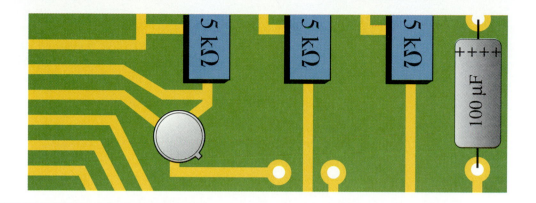

■ SYSTEM APPLICATION

The system application in Section 8–8 involves the electronic control circuits for a waste water treatment system. In particular, you will focus on the application of field-effect transistors in the sensing circuits for chemical measurements.

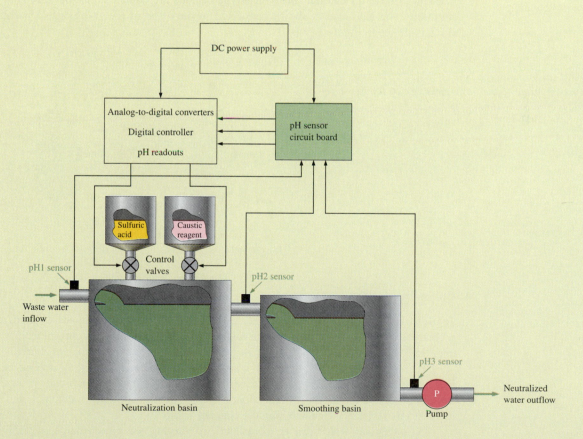

8–1 ■ THE JUNCTION FIELD-EFFECT TRANSISTOR (JFET)

The junction field-effect transistor (JFET) is a type of FET that operates with a reverse-biased junction to control current in a channel. Depending on their structure, JFETs fall into either of two categories, n channel or p channel.

After completing this section, you should be able to

■ **Explain the operation of JFETs**
 ☐ Identify the three terminals of a JFET
 ☐ Explain what a channel is
 ☐ Describe the structural difference between an *n*-channel JFET and a *p*-channel JFET
 ☐ Discuss how voltage controls the current in a JFET
 ☐ Identify the symbols for *n*-channel and *p*-channel JFETs

Figure 8–1(a) shows the basic structure of an *n*-channel **junction field-effect transistor (JFET)**. Wire leads are connected to each end of the *n*-channel; the **drain** is at the upper end, and the **source** is at the lower end. Two *p*-type regions are diffused in the *n*-type material to form a **channel,** and both *p*-type regions are connected to the **gate** lead. For simplicity, the gate lead is shown connected to only one of the *p* regions. A *p*-channel JFET is shown in Figure 8–1(b).

FIGURE 8–1

A representation of the basic structure of the two types of JFET.

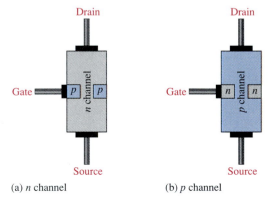

(a) *n* channel

(b) *p* channel

Basic Operation

To illustrate the operation of a JFET, Figure 8–2 shows bias voltages applied to an *n*-channel device. V_{DD} provides a drain-to-source voltage and supplies current from source to drain. V_{GG} sets the reverse-bias voltage between the gate and the source, as shown.

The JFET is *always* operated with the gate-source *pn* junction reverse-biased. Reverse-biasing of the gate-source junction with a negative gate voltage produces a depletion region along the *pn* junction, which extends into the *n* channel and thus increases its resistance by restricting the channel width.

The channel width can be controlled by varying the gate voltage, whereby the amount of drain current, I_D, can also be controlled. Figure 8–3 illustrates this concept. The white areas represent the depletion region created by the reverse bias. It is wider

FIGURE 8–2
A biased n-channel JFET.

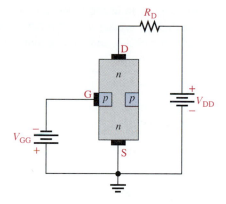

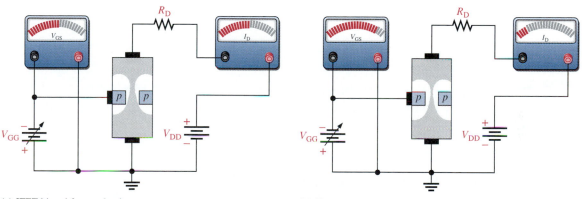

(a) JFET biased for conduction

(b) Greater V_{GG} narrows the channel and decreases I_D.

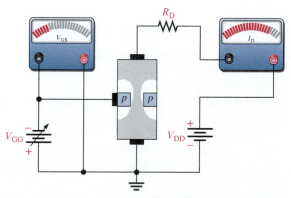

(c) Less V_{GG} widens the channel and increases I_D.

FIGURE 8–3
Effects of V_{GS} on channel width and on drain current ($V_{GG} = V_{GS}$).

toward the drain end of the channel because the reverse-bias voltage between the gate and the drain is greater than that between the gate and the source. We will discuss JFET characteristic curves and some important parameters in Section 8–2.

JFET Symbols

The schematic symbols for both *n*-channel and *p*-channel JFETs are shown in Figure 8–4. Notice that the arrow on the gate points "in" for *n* channel and "out" for *p* channel.

FIGURE 8–4
JFET schematic symbols.

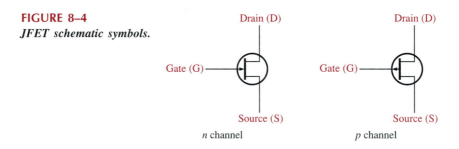

n channel   p channel

SECTION 8–1 REVIEW

1. Name the three terminals of a JFET.
2. Does an *n*-channel JFET require a positive or negative value for V_{GS}?
3. How is the drain current controlled in a JFET?

8–2 ■ JFET CHARACTERISTICS AND PARAMETERS

In this section, you will see how the JFET operates as a voltage-controlled, constant-current device. You will also learn about cutoff and pinch-off as well as JFET transfer characteristics.

After completing this section, you should be able to

■ **Define, discuss, and apply important JFET parameters**
  □ Explain ohmic region, constant-current region, and breakdown
  □ Define *pinch-off voltage*
  □ Describe how gate-to-source voltage controls the drain current
  □ Define *cutoff voltage*
  □ Compare pinch-off and cutoff
  □ Analyze a JFET transfer characteristic curve
  □ Use the equation for the transfer characteristic to calculate I_D
  □ Use a JFET data sheet
  □ Define *transconductance*
  □ Explain and determine input resistance and capacitance
  □ Determine drain-to-source resistance

First, let's consider the case where the gate-to-source voltage is zero ($V_{GS} = 0$ V). This is produced by shorting the gate to the source, as in Figure 8–5(a) where both are grounded. As V_{DD} (and thus V_{DS}) is increased from zero, I_D will increase proportionally through the *n*-type material, as shown in the graph of Figure 8–5(b) between points *A* and *B*. In this region, the channel resistance is essentially constant because the depletion region is not large enough to have significant effect. This is called the *ohmic region* because V_{DS} and I_D are related by Ohm's law.

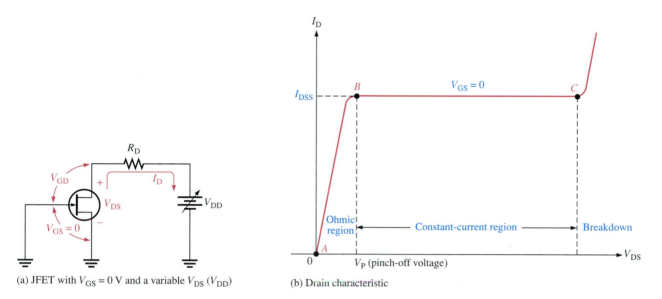

(a) JFET with $V_{GS} = 0$ V and a variable V_{DS} (V_{DD})

(b) Drain characteristic

FIGURE 8–5

The drain characteristic curve of a JFET for $V_{GS} = 0$ showing pinch-off.

At point *B* in Figure 8–5(b), the curve levels off and I_D becomes essentially constant. As V_{DS} increases from point *B* to point *C*, the reverse-bias voltage from gate to drain (V_{GD}) produces a depletion region large enough to offset the increase in V_{DS}, thus keeping I_D relatively constant.

Pinch-Off Voltage

For $V_{GS} = 0$ V, the value of V_{DS} at which I_D becomes essentially constant (point *B* on the curve in Figure 8–5(b)) is the **pinch-off voltage,** V_P. For a given JFET, V_P has a fixed value. As you can see, a continued increase in V_{DS} above the pinch-off voltage produces an almost constant drain current. This value of drain current is I_{DSS} (*D*rain to *S*ource current with gate *S*horted) and is always specified on JFET data sheets. I_{DSS} is the *maximum* drain current that a specific JFET can produce regardless of the external circuit, and it is always specified for the condition, $V_{GS} = 0$ V.

As shown in the graph in Figure 8–5(b), *breakdown* occurs at point *C* when I_D begins to increase very rapidly with any further increase in V_{DS}. Breakdown can result in irreversible damage to the device, so JFETs are always operated below breakdown and

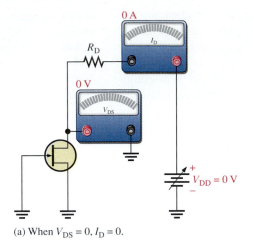

(a) When $V_{DS} = 0$, $I_D = 0$.

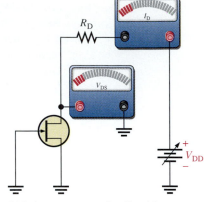

(b) I_D increases proportionally with V_{DS} in the ohmic region.

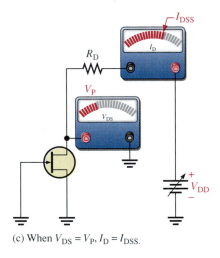

(c) When $V_{DS} = V_P$, $I_D = I_{DSS}$.

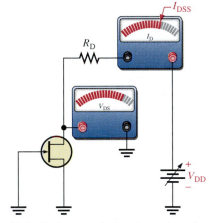

(d) As V_{DS} increases further, I_D remains at I_{DSS} unless breakdown occurs.

FIGURE 8–6

JFET action that produces the characteristic curve for $V_{GS} = 0$ V.

within the *constant-current region* (between points B and C on the graph). The JFET action that produces the drain characteristic curve to the point of breakdown for $V_{GS} = 0$ V is illustrated in Figure 8–6.

V_{GS} Controls I_D

Now, let's connect a bias voltage, V_{GG}, from gate to source as shown in Figure 8–7(a). As V_{GS} is set to increasingly more negative values by adjusting V_{GG}, a family of drain characteristic curves is produced, as shown in Figure 8–7(b). Notice that I_D decreases as the magnitude of V_{GS} is increased to larger negative values because of the narrowing of the channel. Also notice that, for each increase in V_{GS}, the JFET reaches pinch-off (where constant current begins) at values of V_{DS} less than V_P. So, the amount of drain current is controlled by V_{GS}, as illustrated in Figure 8–8.

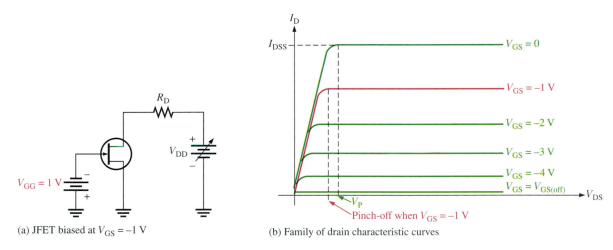

(a) JFET biased at $V_{GS} = -1$ V

(b) Family of drain characteristic curves

FIGURE 8–7

Pinch-off occurs at a lower V_{DS} as V_{GS} is increased to more negative values.

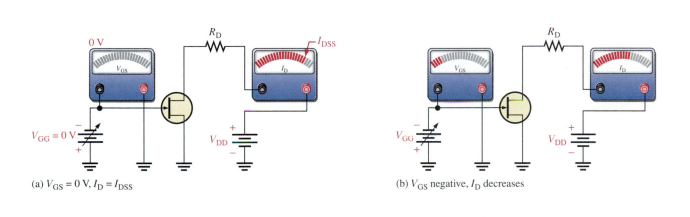

(a) $V_{GS} = 0$ V, $I_D = I_{DSS}$

(b) V_{GS} negative, I_D decreases

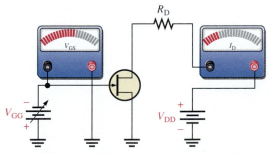

(c) I_D continues to decrease as V_{GS} is made more negative.

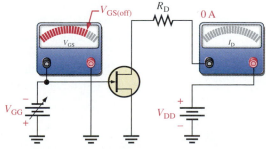

(d) I_D continues to decrease until $V_{GS} = -V_{GS(off)}$, where it becomes zero.

FIGURE 8–8

V_{GS} controls I_D.

Cutoff Voltage

The value of V_{GS} that makes I_D approximately zero is the **cutoff voltage,** $V_{GS(off)}$. The JFET must be operated between $V_{GS} = 0$ V and $V_{GS(off)}$. For this range of gate-to-source voltages, I_D will vary from a maximum of I_{DSS} to a minimum of almost zero.

As you have seen, for an *n*-channel JFET, the more negative V_{GS} is, the smaller I_D becomes in the constant-current region. When V_{GS} has a sufficiently large negative value, I_D is reduced to zero. This cutoff effect is caused by the widening of the depletion region to a point where it completely closes the channel as shown in Figure 8–9.

The basic operation of a *p*-channel JFET is the same as for an *n*-channel device except that a *p*-channel JFET requires a negative V_{DD} and a positive V_{GS}, as illustrated in Figure 8–10.

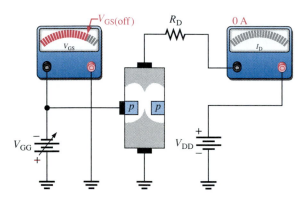

FIGURE 8–9
JFET at cutoff.

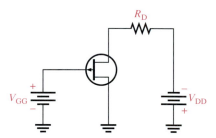

FIGURE 8–10
A biased p-channel JFET.

Comparison of Pinch-Off and Cutoff

As you have seen, there is definitely a difference between pinch-off and cutoff. There is also a connection. V_P is the value of V_{DS} at which the drain current becomes constant and is always measured at $V_{GS} = 0$ V. However, pinch-off occurs for V_{DS} values less than V_P when V_{GS} is nonzero. So, although V_P is a constant, the minimum value of V_{DS} at which I_D becomes constant varies with V_{GS}.

$V_{GS(off)}$ and V_P are always equal in magnitude but opposite in sign. A data sheet usually will give either $V_{GS(off)}$ or V_P, but not both. However, when you know one, you have the other. For example, if $V_{GS(off)} = -5$ V, then $V_P = +5$ V.

EXAMPLE 8–1   For the JFET in Figure 8–11, $V_{GS(off)} = -4$ V and $I_{DSS} = 12$ mA. Determine the *minimum* value of V_{DD} required to put the device in the constant-current region of operation.

FIGURE 8–11

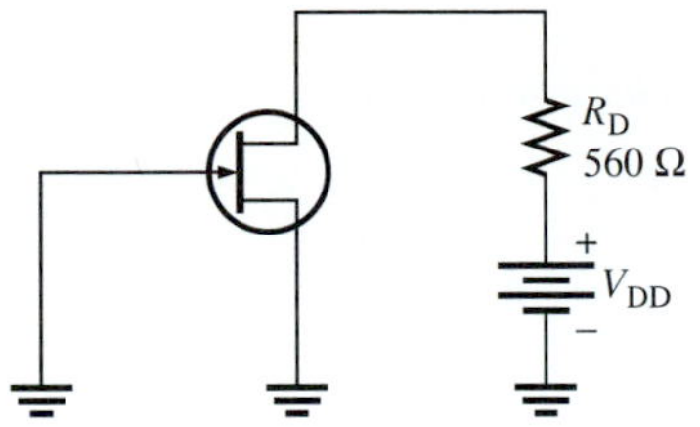

Solution Since $V_{GS(off)} = -4$ V, $V_P = 4$ V. The minimum value of V_{DS} for the JFET to be in its constant-current region is

$$V_{DS} = V_P = 4 \text{ V}$$

In the constant-current region with $V_{GS} = 0$ V,

$$I_D = I_{DSS} = 12 \text{ mA}$$

The drop across the drain resistor is

$$V_{RD} = I_D R_D = (12 \text{ mA})(560 \text{ } \Omega) = 6.72 \text{ V}$$

Apply Kirchhoff's law around the drain circuit.

$$V_{DD} = V_{DS} + V_{RD} = 4 \text{ V} + 6.72 \text{ V} = 10.7 \text{ V}$$

This is the value of V_{DD} to make $V_{DS} = V_P$ and put the device in the constant-current region.

Related Exercise If V_{DD} is increased to 15 V, what is the drain current?

EXAMPLE 8–2

A particular *p*-channel JFET has a $V_{GS(off)} = +4$ V. What is I_D when $V_{GS} = +6$ V?

Solution The *p*-channel JFET requires a positive gate-to-source voltage. The more positive the voltage, the less the drain current. When $V_{GS} = 4$ V, I_D is 0. Any further increase in V_{GS} keeps the JFET cut off, so I_D remains 0.

Related Exercise What is V_P for the JFET described in this example?

JFET Transfer Characteristic

You have learned that a range of V_{GS} values from zero to $V_{GS(off)}$ controls the amount of drain current. For an *n*-channel JFET, $V_{GS(off)}$ is negative, and for a *p*-channel JFET, $V_{GS(off)}$ is positive. Because V_{GS} does control I_D, the relationship between these two quantities is very important. Figure 8–12 is a typical transfer characteristic curve that illustrates graphically the relationship between V_{GS} and I_D.

Notice that the bottom end of the curve is at a point on the V_{GS} axis equal to $V_{GS(off)}$, and the top end of the curve is at a point on the I_D axis equal to I_{DSS}. This curve, of course, shows that the operating limits of a JFET are

$$I_D = 0 \quad \text{when } V_{GS} = V_{GS(off)}$$

FIGURE 8–12

JFET transfer characteristic curve (n-channel).

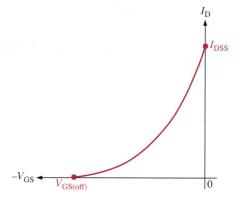

and

$$I_D = I_{DSS} \quad \text{when } V_{GS} = 0$$

The transfer characteristic curve can be developed from the drain characteristic curves by plotting values of I_D for the values of V_{GS} taken from the family of drain curves in the pinch-off region, as illustrated in Figure 8–13 for a specific set of curves. Each point on the transfer characteristic curve corresponds to specific values of V_{GS} and I_D on the drain curves. For example, when $V_{GS} = -2$ V, $I_D = 4.32$ mA. Also, for this specific JFET, $V_{GS(off)} = -5$ V and $I_{DSS} = 12$ mA.

A JFET transfer characteristic curve is nearly parabolic in shape and can therefore be expressed approximately as

$$I_D = I_{DSS}\left(1 - \frac{V_{GS}}{V_{GS(off)}}\right)^2 \tag{8-1}$$

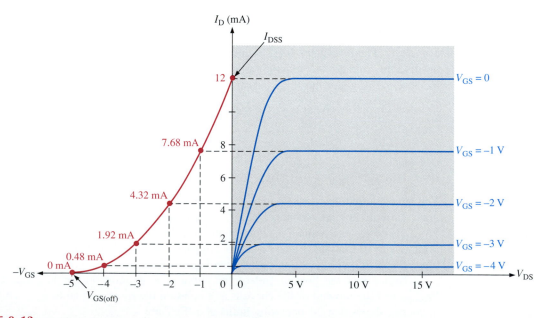

FIGURE 8–13

Example of the development of an n-channel JFET transfer characteristic curve (left) from the JFET drain characteristic curves (right).

With Equation (8–1), I_D can be determined for any V_{GS} if $V_{GS(off)}$ and I_{DSS} are known. These quantities are usually available from the data sheet for a given JFET. Notice the squared term in the equation. Because of its form, a parabolic relationship is known as a *square law*, and therefore, JFETs and MOSFETs are often referred to as *square-law devices*.

The data sheet for a typical JFET series is shown in Figure 8–14.

FIGURE 8–14

JFET data sheet.

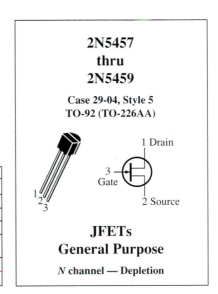

**2N5457
thru
2N5459**

**Case 29-04, Style 5
TO-92 (TO-226AA)**

1 Drain

3 Gate

2 Source

**JFETs
General Purpose**

N channel — Depletion

Maximum Ratings

Rating	Symbol	Value	Unit
Drain-Source voltage	V_{DS}	25	V dc
Drain-Gate voltage	V_{DG}	25	V dc
Reverse gate-source voltage	V_{GSR}	−25	V dc
Gate current	I_G	10	mA dc
Total device dissipation @ $T_A = 25°C$ Derate above 25°C	P_D	310 2.82	mW mW/C°
Junction temperature	T_J	125	°C
Storage channel temperature range	T_{stg}	−65 to +150	°C

Electrical Characteristics ($T_A = 25°C$ unless otherwise noted.)

Characteristic		Symbol	Min	Typ	Max	Unit
OFF Characteristics						
Gate-Source breakdown voltage ($I_G = −10$ μA dc, $V_{DS} = 0$)		$V_{(BR)GSS}$	−25	–	–	V dc
Gate reverse current		I_{GSS}				nA dc
($V_{GS} = −15$ V dc, $V_{DS} = 0$)			–	–	−1.0	
($V_{GS} = −15$ V dc, $V_{DS} = 0, T_A = 100°C$)			–	–	−200	
Gate-Source cutoff voltage		$V_{GS(off)}$				V dc
($V_{DS} = 15$ V dc, $I_D = 10$ nA dc)	2N5457		−0.5	–	−6.0	
	2N5458		−1.0	–	−7.0	
	2N5459		−2.0	–	−8.0	
Gate-Source voltage		V_{GS}				V dc
($V_{DS} = 15$ V dc, $I_D = 100$ μA dc)	2N5457		–	−2.5	–	
($V_{DS} = 15$ V dc, $I_D = 200$ μA dc)	2N5458		–	−3.5	–	
($V_{DS} = 15$ V dc, $I_D = 400$ μA dc)	2N5459		–	−4.5	–	
ON Characteristics						
Zero-Gate-Voltage drain current		I_{DSS}				mA dc
($V_{DS} = 15$ V dc, $V_{GS} = 0$)	2N5457		1.0	3.0	5.0	
	2N5458		2.0	6.0	9.0	
	2N5459		4.0	9.0	16	
Small-signal Characteristics						
Forward transfer admittance common source		$|y_{fs}|$				μmhos or μS
($V_{DS} = 15$ V dc, $V_{GS} = 0, f = 1.0$ kHz)	2N5457		1000	–	5000	
	2N5458		1500	–	5500	
	2N5459		2000	–	6000	
Output admittance common source ($V_{DS} = 15$ V dc, $V_{GS} = 0, f = 1.0$ kHz)		$|y_{os}|$	–	10	50	μmhos or μS
Input capacitance ($V_{DS} = 15$ V dc, $V_{GS} = 0, f = 1.0$ MHz)		C_{iss}	–	4.5	7.0	pF
Reverse transfer capacitance ($V_{DS} = 15$ V dc, $V_{GS} = 0, f = 1.0$ MHz)		C_{rss}	–	1.5	3.0	pF

EXAMPLE 8–3

The data sheet in Figure 8–14 for a 2N5459 JFET indicates that typically $I_{DSS} = 9$ mA and $V_{GS(off)} = -8$ V (maximum). Determine the drain current for $V_{GS} = 0$ V, -1 V, and -4 V.

Solution   For $V_{GS} = 0$ V, $I_D = I_{DSS} = 9$ mA. For $V_{GS} = -1$ V, use Equation (8–1).

$$I_D = I_{DSS}\left(1 - \frac{V_{GS}}{V_{GS(off)}}\right)^2 = (9 \text{ mA})\left(1 - \frac{-1 \text{ V}}{-8 \text{ V}}\right)^2$$

$$= (9 \text{ mA})(1 - 0.125)^2 = (9 \text{ mA})(0.766) = 6.89 \text{ mA}$$

For $V_{GS} = -4$ V:

$$I_D = (9 \text{ mA})\left(1 - \frac{-4 \text{ V}}{-8 \text{ V}}\right)^2 = (9 \text{ mA})(1 - 0.5)^2 = (9 \text{ mA})(0.25) = 2.25 \text{ mA}$$

Related Exercise   Determine I_D for $V_{GS} = -3$ V for the 2N5459 JFET.

JFET Forward Transconductance

The forward transfer conductance (**transconductance**), g_m, is the change in drain current (ΔI_D) for a given change in gate-to-source voltage (ΔV_{GS}) with the drain-to-source voltage constant. It is expressed as a ratio and has the unit of siemens (S).

$$g_m = \frac{\Delta I_D}{\Delta V_{GS}}$$

Other common designations for this parameter are g_{fs} and y_{fs} (forward transfer admittance). As you will see in Chapter 9, g_m is important in FET amplifiers as a major factor in determining the voltage gain.

Because the transfer characteristic curve for a JFET is nonlinear, g_m varies in value depending on the location on the curve as set by V_{GS}. The value for g_m is greater near the top of the curve (near $V_{GS} = 0$) than it is near the bottom (near $V_{GS(off)}$), as illustrated in Figure 8–15. A data sheet normally gives the value of g_m measured at $V_{GS} = 0$ V (g_{m0}). For example, the data sheet for the 2N5457 JFET specifies a minimum g_{m0} (y_{fs}) of 1000 μS with $V_{DS} = 15$ V.

Given g_{m0}, you can calculate an approximate value for g_m at any point on the transfer characteristic curve using the following formula:

$$g_m = g_{m0}\left(1 - \frac{V_{GS}}{V_{GS(off)}}\right) \tag{8–2}$$

When a value of g_{m0} is not available, you can calculate it using values of I_{DSS} and $V_{GS(off)}$. The vertical lines indicate an absolute value (no sign).

$$g_{m0} = \frac{2I_{DSS}}{|V_{GS(off)}|} \tag{8–3}$$

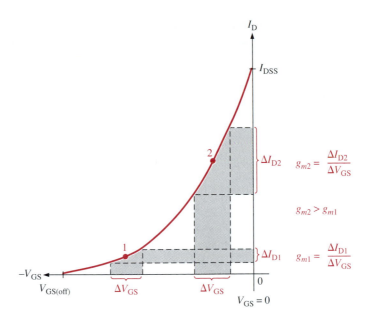

FIGURE 8–15

g_m varies depending on the bias point (V_{GS}).

EXAMPLE 8–4

The following information is included on the data sheet in Figure 8–14 for a 2N5457 JFET: typically, $I_{DSS} = 3.0$ mA, $V_{GS(off)} = -6$ V maximum, and $y_{fs(max)} = 5000$ μS. Determine the forward transconductance for $V_{GS} = -4$ V, and find I_D at this point.

Solution $g_{m0} = y_{fs} = 5000$ μS. Use Equation (8–2) to calculate g_m.

$$g_m = g_{m0}\left(1 - \frac{V_{GS}}{V_{GS(off)}}\right) = (5000\ \mu S)\left(1 - \frac{-4\ V}{-6\ V}\right) = 1667\ \mu S$$

Next, use Equation (8–1) to calculate I_D at $V_{GS} = -4$ V.

$$I_D = I_{DSS}\left(1 - \frac{V_{GS}}{V_{GS(off)}}\right)^2 = (3.0\ mA)\left(1 - \frac{-4\ V}{-6\ V}\right)^2 = 333\ \mu A$$

Related Exercise   A given JFET has the following characteristics: $I_{DSS} = 12$ mA, $V_{GS(off)} = -5$ V, and $g_{m0} = 3000$ μS. Find g_m and I_D when $V_{GS} = -2$ V.

Input Resistance and Capacitance

A JFET operates with its gate-source junction reverse-biased. Therefore, the input resistance at the gate is very high. This high input resistance is one advantage of the JFET over the bipolar transistor. (Recall that a bipolar transistor operates with a forward-biased base-emitter junction.) JFET data sheets often specify the input resistance by giving a value for the gate reverse current I_{GSS} at a certain gate-to-source voltage. The input resistance can then be determined using the following equation:

$$R_{IN} = \left| \frac{V_{GS}}{I_{GSS}} \right|$$ (8–4)

For example, the 2N5457 data sheet in Figure 8–14 lists a maximum I_{GSS} of 1 nA for $V_{GS} = -15$ V at 25°C. I_{GSS} increases with temperature, so the input resistance decreases.

The input capacitance, C_{iss}, is a result of the JFET operating with a reverse-biased *pn* junction. Recall that a reverse-biased *pn* junction acts as a capacitor whose capacitance depends on the amount of reverse voltage. For example, the 2N5457 has a maximum C_{iss} of 7 pF for $V_{GS} = 0$.

EXAMPLE 8–5

A certain JFET has an I_{GSS} of 2 nA for $V_{GS} = -20$ V. Determine the input resistance.

Solution

$$R_{IN} = \left| \frac{V_{GS}}{I_{GSS}} \right| = \frac{20 \text{ V}}{2 \text{ nA}} = 10,000 \text{ M}\Omega$$

Related Exercise Determine the minimum input resistance for the 2N5458 from the data sheet in Figure 8–14.

Drain-to-Source Resistance

You learned from the drain characteristic curve that, above pinch-off, the drain current is relatively constant over a range of drain-to-source voltages. Therefore, a large change in V_{DS} produces only a very small change in I_D. The ratio of these changes is the drain-to-source resistance of the device, r'_{ds}.

$$r'_{ds} = \frac{\Delta V_{DS}}{\Delta I_D}$$

Data sheets often specify this parameter as output conductance, g_{os}, or output admittance, y_{os}.

SECTION 8–2 REVIEW

1. The drain-to-source voltage at the pinch-off point of a particular JFET is 7 V. If the gate-to-source voltage is zero, what is V_P?
2. The V_{GS} of a certain *n*-channel JFET is increased negatively. Does the drain current increase or decrease?
3. What value must V_{GS} have to produce cutoff in a *p*-channel JFET with a $V_P = -3$ V?

8–3 ■ JFET BIASING

Using some of the FET parameters discussed in the previous sections, we will now see how to dc-bias JFETs. The purpose of biasing is to select the proper dc gate-to-source voltage to establish a desired value of drain current and, thus, a proper Q-point. You will learn about two major types of bias circuits, self-bias and voltage-divider bias.

After completing this section, you should be able to

■ **Discuss and analyze JFET bias circuits**
  ☐ Describe self-bias
  ☐ Analyze a self-biased JFET circuit
  ☐ Set the self-biased Q-point
  ☐ Analyze a voltage-divider-biased JFET circuit
  ☐ Use transfer characteristic curves to analyze JFET bias circuits
  ☐ Discuss Q-point stability

Self-Bias

Recall that a JFET must be operated such that the gate-source junction is always reverse-biased. This condition requires a negative V_{GS} for an *n*-channel JFET and a positive V_{GS} for a *p*-channel JFET. This can be achieved using the self-bias arrangements shown in Figure 8–16. The gate resistor, R_G, does not affect the bias because it has essentially no voltage drop across it; and therefore the gate remains at 0 V. R_G is necessary only to isolate an ac signal from ground in amplifier applications, as you will see later.

For the *n*-channel JFET in Figure 8–16(a), I_S produces a voltage drop across R_S and makes the source positive with respect to ground. Since $I_S = I_D$ and $V_G = 0$, then $V_S = I_D R_S$. The gate-to-source voltage is

$$V_{GS} = V_G - V_S = 0 - I_D R_S$$

so

$$V_{GS} = -I_D R_S \tag{8–5}$$

FIGURE 8–16
Self-biased JFETs ($I_S = I_D$ in all FETs).

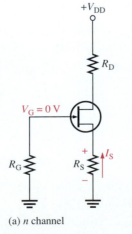

(a) *n* channel

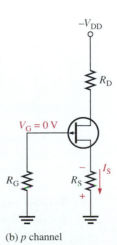

(b) *p* channel

For the p-channel JFET shown in Figure 8–16(b), the current through R_S produces a negative voltage at the source and therefore

$$V_{GS} = +I_DR_S \qquad\qquad (8\text{–}6)$$

In the following analysis, the n-channel JFET in Figure 8–16(a) is used for illustration. Keep in mind that analysis of the p-channel JFET is the same except for opposite-polarity voltages. The drain voltage with respect to ground is determined as follows:

$$V_D = V_{DD} - I_DR_D$$

Since $V_S = I_DR_S$, the drain-to-source voltage is

$$V_{DS} = V_D - V_S = V_{DD} - I_D(R_D + R_S)$$

EXAMPLE 8–6   Find V_{DS} and V_{GS} in Figure 8–17, given that $I_D = 5$ mA.

FIGURE 8–17

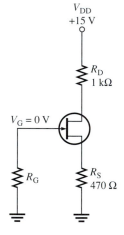

Solution

$$V_S = I_DR_S = (5\text{ mA})(470\ \Omega) = 2.35\text{ V}$$

$$V_D = V_{DD} - I_DR_D = 15\text{ V} - (5\text{ mA})(1\text{ k}\Omega) = 15\text{ V} - 5\text{ V} = 10\text{ V}$$

Therefore,

$$V_{DS} = V_D - V_S = 10\text{ V} - 2.35\text{ V} = 7.65\text{ V}$$

Since $V_G = 0$ V,

$$V_{GS} = V_G - V_S = 0\text{ V} - 2.35\text{ V} = -2.35\text{ V}$$

Related Exercise   Determine V_{DS} and V_{GS} in Figure 8–17 when $I_D = 8$ mA. Assume that $R_D = 860\ \Omega$, $R_S = 390\ \Omega$, and $V_{DD} = 12$ V.

Setting the Q-Point of a Self-Biased JFET

The basic approach to establishing a JFET bias point is to determine I_D for a desired value of V_{GS} or vice versa. Then calculate the required value of R_S using the relationship derived from Equation (8–5) and stated in Equation (8–7). The vertical lines indicate an absolute value.

$$R_S = \left| \frac{V_{GS}}{I_D} \right| \qquad\qquad (8\text{–}7)$$

For a desired value of V_{GS}, I_D can be determined in either of two ways: from the transfer characteristic curve for the particular JFET, or more practically, from Equation (8–1) using I_{DSS} and $V_{GS(off)}$ from the JFET data sheet. The next two examples illustrate these procedures.

EXAMPLE 8–7

Determine the value of R_S required to self-bias an n-channel JFET having the transfer characteristic curve shown in Figure 8–18 at $V_{GS} = -5$ V.

FIGURE 8–18

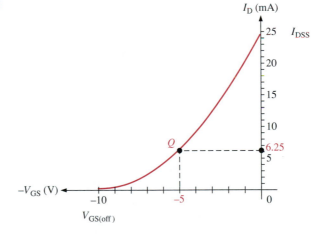

Solution From the graph, $I_D = 6.25$ mA when $V_{GS} = -5$ V. Calculate R_S:

$$R_S = \left| \frac{V_{GS}}{I_D} \right| = \frac{5 \text{ V}}{6.25 \text{ mA}} = 800 \ \Omega$$

Related Exercise Find R_S for $V_{GS} = -3$ V.

EXAMPLE 8–8

Determine the value of R_S required to self-bias a p-channel JFET with $I_{DSS} = 25$ mA and $V_{GS(off)} = 15$ V. V_{GS} is to be 5 V.

Solution   Use Equation (8–1) to calculate I_D.

$$I_D = I_{DSS}\left(1 - \frac{V_{GS}}{V_{GS(off)}}\right)^2 = (25 \text{ mA})\left(1 - \frac{5 \text{ V}}{15 \text{ V}}\right)^2 = (25 \text{ mA})(1 - 0.333)^2 = 11.1 \text{ mA}$$

Now, determine R_S.

$$R_S = \left|\frac{V_{GS}}{I_D}\right| = \frac{5 \text{ V}}{11.1 \text{ mA}} = 450 \text{ }\Omega$$

Since 450 Ω is not a standard value, use a 470 Ω resistor.

Related Exercise   Find the value of R_S required to self-bias a p-channel JFET with $I_{DSS} = 18$ mA and $V_{GS(off)} = 8$ V. $V_{GS} = 4$ V.

Midpoint Bias   It is often desirable to bias a JFET near the midpoint of its transfer characteristic curve where $I_D = I_{DSS}/2$. Under signal conditions, midpoint bias allows a maximum amount of drain current swing between I_{DSS} and 0. Using Equation (8–1), it is shown in Appendix B that I_D is approximately one-half of I_{DSS} when $V_{GS} = V_{GS(off)}/3.4$.

$$I_D = I_{DSS}\left(1 - \frac{V_{GS}}{V_{GS(off)}}\right)^2 = I_{DSS}\left(1 - \frac{V_{GS(off)}/3.4}{V_{GS(off)}}\right)^2 = 0.5 I_{DSS}$$

So, by selecting $V_{GS} = V_{GS(off)}/3.4$, you get a midpoint bias in terms of I_D.

To set the drain voltage at midpoint ($V_D = V_{DD}/2$), select a value of R_D to produce the desired voltage drop. Choose R_G arbitrarily large to prevent loading on the driving stage in a cascaded amplifier arrangement. Example 8–9 illustrates these concepts.

EXAMPLE 8–9

Select resistor values in Figure 8–19 to set up an approximate midpoint bias. The JFET parameters are $I_{DSS} = 15$ mA and $V_{GS(off)} = -8$ V. V_D should be 6 V (one-half of V_{DD}).

FIGURE 8–19

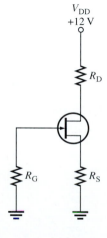

Solution   For midpoint bias,

$$I_D \cong \frac{I_{DSS}}{2} = 7.5 \text{ mA}$$

and

$$V_{GS} \cong \frac{V_{GS(off)}}{3.4} = \frac{-8 \text{ V}}{3.4} = -2.35 \text{ V}$$

Then

$$R_S = \left| \frac{V_{GS}}{I_D} \right| = \frac{2.35 \text{ V}}{7.5 \text{ mA}} = 313 \text{ } \Omega$$
$$V_D = V_{DD} - I_D R_D$$
$$I_D R_D = V_{DD} - V_D$$
$$R_D = \frac{V_{DD} - V_D}{I_D} = \frac{12 \text{ V} - 6 \text{ V}}{7.5 \text{ mA}} = 800 \text{ } \Omega$$

Use the nearest standard values of 330 Ω and 820 Ω.

Related Exercise   Select resistor values in Figure 8–19 to set up an approximate midpoint bias. The JFET parameters are $I_{DSS} = 10$ mA and $V_{GS(off)} = -10$ V. $V_{DD} = 15$ V.

Graphical Analysis of a Self-Biased JFET

You can use the transfer characteristic curve of a JFET and certain parameters to determine the Q-point (I_D and V_{GS}) of a self-biased circuit. A circuit is shown in Figure 8–20(a), and a transfer characteristic curve is shown in Figure 8–20(b). If a curve is not available from the data sheet, you can plot it from Equation (8–1) using data sheet values for I_{DSS} and $V_{GS(off)}$.

   To determine the Q-point of the circuit in Figure 8–20(a), a self-bias dc load line is established as follows. First, calculate V_{GS} when I_D is zero.

$$V_{GS} = -I_D R_S = (0)(470 \text{ } \Omega) = 0 \text{ V}$$

This establishes a point at the origin on the graph ($I_D = 0$, $V_{GS} = 0$). Next, get I_{DSS} from the data sheet and calculate V_{GS} when $I_D = I_{DSS}$. From the curve, $I_{DSS} = 10$ mA for the JFET in Figure 8–20.

$$V_{GS} = -I_D R_S = -(10 \text{ mA})(470 \text{ } \Omega) = -4.7 \text{ V}$$

This establishes a second point on the graph ($I_D = 10$ mA, $V_{GS} = -4.7$ V). Now, with two points, the load line can be drawn on the graph of the transfer characteristic curve as shown in Figure 8–21. The point where the line intersects the transfer characteristic curve is the Q-point of the circuit.

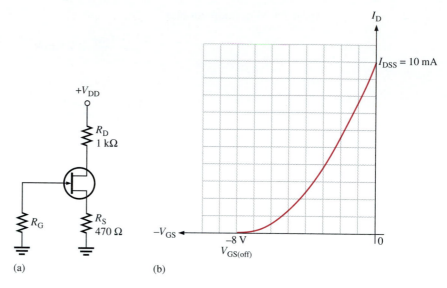

(a) (b)

FIGURE 8–20

A self-biased JFET and its transfer characteristic curve.

FIGURE 8–21

The intersection of the self-bias dc load line and the transfer characteristic curve is the Q-point.

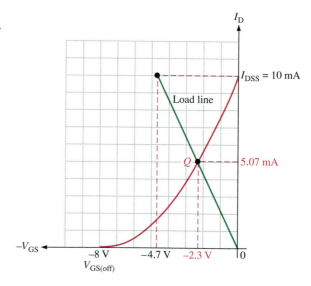

EXAMPLE 8–10 Determine the Q-point for the JFET circuit in Figure 8–22(a). The transfer characteristic curve is given in Figure 8–22(b).

Solution For $I_D = 0$,

$$V_{GS} = -I_D R_S = (0)(680 \ \Omega) = 0 \text{ V}$$

This gives a point at the origin. From the curve, $I_{DSS} = 4$ mA. So for $I_D = I_{DSS} = 4$ mA.

$$V_{GS} = -I_D R_S = -(4 \text{ mA})(680 \text{ }\Omega) = -2.72 \text{ V}$$

This gives a second point at 4 mA and −2.72 V. A line is now drawn between the two points, and the values of I_D and V_{GS} at the intersection of the line and the curve are taken from the graph, as illustrated in Figure 8–22(b). The Q-point values from the graph are $I_D = 2.25$ mA and $V_{GS} = -1.5$ V.

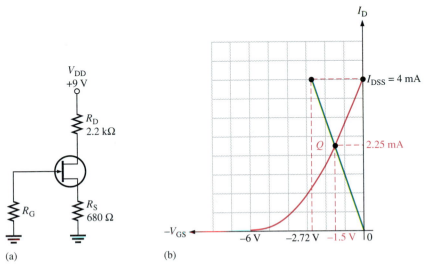

(a)   (b)

FIGURE 8–22

Related Exercise If R_S is increased to 1 kΩ in Figure 8–22(a), what is the new Q-point?

Voltage-Divider Bias

An *n*-channel JFET with voltage-divider bias is shown in Figure 8–23. The voltage at the source of the JFET must be more positive than the voltage at the gate in order to keep the gate-source junction reverse-biased.

FIGURE 8–23
***An n-channel JFET with voltage-divider
bias ($I_S = I_D$).***

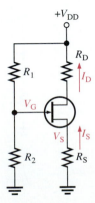

The source voltage is

$$V_S = I_D R_S$$

The gate voltage is set by resistors R_1 and R_2 as expressed by the following equation using the voltage-divider formula:

$$V_G = \left(\frac{R_2}{R_1 + R_2} \right) V_{DD} \qquad (8\text{–}8)$$

The gate-to-source voltage is

$$V_{GS} = V_G - V_S$$

and the source voltage is

$$V_S = V_G - V_{GS}$$

The drain current can be expressed as

$$I_D = \frac{V_S}{R_S}$$

Substituting for V_S,

$$I_D = \frac{V_G - V_{GS}}{R_S} \qquad (8\text{–}9)$$

EXAMPLE 8–11

Determine I_D and V_{GS} for the JFET with voltage-divider bias in Figure 8–24, given that $V_D = 7$ V.

FIGURE 8–24

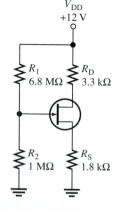

V_{DD}
+12 V

R_1 6.8 MΩ R_D 3.3 kΩ

R_2 1 MΩ R_S 1.8 kΩ

Solution

$$I_D = \frac{V_{DD} - V_D}{R_D} = \frac{12\ V - 7\ V}{3.3\ k\Omega} = \frac{5\ V}{3.3\ k\Omega} = 1.52\ mA$$

Calculate the gate-to-source voltage as follows:

$$V_S = I_D R_S = (1.52 \text{ mA})(1.8 \text{ k}\Omega) = 2.74 \text{ V}$$

$$V_G = \left(\frac{R_2}{R_1 + R_2}\right)V_{DD} = \left(\frac{1 \text{ M}\Omega}{7.8 \text{ M}\Omega}\right)12 \text{ V} = 1.54 \text{ V}$$

$$V_{GS} = V_G - V_S = 1.54 \text{ V} - 2.74 \text{ V} = -1.2 \text{ V}$$

If V_D had not been given in this example, the Q-point values could not have been found without the transfer characteristic curve.

Related Exercise Given that $V_D = 6$ V in Figure 8–24, determine the Q-point.

Graphical Analysis of a JFET with Voltage-Divider Bias

An approach similar to the one used for self-bias can be used with voltage-divider bias to graphically determine the Q-point of a circuit on the transfer characteristic curve.

In a JFET with voltage-divider bias when $I_D = 0$, V_{GS} is not zero, as in the self-biased case, because the voltage divider produces a voltage at the gate independent of the drain current. The voltage-divider dc load line is determined as follows.

For $I_D = 0$,

$$V_S = I_D R_S = (0)R_S = 0 \text{ V}$$

$$V_{GS} = V_G - V_S = V_G - 0 \text{ V} = V_G$$

Therefore, one point on the line is at $I_D = 0$ and $V_{GS} = V_G$.

For $V_{GS} = 0$,

$$I_D = \frac{V_G - V_{GS}}{R_S} = \frac{V_G}{R_S}$$

A second point on the line is at $I_D = V_G/R_S$ and $V_{GS} = 0$. The dc load line is shown in Figure 8–25.

FIGURE 8–25

DC load line for a JFET with voltage-divider bias.

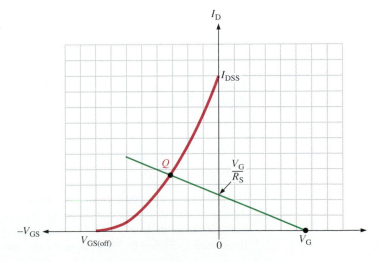

EXAMPLE 8–12

Determine the Q-point for the JFET with voltage-divider bias in Figure 8–26(a), given the transfer characteristic curve in Figure 8–26(b).

Solution   First, establish the two points for the bias line.
For $I_D = 0$,

$$V_{GS} = V_G = \left(\frac{R_2}{R_1 + R_2}\right)V_{DD} = \left(\frac{2.2 \text{ M}\Omega}{4.4 \text{ M}\Omega}\right)8 \text{ V} = 4 \text{ V}$$

The first point is at $I_D = 0$ and $V_{GS} = 4$ V.
For $V_{GS} = 0$,

$$I_D = \frac{V_G - V_{GS}}{R_S} = \frac{V_G}{R_S} = \frac{4 \text{ V}}{3.3 \text{ k}\Omega} = 1.2 \text{ mA}$$

The second point is at $I_D = 1.2$ mA and $V_{GS} = 0$.
The load line is drawn in Figure 8–26(b), and the Q-point values of $I_D = 1.7$ mA and $V_{GS} = -2$ V are picked off the graph, as indicated.

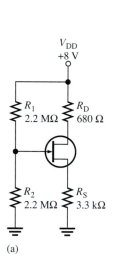

(a)

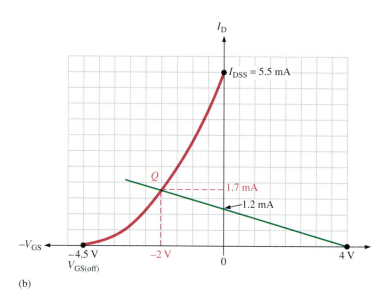

(b)

FIGURE 8–26

Related Exercise   Change R_S to 4.7 kΩ and determine the Q-point for the circuit in Figure 8–26(a).

Q-Point Stability

Unfortunately, the transfer characteristic of a JFET can differ considerably from one device to another of the same type. If, for example, a 2N5459 JFET is replaced in a given bias circuit with another 2N5459, the transfer characteristic curve can vary greatly as illustrated in Figure 8–27(a). In this case, the maximum I_{DSS} is 16 mA and the minimum

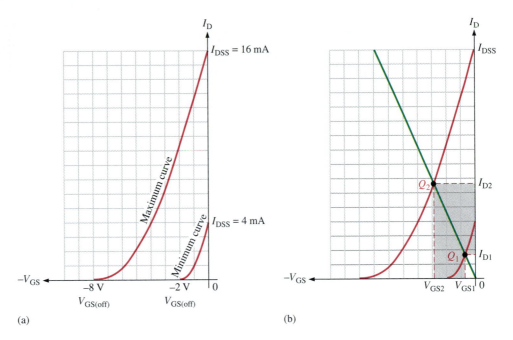

FIGURE 8–27

Variation in the transfer characteristic of 2N5459 JFETs and the effect on the Q-point.

I_{DSS} is 4 mA. Likewise, the maximum $V_{GS(off)}$ is −8 V and the minimum $V_{GS(off)}$ is −2 V. This means that if you have a selection of 2N5459s and you randomly pick one out, it can have values anywhere within these ranges.

If a self-bias dc load line is drawn as illustrated in Figure 8–27(b), the same circuit using a 2N5459 can have a Q-point anywhere along the line from Q_1, the minimum bias point, to Q_2, the maximum bias point. Accordingly, the drain current can be any value between I_{D1} and I_{D2}, as shown. This means that the dc voltage at the drain can have a range of values depending on I_D. Also, the gate-to-source voltage can be any value between V_{GS1} and V_{GS2}, as indicated.

With voltage-divider bias, the dependency of I_D on the range of Q-points is reduced because the slope of the bias line is less than for self-bias. Although V_{GS} varies quite a bit for both self-bias and voltage-divider bias, I_D is much more stable with voltage-divider bias, as illustrated in Figure 8–28.

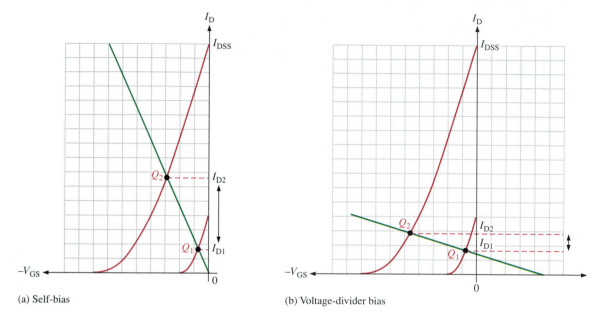

FIGURE 8–28

The change in I_D between the minimum and the maximum Q-points is much less for a JFET with voltage-divider bias than for a self-biased JFET.

<table>
<tr>
<td>

SECTION 8–3 REVIEW

</td>
<td>

1. Should a *p*-channel JFET have a positive or a negative V_{GS}?

2. In a certain self-biased *n*-channel JFET circuit, $I_D = 8$ mA and $R_S = 1$ kΩ. Determine V_{GS}.

3. An *n*-channel JFET with voltage-divider bias has a gate voltage of 3 V and a source voltage of 5 V. Calculate V_{GS}.

</td>
</tr>
</table>

8–4 ■ THE METAL OXIDE SEMICONDUCTOR FET (MOSFET)

The metal oxide semiconductor field-effect transistor (MOSFET) is the second category of field-effect transistor. The MOSFET differs from the JFET in that it has no pn junction structure; instead, the gate of the MOSFET is insulated from the channel by a silicon dioxide (SiO_2) layer. The two basic types of MOSFETs are depletion (D) and enhancement (E). Because of the insulated gate, these devices are sometimes called IGFETs.

After completing this section, you should be able to

■ **Explain the operation of MOSFETs**
  ☐ Describe the structural difference between an *n*-channel and a *p*-channel depletion MOSFET (D-MOSFET)
  ☐ Explain the depletion mode
  ☐ Explain the enhancement mode
  ☐ Identify the symbols for *n*-channel and *p*-channel D-MOSFETs

☐ Describe the structural difference between an *n*-channel and a *p*-channel enhance-ment MOSFET (E-MOSFET)
☐ Identify the symbols for *n*-channel and *p*-channel E-MOSFETs
☐ Explain how D-MOSFETs and E-MOSFETs differ
☐ Discuss power MOSFETs
☐ Discuss dual-gate MOSFETs

Depletion MOSFET (D-MOSFET)

One type of **MOSFET** is the depletion MOSFET (D-MOSFET) and Figure 8–29 illus-trates its basic structure. The drain and source are diffused into the substrate material and then connected by a narrow channel adjacent to the insulated gate. Both *n*-channel and *p*-channel devices are shown in the figure. We will use the *n*-channel device to describe the basic operation. The *p*-channel operation is the same, except the voltage polarities are opposite those of the *n*-channel.

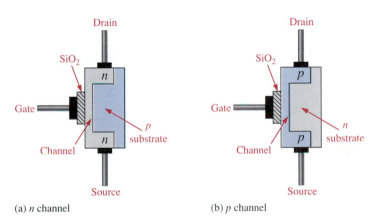

(a) *n* channel   (b) *p* channel

FIGURE 8–29
Representation of the basic structure of D-MOSFETs.

The D-MOSFET can be operated in either of two modes—the depletion mode or the enhancement mode—and is sometimes called a *depletion/enhancement MOSFET*. Since the gate is insulated from the channel, either a positive or a negative gate voltage can be applied. The *n*-channel MOSFET operates in the **depletion** mode when a negative gate-to-source voltage is applied and in the **enhancement** mode when a positive gate-to-source voltage is applied. These devices are generally operated in the depletion mode.

Depletion Mode   Visualize the gate as one plate of a parallel-plate capacitor and the channel as the other plate. The silicon dioxide insulating layer is the dielectric. With a negative gate voltage, the negative charges on the gate repel conduction electrons from the channel, leaving positive ions in their place. Thereby, the *n* channel is depleted of some of its electrons, thus decreasing the channel conductivity. The greater the negative voltage on the gate, the greater the depletion of *n*-channel electrons. At a sufficiently neg-ative gate-to-source voltage, $V_{GS(off)}$, the channel is totally depleted and the drain current

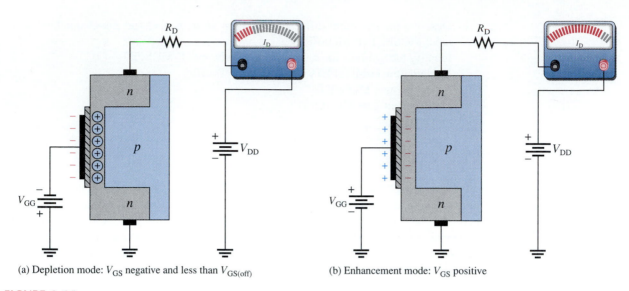

(a) Depletion mode: V_{GS} negative and less than $V_{GS(off)}$ (b) Enhancement mode: V_{GS} positive

FIGURE 8–30
Operation of n-channel D-MOSFET.

is zero. This depletion mode is illustrated in Figure 8–30(a). Like the *n*-channel JFET, the *n*-channel D-MOSFET conducts drain current for gate-to-source voltages between $V_{GS(off)}$ and zero. In addition, the D-MOSFET conducts for values of V_{GS} above zero.

Enhancement Mode With a positive gate voltage, more conduction electrons are attracted into the channel, thus increasing (enhancing) the channel conductivity, as illustrated in Figure 8–30(b).

D-MOSFET Symbols The schematic symbols for both the *n*-channel and the *p*-channel depletion MOSFETs are shown in Figure 8–31. The substrate, indicated by the arrow, is normally (but not always) connected internally to the source. Sometimes, there is a separate substrate pin. An inward substrate arrow is for *n* channel, and an outward arrow is for *p* channel.

FIGURE 8–31
D-MOSFET schematic symbols.

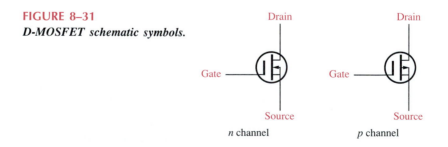

Enhancement MOSFET (E-MOSFET)

The E-MOSFET operates *only* in the enhancement mode and has no depletion mode. It differs in construction from the D-MOSFET in that it has no structural channel. Notice in Figure 8–32(a) that the substrate extends completely to the SiO_2 layer. For an *n*-channel

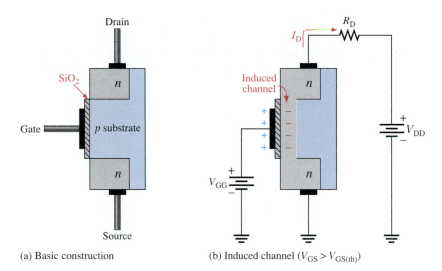

(a) Basic construction   (b) Induced channel ($V_{GS} > V_{GS(th)}$)

FIGURE 8–32
E-MOSFET construction and operation (n-channel).

device, a positive gate voltage above a threshold value *induces* a channel by creating a thin layer of negative charges in the substrate region adjacent to the SiO$_2$ layer, as shown in Figure 8–32(b). The conductivity of the channel is enhanced by increasing the gate-to-source voltage and thus pulling more electrons into the channel area. For any gate voltage below the threshold value, there is no channel.

The schematic symbols for the *n*-channel and *p*-channel E-MOSFETs are shown in Figure 8–33. The broken lines symbolize the absence of a physical channel. Like the D-MOSFET, some devices have a separate substrate connection.

Power MOSFETs

The conventional enhancement MOSFETs have a long thin lateral channel as shown in the structural view in Figure 8–34. This results in a relatively high drain-to-source resistance and limits the E-MOSFET to low power applications. When the gate is positive, the channel is formed close to the gate between the source and the drain, as shown.

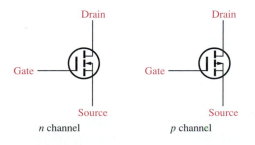

n channel   *p* channel

FIGURE 8–33
E-MOSFET schematic symbols.

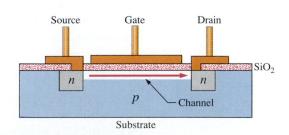

FIGURE 8–34
Cross section of conventional E-MOSFET structure.

Lateral Double Diffused MOSFET (LDMOSFET) The LDMOSFET is a type of enhancement MOSFET designed for power applications. This device has a shorter channel between drain and source than does the conventional E-MOSFET. The shorter channel results in lower resistance which allows higher current and voltage.

Figure 8–35 shows the basic structure of an LDMOSFET. When the gate is positive, a very short n channel is induced in the p layer between the lightly doped source and the n^- region. Current flows from the source through the n regions and the induced channel to the drain, as indicated.

VMOSFET The V-groove MOSFET is another variation of the conventional E-MOSFET designed to achieve higher power capability by creating a shorter and wider channel with less resistance between the drain and source. The shorter, wider channels allow for higher currents and, thus, greater power dissipation. Frequency response is also improved.

The VMOSFET has two source connections, the gate connection on top, and the drain connection on the bottom, as shown in Figure 8–36. The channel is induced vertically along both sides of the V-shaped groove between the drain (n^+ substrate where n^+ means a higher doping level than n^-) and the source connections. The channel length is set by the thickness of the layers, which is controlled by doping densities and diffusion time rather than by mask dimensions.

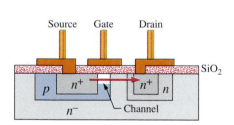

FIGURE 8–35
Cross section of LDMOSFET structure.

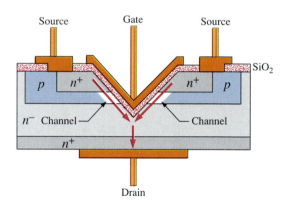

FIGURE 8–36
Cross section of VMOS structure.

TMOS TMOS is similar to VMOS except that it doesn't use a V-shaped groove and is, therefore, easier to manufacture. The structure of TMOS is illustrated in Figure 8–37. The gate structure is embedded in a silicon dioxide layer, and the source contact is continuous over the entire surface area. The drain is on the bottom. TMOS achieves greater packing density than VMOS, while retaining the short vertical channel advantage.

Dual-Gate MOSFETs

The dual-gate MOSFET can be either a depletion or an enhancement type. The only difference is that it has two gates, as shown in Figure 8–38. As previously mentioned, one drawback of an FET is its high input capacitance, which restricts its use at higher frequencies. By using a dual-gate device, the input capacitance is reduced, thus making the device useful in high-frequency RF amplifier applications. Another advantage of the dual-gate arrangement is that it allows for an automatic gain control (AGC) input in certain RF amplifiers.

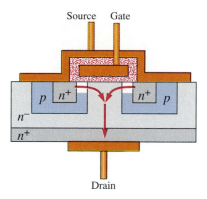

FIGURE 8–37
Cross section of TMOS structure.

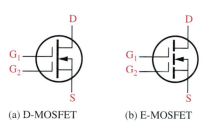

(a) D-MOSFET   (b) E-MOSFET

FIGURE 8–38
Dual-gate n-channel MOSFET symbols.

**SECTION 8–4
REVIEW**

  1. Name the two basic types of MOSFETs.

  2. If the gate-to-source voltage in a depletion MOSFET is zero, what is the current from drain to source?

  3. If the gate-to-source voltage in an E-MOSFET is zero, what is the current from drain to source?

8–5 ■ MOSFET CHARACTERISTICS AND PARAMETERS

Much of the discussion concerning JFET characteristics and parameters applies equally to MOSFETs. In this section, the differences are highlighted.

After completing this section, you should be able to

■ **Define, discuss, and apply important MOSFET parameters**
  ☐ Analyze a D-MOSFET transfer characteristic curve
  ☐ Use the equation for the D-MOSFET transfer characteristic to calculate I_D
  ☐ Analyze an E-MOSFET transfer characteristic curve
  ☐ Use the equation for the E-MOSFET transfer characteristic to calculate I_D
  ☐ Use a MOSFET data sheet
  ☐ Discuss handling precautions for MOS devices

D-MOSFET Transfer Characteristic

As previously discussed, the D-MOSFET can operate with either positive or negative gate voltages. This is indicated on the transfer characteristic curves in Figure 8–39 for both *n*-channel and *p*-channel MOSFETs. The point on the curves where $V_{GS} = 0$ corresponds to I_{DSS}. The point where $I_D = 0$ corresponds to $V_{GS(off)}$. As with the JFET, $V_{GS(off)} = -V_P$.

    The square-law expression in Equation (8–1) for the JFET curve also applies to the D-MOSFET curve, as the following example demonstrates.

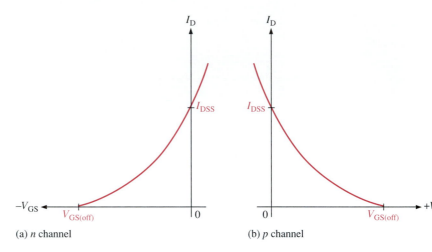

FIGURE 8–39
D-MOSFET transfer characteristic curves.

EXAMPLE 8–13

For a certain D-MOSFET, $I_{DSS} = 10$ mA and $V_{GS(off)} = -8$ V.

(a) Is this an *n*-channel or a *p*-channel?

(b) Calculate I_D at $V_{GS} = -3$ V.

(c) Calculate I_D at $V_{GS} = +3$ V.

Solution

(a) The device has a negative $V_{GS(off)}$; therefore, it is an *n*-channel MOSFET.

(b) $I_D = I_{DSS}\left(1 - \dfrac{V_{GS}}{V_{GS(off)}}\right)^2 = (10 \text{ mA})\left(1 - \dfrac{-3 \text{ V}}{-8 \text{ V}}\right)^2 = 3.91 \text{ mA}$

(c) $I_D = (10 \text{ mA})\left(1 - \dfrac{+3 \text{ V}}{-8 \text{ V}}\right)^2 = 18.9 \text{ mA}$

Related Exercise For a certain D-MOSFET, $I_{DSS} = 18$ mA and $V_{GS(off)} = +10$ V.

(a) Is this an *n* channel or a *p* channel?

(b) Determine I_D at $V_{GS} = +4$ V.

(c) Determine I_D at $V_{GS} = -4$ V.

E-MOSFET Transfer Characteristic

The E-MOSFET uses only channel enhancement. Therefore, an *n*-channel device requires a positive gate-to-source voltage, and a *p*-channel device requires a negative gate-to-source voltage. Figure 8–40 shows the transfer characteristic curves for both types of E-MOSFETs. As you can see, there is no drain current when $V_{GS} = 0$. Therefore, the E-MOSFET does not have a significant I_{DSS} parameter, as do the JFET and the D-MOSFET. Notice also that there is ideally no drain current until V_{GS} reaches a certain nonzero value called the *threshold voltage*, $V_{GS(th)}$.

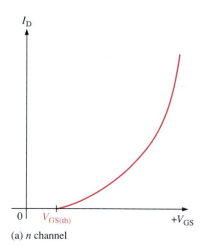
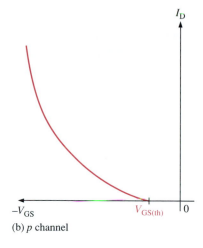

(a) *n* channel   (b) *p* channel

FIGURE 8–40
E-MOSFET transfer characteristic curves.

The equation for the parabolic transfer characteristic curve of the E-MOSFET differs from that of the JFET and the D-MOSFET because the curve starts at $V_{GS(th)}$ rather than $V_{GS(off)}$ on the horizontal axis and never intersects the vertical axis. The equation for the E-MOSFET transfer characteristic curve is

$$I_D = K(V_{GS} - V_{GS(th)})^2 \qquad\qquad \textbf{(8–10)}$$

The constant K depends on the particular MOSFET and can be determined from the data sheet by taking the specified value of I_D, called $I_{D(on)}$, at the given value of V_{GS} and substituting the values into Equation (8–10). A typical E-MOSFET data sheet is given in Figure 8–41.

EXAMPLE 8–14

The data sheet in Figure 8–41 for a 2N7008 E-MOSFET gives $I_{D(on)}$ = 500 mA (minimum) at V_{GS} = 10 V and $V_{GS(th)}$ = 1 V. Determine the drain current for V_{GS} = 5 V.

Solution   First, solve for K using Equation (8–10).

$$K = \frac{I_{D(on)}}{(V_{GS} - V_{GS(th)})^2} = \frac{500 \text{ mA}}{(10 \text{ V} - 1 \text{ V})^2} = \frac{500 \text{ mA}}{81 \text{ V}^2} = 6.17 \text{ mA/V}^2$$

Next, using the value of K, calculate I_D for V_{GS} = 5 V.

$$I_D = K(V_{GS} - V_{GS(th)})^2 = 6.17 \text{ mA/V}^2(5 \text{ V} - 1 \text{ V})^2 = 98.7 \text{ mA}$$

Related Exercise   The data sheet for an E-MOSFET gives $I_{D(on)}$ = 100 mA at V_{GS} = 8 V and $V_{GS(th)}$ = 4 V. Find I_D when V_{GS} = 6 V.

Maximum Ratings

Rating	Symbol	Value	Unit
Drain-Source voltage	V_{DSS}	60	V dc
Drain-Gate voltage ($R_{GS} = 1$ MΩ)	V_{DGR}	-60	V dc
Gate-Source voltage	V_{GS}	±40	V dc
Drain current			mA dc
Continuous	I_D	150	
Pulsed	I_{DM}	1000	
Total power dissipation @ $T_A = 25°C$	P_D	400	mW
Derate above 25°C		3.2	mW/C°
Operating and storage temperature range	T_J, T_{stg}	-55 to $+150$	°C

Thermal Characteristics

Thermal resistance junction to ambient	$R_{\theta JA}$	312.5	C°/W
Maximum lead temperature for soldering purposes, 1/16" from case for 10 seconds	T_L	300	°C

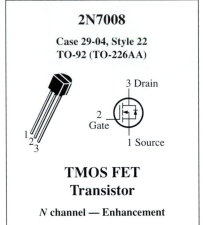

2N7008

Case 29-04, Style 22
TO-92 (TO-226AA)

3 Drain

2 Gate

1 Source

TMOS FET
Transistor

N **channel — Enhancement**

Electrical Characteristics ($T_C = 25°C$ unless otherwise noted.)

Characteristic	Symbol	Min	Max	Unit
OFF Characteristics				
Drain-Source breakdown voltage ($V_{GS} = 0$, $I_D = 100$ μA)	$V_{(BR)DSS}$	60	–	V dc
Zero gate voltage drain current	I_{DSS}			μA dc
($V_{DS} = 50$ V, $V_{GS} = 0$)		–	1.0	
($V_{DS} = 50$ V, $V_{GS} = 0$, $T_J = 125°C$)		–	500	
Gate-Body leakage current, forward ($V_{GSF} = 30$ V dc, $V_{DS} = 0$)	I_{GSSF}	–	-100	nA dc
ON Characteristics				
Gate threshold voltage ($V_{DS} = V_{GS}$, $I_D = 250$ μA)	$V_{GS(th)}$	1.0	2.5	V dc
Static drain-source on-resistance	$r_{DS(on)}$			Ohm
($V_{GS} = 5.0$ V dc, $I_D = 50$ A dc)		–	7.5	
($V_{GS} = 10$ V dc, $I_D = 500$ mA dc, $T_C = 125°C$)		–	13.5	
Drain-Source on-voltage	$V_{DS(on)}$			V dc
($V_{GS} = 5.0$ V, $I_D = 50$ mA)		–	1.5	
($V_{GS} = 10$ V, $I_D = 500$ mA)		–	3.75	
On-state drain current ($V_{GS} = 10$ V, $V_{DS} \geq 2.0V_{D(on)}$)	$I_{D(on)}$	500	–	mA
Forward transconductance ($V_{DS} \geq 2.0V_{DS(on)}$, $I_D = 200$ mA)	g_{fs}	80	–	μmhos or μS
Dynamic Characteristics				
Input capacitance	C_{iss}	–	50	pF
Output capacitance	($V_{DS} = 25$ V, $V_{GS} = 0$ $f = 1.0$ MHz) C_{oss}	–	25	
Reverse transfer capacitance	C_{rss}	–	5.0	
Switching Characteristics				
Turn-On delay time	($V_{DD} = 30$ V, $I_D = 200$ mA t_{on}	–	20	ns
Turn-Off delay time	$R_{gen} = 25$ ohms, $R_L = 150$ ohms) t_{off}	–	20	

FIGURE 8–41
Data sheet for the 2N7008 n-channel E-MOSFET (TMOS construction).

Handling Precautions

All MOS devices are subject to damage from electrostatic discharge (ESD). Because the gate of a MOSFET is insulated from the channel, the input resistance is extremely high (ideally infinite). The gate leakage current I_{GSS} for a typical MOSFET is in the pA range, whereas the gate reverse current for a typical JFET is in the nA range. The input capacitance results from the insulated gate structure. Excess static charge can be accumulated because the input capacitance combines with the very high input resistance and can result in damage to the device. To avoid damage from ESD, certain precautions should be taken when handling MOSFETs:

1. MOS devices should be shipped and stored in conductive foam.

2. All instruments and metal benches used in assembly or test should be connected to earth ground (round or third prong of 110 V wall outlets).

3. The assembler's or handler's wrist should be connected to earth ground with a length of wire and a high-value series resistor.

4. Never remove an MOS device (or any other device, for that matter) from the circuit while the power is on.

5. Do not apply signals while the dc power supply is off.

SECTION 8–5 REVIEW

1. What is the major difference in construction of the D-MOSFET and the E-MOS-FET?

2. Name two parameters of an E-MOSFET that are not specified for D-MOSFETs?

3. What is ESD?

8–6 ■ MOSFET BIASING

In this section, we will look at three ways to bias a MOSFET: zero-bias, voltage-divider bias, and drain-feedback bias. As you know, biasing is very important in amplifier applications, which you will study in the next chapter.

After completing this section, you should be able to

■ **Discuss and analyze MOSFET bias circuits**
   □ Describe zero-bias of a D-MOSFET
   □ Analyze a zero-biased MOSFET circuit
   □ Describe voltage-divider bias of an E-MOSFET
   □ Describe drain-feedback bias of an E-MOSFET
   □ Analyze MOSFET bias circuits

D-MOSFET Bias

Recall that D-MOSFETs can be operated with either positive or negative values of V_{GS}. A simple bias method is to set $V_{GS} = 0$ so that an ac signal at the gate varies the gate-to-source voltage above and below this bias point. A MOSFET with zero bias is shown in Figure 8–42(a). Since $V_{GS} = 0$, $I_D = I_{DSS}$ as indicated. The drain-to-source voltage is expressed as follows.

$$V_{DS} = V_{DD} - I_{DSS}R_D \qquad (8–11)$$

The purpose of R_G is to accommodate an ac signal input by isolating it from ground, as shown in Figure 8–42(b). Since there is no dc gate current, R_G does not affect the zero gate-to-source bias.

FIGURE 8–42

A zero-biased D-MOSFET.

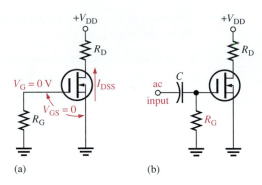

(a)   (b)

EXAMPLE 8–15

Determine the drain-to-source voltage in the circuit of Figure 8–43. The MOSFET data sheet gives $V_{GS(off)} = -8$ V and $I_{DSS} = 12$ mA.

FIGURE 8–43

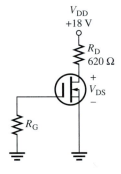

Solution   Since $I_D = I_{DSS} = 12$ mA, the drain-to-source voltage is calculated as follows:

$$V_{DS} = V_{DD} - I_{DSS}R_D = 18\ V - (12\ mA)(620\ \Omega) = 10.6\ V$$

Related Exercise   Find V_{DS} in Figure 8–43 when $V_{GS(off)} = -10$ V and $I_{DSS} = 20$ mA.

E-MOSFET Bias

Recall that E-MOSFETs must have a V_{GS} greater than the threshold value, $V_{GS(th)}$, so zero bias cannot be used. Figure 8–44 shows two ways to bias an E-MOSFET (D-MOSFETs can also be biased using these methods). An *n*-channel device is used for purposes of illustration. In either the voltage-divider or drain-feedback bias arrangement, the purpose is to make the gate voltage more positive than the source by an amount exceeding $V_{GS(th)}$. Equations for the analysis of the voltage-divider bias in Figure 8–44(a) are as follows:

$$V_{GS} = \left(\frac{R_2}{R_1 + R_2} \right) V_{DD}$$

$$V_{DS} = V_{DD} - I_D R_D$$

where $I_D = K(V_{GS} - V_{GS(th)})^2$ from Equation (8–10).

FIGURE 8–44
Common E-MOSFET biasing arrangements.

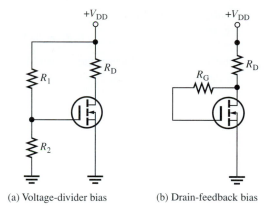

(a) Voltage-divider bias   (b) Drain-feedback bias

In the drain-feedback bias circuit in Figure 8–44(b), there is negligible gate current and, therefore, no voltage drop across R_G. This makes $V_{GS} = V_{DS}$.

EXAMPLE 8–16

Determine V_{GS} and V_{DS} for the E-MOSFET circuit in Figure 8–45. The data sheet for this particular MOSFET gives minimum values of $I_{D(on)} = 500$ mA at $V_{GS} = 10$ V and $V_{GS(th)} = 1$ V.

FIGURE 8–45

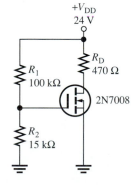

Solution

$$V_{GS} = \left(\frac{R_2}{R_1 + R_2} \right) V_{DD} = \left(\frac{15 \text{ k}\Omega}{115 \text{ k}\Omega} \right) 24 \text{ V} = 3.13 \text{ V}$$

You found K in Example 8–14 using Equation (8–10) as follows:

$$K = \frac{I_{D(on)}}{(V_{GS} - V_{GS(th)})^2} = \frac{500 \text{ mA}}{(10 \text{ V} - 1 \text{ V})^2} = \frac{500 \text{ mA}}{81 \text{ V}^2} = 6.17 \text{ mA/V}^2$$

Now calculate I_D for $V_{GS} = 3.13$ V.

$$I_D = K(V_{GS} - V_{GS(th)})^2 = (6.17 \text{ mA/V}^2)(3.13 \text{ V} - 1 \text{ V})^2$$
$$= (6.17 \text{ mA/V}^2)(2.13 \text{ V})^2 = 28 \text{ mA}$$

Finally, calculate V_{DS}.

$$V_{DS} = V_{DD} - I_D R_D = 24 \text{ V} - (28 \text{ mA})(470 \text{ }\Omega) = 10.8 \text{ V}$$

Related Exercise   Determine V_{GS} and V_{DS} for the 2N7008 E-MOSFET in Figure 8–45 for the maximum specified value of $V_{GS(th)}$. Refer to the data sheet in Figure 8–41.

EXAMPLE 8–17

Determine the amount of drain current in Figure 8–46. The MOSFET has a $V_{GS(th)} = 3$ V.

FIGURE 8–46

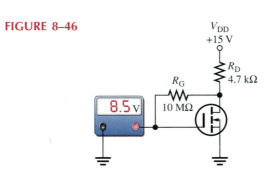

Solution   The meter indicates $V_{GS} = 8.5$ V. Since this is a drain-feedback configuration, $V_{DS} = V_{GS} = 8.5$ V.

$$I_D = \frac{V_{DD} - V_{DS}}{R_D} = \frac{15 \text{ V} - 8.5 \text{ V}}{4.7 \text{ k}\Omega} = 1.38 \text{ mA}$$

Related Exercise   Determine I_D if the meter in Figure 8–46 reads 5 V.

**SECTION 8–6
REVIEW**

1. For a D-MOSFET biased at $V_{GS} = 0$, is the drain current equal to zero, I_{GSS}, or I_{DSS}?

2. For an *n*-channel E-MOSFET with $V_{GS(th)} = 2$ V, V_{GS} must be in excess of what value in order to conduct?

SUMMARY OF FIELD-EFFECT TRANSISTORS

JFETs

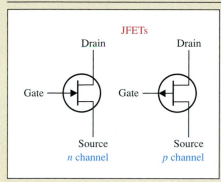

- Gate-source junction must be reverse-biased.
- V_{GS} controls I_D.
- Value of V_{DS} at which I_D becomes constant is the pinch-off voltage.
- Value of V_{GS} at which I_D becomes zero is the cutoff voltage, $V_{GS(off)}$.
- I_{DSS} is drain current when $V_{GS} = 0$.
- Transfer characteristic:

$$I_D = I_{DSS}\left(1 - \frac{V_{GS}}{V_{GS(off)}}\right)^2$$

- Forward transconductance:

$$g_m = g_{m0}\left(1 - \frac{V_{GS}}{V_{GS(off)}}\right)$$

$$g_{m0} = \frac{2I_{DSS}}{|V_{GS(off)}|}$$

D-MOSFETs

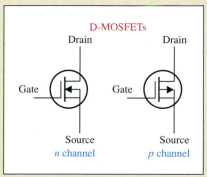

Except that it can be operated in enhancement mode, the D-MOSFET characteristics are the same as JFET.

- *Depletion mode:*
  n channel: V_{GS} negative
  p channel: V_{GS} positive
- *Enhancement mode:*
  n channel: V_{GS} positive
  p channel: V_{GS} negative
- V_{GS} controls I_D.
- Value of V_{GS} at which I_D becomes zero is the cutoff voltage, $V_{GS(off)}$.
- I_{DSS} is drain current when $V_{GS} = 0$.
- Transfer characteristic:

$$I_D = I_{DSS}\left(1 - \frac{V_{GS}}{V_{GS(off)}}\right)^2$$

SUMMARY OF FIELD-EFFECT TRANSISTORS, *continued*

E-MOSFETs

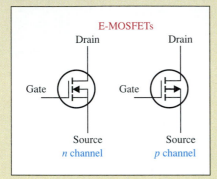

E-MOSFETs

Drain Drain

Gate Gate

Source Source

n channel *p* channel

There is no depletion mode and characteristics differ from D-MOSFET.

- *Enhancement mode:*

  n channel: V_{GS} positive

  p channel: V_{GS} negative

- V_{GS} controls I_D.

- Value of V_{GS} at which I_D starts to flow is the threshold voltage, $V_{GS(th)}$.

- Transfer characteristic:

$$I_D = K(V_{GS} - V_{GS(th)})^2$$

- K in formula can be calculated by substituting data sheet values $I_{D(on)}$ for I_D and V_{GS} at which $I_{D(on)}$ is specified for V_{GS}.

FET BIASING (voltage polarities and current directions reverse for *p* channel)

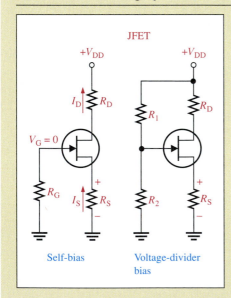

JFET

Self-bias Voltage-divider bias

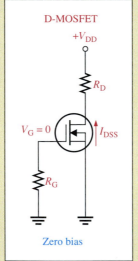

D-MOSFET

Zero bias

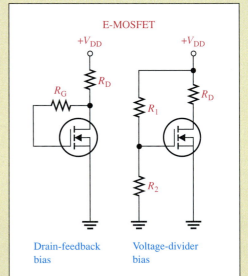

E-MOSFET

Drain-feedback bias Voltage-divider bias

8–7 ■ TROUBLESHOOTING

In this section, we discuss some common faults that may be encountered in several FET circuits and discuss probable causes for each fault.

After completing this section, you should be able to

■ **Troubleshoot FET circuits**
   ☐ Troubleshoot self-biased JFET circuits
   ☐ Troubleshoot zero-biased MOSFET circuits

Faults in Self-Biased JFET Circuits

Symptom 1: $V_D = V_{DD}$ For this condition, the drain current must be zero because there is no voltage drop across R_D, as illustrated in Figure 8–47. As in any circuit, it is good troubleshooting practice to first check for obvious problems such as open or poor connections, as well as charred resistors. Next, disconnect power and measure suspected resistors for opens. If these are OK, the JFET is probably bad. A list of faults that can produce this symptom is as follows:

1. No ground connection at R_S
2. R_S open
3. Open drain lead connection
4. Open source lead connection
5. FET internally open between drain and source

Symptom 2: V_D Significantly Less Than Normal For this condition, unless the supply voltage is lower than it should be, the drain current must be larger than normal because the drop across R_D is too much. Figure 8–47(b) indicates this situation. This symptom can be caused by any one of the following:

1. Open R_G
2. Open gate lead
3. FET internally open at gate

FIGURE 8–47
Two symptoms in a self-biased JFET circuit.

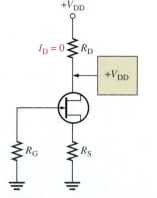
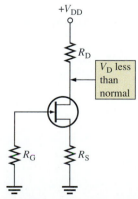

(a) Drain voltage equal to supply voltage   (b) Drain voltage less than normal

Any one of these faults will cause the depletion region in the JFET to disappear and the channel to widen, so that the drain current is limited only by R_D, R_S, and the small channel resistance.

Faults in D-MOSFET and E-MOSFET Circuits

One fault that is difficult to detect is when the gate opens in a zero-biased D-MOSFET or an E-MOSFET with drain-feedback bias. In a zero-biased D-MOSFET, the gate-to-source voltage remains zero when an open occurs in the gate circuit; thus, the drain current doesn't change, and the bias appears normal, as indicated in Figure 8–48.

In an E-MOSFET circuit with voltage-divider bias, an open R_1 makes the gate voltage zero. This causes the transistor to be off and act like an open switch because a gate-to-source threshold voltage greater than zero is required to turn the device on. This condition is illustrated in Figure 8–49(a).

If R_2 opens, the gate is at $+V_{DD}$ and the channel resistance is very low so the device approximates a closed switch. The drain current is limited only by R_D. This condition is illustrated in Figure 8–49(b).

FIGURE 8–48

An open fault in the gate circuit of a D-MOSFET causes no change in I_D.

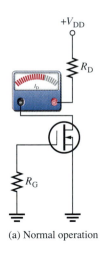
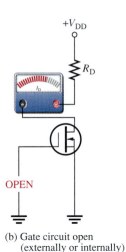

(a) Normal operation

(b) Gate circuit open
(externally or internally)

FIGURE 8–49

Failures in an E-MOSFET circuit with voltage-divider bias.

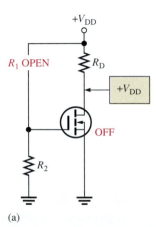
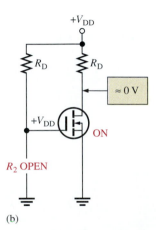

(a)

(b)

SECTION 8–7 REVIEW

1. In a self-biased JFET circuit, the drain voltage equals V_{DD}. If the JFET is OK, what are other possible faults?

2. Why doesn't the drain current change when an open occurs in the gate circuit of a zero-biased D-MOSFET circuit?

3. If the gate of an E-MOSFET becomes shorted to ground in a circuit with voltage-divider bias, what is the drain voltage?

8–8 ■ SYSTEM APPLICATION

This application involves electronic instrumentation for waste water treatment facilities. You have been given responsibility for evaluating the circuits used in a waste water neutralization system. Although both digital and analog circuits are used, you will begin by focusing on the pH sensor circuit board and apply the knowledge you have gained in this chapter in completing your assignment.

The Waste Water Neutralization System

Basic Operation   The diagram of the waste water neutralization system is shown in Figure 8–50. The system measures and controls the pH of waste water. The pH is a measure of the degree of acidity or alkalinity of a solution. Values of pH range from 0 for the strongest acids, through 7 for neutral solutions, on up to 14 for the strongest bases. Typically waste water is not a strong acid or a strong base, so the range of pH values is typically greater than 2 and less than 11. The pH of the water is measured by sensors at the inlet and outlet of the neutralization basin and at the outlet of the smoothing basin where the pH should be 7, indicating a neutral solution.

The pH sensor produces a small voltage proportional to the pH of the liquid in which it is immersed. The output voltage of each pH sensor is fed to the gate of a MOSFET on the pH sensor circuit board. The small gate voltage from the sensor controls the drain current which produces an output voltage at the drain that is inversely proportional to the gate voltage but with a larger magnitude. Rheostats are used as drain resistors for calibrating each circuit individually so that, for a given pH, all the output voltages are equal. This is necessary because of variation in the MOSFET characteristics from one device to another.

The MOSFET output voltages go to the analog-to-digital converters and the digital controller. Based on the digitized pH values, the controller determines whether to add sulfuric acid or caustic reagent to the water and the amount that should be added. The digital controller activates the control valves for the correct amount of chemical to properly adjust the pH level. Also, the digitized pH values are sent to a display panel for visual monitoring.

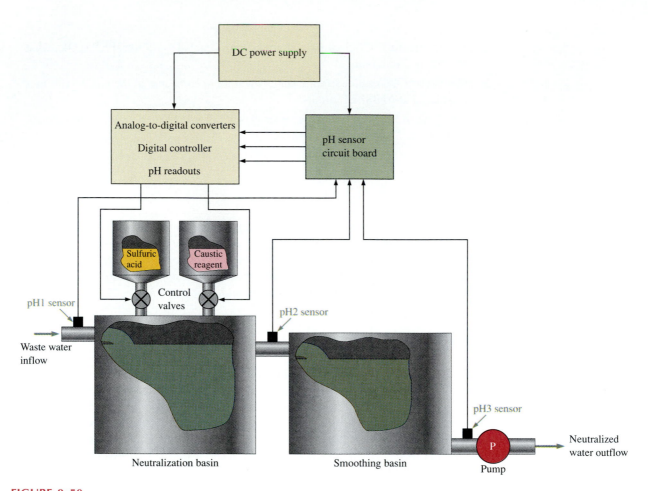

FIGURE 8–50

Diagram of the waste water neutralization system.

The Printed Circuit Board

☐ The system documentation is incomplete and the schematic for the pH sensor circuit board is missing. The transistors are 2N3797.

☐ From the circuit board in Figure 8–51, create a schematic and label all components. There are two feedthrough interconections on the back side.

☐ Label a copy of the board with component and input/output designations in agreement with the schematic.

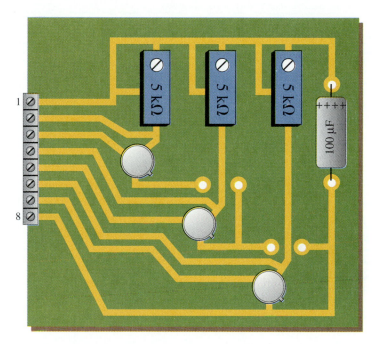

FIGURE 8–51
pH sensor circuit board.

Maximum Ratings

Rating	Symbol	Value	Unit
Drain-Source voltage 2N3796 2N3797	V_{DS}	 25 20	V dc
Gate-Source voltage	V_{GS}	±10	V dc
Drain current	I_D	20	mA dc
Total device dissipation @ $T_A = 25°C$ Derate above 25°C	P_D	200 1.14	mW mW/C°
Junction temperature range	T_J	+175	°C
Storage channel temperature range	T_{stg}	−65 to +200	°C

2N3796
2N3797
Case 22-03, Style 2
TO-18 (TO-206AA)
3 Drain
Gate 2
3 2 1   1 Source
MOSFETs
Low Power Audio
N channel — Depletion

Electrical Characteristics ($T_A = 25°C$ unless otherwise noted.)

Characteristic	Symbol	Min	Typ	Max	Unit
OFF Characteristics					
Drain-Source breakdown voltage ($V_{GS} = -4.0$ V, $I_D = 5.0$ μA)　　2N3796 ($V_{GS} = -7.0$ V, $I_D = 5.0$ μA)　　2N3797	$V_{(BR)DSX}$	 25 20	 30 25	 – –	V dc
Gate reverse current ($V_{GS} = -10$ V, $V_{DS} = 0$) ($V_{GS} = -10$ V, $V_{DS} = 0$, $T_A = 150°C$)	I_{GSS}	 – –	 – –	 · 1.0 200	pA dc
Gate-Source cutoff voltage ($I_D = 0.5$ μA, $V_{DS} = 10$ V)　　2N3796 ($I_D = 2.0$ μA, $V_{DS} = 10$ V)　　2N3797	$V_{GS(off)}$	 – –	 −3.0 −5.0	 −4.0 −7.0	V dc
Drain-Gate reverse current ($V_{DG} = 10$ V, $I_S = 0$)	I_{DGO}	–	–	1.0	pA dc
ON Characteristics					
Zero-Gate-Voltage drain current　　2N3796 ($V_{DS} = 10$ V, $V_{GS} = 0$)　　2N3797	I_{DSS}	 0.5 2.0	 1.5 2.9	 3.0 6.0	mA dc
On-State drain current　　2N3796 ($V_{DS} = 10$ V, $V_{GS} = +3.5$ V)　　2N3797	$I_{D(on)}$	 7.0 9.0	 8.3 14	 14 18	mA dc
Small-Signal Characteristics					
Forward-transfer admittance ($V_{DS} = 10$ V, $V_{GS} = 0$, $f = 1.0$ kHz)　　2N3796 2N3797 ($V_{DS} = 10$ V, $V_{GS} = 0$, $f = 1.0$ MHz)　　2N3796 2N3797	$|y_{fs}|$	 900 1500 900 1500	 1200 2300 – –	 1800 3000 – –	μmhos or μS
Output admittance ($V_{DS} = 10$ V, $V_{GS} = 0$, $f = 1.0$ kHz)　　2N3796 2N3797	$|y_{os}|$	 – –	 12 27	 25 60	μmhos or μS
Input capacitance ($V_{DS} = 10$ V, $V_{GS} = 0$, $f = 1.0$ MHz)　　2N3796 2N3797	C_{iss}	 – –	 5.0 6.0	 7.0 8.0	pF
Reverse transfer capacitance ($V_{DS} = 10$ V, $V_{GS} = 0$, $f = 1.0$ MHz)	C_{rss}	–	0.5	0.8	pF
Functional Characteristics					
Noise figure ($V_{DS} = 10$ V, $V_{GS} = 0$, $f = 1.0$ kHz, $R_S = 3$ megohms)	NF	–	3.8	–	dB

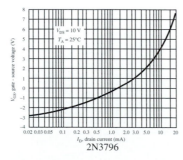

2N3796

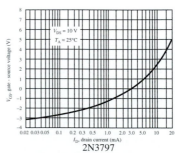

2N3797

FIGURE 8–52

Partial data sheet for the 2N3797 D-MOSFET.

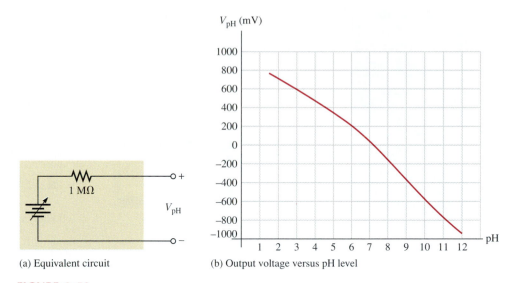

(a) Equivalent circuit   (b) Output voltage versus pH level

FIGURE 8–53
Equivalent circuit and output characteristic for the pH sensor.

Analysis of the pH Sensor Circuits

Refer to the partial data sheet in Figure 8–52 and the pH sensor graph in Figure 8–53.

☐ Determine the input resistance of the D-MOSFET in each circuit.

☐ Determine the minimum, typical, and maximum resistance values to which the rheostat must be set in each circuit to provide a dc drain voltage of +7 V for a neutral solution (pH = 7). The regulated dc supply voltage is +15 V.

☐ Determine the range of output voltage (drain voltage) for a change in the pH sensor voltage of from −500 mV to +500 mV. What is the range of pH values represented? Use typical values from the MOSFET data sheet.

Test Procedure

☐ Develop a step-by-step set of instructions on how to check the pH sensor circuit board for proper operation using the test points (circled numbers) indicated in the test bench setup of Figure 8–54. Assume the test solutions for pH values from 2 to 11 are available.

☐ Specify voltage values for all the measurements to be made. Take into account the loading effect of the 10 MΩ input resistance of the DMM.

☐ Provide a fault analysis for all possible component failures.

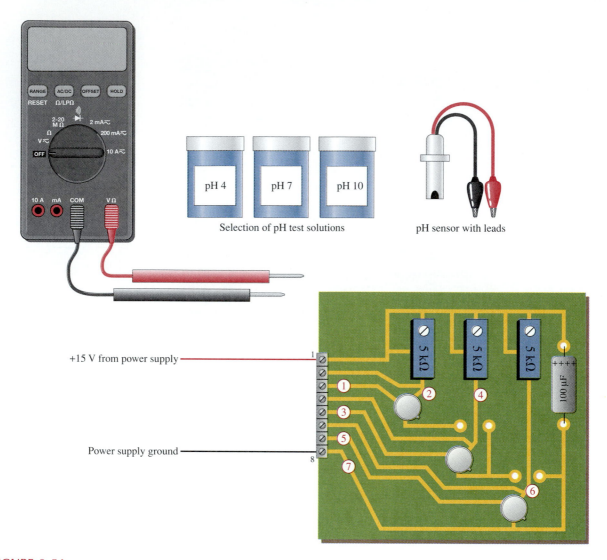

+15 V from power supply

Power supply ground

FIGURE 8–54
Test bench setup for the pH sensor circuit board.

Troubleshooting

Problems have developed in two boards. Based on the sequence of test bench measurements for each board indicated in Figure 8–55, determine the most likely fault in each case. The circled numbers indicate test point connections to the circuit board. Assume typical data sheet values for the MOSFETs. The circuits are supposed to be calibrated, but don't rely on it.

Final Report

Submit a final written report on the pH sensor circuit board using an organized format that includes the following:

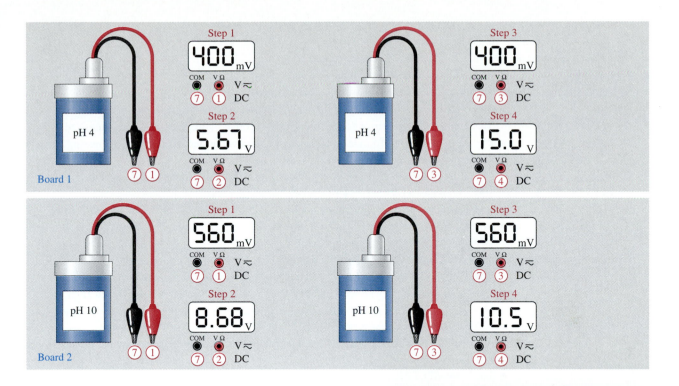

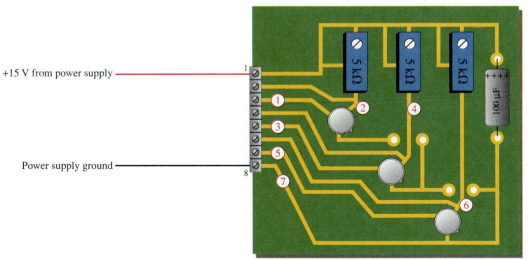

FIGURE 8–55
Test results for two faulty circuit boards.

1. A physical description of the circuits.
2. A discussion of the operation of the circuits.
3. A list of the specifications.
4. A list of parts with part numbers if available.
5. A list of the types of problems on the two faulty circuit boards.
6. A complete description of how you determined the problem on each of the faulty circuit boards.

■ CHAPTER SUMMARY

- Field-effect transistors are unipolar devices (one-charge carrier).
- The three FET terminals are source, drain, and gate.
- The JFET operates with a reverse-biased *pn* junction (gate-to-source).
- The high input resistance of a JFET is due to the reverse-biased gate-source junction.
- Reverse bias of a JFET produces a depletion region within the channel, thus increasing channel resistance.
- For an *n*-channel JFET, V_{GS} can vary from zero negatively to cutoff, $V_{GS(off)}$. For a *p*-channel JFET, V_{GS} can vary from zero positively to $V_{GS(off)}$.
- I_{DSS} is the constant drain current when $V_{GS} = 0$. This is true for both JFETs and D-MOSFETs.
- An FET is called a *square-law device* because of the relationship of I_D to the square of a term containing V_{GS}.
- Unlike JFETs and D-MOSFETs, the E-MOSFET cannot operate with $V_{GS} = 0$ V.
- Midpoint bias for a JFET is $I_D = I_{DSS}/2$, obtained by setting $V_{GS} \cong V_{GS(off)}/3.4$.
- The Q-point in a JFET with voltage-divider bias is more stable than in a self-biased JFET.
- MOSFETs differ from JFETs in that the gate of a MOSFET is insulated from the channel by an SiO_2 layer, whereas the gate and channel in a JFET are separated by a *pn* junction.
- A depletion MOSFET (D-MOSFET) can operate with a zero, positive, or negative gate-to-source voltage.
- The D-MOSFET has a physical channel between drain and source.
- For an *n*-channel D-MOSFET, negative values of V_{GS} produce the depletion mode and positive values produce the enhancement mode.
- The enhancement MOSFET (E-MOSFET) has no physical channel.
- A channel is induced in an E-MOSFET by the application of a V_{GS} greater than the threshold value, $V_{GS(th)}$.
- Midpoint bias for a D-MOSFET is $I_D = I_{DSS}$, obtained by setting $V_{GS} = 0$.
- An E-MOSFET has no I_{DSS} parameter.
- An *n*-channel E-MOSFET has a positive $V_{GS(th)}$. A *p*-channel E-MOSFET has a negative $V_{GS(th)}$.
- LD MOSFET, VMOSFET, and TMOS are E-MOSFET technologies developed for higher power dissipation than a conventional E-MOSFET.

■ GLOSSARY

Channel   The conductive path between the drain and source in an FET.

Cutoff voltage   The value of the gate-to-source voltage that makes the drain current approximately zero.

Depletion   In a MOSFET, the process of removing or depleting the channel of charge carriers and thus decreasing the channel conductivity.

Drain   One of the three terminals of an FET analogous to the collector of a BJT.

Enhancement   In a MOSFET, the process of creating a channel or increasing the conductivity of the channel by the addition of charge carriers.

Field-Effect Transistor (FET)   A type of unipolar, voltage-controlled transistor that uses an induced electric field to control current.

Gate   One of the three terminals of an FET analogous to the base of a BJT.

Junction Field-Effect Transistor (JFET)   One of two major types of field-effect transistor.

MOSFET   Metal oxide semiconductor field-effect transistor; one of two major types of FET; sometimes called IGFET for insulated-gate FET.

Pinch-off voltage   The value of the drain-to-source voltage of an FET at which the drain current becomes constant when the gate-to-source voltage is zero.

Source   One of the three terminals of an FET analogous to the emitter of a BJT.

Transconductance   The ratio of a change in drain current for a change in gate-to-source voltage in an FET.

■ FORMULAS

(8–1) $I_D = I_{DSS}\left(1 - \dfrac{V_{GS}}{V_{GS(off)}}\right)^2$ JFET/D-MOSFET transfer characteristic

(8–2) $g_m = g_{m0}\left(1 - \dfrac{V_{GS}}{V_{GS(off)}}\right)$ Transconductance

(8–3) $g_{m0} = \dfrac{2I_{DSS}}{|V_{GS(off)}|}$ Transconductance at $V_{GS} = 0$

(8–4) $R_{IN} = \left|\dfrac{V_{GS}}{I_{GSS}}\right|$ Input resistance

(8–5) $V_{GS} = -I_D R_S$ n-channel JFET

(8–6) $V_{GS} = +I_D R_S$ p-channel JFET

(8–7) $R_S = \left|\dfrac{V_{GS}}{I_D}\right|$ Source resistance

(8–8) $V_G = \left(\dfrac{R_2}{R_1 + R_2}\right)V_{DD}$ Voltage-divider bias gate voltage

(8–9) $I_D = \dfrac{V_G - V_{GS}}{R_S}$ Voltage-divider bias drain current

(8–10) $I_D = K(V_{GS} - V_{GS(th)})^2$ E-MOSFET transfer characteristic

(8–11) $V_{DS} = V_{DD} - I_{DSS}R_D$ Drain-to-source voltage (zero bias)

■ SELF-TEST

1. The JFET is
   (a) a unipolar device   (b) a voltage-controlled device
   (c) a current-controlled device   (d) answers (a) and (c)
   (e) answers (a) and (b)

2. The channel of a JFET is between the
   (a) gate and drain   (b) drain and source
   (c) gate and source   (d) input and output

3. A JFET always operates with
   (a) the gate-to-source pn junction reverse-biased
   (b) the gate-to-source pn junction forward-biased
   (c) the drain connected to ground
   (d) the gate connected to the source

4. For $V_{GS} = 0$ V, the drain current becomes constant when V_{DS} exceeds
   (a) cutoff   (b) V_{DD} (c) V_P (d) 0 V

5. The constant-current region of an FET lies between
   (a) cutoff and saturation   (b) cutoff and pinch-off
   (c) 0 and I_{DSS} (d) pinch-off and breakdown

6. I_{DSS} is
   (a) the drain current with the source shorted   (b) the drain current at cutoff
   (c) the maximum possible drain current   (d) the midpoint drain current

7. Drain current in the constant-current region increases when
   (a) the gate-to-source bias voltage decreases   (b) the gate-to-source bias voltage increases
   (c) the drain-to-source voltage increases   (d) the drain-to-source voltage decreases

8. In a certain FET circuit, $V_{GS} = 0$ V, $V_{DD} = 15$ V, $I_{DSS} = 15$ mA, and $R_D = 470$ Ω. If R_D is decreased to 330 Ω, I_{DSS} is
   (a) 19.5 mA   (b) 10.5 mA   (c) 15 mA   (d) 1 mA

9. At cutoff, the JFET channel is
   (a) at its widest point   (b) completely closed by the depletion region
   (c) extremely narrow   (d) reverse-biased

10. A certain JFET data sheet gives $V_{GS(off)} = -4$ V. The pinch-off voltage, V_P,
   (a) cannot be determined   (b) is -4 V   (c) depends on V_{GS} (d) is $+4$ V

11. The JFET in Question 10
   (a) is an n channel   (b) is a p channel   (c) can be either

12. For a certain JFET, $I_{GSS} = 10$ nA at $V_{GS} = 10$ V. The input resistance is
   (a) 100 MΩ   (b) 1 MΩ   (c) 1000 MΩ   (d) 1000 mΩ

13. For a certain p-channel JFET, $V_{GS(off)} = 8$ V. The value of V_{GS} for an approximate midpoint bias is
   (a) 4 V   (b) 0 V   (c) 1.25 V   (d) 2.34 V

14. A MOSFET differs from a JFET mainly because
   (a) of the power rating   (b) the MOSFET has two gates
   (c) the JFET has a pn junction   (d) MOSFETs do not have a physical channel

15. A certain D-MOSFET is biased at $V_{GS} = 0$ V. Its data sheet specifies $I_{DSS} = 20$ mA and $V_{GS(off)} = -5$ V. The value of the drain current
   (a) is 0 A   (b) cannot be determined   (c) is 20 mA

16. An n-channel D-MOSFET with a positive V_{GS} is operating in
   (a) the depletion mode   (b) the enhancement mode
   (c) cutoff   (d) saturation

17. A certain p-channel E-MOSFET has a $V_{GS(th)} = -2$ V. If $V_{GS} = 0$ V, the drain current is
   (a) 0 A   (b) $I_{D(on)}$ (c) maximum   (d) I_{DSS}

18. A TMOSFET is a special type of
   (a) D-MOSFET   (b) JFET   (c) E-MOSFET   (d) answers (a) and (c)

▪ BASIC PROBLEMS

SECTION 8–1   The Junction Field-Effect Transistor (JFET)

1. The V_{GS} of a p-channel JFET is increased from 1 V to 3 V.
   (a) Does the depletion region narrow or widen?
   (b) Does the resistance of the channel increase or decrease?

2. Why must the gate-to-source voltage of an n-channel JFET always be either 0 or negative?

3. Sketch the schematic diagrams for a p-channel and an n-channel JFET. Label the terminals.

4. Show how to connect bias voltages between the gate and source of the JFETs in Figure 8–56.

FIGURE 8–56

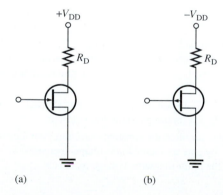

(a)   (b)

SECTION 8–2   JFET Characteristics and Parameters

5. A JFET has a specified pinch-off voltage of 5 V. When $V_{GS} = 0$, what is V_{DS} at the point where the drain current becomes constant?

6. A certain n-channel JFET is biased such that $V_{GS} = -2$ V. What is the value of $V_{GS(off)}$ if V_P is specified to be 6 V? Is the device on?

7. A certain JFET data sheet gives $V_{GS(off)} = -8$ V and $I_{DSS} = 10$ mA. When $V_{GS} = 0$, what is I_D for values of V_{DS} above pinch off? $V_{DD} = 15$ V.

8. A certain p-channel JFET has a $V_{GS(off)} = 6$ V. What is I_D when $V_{GS} = 8$ V?

9. The JFET in Figure 8–57 has a $V_{GS(off)} = -4$ V. Assume that you increase the supply voltage, V_{DD}, beginning at zero until the ammeter reaches a steady value. What does the voltmeter read at this point?

FIGURE 8–57

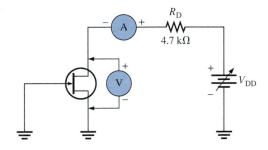

10. The following parameters are obtained from a certain JFET data sheet: $V_{GS(off)} = -8$ V and $I_{DSS} = 5$ mA. Determine the values of I_D for each value of V_{GS} ranging from 0 V to -8 V in 1 V steps. Plot the transfer characteristic curve from these data.

11. For the JFET in Problem 10, what value of V_{GS} is required to set up a drain current of 2.25 mA?

12. For a particular JFET, $g_{m0} = 3200$ µS. What is g_m when $V_{GS} = -4$ V, given that $V_{GS(off)} = -8$ V?

13. Determine the forward transconductance of a JFET biased at $V_{GS} = -2$ V. From the data sheet, $V_{GS(off)} = -7$ V and $g_m = 2000$ µS at $V_{GS} = 0$ V. Also determine the forward transfer admittance, y_{fs}.

14. A p-channel JFET data sheet shows that $I_{GSS} = 5$ nA at $V_{GS} = 10$ V. Determine the input resistance.

15. Using Equation (8–1), plot the transfer characteristic curve for a JFET with $I_{DSS} = 8$ mA and $V_{GS(off)} = -5$ V using at least four points.

SECTION 8–3   JFET Biasing

16. An n-channel self-biased JFET has a drain current of 12 mA and a 100 Ω source resistor. What is the value of V_{GS}?

17. Determine the value of R_S required for a self-biased JFET to produce a V_{GS} of -4 V when $I_D = 5$ mA.

18. Determine the value of R_S required for a self-biased JFET to produce $I_D = 2.5$ mA when $V_{GS} = -3$ V.

19. $I_{DSS} = 20$ mA and $V_{GS(off)} = -6$ V for a particular JFET.
    (a) What is I_D when $V_{GS} = 0$ V?
    (b) What is I_D when $V_{GS} = V_{GS(off)}$?
    (c) If V_{GS} is increased from -4 V to -1 V, does I_D increase or decrease?

FIGURE 8–58

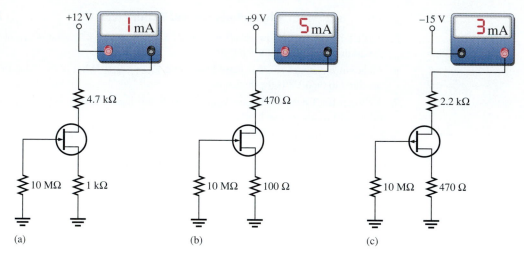

(a)　　　　　(b)　　　　　(c)

20. For each circuit in Figure 8–58, determine V_{DS} and V_{GS}.

21. Using the curve in Figure 8–59, determine the value of R_S required for a 9.5 mA drain current.

22. Set up a midpoint bias for a JFET with I_{DSS} = 14 mA and $V_{GS(off)}$ = −10 V. Use a 24 V dc source as the supply voltage. Show the circuit and resistor values. Indicate the values of I_D, V_{GS}, and V_{DS}.

23. Determine the total input resistance in Figure 8–60. I_{GSS} = 20 nA at V_{GS} = −10 V.

24. Graphically determine the Q-point for the circuit in Figure 8–61(a) using the transfer characteristic curve in Figure 8–61(b).

25. Find the Q-point for the p-channel JFET circuit in Figure 8–62.

26. Given that the drain-to-ground voltage in Figure 8–63 is 5 V, determine the Q-point of the circuit.

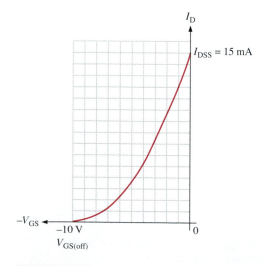

FIGURE 8–59

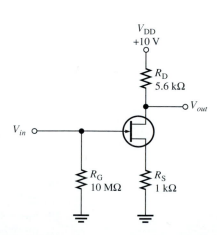

FIGURE 8–60

FIGURE 8–61

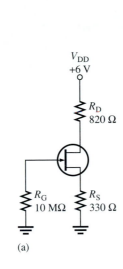

(a)

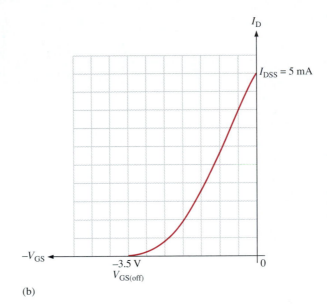

(b)

FIGURE 8–62

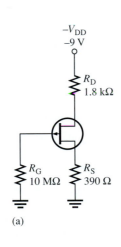

(a)

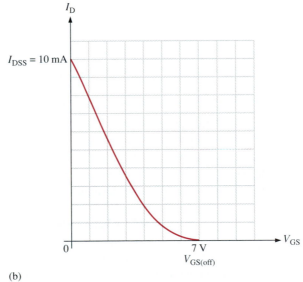

(b)

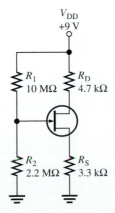

FIGURE 8–63

FIGURE 8–64

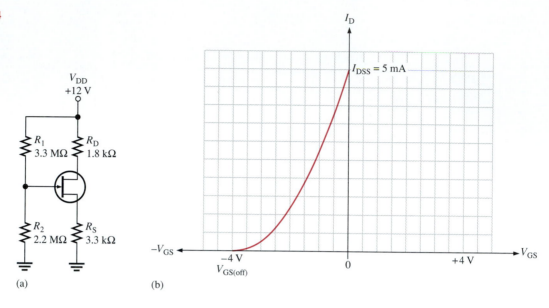

(a)

(b)

27. Find the Q-point values for the JFET with voltage-divider bias in Figure 8–64.

SECTION 8–4 The Metal Oxide Semiconductor FET (MOSFET)

28. Sketch the schematic symbols for *n*-channel and *p*-channel D-MOSFETs and E-MOSFETs. Label the terminals.

29. In what mode is an *n*-channel D-MOSFET with a positive V_{GS} operating?

30. Describe the basic difference between a D-MOSFET and an E-MOSFET.

31. Explain why both types of MOSFETs have an extremely high input resistance at the gate.

SECTION 8–5 MOSFET Characteristics and Parameters

32. The data sheet for a certain D-MOSFET gives $V_{GS(off)} = -5$ V and $I_{DSS} = 8$ mA.
  (a) Is this device *p* channel or *n* channel?
  (b) Determine I_D for values of V_{GS} ranging from -5 V to $+5$ V in increments of 1 V.
  (c) Plot the transfer characteristic curve using the data from part (b).

33. Determine I_{DSS}, given $I_D = 3$ mA, $V_{GS} = -2$ V, and $V_{GS(off)} = -10$ V.

34. The data sheet for an E-MOSFET reveals that $I_{D(on)} = 10$ mA at $V_{GS} = -12$ V and $V_{GS(th)} = -3$ V. Find I_D when $V_{GS} = -6$ V.

SECTION 8–6 MOSFET Biasing

35. Determine in which mode (depletion, enhancement or neither) each D-MOSFET in Figure 8–65 is biased.

36. Each E-MOSFET in Figure 8–66 has a $V_{GS(th)}$ of $+5$ V or -5 V, depending on whether it is an *n*-channel or a *p*-channel device. Determine whether each MOSFET is on or off.

37. Determine V_{DS} for each circuit in Figure 8–67. $I_{DSS} = 8$ mA.

38. Find V_{GS} and V_{DS} for the E-MOSFETs in Figure 8–68. Data sheet information is listed with each circuit.

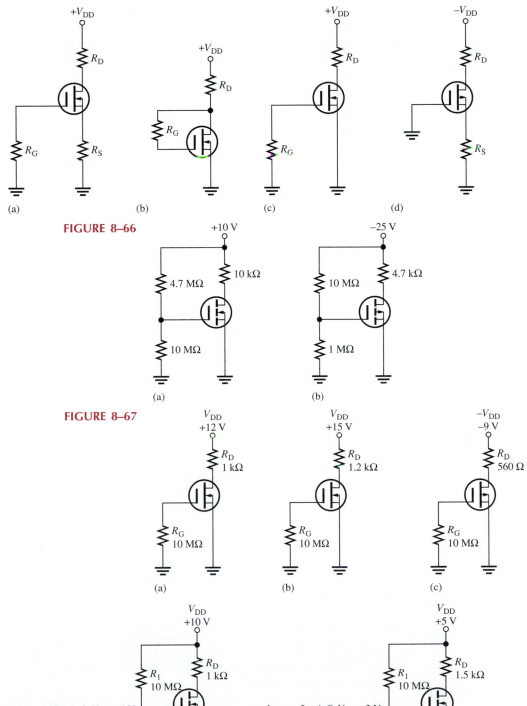

FIGURE 8–65

(a) (b) (c) (d)

FIGURE 8–66

(a) (b)

FIGURE 8–67

(a) (b) (c)

FIGURE 8–68

$I_{D(on)} = 3$ mA @ $V_{GS} = 4$ V
$V_{GS(th)} = 2$ V

$I_{D(on)} = 2$ mA @ $V_{GS} = 3$ V
$V_{GS(th)} = 1.5$ V

(a) (b)

39. Based on the identical V_{GS} measurements, determine the drain current and drain-to-source voltage for each circuit in Figure 8–69.

40. Determine the actual gate-to-source voltage in Figure 8–70 by taking into account the gate leakage current I_{GSS}. Assume that I_{GSS} is 50 pA and I_D is 1 mA under the existing bias conditions.

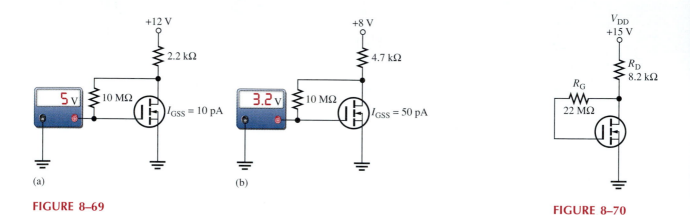

(a)

(b)

FIGURE 8–69

FIGURE 8–70

■ **TROUBLE-SHOOTING PROBLEMS**

SECTION 8–7 Troubleshooting

41. The current reading in Figure 8–58(a) suddenly goes to zero. What are the possible faults?

42. The current reading in Figure 8–58(b) suddenly jumps to approximately 16 mA. What are the possible faults?

43. If the supply voltage in Figure 8–58(c) is accidentally changed to −20 V, what would you see on the ammeter?

44. You measure +10 V at the drain of the MOSFET in Figure 8–66(a). The transistor checks good and the ground connections are OK. What can be the problem?

45. You measure approximately 0 V at the drain of the MOSFET in Figure 8–66(b). You can find no shorts and the transistor checks good. What is the most likely problem?

SECTION 8–8 System Application

46. The 100 µF capacitor on the pH sensor circuit board in Figure 8–51 has opened. What effect could this have on the circuit operation? Explain.

47. Refer to Figure 8–53. What should be the output voltage of the pH sensor for a pH of 5? For a pH of 9?

48. Refer to the test bench setup in Figure 8–54. When measuring the voltage inputs to the pH sensor circuits from the pH sensor in a test solution, you notice that the voltmeter indicates values that are approximately half of what they should be for each circuit. After trying a new sensor, the voltages are still half the expected value. What do you think is wrong?

49. Determine the voltage that would be measured at test point 6 on the circuit board in Figure 8–55 for a pH of 7 if the rheostat is incorrectly set to a value of 1 kΩ. Assume typical values.

■ **DATA SHEET PROBLEMS**

50. What type of FET is the 2N5457?

51. Referring to the data sheet in Figure 8–14, determine the following:
(a) Minimum $V_{GS(off)}$ for the 2N5457.
(b) Maximum drain-to-source voltage for the 2N5457.
(c) Maximum power dissipation for the 2N5458 at an ambient temperature of 25°C.
(d) Maximum reverse gate-to-source voltage for the 2N5459.

52. Referring to Figure 8–14, determine the maximum power dissipation for a 2N5457 at an ambient temperature of 65°C.

53. Referring to Figure 8–14, determine the minimum g_{m0} for the 2N5459 at a frequency of 1 kHz.

54. Referring to Figure 8–14, what is the typical drain current in a 2N5459 for $V_{GS} = 0$ V?

55. Referring to the data sheet in Figure 8–41, determine the minimum gate-to-source voltage at which the MOSFET begins to conduct current.

56. Referring to Figure 8–41, what is the drain current when $V_{GS} = 10$ V?

57. Referring to the data sheet in Figure 8–52, determine I_D in a 2N3797 when $V_{GS} = +3$ V. When $V_{GS} = -2$ V.

58. Referring to Figure 8–52, how much does the maximum forward transconductance of a 2N3796 change over a range of signal frequencies from 1 kHz to 1 MHz?

59. Referring to Figure 8–52, determine the typical value of gate-to-source voltage at which the 2N3796 will go into cutoff.

■ **ADVANCED PROBLEMS**

60. Find V_{DS} and V_{GS} in Figure 8–71 using minimum data sheet values.

61. Determine the maximum I_D and V_{GS} for the circuit in Figure 8–72.

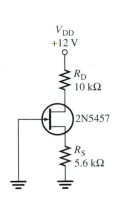

FIGURE 8–71

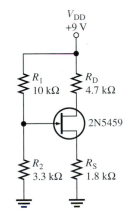

FIGURE 8–72

62. Determine the range of possible Q-point values from minimum to maximum for the circuit in Figure 8–71.

63. Find the typical drain-to-source voltage for the pH sensor circuits in Figure 8–54 when a pH of 5 is measured. Assume the rheostats are set for 7 V at the drains when a pH of 7 is measured.

64. Design a MOSFET circuit with zero bias using a 2N3797 that operates from a +9 V dc supply and produces a V_{DS} of 4.5 V. The maximum current drawn for the source is to be 1 mA.

65. Design a circuit using a 2N7008 MOSFET and a +12 V dc supply voltage with voltage-divider bias that will produce +8 V at the drain and draw a maximum current from the supply of 20 mA.

■ **ANSWERS TO SECTION REVIEWS**

Section 8–1

1. Drain, source, and gate

2. An n-channel JFET requires a negative V_{GS}.

3. I_D is controlled by V_{GS}.

Section 8–2

1. When $V_{DS} = 7$ V at pinch-off and $V_{GS} = 0$ V, $V_P = -7$ V.

2. As V_{GS} increases negatively, I_D decreases.

3. For $V_P = -3$ V, $V_{GS(off)} = +3$ V.

Section 8–3

1. A p-channel JFET requires a positive V_{GS}.

2. $V_{GS} = V_G - V_S = 0$ V $- (8$ mA$)(1$ k$\Omega) = -8$ V

3. $V_{GS} = V_G - V_S = 3$ V $- 5$ V $= -2$ V

Section 8–4

1. Depletion MOSFET (D-MOSFET) and enhancement MOSFET (E-MOSFET)

2. When $V_{GS} = 0$ V for a D-MOSFET, $I_D = I_{DSS}$.

3. When $V_{GS} = 0$ V for an E-MOSFET, $I_D \cong 0$ A.

Section 8–5

1. The D-MOSFET has a structural channel; the E-MOSFET does not.

2. $V_{GS(th)}$ and K are not specified for D-MOSFETs.

3. ESD is ElectroStatic Discharge.

Section 8–6

1. When $V_{GS} = 0$ V, the drain current is equal to I_{DSS}.

2. V_{GS} must exceed $V_{GS(th)} = 2$ V for conduction to occur.

Section 8–7

1. R_S open, no ground connection

2. Because V_{GS} remains at approximately zero

3. The device is off and $V_D = V_{DD}$.

■ **ANSWERS TO RELATED EXERCISES FOR EXAMPLES**

8–1 I_D remains at approximately 12 mA.

8–2 $V_P = -4$ V

8–3 $I_D = 3.51$ mA

8–4 $g_m = 1800$ μS; $I_D = 4.32$ mA

8–5 $R_{IN} = 150,000$ MΩ

8–6 $V_{DS} = 5$ V; $V_{GS} = -3.12$ V

8–7 $R_S = 231$ Ω

8–8 $R_S = 889$ Ω

8–9 $R_S = 586$ Ω; $R_D = 1500$ Ω

8–10 $V_{GS} \cong -1.8$ V, $I_D \cong 1.8$ mA

8–11 $I_D = 1.81$ mA, $V_{GS} = -1.99$ V

8–12 $I_D \cong 1.25$ mA, $V_{GS} \cong -2.25$ V

8–13 **(a)** p channel   **(b)** 6.48 mA   **(c)** 35.3 mA

8–14 $I_D = 25$ mA

8–15 $V_{DS} = 5.6$ V

8–16 $V_{GS} = 3.13$ V; $V_{DS} = 22.3$ V

8–17 $I_D = 2.13$ mA

9

SMALL-SIGNAL FET AMPLIFIERS

■ CHAPTER OBJECTIVES

☐ Explain the operation of FET small-signal amplifiers

☐ Describe the amplification properties of an FET

☐ Explain and analyze the operation of common-source FET amplifiers

☐ Explain and analyze the operation of common-drain FET amplifiers

☐ Explain and analyze the operation of common-gate FET amplifiers

☐ Troubleshoot FET amplifiers

The things you learned about biasing FETs in the last chapter are carried forward into this chapter where FET circuits are used as small-signal amplifiers. Because of their high input resistance and other characteristics, FETs are often preferred over bipolar transistors for certain applications.

Many of the concepts that relate to amplifiers using BJTs apply as well to FET amplifiers. The three FET amplifier configurations are common-source, common-drain, and common-gate. These are analogous to the common-emitter, common-collector, and common-base configurations in bipolar amplifiers.

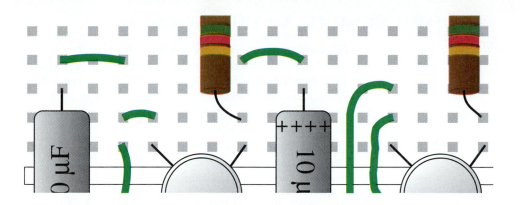

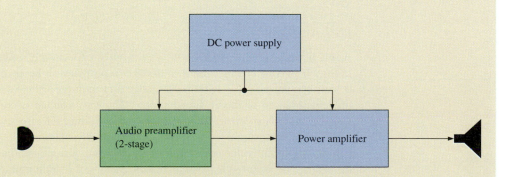

■ SYSTEM APPLICATION

For the system application in Section 9–7, you will be evaluating the feasibility of replacing the BJT amplifier in a public address system with an FET amplifier.

9–1 ■ SMALL-SIGNAL FET AMPLIFIER OPERATION

The concept of small-signal amplifiers was covered in Chapter 6 in relation to bipolar transistors. The concept applies also to small-signal amplifiers using FETs. As you have learned, significant differences exist between BJTs and FETs in terms of their parameters and characteristics. However, when used in amplifier circuits, BJTs and FETs have the same purpose, that is to amplify a small input signal by a desired amount. Because of their high input resistance, FETs are better suited in certain applications. The common-emitter, common-collector, and common-base bipolar amplifier configurations have FET counterparts. These are the common-source (CS), the common-drain (CD), and the common-gate (CG) amplifiers.

After completing this section, you should be able to

- ■ Explain the operation of FET small-signal amplifiers
  - ☐ Describe a JFET common-source amplifier
  - ☐ Describe a D-MOSFET common-source amplifier
  - ☐ Describe an E-MOSFET common-source amplifier
  - ☐ Analyze amplifier operation from a graphical approach

JFET Amplifier

A self-biased *n*-channel JFET with an ac source capacitively coupled to the gate is shown in Figure 9–1(a). The resistor, R_G, serves two purposes: It keeps the gate at approximately 0 V dc (because I_{GSS} is extremely small), and its large value (usually several megohms) prevents loading of the ac signal source. The bias voltage is created by the drop across R_S. The bypass capacitor, C_2, keeps the source of the FET effectively at ac ground.

The input signal voltage causes the gate-to-source voltage to swing above and below its Q-point value, causing a swing in drain current. As the drain current increases, the voltage drop across R_D also increases, causing the drain voltage to decrease. The drain current swings above and below its Q-point value in phase with the gate-to-source volt-

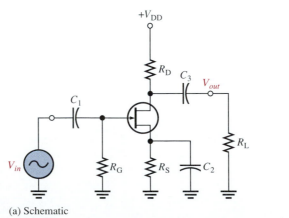

(a) Schematic

(b) Voltage wave form relationship

FIGURE 9–1
JFET common-source amplifier.

age. The drain-to-source voltage swings above and below its Q-point value 180° out of phase with the gate-to-source voltage, as illustrated in Figure 9–1(b).

A Graphical Picture   The operation just described for an *n*-channel JFET can be illustrated graphically on both the transfer characteristic curve and the drain characteristic curve in Figure 9–2. Figure 9–2(a) shows how a sinusoidal variation, V_{gs}, produces a corresponding variation in I_d. As V_{gs} swings from the Q-point to a more negative value, I_d decreases from its Q-point value. As V_{gs} swings to a less negative value, I_d increases. Figure 9–2(b) shows a view of the same operation using the drain curves. The signal at the gate drives the drain current equally above and below the Q-point on the load line, as indicated by the arrows. Lines projected from the peaks of the gate voltage across to the I_D axis and down to the V_{DS} axis indicate the peak-to-peak variations of the drain current and drain-to-source voltage, as shown.

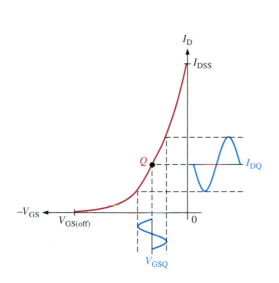

(a) JFET (*n*-channel) transfer characteristic curve showing signal operation

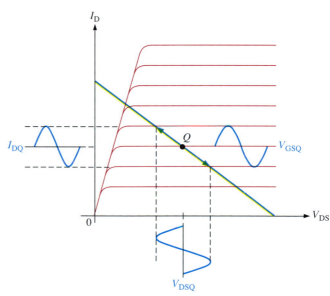

(b) JFET (*n*-channel) drain curves showing signal operation

FIGURE 9–2
JFET characteristic curves.

D-MOSFET Amplifier

A zero-biased *n*-channel D-MOSFET with an ac source capacitively coupled to the gate is shown in Figure 9–3. The gate is at approximately 0 V dc and the source terminal is at ground, thus making $V_{GS} = 0$ V.

The signal voltage causes V_{gs} to swing above and below its zero value, producing a swing in I_d as shown in Figure 9–4. The negative swing in V_{gs} produces the depletion mode, and I_d decreases. The positive swing in V_{gs} produces the enhancement mode, and I_d increases. Note that the enhancement mode is to the right of the vertical axis ($V_{GS} = 0$), and the depletion mode is to the left.

FIGURE 9–3

Zero-biased D-MOSFET common-source amplifier.

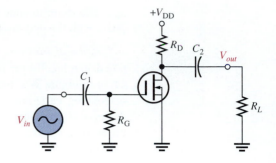

FIGURE 9–4

Depletion-enhancement operation of D-MOSFET shown on transfer characteristic curve.

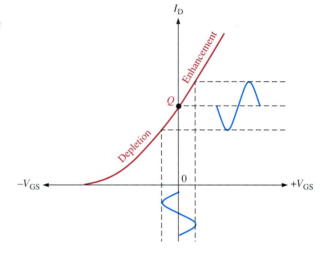

E-MOSFET Amplifier

An *n*-channel E-MOSFET with voltage-divider bias with an ac source capacitively coupled to the gate is shown in Figure 9–5. The gate is biased with a positive voltage such that $V_{GS} > V_{GS(th)}$.

As with the JFET and D-MOSFET, the signal voltage produces a swing in V_{gs} above and below its Q-point value. This, in turn, causes a swing in I_d, as illustrated in Figure 9–6. Operation is entirely in the enhancement mode.

FIGURE 9–5

Common-source E-MOSFET amplifier with voltage-divider bias.

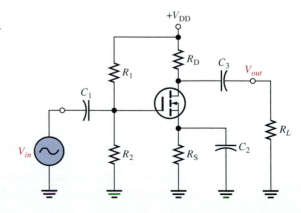

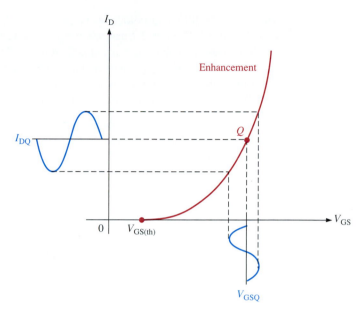

FIGURE 9–6

E-MOSFET (n-channel) operation shown on transfer characteristic curve.

EXAMPLE 9–1

Transfer characteristic curves for an *n*-channel JFET, D-MOSFET, and E-MOSFET are shown in Figure 9–7. Determine the peak-to-peak variation in I_d when V_{gs} is varied ±1 V about its Q-point value for each curve.

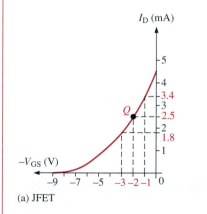

(a) JFET

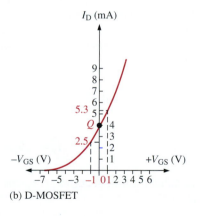

(b) D-MOSFET

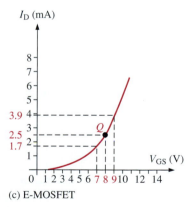

(c) E-MOSFET

FIGURE 9–7

Solution

(a) The JFET Q-point is at $V_{GS} = -2$ V and $I_D = 2.5$ mA. From the graph in Figure 9–7(a), $I_D = 3.4$ mA when $V_{GS} = -1$ V, and $I_D = 1.8$ mA when $V_{GS} = -3$ V. The peak-to-peak drain current is therefore 1.6 mA.

(b) The D-MOSFET Q-point is at $V_{GS} = 0$ V and $I_D = I_{DSS} = 4$ mA. From the graph in Figure 9–7(b), $I_D = 2.7$ mA when $V_{GS} = -1$ V, and $I_D = 5.4$ mA when $V_{GS} = +1$ V. The peak-to-peak drain current is therefore 2.7 mA.

(c) The E-MOSFET Q-point is at $V_{GS} = +8$ V and $I_D = 2.5$ mA. From the graph in Figure 9–7(c), $I_D = 3.9$ mA when $V_{GS} = +9$ V, and $I_D = 1.7$ mA when $V_{GS} = +7$ V. The peak-to-peak drain current is therefore 2.2 mA.

Related Exercise As the Q-point is moved toward the bottom end of the curves in Figure 9–7, does the variation in I_D increase or decrease for the same ±1 V variation in V_{GS}? In addition to the change in the amount that I_D varies, what else will happen?

SECTION 9–1 REVIEW

1. When V_{gs} is at its positive peak, at what points are I_d and V_{ds}?

2. What is the difference between V_{gs} and V_{GS}?

3. Which of the three types of FETs can operate with a gate-to-source Q-point value of 0 V?

9–2 ■ FET AMPLIFICATION

In this section, you will learn about the amplification properties of FETs and how the gain is affected by certain parameters and circuit components. We will simplify the FET to an equivalent circuit to get to the essence of its operation.

After completing this section, you should be able to

■ **Describe the amplification properties of an FET**
   □ Discuss FET parameters using an equivalent circuit
   □ Discuss the voltage gain of an FET
   □ Describe the effect of internal drain-to-source resistance on the voltage gain
   □ Describe the effect of external source resistance on the voltage gain

The transconductance is defined as $g_m = \Delta I_D / \Delta V_{GS}$. In ac quantities, $g_m = I_d / V_{gs}$. By rearranging the terms, we get

$$I_d = g_m V_{gs} \tag{9–1}$$

This equation says that the output current I_d equals the input voltage V_{gs} multiplied by the transconductance g_m.

Equivalent Circuit

An FET equivalent circuit representing the relationship in Equation (9–1) is shown in Figure 9–8. In part (a), the internal resistance, r'_{gs}, appears between gate and source, and a current source equal to $g_m V_{gs}$ appears between drain and source. Also, the internal drain-to-source resistance, r'_{ds}, is included. In part (b), a simplified ideal model is shown. The

FIGURE 9–8

Internal FET equivalent circuits.

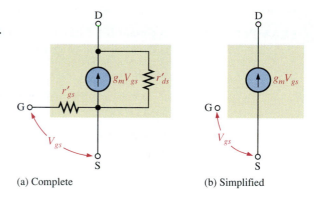

(a) Complete   (b) Simplified

resistance, r'_{gs}, is assumed to be infinitely large so that there is an open circuit between gate and source. Also, r'_{ds} is assumed large enough to neglect.

Voltage Gain

An FET ideal equivalent circuit with an external ac drain resistance is shown in Figure 9–9. The ac voltage gain of this circuit is V_{out}/V_{in}, where $V_{in} = V_{gs}$. The voltage gain expression is therefore

$$A_v = \frac{V_{ds}}{V_{gs}} \qquad\qquad (9\text{–}2)$$

From the equivalent circuit, we see that

$$V_{ds} = I_d R_d$$

From the definition of transconductance in Chapter 8,

$$V_{gs} = \frac{I_d}{g_m}$$

Substituting the two preceding expressions into Equation (9–2) yields

$$A_v = \frac{I_d R_d}{I_d/g_m} = \frac{g_m I_d R_d}{I_d}$$

$$A_v = g_m R_d \qquad\qquad (9\text{–}3)$$

FIGURE 9–9

Simplified FET equivalent circuit with external drain resistor.

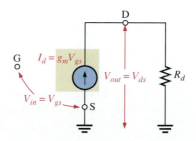

EXAMPLE 9–2

A certain JFET has a $g_m = 4$ mS. With an external ac drain resistance of 1.5 kΩ, what is the ideal voltage gain?

Solution
$$A_v = g_m R_d = (4 \text{ mS})(1.5 \text{ k}\Omega) = 6$$

Related Exercise What is the ideal voltage gain when $g_m = 6000$ μS and $R_d = 2.2$ kΩ?

Effect of r'_{ds} on Gain

If the internal drain-to-source resistance of the FET is taken into account, it appears in parallel with R_d, as indicated in Figure 9–10. The resulting voltage gain expression is stated in Equation (9–4). As you can see, if r'_{ds} is not sufficiently greater than R_d ($r'_{ds} \geqq 10R_d$), the gain is reduced from the ideal case of Equation (9–3) to the following:

$$A_v = g_m \left(\frac{R_d r'_{ds}}{R_d + r'_{ds}} \right) \tag{9–4}$$

FIGURE 9–10
FET equivalent circuit including the internal drain-to-source resistance, r'_{ds}.

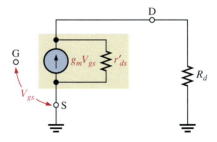

EXAMPLE 9–3

The JFET in Example 9–2 has an $r'_{ds} = 10$ kΩ. Determine the voltage gain when r'_{ds} is taken into account.

Solution The r'_{ds} is effectively in parallel with the external ac drain resistance R_d. Therefore, use Equation (9–4).

$$A_v = g_m \left(\frac{R_d r'_{ds}}{R_d + r'_{ds}} \right) = (4 \text{ mS}) \left(\frac{(1.5 \text{ k}\Omega)(10 \text{ k}\Omega)}{1.5 \text{ k}\Omega + 10 \text{ k}\Omega} \right)$$
$$= (4 \text{ mS})(1.3 \text{ k}\Omega) = 5.2$$

The voltage gain is reduced from a value of 6 (Example 9–2) because r'_{ds} is in parallel with R_d.

Related Exercise A JFET has a $g_m = 6$ mS, an $r'_{ds} = 5$ kΩ, and an external ac drain resistance of 1 kΩ. What is the voltage gain?

Effect of External Source Resistance on Gain

Including an external resistance from the FET's source terminal to ground results in the equivalent circuit of Figure 9–11. Examination of this circuit shows that the total input voltage between the gate and ground is

$$V_{in} = V_{gs} + I_d R_s$$

The output voltage taken across R_d is

$$V_{out} = I_d R_d$$

Therefore, the formula for voltage gain is developed as follows:

$$A_v = \frac{V_{out}}{V_{in}} = \frac{I_d R_d}{V_{gs} + I_d R_s} = \frac{g_m V_{gs} R_d}{V_{gs} + g_m V_{gs} R_s} = \frac{g_m V_{gs} R_d}{V_{gs}(1 + g_m R_s)}$$

$$A_v = \frac{g_m R_d}{1 + g_m R_s} \qquad\qquad (9\text{–}5)$$

FIGURE 9–11

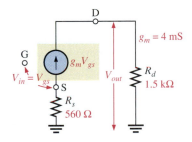

EXAMPLE 9–4

An FET equivalent circuit is shown in Figure 9–11. Determine the voltage gain when the output is taken across R_d. Neglect r'_{ds}.

Solution   There is an external source resistor, so the voltage gain is

$$A_v = \frac{g_m R_d}{1 + g_m R_s} = \frac{(4 \text{ mS})(1.5 \text{ k}\Omega)}{1 + (4 \text{ mS})(560 \text{ }\Omega)} = \frac{6}{1 + 2.24} = \frac{6}{3.24} = 1.85$$

This is the same circuit as in Example 9–2 except for R_s. As you can see, R_s reduces the voltage gain from 6 (Example 9–2) to 1.85.

Related Exercise   For the circuit in Figure 9–11, $g_m = 3.5$ mS, $R_s = 330 \text{ }\Omega$, and $R_d = 1.8 \text{ k}\Omega$. Find the voltage gain when the output is taken across R_d. Neglect r'_{ds}.

SECTION 9–2
REVIEW

1. One FET has a transconductance of 3000 μS and another has a transconductance of 3.5 mS. Which one can produce the higher voltage gain, with all other circuit components the same?

2. An FET circuit has a g_m = 2500 μS and an R_d = 10 kΩ. Ideally, what voltage gain can it produce?

3. Two FETs have the same g_m. One has an r'_{ds} = 50 kΩ and the other has an r'_{ds} = 100 kΩ under the same conditions. Which FET can produce the higher voltage gain when used in a circuit with R_d = 10 kΩ?

9–3 ■ COMMON-SOURCE AMPLIFIERS

Now that you have an idea of how an FET functions as a voltage amplifying device, we will look at a complete amplifier circuit. The common-source (CS) amplifier discussed in this section is comparable to the common-emitter BJT amplifier that you studied in Chapter 6.

After completing this section, you should be able to

■ **Explain and analyze the operation of common-source FET amplifiers**
  ☐ Analyze JFET and MOSFET CS amplifiers
  ☐ Determine the dc values of a CS amplifier
  ☐ Develop an ac equivalent circuit and determine the voltage gain of a CS amplifier
  ☐ Describe the effect of an ac load on the voltage gain
  ☐ Discuss phase inversion in a CS amplifier
  ☐ Determine the input resistance of a CS amplifier

Common-Source JFET Amplifiers

Figure 9–12 shows a **common-source** amplifier with a self-biased *n*-channel JFET. There are coupling capacitors on the input and output in addition to the source bypass capacitor. The circuit has a combination of dc and ac operations.

DC Analysis

To analyze the amplifier in Figure 9–12, you must first determine the dc bias values. To do this, develop a dc equivalent circuit by replacing all capacitors with opens, as shown in Figure 9–13. First, you must determine I_D before you can do any analysis. If the circuit is biased at the midpoint of the load line, you can calculate I_D using I_{DSS} from the FET data sheet as follows:

$$I_D = \frac{I_{DSS}}{2}$$

(9–6)

Otherwise, you must know I_D before you can make any other dc calculations. Determination of I_D from circuit parameter values is tedious because Equation (9–7) must be solved

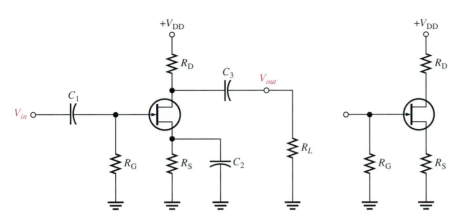

FIGURE 9–12
JFET common-source amplifier.

FIGURE 9–13
DC equivalent circuit for the amplifier in Figure 9–12.

for I_D. (This equation is derived by substitution of $V_{GS} = I_D R_S$ into Equation (9–1).) Solution of the equation for I_D involves expanding it into a quadratic form and then finding the root of the quadratic, as developed in Appendix B.

$$I_D = I_{DSS}\left(1 - \frac{I_D R_S}{V_{GS(off)}}\right)^2 \tag{9–7}$$

The following BASIC program or a programmable calculator such as the TI-85 can be used to solve for I_D in self-biased JFETs using Equation (9–7).

```
10    CLS
20    PRINT "THIS PROGRAM COMPUTES JFET DRAIN CURRENT"
30    PRINT
40    PRINT "THE REQUIRED INPUTS ARE AS FOLLOWS"
50    PRINT "(1) IDSS FROM DATA SHEET"
60    PRINT "(2) VGS(OFF) FROM DATA SHEET"
70    PRINT "(3) RS FROM CIRCUIT DIAGRAM"
80    PRINT:PRINT:PRINT
90    INPUT "TO CONTINUE PRESS 'ENTER'";X
100   CLS
110   INPUT "VALUE OF IDSS IN AMPS";IDSS
120   INPUT "VALUE OF VGS(OFF) IN VOLTS";VGSOFF
130   INPUT "VALUE OF RS IN OHMS";RS
140   CLS
150   A=RS [2*IDSS/VGSOFF[2
160   B=-(1+2*RS*IDSS/ABS(VGSOFF))
170   C=IDSS
180   DI=(-B-SQR(B[2-4*A*C))/(2*A)
190   PRINT "ID=";ABS(DI);"A"
```

Once I_D is determined, the dc analysis can proceed using the following relationships:

$$V_S = V_{GS} = I_D R_S$$

$$V_D = V_{DD} - I_D R_D$$

$$V_{DS} = V_D - V_S$$

AC Equivalent Circuit

To analyze the signal operation of the amplifier in Figure 9–12, develop an ac equivalent circuit as follows. Replace the capacitors by effective shorts, based on the simplifying assumption that $X_C \cong 0$ at the signal frequency. Replace the dc source by a ground, based on the assumption that the voltage source has a zero internal resistance. The V_{DD} terminal is at a zero-volt ac potential and therefore acts as an ac ground.

The ac equivalent circuit is shown in Figure 9–14(a). Notice that the $+V_{DD}$ end of R_d and the source terminal are both effectively at ac ground. Recall that in ac analysis, the ac ground and the actual circuit ground are treated as the same point.

FIGURE 9–14

AC equivalent for the amplifier in Figure 9–12.

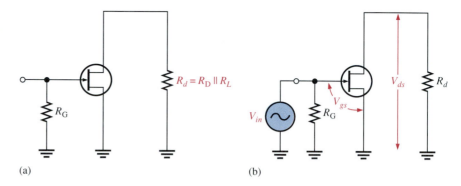

(a)   (b)

Signal Voltage at the Gate An ac voltage source is shown connected to the input in Figure 9–14(b). Since the input resistance to the FET is extremely high, practically all of the input voltage from the signal source appears at the gate with very little voltage dropped across the internal source resistance.

$$V_{gs} = V_{in}$$

Voltage Gain The expression for FET voltage gain that was given in Equation (9–3) applies to the common-source amplifier.

$$\boxed{A_v = g_m R_d} \qquad (9\text{–}8)$$

The output signal voltage V_{ds} at the drain is

$$V_{out} = V_{ds} = A_v V_{gs}$$

or

$$V_{out} = g_m R_d V_{in}$$

where $R_d = R_D \parallel R_L$ and $V_{in} = V_{gs}$.

EXAMPLE 9–5 What is the total output voltage of the unloaded amplifier in Figure 9–15? I_{DSS} is 8 mA and $V_{GS(off)}$ is −10 V.

FIGURE 9–15

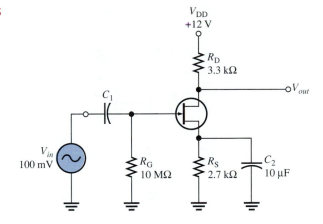

Solution First, find the dc output voltage using the BASIC computer program or a programmable calculator. When Equation (9–7) is executed with the parameter values given, $I_D \cong 1.90$ mA. From this, calculate V_D.

$$V_D = V_{DD} - I_D R_D = 12 \text{ V} - (1.90 \text{ mA})(3.3 \text{ k}\Omega) = 5.73 \text{ V}$$

Next, calculate g_m as follows:

$$V_{GS} = -I_D R_S = -(1.90 \text{ mA})(2.7 \text{ k}\Omega) = -5.13 \text{ V}$$

$$g_{m0} = \frac{2I_{DSS}}{|V_{GS(off)}|} = \frac{2(8 \text{ mA})}{10 \text{ V}} = 1.6 \text{ mS}$$

$$g_m = g_{m0}\left(1 - \frac{V_{GS}}{V_{GS(off)}}\right) = 1.6 \text{ mS}\left(1 - \frac{-5.13 \text{ V}}{-10 \text{ V}}\right) = 779 \text{ }\mu\text{S}$$

Finally, find the ac output voltage.

$$V_{out} = A_v V_{in} = g_m R_D V_{in} = (779 \text{ }\mu\text{S})(3.3 \text{ k}\Omega)(100 \text{ mV}) = 257 \text{ mV rms}$$

The total output voltage is an ac signal with a peak-to-peak value of 257 mV × 2.828 = 727 mV, riding on a dc level of 5.73 V.

Related Exercise What will happen in the amplifier of Figure 9–15 if a transistor with $V_{GS(off)} = -2$ V is used? Assume the other parameters are the same.

Effect of an AC Load on Voltage Gain

When the load is connected to the amplifier's output through a coupling capacitor, as shown in Figure 9–16(a), the ac drain resistance is effectively R_D in parallel with R_L. Remember that the upper end of R_D is at ac ground. The ac equivalent circuit is shown in Figure 9–16(b). The total ac drain resistance is

$$R_d = \frac{R_D R_L}{R_D + R_L}$$

The effect of R_L is to reduce the unloaded voltage gain, as Example 9–6 illustrates.

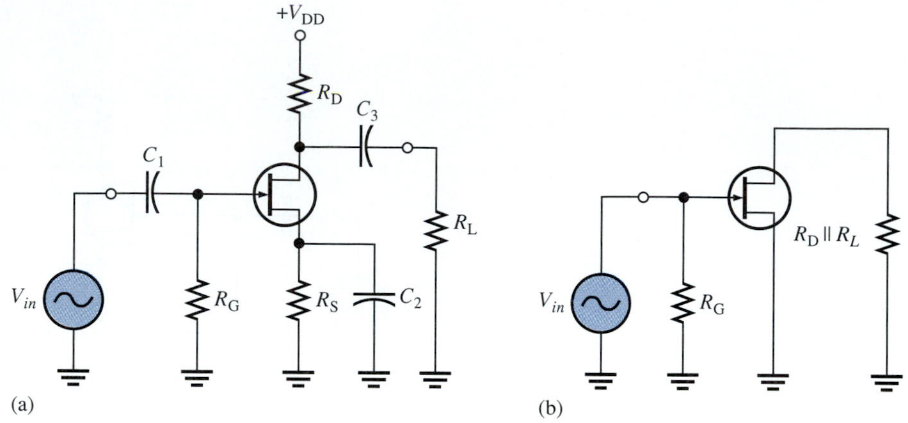

FIGURE 9–16
JFET amplifier and its ac equivalent.

EXAMPLE 9–6

If a 4.7 kΩ load resistor is ac coupled to the output of the amplifier in Example 9–5, what is the resulting rms output voltage?

Solution   The ac drain resistance is

$$R_d = \frac{R_D R_L}{R_D + R_L} = \frac{(3.3 \text{ k}\Omega)(4.7 \text{ k}\Omega)}{8 \text{ k}\Omega} = 1.94 \text{ k}\Omega$$

Calculation of V_{out} yields

$$V_{out} = g_m R_d V_{in} = (779 \text{ μS})(1.94 \text{ k}\Omega)(100 \text{ mV}) = 151 \text{ mV rms}$$

The unloaded ac output voltage was 257 mV rms in Example 9–5.

Related Exercise   If a 10 kΩ load resistor is ac coupled to the output of the amplifier in Example 9–5 and the JFET is replaced with one having a $g_m = 3000$ μS, what is the resulting rms output voltage?

Phase Inversion

The output voltage (at the drain) is 180° out of phase with the input voltage (at the gate). The phase inversion can be designated by a negative voltage gain, $-A_v$. Recall that the common-emitter bipolar amplifier also exhibited a phase inversion.

Input Resistance

Because the input to a common-source amplifier is at the gate, the input resistance is extremely high. Ideally, it approaches infinity and can be neglected. As you know, the high input resistance is produced by the reverse-biased *pn* junction in a JFET and by the insulated gate structure in a MOSFET. The actual input resistance seen by the signal source is the gate-to-ground resistor R_G in parallel with the FET's input resistance,

V_{GS}/I_{GSS}. The reverse leakage current I_{GSS} is typically given on the data sheet for a specific value of V_{GS} so that the input resistance of the device can be calculated.

$$R_{in} = R_G \parallel \left(\frac{V_{GS}}{I_{GSS}} \right) \qquad (9\text{–}9)$$

EXAMPLE 9–7

What input resistance is seen by the signal source in Figure 9–17? $I_{GSS} = 30$ nA at $V_{GS} = 10$ V.

FIGURE 9–17

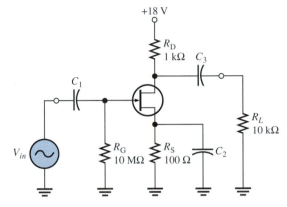

Solution   The input resistance at the gate of the JFET is

$$R_{IN(gate)} = \frac{V_{GS}}{I_{GSS}} = \frac{10 \text{ V}}{30 \text{ nA}} = 333 \text{ M}\Omega$$

The input resistance seen by the signal source is

$$R_{in} = R_G \parallel R_{IN(gate)} = 10 \text{ M}\Omega \parallel 333 \text{ M}\Omega = 9.7 \text{ M}\Omega$$

Related Exercise   How much is the total input resistance if $I_{GSS} = 1$ nA at $V_{GS} = 10$ V?

Common-Source MOSFET Amplifiers

In the previous discussion, we used the JFET to illustrate common-source amplifier analysis. The considerations for MOSFET amplifiers are similar, with the exception of the biasing arrangements required. A D-MOSFET is usually biased at $V_{GS} = 0$, and an E-MOSFET is usually biased at a V_{GS} greater than the threshold value.

D-MOSFET   Figure 9–18 shows a common-source amplifier using a D-MOSFET. The dc analysis of this amplifier is somewhat easier than for a JFET because $I_D = I_{DSS}$ at $V_{GS} = 0$. Once I_D is known, the analysis involves calculating only V_D.

$$V_D = V_{DD} - I_D R_D$$

The ac analysis is the same as for the JFET amplifier.

FIGURE 9–18
D-MOSFET common-source amplifier.

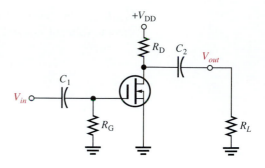

EXAMPLE 9–8

The D-MOSFET used in the amplifier of Figure 9–19 has an I_{DSS} of 12 mA and a g_m of 3.2 mS. Determine both the dc drain voltage and ac output voltage. $V_{in} = 500$ mV.

FIGURE 9–19

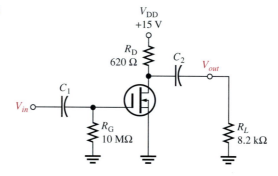

Solution Since the amplifier is zero-biased,

$$I_D = I_{DSS} = 12 \text{ mA}$$

and, therefore,

$$V_D = V_{DD} - I_D R_D = 15 \text{ V} - (12 \text{ mA})(620 \text{ Ω}) = 7.56 \text{ V}$$
$$R_d = R_D \parallel R_L = 620 \text{ Ω} \parallel 8.2 \text{ kΩ} = 576 \text{ Ω}$$

The signal voltage output is

$$V_{out} = g_m R_d V_{in} = (3.2 \text{ mS})(576 \text{ Ω})(500 \text{ mV}) = 922 \text{ mV}$$

Related Exercise If a D-MOSFET with $g_m = 5$ mS and $I_{DSS} = 10$ mA replaces the one in this example, what is the ac output voltage when $V_{in} = 500$ mV?

E-MOSFET Figure 9–20 shows a common-source amplifier using an E-MOSFET. This circuit uses voltage-divider bias to achieve a V_{GS} above threshold. The general dc analysis proceeds as follows using the E-MOSFET characteristic equation (Equation 8–10) to solve for I_D.

$$V_{GS} = \left(\frac{R_2}{R_1 + R_2} \right) V_{DD}$$
$$I_D = K(V_{GS} - V_{GS(th)})^2$$
$$V_{DS} = V_{DD} - I_D R_D$$

FIGURE 9–20

E-MOSFET common-source amplifier.

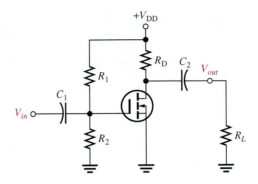

The voltage gain expression is the same as for the JFET and D-MOSFET circuits. The ac input resistance is

$$R_{in} = R_1 \| R_2 \| R_{\text{IN(gate)}}$$

(9–10)

where $R_{\text{IN(gate)}} = V_{GS}/I_{GSS}$.

EXAMPLE 9–9

A common-source amplifier using an E-MOSFET is shown in Figure 9–21. Find V_{GS}, I_D, V_{DS}, and the ac output voltage. $I_{D(on)} = 5$ mA at $V_{GS} = 10$ V, $V_{GS(th)} = 4$ V, and $g_m = 5.5$ mS. $V_{in} = 50$ mV.

FIGURE 9–21

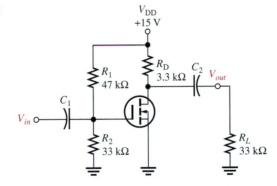

Solution

$$V_{GS} = \left(\frac{R_2}{R_1 + R_2} \right) V_{DD} = \left(\frac{33 \text{ k}\Omega}{80 \text{ k}\Omega} \right) 15 \text{ V} = 6.19 \text{ V}$$

For $V_{GS} = 10$ V,

$$K = \frac{I_{D(on)}}{(V_{GS} - V_{GS(th)})^2} = \frac{5 \text{ mA}}{(10 \text{ V} - 4 \text{ V})^2} = 0.139 \text{ mA/V}^2$$

Therefore,

$$I_D = K(V_{GS} - V_{GS(th)})^2 = 0.139 \text{ mA/V}^2 (6.19 \text{ V} - 4 \text{ V})^2 = 667 \text{ μA}$$

$$V_{DS} = V_{DD} - I_D R_D = 15 \text{ V} - (667 \text{ μA})(3.3 \text{ k}\Omega) = 12.8 \text{ V}$$

$$R_d = R_D \| R_L = 3.3 \text{ k}\Omega \| 33 \text{ k}\Omega = 3 \text{ k}\Omega$$

The ac output voltage is

$$V_{out} = A_v V_{in} = g_m R_d V_{in} = (5.5 \text{ mS})(3 \text{ k}\Omega)(50 \text{ mV}) = 825 \text{ mV}$$

Related Exercise   In Figure 9–21, $I_{D(on)} = 8$ mA at $V_{GS} = 12$ V, $V_{GS(th)} = 3$ V, and $g_m = 3.5$ mS. Find V_{GS}, I_D, V_{DS}, and the ac output voltage. $V_{in} = 50$ mV.

SECTION 9–3 REVIEW

1. What factors determine the voltage gain of a common-source FET amplifier?
2. A certain amplifier has an $R_D = 1$ kΩ. When a load resistance of 1 kΩ is capacitively coupled to the drain, how much does the gain change?

9–4 ■ COMMON-DRAIN AMPLIFIERS

The common-drain (CD) amplifier covered in this section is comparable to the common-collector bipolar amplifier. Recall that the CC amplifier is called an emitter-follower. The common-drain amplifier is sometimes called a source-follower because the voltage at the source is approximately the same amplitude as the input (gate) voltage and in phase with it. In other words, the source voltage follows the gate input voltage.

After completing this section, you should be able to

■ **Explain and analyze the operation of common-drain FET amplifiers**
  □ Analyze a CD amplifier
  □ Determine the voltage gain of a CD amplifier
  □ Determine the input resistance of a CD amplifier

A **common-drain** JFET amplifier is shown in Figure 9–22. Self-biasing is used in this circuit. The input signal is applied to the gate through a coupling capacitor, and the output is at the source terminal. There is no drain resistor.

FIGURE 9–22
JFET common-drain amplifier (source-follower).

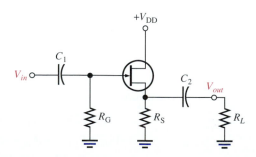

Voltage Gain

As in all amplifiers, the voltage gain is $A_v = V_{out}/V_{in}$. For the **source-follower,** V_{out} is $I_d R_s$ and V_{in} is $V_{gs} + I_d R_s$, as shown in Figure 9–23. Therefore, the gate-to-source voltage gain is $I_d R_s/(V_{gs} + I_d R_s)$. Substituting $I_d = g_m V_{gs}$ into the expression gives the following result:

$$A_v = \frac{g_m V_{gs} R_s}{V_{gs} + g_m V_{gs} R_s}$$

The V_{gs} terms cancel, so

$$A_v = \frac{g_m R_s}{1 + g_m R_s} \qquad \textbf{(9–11)}$$

Notice here that the gain is always slightly less than one. If $g_m R_s \gg 1$, then a good approximation is $A_v \cong 1$. Since the output voltage is at the source, it is in phase with the gate (input) voltage.

FIGURE 9–23
Voltages in a common-drain amplifier.

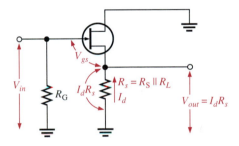

Input Resistance

Because the input signal is applied to the gate, the input resistance seen by the input signal source is extremely high, just as in the common-source amplifier configuration. The gate resistor R_G, in parallel with the input resistance looking in at the gate, is the total input resistance.

$$R_{in} = R_G \| R_{IN(gate)} \qquad \textbf{(9–12)}$$

where $R_{IN(gate)} = V_{GS}/I_{GSS}$.

EXAMPLE 9–10   Determine the voltage gain of the amplifier in Figure 9–24 using the data sheet information in Figure 9–25. Also, determine the input resistance. Use minimum data sheet values where available.

FIGURE 9–24

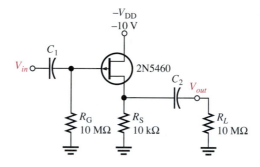

Electrical Characteristics (T_A = 25°C unless otherwise noted.)

Characteristic		Symbol	Min	Typ	Max	Unit
OFF Characteristics						
Gate-Source breakdown voltage	2N5460, 2N5461, 2N5462	$V_{(BR)GSS}$	40	–	–	V dc
(I_G = 10 μA dc, V_{DS} = 0)	2N5463, 2N5464, 2N5465		60	–	–	
Gate reverse current		I_{GSS}				
(V_{GS} = 20 V dc, V_{DS} = 0)	2N5460, 2N5461, 2N5462		–	–	5.0	nA dc
(V_{GS} = 30 V dc, V_{DS} = 0)	2N5463, 2N5464, 2N5465		–	–	5.0	
(V_{GS} = 20 V dc, V_{DS} = 0, T_A = 100°C)	2N5460, 2N5461, 2N5462		–	–	1.0	μA dc
(V_{GS} = 30 V dc, V_{DS} = 0, T_A = 100°C)	2N5463, 2N5464, 2N5465		–	–	1.0	
Gate-Source cutoff voltage		$V_{GS(off)}$				V dc
(V_{DS} = 15 V dc, I_D = 1.0 μA dc)	2N5460, 2N5463		0.75	–	6.0	
	2N5461, 2N5464		1.0	–	7.5	
	2N5462, 2N5465		1.8	–	9.0	
Gate-Source voltage		V_{GS}				
(V_{DS} = 15 V dc, I_D = 0.1 mA dc)	2N5460, 2N5463		0.5	–	4.0	V dc
(V_{DS} = 15 V dc, I_D = 0.2 mA dc)	2N5461, 2N5464		0.8	–	4.5	
(V_{DS} = 15 V dc, I_D = 0.4 mA dc)	2N5462, 2N5465		1.5	–	6.0	
ON Characteristics						
Zero-gate-voltage drain current		I_{DSS}				mA dc
(V_{DS} = 15 V dc, V_{GS} = 0,	2N5460, 2N5463		– 1.0	–	– 5.0	
f = 1.0 kHz)	2N5461, 2N5464		– 2.0	–	– 9.0	
	2N5462, 2N5465		– 4.0	–	– 16	
Small-Signal Characteristics						
Forward transfer admittance		$|Y_{fs}|$				μmhos
(V_{DS} = 15 V dc, V_{GS} = 0, f = 1.0 kHz)	2N5460, 2N5463		1000	–	4000	or
	2N5461, 2N5464		1500	–	5000	μS
	2N5462, 2N5465		2000	–	6000	
Output admittance		$|Y_{os}|$	–	–	75	μmhos or
(V_{DS} = 15 V dc, V_{GS} = 0, f = 1.0 kHz)						μS
Input capacitance		C_{iss}	–	5.0	7.0	pF
(V_{DS} = 15 V dc, V_{GS} = 0, f = 1.0 MHz)						
Reverse transfer capacitance		C_{rss}	–	1.0	2.0	pF
(V_{DS} = 15 V dc, V_{GS} = 0, f = 1.0 MHz)						

FIGURE 9–25

Partial data sheet for the 2N5460 p-channel JFET.

Solution Since $R_L \gg R_S$, $R_s \cong R_S$. From the partial data sheet in Figure 9–25, $g_m = y_{fs} = 1000\ \mu S$ (minimum). The voltage gain is

$$A_v = \frac{g_m R_S}{1 + g_m R_S} = \frac{(1000\ \mu S)(10\ k\Omega)}{1 + (1000\ \mu S)(10\ k\Omega)} = 0.909$$

From the data sheet, $I_{GSS} = 5\ nA$ (maximum) at $V_{GS} = 20\ V$. Therefore,

$$R_{IN(gate)} = \frac{V_{GS}}{I_{GSS}} = \frac{20\ V}{5\ nA} = 4000\ M\Omega$$

$$R_{IN} = R_G \parallel R_{IN(gate)} = 10\ M\Omega \parallel 4000\ M\Omega \cong 10\ M\Omega$$

Related Exercise If the maximum value of g_m of the 2N5460 JFET in the source-follower of Figure 9–24 is used, what is the voltage gain?

<table>
<tr><td>**SECTION 9–4
REVIEW**</td><td>1. What is the ideal maximum voltage gain of a common-drain amplifier?

2. What factors influence the voltage gain of a common-drain amplifier?</td></tr>
</table>

9–5 ■ COMMON-GATE AMPLIFIERS

The common-gate FET amplifier configuration introduced in this section is comparable to the common-base BJT amplifier. Like the CB, the common-gate (CG) amplifier has a low input resistance. This is different from the CS and CD configurations, which have very high input resistances.

After completing this section, you should be able to

■ **Explain and analyze the operation of common-gate FET amplifiers**
  □ Analyze a CG amplifier
  □ Determine the voltage gain of a CG amplifier
  □ Determine the input resistance of a CG amplifier

A self-biased **common-gate** amplifier is shown in Figure 9–26. The gate is connected directly to ground. The input signal is applied at the source terminal through C_1. The output is coupled through C_2 from the drain terminal.

Voltage Gain

The voltage gain from source to drain is developed as follows.

$$A_v = \frac{V_{out}}{V_{in}} = \frac{V_d}{V_{gs}} = \frac{I_d R_d}{V_{gs}} = \frac{g_m V_{gs} R_d}{V_{gs}}$$

$$A_v = g_m R_d \tag{9–13}$$

where $R_d = R_D \parallel R_L$. Notice that the gain expression is the same as for the common-source JFET amplifier.

FIGURE 9–26
JFET common-gate amplifier.

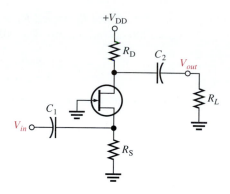

Input Resistance

As you have seen, both the common-source and common-drain configurations have extremely high input resistances because the gate is the input terminal. In contrast, the common-gate configuration where the source is the input terminal has a low input resistance, as shown in the following steps.

First, the input current (source current) is equal to the drain current.

$$I_{in} = I_s = I_d = g_m V_{gs}$$

Second, the input voltage equals V_{gs}.

$$V_{in} = V_{gs}$$

Therefore, the input resistance at the source terminal is

$$R_{in(source)} = \frac{V_{in}}{I_{in}} = \frac{V_{gs}}{g_m V_{gs}}$$

$$R_{in(source)} = \frac{1}{g_m} \qquad\qquad (9\text{–}14)$$

If, for example, g_m has a value of 4000 μS, then

$$R_{in(source)} = \frac{1}{4000 \ \mu S} = 250 \ \Omega$$

EXAMPLE 9–11

Determine the minimum voltage gain and input resistance of the amplifier in Figure 9–27.

FIGURE 9–27

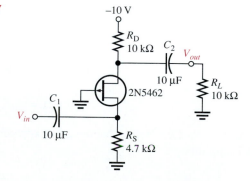

Solution From the data sheet in Figure 9–25 $g_m = 2000\ \mu S$ minimum. This common-gate amplifier has a load resistor, so the effective drain resistance is $R_D \parallel R_L$ and the minimum voltage gain is

$$A_v = g_m(R_D \parallel R_L) = (2000\ \mu S)(10\ k\Omega \parallel 10\ k\Omega) = 10$$

The input resistance at the source terminal is

$$R_{in(source)} = \frac{1}{g_m} = \frac{1}{2000\ \mu S} = 500\ \Omega$$

The signal source actually sees R_S in parallel with $R_{in(source)}$, so the total input resistance is

$$R_{in} = R_{in(source)} \parallel R_S = 500\ \Omega \parallel 4.7\ k\Omega = 452\ \Omega$$

Related Exercise What is the input resistance in Figure 9–27 if R_S is changed to 10 kΩ?

SECTION 9–5 REVIEW

1. What is a major difference between a common-gate amplifier and the other two configurations?
2. What common factor determines the voltage gain and the input resistance of a common-gate amplifier?

SUMMARY OF FET AMPLIFIERS

N channels are shown. V_{DD} is negative for p channel.

COMMON-SOURCE AMPLIFIERS

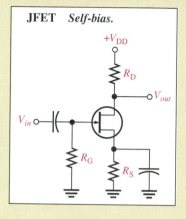

JFET *Self-bias.*

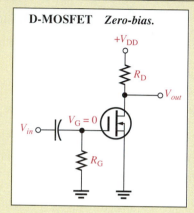

D-MOSFET *Zero-bias.*

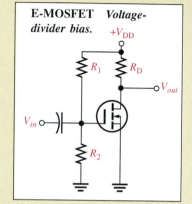

E-MOSFET *Voltage-divider bias.*

■ $I_D = I_{DSS}\left(1 - \dfrac{I_D R_S}{V_{GS(off)}}\right)^2$

■ $A_v = g_m R_d$

■ $R_{in} = R_G \parallel \left(\dfrac{V_{GS}}{I_{DSS}}\right)$

■ $I_D = I_{DSS}$

■ $A_v = g_m R_d$

■ $R_{in} = R_G \parallel \left(\dfrac{V_{GS}}{I_{DSS}}\right)$

■ $I_D = K(V_{GS} - V_{GS(th)})^2$

■ $A_v = g_m R_d$

■ $R_{in} = R_1 \parallel R_2 \parallel \left(\dfrac{V_{GS}}{I_{DSS}}\right)$

COMMON-DRAIN AMPLIFIER

COMMON-GATE AMPLIFIER

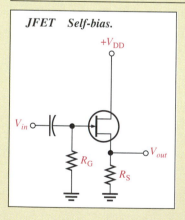

JFET *Self-bias.*

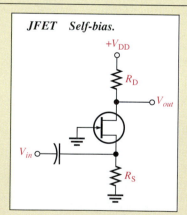

JFET *Self-bias.*

■ $I_D = I_{DSS}\left(1 - \dfrac{I_D R_S}{V_{GS(off)}}\right)^2$

■ $A_v = \dfrac{g_m R_s}{1 + g_m R_s}$

■ $R_{in} = R_G \parallel \left(\dfrac{V_{GS}}{I_{DSS}}\right)$

■ $I_D = I_{DSS}\left(1 - \dfrac{I_D R_S}{V_{GS(off)}}\right)^2$

■ $A_v = g_m R_d$

■ $R_{in} = \dfrac{1}{g_m}$

9–6 ■ TROUBLESHOOTING

A technician who understands the basics of circuit operation and who can, if necessary, perform basic analysis on a given circuit is much more valuable than one who is limited to carrying out routine test procedures. In this section, you will see how to test a circuit board that has only a schematic with no specified test procedure or voltage levels. In this case, basic knowledge of how the circuit operates and the ability to do a quick circuit analysis are useful.

After completing this section, you should be able to

■ **Troubleshoot FET amplifiers**
  □ Troubleshoot a two-stage CS amplifier
  □ Relate a schematic to a circuit board

Assume that you are given a circuit board pulled from the audio amplifier section of a sound system and told simply that it is not working properly. The first step is to obtain the system schematic and locate this particular circuit on it. The circuit is a two-stage FET amplifier, as shown in Figure 9–28. The problem is approached in the following sequence.

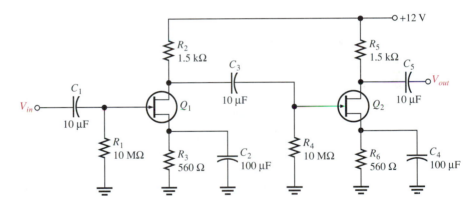

FIGURE 9–28
A two-stage FET amplifier circuit.

Step 1: Determine what the voltage levels in the circuit should be so that you know what to look for. First, pull a data sheet on the particular transistor (assume both Q_1 and Q_2 are found to be the same type of transistor) and determine the g_m so that you can calculate the voltage gain. Assume that for this particular device, a typical g_m of 5000 μS is specified. Calculate the expected typical voltage gain of each stage (notice they are identical). Because the input resistance is very high, the second stage does not significantly load the first stage, as in a bipolar amplifier. So, the unloaded voltage gain for each stage is

$$A_v = g_m R_2 = (5000 \text{ μS})(1.5 \text{ kΩ}) = 7.5$$

Since the stages are identical, the typical overall gain should be

$$A'_v = (7.5)(7.5) = 56.3$$

We will ignore dc levels at this time and concentrate on signal tracing.

Step 2: Arrange a test setup to permit connection of an input test signal, a dc supply voltage, and ground to the circuit. The schematic shows that the dc supply voltage must be +12 V. Choose 10 mV rms as an input test signal. This value is arbitrary (although the capability of your signal source is a factor), but small enough that the expected output signal voltage is well below the absolute peak-to-peak limit of 12 V set by the supply voltage and ground (you know that the output voltage swing cannot go higher than 12 V or lower than 0 V). Set the frequency of the sine wave signal source to an arbitrary value in the audio range (say 10 kHz), since you know this is an audio amplifier. The audio frequency range is generally accepted as 20 Hz to 20 kHz.

Step 3: Check the input signal at the gate of Q_1 and the output signal at the drain of Q_2 with an oscilloscope. The results are shown in Figure 9–29. The measured output voltage has a peak value of 226 mV. The expected typical peak output voltage is

$$V_{out} = V_{in}A'_v = (14.14 \text{ mV})(56.3) = 796 \text{ mV peak}$$

The output is much less than it should be.

Step 4: Trace the signal from the output back toward the input to determine the fault. Figure 9–29 shows the oscilloscope displays of the measured signal voltages. The voltage at the gate of Q_2 is 106 mV peak, as expected (14.14 mV × 7.5 = 106 mV). This signal is properly coupled from the drain of Q_1. Therefore, the problem lies in the second stage. From the oscilloscope displays, the gain of Q_2 is much lower than it should be (213 mV/100 mV = 2.13 instead of 7.5).

Step 5: Analyze the possible causes of the observed malfunction. There are three possible reasons the gain is low:

1. Q_2 has a lower transconductance (g_m) than the specified typical value. Check the data sheet to see if the minimum g_m accounts for the lower measured gain.

2. R_5 has a lower value than shown on the schematic.

3. The bypass capacitor C_4 is open.

The only way to check the g_m is by replacing Q_2 with a new transistor of the same type and rechecking the output signal. You can make certain that R_5 is the proper value by removing one end of the resistor from the circuit board and measuring the resistance with an ohmmeter. To avoid having to unsolder a component, the best way to start isolating the fault is by checking the signal voltage at the source of Q_2. If the capacitor is working properly, there will be only a dc voltage at the source. The presence of a signal voltage at the source indicates that C_4 is open. With R_6 unbypassed, the gain expression is $g_mR_d/(1 + g_mR_d)$ rather than simply g_mR_d, thus resulting in less gain.

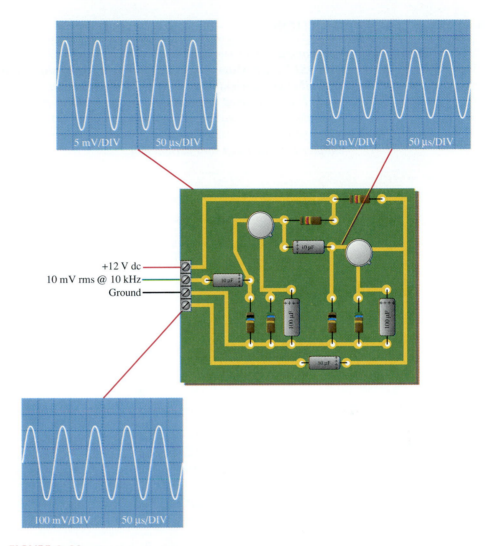

FIGURE 9–29

Oscilloscope displays of signals in the two-stage FET amplifier.

SECTION 9–6 REVIEW

1. What is the prerequisite to effective troubleshooting?

2. Assume that C_2 in the amplifier of Figure 9–28 opened. What symptoms would indicate this failure?

3. If C_3 opened in the amplifier, would the voltage gain of the first stage be affected?

9–7 ▪ SYSTEM APPLICATION

The company you work for has the opportunity to purchase a quantity of 2N3797 D-MOSFET transistors for a very low price. Your supervisor wants to investigate the possibility of redesigning the audio preamplifier used in the public address system currently being produced to replace the bipolar transistors with the FETs. You are asked to evaluate a design and breadboard the circuit to evaluate its performance compared to the existing bipolar amplifier. You will apply the knowledge you have gained in this chapter in completing your assignment.

Review of the Bipolar Amplifier

A schematic of the bipolar amplifier currently used in the public address/audio paging system is shown in Figure 9–30.

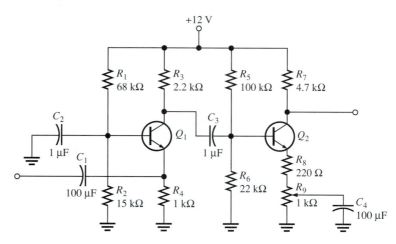

FIGURE 9–30
Bipolar transistor audio preamplifier. Both transistors are 2N3904.

Performance of the Bipolar Amplifier

The overall voltage gain of the two-stage bipolar amplifier was determined to be adjustable over a range from 145 to 733 as follows:

Stage 1:

$$V_B = \left(\frac{R_2 \parallel \beta_{DC}R_4}{R_1 + R_2 \parallel \beta_{DC}R_4} \right) V_{CC} = 1.93 \text{ V}$$

$$V_E = V_B - 0.7 \text{ V} = 1.23 \text{ V}$$

$$I_E = \frac{V_E}{R_E} = 1.23 \text{ mA}$$

$$r'_e = 20.3 \text{ }\Omega$$

$$A_v = \frac{R_c}{r'_e} = \frac{1.93 \text{ k}\Omega}{20.3 \text{ }\Omega} = 95$$

Stage 2:

$$V_B = \left(\frac{R_6 \parallel \beta_{DC}(R_8 + R_9)}{R_5 + R_2 \parallel \beta_{DC}(R_8 + R_9)} \right) V_{CC} = 1.88 \text{ V}$$

$$V_E = V_B - 0.7 \text{ V} = 1.18 \text{ V}$$

$$I_E = \frac{V_E}{R_E} = 0.97 \text{ mA}$$

$$r'_e = 25.8 \text{ }\Omega$$

$$A_{v(max)} = \frac{R_c}{r'_e + R_8} = \frac{4.7 \text{ k}\Omega}{245.8 \text{ }\Omega} = 19.1$$

$$A_{v(min)} = \frac{R_c}{r'_e + R_8 + R_9} = \frac{4.7 \text{ k}\Omega}{1245.8 \text{ }\Omega} = 3.77$$

Attenuation of Input: The microphone input has a 30 Ω resistance.

$$\text{Attenuation} = \frac{20.3 \text{ }\Omega}{20.3 \text{ }\Omega + 30 \text{ }\Omega} = 0.404$$

Overall Voltage Gain:

$$A_{v(tot)} = (0.404)(95)(19.1) = 733 \qquad \text{maximum}$$
$$A_{v(tot)} = (0.404)(95)(3.77) = 145 \qquad \text{minimum}$$

Basic MOSFET Amplifier Design

Your boss has given you a basic design for one stage of a MOSFET amplifier shown in Figure 9–31 and asked you to evaluate it to determine if it can be used in a multistage amplifier to provide the same voltage gain as the bipolar version.

FIGURE 9–31

Schematic of one MOSFET amplifier stage.

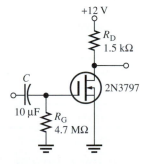

☐ Using the data sheet in Figure 9–32, determine the minimum and maximum dc drain-to-source voltage. Remember, for a zero-biased D-MOSFET, the drain current equals I_{DSS}.

☐ Using the data sheet in Figure 9–32, determine the minimum and maximum voltage gains.

☐ Discuss the problem presented by the variation in I_{DSS} from one device to the next and recommend an approach to minimize the problem.

☐ Discuss the problem presented by the fact that the voltage gain of the MOSFET amplifier stage is dependent on g_m, and recommend an approach to minimize the problem.

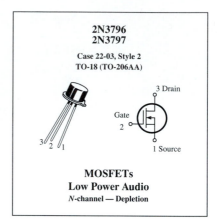

Maximim Ratings

Rating	Symbol	Value	Unit

Drain-Source voltage
    2N3796
    2N3797
	V_{DS}	25
		20
			V dc

Gate-Source voltage	V_{GS}	±10	V dc
Drain current	I_D	20	mA dc
Total device dissipation @ $T_A = 25°C$ Derate above 25°C	P_D	200 1.14	mW mW/C°
Junction temperature	T_J	+175	°C
Storage channel temperature range	T_{stg}	− 65 to + 200	°C

Electrical Characteristics ($T_A = 25°C$ unless otherwise noted.)

Characteristic		Symbol	Min	Typ	Max	Unit
OFF Characteristics						

Drain-Source breakdown voltage
    ($V_{GS} = -4.0$ V, $I_D = 5.0$ μA)
    ($V_{GS} = -7.0$ V, $I_D = 5.0$ μA)
	2N3796
	2N3797
		$V_{(BR)DSX}$	25
			20
				30
				25
					–
					–
						V dc

Gate reverse current
    ($V_{GS} = -10$ V, $V_{DS} = 0$)
    ($V_{GS} = -10$ V, $V_{DS} = 0, T_A = 150°C$)
		I_{GSS}	–
			–
				–
				–
					1.0
					200
						pA dc

Gate-Source cutoff voltage
    ($I_D = 0.5$ μA, $V_{DS} = 10$ V)
    ($I_D = 2.0$ μA, $V_{DS} = 10$ V)
	2N3796
	2N3797
		$V_{GS(off)}$	–
			–
				− 3.0
				− 5.0
					− 4.0
					− 7.0
						V dc

Drain-Gate reverse current ($V_{DG} = 10$ V, $I_S = 0$)		I_{DGO}	–	–	1.0	pA dc
ON Characteristics						
Zero-gate-voltage drain current ($V_{DS} = 10$ V, $V_{GS} = 0$)	2N3796 2N3797	I_{DSS}	0.5 2.0	1.5 2.9	3.0 6.0	mA dc
On-State drain current ($V_{DS} = 10$ V, $V_{GS} = +3.5$ V)	2N3796 2N3797	$I_{D(on)}$	7.0 9.0	8.3 14	14 18	mA dc
Small-Signal Characteristics						
Forward transfer admittance ($V_{DS} = 10$ V, $V_{GS} = 0, f = 1.0$ kHz)	2N3796 2N3797	$|Y_{fs}|$	900 1500	1200 2300	1800 3000	μmhos or μS
($V_{DS} = 10$ V, $V_{GS} = 0, f = 1.0$ MHz)	2N3796 2N3797		900 1500	– –	– –	
Output admittance ($V_{DS} = 10$ V, $V_{GS} = 0, f = 1.0$ kHz)	2N3796 2N3797	$|Y_{os}|$	– –	12 27	25 60	μmhos or μS
Input capacitance ($V_{DS} = 10$ V, $V_{GS} = 0, f = 1.0$ MHz)	2N3796 2N3797	C_{iss}	– –	5.0 6.0	7.0 8.0	pF
Reverse transfer capacitance ($V_{DS} = 10$ V, $V_{GS} = 0, f = 1.0$ MHz)		C_{rss}	–	0.5	0.8	pF

FIGURE 9–32

Partial data sheet for the 2N3797 D-MOSFET.

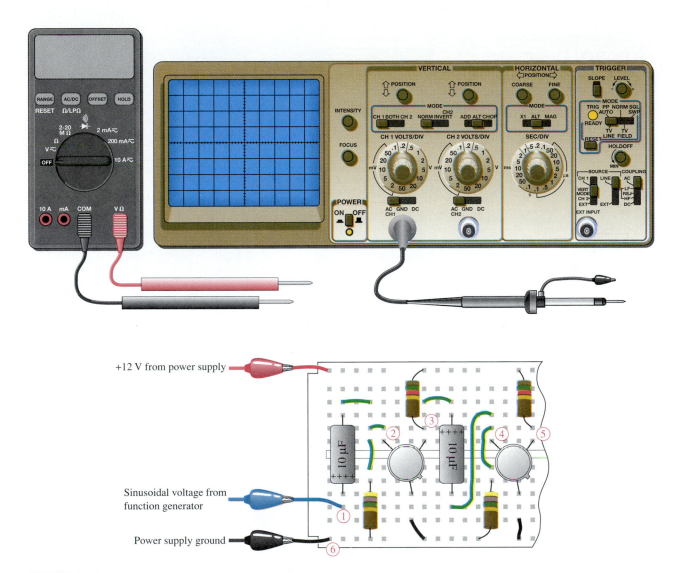

FIGURE 9–33
Breadboard test bench for a two-stage MOSFET amplifier.

The Amplifier

The breadboard test bench for the MOSFET amplifier is shown in Figure 9–33. Two stages have been wired on the connector board for testing. The two 2N3797 transistors in the circuit have been randomly picked from a large selection of the devices.

Verify that the circuit is properly wired on the connector board.

Amplifier Performance on the Test Bench

☐ Measurements taken on the breadboarded circuit are shown in Figure 9–34 where each set of measurements is made with different transistors. That is, the first set of measurements is made with two randomly selected 2N3797s and then the transistors are

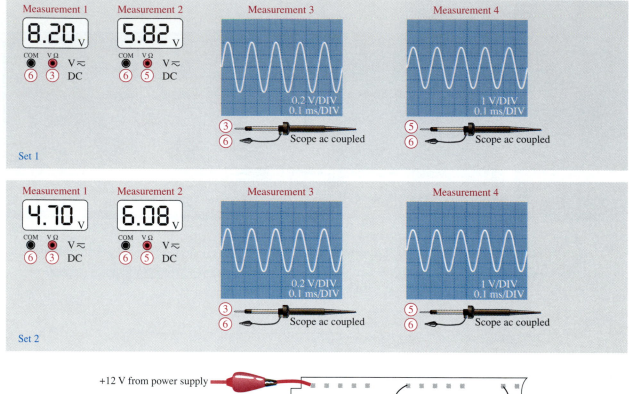

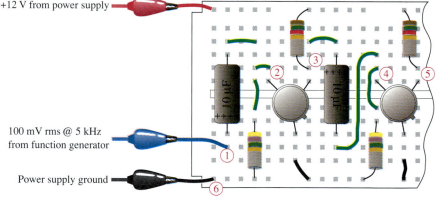

FIGURE 9–34

Measurements of amplifier voltages for two sets of MOSFETs.

replaced with two different randomly selected devices for the second set of measurements. The circled numbers indicate test point connections to the circuit.

☐ Explain the difference in the two sets of dc voltage readings.

☐ Explain the difference in the two sets of ac voltage readings.

☐ Can I_{DSS} and g_m be determined for each transistor from the measurements in Figure 9–34? If so, determine the values for each of the four different MOSFETs.

☐ Determine the gain of each stage for each set of measurements in Figure 9–34.

☐ Assuming the maximum theoretical gain can be achieved for V_{DS} centered at 6 V by careful selection of the MOSFETs for maximum g_m and typical I_{DSS}, how many stages will be required to match the maximum gain of the bipolar amplifier in Figure 9–30? Remember to take into account the attenuation, if any, of the input circuit.

Recommendation

Submit a recommendation to your supervisor containing the following items:

1. A summary of your analysis and test results for the MOSFET amplifier.

2. A comparison of the performance of the MOSFET amplifier to the bipolar amplifier.

3. A statement of recommendation to either replace the bipolar amplifier with a MOSFET amplifier or to retain the bipolar amplifier in the system.

4. Reasons justifying your recommendation.

■ CHAPTER SUMMARY

- The drain of an FET is analogous to the collector of a BJT, the source of an FET is analogous to the emitter of a BJT, and the gate of an FET is analogous to the base of a BJT.
- The transconductance, g_m, of an FET relates the output current I_d to the input voltage V_{gs}.
- The voltage gain of a common-source amplifier is determined largely by the transconductance g_m and the drain resistance R_d.
- The internal drain-to-source resistance r'_{ds} of the FET influences (reduces) the gain if it is not sufficiently greater than R_d so that it can be neglected.
- An unbypassed resistance between source and ground (R_S) reduces the voltage gain of an FET amplifier.
- A load resistance connected to the drain of a common-source amplifier reduces the voltage gain.
- There is a 180° phase inversion between gate and drain voltages.
- The input resistance at the gate of an FET is extremely high.
- The voltage gain of a common-drain amplifier (source-follower) is always slightly less than 1.
- There is no phase inversion between gate and source in a source-follower.
- The input resistance of a common-gate amplifier is the reciprocal of g_m.
- The total voltage gain of a multistage amplifier is the product of the individual voltage gains (sum of dB gains).
- Generally, higher voltage gains can be achieved with bipolar amplifiers than with FET amplifiers.

■ GLOSSARY

Common-drain An FET amplifier configuration in which the drain is the grounded terminal.

Common-gate An FET amplifier configuration in which the gate is the grounded terminal.

Common-source An FET amplifier configuration in which the source is the grounded terminal.

Source-follower The common-drain amplifier.

■ FORMULAS

FET Amplification

(9–1)	$I_d = g_m V_{gs}$	Drain current
(9–2)	$A_v = \dfrac{V_{ds}}{V_{gs}}$	General voltage gain

(9–3) $\quad A_v = g_m R_d$ $\qquad\qquad\qquad$ Voltage gain

(9–4) $\quad A_v = g_m\left(\dfrac{R_d r'_{ds}}{R_d + r'_{ds}}\right)$ $\qquad$ Voltage gain considering r'_{ds}

(9–5) $\quad A_v = \dfrac{g_m R_d}{1 + g_m R_s}$ $\qquad\qquad$ Voltage gain with R_s

Common-Source Amplifier

(9–6) $\quad I_D = \dfrac{I_{DSS}}{2}$ $\qquad\qquad\qquad$ For centered Q-point

(9–7) $\quad I_D = I_{DSS}\left(1 - \dfrac{I_D R_S}{V_{GS(off)}}\right)^2$ $\qquad$ Self-biased JFET current

(9–8) $\quad A_v = g_m R_d$ $\qquad\qquad\qquad$ Voltage gain

(9–9) $\quad R_{in} = R_G \left\|\left(\dfrac{V_{GS}}{I_{GSS}}\right)\right.$ $\qquad$ Input resistance, self-bias and zero-bias

(9–10) $\quad R_{in} = R_1 \| R_2 \| R_{IN(gate)}$ $\qquad$ Input resistance, voltage-divider bias

Common-Drain Amplifier

(9–11) $\quad A_v = \dfrac{g_m R_s}{1 + g_m R_s}$ $\qquad\qquad$ Voltage gain

(9–12) $\quad R_{in} = R_G \| R_{IN(gate)}$ $\qquad\qquad$ Input resistance

Common-Gate Amplifier

(9–13) $\quad A_v = g_m R_d$ $\qquad\qquad\qquad$ Voltage gain

(9–14) $\quad R_{in(source)} = \dfrac{1}{g_m}$ $\qquad\qquad$ Input resistance

---

■ **SELF-TEST**

1. In a common-source amplifier, the output voltage is
   (a) 180° out of phase with the input (b) in phase with the input
   (c) taken at the source (d) taken at the drain
   (e) answers (a) and (c) (f) answers (a) and (d)

2. In a certain common-source (CS) amplifier, $V_{ds} = 3.2$ V rms and $V_{gs} = 280$ mV rms. The voltage gain is
   (a) 1 (b) 11.4 (c) 8.75 (d) 3.2

3. In a certain CS amplifier, $R_D = 1$ kΩ, $R_S = 560$ Ω, $V_{DD} = 10$ V, and $g_m = 4500$ μS. If the source resistor is completely bypassed, the voltage gain is
   (a) 450 (b) 45 (c) 4.5 (d) 2.52

4. Ideally, the equivalent circuit of an FET contains
   (a) a current source in series with a resistance
   (b) a resistance between drain and source terminals
   (c) a current source between gate and source terminals
   (d) a current source between drain and source terminals

5. The value of the current source in Question 4 is dependent on the
   (a) transconductance and gate-to-source voltage
   (b) dc supply voltage
   (c) external drain resistance
   (d) answers (b) and (c)

6. A certain common-source amplifier has a voltage gain of 10. If the source bypass capacitor is removed,
   (a) the voltage gain will increase (b) the transconductance will increase
   (c) the voltage gain will decrease (d) the Q-point will shift

7. A CS amplifier has a load resistance of 10 kΩ and $R_D = 820\ \Omega$. If $g_m = 5$ mS and $V_{in} = 500$ mV, the output signal voltage is
   (a) 1.89 V (b) 2.05 V (c) 25 V (d) 0.5 V

8. If the load resistance in Question 7 is removed, the output voltage will
   (a) stay the same (b) decrease (c) increase (d) be zero

9. A certain common-drain (CD) amplifier with $R_S = 1$ kΩ has a transconductance of 6000 μS. The voltage gain is
   (a) 1 (b) 0.86 (c) 0.98 (d) 6

10. The data sheet for the transistor used in a CD amplifier specifies $I_{GSS} = 5$ nA at $V_{GS} = 10$ V. If the resistor from gate to ground, R_G, is 50 MΩ, the total input resistance is approximately
    (a) 50 MΩ (b) 200 MΩ (c) 40 MΩ (d) 20.5 MΩ

11. The common-gate (CG) amplifier differs from both the CS and CD configurations in that it has a
    (a) much higher voltage gain (b) much lower voltage gain
    (c) much higher input resistance (d) much lower input resistance

12. If you are looking for both good voltage gain and high input resistance, you must use a
    (a) CS amplifier (b) CD amplifier (c) CG amplifier

13. For small-signal operation, an n-channel JFET must be biased at
    (a) $V_{GS} = 0$ V (b) $V_{GS} = V_{GS(off)}$
    (c) $-V_{GS(off)} < V_{GS} < 0$ V (d) 0 V $< V_{GS} < +V_{GS(off)}$

14. Two FET amplifiers are cascaded. The first stage has a voltage gain of 5 and the second stage has a voltage gain of 7. The overall voltage gain is
    (a) 35 (b) 12 (c) dependent on the second stage loading

15. If there is an internal open between the drain and source in a CS amplifier, the drain voltage is equal to
    (a) 0 V (b) V_{DD} (c) a value less than normal (d) V_{GS}

■ **BASIC PROBLEMS**

SECTION 9–1 Small-Signal FET Amplifier Operation

1. Identify the type of FET and its bias arrangement in Figure 9–35. Ideally, what is V_{GS}?
2. Calculate the dc voltages from each terminal to ground for the FETs in Figure 9–35.

FIGURE 9–35

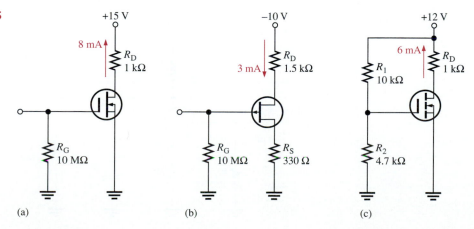

(a) (b) (c)

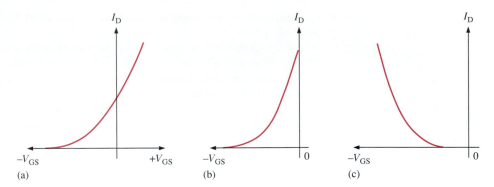

(a)   (b)   (c)

FIGURE 9–36

3. Identify each characteristic curve in Figure 9–36 by the type of FET that it represents.

4. Refer to the JFET transfer characteristic curve in Figure 9–7(a) and determine the peak-to-peak value of I_d when V_{gs} is varied ±1.5 V about its Q-point value.

5. Repeat Problem 4 for the curves in Figure 9–7(b) and Figure 9–7(c).

SECTION 9–2   FET Amplification

6. An FET has a $g_m = 6000\ \mu S$. Determine the rms drain current for each of the following rms values of V_{gs}.
   (a) 10 mV   **(b)** 150 mV   **(c)** 0.6 V   **(d)** 1 V

7. The gain of a certain JFET amplifier with a source resistance of zero is 20. Determine the drain resistance if the g_m is 3500 μS.

8. A certain FET amplifier has a g_m of 4.2 mS, $r'_{ds} = 12\ k\Omega$, and $R_D = 4.7\ k\Omega$. What is the voltage gain? Assume the source resistance is 0 Ω.

9. What is the gain for the amplifier in Problem 8 if the source resistance is 1 kΩ?

SECTION 9–3   Common-Source Amplifiers

10. Given that $I_D = 2.83$ mA in Figure 9–37, find V_{DS} and V_{GS}. $V_{GS(off)} = -7$ V and $I_{DSS} = 8$ mA.

11. If a 50 mV rms input signal is applied to the amplifier in Figure 9–37, what is the peak-to-peak output voltage? $g_m = 5000\ \mu S$.

12. If a 1500 Ω load is ac coupled to the output in Figure 9–37, what is the resulting output voltage (rms) when a 50 mV rms input is applied? $g_m = 5000\ \mu S$.

FIGURE 9–37

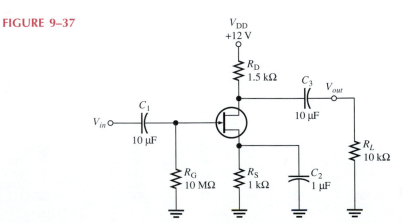

13. Determine the voltage gain of each common-source amplifier in Figure 9–38.

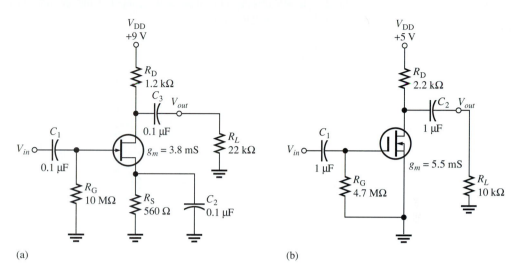

(a)   (b)

FIGURE 9–38

14. Draw the dc and ac equivalent circuits for the amplifier in Figure 9–39.

15. Determine the drain current in Figure 9–39 given that $I_{DSS} = 12.7$ mA and $V_{GS(off)} = -4$ V. The Q-point is centered.

16. What is the gain of the amplifier in Figure 9–39 if C_2 is removed?

17. A 4.7 kΩ resistor is connected in parallel with R_L in Figure 9–39. What is the voltage gain?

FIGURE 9–39

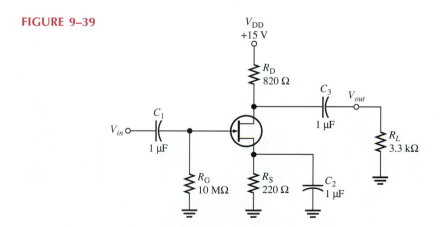

18. For the common-source amplifier in Figure 9–40 determine I_D, V_{GS}, and V_{DS} for a centered Q-point. $I_{DSS} = 9$ mA, and $V_{GS(off)} = -3$ V.

19. If a 10 mV rms signal is applied to the input of the amplifier in Figure 9–40, what is the rms value of the output signal?

FIGURE 9–40

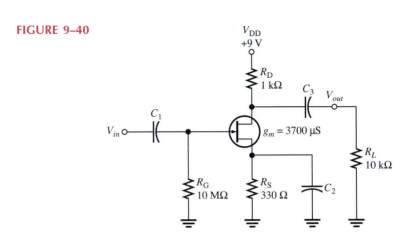

20. Determine V_{GS}, I_D, and V_{DS} for the amplifier in Figure 9–41. $I_{D(on)} = 18$ mA at $V_{GS} = 10$ V, $V_{GS(th)} = 2.5$ V, and $g_m = 3000$ μS.

21. Determine the input resistance seen by the signal source in Figure 9–42. Assume $I_{GSS} = 25$ nA at $V_{GS} = -15$ V.

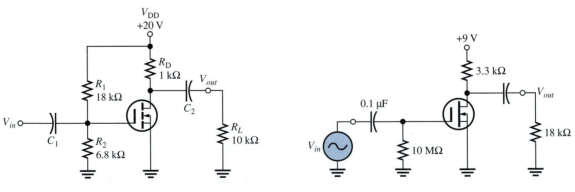

FIGURE 9–41

FIGURE 9–42

22. Determine the total drain voltage waveform (dc and ac) and the V_{out} waveform in Figure 9–43. $g_m = 4.8$ mS and $I_{DSS} = 15$ mA. Observe that $V_{GS} = 0$.

23. For the unloaded amplifier in Figure 9–44, find V_{GS}, I_D, V_{DS}, and the rms output voltage V_{ds}. $I_{D(on)} = 8$ mA at $V_{GS} = 12$ V, $V_{GS(th)} = 4$ V, and $g_m = 4500$ μS.

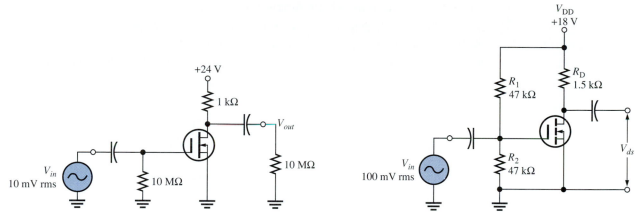

FIGURE 9–43   **FIGURE 9–44**

SECTION 9–4   Common-Drain Amplifiers

24. For the source-follower in Figure 9–45, determine the voltage gain and input resistance. $I_{GSS} =$ 50 pA at $V_{GS} = -15$ V and $g_m = 5500$ μS.

25. If the JFET in Figure 9–45 is replaced with one having a g_m of 3000 μS, what are the gain and the input resistance with all other conditions the same?

FIGURE 9–45

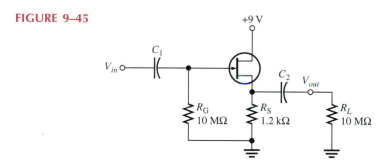

26. Find the gain of each amplifier in Figure 9–46.

27. Determine the voltage gain of each amplifier in Figure 9–46 when the capacitively coupled load is changed to 10 kΩ.

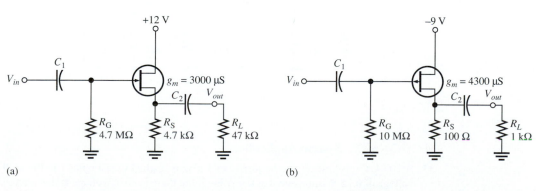

(a)   (b)

FIGURE 9–46

SECTION 9–5   Common-Gate Amplifiers

28. A common-gate amplifier has a $g_m = 4000 \ \mu S$ and $R_d = 1.5 \ k\Omega$. What is its gain?

29. What is the input resistance of the amplifier in Problem 28?

30. Determine the voltage gain and input resistance of the common-gate amplifier in Figure 9–47.

FIGURE 9–47

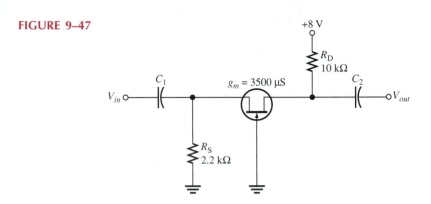

■ **TROUBLE-SHOOTING PROBLEMS**

SECTION 9–6   Troubleshooting

31. What symptom(s) would indicate each of the following failures when a signal voltage is applied to the input in Figure 9–48?

   (a) Q_1 open from drain to source   **(b)** R_3 open
   (c) C_2 shorted   **(d)** C_3 open
   (e) Q_2 open from drain to source

32. If $V_{in} = 10 \ mV$ rms in Figure 9–48, what is V_{out} for each of the following faults?

   (a) C_1 open   **(b)** C_4 open
   (c) a short from the source of Q_2 to ground   **(d)** Q_2 has an open gate

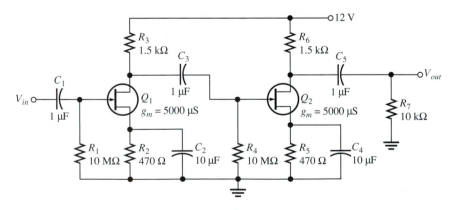

FIGURE 9–48

SECTION 9–7   System Application

33. A 100 mV rms voltage at a frequency of 1 kHz is applied to test point 1 in Figure 9–49. A dc voltage of 6.75 V is measured at test point 5, but there is no ac voltage at this point. Determine the one and only fault in the circuit.

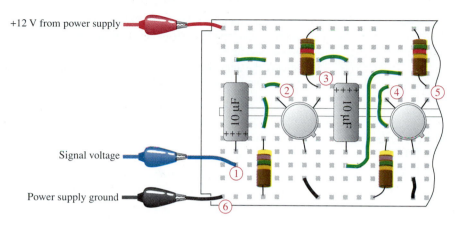

FIGURE 9–49

34. Assume that the fault in Problem 33 for the circuit of Figure 9–49 has been corrected and the input voltage is increased to 250 mV rms. If the following voltages are measured at the specified test points, determine the possible fault(s) and state what you would do to repair the circuit.

   Test point 2: 250 mV rms
   Test point 3: 800 mV rms
   Test point 4: 530 mV rms
   Test point 5: 2.12 V rms

35. Assume the previous faults in the circuit of Figure 9–49 have been corrected. Determine the dc and ac voltages that should be measured at test point 5 if the first stage transistor has an $I_{DSS} = 2.85$ mA and a $g_m = 2200$ μS and the second stage transistor has an $I_{DSS} = 5.10$ mA and a $g_m = 2600$ μS. The input signal is 100 mV rms.

■ **DATA SHEET PROBLEMS**

36. What type of FET is the 2N3796?

37. Referring to the data sheet in Figure 9–32, determine the following:
   (a) typical $V_{GS(off)}$ for the 2N3796
   (b) maximum drain-to-source voltage for the 2N3797
   (c) maximum power dissipation for the 2N3797 at an ambient temperature of 25°C
   (d) maximum gate-to-source voltage for the 2N3797

38. Referring to Figure 9–32, determine the maximum power dissipation for a 2N3796 at an ambient temperature of 55°C.

39. Referring to Figure 9–32, determine the minimum g_{m0} for the 2N3796 at a frequency of 1 kHz.

40. What is the drain current when $V_{GS} = +3.5$ V for the 2N3797?

41. Typically, what is the drain current for a zero-biased 2N3796?

42. What is the maximum possible voltage gain for a 2N3796 common-source amplifier with $R_d = 2.2$ kΩ?

■ **ADVANCED PROBLEMS**

43. The MOSFET in a certain single-stage common-source amplifier has a range of forward transconductance values from 2.5 mS to 7.5 mS. If the amplifier is capacitively coupled to a variable load that ranges from 4 kΩ to 10 kΩ and the dc drain resistance is 1 kΩ, determine the minimum and maximum voltage gains.

44. Design an amplifier using a 2N3797 that operates from a 24 V supply voltage. The typical dc drain-to-source voltage should be approximately 12 V and the typical voltage gain should be approximately 9.

45. Modify the amplifier you designed in Problem 43 so that the voltage gain can be set at 9 for any randomly selected 2N3797.

■ **ANSWERS TO SECTION REVIEWS**

Section 9–1

1. I_d is at its positive peak and V_{ds} is at its negative peak when V_{gs} is at its positive peak.

2. V_{gs} is an ac quantity, V_{GS} is a dc quantity.

3. The D-MOSFET can operate with $V_{GS} = 0$ V at the Q-point.

Section 9–2

1. The FET with $g_m = 3.5$ mS can produce the higher gain.

2. $A_v = g_m R_d = (2500 \ \mu S)(10 \ k\Omega) = 25$

3. The FET with $R_{ds} = 100 \ k\Omega$ can produce the higher gain.

Section 9–3

1. Voltage gain of a CS amplifier is determined by g_m and R_d.

2. The gain is halved because $R_d = R_D/2$.

Section 9–4

1. The ideal maximum voltage gain of a CD amplifier is 1.

2. The voltage gain of a CD amplifier is determined by g_m and R_s.

Section 9–5

1. The CG amplifier has a low input resistance ($1/g_m$).

2. g_m affects both voltage gain and input resistance.

Section 9–6

1. To be a good troubleshooter, you must understand the circuit.

2. There would be a lower than normal first-stage gain if C_2 opens.

3. No, but there would be a loss of signal to the second stage.

■ **ANSWERS TO RELATED EXERCISES FOR EXAMPLES**

9–1 ΔI_D decreases; distortion and clipping at cutoff

9–2 13.2

9–3 5

9–4 2.92

9–5 I_D will be 547 μA. V_D will increase to 10.2 V.

9–6 744 mV

9–7 $R_{in} = 9.99 \ M\Omega$

9–8 1.44 V

9–9 $V_{GS} = 6.19$ V; $I_D = 1$ mA; $V_{DS} = 11.7$ V; $V_{out} = 525$ mV

9–10 0.976

9–11 $R_{in} = 476 \ \Omega$

10

AMPLIFIER FREQUENCY RESPONSE

■ CHAPTER OBJECTIVES

☐ Discuss the frequency response of an amplifier

☐ Express the gain of an amplifier in decibels (dB)

☐ Analyze the low-frequency response of a BJT amplifier

☐ Use Miller's theorem to determine amplifier capacitances

☐ Analyze the high-frequency response of a BJT amplifier

☐ Analyze an amplifier for total frequency response

☐ Analyze the frequency response of an FET amplifier

☐ Analyze the frequency response of multistage amplifiers

☐ Measure the frequency response of an amplifier

In the previous chapters on amplifiers, the effects of the input frequency on the amplifier's operation were neglected in order to focus on other concepts. The coupling and bypass capacitors were considered to be ideal shorts and the internal transistor capacitances were considered to be ideal opens. This treatment is valid when the frequency is in the amplifier's midrange.

As you know, capacitive reactance decreases with increasing frequency and vice versa. When the frequency is low enough, the coupling and bypass capacitors can no longer be considered as shorts because their reactances are large enough to have a significant effect. Also, when the frequency is high enough, the internal transistor capacitances can no longer be considered as opens because their reactances become small enough to have a significant effect on the amplifier operation. A complete picture of an amplifier's response must take into account the full range of frequencies over which the amplifier may operate.

In this chapter, you will study the frequency effects on amplifier gain and phase shift. The coverage applies to both BJT and FET amplifiers, and a mix of both are included to illustrate the concepts.

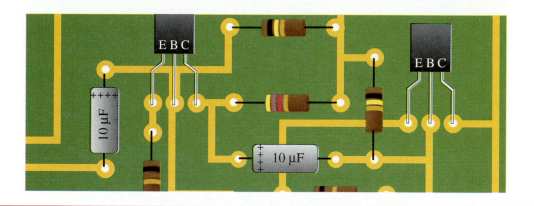

■ SYSTEM APPLICATION

For the system application in Section 10–10, you will apply knowledge gained in this chapter to determine how the voltage gain of a two-stage amplifier circuit varies with the frequency of the input voltage.

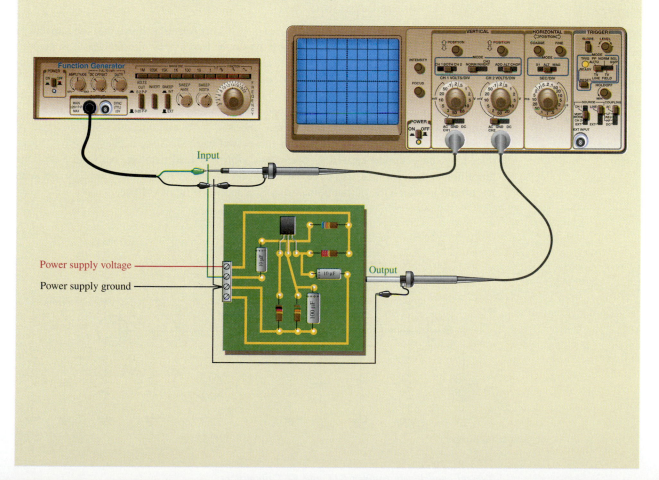

10–1 ■ GENERAL CONCEPTS

In the previous coverage of amplifiers, the capacitive reactance of the coupling and bypass capacitors was assumed to be 0 Ω at the signal frequency and, therefore, had no effect on the amplifier's gain or phase shift. Also, the internal transistor capacitances were assumed to be small enough to neglect at the operating frequency. All of these simplifying assumptions are valid and necessary for studying amplifier theory. However, they do give a limited picture of an amplifier's total operation, so in this section you begin to study the frequency effects of these capacitances. The frequency response of an amplifier is the change in gain or phase shift over a specified range of input signal frequencies.

After completing this section, you should be able to

■ **Discuss the frequency response of an amplifier**
  ☐ Explain the effect of coupling capacitors
  ☐ Explain the effect of bypass capacitors
  ☐ Discuss the internal transistor capacitances and explain their effects

Effect of Coupling Capacitors

Recall from basic circuit theory that $X_C = 1/2\pi fC$. This formula shows that the capacitive reactance varies inversely with frequency. At lower frequencies the reactance is greater, and it decreases as the frequency increases. At lower frequencies—for example, audio frequencies below 10 Hz—capacitively coupled amplifiers such as those in Figure 10–1 have less voltage gain than they have at higher frequencies. The reason is that at lower frequencies more signal voltage is dropped across C_1 and C_3 because their reactances are higher. This is the result of the limitation imposed by availability and physical size of large values of capacitance required in these applications. Also, a phase shift is introduced by the coupling capacitors because C_1 forms a lead network with the R_{in} of the amplifier, and C_3 forms a lead circuit with R_L and R_C or R_D. Recall that a *lead circuit* is an *RC* circuit in which the output voltage across R leads the input voltage.

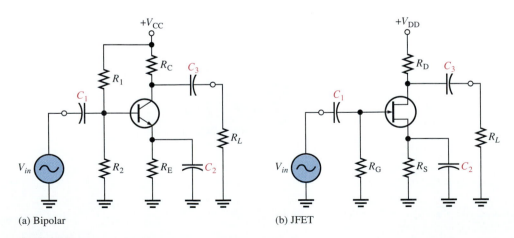

(a) Bipolar   (b) JFET

FIGURE 10–1
Typical capacitively coupled BJT and FET amplifiers.

Effect of Bypass Capacitors

At lower frequencies, the reactance of the emitter bypass capacitor C_2 becomes significant and the emitter (or FET source terminal) is no longer at ac ground. X_{C2} in parallel with R_E (or R_S) creates an impedance that reduces the gain. This is illustrated in Figure 10–2.

When the frequency is sufficiently high, $X_C \cong 0 \ \Omega$ and the voltage gain is $A_v = R_C/r'_e$. At lower frequencies, $X_C \gg 0 \ \Omega$ and the voltage gain is $A_v = R_C/(r'_e + Z_e)$.

FIGURE 10–2

Nonzero reactance of the bypass capacitor in parallel with R_E creates an emitter impedance, (Z_e), which reduces the voltage gain.

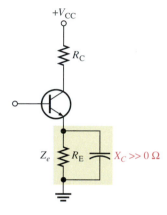

Effect of Internal Transistor Capacitances

At high frequencies, the coupling and bypass capacitors become effective ac shorts and do not affect the amplifier's response. Internal device capacitances, however, do come into play, reducing the amplifier's gain and introducing phase shift as the signal frequency increases.

Figure 10–3 shows the internal capacitances for both a bipolar transistor and a JFET. In the case of the bipolar transistor, C_{be} is the base-emitter junction capacitance and C_{bc} is the base-collector junction capacitance. In the case of the JFET, C_{gs} is the internal capacitance between gate and source and C_{gd} is the internal capacitance between gate and drain.

FIGURE 10–3

Internal transistor capacitances.

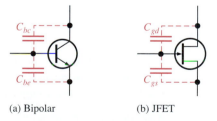

(a) Bipolar   (b) JFET

Data sheets often refer to the bipolar transistor capacitance C_{bc} as the output capacitance, often designated C_{ob}. The capacitance C_{be} is often designated as the input capacitance C_{ib}. Data sheets for FETs normally specify input capacitance C_{iss} and reverse transfer capacitance C_{rss}. From these, C_{gs} and C_{gd} can be calculated, as you will see in Section 10–7.

At lower frequencies, the internal capacitances have a very high reactance because of their low capacitance value (usually only a few picofarads). Therefore, they look like opens and have no effect on the transistor's performance. As the frequency goes up, the internal capacitive reactances go down, and at some point they begin to have a significant effect on the transistor's gain. When the reactance of C_{be} (or C_{gs}) becomes small enough, a significant amount of the signal voltage is lost due to a voltage-divider effect of the source resistance and the capacitive reactance of C_{be}, as illustrated in Figure 10–4(a). When the reactance of C_{bc} (or C_{gd}) becomes small enough, a significant amount of output signal voltage is fed back out of phase with the input (negative feedback), thus effectively reducing the voltage gain. This is illustrated in Figure 10–4(b).

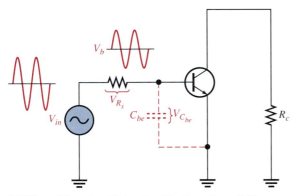

(a) Effect of C_{be}, where V_b is reduced by the voltage-divider action of R_s and $X_{C_{be}}$.

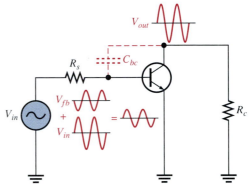

(b) Effect of C_{bc}, where part of V_{out} (V_{fb}) goes back through C_{bc} to the base and reduces the input signal because it is approximately 180° out of phase with V_{in}.

FIGURE 10–4

AC equivalent circuit for a bipolar amplifier showing effects of the internal capacitances C_{be} and C_{bc}.

SECTION 10–1
REVIEW

1. In an ac amplifier, which capacitors affect the low-frequency gain?
2. How is the high-frequency gain of an amplifier limited?
3. When can coupling and bypass capacitors be neglected?

10–2 ■ THE DECIBEL

Before we get into the analysis of amplifier frequency response, the important topic of decibels must be discussed. Decibels are a common form of gain measurement and are used in expressing amplifier response.

After completing this section, you should be able to

■ **Express the gain of an amplifier in decibels (dB)**
  ☐ Express power gain in dB
  ☐ Express voltage gain in dB
  ☐ State the distinction between a positive and a negative dB gain
  ☐ Define *critical frequency*
  ☐ Express power in terms of dBm

In Chapter 6 we introduced the use of decibels in expressing gain. The decibel unit is important in amplifier measurements. The basis for the decibel unit stems from the logarithmic response of the human ear to the intensity of sound. The *decibel* is a measurement of the ratio of one power to another or one voltage to another. Power gain is expressed in decibels (dB) by the following formula:

$$A_{p(\text{dB})} = 10 \log A_p \qquad\qquad (10\text{--}1)$$

where A_p is the actual power gain, P_{out}/P_{in}. Voltage gain is expressed in decibels by the following formula:

$$A_{v(\text{dB})} = 20 \log A_v \qquad\qquad (10\text{--}2)$$

If A_v is greater than one, the dB gain is positive. If A_v is less than one, the dB gain is negative and is usually called *attenuation*.

EXAMPLE 10–1

Express each of the following ratios in dB:

(a) $\dfrac{P_{out}}{P_{in}} = 250$ (b) $\dfrac{P_{out}}{P_{in}} = 100$ (c) $A_v = 10$

(d) $A_p = 0.5$ (e) $\dfrac{V_{out}}{V_{in}} = 0.707$

Solution

(a) $A_{p(\text{dB})} = 10 \log(250) = 24$ dB   (b) $A_{p(\text{dB})} = 10 \log(100) = 20$ dB
(c) $A_{v(\text{dB})} = 20 \log(10) = 20$ dB   (d) $A_{p(\text{dB})} = 10 \log(0.5) = -3$ dB
(e) $A_{v(\text{dB})} = 20 \log(0.707) = -3$ dB

Related Exercise   Express each of the following gains in dB: (a) $A_v = 1200$, (b) $A_p = 50$, (c) $A_v = 125{,}000$.

---

0 dB Reference

It is often convenient in amplifier analysis to assign a certain value of gain as the 0 dB reference. This does not mean that the actual voltage gain is 1 (which is 0 dB); it means that the reference gain, no matter what its actual value, is used as a reference with which to compare other values of gain and is therefore assigned a 0 dB value.

Many amplifiers exhibit a maximum gain over a certain range of frequencies and a reduced gain at frequencies below and above this range. The maximum gain is called the *midrange gain* in this case and is assigned a 0 dB value. Any value of gain below **midrange** can be referenced to 0 dB and expressed as a negative dB value. For example, if the midrange voltage gain of a certain amplifier is 100 and the gain at a certain frequency below midrange is 50, then this reduced gain can be expressed as 20 $\log(50/100) = 20 \log(0.5) = -6$ dB. This indicates that it is 6 dB *below* the 0 dB reference. Halving the output voltage for a steady input voltage is a 6 dB reduction in the gain. Correspondingly, a doubling of the output voltage is a 6 dB increase in the gain. Figure 10–5 illustrates a gain-versus-frequency curve showing several dB points.

FIGURE 10–5

Normalized gain-versus-frequency curve.

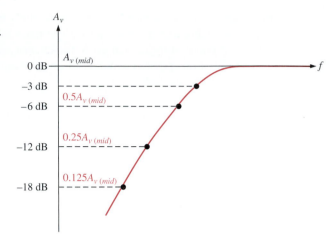

Table 10–1 shows how doubling or halving voltage gains translates into dB values. Notice that each step in the doubling or halving of the voltage gain increases or decreases the dB value by 6 dB. Notice in the table that every time the voltage gain is doubled, the dB value increases by 6 dB, and every time the gain is halved, the dB value decreases by 6 dB.

TABLE 10–1

Decibel values corresponding to doubling and halving of the voltage gain

Voltage Gain (A_v)	dB (with Respect to Zero Reference)
32	$20 \log(32) = 30$ dB
16	$20 \log(16) = 24$ dB
8	$20 \log(8) = 18$ dB
4	$20 \log(4) = 12$ dB
2	$20 \log(2) = 6$ dB
1	$20 \log(1) = 0$ dB
0.707	$20 \log(0.707) = -3$ dB
0.5	$20 \log(0.5) = -6$ dB
0.25	$20 \log(0.25) = -12$ dB
0.125	$20 \log(0.125) = -18$ dB
0.0625	$20 \log(0.0625) = -24$ dB
0.03125	$20 \log(0.03125) = -30$ dB

The Critical Frequency

The **critical frequency** (also known as **cutoff frequency** or *corner frequency*) is the frequency at which the output power drops to one-half of its midrange value. This corresponds to a 3 dB reduction in the power gain, as expressed by the following formula:

$$A_{p(\text{dB})} = 10 \log(0.5) = -3 \text{ dB}$$

Also, at the critical frequency the output voltage is 70.7 percent of its midrange value and is expressed in dB as

$$A_{v(\text{dB})} = 20 \log(0.707) = -3 \text{ dB}$$

At the critical frequency, the voltage gain is down 3 dB or is 70.7% of its midrange value. At this same frequency, the power is one-half of its midrange value.

EXAMPLE 10–2

A certain amplifier has a midrange rms output voltage of 10 V. What is the rms output voltage for each of the following dB gain reductions with a constant rms input voltage?

(a) −3 dB   **(b)** −6 dB   **(c)** −12 dB   **(d)** −24 dB

Solution   Multiply the midrange output voltage by the voltage gain corresponding to the specified dB value in Table 10–1.

(a) At −3 dB, $V_{out} = 0.707(10 \text{ V}) = 7.07 \text{ V}$
(b) At −6 dB, $V_{out} = 0.5(10 \text{ V}) = 5 \text{ V}$
(c) At −12 dB, $V_{out} = 0.25(10 \text{ V}) = 2.5 \text{ V}$
(d) At −24 dB, $V_{out} = 0.0625(10 \text{ V}) = 0.625 \text{ V}$

Related Exercise   Determine the output voltage at the following dB levels for a midrange value of 50 V:

(a) 0 dB   **(b)** −18 dB   **(c)** −30 dB

dBm Power Measurement

A unit that is often used in measuring power is the dBm. The term *dBm* means decibels referenced to 1 mW of power. When dBm is used, all power measurements are relative to a reference level of 1 mW. A 3 dBm increase corresponds to doubling of the power, and a 3 dBm decrease corresponds to a halving of the power. For example, +3 dBm corresponds to 2 mW (twice 1 mW), and −3 dBm corresponds to 0.5 mW (half of 1 mW). Table 10–2 shows several dBm values.

TABLE 10–2
Power in terms of dBm

Power	dBm
0.03125 mW	−15 dBm
0.0625 mW	−12 dBm
0.125 mW	−9 dBm
0.25 mW	−6 dBm
0.5 mW	−3 dBm
1 mW	0 dBm
2 mW	3 dBm
4 mW	6 dBm
8 mW	9 dBm
16 mW	12 dBm
32 mW	15 dBm

SECTION 10–2 REVIEW

1. How much increase in actual voltage gain corresponds to +12 dB?
2. Convert a power gain of 25 to decibels.
3. What power corresponds to 0 dBm?

10–3 ■ LOW-FREQUENCY AMPLIFIER RESPONSE

In this section, we will examine how the voltage gain and phase shift of a capacitively coupled amplifier are affected by frequencies below which the capacitive reactance becomes too large to neglect. At the end of the section, a comparison is made to direct-coupled amplifiers.

After completing this section, you should be able to

■ **Analyze the low-frequency response of a BJT amplifier**
  □ Determine midrange voltage gain
  □ Generally describe how *RC* circuits affect the gain
  □ Identify the input *RC* circuit
  □ Determine the lower critical frequency of the input *RC* circuit
  □ Discuss gain roll-off in terms of dB/decade and dB/octave
  □ Determine phase shift of the input *RC* circuit
  □ Identify the output *RC* circuit
  □ Determine the lower critical frequency of the output *RC* circuit
  □ Determine phase shift of the output *RC* circuit
  □ Identify the bypass *RC* circuit
  □ Determine the lower critical frequency of the bypass *RC* circuit
  □ Describe a Bode plot
  □ Analyze an amplifier for total low-frequency response
  □ Compare direct-coupled amplifiers to capacitively coupled amplifiers

A typical capacitively coupled small-signal common-emitter amplifier is shown in Figure 10–6 (the approach for FETs is similar). Assuming that the coupling and bypass capacitors are ideal shorts at the midrange signal frequency, you can calculate the voltage gain as you have done before. You can calculate the midrange voltage gain using Equation (10–3), where $R_c = R_C \parallel R_L$.

$$A_{v(mid)} = \frac{R_c}{r'_e}$$

(10–3)

FIGURE 10–6
Typical capacitively coupled amplifier.

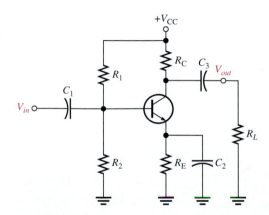

The amplifier in Figure 10–6 has three high-pass *RC* circuits that affect its gain as the frequency is reduced below midrange. These are shown in the low-frequency equivalent circuit in Figure 10–7. Unlike the ac equivalent circuit used in previous chapters, which represented midrange response ($X_C \cong 0 \ \Omega$), the low-frequency equivalent retains the coupling and bypass capacitors because X_C is not small enough to neglect when the signal frequency is sufficiently low.

FIGURE 10–7

The low-frequency equivalent of the amplifier in Figure 10–6 consists of three high-pass RC circuits.

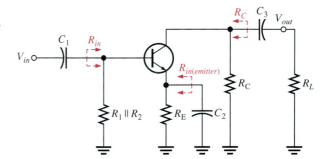

One *RC* circuit is formed by the input coupling capacitor C_1 and the input resistance of the amplifier. The second *RC* circuit is formed by the output coupling capacitor C_3, the resistance looking in at the collector, and the load resistance. The third *RC* circuit that affects the low-frequency response is formed by the emitter-bypass capacitor C_2 and the resistance looking in at the emitter.

The Input *RC* Circuit

The *RC* circuit for the amplifier in Figure 10–6 is formed by C_1 and the amplifier's input resistance and is shown in Figure 10–8. (Input resistance was discussed in Chapter 6.) As the signal frequency decreases, X_{C1} increases. This causes less voltage across the input resistance of the amplifier at the base because more is dropped across X_{C1}. As you can see, the overall voltage gain of the amplifier is reduced. The voltage relationship for the *RC* circuit in Figure 10–8 (neglecting the internal resistance of the input signal source) can be stated as

$$V_{base} = \left(\frac{R_{in}}{\sqrt{R_{in}^2 + X_{C1}^2}} \right) V_{in} \qquad (10\text{–}4)$$

FIGURE 10–8

RC circuit formed by the input coupling capacitor and the amplifier's input resistance.

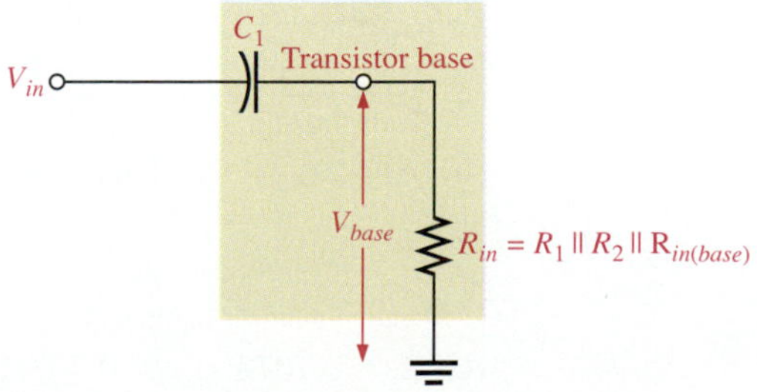

As previously mentioned, a critical point in the amplifier's response is generally accepted to occur when the output voltage is 70.7 percent of its midrange value. This condition occurs in the input RC circuit when $X_{C1} = R_{in}$, as shown in the following steps using Equation (10–4).

$$V_{base} = \left(\frac{R_{in}}{\sqrt{R_{in}^2 + R_{in}^2}}\right)V_{in} = \left(\frac{R_{in}}{\sqrt{2R_{in}^2}}\right)V_{in} = \left(\frac{R_{in}}{\sqrt{2}\,R_{in}}\right)V_{in} = \left(\frac{1}{\sqrt{2}}\right)V_{in}$$
$$= 0.707V_{in}$$

In terms of measurement in decibels,

$$20\log\left(\frac{V_{base}}{V_{in}}\right) = 20\log(0.707) = -3\text{ dB}$$

Lower Critical Frequency The condition where the gain is down 3 dB is logically called the *−3 dB point* of the amplifier response; the overall gain is 3 dB less than at midrange frequencies because of the attenuation of the RC circuit. The frequency, f_c, at which this condition occurs is called the *lower critical frequency* (also known as the *lower cutoff frequency, lower corner frequency,* or *lower break frequency*) and can be calculated as follows.

$$X_{C1} = \frac{1}{2\pi f_c C_1} = R_{in}$$

$$f_c = \frac{1}{2\pi R_{in}C_1} \qquad\qquad \textbf{(10–5)}$$

If the resistance of the input source is taken into account, Equation (10–5) becomes

$$f_c = \frac{1}{2\pi(R_s + R_{in})C_1}$$

EXAMPLE 10–3

For an input RC circuit in a certain amplifier, $R_{in} = 1\text{ k}\Omega$ and $C_1 = 1\text{ μF}$. Neglect the source resistance.

(a) Determine the lower critical frequency.
(b) What is the attenuation of the RC circuit at the lower critical frequency?
(c) If the midrange voltage gain of the amplifier is 100, what is the gain at the lower critical frequency?

Solution

(a) $f_c = \dfrac{1}{2\pi R_{in}C_1} = \dfrac{1}{2\pi(1\text{ k}\Omega)(1\text{ μF})} = 159\text{ Hz}$

(b) At f_c, $X_{C1} = R_{in}$. Therefore, from Equation (10–4),

$$\frac{V_{base}}{V_{in}} = 0.707$$

(c) $A_v = 0.707A_{v(mid)} = 0.707(100) = 70.7$

Related Exercise   For an input RC circuit in a certain amplifier, $R_{in} = 10$ kΩ and $C_1 = 2.2$ μF.

(a) What is f_c?   **(b)** What is the attenuation at f_c?

(c) If $A_{v(mid)} = 500$, what is A_v at f_c?

Voltage Gain Roll-Off at Low Frequencies   As you have seen, the input RC circuit reduces the overall voltage gain of an amplifier by 3 dB when the frequency is reduced to the critical value f_c. As the frequency continues to decrease below f_c, the overall voltage gain also continues to decrease. The decrease in voltage gain with frequency is called **roll-off.** *For each ten times reduction in frequency below f_c, there is a 20 dB reduction in voltage gain.*

Let's take a frequency that is one-tenth of the critical frequency ($f = 0.1f_c$). Since $X_{C1} = R_{in}$ at f_c, then $X_{C1} = 10R_{in}$ at $0.1f_c$ because of the inverse relationship of X_{C1} and f. The attenuation of the RC circuit is therefore

$$\frac{V_{base}}{V_{in}} = \frac{R_{in}}{\sqrt{R_{in}^2 + X_{C1}^2}} = \frac{R_{in}}{\sqrt{R_{in}^2 + (10R_{in})^2}} = \frac{R_{in}}{\sqrt{R_{in}^2 + 100R_{in}^2}}$$

$$= \frac{R_{in}}{\sqrt{R_{in}^2(1 + 100)}} = \frac{R_{in}}{R_{in}\sqrt{101}} = \frac{1}{\sqrt{101}} \cong \frac{1}{10} = 0.1$$

The dB attenuation is

$$20 \log\left(\frac{V_{base}}{V_{in}}\right) = 20 \log(0.1) = -20 \text{ dB}$$

dB/Decade   A ten-times change in frequency is called a **decade.** So, for the input RC circuit, the attenuation is reduced by 20 dB for each decade that the frequency decreases below the critical frequency. This causes the overall voltage gain to drop 20 dB per decade. For example, if the frequency is reduced to one-hundredth of f_c (a two-decade decrease), the amplifier voltage gain drops 20 dB for each decade, giving a total decrease in voltage gain of -20 dB $+ (-20$ dB$) = -40$ dB. This is illustrated in Figure 10–9, which is a graph of dB voltage gain versus frequency. This graph is the low-frequency response curve for the amplifier showing the effect of the input RC circuit on the voltage gain.

FIGURE 10–9

dB voltage gain versus frequency for the input RC circuit.

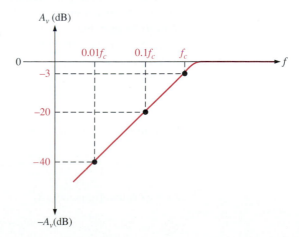

dB/Octave Sometimes, the voltage gain roll-off of an amplifier is expressed in dB/octave rather than dB/decade. An **octave** corresponds to a doubling or halving of the frequency. For example, an increase in frequency from 100 Hz to 200 Hz is an octave. Likewise, a decrease in frequency from 100 kHz to 50 kHz is also an octave. A rate of −20 dB/decade is approximately equivalent to −6 dB/octave, a rate of −40 dB/decade is approximately equivalent to −12 dB/octave, and so on.

EXAMPLE 10–4

The midrange voltage gain of a certain amplifier is 100. The input RC circuit has a lower critical frequency of 1 kHz. Determine the actual voltage gain at $f = 1$ kHz, $f = 100$ Hz, and $f = 10$ Hz.

Solution When $f = 1$ kHz, the voltage gain is 3 dB less than at midrange. At −3 dB, the voltage gain is reduced by a factor of 0.707.

$$A_v = (0.707)(100) = 70.7$$

When $f = 100$ Hz $= 0.1f_c$, the voltage gain is 20 dB less than at f_c. The voltage gain at −20 dB is one-tenth of that at the midrange frequencies.

$$A_v = (0.1)(100) = 10$$

When $f = 10$ Hz $= 0.01f_c$, the voltage gain is 20 dB less than at $f = 0.1f_c$ or −40 dB. The voltage gain at −40 dB is one-tenth of that at −20 dB or one-hundredth that at the midrange frequencies.

$$A_v = (0.01)(100) = 1$$

Related Exercise The midrange voltage gain of an amplifier is 300. The lower critical frequency of the input RC circuit is 400 Hz. Determine the actual voltage gain at 400 Hz, 40 Hz, and 4 Hz.

Phase Shift in the Input RC Circuit In addition to reducing the voltage gain, the input RC circuit also causes an increasing phase shift through the amplifier as the frequency decreases. At midrange frequencies, the phase shift through the RC circuit is approximately zero because $X_{C1} \cong 0 \ \Omega$. At lower frequencies, higher values of X_{C1} cause a phase shift to be introduced, and the output voltage of the RC circuit leads the input voltage. As you learned in ac circuit theory, the phase angle in an RC circuit is expressed as

$$\theta = \tan^{-1}\left(\frac{X_{C1}}{R_{in}}\right) \tag{10–6}$$

For midrange frequencies, $X_{C1} \cong 0 \ \Omega$, so

$$\theta = \tan^{-1}\left(\frac{0 \ \Omega}{R_{in}}\right) = \tan^{-1}(0) = 0°$$

At the critical frequency, $X_{C1} = R_{in}$, so

$$\theta = \tan^{-1}\left(\frac{R_{in}}{R_{in}}\right) = \tan^{-1}(1) = 45°$$

A decade below the critical frequency, $X_{C1} = 10R_{in}$, so

$$\theta = \tan^{-1}\left(\frac{10R_{in}}{R_{in}}\right) = \tan^{-1}(10) = 84.3°$$

A continuation of this analysis will show that the phase shift through the input RC circuit approaches 90° as the frequency approaches zero. A plot of phase angle versus frequency is shown in Figure 10–10. The result is that the voltage at the base of the transistor *leads* the input signal voltage in phase below midrange, as shown in Figure 10–11.

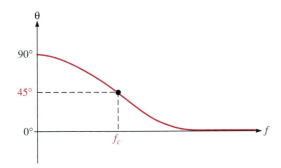

FIGURE 10–10
Phase angle versus frequency for the input RC circuit.

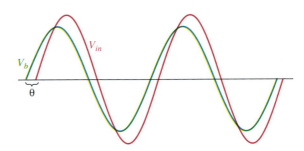

FIGURE 10–11
Input RC circuit causes the base voltage to lead the input voltage below midrange by an amount equal to the circuit phase angle θ.

The Output *RC* Circuit

The second high-pass RC circuit in the amplifier of Figure 10–6 is formed by the coupling capacitor C_3, the resistance looking in at the collector, and the load resistance R_L, as shown in Figure 10–12(a). In determining the output resistance, looking in at the collector, the transistor is treated as an ideal current source (with infinite internal resistance), and the upper end of R_C is effectively at ac ground, as shown in Figure 10–12(b). Therefore, Thevenizing the circuit to the left of the capacitor C_3 produces an equivalent voltage source equal to the collector voltage and a series resistance equal to R_C, as shown in Figure 10–12(c). The critical frequency of this RC circuit is

$$f_c = \frac{1}{2\pi(R_C + R_L)C_3} \tag{10–7}$$

The effect of the output RC circuit on the amplifier voltage gain is similar to that of the input RC circuit. As the signal frequency decreases, X_{C3} increases. This causes less voltage across the load resistance because more is dropped across X_{C3}. The signal voltage is reduced by a factor of 0.707 when the frequency is reduced to the lower critical value, f_c, for the circuit. This corresponds to a 3 dB reduction in voltage gain.

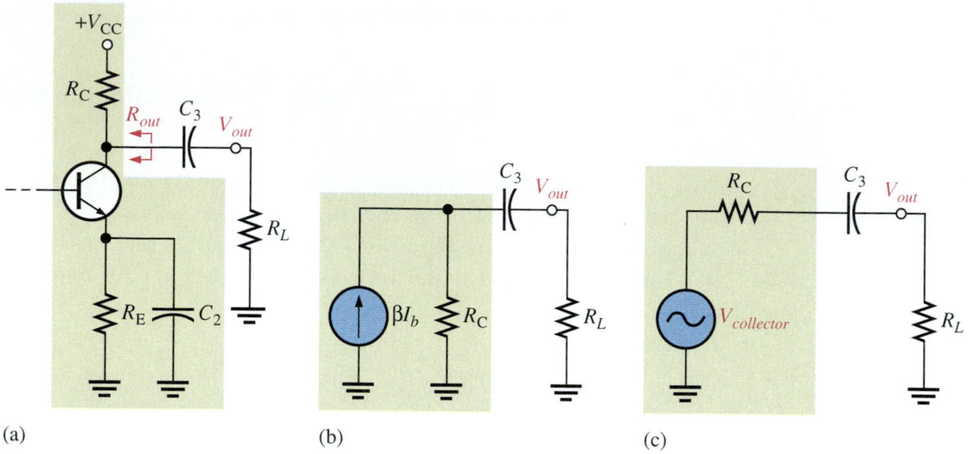

FIGURE 10–12

Development of the low-frequency output RC circuit.

EXAMPLE 10–5

For an output RC circuit in a certain amplifier, $R_C = 10$ kΩ, $C_3 = 0.1$ μF, and $R_L = 10$ kΩ.

(a) Determine the critical frequency.

(b) What is the attenuation of the RC circuit at the critical frequency?

(c) If the midrange voltage gain of the amplifier is 50, what is the gain at the critical frequency?

Solution

(a) $f_c = \dfrac{1}{2\pi(R_C + R_L)C_3} = \dfrac{1}{2\pi(20\ \text{k}\Omega)(0.1\ \mu\text{F})} = 79.6$ Hz

(b) For the midrange frequencies, $X_{C3} \cong 0$ Ω; thus, the attenuation of the circuit is

$$\frac{V_{out}}{V_{collector}} = \frac{R_L}{R_C + R_L} = \frac{10\ \text{k}\Omega}{20\ \text{k}\Omega} = 0.5$$

or, in dB, $V_{out}/V_{collector} = 20 \log(0.5) = -6$ dB. This shows that, in this case, the midrange voltage gain is reduced by 6 dB because of the load resistor. At the critical frequency, $X_{C3} = R_C + R_L$.

$$\frac{V_{out}}{V_{in}} = \frac{R_L}{\sqrt{(R_C + R_L)^2 + X_{C3}^2}} = \frac{10\ \text{k}\Omega}{\sqrt{(20\ \text{k}\Omega)^2 + (20\ \text{k}\Omega)^2}} = 0.354$$

or, in dB, $V_{out}/V_{collector} = 20 \log(0.354) = -9$ dB. As you can see, the gain at f_c is 3 dB less than the gain at midrange.

(c) $A_v = 0.707A_{v(mid)} = 0.707(50) = 35.4$

Related Exercise The output RC circuit in a certain amplifier has the following values: $R_C = 3.9$ kΩ, $C_3 = 1$ μF, and $R_L = 8.2$ kΩ.

(a) Find the critical frequency.   (b) What is the attenuation at f_c?

(c) If $A_{v(mid)}$ is 100, what is the gain at f_c?

Phase Shift in the Output RC Circuit   The phase shift in the output *RC* circuit is

$$\theta = \tan^{-1}\left(\frac{X_{C3}}{R_C + R_L}\right) \tag{10–8}$$

$\theta \cong 0°$ for the midrange frequencies and approaches 90° as the frequency approaches zero (X_{C3} approaches infinity). At the critical frequency f_c, the phase shift is 45°.

The Bypass *RC* Circuit

The third *RC* circuit that affects the low-frequency gain of the amplifier in Figure 10–6 includes the bypass capacitor C_2. For midrange frequencies, it is assumed that $X_{C2} \cong 0\ \Omega$, effectively shorting the emitter to ground so that the amplifier gain is R_c/r'_e, as you already know. As the frequency is reduced, X_{C2} increases and no longer provides a sufficiently low reactance to effectively place the emitter at ac ground. This is illustrated in Figure 10–13. Because the impedance from emitter to ground increases, the gain decreases. In this case, R_e in the formula, $A_v = R_c/(r'_e + R_e)$, is replaced by an impedance formed by R_E in parallel with X_{C2}.

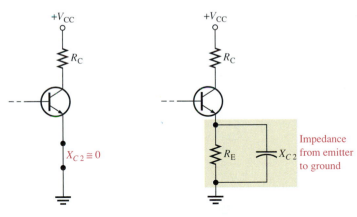

(a) For midrange frequencies, C_2 effectively shorts the emitter to ground.

(b) Below f_c, X_{C2} and R_E form an impedance between the emitter and ground.

FIGURE 10–13
At low frequencies, X_{C2} in parallel with R_E creates an impedance that reduces the voltage gain.

The bypass *RC* circuit is formed by C_2 and the resistance looking in at the emitter $R_{in(emitter)}$, as shown in Figure 10–14(a). The resistance looking in at the emitter is derived as follows. First, Thevenin's theorem is applied looking from the base of the transistor toward the input source V_{in}, as shown in Figure 10–14(b). This results in an equivalent resistance R_{th} in series with the base, as shown in Figure 10–14(c). The resistance looking in at the emitter is determined with the input source shorted, as shown in Figure 10–14(d), and is expressed as follows:

$$R_{in(emitter)} = \frac{V_e}{I_e} + r'_e \cong \frac{V_b}{\beta_{ac}I_b} + r'_e = \frac{I_b R_{th}}{\beta_{ac}I_b} + r'_e$$

$$R_{in(emitter)} = \frac{R_{th}}{\beta_{ac}} + r'_e \qquad\qquad\textbf{(10–9)}$$

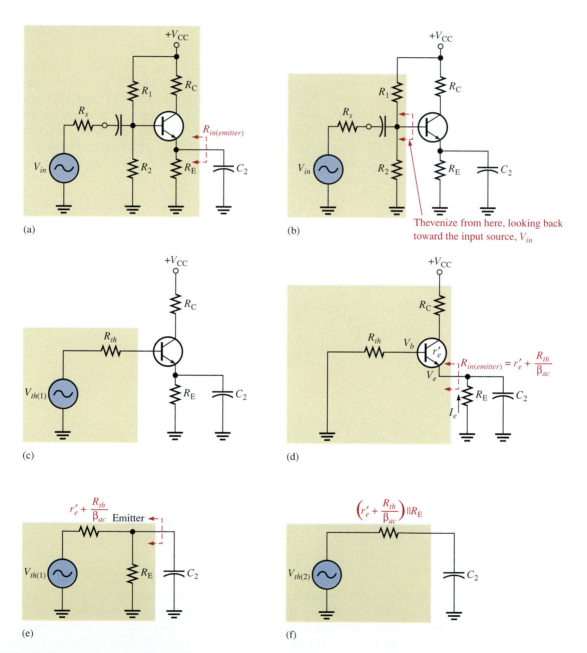

FIGURE 10–14
Development of the bypass RC circuit.

Looking from the capacitor C_2, $R_{th}/\beta_{ac} + r'_e$ is in parallel with R_E, as shown in Figure 10–14(e). Thevenizing again, we get the equivalent RC circuit shown in Figure 10–14(f). The critical frequency for this bypass circuit is

$$f_c = \frac{1}{2\pi[(r'_e + R_{th}/\beta_{ac}) \parallel R_E]C_2} \qquad (10\text{–}10)$$

EXAMPLE 10–6

Determine the critical frequency of the bypass circuit for the amplifier in Figure 10–15 ($r'_e = 16\ \Omega$).

FIGURE 10–15

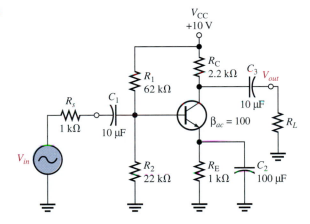

Solution Thevenize the base circuit (looking from the base toward the input source).

$$R_{th} = R_1 \parallel R_2 \parallel R_s = 62\ \text{k}\Omega \parallel 22\ \text{k}\Omega \parallel 1\ \text{k}\Omega \cong 942\ \Omega$$

The resistance looking in at the emitter is

$$R_{in(emitter)} = r'_e + \frac{R_{th}}{\beta_{ac}} = 16\ \Omega + 9.42\ \Omega = 25.4\ \Omega$$

The resistance of the equivalent RC bypass circuit is $R_{in(emitter)} \parallel R_E$.

$$R_{in(emitter)} \parallel R_E = 25.4\ \Omega \parallel 1000\ \Omega = 24.8\ \Omega$$

The critical frequency of the bypass circuit is

$$f_c = \frac{1}{2\pi(R_{in(emitter)} \parallel R_E)C_2} = \frac{1}{2\pi(24.8\ \Omega)(100\ \mu\text{F})} = 64.2\ \text{Hz}$$

Related Exercise In Figure 10–15, the source resistance, R_s, is 50 Ω, and the transistor ac beta is 150. Determine the critical frequency of the bypass circuit.

The Bode Plot

A plot of dB voltage gain versus frequency on semilog graph paper is called a **Bode plot.**
A generalized Bode plot for an RC circuit like that shown in Figure 10–16(a) appears
in part (b) of the figure. The ideal response curve is drawn with a solid line. Notice that
it is a flat (0 dB) down to the critical frequency, at which point the gain drops at
−20 dB/decade as shown. Above f_c are the midrange frequencies. The actual response
curve is shown with the dashed line. Notice that it decreases gradually in midrange and is
down to −3 dB at the critical frequency. Often, the ideal response is used to simplify
amplifier analysis. As previously mentioned, the critical frequency at which the curve
"breaks" into a −20 dB/decade drop is sometimes called the *lower break frequency.*

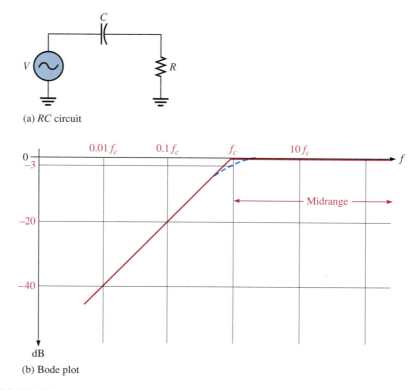

(a) *RC* circuit

(b) Bode plot

FIGURE 10–16
RC circuit and its low-frequency response.

Total Low-Frequency Response of an Amplifier

Now that we have individually examined the three high-pass RC circuits that affect the
amplifier's voltage gain at low frequencies, let's look at the combined effect of all three
circuits. Each circuit has a critical frequency determined by the R and C values. The critical frequencies of the three RC circuits are not necessarily all equal. If one of the RC circuits has a critical (break) frequency higher than the other two, then it is the *dominant RC*
circuit. The dominant circuit determines the frequency at which the overall voltage gain
of the amplifier begins to drop at −20 dB/decade. The other circuits cause an additional
−20 dB/decade roll-off below their respective critical (break) frequencies.

To get a better picture of what happens at low frequencies, refer to the Bode plot in Figure 10–17, which shows the superimposed ideal responses for the three *RC* circuits (blue lines). In this example, each *RC* circuit has a different critical frequency. The input *RC* circuit is dominant (highest f_c) in this case, and the bypass circuit has the lowest f_c. The overall response is shown as the red line.

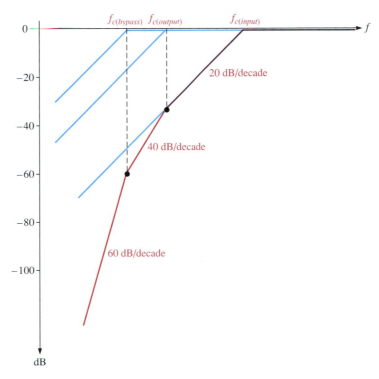

FIGURE 10–17

Composite Bode plot of an amplifier response for three low-frequency RC circuits with different critical frequencies. Total response is shown by the red curve.

Here is what happens. As the frequency is reduced from midrange, the first "break point" occurs at the critical frequency of the input *RC* circuit, $f_{c(input)}$, and the gain begins to drop at −20 dB/decade. This constant roll-off rate continues until the critical frequency of the output *RC* circuit, $f_{c(output)}$, is reached. At this break point, the output *RC* circuit adds another −20 dB/decade to make a total roll-off of −40 dB/decade. This constant −40 dB/decade roll-off continues until the critical frequency of the bypass *RC* circuit, $f_{c(bypass)}$, is reached. At this break point, the bypass *RC* circuit adds still another −20 dB/decade, making the gain roll-off at −60 dB/decade.

If all three *RC* circuits have the same critical frequency, the response curve has one break point at that value of f_c, and the voltage gain rolls off at −60 dB/decade below that value, as shown in Figure 10–18. Keep in mind that the ideal response curves have been used. Actually, the midrange voltage gain does not extend down to the dominant critical frequency but is really at −9 dB below the midrange voltage gain at that point (−3 dB for each *RC* circuit).

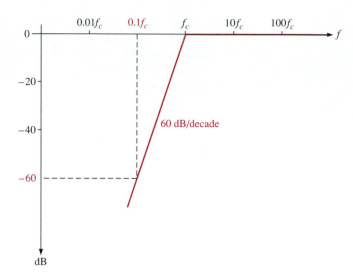

FIGURE 10–18

Composite Bode plot of an amplifier response where all three RC circuits have same f_c.

EXAMPLE 10–7

Determine the total low-frequency response of the amplifier in Figure 10–19. $\beta_{ac} = 100$ and $r'_e = 16\ \Omega$.

FIGURE 10–19

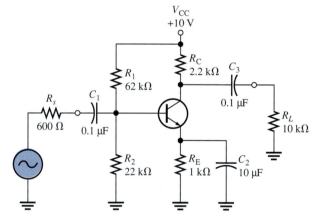

Solution   Each RC circuit is analyzed to determine its critical frequency. For the input RC circuit with the source resistance R_s taken into account:

$$R_{in} = R_1 \parallel R_2 \parallel \beta_{ac}r'_e = 62\ \text{k}\Omega \parallel 22\ \text{k}\Omega \parallel 1.6\ \text{k}\Omega = 1.46\ \text{k}\Omega$$

$$f_{c(input)} = \frac{1}{2\pi(R_s + R_{in})C_1} = \frac{1}{2\pi(600\ \Omega + 1.46\ \text{k}\Omega)(0.1\ \mu\text{F})} = 733\ \text{Hz}$$

For the bypass RC circuit:

$$R_{th} = R_1 \parallel R_2 \parallel R_s = 62\ \text{k}\Omega \parallel 22\ \text{k}\Omega \parallel 600\ \Omega = 579\ \Omega$$

$$R_{in(emitter)} = \frac{R_{th}}{\beta_{ac}} + r'_e = \frac{579\ \Omega}{100} + 16\ \Omega = 21.8\ \Omega$$

$$f_{c(bypass)} = \frac{1}{2\pi(R_{in(emitter)} \parallel R_E C_2} = \frac{1}{2\pi(21.8\ \Omega \parallel 1\ k\Omega)(10\ \mu F)}$$

$$= \frac{1}{2\pi(21.3\ \Omega)(10\ \mu F)} = 730\ Hz$$

For the output *RC* circuit:

$$f_{c(output)} = \frac{1}{2\pi(R_C + R_L)C_3} = \frac{1}{2\pi(2.2\ k\Omega + 10\ k\Omega)(0.1\ \mu F)} = 130.5\ Hz$$

The above analysis shows that the input circuit produces the dominant (highest) lower critical frequency. The midrange voltage gain of the amplifier is

$$A_{v(mid)} = \frac{R_c}{r'_e} = \frac{2.2\ k\Omega \parallel 10\ k\Omega}{16\ \Omega} = 113$$

The midrange attenuation of the input circuit is

$$\frac{R_1 \parallel R_2 \parallel \beta_{ac}r'_e}{R_s + R_1 \parallel R_2 \parallel \beta_{ac}r'_e} = \frac{62\ k\Omega \parallel 22\ k\Omega \parallel 1600\ \Omega}{600\ \Omega + 62\ k\Omega \parallel 22\ k\Omega \parallel 1600\ \Omega} = \frac{1456}{2056} = 0.708$$

The overall voltage gain is

$$A'_{v(mid)} = 0.708(113) = 80$$

and is expressed in dB as

$$A'_{v(mid)} = 20\ \log(80) = 38.1\ dB$$

The Bode plot of the low-frequency response of this amplifier is shown in Figure 10–20.

FIGURE 10–20
Bode plot for the low-frequency response of the amplifier in Figure 10–19.

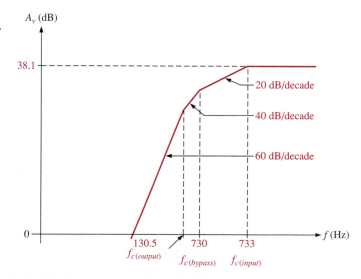

Related Exercise If the overall voltage gain of the amplifier is reduced, how are the lower critical frequencies affected?

Low-Frequency Response of Direct-Coupled Amplifiers

Recall from Chapter 6 that direct-coupled amplifiers have no coupling or bypass capacitors, which allows the frequency response to extend down to dc (0 Hz). Direct-coupled amplifiers are commonly used in linear ICs because of their simplicity and because they can effectively amplify signals with frequencies less than 10 Hz (the frequency to which capacitively coupled amplifiers are generally limited because of the size of the capacitors required).

A direct-coupled amplifier has the same gain for a dc input voltage as it does for a signal with a midrange frequency. The gain of a direct-coupled amplifier such as the one in Figure 10–21 is determined by the ratio R_C/R_E and remains constant at lower frequencies because there are no frequency-dependent components.

FIGURE 10–21
Basic two-stage direct-coupled amplifier stage.

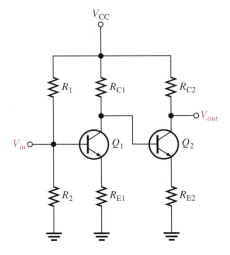

SECTION 10–3 REVIEW

1. A certain amplifier exhibits three critical frequencies in its low-frequency response: $f_{c1} = 130$ Hz, $f_{c2} = 167$ Hz, and $f_{c3} = 75$ Hz. Which is the dominant critical frequency?

2. If the midrange voltage gain of the amplifier in Question 1 is 50 dB, what is the gain at the dominant f_c?

3. A certain RC circuit has an $f_c = 235$ Hz, above which the attenuation is 0 dB. What is the dB attenuation at 23.5 Hz?

10–4 ■ MILLER CAPACITANCE

Miller's theorem shows how the internal capacitance that is inherent in the structure of a transistor is affected by the voltage gain of the amplifier in which the transistor is used. The internal capacitance between the input and output is greatly magnified when looking from the input and magnified slightly when looking from the output of an amplifier.

After completing this section, you should be able to

■ **Use Miller's theorem to determine amplifier capacitances**
  ☐ Determine the Miller input capacitance of a BJT or an FET
  ☐ Determine the Miller output capacitance of a BJT or an FET

Miller's Theorem

Miller's theorem can be used to simplify the analysis of inverting amplifiers at high frequencies where the internal transistor capacitances are important. The capacitance C_{bc} in bipolar transistors (C_{gd} in FETs) between the input (base or gate) and the output (collector or drain) is shown in Figure 10–22(a) in a generalized form. A_v is the absolute voltage gain of the amplifier at midrange frequencies, and C represents either C_{bc} or C_{gd}.

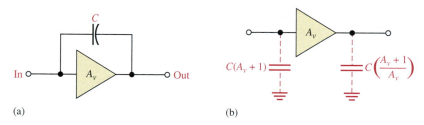

(a)   (b)

FIGURE 10–22
General case of Miller input and output capacitances. C represents C_{bc} or C_{gd}.

Miller's theorem states that C effectively appears as a capacitance from input to ground, as shown in Figure 10–22(b), that can be expressed as follows:

$$C_{in(Miller)} = C(A_v + 1) \tag{10–11}$$

This formula shows that C_{bc} (or C_{gd}) has a much greater impact on input capacitance than its actual value. For example, if $C_{bc} = 6$ pF and the amplifier gain is 50, then $C_{in(Miller)} = 306$ pF. Figure 10–23 shows how this effective input capacitance appears in the actual ac equivalent circuit in parallel with C_{be} (or C_{gs}).

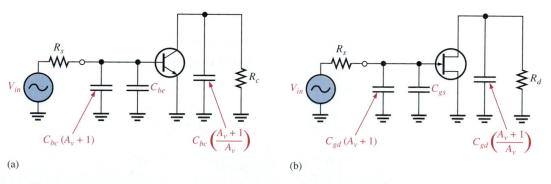

(a)   (b)

FIGURE 10–23
Amplifier ac equivalent circuits showing internal capacitances and effective Miller capacitances.

Miller's theorem also states that C effectively appears as a capacitance from output to ground, as shown in Figure 10–22(b), that can be expressed as follows:

$$C_{out(Miller)} = C\left(\frac{A_v + 1}{A_v}\right)$$

(10–12)

This formula indicates that if the voltage gain is 10 or greater, $C_{out(Miller)}$ is approximately equal to C_{bc} or C_{gd} because $(A_v + 1)/A_v$ is approximately equal to 1. Figure 10–23 also shows how this effective output capacitance appears in the ac equivalent circuit for bipolar transistors and FETs. Equations (10–11) and (10–12) are derived in Appendix B.

EXAMPLE 10–8   Apply Miller's theorem to the small-signal amplifier in Figure 10–24 to determine a high-frequency equivalent circuit, given that $C_{bc} = 3$ pF and $C_{be} = 1$ pF.

FIGURE 10–24

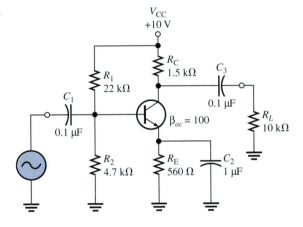

Solution   You must find the voltage gain before you can apply Miller's theorem. Recall that the voltage gain is expressed as

$$A_v = \frac{R_c}{r'_e}$$

The ac collector resistance, R_c, equals R_C in parallel with R_L.

$$R_c = 1.5 \text{ k}\Omega \parallel 10 \text{ k}\Omega = 1.3 \text{ k}\Omega$$

To find r'_e, you must know I_E.

$$V_B = \left(\frac{R_2}{R_1 + R_2}\right)V_{CC} = \left(\frac{4.7 \text{ k}\Omega}{26.7 \text{ k}\Omega}\right)10 \text{ V} = 1.76 \text{ V}$$

$$I_E = \frac{V_E}{R_E} = \frac{1.76 \text{ V} - 0.7 \text{ V}}{560 \text{ }\Omega} = 1.89 \text{ mA}$$

$$r'_e = \frac{25 \text{ mV}}{I_E} = \frac{25 \text{ mV}}{1.89 \text{ mA}} \cong 13.2 \text{ }\Omega$$

Therefore,

$$A_v = \frac{1.3 \text{ k}\Omega}{13.2 \text{ }\Omega} = 98$$

Apply Miller's theorem:

$$C_{in(Miller)} = C_{bc}(A_v + 1) = (3 \text{ pF})(99) = 297 \text{ pF}$$

and

$$C_{out(Miller)} = C_{bc}\left(\frac{A_v + 1}{A_v}\right) = (3 \text{ pF})\left(\frac{99}{98}\right) \cong 3 \text{ pF}$$

The high-frequency equivalent circuit for the amplifier in Figure 10–24 is shown in Figure 10–25.

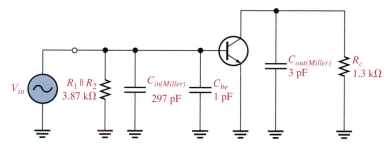

FIGURE 10–25
High-frequency equivalent circuit for the amplifier in Figure 10–24.

Related Exercise   If the voltage gain of the amplifier in Figure 10–24 is reduced to 50, what are the Miller input and output capacitances?

SECTION 10–4 REVIEW

1. Determine $C_{in(Miller)}$ if $A_v = 50$ and $C_{bc} = 5$ pF.
2. Determine $C_{out(Miller)}$ if $A_v = 25$ and $C_{bc} = 3$ pF.

10–5 ■ HIGH-FREQUENCY AMPLIFIER RESPONSE

You have seen how the coupling and bypass capacitors affect the voltage gain of an amplifier at lower frequencies where the capacitive reactances are significant. In the midrange of the amplifier, the effects of the capacitors are minimal and can be neglected. If the frequency is increased sufficiently, a point is reached where the transistor's internal capacitances begin to have a significant effect on the gain. In this section, we use a bipolar transistor amplifier to illustrate the principles, but the approach for FET amplifiers is similar. The basic differences are the specifications of the internal capacitances and the input resistance. This coverage applies to both capacitively coupled and direct-coupled amplifiers.

After completing this section, you should be able to

■ **Analyze the high-frequency response of a BJT amplifier**
  ☐ State and apply Miller's theorem
  ☐ Generally describe how internal capacitances affect the gain
  ☐ Identify the input *RC* circuit
  ☐ Determine the upper critical frequency of the input *RC* circuit
  ☐ Determine phase shift of the input *RC* circuit
  ☐ Identify the output *RC* circuit
  ☐ Determine the upper critical frequency of the output *RC* circuit
  ☐ Determine phase shift of the output *RC* circuit
  ☐ Analyze an amplifier for total high-frequency response

A high-frequency ac equivalent circuit for the amplifier in Figure 10–26(a) is shown in Figure 10–26(b). Notice that the coupling and bypass capacitors are treated as effective shorts, and the internal capacitances, C_{be} and C_{bc}, which are significant only at high frequencies, appear in the diagram. As previously mentioned, C_{be} is sometimes called the input capacitance C_{ib} and C_{bc} is sometimes called the output capacitance C_{ob}. C_{be} is specified on data sheets at a certain value of V_{BE}. Often, a data sheet will list C_{ib} as C_{ibo} and C_{ob} as C_{obo}. The *o* indicates the capacitance is measured with the base open. For example, a 2N2222A has a C_{be} of 25 pF at $V_{BE} = 0.5$ V dc, $I_C = 0$, and $f = 100$ kHz. Also, C_{bc} is specified at a certain value of V_{CB}. The 2N2222A has a maximum C_{bc} of 8 pF at $V_{CB} = 10$ V dc.

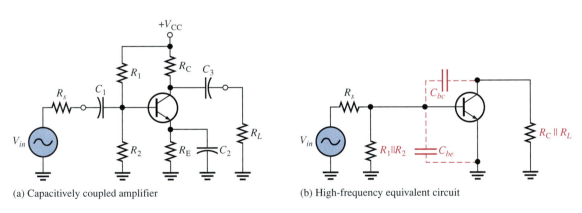

(a) Capacitively coupled amplifier

(b) High-frequency equivalent circuit

FIGURE 10–26

Capacitively coupled amplifier and its high-frequency equivalent circuit.

Miller's Theorem in High-Frequency Analysis

By applying Miller's theorem to the circuit in Figure 10–26(b) and using the midrange voltage gain, you have a circuit that can be analyzed for high-frequency response. Looking in from the signal source, the capacitance C_{bc} appears in the Miller input capacitance from base to ground.

$$C_{in(Miller)} = C_{bc}(A_v + 1)$$

C_{be} simply appears as a capacitance to ac ground, as shown in Figure 10–27, in parallel with $C_{in(Miller)}$. Looking in at the collector, C_{bc} appears in the Miller output capacitance from collector to ground. As shown in Figure 10–27, it appears in parallel with R_c.

$$C_{out(Miller)} = C_{bc}\left(\frac{A_v + 1}{A_v}\right)$$

These two Miller capacitances create a high-frequency input RC circuit and a high-frequency output RC circuit. These two circuits differ from the low-frequency input and output circuits, which act as high-pass filters, because the capacitances go to ground and therefore act as low-pass filters. The equivalent circuit in Figure 10–27 is an ideal model because stray capacitances that are due to circuit interconnections are neglected.

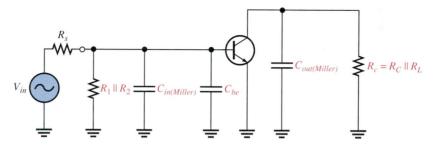

FIGURE 10–27
High-frequency equivalent circuit after applying Miller's theorem.

The Input *RC* Circuit

At high frequencies, the input circuit appears as in Figure 10–28(a), where $\beta_{ac}r'_e$ is the input resistance at the base of the transistor because the bypass capacitor effectively shorts the emitter to ground. By combining C_{be} and $C_{in(Miller)}$ in parallel and repositioning, you get the simplified circuit shown in Figure 10–28(b). Next, by Thevenizing the circuit to the left of the capacitor, as indicated, the input RC circuit is reduced to the equivalent form shown in Figure 10–28(c).

As the frequency increases, the capacitive reactance becomes smaller. This causes the signal voltage at the base to decrease, so the amplifier's voltage gain decreases. The reason for this is that the capacitance and resistance act as a voltage divider and, as the frequency increases, more voltage is dropped across the resistance and less across the capacitance. At the critical frequency, the gain is 3 dB less than its midrange value. Just as with the low-frequency response, the critical frequency, f_c, is the frequency at which the capacitive reactance is equal to the total resistance.

$$X_{Ctot} = R_s \| R_1 \| R_2 \| \beta_{ac}r'_e$$

Therefore,

$$\frac{1}{2\pi f_c C_{tot}} = R_s \| R_1 \| R_2 \| \beta_{ac}r'_e$$

and

$$f_c = \frac{1}{2\pi(R_s \parallel R_1 \parallel R_2 \parallel \beta_{ac}r'_e)C_{tot}}$$ (10–13)

where R_s is the resistance of the signal source and $C_{tot} = C_{be} + C_{in(Miller)}$. As the frequency goes above f_c, the input RC circuit causes the gain to roll off at a rate of −20 dB/decade just as with the low-frequency response.

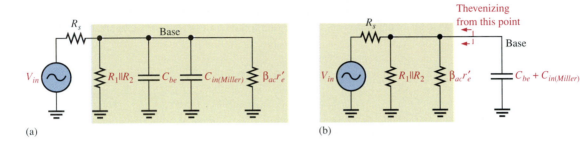

(a)   (b)

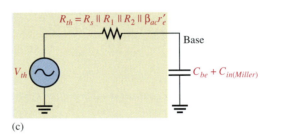

(c)

FIGURE 10–28
Development of the high-frequency input RC circuit.

EXAMPLE 10–9   Derive the high-frequency input RC circuit for the amplifier in Figure 10–29. Also determine the critical frequency. The transistor's data sheet provides the following: $\beta_{ac} = 125$, $C_{be} = 20$ pF, and $C_{bc} = 3$ pF.

FIGURE 10–29

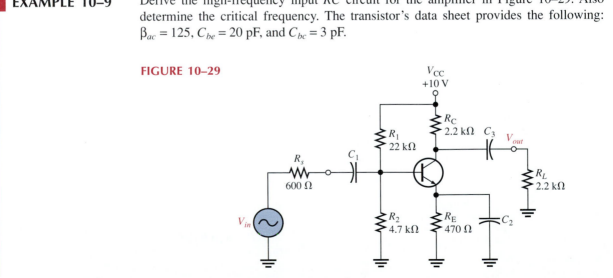

Solution First, find r'_e as follows:

$$V_B = \left(\frac{R_2}{R_1 + R_2}\right)V_{CC} = \left(\frac{4.7 \text{ k}\Omega}{26.7 \text{ k}\Omega}\right)10 \text{ V} = 1.76 \text{ V}$$

$$V_E = V_B - 0.7 \text{ V} = 1.06 \text{ V}$$

$$I_E = \frac{V_E}{R_E} = \frac{1.06 \text{ V}}{470 \ \Omega} = 2.26 \text{ mA}$$

$$r'_e = \frac{25 \text{ mV}}{I_E} = 11.1 \ \Omega$$

The total resistance of the input circuit is

$$R_s \parallel R_1 \parallel R_2 \parallel \beta_{ac}r'_e = 600 \ \Omega \parallel 22 \text{ k}\Omega \parallel 4.7 \text{ k}\Omega \parallel 125(11.1 \ \Omega) = 378 \ \Omega$$

Next, in order to determine the capacitance, you must calculate the midrange gain of the amplifier so that you can apply Miller's theorem.

$$A_{v(mid)} = \frac{R_c}{r'_e} = \frac{R_C \parallel R_L}{r'_e} = \frac{1.1 \text{ k}\Omega}{11.1 \ \Omega} = 99$$

Apply Miller's theorem:

$$C_{in(Miller)} = C_{bc}(A_{v(mid)} + 1) = (3 \text{ pF})(100) = 300 \text{ pF}$$

The total input capacitance is $C_{in(Miller)}$ in parallel with C_{be}.

$$C_{in(tot)} = C_{in(Miller)} + C_{be} = 300 \text{ pF} + 20 \text{ pF} = 320 \text{ pF}$$

The resulting high-frequency input RC circuit is shown in Figure 10–30. The critical frequency is

$$f_c = \frac{1}{2\pi(378 \ \Omega)(320 \text{ pF})} = 1.32 \text{ MHz}$$

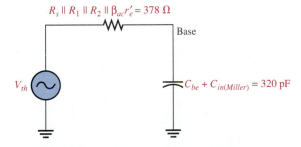

FIGURE 10–30
High-frequency input RC circuit for the amplifier in Figure 10–29.

Related Exercise Determine the high-frequency input RC circuit for Figure 10–29 and find its critical frequency if a transistor with the following specifications is used: $\beta_{ac} = 75$, $C_{be} = 15 \text{ pF}$, $C_{bc} = 2 \text{ pF}$.

Phase Shift of the Input RC Circuit Because the output voltage of a high-frequency input *RC* circuit is across the capacitor, the output of the circuit lags the input. The phase angle is expressed as

$$\theta = \tan^{-1}\left(\frac{R_s \parallel R_1 \parallel R_2 \parallel \beta_{ac} r'_e}{X_{C_{tot}}} \right) \tag{10–14}$$

At the critical frequency, the phase angle is 45° with the signal voltage at the base of the transistor lagging the input signal. As the frequency increases above f_c, the phase angle increases above 45° and approaches 90° when the frequency is sufficiently high.

The Output *RC* Circuit

The high-frequency output *RC* circuit is formed by the Miller output capacitance and the resistance looking in at the collector, as shown in Figure 10–31(a). In determining the output resistance, the transistor is treated as a current source (open) and one end of R_C is effectively ac ground, as shown in Figure 10–31(b). By rearranging the position of the capacitance in the diagram and Thevenizing the circuit to the left, as shown in Figure 10–31(c), you get the equivalent circuit in Figure 10–31(d). The equivalent output *RC* circuit consists of a resistance equal to R_C and R_L in parallel in series with a capacitance which is determined by the Miller formula:

$$C_{out(Miller)} = C_{bc}\left(\frac{A_v + 1}{A_v} \right)$$

If the voltage gain is at least 10, this formula can be approximated as

$$C_{out(Miller)} \cong C_{bc}$$

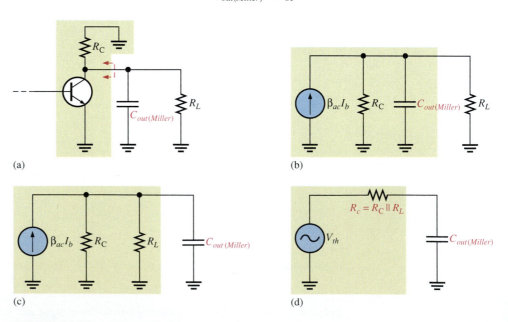

(a)   (b)

(c)   (d)

FIGURE 10–31
Development of high-frequency output RC circuit.

The critical frequency is determined with the following equation, where $R_c = R_C \| R_L$.

$$f_c = \frac{1}{2\pi R_c C_{out(Miller)}}$$ (10–15)

Just as in the input RC circuit, the output RC circuit reduces the gain by 3 dB at the critical frequency. When the frequency goes above the critical value, the gain drops at a −20 dB/decade rate. The phase shift introduced by the output RC circuit is

$$\theta = \tan^{-1}\left(\frac{R_c}{X_{C_{out(Miller)}}}\right)$$ (10–16)

EXAMPLE 10–10 Determine the critical frequency of the amplifier in Example 10–9 (Figure 10–29) due to its output RC circuit.

Solution The Miller output capacitance is calculated as follows:

$$C_{out(Miller)} = C_{bc}\left(\frac{A_v + 1}{A_v}\right) = (3 \text{ pF})\left(\frac{99 + 1}{99}\right) \cong 3 \text{ pF}$$

The equivalent resistance is

$$R_c = R_C \| R_L = 2.2 \text{ k}\Omega \| 2.2 \text{ k}\Omega = 1.1 \text{ k}\Omega$$

The equivalent circuit is shown in Figure 10–32, and the critical frequency is determined as follows ($C_{out(Miller)} \cong C_{bc}$).

$$f_c = \frac{1}{2\pi R_c C_{bc}} = \frac{1}{2\pi(1.1 \text{ k}\Omega)(3 \text{ pF})} = 48.2 \text{ MHz}$$

FIGURE 10–32

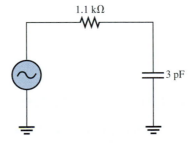

1.1 kΩ

3 pF

Related Exercise If another transistor with $C_{bc} = 5$ pF is used in the amplifier, what is f_c?

Total High-Frequency Response of an Amplifier

As you have seen, the two RC circuits created by the internal transistor capacitances influence the high-frequency response of an amplifier. As the frequency increases and reaches the high end of its midrange values, one of the RC circuits will cause the amplifier's gain to begin dropping off. The frequency at which this occurs is the dominant critical frequency; it is the lower of the two critical high frequencies. An ideal high-frequency Bode plot is shown in Figure 10–33(a). It shows the first break point at $f_{c(input)}$ where the voltage gain begins to roll off at −20 dB/decade. At $f_{c(output)}$, the gain begins dropping at −40 dB/decade because each RC circuit is providing a −20 dB/decade roll-off. Figure 10–33(b) shows a nonideal Bode plot where the voltage gain is actually −3 dB below midrange at $f_{c(input)}$. Other possibilities are that the output RC circuit is dominant or that both circuits have the same critical frequency.

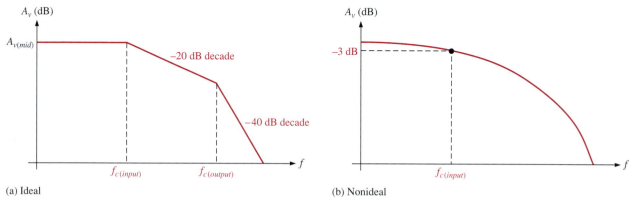

(a) Ideal   (b) Nonideal

FIGURE 10–33
High-frequency Bode plots.

SECTION 10–5 REVIEW	1. What determines the high-frequency response of an amplifier?

SECTION 10–5 REVIEW

1. What determines the high-frequency response of an amplifier?

2. If an amplifier has a midrange voltage gain of 80, the transistor's C_{bc} is 4 pF, and $C_{be} = 8$ pF what is the total input capacitance?

3. A certain amplifier has $f_{c(input)} = 3.5$ MHz and $f_{c(output)} = 8.2$ MHz. Which circuit dominates the high-frequency response?

10–6 ■ TOTAL AMPLIFIER FREQUENCY RESPONSE

In the previous sections, you learned how each RC circuit in an amplifier affects the frequency response. In this section, we will bring these concepts together and examine the total response of typical amplifiers and the specifications relating to their performance.

After completing this section, you should be able to

■ **Analyze an amplifier for total frequency response**
  □ Determine the bandwidth
  □ Define *gain-bandwidth product*
  □ Explain the half-power points

Figure 10–34(b) shows a generalized ideal response curve (Bode plot) for an amplifier of the type shown in Figure 10–34(a). As previously discussed, the three break points at the lower critical frequencies (f_{c1}, f_{c2}, and f_{c3}) are produced by the three low-frequency RC circuits formed by the coupling and bypass capacitors. The break points at the upper critical frequencies, f_{c4} and f_{c5}, are produced by the two high-frequency RC circuits formed by the transistor's internal capacitances.

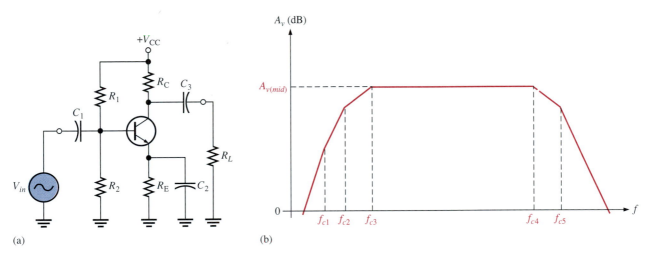

(a)   (b)

FIGURE 10–34
An amplifier and its generalized ideal response curve (Bode plot).

Of particular interest are the two dominant critical frequencies f_{c3} and f_{c4} in Figure 10–34(b). These two frequencies are where the voltage gain of the amplifier is 3 dB below its midrange value. These dominant frequencies are referred to as the *lower critical frequency, f_{cl},* and the *upper critical frequency, f_{cu}.*

Bandwidth

An amplifier normally operates with frequencies between f_{cl} and f_{cu}. As you know, when the input signal frequency is at f_{cl} or f_{cu}, the output signal voltage level is 70.7 percent of its midrange value or −3 dB. If the signal frequency drops below f_{cl}, the gain and thus the output signal level drops at 20 dB/decade until the next critical frequency is reached. The same occurs when the signal frequency goes above f_{cu}.

The range (band) of frequencies lying between f_{cl} and f_{cu} is defined as the **bandwidth** of the amplifier, as illustrated in Figure 10–35. Only the dominant critical frequencies appear in the response curve because they determine the bandwidth. Also, sometimes the other critical frequencies are far enough away from the dominant frequencies that they play no significant role in the total amplifier response and can be neglected. The amplifier's bandwidth is expressed in units of hertz as

$$BW = f_{cu} - f_{cl}$$

(10–17)

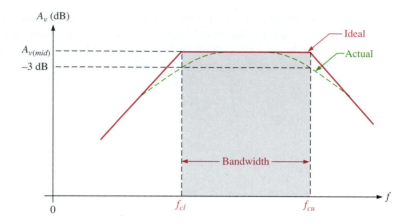

FIGURE 10–35
Response curve illustrating the bandwidth of an amplifier.

Ideally, all signal frequencies lying in an amplifier's bandwidth are amplified equally. For example, if a 10 mV rms signal is applied to an amplifier with a gain of 20, it is amplified to 200 mV rms for all frequencies in the bandwidth. Actually, the gain is down 3 dB at f_{cl} and f_{cu}.

EXAMPLE 10–11   What is the bandwidth of an amplifier having an f_{cl} of 200 Hz and an f_{cu} of 2 kHz?

Solution $BW = f_{cu} - f_{cl} = 2000 \text{ Hz} - 200 \text{ Hz} = 1800 \text{ Hz}$

Notice that bandwidth has the unit of hertz.

Related Exercise   If f_{cl} is increased, does the bandwidth increase or decrease? If f_{cu} is increased, does the bandwidth increase or decrease?

Gain-Bandwidth Product

One characteristic of amplifiers is that the product of the voltage gain and the bandwidth is always constant when the roll-off is −20 dB/decade. This characteristic is called the **gain-bandwidth product.** Let's assume that the lower critical frequency of a particular amplifier is much less than the upper critical frequency.

$$f_{cl} \ll f_{cu}$$

The bandwidth can then be approximated as

$$BW = f_{cu} - f_{cl} \cong f_{cu}$$

Unity-Gain Frequency   The simplified Bode plot for this condition is shown in Figure 10–36. Notice that f_{cl} is neglected because it is so much smaller than f_{cu}, and the bandwidth approximately equals f_{cu}. Beginning at f_{cu}, the gain rolls off until unity gain (0 dB) is reached. The frequency at which the amplifier's gain is one is called the *unity-gain fre-*

quency, f_T. The significance of f_T is that it always equals the product of the midrange voltage gain times the bandwidth and is constant for a given transistor.

$$f_T = A_{v(mid)}BW \qquad (10\text{–}18)$$

For the case shown in Figure 10–36, $f_T = A_{v(mid)}f_{cu}$. For example, if a transistor data sheet specifies $f_T = 100$ MHz, this means that the transistor is capable of producing a voltage gain of one up to 100 MHz, or a gain of 100 up to 1 MHz, or any combination of gain and bandwidth that produces a product of 100 MHz.

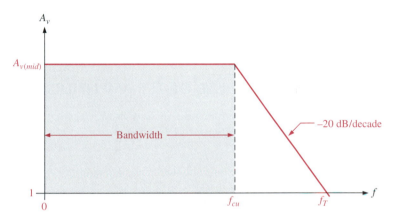

FIGURE 10–36
Simplified response curve where f_{cl} is negligible (assumed to be zero) compared to f_{cu}.

EXAMPLE 10–12

A certain transistor has an f_T of 175 MHz. When this transistor is used in an amplifier with a midrange voltage gain of 50, what bandwidth can be achieved ideally?

Solution

$$f_T = A_{v(mid)}BW$$

$$BW = \frac{f_T}{A_{v(mid)}} = \frac{175 \text{ MHz}}{50} = 3.5 \text{ MHz}$$

Related Exercise An amplifier has a midrange voltage gain of 20 and a bandwidth of 1 MHz. What is the f_T of the transistor?

Half-Power Points

The upper and lower critical frequencies are sometimes called the *half-power frequencies.* This term is derived from the fact that the output power of an amplifier at its critical frequencies is one-half of its midrange power, as previously mentioned. This can be shown as follows, starting with the fact that the output voltage is 0.707 of its midrange value at the critical frequencies.

$$V_{out(f_c)} = 0.707V_{out(mid)}$$

$$P_{out(f_c)} = \frac{V_{out(f_c)}^2}{R_{out}} = \frac{(0.707V_{out(mid)})^2}{R_{out}} = \frac{0.5V_{out(mid)}^2}{R_{out}} = 0.5P_{out(mid)}$$

SECTION 10–6 REVIEW

1. What is the gain of an amplifier at f_T?
2. What is the bandwidth of an amplifier when $f_{cu} = 25$ kHz and $f_{cl} = 100$ Hz?
3. The f_T of a certain transistor is 130 MHz. What gain can be achieved with a bandwidth of 50 MHz?

10–7 ■ FREQUENCY RESPONSE OF FET AMPLIFIERS

In this section, we will look at the frequency response of FET amplifiers using both D-MOSFETs and JFETs. As you will see, the approach is basically the same as for BJTs.

After completing this section, you should be able to

■ **Analyze the frequency response of an FET amplifier**
  □ Identify the low-frequency input RC circuit
  □ Determine the lower critical frequency and phase shift of the input RC circuit
  □ Identify the low-frequency output RC circuit
  □ Determine the lower critical frequency and phase shift of the output RC circuit
  □ Determine values for the internal FET capacitances.
  □ Apply Miller's theorem
  □ Identify the high-frequency input RC circuit
  □ Determine the upper critical frequency and phase shift of the input RC circuit
  □ Identify the high-frequency output RC circuit
  □ Determine the upper critical frequency and phase shift of the output RC circuit

A zero-biased D-MOSFET amplifier with capacitive coupling on the input and output is shown in Figure 10–37.

FIGURE 10–37
Zero-biased D-MOSFET amplifier.

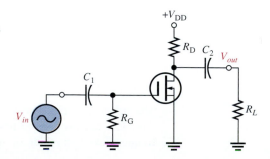

Low-Frequency Response

As you learned in Chapter 9, the midrange voltage gain of a zero-biased amplifier is

$$A_{v(mid)} = g_m R_d$$

This is the gain at frequencies high enough so that the capacitive reactances are approximately zero. The amplifier in Figure 10–37 has only two RC circuits that influence its low-frequency response. One RC circuit is formed by the input coupling capacitor C_1 and the input resistance, as shown in Figure 10–38. The other circuit is formed by the output coupling capacitor C_2 and the output resistance looking in at the drain.

FIGURE 10–38
Input RC circuit for the amplifier in Figure 10–37.

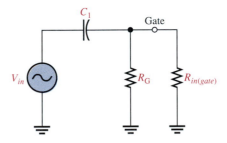

The Low-Frequency Input RC Circuit Just as in the previous case for the bipolar transistor amplifier, the reactance of the input coupling capacitor increases as the frequency decreases. When $X_{C1} = R_{in}$, the gain is down 3 dB below its midrange value. The lowest critical frequency is

$$f_c = \frac{1}{2\pi R_{in} C_1}$$

The input resistance is

$$R_{in} = R_G \parallel R_{in(gate)}$$

where $R_{in(gate)}$ is determined from data sheet information as

$$R_{in(gate)} = \left| \frac{V_{GS}}{I_{GSS}} \right|$$

So, the critical frequency is

$$f_c = \frac{1}{2\pi (R_G \parallel R_{in(gate)}) C_1} \tag{10–19}$$

The gain rolls off below f_c at 20 dB/decade, as previously shown. The phase shift in the low-frequency input RC circuit is

$$\theta = \tan^{-1}\left(\frac{X_{C1}}{R_{in}} \right) \tag{10–20}$$

EXAMPLE 10–13   What is the critical frequency of the input RC circuit in the amplifier of Figure 10–39?

FIGURE 10–39

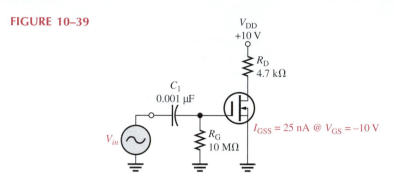

Solution   First determine R_{in} and then calculate f_c.

$$R_{in(gate)} = \left| \frac{V_{GS}}{I_{GSS}} \right| = \frac{10 \text{ V}}{25 \text{ nA}} = 400 \text{ M}\Omega$$

$$R_{in} = R_G \| R_{in(gate)} = 10 \text{ M}\Omega \| 400 \text{ M}\Omega = 9.8 \text{ M}\Omega$$

$$f_c = \frac{1}{2\pi R_{in}C_1} = \frac{1}{2\pi(9.8 \text{ M}\Omega)(0.001 \text{ μF})} = 16.2 \text{ Hz}$$

The critical frequency of the input RC circuit of an FET amplifier is usually very low because of the very high input resistance.

Related Exercise   How much does the critical frequency of the input RC circuit change if the FET in Figure 10–39 is replaced by one with $I_{GSS} = 10$ nA @ $V_{GS} = -8$ V?

The Low-Frequency Output RC Circuit   The second RC circuit that affects the low-frequency response of the amplifier in Figure 10–37 is formed by a coupling capacitor C_2 and the output resistance looking in at the drain, as shown in Figure 10–40(a). R_L is also included. As in the case of the bipolar transistor, the FET is treated as a current source, and the upper end of R_D is effectively ac ground, as shown in Figure 10–40(b). The Thevenin equivalent of the circuit to the left of C_2 is shown in Figure 10–40(c). The critical frequency for this circuit is

$$f_c = \frac{1}{2\pi(R_D + R_L)C_2} \tag{10–21}$$

The effect of the output RC circuit on the amplifier's voltage gain below the midrange is similar to that of the input RC circuit. The circuit with the highest critical frequency dominates because it is the one that first causes the gain to roll off as the frequency drops below its midrange values. The phase shift in the low-frequency output RC circuit is

$$\theta = \tan^{-1}\left(\frac{X_{C2}}{R_D + R_L} \right) \tag{10–22}$$

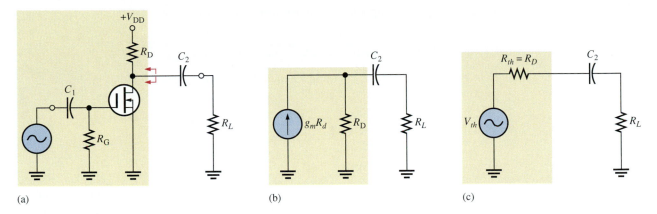

(a)   (b)   (c)

FIGURE 10–40
Development of the output RC circuit.

Again, at the critical frequency, the phase angle is 45° and approaches 90° as the frequency approaches zero. However, starting at the critical frequency, the phase angle decreases from 45° and becomes very small as the frequency goes higher.

EXAMPLE 10–14   Determine the total low-frequency response of the FET amplifier in Figure 10–41. Assume that the load is another identical amplifier with the same R_{in}. The data sheet shows $I_{GSS} = 100$ nA at $V_{GS} = -12$ V.

FIGURE 10–41

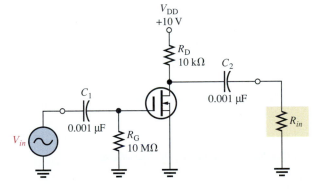

Solution   First, find the critical frequency for the low-frequency input RC circuit.

$$R_{in(gate)} = \left| \frac{V_{GS}}{I_{GSS}} \right| = \frac{12 \text{ V}}{100 \text{ nA}} = 120 \text{ M}\Omega$$

$$R_{in} = R_G \| R_{in(gate)} = 10 \text{ M}\Omega \| 120 \text{ M}\Omega = 9.2 \text{ M}\Omega$$

$$f_{c(input)} = \frac{1}{2\pi R_{in} C_1} = \frac{1}{2\pi (9.2 \text{ M}\Omega)(0.001 \text{ μF})} = 17.3 \text{ Hz}$$

The low-frequency output *RC* circuit has a critical frequency of

$$f_{c(output)} = \frac{1}{2\pi(R_D + R_L)C_2} = \frac{1}{2\pi(9.2 \text{ M}\Omega)(0.001 \text{ μF})} = 17.3 \text{ Hz}$$

Related Exercise If the circuit in Figure 10–41 were operated with no load, how is the low-frequency response affected?

High-Frequency Response

The approach to the high-frequency analysis of an FET amplifier is very similar to that of a bipolar amplifier. The basic differences are the specifications of the internal FET capacitances and the determination of the input resistance.

Figure 10–42(a) shows a JFET common-source amplifier that will be used to illustrate high-frequency analysis. A high-frequency equivalent circuit for the amplifier is shown in Figure 10–42(b). Notice that the coupling and bypass capacitors are assumed to have negligible reactances and are considered to be shorts. The internal capacitances C_{gs} and C_{gd} appear in the equivalent circuit because their reactances are significant at high frequencies.

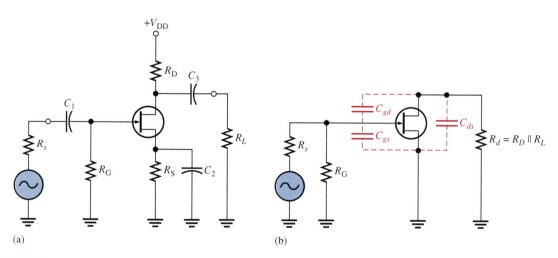

(a)   (b)

FIGURE 10–42

A JFET amplifier and its high-frequency equivalent circuit.

Values of C_{gs}, C_{gd}, and C_{ds} FET data sheets do not normally provide values for C_{gs}, C_{gd}, or C_{ds}. Instead, three other values are usually specified because they are easier to measure. These are C_{iss}, the input capacitance; C_{rss}, the reverse transfer capacitance; and C_{oss}, the output capacitance. Because of the manufacturer's method of measurement, the following relationships allow you to determine the capacitor values needed for analysis.

$$C_{gd} = C_{rss} \qquad\qquad (10\text{–}23)$$

$$C_{gs} = C_{iss} - C_{rss} \qquad\qquad \textbf{(10–24)}$$

$$C_{ds} = C_{oss} - C_{rss} \qquad\qquad \textbf{(10–25)}$$

C_{oss} is not specified as often as the other values on data sheets. Sometimes, it is designated as $C_{d(sub)}$, the drain-to-substrate capacitance. In cases where a value is not available, you must either assume a value or neglect C_{ds}.

EXAMPLE 10–15   The data sheet for a 2N3823 JFET gives C_{iss} = 6 pF and C_{rss} = 2 pF. Determine C_{gd} and C_{gs}.

Solution
$$C_{gd} = C_{rss} = 2 \text{ pF}$$
$$C_{gs} = C_{iss} - C_{rss} = 6 \text{ pF} - 2 \text{ pF} = 4 \text{ pF}$$

Related Exercise   Although C_{oss} is not specified on the data sheet for the 2N3823 JFET, assume a value of 3 pF and determine C_{ds}.

Using Miller's Theorem   Miller's theorem is applied the same way in FET amplifier high-frequency analysis as was done in bipolar transistor amplifiers. Looking in from the signal source in Figure 10–42(b), C_{gd} effectively appears in the Miller input capacitance as follows.

$$C_{in(Miller)} = C_{gd}(A_v + 1) \qquad\qquad \textbf{(10–26)}$$

C_{gs} simply appears as a capacitance to ac ground in parallel with $C_{in(Miller)}$, as shown in Figure 10–43. Looking in at the drain, C_{gd} effectively appears in the Miller output capacitance from drain to ground in parallel with R_d, as shown in Figure 10–43.

$$C_{out(Miller)} = C_{gd}\left(\frac{A_v + 1}{A_v}\right) \qquad\qquad \textbf{(10–27)}$$

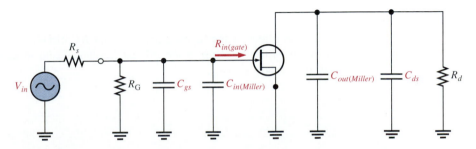

FIGURE 10–43
High-frequency equivalent circuit after applying Miller's theorem.

These two Miller capacitances contribute to a high-frequency input RC circuit and a high-frequency output RC circuit. Both are low-pass filters which produce phase lag.

The High-Frequency Input RC Circuit   The high-frequency input circuit forms a low-pass type of filter and is shown in Figure 10–44(a). Because both R_G and the input resistance at the gate of FETs are extremely high, the controlling resistance for the input circuit is the resistance of the input source as long as $R_s \ll R_{in}$. This is because R_s appears in parallel with R_{in} when Thevenin's theorem is applied. The simplified input RC circuit appears in Figure 10–44(b). The critical frequency is

$$f_c = \frac{1}{2\pi R_s C_{tot}}$$

(10–28)

where $C_{tot} = C_{gs} + C_{in(Miller)}$. The input circuit produces a phase shift of

$$\theta = \tan^{-1}\left(\frac{R_s}{X_{C_{tot}}}\right)$$

(10–29)

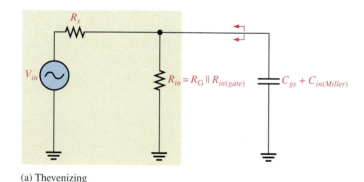

(a) Thevenizing

(b) Thevenin equivalent input circuit, neglecting R_{in}

FIGURE 10–44
High-frequency input RC circuit.

    The effect of the high-frequency input RC circuit is to reduce the midrange gain of the amplifier by 3 dB at the critical frequency and to cause the gain to decrease at −20 dB/decade above f_c.

EXAMPLE 10–16   Find the critical frequency of the high-frequency input RC circuit for the amplifier in Figure 10–45. $C_{iss} = 8$ pF and $C_{rss} = 3$ pF. $g_m = 6500$ μS.

FIGURE 10–45

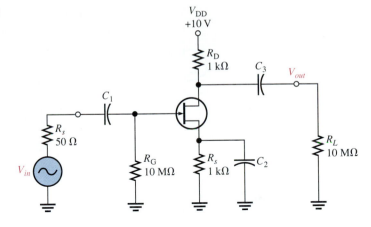

Solution   Determine C_{gd} and C_{gs}.

$$C_{gd} = C_{rss} = 3 \text{ pF}$$
$$C_{gs} = C_{iss} - C_{rss} = 8 \text{ pF} - 3 \text{ pF} = 5 \text{ pF}$$

Determine the critical frequency for the input RC circuit as follows:

$$A_v = g_m R_d = g_m(R_D \parallel R_L) \cong (6500 \text{ μS})(1 \text{ kΩ}) = 6.5$$
$$C_{in(Miller)} = C_{gd}(A_v + 1) = (3 \text{ pF})(7.5) = 22.5 \text{ pF}$$

The total input capacitance is

$$C_{tot} = C_{gs} + C_{in(Miller)} = 5 \text{ pF} + 22.5 \text{ pF} = 27.5 \text{ pF}$$

The critical frequency is

$$f_c = \frac{1}{2\pi R_s C_{tot}} = \frac{1}{2\pi(50 \text{ Ω})(27.5 \text{ pF})} = 116 \text{ MHz}$$

Related Exercise   If the gain of the amplifier in Figure 10–45 is increased to 10, what happens to f_c?

The High-Frequency Output *RC* Circuit   The high-frequency output *RC* circuit is formed by the Miller output capacitance and the output resistance looking in at the drain, as shown in Figure 10–46(a). As in the case of the bipolar transistor, the FET is treated as a current source. When you apply Thevenin's theorem, you get an equivalent output *RC* circuit consisting of R_D in parallel with R_L and an equivalent output capacitance as follows:

$$C_{out(Miller)} = C_{gd}\left(\frac{A_v + 1}{A_v}\right)$$

This equivalent output circuit is shown in Figure 10–46(b). The critical frequency of the output *RC* lag circuit is

$$f_c = \frac{1}{2\pi R_d C_{out(Miller)}}$$

(10–30)

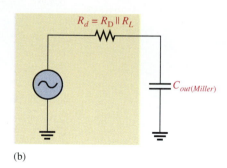

(a) (b)

FIGURE 10–46
High-frequency output RC circuit.

The output circuit produces a phase shift of

$$\theta = \tan^{-1}\left(\frac{R_d}{X_{C_{out(Miller)}}}\right) \qquad \textbf{(10–31)}$$

EXAMPLE 10–17 Determine the critical frequency of the high-frequency output *RC* circuit for the amplifier in Figure 10–45. What is the phase shift introduced by this circuit at the critical frequency?

Solution   Since R_L is very large compared to R_D, it can be neglected, and the equivalent output resistance is

$$R_d \cong R_D = 1 \text{ k}\Omega$$

The equivalent output capacitance is

$$C_{out(Miller)} = C_{gd}\left(\frac{A_v + 1}{A_v}\right) = (3 \text{ pF})\left(\frac{7.5}{6.5}\right) = 3.46 \text{ pF}$$

Therefore, the critical frequency is

$$f_c = \frac{1}{2\pi R_d C_{out\,(Miller)}} = \frac{1}{2\pi(1 \text{ k}\Omega)(3.46 \text{ pF})} = 46 \text{ MHz}$$

Although it has been neglected, any stray wiring capacitance could significantly affect the frequency response because $C_{out(Miller)}$ is very small.

The phase angle is always 45° at f_c for an *RC* circuit and the output lags.

In Example 10–16, the critical frequency of the high-frequency input *RC* circuit was found to be 116 MHz. Therefore, the critical frequency for the output circuit is dominant since it is the lesser of the two.

Related Exercise   If A_v of the amplifier in Figure 10–45 is increased to 10, what is the f_c of the output circuit?

**SECTION 10–7
REVIEW**
1. What is the amount of phase shift contributed by an input circuit when $X_C = 0.5R_{in}$ at a certain frequency below f_{cl}?
2. What is the critical frequency when $R_D = 1.5\ \text{k}\Omega$, $R_L = 5\ \text{k}\Omega$, and $C_3 = 0.002\ \mu\text{F}$ in Figure 10–45?
3. What are the capacitances that are usually specified on an FET data sheet?
4. If $C_{gs} = 4\ \text{pF}$ and $C_{gd} = 3\ \text{pF}$, what is the total input capacitance of an FET amplifier whose voltage gain is 25?

10–8 ■ FREQUENCY RESPONSE OF MULTISTAGE AMPLIFIERS

To this point, you have seen how the voltage gain of a single-stage amplifier changes over frequency. When two or more stages are cascaded to form a multistage amplifier, the overall frequency response is determined by the frequency response of each stage depending on the relationships of the critical frequencies.

After completing this section, you should be able to

■ **Analyze multistage amplifiers for frequency response**
  ☐ Determine the dominant critical frequencies when the critical frequencies of each stage are the same
  ☐ Determine the dominant critical frequencies when the critical frequencies of each stage differ
  ☐ Determine the bandwidth of a multistage amplifier

When amplifier stages are cascaded to form a multistage amplifier, the dominant frequency response is determined by the responses of the individual stages. There are two cases to consider:

1. Each stage has a different lower critical frequency and a different upper critical frequency.
2. Each stage has the same lower critical frequency and the same upper critical frequency.

Different Critical Frequencies

When the lower critical frequency, f_{cl}, of each amplifier stage is different, the dominant lower critical frequency, f'_{cl}, equals the critical frequency of the stage with the highest f_{cl}.

When the upper critical frequency, f_{cu}, of each amplifier stage is different, the dominant upper critical frequency, f'_{cu}, equals the critical frequency of the stage with the lowest f_{cu}.

Overall Bandwidth The bandwidth of a multistage amplifier is the difference between the dominant lower critical frequency and the dominant upper critical frequency.

$$BW = f'_{cu} - f'_{cl}$$

EXAMPLE 10–18

In a certain 2-stage amplifier, one stage has a lower critical frequency of 850 Hz and an upper critical frequency of 100 kHz. The other has a lower critical frequency of 1 kHz and an upper critical frequency of 230 kHz. Determine the overall bandwidth of the 2-stage amplifier.

Solution

$$f'_{cl} = 1 \text{ kHz}$$

$$f'_{cu} = 100 \text{ kHz}$$

$$BW = f'_{cu} - f'_{cl} = 100 \text{ kHz} - 1 \text{ kHz} = 99 \text{ kHz}$$

Related Exercise   A certain 3-stage amplifier has the following critical frequencies for each stage: $f_{cl(1)} = 1500$ Hz, $f_{cl(2)} = 980$ Hz, and $f_{cl(3)} = 130$ Hz. What is the dominant lower critical frequency?

Equal Critical Frequencies

When each amplifier stage in a multistage arrangement has equal critical frequencies, you may think that the dominant critical frequency is equal to the critical frequency of each stage. This is not the case, however.

When the lower critical frequencies of each stage in a multistage amplifier are all the same, the dominant lower critical frequency is increased by a factor of $1/\sqrt{2^{1/n} - 1}$ as shown by the following formula (n is the number of stages in the multistage amplifier):

$$f'_{cl} = \frac{f_{cl}}{\sqrt{2^{1/n} - 1}} \tag{10–32}$$

When the upper critical frequencies of each stage are all the same, the dominant upper critical frequency is reduced by a factor of $\sqrt{2^{1/n} - 1}$, as shown by the following formula:

$$f'_{cu} = f_{cu}\sqrt{2^{1/n} - 1} \tag{10–33}$$

The proofs of these formulas are quite involved and are reserved for Appendix B.

EXAMPLE 10–19

Both stages in a certain 2-stage amplifier have a lower critical frequency of 500 Hz and an upper critical frequency of 80 kHz. Determine the overall bandwidth.

Solution

$$f'_{cl} = \frac{f_{cl}}{\sqrt{2^{1/n} - 1}} = \frac{500 \text{ Hz}}{\sqrt{2^{0.5} - 1}} = \frac{500 \text{ Hz}}{0.644} = 776 \text{ Hz}$$

$$f'_{cu} = f_{cu}\sqrt{2^{1/n} - 1} = (80 \text{ kHz})(0.644) = 51.5 \text{ kHz}$$

$$BW = f'_{cu} - f'_{cl} = 51.5 \text{ kHz} - 776 \text{ Hz} = 50.7 \text{ kHz}$$

Related Exercise   If a third identical stage is connected in cascade to the 2-stage amplifier in this example, what is the resulting overall bandwidth?

1. One stage in an amplifier has $f_{cl} = 1$ kHz and the other stage has $f_{cl} = 325$ Hz. What is the dominant lower critical frequency?

2. In a certain 3-stage amplifier $f_{cu(1)} = 50$ kHz, $f_{cu(2)} = 55$ kHz, and $f_{cu(3)} = 49$ kHz. What is the dominant upper critical frequency?

3. When more identical stages are added to a multistage amplifier with each stage having the same critical frequency, does the bandwidth increase or decrease?

10–9 ■ FREQUENCY RESPONSE MEASUREMENT

In this section, two basic methods of measuring the frequency response of an amplifier are covered. You will concentrate on determining the two dominant critical frequencies. From these values, you can get the bandwidth.

After completing this section, you should be able to

■ **Measure the frequency response of an amplifier**
  ☐ Use frequency and amplitude measurement to determine the critical frequencies of an amplifier
  ☐ Relate pulse characteristics to frequency
  ☐ Identify the effects of frequency response on pulse shape
  ☐ Use step response measurement to determine the critical frequencies of an amplifier

Frequency and Amplitude Measurement

Figure 10–47(a) shows the test setup for an amplifier circuit board. The amplifier is driven by a sinusoidal voltage source with a dual-channel oscilloscope connected to the input and to the output. The input frequency is set to a midrange value, and its amplitude is adjusted to establish an output signal reference level, as shown in Figure 10–47(b). This output voltage reference level for midrange should be set at a convenient value within the linear operation of the amplifier, for example, 100 mV, 1 V, 10 V, and so on. In this case, set the output signal to a peak value of 1 V.

Next, the frequency of the input voltage is decreased until the peak value of the output drops to 0.707 V. The amplitude of the input voltage must be kept constant as the frequency is reduced. Readjustment may be necessary because of changes in loading of the voltage source with frequency. When the output is 0.707 V, the frequency is measured, and you have the value for f_{cl} as indicated in Figure 10–47(c).

Next, the input frequency is increased back up through midrange and beyond until the peak value of the output voltage again drops to 0.707 V. Again, the amplitude of the input must be kept constant as the frequency is increased. When the output is 0.707 V, the frequency is measured and you have the value for f_{cu} as indicated in Figure 10–47(d). From these two frequency measurements, you can find the bandwidth by the formula $BW = f_{cu} - f_{cl}$.

Response of Linear Amplifiers to Pulse Inputs

As you know, pulse waveforms contain **harmonics** and are a composite of frequencies from dc (0 Hz) on up to very high frequencies. The flat parts of an ideal pulse waveform are constant, nonchanging voltages during the pulse interval and between pulses and rep-

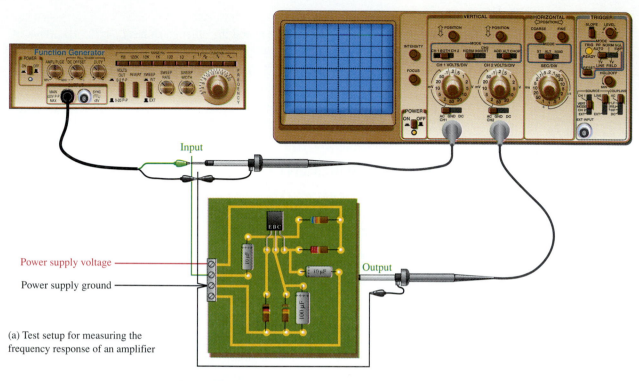

Input

Power supply voltage

Power supply ground

Output

(a) Test setup for measuring the frequency response of an amplifier

Input frequency control on function generator

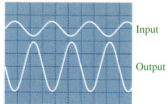

Input

Output

Amplifier input and output voltages
50 µs/DIV
Input: 50 mV/DIV
Output: 0.5 V/DIV

(b) Frequency set to a midrange value (6.67 kHz in this case). Input voltage adjusted for an output of 1 V peak.

Input frequency control on function generator

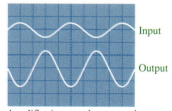

Input

Output

Amplifier input and output voltages
2 ms/DIV
Input: 50 mV/DIV
Output: 0.5 V/DIV

(c) Frequency is reduced until the output is 0.707 V peak. This is the lower critical frequency.

Input frequency control on function generator

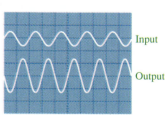

Input

Output

Amplifier input and output voltages
1 µs/DIV
Input: 50 mV/DIV
Output: 0.5 V/DIV

(d) Frequency is increased until the output is again 0.707 V peak. This is the upper critical frequency.

FIGURE 10–47

A general procedure for measuring an amplifier's frequency response.

resent the dc component of the waveform. The rising and falling edges of the pulses, because they are fast-changing, contain the high-frequency components. A pulse consists of two voltage **steps**: one at the rising edge when the pulse goes from low to high and the other at the falling edge when it goes from high to low. When you apply a pulse waveform to the input of an amplifier, you are actually applying a dc voltage plus all the harmonic frequencies that make up the pulse waveform. Suppose an amplifier has a bandwidth that extends from dc (0 Hz) to an infinitely high frequency. The output of the amplifier, in this case, would have the same shape as the input waveform (except it would be amplified) as illustrated in Figure 10–48(a) because all of the input frequency components would pass through.

Practical amplifiers, however, do not have infinite bandwidths and the limitation of the frequency response will affect the shape of the output pulse waveform as you will see.

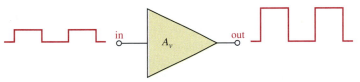

(a) Pulse response of an amplifier with an infinite bandwidth (0 Hz to infinite Hz)

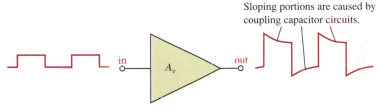

(b) Low-frequency pulse response of amplifier with limited bandwidth

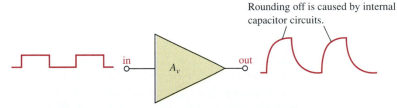

(c) High-frequency pulse response of amplifier with limited bandwidth

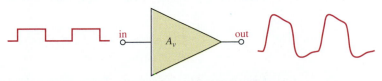

(d) Overall pulse response of amplifier with limited bandwidth

FIGURE 10–48

Illustration of the general response of a linear amplifier to a pulse input.

Effect of the Low-Frequency Response on the Pulse Shape   If a pulse waveform is applied to a noninverting capacitively coupled amplifier with a low-frequency response that does not extend down to dc, the flat portions of the pulse waveform are affected, as shown in Figure 10–48(b). The limitation of the high-frequency response is neglected for the time being. The "sloping" of the flat portions is caused by the coupling capacitors in the amplifier. Let's look at the input *RC* circuit. The input coupling capacitor and the input resistance form a differentiator. The time constant of this differentiator determines how fast the capacitor charges and discharges with a pulse input and thereby adds a "slope" to the flat portions of the input pulse. The "slope" is actually an exponential curve. The coupling capacitor determines at what frequency the sloping effect becomes significant.

Effect of the High-Frequency Response on the Pulse Shape   At high frequencies, the rising and falling edges of the pulses are affected as shown in Figure 10–48(c). This discussion assumes a low-frequency response down to dc. The rounding off of the edges of the pulses is caused by the charging and discharging of the internal input and output capacitances that come into play only at higher frequencies. These capacitances form integrators with the circuit resistances. The value of these capacitances determine at what frequency the rounding off becomes significant.

The combined low-frequency and high-frequency responses of an amplifier cause both rounding of the pulse edges and a sloping of the flat portions, as indicated in Figure 10–48(d).

Step Response Measurement

The lower and upper critical frequencies of an amplifier can be determined using the *step response method* by applying a voltage step to the input of the amplifier and measuring the rise and fall times of the resulting output voltage. The basic test setup shown in Figure 10–47(a) is used except that the pulse output of the function generator is selected. The input step is created by the rising edge of a pulse that has a long duration compared to the rise and fall times to be measured.

High-Frequency Measurement   When a step input is applied, the amplifier's high-frequency *RC* circuits (internal capacitances) prevent the output from responding immediately to the step input. As a result, the output voltage has a rise time (t_r) associated with it, as shown in Figure 10–49(a). In fact, the rise time is inversely related to the upper critical frequency (f_{cu}) of the amplifier. As f_{cu} becomes lower, the rise time of the output becomes greater. The oscilloscope display illustrates how the rise time is measured from the 10% amplitude point to the 90% amplitude point. The scope must be set on a short time base so the relatively short interval of the rise time can be accurately observed. Once this measurement is made, f_{cu} can be calculated with the following formula:

$$f_{cu} = \frac{0.35}{t_r} \qquad\qquad \textbf{(10–34)}$$

Low-Frequency Measurement   To determine the lower critical frequency (f_{cl}) of the amplifier, the step input must be of sufficiently long duration to observe the full charging time of the low-frequency *RC* circuits (coupling capacitances), which cause the "sloping" of the output and which we will refer to as the fall time (t_f). This is illustrated in Figure

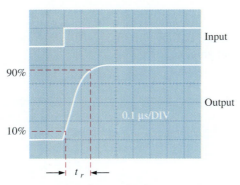
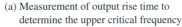

(a) Measurement of output rise time to
determine the upper critical frequency

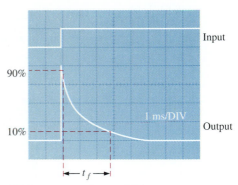

(b) Measurement of output fall time to
determine the lower critical frequency

FIGURE 10–49
Measurement of the rise and fall times associated with the amplifier's step response. The outputs are inverted.

10–49(b). The fall time is inversely related to the low critical frequency of the amplifier. As f_{cl} becomes higher, the fall time of the output becomes less. The scope display illustrates how the fall time is measured from the 90% point to the 10% point. The scope must be set on a long time base so the complete interval of the fall time can be observed. Once this measurement is made, f_{cl} can be determined with the following formula. The derivations of Equations 10–34 and 10–35 are in Appendix B.

$$f_{cl} = \frac{0.35}{t_f} \qquad (10\text{–}35)$$

SECTION 10–9 REVIEW

1. In Figure 10–47, what are the lower and upper critical frequencies?
2. The rise time and the fall time of an amplifier's output voltage are measured between what two points on the voltage transition?
3. In Figure 10–49, what is the rise time?
4. In Figure 10–49, what is the fall time?
5. What is the bandwidth of the amplifier whose step response is measured in Figure 10–49?

10–10 ■ SYSTEM APPLICATION

An audio amplifier circuit board is being finalized for production and will be used in a new intercom system that your company is planning to market. Your supervisor gives you the board with no documentation and asks you to check the frequency response of the amplifier. You will apply the knowledge you have gained in this chapter in completing your assignment.

The Schematic

The amplifier circuit board is shown in Figure 10–50. Draw the schematic of the circuit and label all component values.

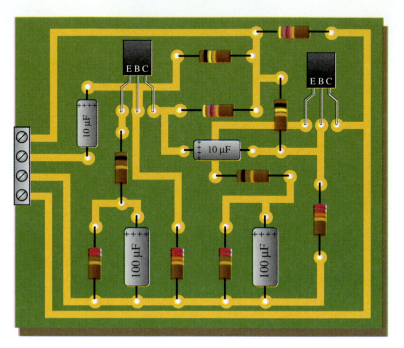

FIGURE 10–50
Amplifier circuit board. Both transistors are 2N3904.

Analysis of the Amplifier Circuit

Refer to the partial data sheet in Figure 10–51 when necessary. V_{CC} = 12V. Use a 10 kΩ load resistor on the final output.

☐ Determine the lower critical frequency of the first stage.
☐ Determine the upper critical frequency of the first stage.
☐ Determine the lower critical frequency of the second stage.
☐ Determine the upper critical frequency of the second stage.
☐ Calculate the dominant lower critical frequency.
☐ Calculate the dominant upper critical frequency.
☐ Calculate the overall bandwidth.

Test Procedure

☐ Specify a procedure for measuring the frequency response of the amplifier board.
☐ Indicate the point-to-point hookup with the circled numbers for the circuit board and measuring instruments on the test bench setup of Figure 10–52.

Electrical Characteristics (T_A = 25°C unless otherwise noted.)

Characteristic		Symbol	Min	Max	Unit
Output capacitance (V_{CB} = 5.0 V dc, I_E = 0, f = 1.0 MHz)		C_{obo}	–	4.0	pF
Input capacitance (V_{BE} = 0.5 V dc, I_C = 0, f = 1.0 MHz)		C_{ibo}	–	8.0	pF
Input impedance (I_C = 1.0 mA dc, V_{CE} = 10 V dc, f = 1.0 kHz)	2N3903 2N3904	h_{ie}	1.0 1.0	8.0 10	k ohms
Voltage feedback ratio (I_C = 1.0 mA dc, V_{CE} = 10 V dc, f = 1.0 kHz)	2N3903 2N3904	h_{re}	0.1 0.5	5.0 8.0	$\times 10^{-4}$
Small-signal current gain (I_C = 1.0 mA dc, V_{CE} = 10 V dc, f = 1.0 kHz)	2N3903 2N3904	h_{fe}	50 100	200 400	–
Output admittance (I_C = 1.0 mA dc, V_{CE} = 10 V dc, f = 1.0 kHz)		h_{oe}	1.0	40	μmhos or μS
Noise figure (I_C = 100 μA dc, V_{CE} = 5.0 V dc, R_S = 1.0 k ohms, f = 1.0 kHz)	2N3903 2N3904	NF	– –	6.0 5.0	dB
DC current gain (I_C = 0.1 mA dc, V_{CE} = 1.0 V dc)	2N3903 2N3904	h_{FE}	20 40	– –	–
(I_C = 1.0 mA dc, V_{CE} = 1.0 V dc)	2N3903 2N3904		35 70	– –	
(I_C = 10 mA dc, V_{CE} = 1.0 V dc)	2N3903 2N3904		50 100	150 300	
(I_C = 50 mA dc, V_{CE} = 1.0 V dc)	2N3903 2N3904		30 60	– –	
(I_C = 100 mA dc, V_{CE} = 1.0 V dc)	2N3903 2N3904		15 30	– –	
Collector-Emitter saturation voltage (I_C = 10 mA dc, I_B = 1.0 mA dc) (I_C = 50 mA dc, I_B = 5.0 mA dc)		$V_{CE(sat)}$	– –	0.2 0.3	V dc
Base-Emitter saturation voltage (I_C = 10 mA dc, I_B = 1.0 mA dc) (I_C = 50 mA dc, I_B = 5.0 mA dc)		$V_{BE(sat)}$	0.65 –	0.85 0.95	V dc
Small-Signal Characteristics					
Current-gain — Bandwidth product (I_C = 10 mA dc, V_{CE} = 20 V dc, f = 100 MHz)	2N3903 2N3904	f_T	250 300	– –	MHz

FIGURE 10–51

Partial data sheet for the 2N3903 and 2N3904 transistors.

☐ Specify voltage and frequency values for all the measurements to be made with the input voltage adjusted to produce a midrange peak output voltage of 1 V.

The Frequency Response on the Test Bench

Frequency response measurements of the amplifier circuit board are shown in Figure 10–53.

☐ Assume that the input voltage is held at a constant amplitude over the frequency range.

☐ Assume that the frequency response of the oscilloscope is sufficient to accurately measure signals with frequencies well above the upper critical frequency.

☐ Determine the correct SEC/DIV settings for each oscilloscope display.

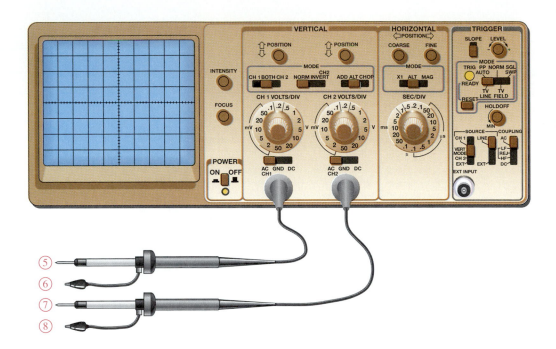

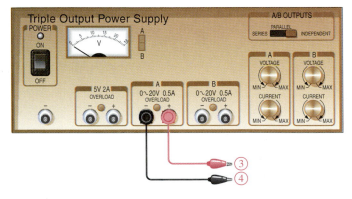

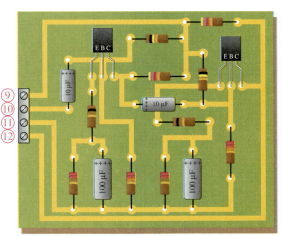

FIGURE 10–52

Test bench setup for frequency response measurements of the amplifier board.

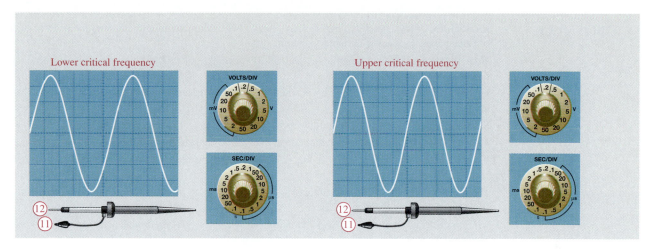

FIGURE 10–53
Amplifier circuit board measurements for frequency response.

Final Report

Submit a final written report on the amplifier circuit board using an organized format that includes the following:

1. A description of the circuit.
2. A discussion of the operation of the circuit.
3. A list of the specifications including gain, critical frequencies, and bandwidth.
4. A list of parts with part numbers if available.

■ CHAPTER SUMMARY

- The coupling and bypass capacitors of an amplifier affect the low-frequency response.
- The internal transistor capacitances affect the high-frequency response.
- Critical frequencies are values of frequency at which the *RC* circuits reduce the voltage gain to 70.7% of its midrange value.
- Each *RC* circuit causes the gain to drop at a rate of 20 dB/decade.
- For the low-frequency *RC* circuits, the *highest* critical frequency is the dominant critical frequency.
- For the high-frequency *RC* circuits, the *lowest* critical frequency is the dominant critical frequency.
- A decade of frequency change is a ten-times change (increase or decrease).
- An octave of frequency change is a two-times change (increase or decrease).
- The bandwidth of an amplifier is the range of frequencies between the lower critical frequency and the upper critical frequency.
- The gain-bandwidth product is a transistor parameter that is constant and equal to the unity-gain frequency.

■ GLOSSARY

Bandwidth The characteristic of certain types of electronic circuits that specifies the usable range of frequencies that pass from input to output.

Bode plot An idealized graph of the gain in dB versus frequency used to graphically illustrate the response of an amplifier or filter.

Critical frequency The frequency at which the response of an amplifier or filter is 3 dB less than at midrange.

Cutoff frequency Another term for critical frequency.

Decade A ten-times increase or decrease in the value of a quantity such as frequency.

Gain-bandwidth product A characteristic of amplifiers whereby the product of the gain and the bandwidth is always constant.

Harmonics The frequencies contained in a composite waveform that are integer multiples of the repetition frequency (fundamental).

Midrange The frequency range of an amplifier lying between the lower and upper critical frequencies.

Octave A two-times increase or decrease in the value of a quantity such as frequency.

Roll-off The decrease in the gain of an amplifier above and below the critical frequencies.

Step A fast voltage transition from one level to another.

■ FORMULAS

The Decibel

(10–1)	$A_{p(dB)} = 10 \log A_p$	Power gain in decibels
(10–2)	$A_{v(dB)} = 20 \log A_v$	Voltage gain in decibels

Low-Frequency Response

(10–3)	$A_{v(mid)} = \dfrac{R_c}{r'_e}$	Midrange voltage gain (bipolar)
(10–4)	$V_{base} = \left(\dfrac{R_{in}}{\sqrt{R_{in}^2 + X_{C1}^2}} \right) V_{in}$	RC input circuit
(10–5)	$f_c = \dfrac{1}{2\pi R_{in} C_1}$	Critical frequency, input circuit
(10–6)	$\theta = \tan^{-1} \left(\dfrac{X_{C1}}{R_{in}} \right)$	Phase angle, input circuit
(10–7)	$f_c = \dfrac{1}{2\pi(R_C + R_L)C_3}$	Critical frequency, output circuit
(10–8)	$\theta = \tan^{-1} \left(\dfrac{X_{C3}}{R_C + R_L} \right)$	Phase angle, output circuit
(10–9)	$R_{in(emitter)} = \dfrac{R_{th}}{\beta_{ac}} + r'_e$	Resistance looking in at emitter
(10–10)	$f_c = \dfrac{1}{2\pi[(r'_e + R_{th}/\beta_{ac}) \parallel R_E]C_2}$	Critical frequency, bypass circuit

Miller's Theorem

(10–11)	$C_{in(Miller)} = C(A_v + 1)$	Miller input capacitance, where $C = C_{bc}$ or C_{gd}
(10–12)	$C_{out(Miller)} = C \left(\dfrac{A_v + 1}{A_v} \right)$	Miller output capacitance, where $C = C_{bc}$ or C_{gd}

High-Frequency Response

(10–13) $\quad f_c = \dfrac{1}{2\pi(R_s \parallel R_1 \parallel R_2 \parallel \beta_{ac}r'_e)C_{tot}}$ Critical frequency, input circuit

(10–14) $\quad \theta = \tan^{-1}\!\left(\dfrac{R_s \parallel R_1 \parallel R_2 \parallel \beta_{ac}r'_e}{X_{C_{tot}}}\right)$ Phase angle, input circuit

(10–15) $\quad f_c = \dfrac{1}{2\pi R_c C_{out(Miller)}}$ Critical frequency, output circuit

(10–16) $\quad \theta = \tan^{-1}\!\left(\dfrac{R_c}{X_{C_{out(Miller)}}}\right)$ Phase angle, output circuit

Total Response

(10–17) $\quad BW = f_{cu} - f_{cl}$ Bandwidth

(10–18) $\quad f_T = A_{v(mid)}BW$ Gain-bandwidth product

FET Frequency Response

(10–19) $\quad f_c = \dfrac{1}{2\pi(R_G \parallel R_{in(gate)})C_1}$ Critical frequency, input circuit

(10–20) $\quad \theta = \tan^{-1}\!\left(\dfrac{X_{C1}}{R_{in}}\right)$ Phase angle, input circuit

(10–21) $\quad f_c = \dfrac{1}{2\pi(R_D + R_L)C_2}$ Critical frequency, output circuit

(10–22) $\quad \theta = \tan^{-1}\!\left(\dfrac{X_{C2}}{R_D + R_L}\right)$ Phase angle, output circuit

(10–23) $\quad C_{gd} = C_{rss}$ Gate-to-drain capacitance

(10–24) $\quad C_{gs} = C_{iss} - C_{rss}$ Gate-to-source capacitance

(10–25) $\quad C_{ds} = C_{oss} - C_{rss}$ Drain-to-source capacitance

(10–26) $\quad C_{in(Miller)} = C_{gd}(A_v + 1)$ Miller input capacitance

(10–27) $\quad C_{out(Miller)} = C_{gd}\!\left(\dfrac{A_v + 1}{A_v}\right)$ Miller output capacitance

(10–28) $\quad f_c = \dfrac{1}{2\pi R_s C_{tot}}$ Critical frequency, input circuit

(10–29) $\quad \theta = \tan^{-1}\!\left(\dfrac{R_s}{X_{C_{tot}}}\right)$ Phase angle, input circuit

(10–30) $\quad f_c = \dfrac{1}{2\pi R_d C_{out(Miller)}}$ Critical frequency, output circuit

(10–31) $\quad \theta = \tan^{-1}\!\left(\dfrac{R_d}{X_{C_{out(Miller)}}}\right)$ Phase angle, output circuit

Multistage Response

(10–32) $\quad f'_{cl} = \dfrac{f_{cl}}{\sqrt{2^{1/n} - 1}}$ Lower critical frequency for case of equal critical frequencies

(10–33) $\quad f'_{cu} = f_{cu}\sqrt{2^{1/n} - 1}$ Upper critical frequency for case of equal critical frequencies

Measurement Techniques

$$(10\text{--}34) \qquad f_{cu} = \frac{0.35}{t_r} \qquad\qquad \text{Upper critical frequency}$$

$$(10\text{--}35) \qquad f_{cl} = \frac{0.35}{t_f} \qquad\qquad \text{Lower critical frequency}$$

■ **SELF-TEST**

1. The low-frequency response of an amplifier is determined in part by
   (a) the voltage gain   (b) the type of transistor
   (c) the supply voltage   (d) the coupling capacitors

2. The high-frequency response of an amplifier is determined in part by
   (a) the gain-bandwidth product   (b) the bypass capacitor
   (c) the internal transistor capacitances   (d) the roll-off

3. The bandwidth of an amplifier is determined by
   (a) the midrange gain   (b) the critical frequencies
   (c) the roll-off rate   (d) the input capacitance

4. The gain of a certain amplifier decreases by 6 dB when the frequency is reduced from 1 kHz to 10 Hz. The roll-off is
   (a) 3 dB/decade   (b) 6 dB/decade
   (c) 3 dB/octave   (d) 6 dB/octave

5. The gain of a particular amplifier at a given frequency decreases by 6 dB when the frequency is doubled. The roll-off is
   (a) 12 dB/decade   (b) 20 dB/decade
   (c) 6 dB/octave   (d) answers (b) and (c)

6. The Miller input capacitance of an amplifier is dependent, in part, on
   (a) the input coupling capacitor   (b) the voltage gain
   (c) the bypass capacitor   (d) none of these

7. An amplifier has the following critical frequencies: 1.2 kHz, 950 Hz, 8 kHz, and 8.5 kHz. The bandwidth is
   (a) 7550 Hz   (b) 7300 Hz   (c) 6800 Hz   (d) 7050 Hz

8. Ideally, the midrange gain of an amplifier
   (a) increases with frequency   (b) decreases with frequency
   (c) remains constant with frequency   (d) depends on the coupling capacitors

9. The frequency at which an amplifier's gain is one is called the
   (a) unity-gain frequency   (b) midrange frequency
   (c) corner frequency   (d) break frequency

10. When the voltage gain of an amplifier is increased, the bandwidth
    (a) is not affected   (b) increases
    (c) decreases   (d) becomes distorted

11. If the f_T of the transistor used in a certain amplifier is 75 MHz and the bandwidth is 10 MHz, the voltage gain must be
    (a) 750   (b) 7.5   (c) 10   (d) 1

12. In the midrange of an amplifier's bandwidth, the peak output voltage is 6 V. At the lower critical frequency, the peak output voltage is
    (a) 3 V   (b) 3.82 V   (c) 8.48 V   (d) 4.24 V

13. At the upper critical frequency, the peak output voltage of a certain amplifier is 10 V. The peak voltage in the midrange of the amplifier is
    (a) 7.07 V   (b) 6.37 V   (c) 14.14 V   (d) 10 V

14. In the step response of a noninverting amplifier, a longer rise time means
    (a) a narrower bandwidth   (b) a lower f_{cl}
    (c) a higher f_{cu} (d) answers (a) and (b)

15. The lower critical frequency of a direct-coupled amplifier with no bypass capacitor is
    (a) variable   (b) 0 Hz
    (c) dependent on the bias   (d) none of these

▪ BASIC PROBLEMS

SECTION 10–1   General Concepts

1. In a capacitively coupled amplifier, the input coupling capacitor and the output coupling capacitor form two of the circuits (along with the respective resistances) that determine the low-frequency response. Assuming that the input and output impedances are the same and neglecting the bypass circuit, which circuit will first cause the gain to drop from its midrange value as the frequency is lowered?

2. Explain why the coupling capacitors do not have a significant effect on gain at sufficiently high-signal frequencies.

3. List the capacitances that affect high-frequency gain in both bipolar and FET amplifiers.

SECTION 10–2   The Decibel

4. A certain amplifier exhibits an output power of 5 W with an input power of 0.5 W. What is the power gain in dB?

5. If the output voltage of an amplifier is 1.2 V rms and its voltage gain is 50, what is the rms input voltage? What is the gain in dB?

6. The midrange voltage gain of a certain amplifier is 65. At a certain frequency beyond midrange, the gain drops to 25. What is the gain reduction in dB?

7. What are the dBm values corresponding to the following power values?
   (a) 2 mW   (b) 1 mW   (c) 4 mW   (d) 0.25 mW

8. Express the midrange voltage gain of the amplifier in Figure 10–54 in decibels. Also express the voltage gain in dB for the critical frequencies.

FIGURE 10–54

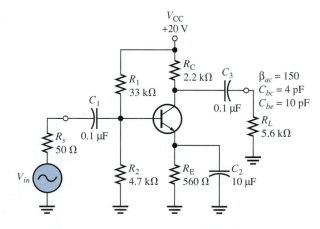

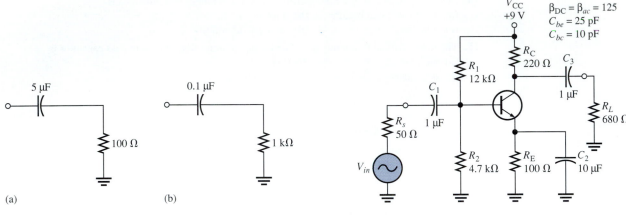

(a)   (b)

FIGURE 10–55   **FIGURE 10–56**

SECTION 10–3   Low-Frequency Amplifier Response

9. Determine the critical frequencies of each RC circuit in Figure 10–55.

10. Determine the critical frequencies associated with the low-frequency response of the amplifier in Figure 10–56. Which is the dominant critical frequency? Sketch the Bode plot.

11. Determine the voltage gain of the amplifier in Figure 10–56 at one-tenth of the dominant critical frequency, at the dominant critical frequency, and at ten times the dominant critical frequency for the low-frequency response.

12. Determine the phase shift at each of the frequencies used in Problem 11.

SECTION 10–4   Miller Capacitance

13. In the amplifier of Figure 10–54 list the capacitors that affect the low-frequency response of the amplifier and those that affect the high-frequency response.

14. Determine the Miller input capacitance in Figure 10–54.

15. Determine the Miller output capacitance in Figure 10–54.

16. Determine the Miller input and output capacitances for the amplifier in Figure 10–57.

17. Determine the high-frequency equivalent circuit for the amplifier in Figure 10–57.

18. Repeat Problems 16 and 17 for the amplifier in Figure 10–58.

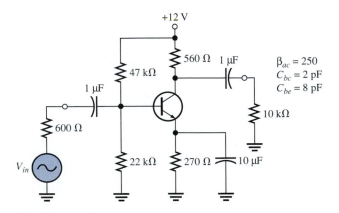

FIGURE 10–57

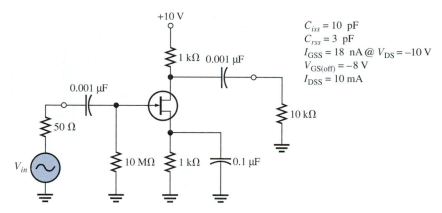

FIGURE 10–58

SECTION 10–5   High-Frequency Amplifier Response

19. Determine the critical frequencies associated with the high-frequency response of the amplifier in Figure 10–56. Identify the dominant critical frequency and sketch the Bode plot.

20. Determine the voltage gain of the amplifier in Figure 10–56 at the following frequencies: $0.1f_c$, f_c, $10f_c$, and $100f_c$, where f_c is the dominant critical frequency in the high-frequency response.

SECTION 10–6   Total Amplifier Frequency Response

21. A particular amplifier has the following low critical frequencies: 25 Hz, 42 Hz, and 136 Hz. It also has high critical frequencies of 8 kHz and 20 kHz. Determine the upper and lower critical frequencies.

22. Determine the bandwidth of the amplifier in Figure 10–56.

23. $f_T = 200$ MHz is taken from the data sheet of a transistor used in a certain amplifier. If the midrange gain is determined to be 38 and if f_{cl} is low enough to be neglected compared to f_{cu}, what bandwidth would you expect? What value of f_{cu} would you expect?

24. If the midrange gain of a given amplifier is 50 dB and therefore 47 dB at f_{cu}, how much gain is there at $2f_{cu}$? At $4f_{cu}$? At $10f_{cu}$?

FIGURE 10–59

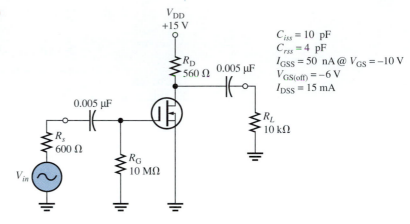

SECTION 10–7   Frequency Response of FET Amplifiers

25. Determine the critical frequencies associated with the low-frequency response of the amplifier in Figure 10–59. Indicate the dominant critical frequency and draw the Bode plot.

26. **(a)** Find the voltage gain of the amplifier in Figure 10–59 at the following frequencies: f_c, $0.1f_c$, and $10f_c$, where f_c is the dominant critical frequency.
   (b) Find the phase shift at each frequency in part (a).

27. The data sheet for the FET in Figure 10–59 gives $C_{rss} = 4$ pF and $C_{iss} = 10$ pF. Determine the critical frequencies associated with the high-frequency response of the amplifier, and indicate the dominant frequency.

28. Determine the voltage gain in dB and the phase shift at each of the following multiples of the dominant critical frequency in Figure 10–59 for the high-frequency response: $0.1f_c$, f_c, $10f_c$, and $100f_c$.

SECTION 10–8   Frequency Response of Multistage Amplifiers

29. In a certain two-stage amplifier, the first stage has critical frequencies of 230 Hz and 1.2 MHz. The second stage has critical frequencies of 195 Hz and 2 MHz. What are the dominant critical frequencies?

30. What is the bandwidth of the two-stage amplifier in Problem 29?

31. Determine the bandwidth of a two-stage amplifier in which each stage has a lower critical frequency of 400 Hz and an upper critical frequency of 800 kHz.

32. What is the dominant lower critical frequency of a three-stage amplifier in which $f_{cl} = 50$ Hz for each stage.

33. In a certain two-stage amplifier, the lower critical frequencies are $f_{cl(1)} = 125$ Hz, $f_{cl(2)} = 125$ Hz, $f_{cu(1)} = 3$ MHz, and $f_{cu(2)} = 2.5$ MHz. Determine the bandwidth.

SECTION 10–9   Frequency Response Measurement

34. In a step response test of a certain amplifier, $t_r = 20$ ns and $t_f = 1$ ms. Determine f_{cl} and f_{cu}.

35. Suppose you are measuring the frequency response of an amplifier with a signal source and an oscilloscope. Assume that the signal level and frequency are set such that the oscilloscope indicates an output voltage level of 5 V rms in the midrange of the amplifier's response. If you wish to determine the upper critical frequency, indicate what you would do and what scope indication you would look for.

36. Determine the approximate bandwidth of an amplifier from the indicated results of the step-response test in Figure 10–60.

FIGURE 10–60

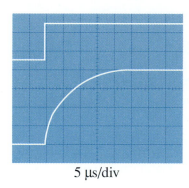 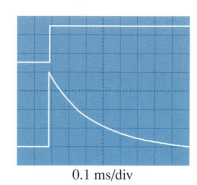

5 μs/div   0.1 ms/div

SECTION 10–10   System Application

37. Determine the dominant lower critical frequency for the amplifier in Figure 10–50 if the coupling capacitors are changed to 1 μF.

38. Does the change in Problems 37 significantly affect the overall bandwidth?

39. How does a change from 10 kΩ to 100 kΩ in load resistance on the final output of the amplifier in Figure 10–50 affect the dominant lower critical frequency?

40. Determine the step response of the amplifier in Figure 10–50 by specifying the rise and fall times on the output.

■ DATA SHEET PROBLEMS

41. Referring to the partial data sheet in Figure 10–51, determine the total input capacitance for a 2N3903 amplifier if the voltage gain is 25.

42. A certain amplifier uses a 2N3904 and has a midrange voltage gain of 50. Referring to the partial data sheet in Figure 10–51, determine its minimum bandwidth.

43. The data sheet for a 2N4351 MOSFET specifies the maximum values of internal capacitances as follows: $C_{iss} = 5$ pF, $C_{rss} = 1.3$ pF, and $C_{d(sub)} = 5$ pF. Determine C_{gd}, C_{gs}, and C_{ds}.

■ ADVANCED PROBLEMS

44. Two single-stage capacitively coupled amplifiers like the one in Figure 10–54 are connected as a two-stage amplifier (R_L is removed from the first stage). Determine whether or not this configuration will operate as a linear amplifier with an input voltage of 10 mV rms. If not, modify the design to achieve maximum gain without distortion.

45. Two stages of the amplifier in Figure 10–59 are connected in cascade. Determine the overall bandwidth.

46. Redesign the amplifier in Figure 10–50 for an adjustable voltage gain of 50 to 500 and a lower critical frequency of 1 kHz.

■ **ANSWERS TO SECTION REVIEWS**

SECTION 10–1

1. The coupling and bypass capacitors affect the low-frequency gain.
2. The high-frequency gain is limited by internal capacitances.
3. Coupling and bypass capacitors can be neglected at frequencies for which their reactances are negligible.

SECTION 10–2

1. +12 dB corresponds to a voltage gain of approximately 4.
2. $A_p = 10 \log(25) = 13.98$ dB
3. 0 dBm corresponds to 1 mW.

SECTION 10–3

1. $f_{c2} = 167$ Hz is dominant.
2. $A_{v(dB)} = 50$ dB $- 3$ dB $= 47$ dB
3. -20 dB attenuation at one decade below f_c.

SECTION 10–4

1. $C_{in(Miller)} = (5 \text{ pF})(51) = 255$ pF
2. $C_{out(Miller)} = (3 \text{ pF})(1.04) = 3.12$ pF

SECTION 10–5

1. The internal transistor capacitances determine the high-frequency response.
2. $C_{in(tot)} = C_{in(Miller)} + C_{ce} = (4 \text{ pF})(81) + 8 \text{ pF} = 342$ pF
3. The input RC circuit dominates.

SECTION 10–6

1. The gain is 1 at f_T.
2. $BW = 25$ kHz $- 100$ Hz $= 24.9$ kHz
3. $A_v = 130$ MHz$/50$ MHz $= 2.6$

SECTION 10–7

1. $\theta = \tan^{-1}(0.5) = 26.6°$
2. $f_c = 1/(2\pi(6500 \ \Omega)(0.002 \ \mu\text{F})) = 12.2$ kHz
3. C_{iss} and C_{rss} are usually specified on the data sheet.
4. $C_{in(total)} = (3 \text{ pF})(26) + 4 \text{ pF} = 82$ pF

SECTION 10–8

1. $f'_{cl} = 1$ kHz
2. $f'_{cu} = 49$ kHz
3. BW decreases.

SECTION 10–9

1. $f_{cl} = 125$ Hz; $f_{cu} = 500$ kHz

2. Rise time is between the 10% and 90% points and fall time is between the 90% and 10% points.

3. $t_r = 140$ ns

4. $t_f = 2.8$

5. Since $f_{cu} \gg f_{cl}$, $BW \cong f_{cu} = 2.5$ MHz.

■ **ANSWERS TO RELATED EXERCISES FOR EXAMPLES**

10–1   **(a)** 61.6 dB   **(b)** 17 dB   **(c)** 102 dB

10–2   **(a)** 50 V   **(b)** 6.25 V   **(c)** 1.56 V

10–3   **(a)** 7.23 Hz   **(b)** 0.707   **(c)** 354

10–4   212 @ 400 Hz; 30 @ 40 Hz; 3 @ 4 Hz

10–5   **(a)** 13.2 Hz   **(b)** 0.479   **(c)** 70.7

10–6   93.4 Hz

10–7   No effect

10–8   153 pF; 3.06 pF

10–9   320 Ω in series with 215 pF, $f_c = 2.31$ MHz

10–10   28.9 MHz

10–11   *BW* decreases; *BW* increases

10–12   20 MHz

10–13 f_{cl} changes from 16.2 Hz to 16.1 Hz.

10–14   Ideally, the low-frequency response is not affected because the input *RC* circuit is unchanged.

10–15   1 pF

10–16 f_c decreases to 83.8 MHz.

10–17   48.2 MHz

10–18   1500 Hz

10–19   39.8 kHz

11

THYRISTORS AND OTHER DEVICES

■ CHAPTER OBJECTIVES

☐ Describe the basic structure and operation of the Shockley diode
☐ Describe the basic structure and operation of an SCR
☐ Discuss several SCR applications
☐ Describe the basic operation of an SCS
☐ Describe the basic structure and operation of diacs and triacs
☐ Describe the basic structure and operation of the UJT
☐ Describe the structure and operation of the PUT
☐ Describe the phototransistor and its operation
☐ Describe the LASCR and its operation
☐ Discuss various types of optical couplers

In this chapter, we examine several devices. Some or all of the topics in this chapter may be considered optional in those electronics programs where these devices will be studied in more depth in a later industrial electronics course. First, we discuss a family of devices that are constructed of four semiconductor layers (*pnpn*) known as thyristors. Thyristors include the Shockley diode, the silicon-controlled rectifier (SCR), the silicon-controlled switch (SCS), the diac, and the triac. These various types of thyristors share certain characteristics in addition to their four-layer construction. They act as open circuits capable of withstanding a certain rated voltage until they are triggered on, they become low-resistance current paths and remain so, even after the trigger is removed, until the current is reduced to a certain level or until they are triggered off, depending on the type of device. Other devices described in this chapter include the unijunction transistor (UJT), the programmable unijunction transistor (PUT), and several optoelectronic devices.

Thyristors can be used to control the amount of ac power to a load and are used in lamp dimmers, motor speed controls, ignition systems, and charging circuits, to name a few. UJTs and PUTs are also used as trigger devices for thyristors and also in oscillators and timing circuits.

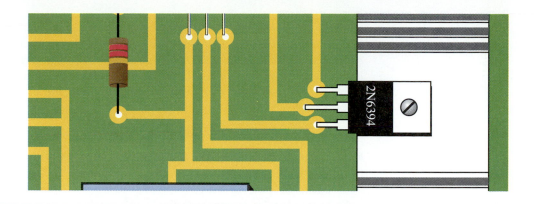

■ SYSTEM APPLICATION

The system application in this chapter is a speed-control system for a production line conveyor. The system senses the number of parts passing a point in a specified period of time and adjusts the rate of movement of the conveyor belt to achieve a desired rate of flow of the parts. The focus is on the conveyor motor speed-control circuit. You will apply the knowledge gained in this chapter to the system application in Section 11–11.

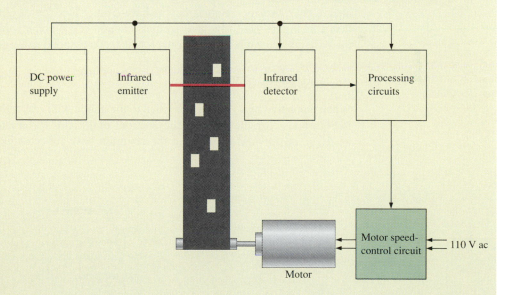

11–1 ■ THE SHOCKLEY DIODE

The Shockley diode is a thyristor with two terminals, the anode and the cathode. It is constructed of four semiconductor layers that form a pnpn structure. The device acts as a switch and remains off until the forward voltage reaches a certain value; then it turns on and conducts. Conduction continues until the current is reduced below a specified value.

After completing this section, you should be able to

■ **Describe the basic structure and operation of the Shockley diode**
   ☐ Identify the Shockley diode symbol
   ☐ Define *forward-breakover voltage*
   ☐ Define *holding current*
   ☐ Define *switching current*
   ☐ Discuss an application

The **Shockley diode** is a **thyristor,** a class of devices constructed of four semiconductor layers. The basic construction of a Shockley diode and its schematic symbol are shown in Figure 11–1. The *pnpn* structure can be represented by an equivalent circuit consisting of a *pnp* transistor and an *npn* transistor, as shown in Figure 11–2(a). The upper *pnp* layers form Q_1 and the lower *npn* layers form Q_2, with the two middle layers shared by both equivalent transistors. Notice that the base-emitter junction of Q_1 corresponds to *pn* junction 1 in Figure 11–1, the base-emitter junction of Q_2 corresponds to *pn* junction 3, and the base-collector junctions of both Q_1 and Q_2 correspond to *pn* junction 2.

Basic Operation

When a positive bias voltage is applied to the anode with respect to the cathode, as shown in Figure 11–2(b), the base-emitter junctions of Q_1 and Q_2 (*pn* junctions 1 and 3 in Figure 11–1(a)) are forward-biased, and the common base-collector junction (*pn* junction 2 in

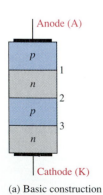

(a) Basic construction

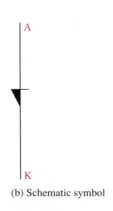

(b) Schematic symbol

FIGURE 11–1
The Shockley diode.

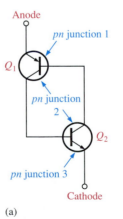

(a)

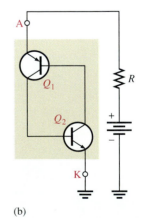

(b)

FIGURE 11–2
Shockley diode equivalent circuit.

FIGURE 11–3

Currents in a basic Shockley diode equivalent circuit.

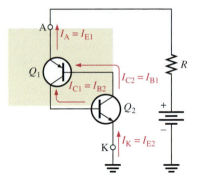

Figure 11–1(a)) is reverse-biased. Therefore, both equivalent transistors are in the linear region.

At low values of forward-bias voltage, an expression for the anode current, I_A, is developed as follows, using the standard transistor relationships and Figure 11–3. In this analysis, the component of leakage current, I_{CBO}, is taken into account.

$$I_{B1} = I_{E1} - I_{C1} - I_{CBO1}$$

Since $I_C = \alpha_{DC} I_E$,

$$I_{B1} = I_{E1} - \alpha_{DC} I_{E1} - I_{CBO1} = (1 - \alpha_{DC1}) I_{E1} - I_{CBO1}$$

The anode current, I_A, is the same as I_{E1}, and therefore

$$I_{B1} = (1 - \alpha_{DC1}) I_A - I_{CBO1} \tag{11–1}$$

$$I_{C2} = \alpha_{DC2} I_{E2} + I_{CBO2}$$

The cathode current, I_K, is the same as I_{E2}, and therefore

$$I_{C2} = \alpha_{DC2} I_K + I_{CBO2} \tag{11–2}$$

Since $I_{C2} = I_{B1}$,

$$\alpha_{DC2} I_K + I_{CBO2} = (1 - \alpha_{DC1}) I_A - I_{CBO1}$$

I_A and I_K are equal. Substituting I_A for I_K in the above equation and solving for I_A, you get

$$\alpha_{DC2} I_A + I_{CBO2} = (1 - \alpha_{DC1}) I_A - I_{CBO1}$$

$$\alpha_{DC2} I_A - (1 - \alpha_{DC1}) I_A = -I_{CBO1} - I_{CBO2}$$

$$I_A[(1 - \alpha_{DC1}) - \alpha_{DC2}] = I_{CBO1} + I_{CBO2}$$

$$I_A = \frac{I_{CBO1} + I_{CBO2}}{1 - (\alpha_{DC1} + \alpha_{DC2})} \tag{11–3}$$

At low-current levels, the transistor alpha (α_{DC}) is very small. Therefore, at low-bias levels, there is very little anode current in the Shockley diode as Equation (11–3) demonstrates, and thus it is in the *off* state or forward-blocking region.

EXAMPLE 11–1

A certain Shockley diode is biased in the forward-blocking region with an anode-to-cathode voltage of 20 V. Under this bias condition, $\alpha_{DC1} = 0.35$ and $\alpha_{DC2} = 0.45$. The leakage currents are 100 nA. Determine the anode current and the forward resistance of the diode.

Solution   Use Equation (11–3) to determine the anode current.

$$I_A = \frac{I_{CBO1} + I_{CBO2}}{1 - (\alpha_{DC1} + \alpha_{DC2})} = \frac{200 \text{ nA}}{1 - 0.8} = \frac{200 \text{ nA}}{0.2} = 1 \text{ μA}$$

This is the forward current when the device is off, but forward-biased with $V_{AK} = +20$ V. The forward resistance is, therefore,

$$R_{AK} = \frac{V_{AK}}{I_A} = \frac{20 \text{ V}}{1 \text{ μA}} = 20 \text{ MΩ}$$

Related Exercise   If the anode current is 2 μA and $V_{AK} = 20$ V, what is the Shockley diode's forward resistance in the forward-blocking region?

Forward-Breakover Voltage   The operation of the Shockley diode may seem strange because it is forward-biased, yet it acts essentially as an open switch. As previously mentioned, there is a region of forward bias, called the *forward-blocking region,* in which the device has a very high forward resistance (ideally an open) and is in the *off* state. The forward-blocking region exists from $V_{AK} = 0$ V up to a value of V_{AK} called the **forward-breakover voltage, $V_{BR(F)}$.** This is indicated on the Shockley diode characteristic curve in Figure 11–4.

FIGURE 11–4
Shockley diode characteristic curve.

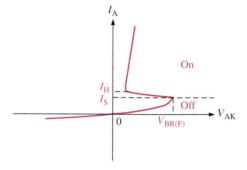

As V_{AK} is increased from 0, the anode current, I_A, gradually increases, as shown on the graph. As I_A increases, so do α_{DC1} and α_{DC2} increase. At the point where $\alpha_{DC1} + \alpha_{DC2} = 1$, the denominator in Equation (11–3) becomes zero, and $I_A = I_S$, the switching current. At this point, $V_{AK} = V_{BR(F)}$, and the internal transistor structures become saturated. When this happens, the forward voltage drop V_{AK} suddenly decreases to a low value approximately equal to $V_{BE} + V_{CE(sat)}$ as I_A increases, and the Shockley diode enters the forward-conduction region as indicated in Figure 11–4. Now, the device is in the *on* state and acts as a closed switch. When the anode current drops back below the holding

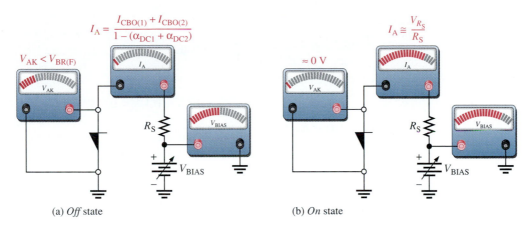

FIGURE 11–5
On/off states of the Shockley diode.

value, I_H, the device turns off. The *on/off* states of the Shockley diode are illustrated in Figure 11–5.

Holding Current   Once the Shockley diode is conducting (in the *on* state), it will continue to conduct until the anode current is reduced below a specified level, called the **holding current, I_H.** This parameter is also indicated on the characteristic curve in Figure 11–4. When I_A falls below I_H, the device rapidly switches back to the *off* state and enters the forward-blocking region.

Switching Current   The value of the anode current at the point where the device switches from the forward-blocking region (off) to the forward-conduction region (on) is called the **switching current, I_S.** This value of current is always less than the holding current, I_H.

EXAMPLE 11–2

(a) Determine the value of anode current in Figure 11–6(a), when the device is off and the dc alphas are 0.4. The leakage currents are 0.08 μA each.

(b) Determine the value of anode current in Figure 11–6(b) when the device is on. $V_{BR(F)} = 40$ V. Assume $V_{BE} = 0.7$ V and $V_{CE(sat)} = 0.1$ V for the internal transistor structure.

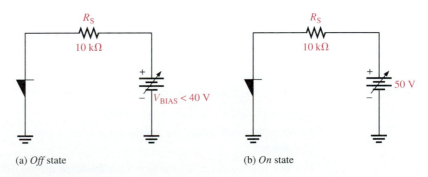

FIGURE 11–6

Solution

(a) $I_A = \dfrac{I_{CBO1} + I_{CBO2}}{1 - (\alpha_{DC1} + \alpha_{DC2})} = \dfrac{0.16\ \mu A}{1 - 0.8} = 0.8\ \mu A$

(b) The voltage at the anode is

$$V_A = V_{BE} + V_{CE(sat)} = 0.7\ V + 0.1\ V = 0.8\ V$$

The voltage across R_S is

$$V_{R_S} = V_{BIAS} - V_A = 50\ V - 0.8\ V = 49.2\ V$$

The anode current is

$$I_A = \frac{V_{R_S}}{R_S} = \frac{49.2\ V}{10\ k\Omega} = 4.92\ mA$$

Related Exercise What is the forward resistance of the Shockley diode in Figure 11–6(b)?

An Application

The circuit in Figure 11–7(a) is a relaxation **oscillator.** The operation is as follows. When the switch is closed, the capacitor charges through R until its voltage reaches the forward-breakover voltage of the Shockley diode. At this point the diode switches into conduction, and the capacitor rapidly discharges through the diode. Discharging continues until the current through the diode falls below the holding value. At this point, the diode switches back to the *off* state, and the capacitor begins to charge again. The result of this action is a voltage waveform across C like that shown in Figure 11–7(b).

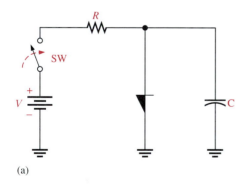

(a)

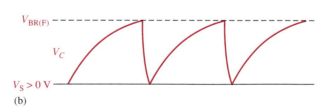

(b)

FIGURE 11–7
Shockley diode relaxation oscillator.

SECTION 11–1 REVIEW

1. Why is the Shockley diode classified as a thyristor?
2. What is the forward-blocking region?
3. What happens when the anode-to-cathode voltage exceeds the forward-breakover voltage?
4. Once it is on, how can the Shockley diode be turned off?

11–2 ■ THE SILICON-CONTROLLED RECTIFIER (SCR)

The silicon-controlled rectifer (SCR) is another four-layer pnpn device similar to the Shockley diode except with three terminals: anode, cathode, and gate. Like the Shockley diode, the SCR has two possible states of operation. In the off state, it acts ideally as an open circuit between the anode and the cathode; actually, rather than an open, there is a very high resistance. In the on state, the SCR acts ideally as a short from the anode to the cathode; actually, there is a small on (forward) resistance. The SCR is used in many applications, including motor controls, time-delay circuits, heater controls, phase controls, and relay controls, to name a few.

After completing this section, you should be able to

■ **Describe the basic structure and operation of an SCR**
- ☐ Identify an SCR by the schematic symbol
- ☐ Draw the bipolar equivalent circuit of an SCR
- ☐ Explain how to turn an SCR on and off
- ☐ Explain the characteristic curves of an SCR
- ☐ Define *forced commutation*
- ☐ Define various SCR parameters

The basic structure of a **silicon-controlled rectifier (SCR)** is shown in Figure 11–8(a) and the schematic symbol is shown in Figure 11–8(b). Typical SCR packages are shown in Figure 11–8(c). Other types of thyristors are found in the same or similar packages.

FIGURE 11–8
The silicon-controlled rectifier (SCR).

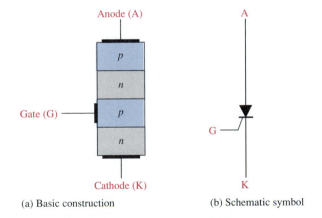

(a) Basic construction   (b) Schematic symbol

(c) Typical packages

SCR Equivalent Circuit

Like the Shockley diode operation, the SCR operation can best be understood by thinking of its internal *pnpn* structure as a two-transistor arrangement, as shown in Figure 11–9. This structure is like that of the Shockley diode except for the gate connection. The upper *pnp* layers act as a transistor, Q_1, and the lower *npn* layers act as a transistor, Q_2. Again, notice that the two middle layers are "shared."

FIGURE 11–9
SCR equivalent circuit.

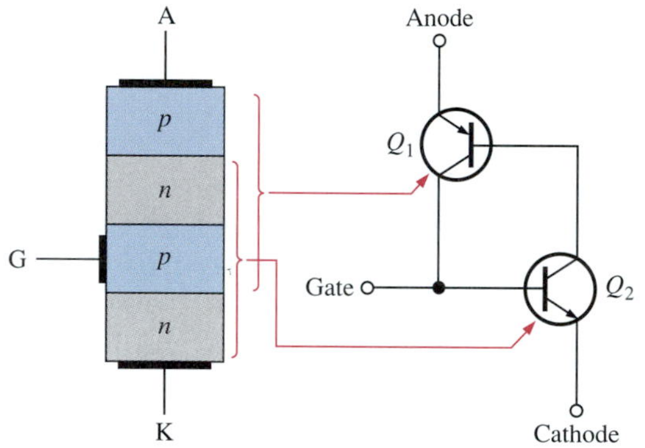

Turning the SCR On

When the gate current, I_G, is zero, as shown in Figure 11–10(a), the device acts as a Shockley diode in the *off* state. In this state, the very high resistance between the anode and cathode can be approximated by an open switch, as indicated. When a positive pulse of current (**trigger**) is applied to the gate, both transistors turn on (the anode must be more positive than the cathode). This action is shown in Figure 11–10(b). I_{B2} turns on Q_2, providing a path for I_{B1} out of the Q_2 collector, thus turning on Q_1. The collector current of Q_1 provides additional base current for Q_2 so that Q_2 stays in conduction after the trigger pulse is removed from the gate. By this regenerative action, Q_2 sustains the saturated conduction of Q_1 by providing a path for I_{B1}; in turn, Q_1 sustains the saturated conduction of Q_2 by providing I_{B2}. Thus, the device stays on (latches) once it is triggered on, as shown in Figure 11–10(c). In this state, the very low resistance between the anode and cathode can be approximated by a closed switch, as indicated.

Like the Shockley diode, an SCR can also be turned on without gate triggering by increasing anode-to-cathode voltage to a value exceeding the forward-breakover voltage $V_{BR(F)}$, as shown on the characteristic curve in Figure 11–11(a). The forward-breakover voltage decreases as I_G is increased above 0 V, as shown by the set of curves in Figure 11–11(b). Eventually, a value of I_G is reached at which the SCR turns on at a very low anode-to-cathode voltage. So, as you can see, the gate current controls the value of forward voltage $V_{BR(F)}$ required for turn-on.

Although anode-to-cathode voltages in excess of $V_{BR(F)}$ will not damage the device if current is limited, this situation should be avoided because the normal control of the SCR is lost. It should normally be triggered on only with a pulse at the gate.

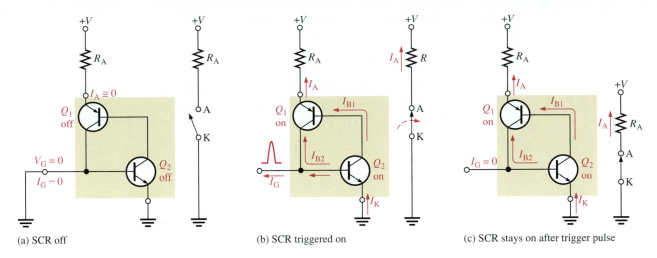

(a) SCR off

(b) SCR triggered on

(c) SCR stays on after trigger pulse

FIGURE 11–10

The SCR turn-on process with the switch equivalents shown.

FIGURE 11–11

SCR characteristic curves.

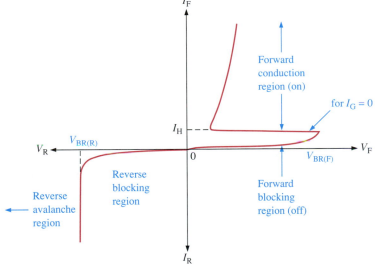

(a) For $I_G = 0$

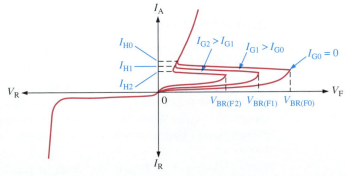

(b) For various I_G values

Turning the SCR Off

When the gate returns to 0 V after the trigger pulse is removed, the SCR cannot turn off; it stays in the forward-conduction region. The anode current must drop below the value of the holding current, I_H, in order for turn-off to occur. The holding current is indicated in Figure 11–11.

There are two basic methods for turning off an SCR: *anode current interruption* and *forced commutation*. The anode current can be interrupted by either a momentary series or parallel switching arrangement, as shown in Figure 11–12. The series switch in part (a) simply reduces the anode current to zero and causes the SCR to turn off. The parallel switch in part (b) routes part of the total current away from the SCR, thereby reducing the anode current to a value less than I_H.

The **forced commutation** method basically requires momentarily forcing current through the SCR in the direction opposite to the forward conduction so that the net forward current is reduced below the holding value. The basic circuit, as shown in Figure 11–13, consists of a switch (normally a transistor switch) and a battery in parallel with the SCR. While the SCR is conducting, the switch is open, as shown in part (a). To turn off the SCR, the switch is closed, placing the battery across the SCR and forcing current through it opposite to the forward current, as shown in part (b). Typically, turn-off times for SCRs range from a few microseconds up to about 30 μs.

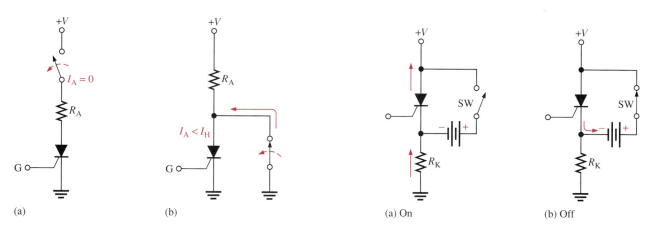

FIGURE 11–12
SCR turn-off by anode current interruption.

FIGURE 11–13
SCR turn-off by forced commutation.

SCR Characteristics and Ratings

Several of the most important SCR characteristics and ratings are defined as follows. Use the curve in Figure 11–11(a) for reference where appropriate.

Forward-breakover voltage, $V_{BR(F)}$ This is the voltage at which the SCR enters the forward-conduction region. The value of $V_{BR(F)}$ is maximum when $I_G = 0$ and is designated $V_{BR(F0)}$. When the gate current is increased, $V_{BR(F)}$ decreases and is designated $V_{BR(F1)}$, $V_{BR(F2)}$, and so on, for increasing steps in gate current (I_{G1}, I_{G2}, and so on).

Holding current, I_H This is the value of anode current below which the SCR switches from the forward-conduction region to the forward-blocking region. The value increases with decreasing values of I_G and is maximum for $I_G = 0$.

Gate trigger current, I_{GT}   This is the value of gate current necessary to switch the SCR from the forward-blocking region to the forward-conduction region under specified conditions.

Average forward current, $I_{F(avg)}$ This is the maximum continuous anode current (dc) that the device can withstand in the conduction state under specified conditions.

Forward-conduction region   This region corresponds to the *on* condition of the SCR where there is forward current from cathode to anode through the very low resistance (approximate short) of the SCR.

Forward- and reverse-blocking regions   These regions correspond to the *off* condition of the SCR where the forward current from cathode to anode is blocked by the effective open circuit of the SCR.

Reverse-breakdown voltage, $V_{BR(R)}$ This parameter specifies the value of reverse voltage from cathode to anode at which the device breaks into the avalanche region and begins to conduct heavily (the same as in a *pn* junction diode).

SECTION 11–2 REVIEW

1. What is an SCR?
2. Name the SCR terminals.
3. How can an SCR be turned on (made to conduct)?
4. How can an SCR be turned off?

11–3 ▪ SCR APPLICATIONS

The SCR has many uses in the areas of power control and switching applications. A few of the basic applications are described in this section.

After completing this section, you should be able to

■ **Discuss several SCR applications**
  ☐ Explain how an SCR is used to control current
  ☐ Describe half-wave power control
  ☐ Explain a basic phase control circuit
  ☐ Discuss the function of an SCR in a lighting system for power interruptions
  ☐ Explain an over-voltage protection or "crowbar" circuit

On-Off Control of Current

Figure 11–14 shows an SCR circuit that permits current to be switched to a load by the momentary closure of switch SW1 and removed from the load by the momentary closure of switch SW2.

Assuming the SCR is initially off, momentary closure of SW1 provides a pulse of current from the gate, thus triggering the SCR on so that it conducts current through R_L. The SCR remains in conduction even after the momentary contact of SW1 is removed. When SW2 is momentarily closed, current is shunted around the SCR, thus reducing its

FIGURE 11–14
On-off SCR control circuit.

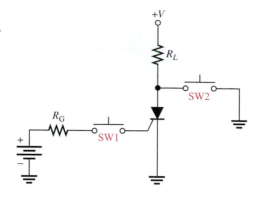

anode current below the holding value I_H. This turns the SCR off and thus reduces the load current to zero.

Half-Wave Power Control

A common application of SCRs is in the control of ac power for lamp dimmers, electric heaters, and electric motors.

A half-wave, variable-resistance, phase-control circuit is shown in Figure 11–15; 120 V ac are applied across terminals A and B; R_L represents the resistance of the load (for example, a heating element or lamp filament). R_1 is a current-limiting resistor, and potentiometer R_2 sets the trigger level for the SCR. By adjusting R_2, the SCR can be made to trigger at any point on the positive half-cycle of the ac waveform between 0° and 90° as shown in Figure 11–16.

FIGURE 11–15
Half-wave, variable-resistance, phase-control circuit.

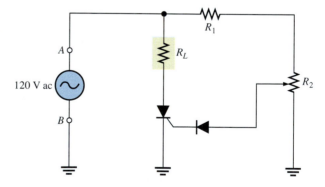

When the SCR triggers near the beginning of the cycle (approximately 0°), as in Figure 11–16(a), it conducts for approximately 180° and maximum power is delivered to the load. When it fires near the peak of the positive half-cycle (90°), as in Figure 11–16(b), the SCR conducts for approximately 90° and less power is delivered to the load. By adjusting R_2, triggering can be made to occur anywhere between these two extremes, and therefore, a variable amount of power can be delivered to the load. Figure 11–16(c) shows triggering at the 45° point as an example. When the ac input goes negative, the SCR turns off and does not conduct again until the trigger point on the next positive half-cycle. The diode prevents the negative ac voltage from being applied to the gate of the SCR.

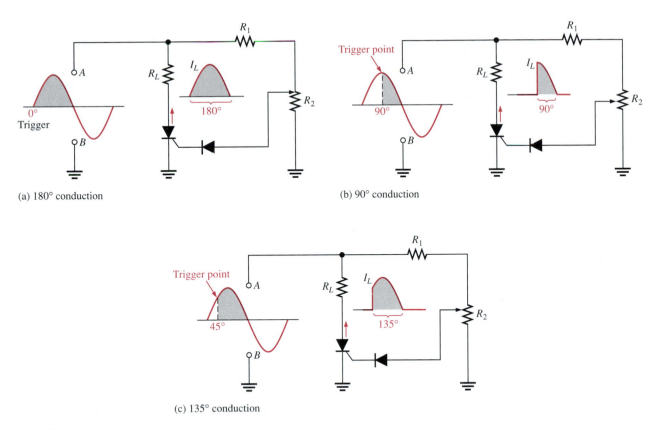

(a) 180° conduction

(b) 90° conduction

(c) 135° conduction

FIGURE 11–16
Operation of the phase-control circuit.

Lighting System for Power Interruptions

As another example of SCR applications, we will examine a circuit that will maintain lighting by using a backup battery when there is an ac power failure. Figure 11–17 shows a center-tapped full-wave rectifier used for providing ac power to a low-voltage lamp. As long as the ac power is available, the battery charges through diode D_3 and R_1.

The SCR's cathode voltage is established when the capacitor charges to the peak value of the full-wave rectified ac (6.3 V rms less the drops across R_2 and D_1). The anode is at the 6 V battery voltage, making it less positive than the cathode, thus preventing conduction. The SCR's gate is at a voltage established by the voltage divider made up of R_2 and R_3. Under these conditions the lamp is illuminated by the ac input power and the SCR is off, as shown in Figure 11–17(a).

When there is an interruption of ac power, the capacitor discharges through the closed path D_3, R_1, and R_3, making the cathode less positive than the anode or the gate. This action establishes a triggering condition, and the SCR begins to conduct. Current from the battery is through the SCR and the lamp, thus maintaining illumination, as shown in Figure 11–17(b). When ac power is restored, the capacitor recharges and the SCR turns off. The battery begins recharging.

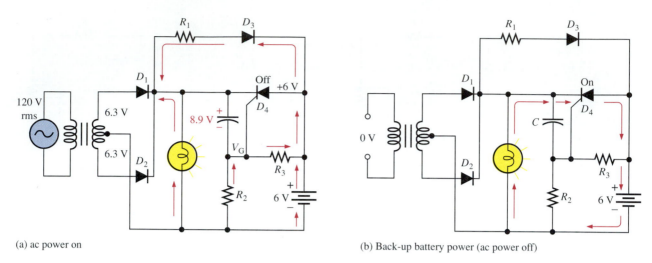

(a) ac power on

(b) Back-up battery power (ac power off)

FIGURE 11–17
Automatic back-up lighting circuit.

An Over-Voltage Protection Circuit

Figure 11–18 shows a simple over-voltage protection circuit, sometimes called a "crow-bar" circuit, in a dc power supply. The dc output voltage from the regulator is monitored by the zener diode D_1 and the resistive voltage divider (R_1 and R_2). The upper limit of the output voltage is set by the zener voltage. If this voltage is exceeded, the zener conducts and the voltage divider produces an SCR trigger voltage. The trigger voltage fires the SCR, which is connected across the line voltage. The SCR current causes the fuse to blow, thus disconnecting the line voltage from the power supply.

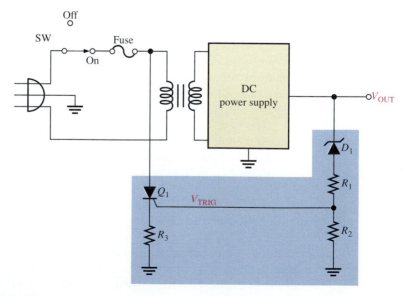

FIGURE 11–18
A basic SCR over-voltage protection circuit.

11–4 ■ THE SILICON-CONTROLLED SWITCH (SCS)

The silicon-controlled switch (SCS) is similar in construction to the SCR. The SCS, however, has two gate terminals, the cathode gate and the anode gate. The SCS can be turned on and off using either gate terminal. Remember that the SCR can be only turned on using its gate terminal. Normally, the SCS is available in power ratings lower than those of the SCR.

After completing this section, you should be able to

■ **Describe the basic operation of an SCS**
  □ Identify an SCS by its schematic symbol
  □ Use a bipolar equivalent circuit to describe SCS operation
  □ Compare the SCS to the SCR

The symbol and terminal identification for the **silicon-controlled switch (SCS)** are shown in Figure 11–19.

As with the previous four-layer devices, the basic operation of the SCS can be understood by referring to the transistor equivalent, shown in Figure 11–20. To start, assume that both Q_1 and Q_2 are off, and therefore that the SCS is not conducting. A positive pulse on the cathode gate drives Q_2 into conduction and thus provides a path for Q_1 base current. When Q_1 turns on, its collector current provides base current for Q_2, thus sustaining the *on* state of the device. This regenerative action is the same as in the turn-on process of the SCR and the Shockley diode and is illustrated in Figure 11–20(a).

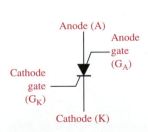

FIGURE 11–19
The silicon-controlled switch (SCS).

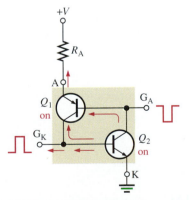

(a) *Turn-on*: Positive pulse on G_K *or* negative pulse on G_A

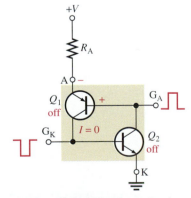

(b) *Turn-off*: Positive pulse on G_A *or* negative pulse on G_K

FIGURE 11–20
SCS operation.

The SCS can also be turned on with a negative pulse on the anode gate, as indicated in Figure 11–20(a). This drives Q_1 into conduction which, in turn, provides base current for Q_2. Once Q_2 is on, it provides a path for Q_1 base current, thus sustaining the *on* state.

To turn the SCS off, a positive pulse is applied to the anode gate. This reverse-biases the base-emitter junction of Q_1 and turns it off. Q_2, in turn, cuts off and the SCS ceases conduction, as shown in Figure 11–20(b). The device can also be turned off with a negative pulse on the cathode gate, as indicated in part (b). The SCS typically has a faster turn-off time than the SCR.

In addition to the positive pulse on the anode gate or the negative pulse on the cathode gate, there are other methods for turning off an SCS. Figure 11–21(a) and (b) shows two switching methods to reduce the anode current below the holding value. In each case, the bipolar transistor acts as a switch.

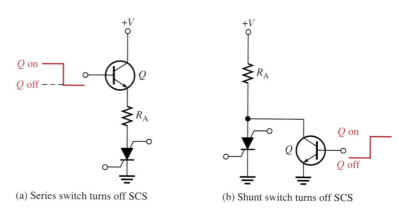

(a) Series switch turns off SCS   (b) Shunt switch turns off SCS

FIGURE 11–21
Transistor switches reduce I_A below I_H and turn off the SCS.

Applications

The SCS and SCR are used in similar applications. The SCS has the advantage of faster turn-off with pulses on either gate terminal; however, it is more limited in terms of maximum current and voltage ratings. Also, the SCS is sometimes used in digital applications such as counters, registers, and timing circuits.

SECTION 11–4 REVIEW	
	1. Explain the difference between an SCS and an SCR.
	2. How can an SCS be turned on?
	3. Describe four ways an SCS can be turned off.

11–5 ■ THE DIAC AND TRIAC

Both the diac and the triac are types of thyristors that can conduct current in both directions (bilateral). The difference between the two devices is that the diac has two terminals, while the triac has a third terminal, which is the gate. The diac functions

basically like two parallel Shockley diodes turned in opposite directions. The triac functions basically like two parallel SCRs turned in opposite directions with a common gate terminal.

After completing this section, you should be able to

▪ **Describe the basic structure and operation of diacs and triacs**
  ☐ Identify a diac or triac by the schematic symbol
  ☐ Discuss the equivalent circuit and bias conditions
  ☐ Explain the characteristic curve
  ☐ Discuss an application

The Diac

The **diac** basic construction and schematic symbol are shown in Figure 11–22. Notice that there are two terminals, labelled A_1 and A_2. Conduction occurs in the diac when the breakover voltage is reached with either polarity across the two terminals. The curve in Figure 11–23 illustrates this characteristic. Once breakover occurs, current is in a direction depending on the polarity of the voltage across the terminals. The device turns off when the current drops below the holding value.

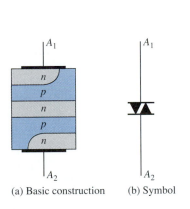

(a) Basic construction   (b) Symbol

FIGURE 11–22
The diac.

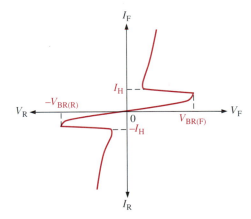

FIGURE 11–23
Diac characteristic curve.

The equivalent circuit of a diac consists of four transistors arranged as shown in Figure 11–24(a). When the diac is biased as in Figure 11–24(b), the *pnpn* structure from A_1 to A_2 provides the four-layer device operation as was described for the Shockley diode. In the equivalent circuit, Q_1 and Q_2 are forward-biased, and Q_3 and Q_4 are reverse-biased. The device operates on the upper right portion of the characteristic curve in Figure 11–23 under this bias condition. When the diac is biased as shown in Figure 11–24(c), the *pnpn* structure from A_2 and A_1 is used. In the equivalent circuit, Q_3 and Q_4 are forward-biased, and Q_1 and Q_2 are reverse-biased. Under this bias condition, the device operates on the lower left portion of the characteristic curve, as shown in Figure 11–23.

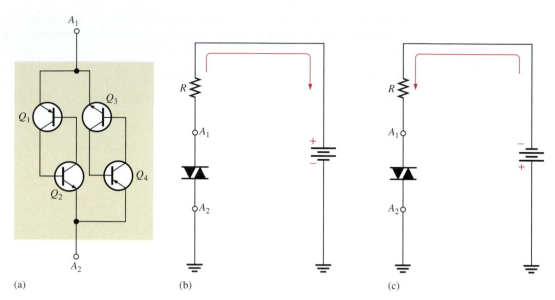

(a) (b) (c)

FIGURE 11–24

Diac equivalent circuit and bias conditions.

The Triac

The **triac** is like a diac with a gate terminal. The triac can be turned on by a pulse of gate current and does not require the breakover voltage to initiate conduction, as does the diac. Basically, the triac can be thought of simply as two SCRs connected in parallel and in opposite directions with a common gate terminal. Unlike the SCR, the triac can conduct current in either direction when it is triggered on, depending on the polarity of the voltage across its A_1 and A_2 terminals. Figure 11–25(a) and (b) shows the basic construction and schematic symbol for the triac. The characteristic curve is shown in Figure 11–26. Notice that the breakover potential decreases as the gate current increases, just as with the SCR.

As with other four-layer devices, the triac ceases to conduct when the anode current drops below the specified value of the holding current, I_H. The only way to turn off the triac is to reduce the current to a sufficiently low level.

Figure 11–27 shows the triac being triggered into both directions of conduction. In part (a), terminal A_1 is biased positive with respect to A_2, so the triac conducts as shown when triggered by a positive pulse at the gate terminal. The transistor equivalent circuit in

FIGURE 11–25

The triac.

(a) Basic construction

(b) Symbol

FIGURE 11–26
Triac characteristic curves.

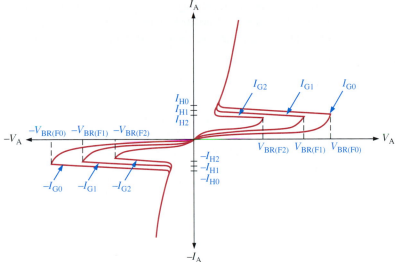

FIGURE 11–27
Bilateral operation of a triac.

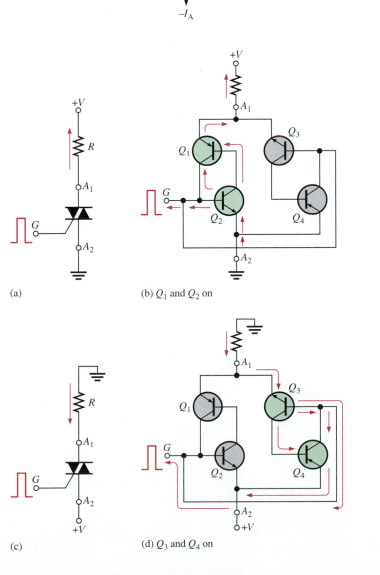

(a)

(b) Q_1 and Q_2 on

(c)

(d) Q_3 and Q_4 on

part (b) shows that Q_1 and Q_2 conduct when a positive trigger pulse is applied. In part (c), terminal A_2 is biased positive with respect to A_1, so the triac conducts as shown. In this case, Q_3 and Q_4 conduct as indicated in part (d) upon application of a positive trigger pulse.

Applications

Like the SCR, triacs are also used to control average power to a load by the method of phase control. The triac can be triggered such that the ac power is supplied to the load for a controlled portion of each half-cycle. During each positive half-cycle of the ac, the triac is off for a certain interval, called the *delay angle* (measured in degrees), and then it is triggered on and conducts current through the load for the remaining portion of the positive half-cycle, called the *conduction angle*. Similar action occurs on the negative half-cycle except that, of course, current is conducted in the opposite direction through the load. Figure 11–28 illustrates this action.

One example of phase control using a triac is illustrated in Figure 11–29(a). Diodes are used to provide trigger pulses to the gate of the triac. Diode D_1 conducts during the positive half-cycle. The value of R_1 sets the point on the positive half-cycle at which the triac triggers. Notice that during this portion of the ac cycle, A_1 and G are positive with respect to A_2.

FIGURE 11–28
Basic triac phase control.

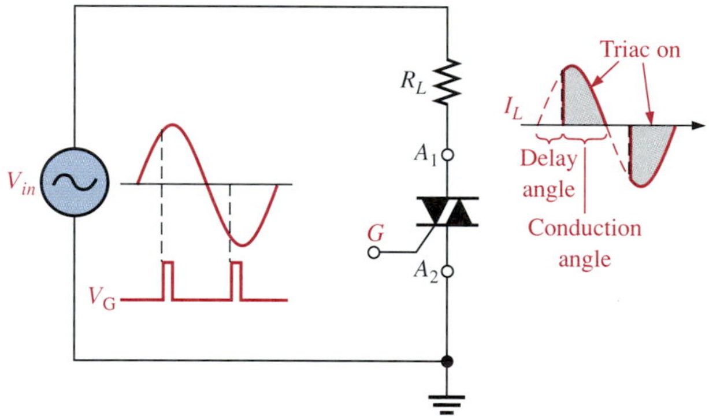

FIGURE 11–29
Triac phase-control circuit.

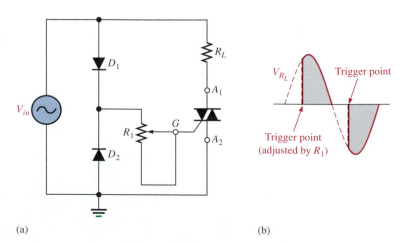

(a)   (b)

Diode D_2 conducts during the negative half-cycle, and R_1 sets the trigger point. Notice that during this portion of the ac cycle, A_2 and G are positive with respect to A_1. The resulting waveform across R_L is shown in Figure 11–29(b).

In the phase-control circuit, it is necessary that the triac turn off at the end of each positive and each negative alternation of the ac. Figure 11–30 illustrates that there is an interval near each 0 crossing where the triac current drops below the holding value, thus turning the device off.

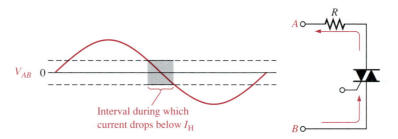

FIGURE 11–30
Triac turn-off interval.

11–6 ■ THE UNIJUNCTION TRANSISTOR (UJT)

The unijunction transistor does not belong to the thyristor family because it does not have a four-layer type of construction. The term unijunction refers to the fact that the UJT has a single pn junction. As you will see in this section, the UJT is useful in certain oscillator applications and as a triggering device in thyristor circuits.

After completing this section, you should be able to

■ **Describe the basic structure and operation of the UJT**
  □ Identify the UJT by its symbol
  □ Draw the equivalent circuit
  □ Explain why a UJT is not a thyristor
  □ Define *standoff ratio*
  □ Analyze the operation of a UJT relaxation oscillator

The **unijunction transistor (UJT)** is a three-terminal device whose basic construction is shown in Figure 11–31(a); the schematic symbol appears in Figure 11–31(b). Notice the terminals are labelled Emitter (E), Base 1 (B_1), and Base 2 (B_2). Do not confuse this symbol with that of a JFET; *the difference is that the arrow is at an angle for the UJT.* The

FIGURE 11–31

The unijunction transistor (UJT).

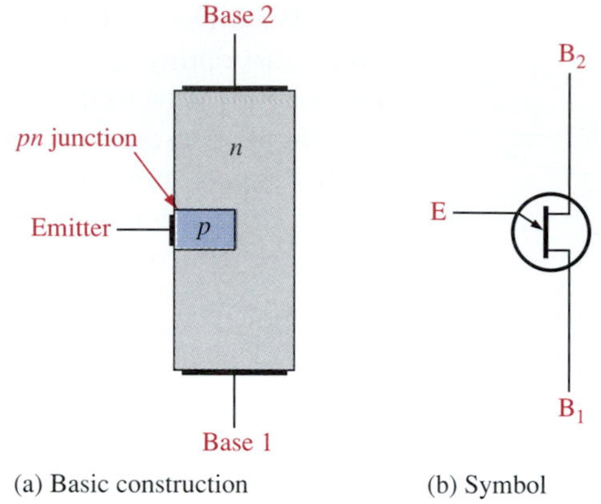

(a) Basic construction   (b) Symbol

UJT has only one *pn* junction, and therefore, the characteristics of this device are very different from those of either the bipolar junction transistor or the FET, as you will see.

Equivalent Circuit

The equivalent circuit for the UJT, shown in Figure 11–32(a), will aid in understanding the basic operation. The diode shown in the figure represents the *pn* junction, r'_{B1} represents the internal dynamic resistance of the silicon bar between the emitter and base 1, and r'_{B2} represents the dynamic resistance between the emitter and base 2. The sum $r'_{B1} + r'_{B2}$ is the total resistance between the base terminals and is called the *interbase resistance*, r'_{BB}.

$$r'_{BB} = r'_{B1} + r'_{B2}$$

The value of r'_{B1} varies inversely with emitter current I_E, and therefore, it is shown as a variable resistor. Depending on I_E, the value of r'_{B1} can vary from several thousand ohms down to tens of ohms. The internal resistances r'_{B1} and r'_{B2} form a voltage divider

FIGURE 11–32

UJT equivalent circuit.

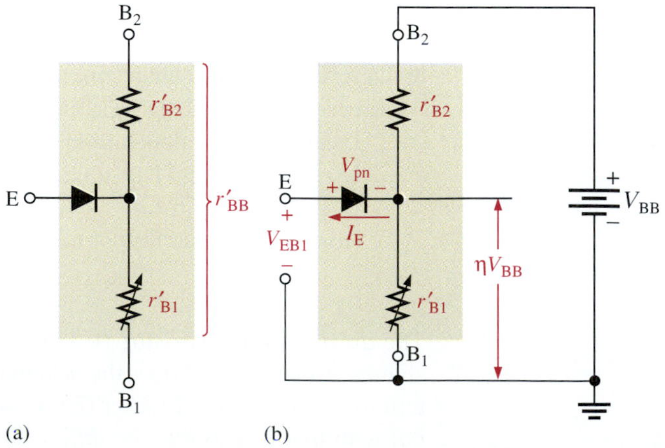

(a)   (b)

when the device is biased as shown in Figure 11–32(b). The voltage across the resistance r'_{B1} can be expressed as

$$V_{r'_{B1}} = \left(\frac{r'_{B1}}{r'_{BB}}\right)V_{BB}$$

Standoff Ratio

The ratio r'_{B1}/r'_{BB} is a UJT characteristic called the intrinsic **standoff ratio** and designated by η (Greek *eta*).

$$\eta = \frac{r'_{B1}}{r'_{BB}} \qquad\qquad \textbf{(11–4)}$$

As long as the applied emitter voltage V_{EB1} is less than $V_{r'_{B1}} + V_{pn}$, there is no emitter current because the *pn* junction is not forward-biased (V_{pn} is the barrier potential of the *pn* junction). The value of emitter voltage that causes the *pn* junction to become forward-biased is called V_P (peak-point voltage) and is expressed as

$$V_P = \eta V_{BB} + V_{pn} \qquad\qquad \textbf{(11–5)}$$

When V_{EB1} reaches V_P, the *pn* junction becomes forward-biased and I_E begins. Holes are injected into the *n*-type bar from the *p*-type emitter. This increase in holes causes an increase in free electrons, thus increasing the conductivity between emitter and B_1 (decreasing r'_{B1}).

After turn-on, the UJT operates in a negative resistance region up to a certain value of I_E, as shown by the characteristic curve in Figure 11–33. As you can see, after the peak point ($V_E = V_P$ and $I_E = I_P$), V_E decreases as I_E continues to increase, thus producing the negative resistance characteristic. Beyond the valley point ($V_E = V_V$ and $I_E = I_V$), the device is in saturation, and V_E increases very little with an increasing I_E.

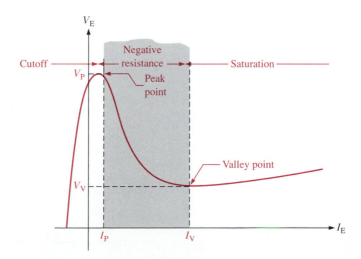

FIGURE 11–33
UJT characteristic curve for a fixed value of V_{BB}.

EXAMPLE 11–3   The data sheet of a certain UJT gives $\eta = 0.6$. Determine the peak-point emitter voltage V_P if $V_{BB} = 20$ V.

Solution $V_P = \eta V_{BB} + V_{pn} = 0.6(20 \text{ V}) + 0.7 \text{ V} = 12.7 \text{ V}$

Related Exercise How can the peak-point emitter voltage of a UJT be increased?

A UJT Application

The UJT can be used as a trigger device for SCRs and triacs. Other applications include nonsinusoidal oscillators, sawtooth generators, phase control, and timing circuits. Figure 11–34 shows a UJT relaxation oscillator as an example of one application.

The operation is as follows. When dc power is applied, the capacitor C charges exponentially through R_1 until it reaches the peak-point voltage V_P. At this point, the *pn* junction becomes forward-biased, and the emitter characteristic goes into the negative resistance region (V_E decreases and I_E increases). The capacitor then quickly discharges through the forward-biased junction, r'_B, and R_2. When the capacitor voltage decreases to the valley-point voltage V_V, the UJT turns off, the capacitor begins to charge again, and the cycle is repeated, as shown in the emitter voltage waveform in Figure 11–35 (top). During the discharge time of the capacitor, the UJT is conducting. Therefore, a voltage is developed across R_2, as shown in the waveform diagram in Figure 11–35 (bottom).

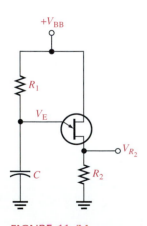

FIGURE 11–34
Relaxation oscillator.

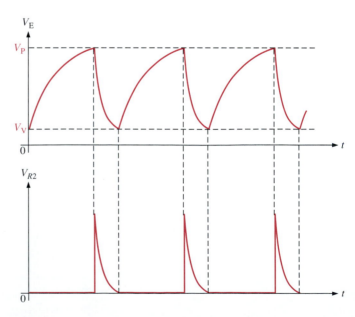

FIGURE 11–35
Waveforms for UJT relaxation oscillator.

Conditions for Turn-On and Turn-Off

In the relaxation oscillator of Figure 11–34, certain conditions must be met for the UJT to reliably turn on and turn off. First, to ensure turn-on, R_1 must not limit I_E at the peak point to less than I_P. To ensure this, the voltage drop across R_1 at the peak point should be greater than $I_P R_1$. Thus, the condition for turn-on is

$$V_{BB} - V_P > I_P R_1$$

or

$$R_1 < \frac{V_{BB} - V_P}{I_P} \qquad (11\text{–}6)$$

To ensure turn-off of the UJT at the valley point, R_1 must be large enough that I_E (at the valley point) can decrease below the specified value of I_V. This means that the voltage across R_1 at the valley point must be less than $I_V R_1$. Thus, the condition for turn-off is

$$V_{BB} - V_V < I_V R_1$$

or

$$R_1 > \frac{V_{BB} - V_V}{I_V} \qquad (11\text{–}7)$$

So, for a proper turn-on and turn-off, R_1 must be in the range

$$\frac{V_{BB} - V_P}{I_P} > R_1 > \frac{V_{BB} - V_V}{I_V}$$

EXAMPLE 11–4

Determine a value of R_1 in Figure 11–36 that will ensure proper turn-on and turn-off of the UJT. The characteristic of the UJT exhibits the following values: $\eta = 0.5$, $V_V = 1$ V, $I_V = 10$ mA, $I_P = 20$ μA, and $V_P = 14$ V.

FIGURE 11–36

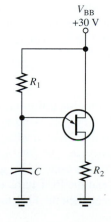

Solution

$$\frac{V_{BB} - V_P}{I_P} > R_1 > \frac{V_{BB} - V_V}{I_V}$$

$$\frac{30 \text{ V} - 14 \text{ V}}{20 \text{ μA}} > R_1 > \frac{30 \text{ V} - 1 \text{ V}}{10 \text{ mA}}$$

$$800 \text{ kΩ} > R_1 > 2.9 \text{ kΩ}$$

As you can see, R_1 has quite a wide range of possible values.

Related Exercise Determine a value of R_1 in Figure 11–36 that will ensure proper turn-on and turn-off for the following values: $\eta = 0.33$, $V_V = 0.8$ V, $I_V = 15$ mA, $I_P = 35$ μA, and $V_P = 18$ V.

SECTION 11–6 REVIEW

1. Name the UJT terminals.

2. What is the intrinsic standoff ratio?

3. In a basic UJT relaxation oscillator such as in Figure 11–34, what three factors determine the period of oscillation?

11–7 ■ THE PROGRAMMABLE UNIJUNCTION TRANSISTOR (PUT)

The programmable unijunction transistor (PUT) is actually a type of thyristor and not like the UJT at all in terms of structure. The only similarity to a UJT is that the PUT can be used in some oscillator applications to replace the UJT. The PUT is more similar to an SCR except that its anode-to-gate voltage can be used to both turn on and turn off the device.

After completing this section, you should be able to

■ **Describe the structure and operation of the PUT**
   □ Compare PUT structure to that of the SCR
   □ Explain the difference between a PUT and a UJT
   □ State how to set the PUT trigger voltage
   □ Discuss an application

The structure of the **programmable unijunction transistor (PUT)** is similar to that of an SCR (four-layer) except that the gate is brought out as shown in Figure 11–37. Notice that the gate is connected to the *n* region adjacent to the anode. This *pn* junction controls the *on* and *off* states of the device. The gate is always biased positive with respect to the cathode. When the anode voltage exceeds the gate voltage by approximately 0.7 V, the *pn* junction is forward-biased and the PUT turns on. The PUT stays on until the anode voltage falls back below this level, then the PUT turns off.

FIGURE 11–37
The programmable unijunction transistor (PUT).

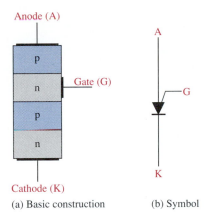

(a) Basic construction   (b) Symbol

Setting the Trigger Voltage

The gate can be biased to a desired voltage with an external voltage divider, as shown in Figure 11–38(a), so that when the anode voltage exceeds this "programmed" level, the PUT turns on.

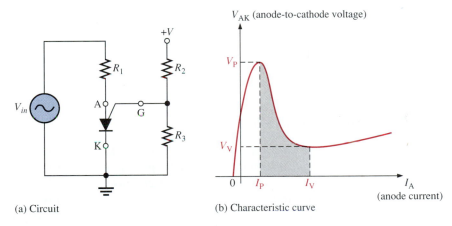

(a) Circuit   (b) Characteristic curve

FIGURE 11–38
PUT biasing.

An Application

A plot of the anode-to-cathode voltage V_{AK} versus anode current I_A in Figure 11–38(b) reveals a characteristic curve similar to that of the UJT. Therefore, the PUT replaces the UJT in many applications. One such application is the relaxation oscillator in Figure 11–39(a). Its basic operation is as follows.

The gate is biased at +9 V by the voltage divider consisting of resistors R_2 and R_3. When dc power is applied, the PUT is off and the capacitor charges toward +18 V through R_1. When the capacitor reaches $V_G + 0.7$ V, the PUT turns on and the capacitor rapidly discharges through the low *on* resistance of the PUT and R_4. A voltage spike is developed across R_4 during the discharge. As soon as the capacitor discharges, the PUT turns off and the charging cycle starts over, as shown by the waveforms in Figure 11–39(b).

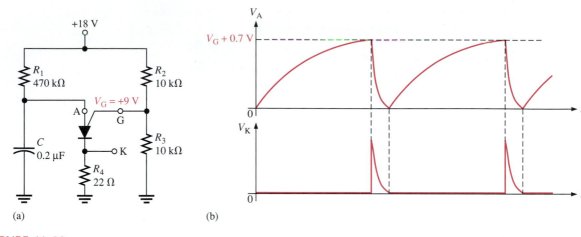

FIGURE 11–39
PUT relaxation oscillator.

SECTION 11–7 REVIEW

1. What does the term *programmable* mean as used in programmable unijunction transistor (PUT)?

2. Compare the structure and the operation of a PUT to those of other devices such as the UJT and SCR.

11–8 ■ THE PHOTOTRANSISTOR

The phototransistor has a light-sensitive, collector-base pn junction. It is exposed to incident light through a lens opening in the transistor package. When there is no incident light, there is a small thermally generated collector-to-emitter leakage current, I_{CEO}; this is called the dark current and is typically in the nA range. When light strikes the collector-base pn junction, a base current, I_λ, is produced that is directly proportional to the light intensity. This action produces a collector current that increases with I_λ. Except for the way base current is generated, the phototransistor behaves as a conventional bipolar transistor. In many cases, there is no electrical connection to the base.

After completing this section, you should be able to

■ **Describe a phototransistor and its operation**
  ☐ Explain how the base current is produced
  ☐ Discuss how phototransistors are used

The relationship between the collector current and the light-generated base current in a phototransistor is

$$I_C = \beta_{DC}I_\lambda$$

(11–8)

The schematic symbol and some typical phototransistors are shown in Figure 11–40. Since the actual photogeneration of base current occurs in the collector-base region, the larger the physical area of this region, the more base current is generated. Thus, a typical phototransistor is designed to offer a large area to the incident light, as the simplified structure diagram in Figure 11–41 illustrates.

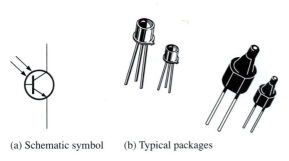

(a) Schematic symbol   (b) Typical packages

FIGURE 11–40
Phototransistor.

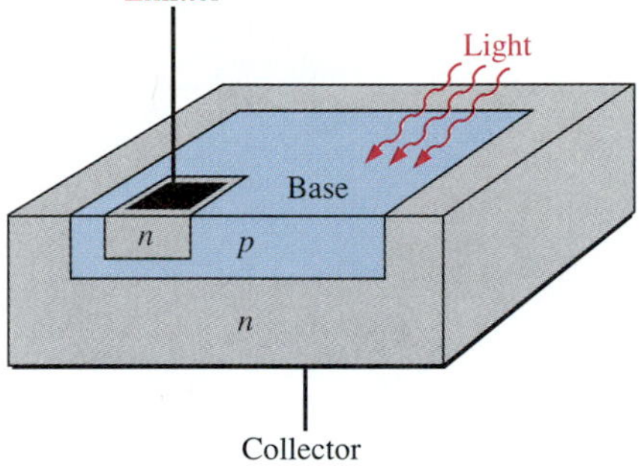

FIGURE 11–41
Typical phototransistor chip structure.

A **phototransistor** can be either a two-lead or a three-lead device. In the three-lead configuration, the base lead is brought out so that the device can be used as a conventional bipolar transistor with or without the additional light-sensitivity feature. In the two-lead configuration, the base is not electrically available, and the device can be used only with light as the input. In many applications, the phototransistor is used in the two-lead version. Figure 11–42 shows a phototransistor with a biasing circuit and typical collector characteristic curves. Notice that each individual curve on the graph corresponds to a certain value of light intensity (in this case, the units are mW/cm^2) and that the collector current increases with light intensity.

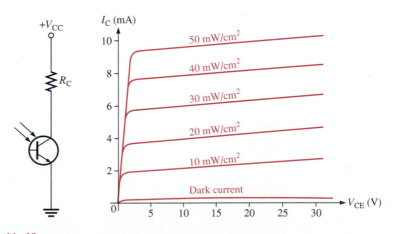

FIGURE 11–42
Phototransistor bias circuit and typical collector characteristic curves.

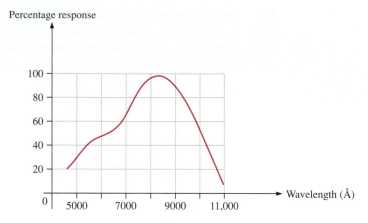

FIGURE 11–43
Typical phototransistor spectral response.

Phototransistors are not sensitive to all light but only to light within a certain range of wavelengths. They are most sensitive to particular wavelengths, as shown by the peak of the spectral response curve in Figure 11–43.

Photodarlington

The photodarlington consists of a phototransistor connected in a darlington arrangement with a conventional transistor, as shown in Figure 11–44. Because of the higher current gain, this device has a much higher collector current and exhibits a greater light sensitivity than does a regular phototransistor.

Applications

Phototransistors are used in a wide variety of applications. A light-operated relay circuit is shown in Figure 11–45. The phototransistor Q_1 drives the bipolar transistor Q_2. When there is sufficient incident light on Q_1, transistor Q_2 is driven into saturation, and collector current through the relay coil energizes the relay.

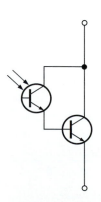

FIGURE 11–44
Photodarlington.

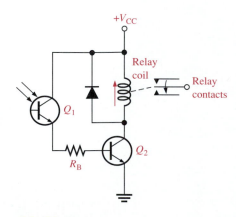

FIGURE 11–45
Light-operated relay circuit.

Figure 11–46 shows a circuit in which a relay is de-energized by incident light on the phototransistor. When there is insufficient light, transistor Q_2 is biased on, keeping the relay energized. When there is sufficient light, phototransistor Q_1 turns on; this pulls the base of Q_2 low, thus turning Q_2 off and de-energizing the relay.

These relay circuits can be used in a variety of applications such as automatic door activators, process counters, and various alarm systems. Another simple application is illustrated in Figure 11–47. The phototransistor is normally on, holding the gate of the SCR low. When the light is interrupted, the phototransistor turns off. The high-going transition on the collector triggers the SCR and sets off the alarm mechanism. The momentary contact switch SW1 provides for resetting the alarm. Smoke and intrusion detection are possible uses for this circuit.

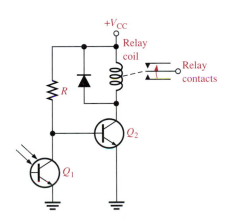

FIGURE 11–46
Darkness-operated relay circuit.

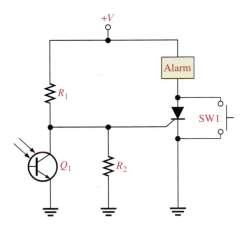

FIGURE 11–47
Light-interruption alarm.

SECTION 11–8 REVIEW

1. How does a phototransistor differ from a conventional bipolar transistor?
2. Some, but not all, phototransistors have external _____-leads.
3. The collector current in a phototransistor circuit depends on what two factors?

11–9 ■ THE LIGHT-ACTIVATED SCR (LASCR)

The light-activated silicon-controlled rectifier (LASCR) operates essentially as does the conventional SCR except that it can also be light-triggered. Most LASCRs have an available gate terminal so that the device can also be triggered by an electrical pulse just as a conventional SCR.

After completing this section, you should be able to

■ **Describe the LASCR and its operation**
   ☐ Compare the LASCR to the conventional SCR
   ☐ Discuss an application

Figure 11–48 shows a **light-activated silicon-controlled rectifier (LASCR)** schematic symbol and typical packages. The LASCR is most sensitive to light when the gate terminal is open. If necessary, a resistor from the gate to the cathode can be used to reduce the sensitivity. Figure 11–49 shows an LASCR used to energize a latching relay. The input source turns on the lamp; and the resulting incident light triggers the LASCR. The anode current energizes the relay and closes the contact. Notice that the input source is electrically isolated from the rest of the circuit.

FIGURE 11–48
Light-activated SCRs.

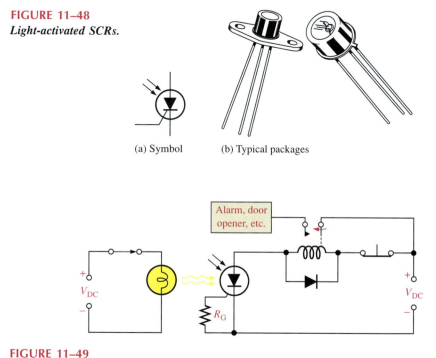

(a) Symbol   (b) Typical packages

FIGURE 11–49
A light-activated SCR circuit.

SECTION 11–9 REVIEW

1. Can most LASCRs be operated as conventional SCRs?
2. What is required in Figure 11–49 to turn off the LASCR and de-energize the relay?

11–10 ■ OPTICAL COUPLERS

Optical couplers are designed to provide complete electrical isolation between an input circuit and an output circuit. The usual purpose of isolation is to provide protection from high-voltage transients, surge voltage, or low-level noise that could possibly result in an erroneous output or damage to the device. Optical couplers also allow interfacing circuits with different voltage levels, different grounds, and so on.

After completing this section, you should be able to

■ **Discuss various types of optical couplers**
  ☐ Define isolation voltage
  ☐ Define dc current transfer ratio
  ☐ Define LED trigger current
  ☐ Define transfer gain
  ☐ Discuss fiber optics
  ☐ Explain refraction and reflection of light

The input circuit of an optical coupler is typically an LED, but the output circuit can take several forms, such as the phototransistor shown in Figure 11–50(a). When the input voltage forward-biases the LED, light transmitted to the phototransistor turns it on, producing current through the external load, as shown in Figure 11–50(b). Some typical devices are shown in Figure 11–50(c).

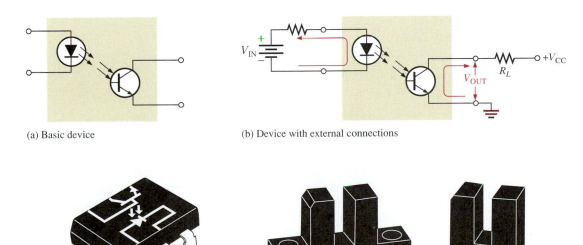

(a) Basic device

(b) Device with external connections

(c) Typical packages

FIGURE 11–50
Phototransistor couplers.

Several other types of couplers are shown in Figure 11–51. The darlington transistor coupler in Figure 11–51(a) can be used when increased output current capability is needed beyond that provided by the phototransistor output. The disadvantage is that the photodarlington has a switching speed less than that of the phototransistor.

An LASCR output coupler is shown in Figure 11–51(b). This device can be used in applications where, for example, a low-level input voltage is required to latch a high-voltage relay for the purpose of activating some type of electromechanical device.

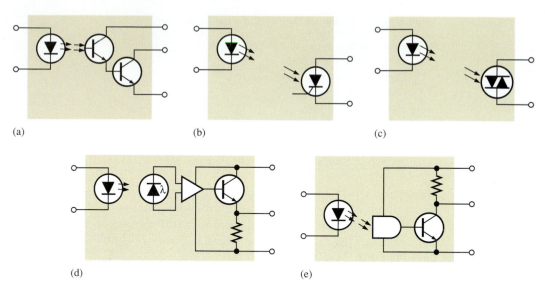

FIGURE 11–51

Common types of optical-coupling devices.

A phototriac output coupler is illustrated in Figure 11–51(c). This device is designed for applications requiring isolated triac triggering, such as switching a 110 V ac line from a low-level input.

Figure 11–51(d) shows an optically isolated ac linear coupler. This device converts an input current variation to an output voltage variation. The output circuit consists of an amplifier with a photodiode across its input terminals. Light variations emitted from the LED are picked up by the photodiode, providing an input signal to the amplifier. The output of the amplifier is buffered with the emitter-follower stage. The optically isolated ac linear coupler can be used for telephone line coupling, peripheral equipment isolation, and audio applications.

Part (e) of Figure 11–51 shows a digital output coupler. This device consists of a high-speed detector circuit followed by a transistor buffer stage. When there is current through the input LED, the detector is light-activated and turns on the output transistor, so that the collector switches to a low-voltage level. When there is no current through the LED, the output is at the high-voltage level. The digital output coupler can be used in applications requiring compatibility with digital circuits, such as interfacing computer terminals with peripheral devices.

Isolation Voltage The *isolation voltage* of an optical coupler is the maximum voltage that can exist between the input and output terminals without dielectric breakdown occurring. Typical values are about 7500 V ac peak.

DC Current Transfer Ratio This parameter is the ratio of the output current to the input current through the LED. It is usually expressed as a percentage. For a phototransistor output, typical values range from 2 percent to 100 percent. For a photodarlington output, typical values range from 50 percent to 500 percent.

LED Trigger Current This parameter applies to the LASCR output coupler and the phototriac output device. The *trigger current* is the value of current required to trigger the thyristor output device. Typically, the trigger current is in the mA range.

Transfer Gain This parameter applies to the optically isolated ac linear coupler. The *transfer gain* is the ratio of output voltage to input current, and a typical value is 200 mV/mA.

Fiber Optics

Fiber optics provides a means for coupling a photo-emitting device to a photodetector via a light-transmitting cable consisting of optical fibers. Typical applications include medical electronics, industrial controls, microprocessor systems, security systems, and communications. Glass fiber cables are used to maximize the optical coupling between optoelectronic devices. Fiber optics is based on the principle of internal reflection.

Every material that conducts light has an **index of refraction, *n*.** A ray of light striking the interface between two materials with different indices of refraction will either be reflected or refracted, depending on the angle at which the light strikes the interface surface, as illustrated in Figure 11–52. If the incident angle, θ_i, is equal to or greater than a certain value called the *critical angle,* θ_c, the light is reflected. If θ_i is less than the critical angle, the light is refracted.

A glass fiber is clad with a layer of glass with a lower index of refraction than the fiber. A ray of light entering the end of the cable will be refracted as shown in Figure 11–53. If, after refraction, it approaches and hits the glass interface at an angle greater than the critical angle, it is reflected within the fiber as shown. Since, according to a law of physics, the angle of reflection must equal the angle of incidence, the light ray will bounce down the fiber and emerge, refracted, at the other end, as shown in the figure.

FIGURE 11–52
Refraction and reflection of light rays.

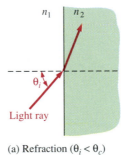 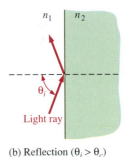

(a) Refraction ($\theta_i < \theta_c$)   (b) Reflection ($\theta_i > \theta_c$)

FIGURE 11–53
A light ray in optical fiber.

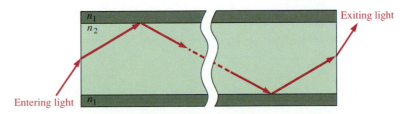

SECTION 11–10 REVIEW	1. What type of input device is normally used in an optical coupler?
	2. List five types of output devices used in optical couplers.

11–11 ■ SYSTEM APPLICATION

Your company is in the process of designing an optical counting and control system for controlling the rate at which objects on an assembly line are moved. This system is a variation of the system in Chapter 3 for counting baseballs for packing control. In this system, objects passing on an assembly line conveyor belt are counted, and the speed of the conveyor is adjusted according to a predetermined rate. The focus of this application is the conveyor motor speed-control circuit. The knowledge you have gained in this chapter will be applied in completing your assignment.

Basic Operation of the System

The system controls the speed of the conveyor so that a preset average number of unequally spaced parts flow past a point on the production line in a specified period of time. This is to allow a proper amount of time for the production line workers to perform certain tasks on each part. A basic diagram of the conveyor speed-control system is shown in Figure 11–54.

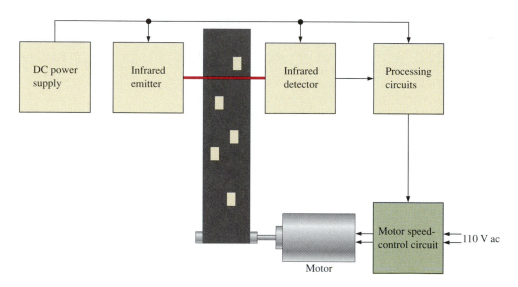

FIGURE 11–54
Diagram of the conveyor control system.

Each time a part on the moving conveyor belt passes the optical detector and interrupts the infrared light beam, a digital counter in the processing circuits is advanced by one. The count of the passing parts is accumulated over a specified period of time and converted to a proportional voltage by the processing circuits based on the desired number of parts in a specified time interval. The more parts that pass the sensor, the higher the voltage. The proportional voltage is applied to the motor speed-control circuit which, in turn, adjusts the speed of the electric motor that drives the conveyor belt in order to maintain the desired number of parts per unit time.

The Motor Speed-Control Circuit The proportional voltage from the processing circuits is applied to the gate of a PUT on the motor speed-control circuit board. This voltage determines the point in the ac cycle that the SCR is triggered on. For a higher PUT gate voltage, the SCR turns on later in the half-cycle and therefore delivers less average power to the motor causing its speed to decrease. For a lower PUT gate voltage, the SCR turns on earlier in the half-cycle delivering more average power to the motor and increasing its speed. This process continually adjusts the motor speed to maintain the required number of objects per unit time. The potentiometer is used for calibration of the SCR trigger point.

The Motor Speed-Control Circuit Board

☐ Make sure that the circuit board shown in Figure 11–55(a) is correctly assembled by comparison with the schematic in part (b). Device pin diagrams are shown.

☐ Label a copy of the board with component and input/output designations in agreement with the schematic.

FIGURE 11–55
Motor speed-control circuit board.

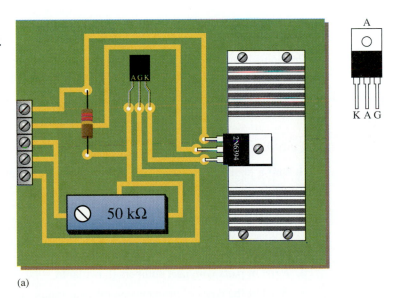

(a)

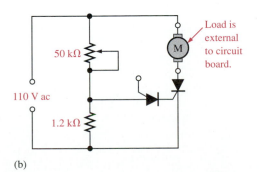

(b)

Analysis of the Motor Speed-Control Circuit

Refer to the schematic in Figure 11–55(b). A 1 kΩ resistor is connected in place of the motor and 110 V, 60 Hz voltage is applied across the input terminals.

☐ Determine the voltage waveforms at the anode, cathode, and gate of the SCR and the PUT with respect to ground for a PUT gate voltage of 0 V and with the potentiometer set at 25 kΩ.

☐ Determine voltage across the 1 kΩ resistor, for each of the following PUT gate voltages: 0 V, 2 V, 4 V, 6 V, 8 V, and 10 V. The potentiometer is still at 25 kΩ.

Test Procedure

☐ Develop a step-by-step set of instructions on how to check the motor speed-control circuit board for proper operation using the numbered test points indicated in the test bench setup of Figure 11–56.

☐ Specify voltage values for all the measurements to be made.

☐ Provide a fault analysis for all possible component failures.

Troubleshooting

Problems have developed in four boards. Based on the test bench measurements for each board indicated in Figure 11–57, determine the most likely fault in each case. The circled numbers indicate test point connections to the circuit board. Assume that each board has 110 V ac applied.

Final Report

Submit a final written report on the motor speed-control circuit board using an organized format that includes the following:

1. A physical description of the circuits.
2. A discussion of the operation of the circuits.
3. A list of the specifications.
4. A list of parts with part numbers if available.
5. A list of the types of problems on the four faulty circuit boards.
6. A description of how you determined the problem on each of the faulty circuit boards.

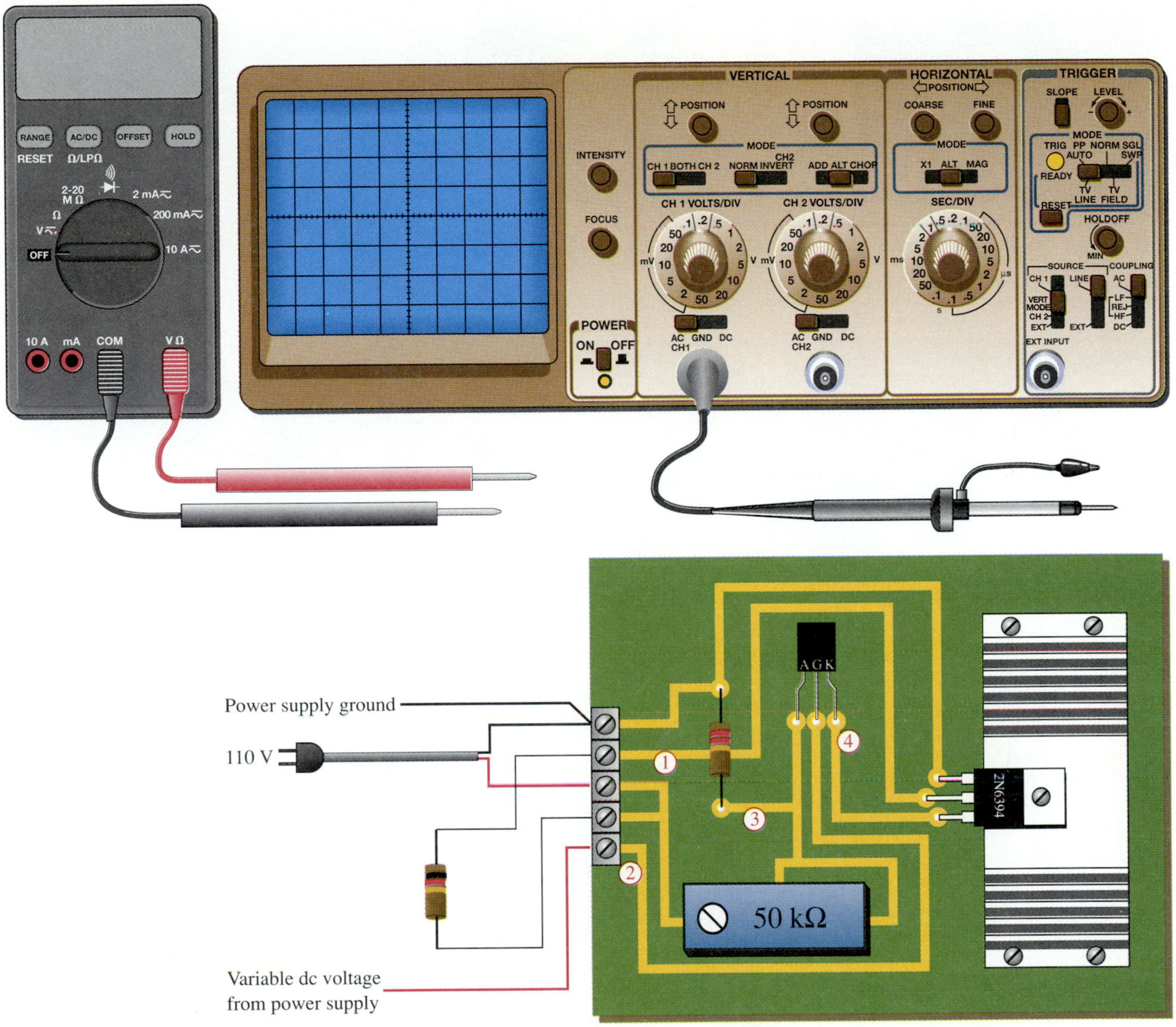

FIGURE 11–56

Test bench setup for the motor speed-control circuit board.

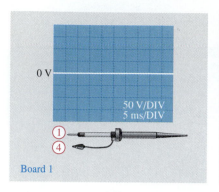

Board 1

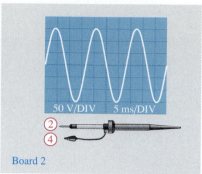

Board 2

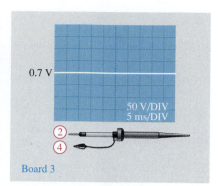

Board 3

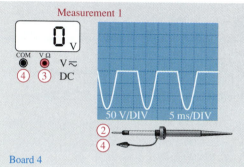

Board 4

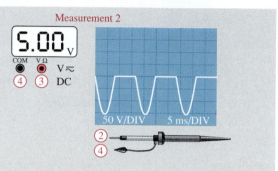

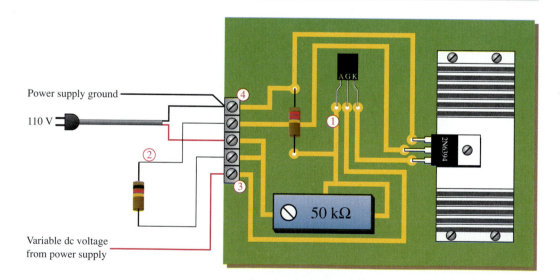

FIGURE 11–57

Test results for four faulty circuit boards.

<hr/>

■ **CHAPTER SUMMARY**

■ Thyristors are devices constructed with four semiconductor layers (*pnpn*).

■ Thyristors include Shockley diodes, SCRs, SCSs, diacs, triacs, and PUTs.

■ The Shockley diode is a thyristor that conducts when the voltage across its terminals exceeds the breakover potential.

- The silicon-controlled rectifier (SCR) can be triggered on by a pulse at the gate and turned off by reducing the anode current below the specified holding value.
- The silicon-controlled switch (SCS) has two gate terminals and can be turned on by a pulse at the cathode gate and turned off by a pulse at the anode gate.
- The diac can conduct current in either direction and is turned on when a breakover voltage is exceeded. It turns off when the current drops below the holding value.
- The triac, like the diac, is a bidirectional device. It can be turned on by a pulse at the gate and conducts in a direction depending on the voltage polarity across the two anode terminals.
- The intrinsic standoff ratio of a unijunction transistor (UJT) determines the voltage at which the device will trigger on.
- The programmable unijunction transistor (PUT) can be externally programmed to turn on at a desired anode-to-gate voltage level.
- In a phototransistor, base current is generated by light input.
- Light acts as the trigger source in light-activated SCRs (LASCRs).
- Optical coupling devices provide electrical isolation between a source and an output circuit.
- Fiber optics provides a light path from a light-emitting device to a light-activated device.
- Device symbols are shown in Figure 11–58.

FIGURE 11–58

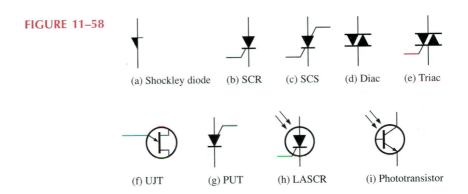

(a) Shockley diode (b) SCR (c) SCS (d) Diac (e) Triac

(f) UJT (g) PUT (h) LASCR (i) Phototransistor

■ GLOSSARY

Diac A two-terminal four-layer semiconductor device (thyristor) that can conduct current in either direction when properly activated.

Forced commutation A method of turning off an SCR.

Forward-breakover voltage ($V_{BR(F)}$) The voltage at which a device enters the forward-blocking region.

Holding current (I_H) The value of the anode current below which a device switches from the forward-conduction region to the forward-blocking region.

Index of refraction A property of light-conducting materials that specifies how much a light ray will bend when passing from one material to another.

Light-activated silicon-controlled rectifier (LASCR) A four-layer semiconductor device (thyristor) that conducts current in one direction when activated by a sufficient amount of light and continues to conduct until the current falls below a specified value.

Oscillator An electronic circuit based on positive feedback that produces a time-varying output signal without an external input signal.

Phototransistor A transistor in which base current is produced when light strikes the photosensitive semiconductor base region.

Programmable unijunction transistor (PUT)   A type of three-terminal thyristor (more like an SCR than a UJT) that is triggered into conduction when the voltage at the anode exceeds the voltage at the gate.

Shockley diode   The type of two-terminal thyristor that conducts current when the anode-to-cathode voltage reaches a specified "breakover" value.

Silicon-controlled rectifier (SCR)   A type of three-terminal thyristor that conducts current when triggered on by a voltage at the single gate terminal and remains on until the anode current falls below a specified value.

Silicon-controlled switch (SCS)   A type of four-terminal thyristor that has two gate terminals that are used to trigger the device on and off.

Standoff ratio   The characteristic of a UJT that determines its turn-on point.

Switching current (I_S)   The value of the anode current at the point where the device switches from the forward-blocking region to the forward-conduction region.

Thyristor   A class of four-layer (*pnpn*) semiconductor devices.

Triac   A three-terminal thyristor that can conduct current in either direction when properly activated.

Trigger   The activating input of some electronic devices and circuits.

Unijunction transistor (UJT)   A three-terminal single *pn* junction device that exhibits a negative resistance characteristic.

■ FORMULAS

Shockley Diode

$$(11–1) \qquad I_{B1} = (1 - \alpha_{DC1})I_A - I_{CBO1} \qquad \text{Shockley diode, } Q_1 \text{ base current}$$

$$(11–2) \qquad I_{C2} = \alpha_{DC2}I_K + I_{CBO2} \qquad \text{Shockley diode, } Q_2 \text{ base current}$$

$$(11–3) \qquad I_A = \frac{I_{CBO1} + I_{CBO2}}{1 - (\alpha_{DC1} + \alpha_{DC2})} \qquad \text{Shockley diode, anode current (or cathode current)}$$

Unijunction Transistor

$$(11–4) \qquad \eta = \frac{r'_{B1}}{r'_{BB}} \qquad \text{UJT intrinsic standoff ratio}$$

$$(11–5) \qquad V_P = \eta V_{BB} + V_{pn} \qquad \text{UJT peak-point voltage}$$

$$(11–6) \qquad R_1 < \frac{V_{BB} - V_P}{I_P} \qquad \text{Turn-on condition, relaxation oscillator}$$

$$(11–7) \qquad R_1 > \frac{V_{BB} - V_V}{I_V} \qquad \text{Turn-off condition, relaxation oscillator}$$

The Phototransistor

$$(11–8) \qquad I_C = \beta_{DC}I_\lambda \qquad \text{Phototransistor collector current}$$

■ SELF-TEST

1. A thyristor has
   - **(a)** two *pn* junctions
   - **(b)** 3 *pn* junctions
   - **(c)** four *pn* junctions
   - **(d)** only two terminals
2. Common types of thyristors include
   - **(a)** BJTs and SCRs
   - **(b)** UJTs and PUTs
   - **(c)** FETs and triacs
   - **(d)** diacs and triacs

3. A Shockley diode turns on when the anode to cathode voltage
   (a) exceeds 0.7 V   (b) exceeds the gate voltage
   (c) exceeds the forward-breakover voltage (d) exceeds the forward-blocking voltage

4. Once it is conducting, a Shockley diode can be turned off by
   (a) reducing the current below a certain value (b) disconnecting the anode voltage
   (c) answers (a) and (b)   (d) none of these

5. An SCR differs from the Shockley diode because
   (a) it has a gate terminal   (b) it is not a thyristor
   (c) it does not have four layers   (d) it cannot be turned on and off

6. An SCR can be turned off by
   (a) forced commutation   (b) a negative pulse on the gate
   (c) anode current interruption   (d) all of the above
   (e) answers (a) and (c)

7. In the forward-blocking region, the SCR is
   (a) reverse-biased   (b) in the *off* state
   (c) in the *on* state (d) at the point of breakdown

8. The specified value of holding current for an SCR means that
   (a) the device will turn on when the anode current exceeds this value
   (b) the device will turn off when the anode current falls below this value
   (c) the device may be damaged if the anode current exceeds this value
   (d) the gate current must equal or exceed this value to turn the device on

9. The SCS differs from the SCR because
   (a) it does not have a gate terminal   (b) its holding current is less
   (c) it can handle much higher currents (d) it has two gate terminals

10. The SCS can be turned on by
    (a) an anode voltage that exceeds forward-breakover voltage
    (b) a positive pulse on the cathode gate
    (c) a negative pulse on the anode gate
    (d) either (b) or (c)

11. The SCS can be turned off by
    (a) a negative pulse on the cathode gate and a positive pulse on the anode gate
    (b) reducing the anode current to below the holding value
    (c) answers (a) and (b)
    (d) a positive pulse on the cathode gate and a negative pulse on the anode gate

12. The diac is
    (a) a thyristor
    (b) a bilateral, two-terminal device
    (c) like two parallel Shockley diodes in reverse directions
    (d) all of these answers

13. The triac is
    (a) like a bidirectional SCR (b) a four-terminal device
    (c) not a thyristor   (d) answers (a) and (b)

14. Which of the following is *not* a characteristic of the UJT?
    (a) intrinsic standoff ratio (b) negative resistance
    (c) peak-point voltage   (d) bilateral conduction

15. The PUT is
    (a) much like the UJT   (b) not a thyristor
    (c) triggered on and off by the gate to anode voltage (d) not a four-layer device

16. In a phototransistor, base current is
    (a) set by a bias voltage   (b) directly proportional to light
    (c) inversely proportional to light (d) not a factor

■ **BASIC PROBLEMS**

SECTION 11–1   The Shockley Diode

1. The Shockley diode in Figure 11–59 is biased such that it is in the forward-blocking region. For this particular bias condition, $\alpha_{DC} = 0.38$ for both internal transistor structures. $I_{CBO1} = 75$ nA and $I_{CBO2} = 80$ nA. Determine the anode current and cathode current.

FIGURE 11–59

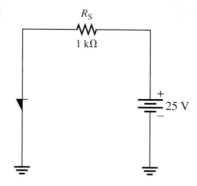

2. **(a)** Determine the forward resistance of the Shockley diode in Figure 11–59 in the forward-blocking region.
   (b) If the forward-breakover voltage is 50 V, how much must V_{AK} be increased to switch the diode into the forward-conduction region?

SECTION 11–2   The Silicon-Controlled Rectifier (SCR)

3. Explain the operation of an SCR in terms of its transistor equivalent.

4. To what value must the variable resistor be adjusted in Figure 11–60 in order to turn the SCR off? Assume $I_H = 10$ mA and $V_{AK} = 0.7$ V.

FIGURE 11–60

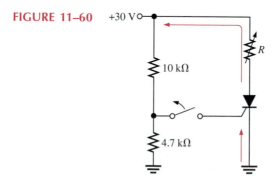

SECTION 11–3   SCR Applications

5. Describe how you would modify the circuit in Figure 11–15 so that the SCR triggers and conducts on the negative half-cycle of the input.

6. What is the purpose of diodes D_1 and D_2 in Figure 11–17?

7. Sketch the V_R waveform for the circuit in Figure 11–61, given the indicated relationship of the input waveforms.

FIGURE 11–61

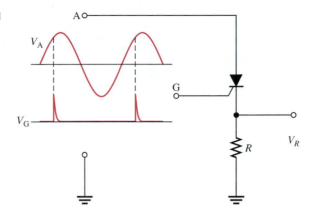

SECTION 11–4 The Silicon-Controlled Switch (SCS)

8. Explain the turn-on and turn-off operation of an SCS in terms of its transistor equivalent.

9. Name the terminals of an SCS.

SECTION 11–5 The Diac and Triac

10. Sketch the current waveform for the circuit in Figure 11–62. The diac has a breakover potential of 20 V. $I_H = 20$ mA.

11. Repeat Problem 10 for the triac circuit in Figure 11–63. The breakover potential is 25 V and $I_H = 1$ mA.

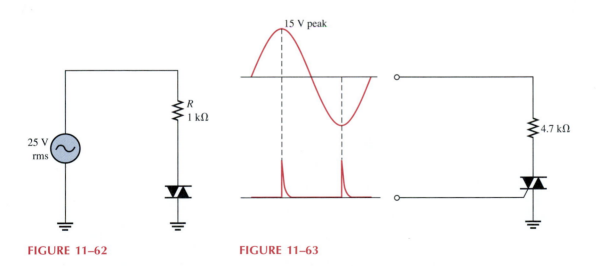

FIGURE 11–62

FIGURE 11–63

SECTION 11–6 The Unijunction Transistor (UJT)

12. In a certain UJT, $r'_{B1} = 2.5$ kΩ and $r'_{B2} = 4$ kΩ. What is the intrinsic standoff ratio?

13. Determine the peak-point voltage for the UJT in Problem 12 if $V_{BB} = 15$ V.

FIGURE 11–64

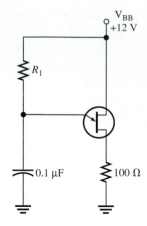

14. Find the range of values of R_1 in Figure 11–64 that will ensure proper turn-on and turn-off of the UJT. $\eta = 0.68$, $V_V = 0.8$ V, $I_V = 15$ mA, $I_P = 10$ μA, and $V_P = 10$ V.

SECTION 11–7 The Programmable Unijunction Transistor (PUT)

15. At what anode voltage (V_A) will each PUT in Figure 11–65 begin to conduct?

16. Draw the current waveform for each circuit in Figure 11–65 when there is a 10 V peak sinusoidal voltage at the anode. Neglect the forward voltage of the PUT.

17. Sketch the voltage waveform across R_1 in Figure 11–66 in relation to the input voltage waveform.

FIGURE 11–65

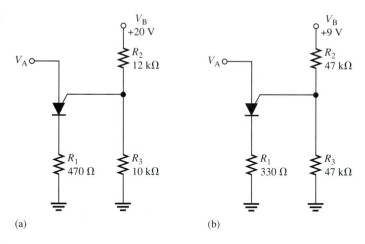

(a)   (b)

FIGURE 11–66

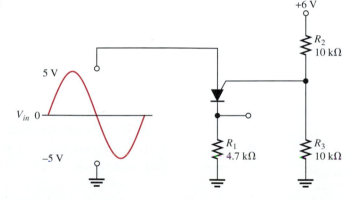

SECTION 11–8   The Phototransistor

18. A certain phototransistor in a circuit has a $\beta_{DC} = 200$. If $I_\lambda = 100 \ \mu A$, what is the collector current?

19. Determine the output voltage in Figure 11–67, **(a)** when the light source is off, and **(b)** when it is on (assuming the transistor saturates).

20. Determine the emitter current in the photodarlington circuit in Figure 11–68 if, for each $1m/m^2$ of light intensity, 1 μA of base current is produced in Q_1.

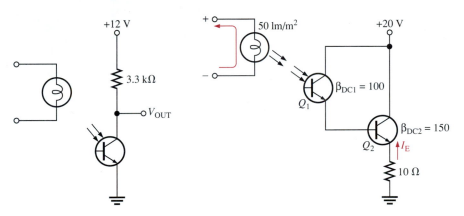

FIGURE 11–67   **FIGURE 11–68**

SECTION 11–9   The Light-Activated SCR (LASCR)

21. By examination of the circuit in Figure 11–69, explain its purpose and basic operation.

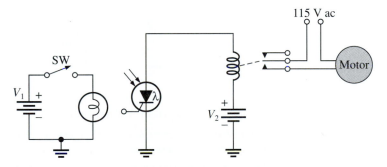

FIGURE 11–69

22. Determine the voltage waveform across R_K in Figure 11–70.

FIGURE 11–70

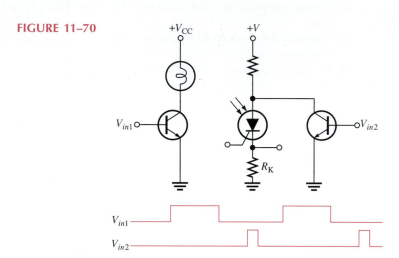

SECTION 11–10 Optical Couplers

23. A particular optical coupler has a current transfer ratio of 30 percent. If the input current is 100 mA, what is the output current?

24. The optical coupler shown in Figure 11–71 is required to deliver at least 10 mA to the external load. If the current transfer ratio is 60 percent, how much current must be supplied to the input?

FIGURE 11–71

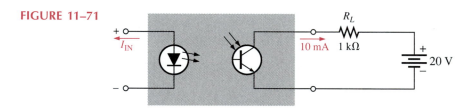

SECTION 11–11 System Application

25. In the motor speed-control circuit of Figure 11–55, at which PUT gate voltage does the electric motor run at the fastest speed: 0 V, 2 V, or 5 V?

26. Does the SCR in the motor speed-control circuit turn on earlier or later in the ac cycle if the resistance of the rheostat is reduced?

27. Describe the SCR action as the PUT gate voltage is increased in the motor speed-control circuit.

■ ADVANCED PROBLEMS

28. Refer to the SCR over-voltage protection circuit in Figure 11–18. For a +12 V output dc power supply, specify the component values that will provide protection for the circuit if the output voltage exceeds +15 V. Assume the fuse is rated at 1 A.

29. Design an SCR crowbar circuit to protect electronic circuits against a voltage from the power supply in excess of 6.2 V.

30. Design a relaxation oscillator to produce a frequency of 2.5 kHz using a UJT with $\eta = 0.75$ and a valley voltage of 1 V. The circuit must operate from a +12 V dc source. Design values of $I_V = 10$ mA and $I_P = 20$ μA are to be used.

■ ANSWERS TO SECTION REVIEWS

Section 11–1

1. The Shockley diode is a thyristor because it is a four-layer device.

2. In the forward-blocking region, the diode is off (nonconducting).

3. The device turns on and conducts when V_{AK} exceeds the forward breakover voltage.

4. When the anode current is reduced below the holding current value, the device turns off.

Section 11–2

1. An SCR (silicon-controlled rectifier) is a three-terminal thyristor.

2. The SCR terminals are anode, cathode, and gate.

3. A positive gate pulse turns the SCR on.

4. Reduce the anode current below I_H (holding current) to turn a conducting SCR off.

Section 11–3

1. The SCR will conduct for more than 90° but less than 180°.

2. To block discharge of the battery through that path

Section 11–4

1. An SCS can be turned off with the application of a gate pulse, but an SCR cannot.

2. A positive pulse on the cathode gate or a negative pulse on the anode gate turns the SCS on.

3. An SCS can be turned off by any of the following:
   (a) positive pulse on anode gate   **(b)** negative pulse on cathode gate
   (c) reduce anode current below holding value   **(d)** completely interrupt anode current

Section 11–5

1. The diac is like two parallel Shockley diodes connected in opposite directions.

2. A triac is like two parallel SCRs having a common gate and connected in opposite directions.

3. A triac has a gate terminal, but a diac does not.

Section 11–6

1. The UJT terminals are base 1, base 2, and emitter.

2. $\eta = r'_{B1}/r'_{BB}$

3. R, C, and η determine the period.

Section 11–7

1. *Programmable* means that the turn-on voltage can be adjusted to a desired value.

2. The PUT is a thyristor, similar in structure to an SCR, but it is turned on by the anode-to-gate voltage. It has a negative resistance characteristic like the UJT.

Section 11–8

1. The base current of a phototransistor is light induced.

2. Base

3. The collector current depends on β_{DC} and I_λ.

Section 11–9

1. Most LASCRs can be operated as conventional SCRs.

2. A series switch to shut off the anode current is required.

Section 11–10

1. An LED

2. Phototransistor, photodarlington, LASCR, phototriac, linear amplifier

---

▪ **ANSWERS TO RELATED EXERCISES FOR EXAMPLES**

11–1 10 MΩ

11–2 163 Ω

11–3 By increasing V_{BB}

11–4 343 kΩ $> R_1 >$ 1.95 kΩ

12

OPERATIONAL AMPLIFIERS

■ CHAPTER OBJECTIVES

☐ Describe the basic op-amp and its characteristics

☐ Discuss the differential amplifier and its operation

☐ Discuss several op-amp parameters

☐ Explain negative feedback in op-amp circuits

☐ Discuss op-amp compensation

☐ Analyze three basic op-amp configurations

☐ Describe impedances of the three op-amp configurations

☐ Troubleshoot op-amp circuits

So far, you have studied a number of important electronic devices. These devices, such as the diode and the transistor, are separate devices that are individually packaged and interconnected in a circuit with other devices to form a complete, functional unit. Such devices are referred to as *discrete components*.

Now you will begin the study of linear integrated circuits, where many transistors, diodes, resistors, and capacitors are fabricated on a single tiny chip of semiconductor material and packaged in a single case to form a functional circuit. An integrated circuit, such as an operational amplifier (op-amp), is treated as a single device. This means that you will be concerned with what the circuit does more from an external viewpoint than from an internal, component-level viewpoint.

In this chapter, you will learn the basics of operational amplifiers, which are the most versatile and widely used of all linear integrated circuits.

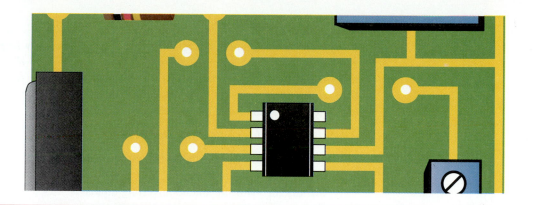

SYSTEM APPLICATION

For the system application in Section 12–9, the op-amp is used in an industrial application. The system is a combination of optics and electronics that is used to determine the chemical content of solutions using a variable wavelength source, a photocell, and an op-amp circuit. You will apply your knowledge of op-amps to this application.

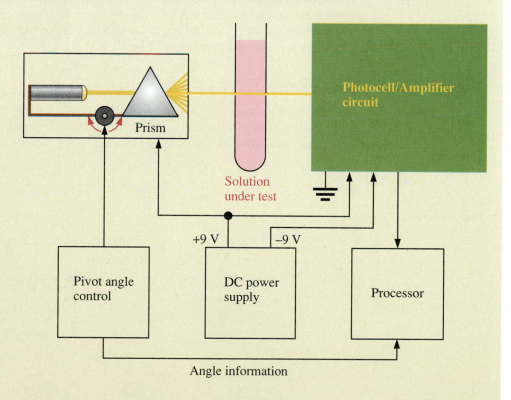

12–1 ■ INTRODUCTION TO OPERATIONAL AMPLIFIERS

Early operational amplifiers (op-amps) were used primarily to perform mathematical operations such as addition, subtraction, integration, and differentiation, hence the term operational. These early devices were constructed with vacuum tubes and worked with high voltages. Today's op-amps are linear integrated circuits that use relatively low dc supply voltages and are reliable and inexpensive.

After completing this section, you should be able to

■ **Describe the basic op-amp and its characteristics**
  □ Identify the op-amp symbol
  □ Identify the inverting and noninverting inputs
  □ Describe the ideal op-amp
  □ Describe the practical op-amp

Symbol and Terminals

The standard op-amp symbol is shown in Figure 12–1(a). It has two input terminals, the inverting input (−) and the noninverting input (+), and one output terminal. The typical **operational amplifier** operates with two dc supply voltages, one positive and the other negative, as shown in Figure 12–1(b). Usually these dc voltage terminals are left off the schematic symbol for simplicity but are always understood to be there. Some typical op-amp IC packages are shown in Figure 12–1(c).

The Ideal Op-Amp

In order to get a concept of what an op-amp is, we will consider its *ideal* characteristics. A practical op-amp, of course, falls short of these ideal standards, but it is much easier to understand and analyze the device from an ideal point of view.

FIGURE 12–1
Op-amp symbols and packages.

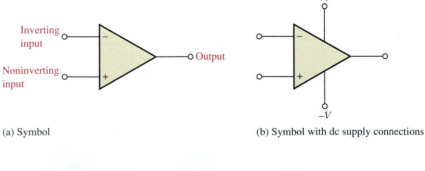

(a) Symbol

(b) Symbol with dc supply connections

(c) Typical packages. Looking from the top, pin 1 always is to the left of the notch or dot on the DIP and SO packages.

FIGURE 12–2
Ideal op-amp representation.

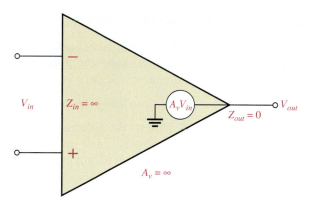

First, the ideal op-amp has infinite voltage gain and infinite bandwidth. Also, it has an infinite input impedance (open), so that it does not load the driving source. Finally, it has a zero output impedance. These characteristics are illustrated in Figure 12–2. The input voltage V_{in} appears between the two input terminals, and the output voltage is $A_v V_{in}$, as indicated by the internal voltage source symbol. The concept of infinite input impedance is a particularly valuable analysis tool for the various op-amp configurations that will be discussed later in the chapter.

The Practical Op-Amp

Although modern integrated circuit op-amps approach parameter values that can be treated as ideal in many cases, the ideal device cannot be made.

Any device has limitations, and the integrated circuit op-amp is no exception. Op-amps have both voltage and current limitations. Peak-to-peak output voltage, for example, is usually limited to slightly less than the two supply voltages. Output current is also limited by internal restrictions such as power dissipation and component ratings.

Characteristics of a practical op-amp are very high voltage gain, very high input impedance, and very low output impedance. Some of these are labelled in Figure 12–3.

FIGURE 12–3
Practical op-amp representation.

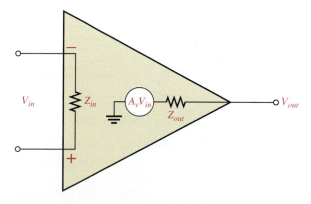

SECTION 12–1 REVIEW	1. What are the connections to a basic op-amp?
	2. Describe some of the characteristics of a practical op-amp.

12–2 ■ THE DIFFERENTIAL AMPLIFIER

The op-amp, in its basic form, typically consists of two or more differential amplifier stages. Because the differential amplifier (diff-amp) is fundamental to the op-amp's internal operation, it is useful to have a basic understanding of this type of circuit.

After completing this section, you should be able to

■ **Discuss the differential amplifier and its operation**
  □ Explain single-ended input operation
  □ Explain differential-input operation
  □ Explain common-mode operation
  □ Define *common-mode rejection ratio*
  □ Discuss the use of differential amplifiers in op-amps

A basic **differential amplifier** circuit and its block symbol are shown in Figure 12–4. The diff-amp stages that make up part of the op-amp provide high voltage gain and common-mode rejection (defined later).

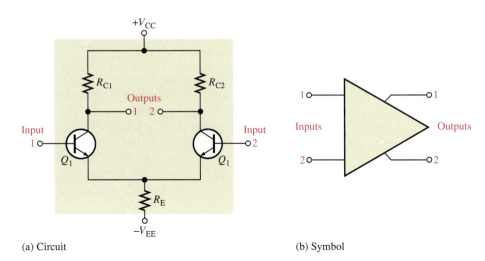

(a) Circuit   (b) Symbol

FIGURE 12–4
Basic differential amplifier.

Basic Operation

Although an op-amp typically has more than one differential amplifier stage, we will use a single diff-amp to illustrate the basic operation. The following discussion is in relation to Figure 12–5 and consists of a basic dc analysis of the diff-amp's operation.

First, when both inputs are grounded (0 V), the emitters are at −0.7 V, as indicated in Figure 12–5(a). It is assumed that the transistors are identically matched by careful process control during manufacturing so that their dc emitter currents are the same when there is no input signal.

$$I_{E1} = I_{E2}$$

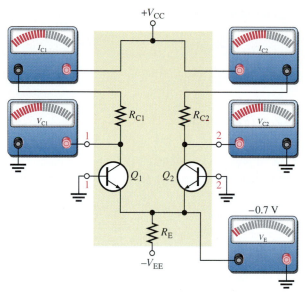

(a) Both inputs grounded

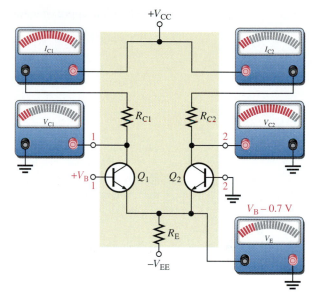

(b) Bias voltage on input 1, input 2 grounded

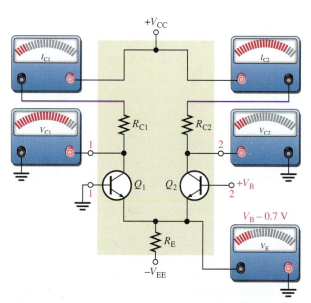

(c) Bias voltage on input 2, input 1 grounded

FIGURE 12–5

Basic operation of a differential amplifier (ground is zero volts).

Since both emitter currents combine through R_E,

$$I_{E1} = I_{E2} = \frac{I_{R_E}}{2}$$

where

$$I_{R_E} = \frac{V_E - V_{EE}}{R_E}$$

Based on the approximation that $I_C \cong I_E$, it can be stated that

$$I_{C1} = I_{C2} \cong \frac{I_{R_E}}{2}$$

Since both collector currents and both collector resistors are equal (when the input voltage is zero),

$$V_{C1} = V_{C2} = V_{CC} - I_{C1}R_{C1}$$

This condition is illustrated in Figure 12–5(a).

Next, input 2 is left grounded, and a positive bias voltage is applied to input 1, as shown in Figure 12–5(b). The positive voltage on the base of Q_1 increases I_{C1} and raises the emitter voltage to

$$V_E = V_B - 0.7 \text{ V}$$

This action reduces the forward bias (V_{BE}) of Q_2 because its base is held at 0 V (ground), thus causing I_{C2} to decrease as indicated in part (b) of the diagram. The net result is that the increase in I_{C1} causes a decrease in $V_{C1,}$ and the decrease in I_{C2} causes an increase in V_{C2}, as shown.

Finally, input 1 is grounded and a positive bias voltage is applied to input 2, as shown in Figure 12–5(c). The positive bias voltage causes Q_2 to conduct more, thus increasing I_{C2}. Also, the emitter voltage is raised. This reduces the forward bias of Q_1, since its base is held at ground, and causes I_{C1} to decrease. The result is that the increase in I_{C2} produces a decrease in V_{C2}, and the decrease in I_{C1} causes V_{C1} to increase, as shown.

Modes of Signal Operation

Single-Ended Input   When a diff-amp is operated in this mode, one input is grounded and the signal voltage is applied only to the other input, as shown in Figure 12–6. In the case where the signal voltage is applied to input 1 as in part (a), an inverted, amplified signal voltage appears at output 1 as shown. Also, a signal voltage appears in phase at the emitter of Q_1. Since the emitters of Q_1 and Q_2 are common, the emitter signal becomes an input to Q_2, which functions as a common-base amplifier. The signal is amplified by Q_2 and appears, noninverted, at output 2. This action is illustrated in part (a).

In the case where the signal is applied to input 2 with input 1 grounded, as in Figure 12–6(b), an inverted, amplified signal voltage appears at output 2. In this situation, Q_1 acts as a common-base amplifier, and a noninverted, amplified signal appears at output 1. This action is illustrated in part (b) of the figure.

Differential Input   In this mode, two opposite-polarity (out-of-phase) signals are applied to the inputs, as shown in Figure 12–7(a). This type of operation is also referred to as *double-ended*. Each input affects the outputs, as you will see in the following discussion. Figure 12–7(b) shows the output signals due to the signal on input 1 acting alone as a single-ended input. Figure 12–7(c) shows the output signals due to the signal on input 2 acting alone as a single-ended input. Notice, in parts (b) and (c), that the signals on output 1 are of the same polarity. The same is also true for output 2. By superimposing both output 1 signals and both output 2 signals, we get the total differential operation, as pictured in Figure 12–7(d).

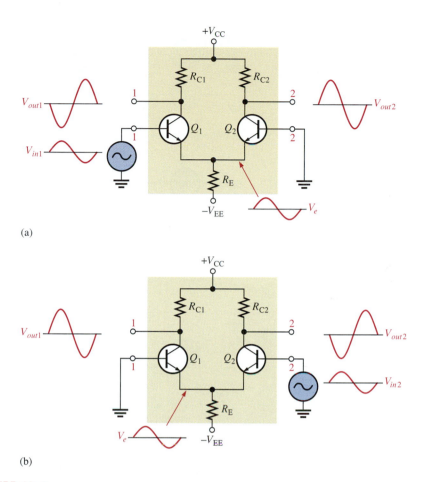

(a)

(b)

FIGURE 12–6

Single-ended input operation of a differential amplifier.

Common-Mode Input One of the most important aspects of the operation of a diff-amp can be seen by considering the **common-mode** condition where two signal voltages of the same phase, frequency, and amplitude are applied to the two inputs, as shown in Figure 12–8(a). Again, by considering each input signal as acting alone, the basic operation can be understood. Figure 12–8(b) shows the output signals due to the signal on only input 1, and Figure 12–8(c) shows the output signals due to the signal on only input 2. Notice that the corresponding signals on output 1 are of the opposite polarity, and so are the ones on output 2. When these are superimposed, they cancel, resulting in a zero output voltage, as shown in Figure 12–8(d).

This action is called *common-mode rejection*. Its importance lies in the situation where an unwanted signal appears commonly on both diff-amp inputs. Common-mode rejection means that this unwanted signal will not appear on the outputs and distort the desired signal. Common-mode signals (noise) generally are the result of the pick-up of radiated energy on the input lines, from adjacent lines, or the 60 Hz power line, or other sources.

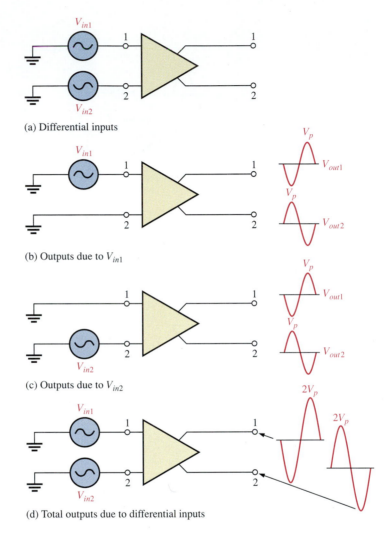

FIGURE 12–7
Differential operation of a differential amplifier.

Common-Mode Rejection Ratio

Desired signals appear on only one input or with opposite polarities on both input lines. These desired signals are amplified and appear on the outputs as previously discussed. Unwanted signals (noise) appearing with the same polarity on both input lines are essentially cancelled by the diff-amp and do not appear on the outputs. The measure of an amplifier's ability to reject common-mode signals is a parameter called the **common-mode rejection ratio (CMRR).**

Ideally, a diff-amp provides a very high gain for desired signals (single-ended or differential), and zero gain for common-mode signals. Practical diff-amps, however, do exhibit a very small common-mode gain (usually much less than one), while providing a high differential voltage gain (usually several thousand). The higher the differential gain with respect to the common-mode gain, the better the performance of the diff-amp in terms of rejection of common-mode signals. This suggests that a good measure of the

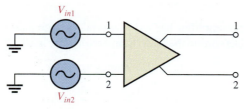

(a) Common-mode inputs

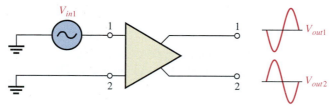

(b) Outputs due to V_{in1}

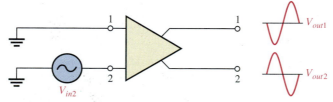

(c) Outputs due to V_{in2}

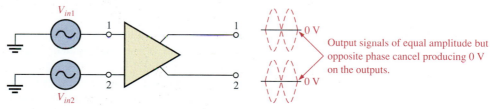

Output signals of equal amplitude but opposite phase cancel producing 0 V on the outputs.

(d) Outputs cancel when common-mode signals are applied.

FIGURE 12–8

Common-mode operation of a differential amplifier.

diff-amp's performance in rejecting unwanted common-mode signals is the ratio of the differential voltage gain $A_{v(d)}$ to the common-mode gain, A_{cm}. This ratio is the common-mode rejection ratio, CMRR.

$$\text{CMRR} = \frac{A_{v(d)}}{A_{cm}} \qquad \textbf{(12–1)}$$

The higher the CMRR, the better. A very high value of CMRR means that the differential gain $A_{v(d)}$ is high and the common-mode gain A_{cm} is low. The CMRR is often expressed in decibels (dB) as

$$\text{CMRR} = 20 \log\left(\frac{A_{v(d)}}{A_{cm}}\right) \qquad \textbf{(12–2)}$$

EXAMPLE 12–1

A certain differential amplifier has a differential voltage gain of 2000 and a common-mode gain of 0.2. Determine the CMRR and express it in dB.

Solution $A_{v(d)} = 2000$, and $A_{cm} = 0.2$. Therefore,

$$\text{CMRR} = \frac{A_{v(d)}}{A_{cm}} = \frac{2000}{0.2} = 10,000$$

Expressed in dB,

$$\text{CMRR} = 20 \log(10,000) = 80 \text{ dB}$$

Related Exercise   Determine the CMRR and express it in dB for an amplifier with a differential voltage gain of 8500 and a common-mode gain of 0.25.

A CMRR of 10,000, for example, means that the desired input signal (differential) is amplified 10,000 times more than the unwanted noise (common-mode). So, as an example, if the amplitudes of the differential input signal and the common-mode noise are equal, the desired signal will appear on the output 10,000 times greater in amplitude than the noise. Thus, the noise or interference has been essentially eliminated.

Example 12–2 illustrates further the idea of common-mode rejection and the general signal operation of the differential amplifier.

EXAMPLE 12–2

The differential amplifier shown in Figure 12–9 has a differential voltage gain of 2500 and a CMRR of 30,000. In part (a), a single-ended input signal of 500 μV rms is applied. At the same time a 1 V, 60 Hz interference signal appears on both inputs as a result of radiated pick-up from the ac power system. In part (b), differential input signals of 500 μV rms each are applied to the inputs. The common-mode interference is the same as in part (a).

(a) Determine the common-mode gain.

(b) Express the CMRR in dB.

(c) Determine the rms output signal for Figure 12–9(a) and (b).

(d) Determine the rms interference voltage on the output.

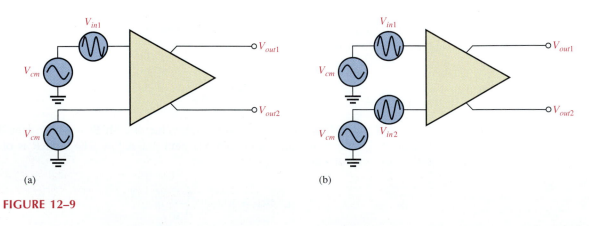

(a)   (b)

FIGURE 12–9

Solution

(a) CMRR $= \dfrac{A_{v(d)}}{A_{cm}}$. Therefore,

$$A_{cm} = \frac{A_{v(d)}}{\text{CMRR}} = \frac{2500}{30{,}000} = 0.083$$

(b) CMRR $= 20 \log(30{,}000) = 89.5$ dB.

(c) In Figure 12–9(a), the differential input voltage, $V_{in(d)}$, is the difference between the voltage on input 1 and that on input 2. Since input 2 is grounded, its voltage is zero. Therefore,

$$V_{in(d)} = V_{in1} - V_{in2} = 500 \ \mu V - 0 \ V = 500 \ \mu V$$

The output signal voltage in this case is taken at output 1.

$$V_{out1} = A_{v(d)}V_{in(d)} = (2500)(500 \ \mu V) = 1.25 \ V \ \text{rms}$$

In Figure 12–9(b), the differential input voltage is the difference between the two opposite-polarity, 500 μV signals.

$$V_{in(d)} = V_{in1} - V_{in2} = 500 \ \mu V - (-500 \ \mu V) = 1000 \ \mu V = 1 \ mV$$

The output voltage signal is

$$V_{out1} = A_{v(d)}V_{in(d)} = (2500)(1 \ mV) = 2.5 \ V \ \text{rms}$$

This shows that a differential input (two opposite-polarity signals) results in a gain that is double that for a single-ended input.

(d) The common-mode input is 1 V rms. The common-mode gain A_{cm} is 0.083. The interference (common-mode) voltage on the output is therefore

$$A_{cm} = \frac{V_{out(cm)}}{V_{in(cm)}}$$

$$V_{out(cm)} = A_{cm}V_{in(cm)} = (0.083)(1 \ V) = 83 \ mV$$

Related Exercise   The amplifier in Figure 12–9 has a differential voltage gain of 4200 and a CMRR of 25,000. For the same single-ended and differential input signals as described in the example: (a) Find A_{cm}. (b) Express the CMRR in dB. (c) Determine the rms output signal for parts (a) and (b) of the figure. (d) Determine the rms interference (common-mode) voltage appearing on the output.

A Simple Op-Amp Arrangement

Figure 12–10 shows two differential amplifier (diff-amp) stages and an emitter-follower connected to form a simple op-amp. The first stage can be used with a single-ended or a differential input. The differential outputs of the first stage are directly coupled into the differential inputs of the second stage. The output of the second stage is single-ended to drive an emitter-follower to achieve a relatively low output impedance. Both differential stages together provide a high voltage gain and a high CMRR.

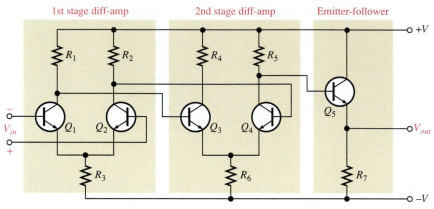

(a) Circuit

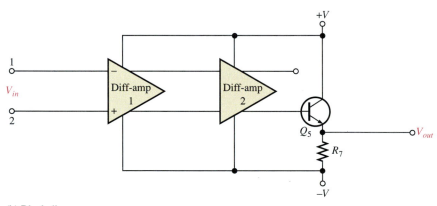

(b) Block diagram

FIGURE 12–10

Simplified internal circuit of a basic op-amp.

SECTION 12–2 REVIEW

1. Distinguish between differential and single-ended inputs.
2. What is common-mode rejection?
3. For a given value of differential gain, does a higher CMRR result in a higher or lower common-mode gain?

12–3 ■ OP-AMP PARAMETERS

In this section, several important op-amp parameters are defined. (These are listed in the objectives that follow.) Also four popular IC op-amps are compared in terms of these parameters.

After completing this section, you should be able to

■ **Discuss several op-amp parameters**
   □ Define *input offset voltage*
   □ Discuss input offset voltage drift with temperature
   □ Define *input bias current*
   □ Define *input impedance*
   □ Define *input offset current*
   □ Define *output impedance*
   □ Discuss common-mode input voltage range
   □ Discuss open-loop voltage gain
   □ Define *common-mode rejection ratio*
   □ Define *slew rate*
   □ Discuss frequency response
   □ Compare op-amp parameters

Input Offset Voltage

The ideal op-amp produces zero volts out for zero volts in. In a practical op-amp, however, a small dc voltage appears at the output when no differential input voltage is applied. Its primary cause is a slight mismatch of the base-to-emitter voltages of the differential input stage, as illustrated in Figure 12–11(a). The output voltage of the differential input stage is expressed as

$$V_{OUT} = I_{C2}R_C - I_{C1}R_C \tag{12-3}$$

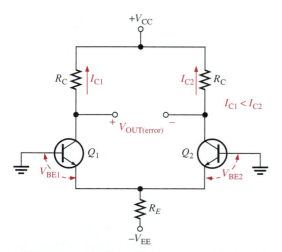

(a) A V_{BE} mismatch ($V_{BE1} > V_{BE2}$) causes a small output error voltage.

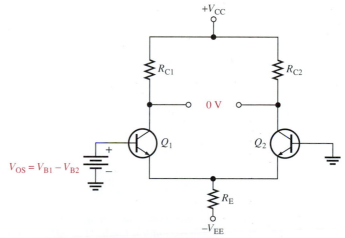

(b) The input offset voltage is the difference in the voltage between the inputs that is necessary to eliminate the output error voltage (makes $V_{OUT} = 0$).

FIGURE 12–11

Illustration of input offset voltage, V_{OS}.

A small difference in the base-to-emitter voltages of Q_1 and Q_2 causes a small difference in the collector currents. This results in a nonzero value of V_{OUT}. (The collector resistors are equal.)

As specified on an op-amp data sheet, the *input offset voltage, V_{OS},* is the differential dc voltage required between the inputs to force the differential output to zero volts. V_{OS} is demonstrated in Figure 12–11(b). Typical values of input offset voltage are in the range of 2 mV or less. In the ideal case, it is 0 V.

Input Offset Voltage Drift with Temperature

The *input offset voltage drift* is a parameter related to V_{OS} that specifies how much change occurs in the input offset voltage for each degree change in temperature. Typical values range anywhere from about 5 µV per degree Celsius to about 50 µV per degree Celsius. Usually, an op-amp with a higher nominal value of input offset voltage exhibits a higher drift.

Input Bias Current

You have seen that the input terminals of a bipolar differential amplifier are the transistor bases and, therefore, the input currents are the base currents.

The *input bias current* is the dc current required by the inputs of the amplifier to properly operate the first stage. By definition, the input bias current is the average of both input currents and is calculated as follows:

$$I_{BIAS} = \frac{I_1 + I_2}{2} \qquad \text{(12–4)}$$

The concept of input bias current is illustrated in Figure 12–12.

FIGURE 12–12
Input bias current is the average of the two op-amp input currents.

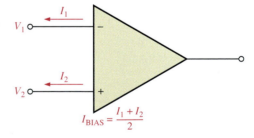

Input Impedance

Two basic ways of specifying the input impedance of an op-amp are the differential and the common mode. The *differential input impedance* is the total resistance between the inverting and the noninverting inputs and is illustrated in Figure 12–13(a). Differential impedance is measured by determining the change in bias current for a given change in differential input voltage. The *common-mode input impedance* is the resistance between each input and ground and is measured by determining the change in bias current for a given change in common-mode input voltage. It is depicted in Figure 12–13(b).

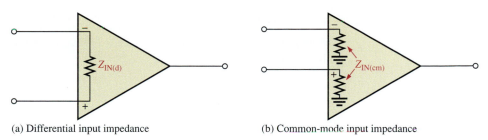

(a) Differential input impedance   (b) Common-mode input impedance

FIGURE 12–13
Op-amp input impedance.

Input Offset Current

Ideally, the two input bias currents are equal, and thus their difference is zero. In a practical op-amp, however, the bias currents are not exactly equal.

The *input offset current* is the difference of the input bias currents, expressed as

$$I_{OS} = |I_1 - I_2| \qquad \qquad \textbf{(12–5)}$$

Actual magnitudes of offset current are usually at least an order of magnitude (ten times) less than the bias current. In many applications, the offset current can be neglected. However, high-gain, high-input impedance amplifiers should have as little I_{OS} as possible, because the difference in currents through large input resistances develops a substantial offset voltage, as shown in Figure 12–14.

The offset voltage developed by the input offset current is

$$V_{OS} = I_1 R_{in} - I_2 R_{in} = (I_1 - I_2)R_{in}$$

$$V_{OS} = I_{OS}R_{in} \qquad \qquad \textbf{(12–6)}$$

FIGURE 12–14
Effect of input offset current.

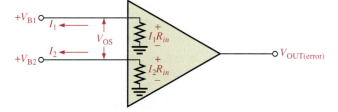

The error created by I_{OS} is amplified by the gain A_v of the op-amp and appears in the output as

$$V_{OUT(error)} = A_v I_{OS} R_{in} \qquad \qquad \textbf{(12–7)}$$

The change in offset current with temperature is often an important consideration. Values of temperature coefficient in the range of 0.5 nA per degree Celsius are common.

Output Impedance

The *output impedance* is the resistance viewed from the output terminal of the op-amp, as indicated in Figure 12–15.

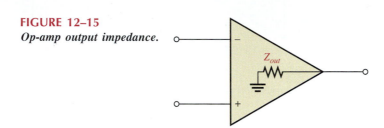

FIGURE 12–15
Op-amp output impedance.

Common-Mode Input Voltage Range

All op-amps have limitations on the range of voltages over which they will operate. The *common-mode input voltage range* is the range of input voltages which, when applied to both inputs, will not cause clipping or other output distortion. Many op-amps have common-mode input voltage ranges of ±10 V with dc supply voltages of ±15 V.

Open-Loop Voltage Gain

The *open-loop voltage gain, A_{ol}*, is the gain of the op-amp without any external feedback from output to input. A good op-amp has a very high **open-loop gain;** 50,000 to 200,000 is typical.

Common-Mode Rejection Ratio

The *common-mode rejection ratio* (CMRR), as discussed in conjunction with the diff-amp, is a measure of an op-amp's ability to reject common-mode signals. An infinite value of CMRR means that the output is zero when the same signal is applied to both inputs (common-mode).

An infinite CMRR is never achieved in practice, but a good op-amp does have a very high value of CMRR. As previously mentioned, common-mode signals are undesired interference voltages such as 60 Hz power-supply ripple and noise voltages due to pick-up of radiated energy. A high CMRR enables the op-amp to virtually eliminate these interference signals from the output.

The accepted definition of CMRR for an op-amp is the open-loop gain (A_{ol}) divided by the common-mode gain.

$$\text{CMRR} = \frac{A_{ol}}{A_{cm}} \qquad\qquad \textbf{(12–8)}$$

It is commonly expressed in decibels as follows:

$$\text{CMRR} = 20 \log\left(\frac{A_{ol}}{A_{cm}}\right) \qquad\qquad \textbf{(12–9)}$$

EXAMPLE 12–3

A certain op-amp has an open-loop gain of 100,000 and a common-mode gain of 0.25. Determine the CMRR and express it in dB.

Solution

$$\text{CMRR} = \frac{A_{ol}}{A_{cm}} = \frac{100,000}{0.25} = 400,000$$

Expressed in dB,

$$\text{CMRR} = 20 \log(400,000) = 112 \text{ dB}$$

Related Exercise If a particular op-amp has a CMRR of 90 dB and a common-mode gain of 0.4, what is the open-loop gain?

Slew Rate

The maximum rate of change of the output voltage in response to a step input voltage is the **slew rate** of an op-amp. The slew rate is dependent upon the high-frequency response of the amplifier stages within the op-amp.

Slew rate is measured with an op-amp connected as shown in Figure 12–16(a). This particular op-amp connection is a unity-gain, noninverting configuration which will be discussed later. It gives a worst-case (slowest) slew rate. Recall that the high-frequency components of a voltage step are contained in the rising edge and that the upper critical frequency of an amplifier limits its response to a step input. The lower the upper critical frequency is, the more slope there is on the output for a step input.

FIGURE 12–16
Slew-rate measurement.

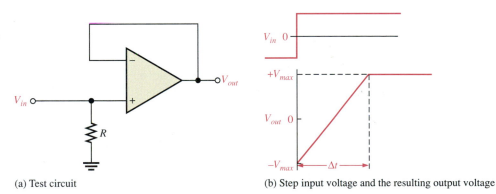

(a) Test circuit

(b) Step input voltage and the resulting output voltage

A pulse is applied to the input as shown, and the ideal output voltage is measured as indicated in Figure12–16(b). The width of the input pulse must be sufficient to allow the output to "slew" from its lower limit to its upper limit, as shown. As you can see, a certain time interval, Δt, is required for the output voltage to go from its lower limit $-V_{max}$ to its upper limit $+V_{max}$, once the input step is applied. The slew rate is expressed as

$$\text{Slew rate} = \frac{\Delta V_{out}}{\Delta t} \qquad \qquad \textbf{(12–10)}$$

where $\Delta V_{out} = +V_{max} - (-V_{max})$. The unit of slew rate is volts per microsecond (V/μs).

EXAMPLE 12–4   The output voltage of a certain op-amp appears as shown in Figure 12–17 in response to a step input. Determine the slew rate.

FIGURE 12–17

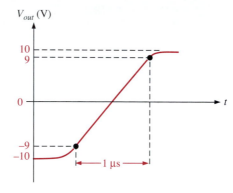

Solution   The output goes from the lower to the upper limit in 1 μs. Since this is not an ideal response, the limits are taken at the 90 percent points, as indicated. So, the upper limit is +9 V and the lower limit is −9 V. The slew rate is

$$\frac{\Delta V_{out}}{\Delta t} = \frac{+9 \text{ V} - (-9 \text{ V})}{1 \text{ μs}} = 18 \text{ V/μs}$$

Related Exercise   When a pulse is applied to an op-amp, the output voltage goes from −8 V to +7 V in 0.75 μs. What is the slew rate?

Frequency Response

The internal amplifier stages that make up an op-amp have voltage gains limited by junction capacitances, as discussed in Chapter 10. Although the differential amplifiers used in op-amps are somewhat different from the basic amplifiers discussed, the same principles apply. An op-amp has no internal coupling capacitors, however; therefore, the low-frequency response extends down to dc. Frequency-related characteristics will be discussed in the next chapter.

Comparison of Op-Amp Parameters

Table 12–1 provides a comparison of values of some of the parameters just described for four common integrated circuit op-amps. Any values not listed were not given on the manufacturer's data sheet. All values are typical at 25°C.

Other Features

Most available op-amps have three important features: short-circuit protection, no latch-up, and input offset nulling. Short-circuit protection keeps the circuit from being damaged if the output becomes shorted, and the no latch-up feature prevents the op-amp from hanging up in one output state (high or low voltage level) under certain input conditions. Input offset nulling is achieved by an external potentiometer that sets the output voltage at precisely zero with zero input.

TABLE 12–1

Parameter	Op-Amp Type			
	741C	LM101A	LM108	LM218
Input offset voltage	1 mV	1 mV	0.7 mV	2 mV
Input bias current	80 nA	120 nA	0.8 nA	120 nA
Input offset current	20 nA	40 nA	0.05 nA	6 nA
Input impedance	2 MΩ	800 kΩ	70 MΩ	3 MΩ
Output impedance	75 Ω	—	—	—
Open-loop gain	200,000	160,000	300,000	200,000
Slew rate	0.5 V/μs	—	—	70 V/μs
CMRR	90 dB	90 dB	100 dB	100 dB

SECTION 12–3 REVIEW

1. List at least ten op-amp parameters.
2. Which two parameters, not including the frequency response, are frequency dependent?

12–4 ▪ NEGATIVE FEEDBACK

Negative feedback is one of the most useful concepts in electronic circuits, particularly in op-amp applications. Negative feedback is the process whereby a portion of the output voltage of an amplifier is returned to the input with a phase angle that opposes (or subtracts from) the input signal.

After completing this section, you should be able to

▪ **Explain negative feedback in op-amp circuits**
  ☐ Discuss why negative feedback is used
  ☐ Describe the effects of negative feedback

Negative feedback is illustrated in Figure 12–18. The inverting input effectively makes the feedback signal 180° out of phase with the input signal.

FIGURE 12–18
Illustration of negative feedback.

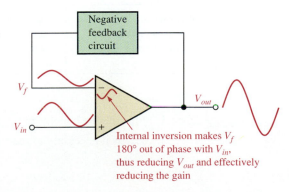

Internal inversion makes V_f 180° out of phase with V_{in}, thus reducing V_{out} and effectively reducing the gain

Why Use Negative Feedback?

As you have seen, the inherent open-loop gain of a typical op-amp is very high (usually greater than 100,000). Therefore, an extremely small input voltage drives the op-amp into its saturated output states. In fact, even the input offset voltage of the op-amp can drive it into saturation. For example, assume $V_{in} = 1$ mV and $A_{ol} = 100,000$. Then,

$$V_{in}A_{ol} = (1 \text{ mV})(100,000) = 100 \text{ V}$$

Since the output level of an op-amp can never reach 100 V, it is driven deep into saturation and the output is limited to its maximum output levels, as illustrated in Figure 12–19 for both a positive and a negative input voltage of 1 mV.

FIGURE 12–19

Without negative feedback, a small input voltage drives the op-amp to its output limits and it becomes nonlinear.

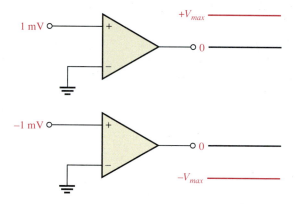

The usefulness of an op-amp operated without negative feedback is severely restricted and is generally limited to comparator and other applications (to be studied in Chapter 14). With negative feedback, the voltage gain (A_{cl}) can be reduced and controlled so that the op-amp can function as a linear amplifier. In addition to providing a controlled, stable voltage gain, negative feedback also provides for control of the input and output impedances and amplifier bandwidth. Table 12–2 summarizes the general effects of negative feedback on op-amp performance.

TABLE 12–2

	Voltage Gain	Input Z	Output Z	Bandwidth
Without negative feedback	A_{ol} is too high for linear amplifier applications	Relatively high (see Table 12–1)	Relatively low (see Table 12–1)	Relatively narrow
With negative feedback	A_{cl} is set by the feedback circuit to desired value	Can be increased or reduced to a desired value depending on type of circuit	Can be reduced to a desired value	Significantly wider

SECTION 12–4 REVIEW

1. What are the benefits of negative feedback in an op-amp circuit?
2. Why is it necessary to reduce the gain of an op-amp from its open-loop value?

12–5 ▪ OP-AMP CONFIGURATIONS WITH NEGATIVE FEEDBACK

In this section, we will discuss several basic ways in which an op-amp can be connected using negative feedback to stabilize the gain and increase frequency response. The extremely high open-loop gain of an op-amp creates an unstable situation because a small noise voltage on the input can be amplified to a point where the amplifier is driven out of its linear region. Also, unwanted oscillations can occur. In addition, the open-loop gain parameter of an op-amp can vary greatly from one device to the next. Negative feedback takes a portion of the output and applies it back out of phase with the input, creating an effective reduction in gain. This closed-loop gain is usually much less than the open-loop gain and independent of it.

After completing this section, you should be able to

▪ **Analyze three basic op-amp configurations**
  ☐ Identify the noninverting amplifier configuration
  ☐ Determine the voltage gain of a noninverting amplifier
  ☐ Identify the voltage-follower configuration
  ☐ Identify the inverting amplifier configuration
  ☐ Determine the voltage gain of an inverting amplifier

Noninverting Amplifier

An op-amp connected in a **closed loop** configuration as a **noninverting amplifier** with a controlled amount of voltage gain is shown in Figure 12–20. The input signal is applied to the noninverting input. The output is applied back to the inverting input through the feedback circuit formed by R_i and R_f. This creates negative feedback as follows. R_i and R_f form a voltage-divider circuit, which reduces V_{out} and connects the reduced voltage V_f to the inverting input. The feedback voltage is expressed as

$$V_f = \left(\frac{R_i}{R_i + R_f} \right) V_{out}$$

The difference of the input voltage, V_{in}, and the feedback voltage, V_f, is the differential input to the op-amp, as shown in Figure 12–21. This differential voltage is

FIGURE 12–20
Noninverting amplifier.

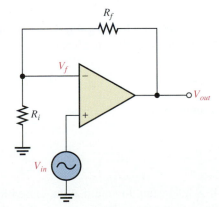

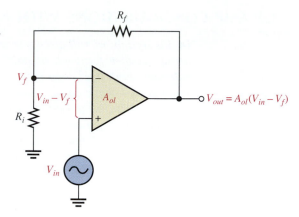

FIGURE 12–21
Differential input, $V_{in} - V_f$.

amplified by the open-loop gain of the op-amp (A_{ol}) and produces an output voltage expressed as

$$V_{out} = A_{ol}(V_{in} - V_f)$$

Let $R_i/(R_i + R_f) = B$, which is the attenuation of the feedback circuit. Then substitute BV_{out} for V_f:

$$V_{out} = A_{ol}(V_{in} - BV_{out})$$

Then apply basic algebra as follows:

$$V_{out} = A_{ol}V_{in} - A_{ol}BV_{out}$$
$$V_{out} + A_{ol}BV_{out} = A_{ol}V_{in}$$
$$V_{out}(1 + A_{ol}B) = A_{ol}V_{in}$$

Since the total voltage gain of the amplifier in Figure 12–20 is V_{out}/V_{in},

$$\frac{V_{out}}{V_{in}} = \frac{A_{ol}}{1 + A_{ol}B} \qquad (12\text{–}11)$$

The product $A_{ol}B$ is typically much greater than 1, so Equation (12–11) simplifies to

$$\frac{V_{out}}{V_{in}} = \frac{A_{ol}}{A_{ol}B} = \frac{1}{B}$$

The **closed-loop gain,** $A_{cl(NI)}$, of the noninverting (NI) amplifier is the reciprocal of the attenuation (B) of the feedback circuit (voltage-divider).

$$A_{cl(NI)} = \frac{V_{out}}{V_{in}} = \frac{1}{B} = \frac{R_i + R_f}{R_i}$$

Therefore,

$$A_{cl(NI)} = 1 + \frac{R_f}{R_i} \qquad (12\text{–}12)$$

It is interesting to note that the closed-loop gain is not at all dependent on the op-amp's open-loop gain under the condition $A_{ol}B \gg 1$. The closed-loop gain can be set by selecting values of R_i and R_f.

EXAMPLE 12–5   Determine the gain of the amplifier in Figure 12–22. The open-loop voltage gain is 100,000.

FIGURE 12–22

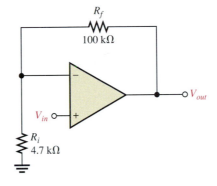

Solution   This is a noninverting op-amp configuration. Therefore, the closed-loop gain is

$$A_{cl(\mathrm{NI})} = 1 + \frac{R_f}{R_i} = 1 + \frac{100\ \mathrm{k\Omega}}{4.7\ \mathrm{k\Omega}} = 22.3$$

Related Exercise   If the open-loop gain of the amplifier in Figure 12–22 is 150,000 and R_f is increased to 150 kΩ, determine the closed-loop gain.

Voltage-Follower

The **voltage-follower** configuration is a special case of the noninverting amplifier where all of the output voltage is fed back to the inverting input, as shown in Figure 12–23. As you can see, the straight feedback connection has a voltage gain of approximately one. The closed-loop voltage gain of a noninverting amplifier is $1/B$ as previously derived. Since $B = 1$ for a voltage-follower, the closed-loop gain of the voltage-follower is

$$A_{cl(\mathrm{VF})} = 1 \qquad\qquad (12\text{–}13)$$

FIGURE 12–23
Op-amp voltage follower.

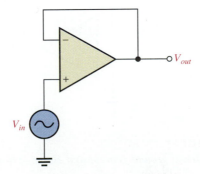

The most important features of the voltage-follower configuration are its very high input impedance and its very low output impedance. These features make it a nearly ideal buffer amplifier for interfacing high-impedance sources and low-impedance loads. This is discussed further in Section 12–6.

Inverting Amplifier

An op-amp connected as an **inverting amplifier** with a controlled amount of voltage gain is shown in Figure 12–24. The input signal is applied through a series input resistor R_i to the inverting input. Also, the output is fed back through R_f to the same input. The noninverting input is grounded.

FIGURE 12–24
Inverting amplifier.

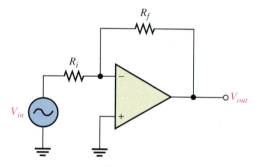

At this point, the ideal op-amp parameters mentioned earlier are useful in simplifying the analysis of this circuit. In particular, the concept of infinite input impedance is of great value. An infinite input impedance implies zero current at the inverting input. If there is zero current through the input impedance, then there must be *no* voltage drop between the inverting and noninverting inputs. This means that the voltage at the inverting (−) input is zero because the other input (+) is grounded. This zero voltage at the inverting input terminal is referred to as *virtual ground*. This condition is illustrated in Figure 12–25(a).

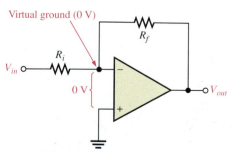

(a) Virtual ground

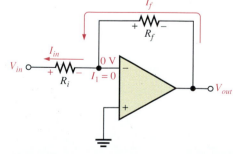

(b) $I_{in} = I_f$ and current into the inverting input (I_1) is 0.

FIGURE 12–25

Virtual ground concept and closed-loop voltage gain development for the inverting amplifier.

Since there is zero current from the inverting input, the current through R_i and the current through R_f are equal, as shown in Figure 12–25(b).

$$I_{in} = I_f$$

The voltage across R_i equals V_{in} because of virtual ground on the other side of the resistor. Therefore,

$$I_{in} = \frac{V_{in}}{R_i}$$

Also, the voltage across R_f equals $-V_{out}$ because of virtual ground, and therefore,

$$I_f = \frac{-V_{out}}{R_f}$$

Since $I_f = I_{in}$,

$$\frac{-V_{out}}{R_f} = \frac{V_{in}}{R_i}$$

Rearranging the terms, you get

$$\frac{V_{out}}{V_{in}} = -\frac{R_f}{R_i}$$

Of course, V_{out}/V_{in} is the gain of the inverting (I) amplifier.

$$A_{cl(I)} = -\frac{R_f}{R_i} \qquad \textbf{(12–14)}$$

Equation (12–14) shows that the closed-loop voltage gain $A_{cl(I)}$ of the inverting amplifier is the ratio of the feedback resistance R_f to the resistance R_i. *The closed-loop gain is independent of the op-amp's internal open-loop gain.* Thus, the negative feedback stabilizes the voltage gain. The negative sign indicates inversion.

EXAMPLE 12–6

Given the op-amp configuration in Figure 12–26, determine the value of R_f required to produce a closed-loop voltage gain of -100.

FIGURE 12–26

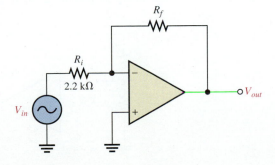

Solution   Knowing that $R_i = 2.2$ kΩ and $|A_{cl(I)}| = 100$, calculate R_f as follows:

$$|A_{cl(I)}| = \frac{R_f}{R_i}$$

$$R_f = |A_{cl(I)}|R_i = (100)(2.2 \text{ k}\Omega) = 220 \text{ k}\Omega$$

Related Exercise   If R_i is changed to 2.7 kΩ in Figure 12–26, what value of R_f is required to produce a closed-loop gain with an absolute value of 25?

SECTION 12–5 REVIEW

1. What is the main purpose of negative feedback?
2. The closed-loop voltage gain of each of the op-amp configurations discussed is dependent on the internal open-loop voltage gain of the op-amp (T or F).
3. The attenuation of the negative feedback circuit of a noninverting op-amp configuration is 0.02. What is the closed-loop gain of the amplifier?

12–6 ■ EFFECTS OF NEGATIVE FEEDBACK ON OP-AMP IMPEDANCES

In this section, you will see how negative feedback affects the input and output impedances of an op-amp. The effects on both inverting and noninverting amplifiers are examined.

After completing this section, you should be able to

■ **Describe impedances of the three op-amp configurations**
  □ Determine input and output impedances of a noninverting amplifier
  □ Determine input and output impedances of a voltage-follower
  □ Determine input and output impedances of an inverting amplifier

Input Impedance of the Noninverting Amplifier

The input impedance of the noninverting amplifier can be developed with the aid of Figure 12–27. For this analysis, assume a small differential voltage, V_d, exists between the

FIGURE 12–27

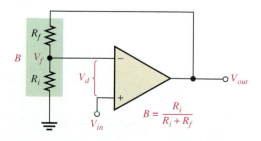

$$B = \frac{R_i}{R_i + R_f}$$

two inputs, as indicated. This means that you cannot assume the op-amp's input imped-ance to be infinite or the input current to be zero. Express the input voltage as

$$V_{in} = V_d + V_f$$

Substituting BV_{out} for V_f, you get

$$V_{in} = V_d + BV_{out}$$

Since $V_{out} \cong A_{ol}V_d$ (A_{ol} is the open-loop gain of the op-amp),

$$V_{in} = V_d + A_{ol}BV_d = (1 + A_{ol}B)V_d$$

Now substituting $I_{in}Z_{in}$ for V_d, you get

$$V_{in} = (1 + A_{ol}B)I_{in}Z_{in}$$

where Z_{in} is the open-loop input impedance of the op-amp (without feedback connec-tions).

$$\frac{V_{in}}{I_{in}} = (1 + A_{ol}B)Z_{in}$$

V_{in}/I_{in} is the overall input impedance of the closed-loop noninverting configuration.

$$Z_{in(NI)} = (1 + A_{ol}B)Z_{in} \qquad (12\text{--}15)$$

This equation shows that the input impedance of this amplifier configuration with nega-tive feedback is much greater than the internal input impedance of the op-amp itself (without feedback).

Output Impedance of the Noninverting Amplifier

You can develop an expression for output impedance with the aid of Figure 12–28.

By applying Kirchhoff's law to the output circuit, you get

$$V_{out} = A_{ol}V_d - Z_{out}I_{out}$$

The differential input voltage is $V_{in} - V_f$; therefore, by assuming that $A_{ol}V_d \gg Z_{out}I_{out}$, you can express the output voltage as

$$V_{out} \cong A_{ol}(V_{in} - V_f)$$

FIGURE 12–28

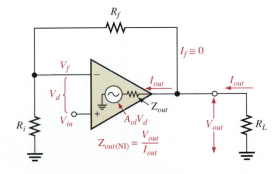

Substituting BV_{out} for V_f, you get

$$V_{out} \cong A_{ol}(V_{in} - BV_{out})$$

Remember, B is the attenuation of the negative feedback circuit. Expanding and factoring, you get

$$V_{out} \cong A_{ol}V_{in} - A_{ol}BV_{out}$$
$$A_{ol}V_{in} \cong V_{out} + A_{ol}BV_{out} \cong (1 + A_{ol}B)V_{out}$$

Since the output impedance of the noninverting configuration is $Z_{out(NI)} = V_{out}/I_{out}$, you can substitute $I_{out}Z_{out(NI)}$ for V_{out}:

$$A_{ol}V_{in} = (1 + A_{ol}B)I_{out}Z_{out(NI)}$$

Dividing both sides of the above expression by I_{out}, you get

$$\frac{A_{ol}V_{in}}{I_{out}} = (1 + A_{ol}B)Z_{out(NI)}$$

The term on the left is the internal output impedance of the op-amp (Z_{out}) because, without feedback, $A_{ol}V_{in} = V_{out}$. Therefore,

$$Z_{out} = (1 + A_{ol}B)Z_{out(NI)}$$

Thus,

$$Z_{out(NI)} = \frac{Z_{out}}{1 + A_{ol}B} \qquad (12\text{--}16)$$

This equation shows that the output impedance of this amplifier configuration with negative feedback is much less than the internal output impedance of the op-amp itself (without feedback).

EXAMPLE 12–7

(a) Determine the input and output impedances of the amplifier in Figure 12–29. The op-amp data sheet gives $Z_{in} = 2$ MΩ, $Z_{out} = 75$ Ω, and $A_{ol} = 200,000$.

(b) Find the closed-loop voltage gain.

FIGURE 12–29

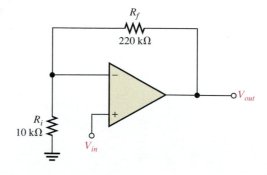

Solution

(a) The attenuation of the feedback circuit is

$$B = \frac{R_i}{R_i + R_f} = \frac{10 \text{ k}\Omega}{230 \text{ k}\Omega} = 0.043$$

$$Z_{in(\text{NI})} = (1 + A_{ol}B)Z_{in} = [1 + (200{,}000)(0.043)](2 \text{ M}\Omega)$$
$$= (1 + 8600)(2 \text{ M}\Omega) = 17{,}202 \text{ M}\Omega$$

$$Z_{out(\text{NI})} = \frac{Z_{out}}{1 + A_{ol}B} = \frac{75 \text{ }\Omega}{1 + 8600} = 8.7 \text{ m}\Omega$$

(b) $A_{cl(\text{NI})} = \dfrac{1}{B} = \dfrac{1}{0.043} \cong 23.3$

Related Exercise

(a) Determine the input and output impedances in Figure 12–29 for op-amp data sheet values of $Z_{in} = 3.5 \text{ M}\Omega$, $Z_{out} = 82 \text{ }\Omega$, and $A_{ol} = 135{,}000$.

(b) Find A_{cl}.

Voltage-Follower Impedances

Since the voltage-follower is a special case of the noninverting configuration, the same impedance formulas are used with $B = 1$.

$$Z_{in(\text{VF})} = (1 + A_{ol})Z_{in} \tag{12–17}$$

$$Z_{out(\text{VF})} = \frac{Z_{out}}{1 + A_{ol}} \tag{12–18}$$

As you can see, the voltage-follower input impedance is greater for a given A_{ol} and Z_{in} than for the noninverting configuration with the voltage-divider feedback circuit. Also, its output impedance is much smaller.

EXAMPLE 12–8

The same op-amp in Example 12–7 is used in a voltage-follower configuration. Determine the input and output impedances.

Solution Since $B = 1$,

$$Z_{in(\text{VF})} = (1 + A_{ol})Z_{in} = (1 + 200{,}000)(2 \text{ M}\Omega) \cong 400{,}000 \text{ M}\Omega$$

$$Z_{out(\text{VF})} = \frac{Z_{out}}{1 + A_{ol}} = \frac{75 \text{ }\Omega}{1 + 200{,}000} = 0.38 \text{ m}\Omega$$

Notice that $Z_{in(\text{VF})}$ is much greater than $Z_{in(\text{NI})}$, and $Z_{out(\text{VF})}$ is much less than $Z_{out(\text{NI})}$ from Example 12–7.

Related Exercise If the op-amp in this example is replaced with one having a higher open-loop gain, how are the input and output impedances affected?

Impedances of an Inverting Amplifier

The input and output impedances of the inverting op-amp configuration are developed with the aid of Figure 12–30. Both the input signal and the negative feedback are applied, through resistors, to the inverting terminal.

FIGURE 12–30
Inverting amplifier.

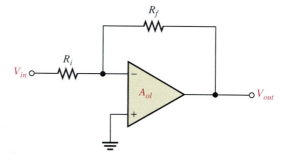

Miller's Theorem Miller's theorem can be applied to the inverting configuration. By Miller's theorem, the effective input impedance of an amplifier with a feedback resistor from output to input as in Figure 12–30 is

$$Z_{in(Miller)} = \frac{R_f}{A_{ol} + 1}$$ (12–19)

and

$$Z_{out(Miller)} = \left(\frac{A_{ol}}{A_{ol} + 1}\right) R_f$$ (12–20)

Applying Miller's theorem to the circuit of Figure 12–30, you get the equivalent circuit of Figure 12–31. As indicated, the Miller input impedance appears in parallel with the internal input impedance of the op-amp, and R_i appears in series with this as follows:

$$Z_{in(I)} = R_i + Z_{in(Miller)} \| Z_{in} = R_i + \frac{R_f}{A_{ol} + 1} \| Z_{in}$$

Typically, $R_f/(A_{ol} + 1)$ is much less than the Z_{in} of an open-loop op-amp; also, $A_{ol} \gg 1$. So the formula simplifies to

$$Z_{in(I)} \cong R_i + \frac{R_f}{A_{ol}}$$

FIGURE 12–31
Miller equivalent for the inverting amplifier in Figure 12–30.

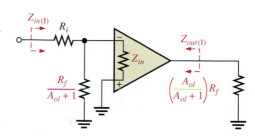

Since R_i appears in series with R_f/A_{ol} and if $R_i \gg R_f/A_{ol}$, $Z_{in(I)}$ reduces to

$$Z_{in(I)} \cong R_i \qquad\qquad\qquad \text{(12–21)}$$

The Miller output impedance appears in parallel with Z_{out} of the op-amp.

$$Z_{out(I)} = Z_{out\,(Miller)} \parallel Z_{out} = \left(\frac{A_{ol}}{A_{ol}+1}\right)R_f \parallel Z_{out}$$

Typically $A_{ol} \gg 1$ and $R_f \gg Z_{out}$, so $Z_{out(I)}$ simplifies to

$$Z_{out(I)} \cong Z_{out} \qquad\qquad\qquad \text{(12–22)}$$

EXAMPLE 12–9

Find the values of the input and output impedances in Figure 12–32. Also, determine the closed-loop voltage gain. The op-amp has the following parameters: $A_{ol} = 50{,}000$; $Z_{in} = 4\text{ M}\Omega$; and $Z_{out} = 50\ \Omega$.

FIGURE 12–32

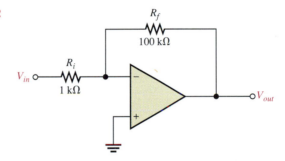

Solution

$$Z_{in(I)} \cong R_i = 1\text{ k}\Omega$$
$$Z_{out(I)} \cong Z_{out} = 50\ \Omega$$
$$A_{cl(I)} = -\frac{R_f}{R_i} = -\frac{100\text{ k}\Omega}{1\text{ k}\Omega} = -100$$

Related Exercise Determine the input and output impedances and the closed-loop voltage gain in Figure 12–32. The op-amp parameters and circuit values are as follows: $A_{ol} = 100{,}000$; $Z_{in} = 5\text{ M}\Omega$; $Z_{out} = 75\ \Omega$; $R_i = 560\ \Omega$; and $R_f = 82\text{ k}\Omega$.

SECTION 12–6 REVIEW

1. How does the input impedance of a noninverting amplifier configuration compare to the input impedance of the op-amp itself?

2. When an op-amp is connected in a voltage-follower configuration, does the input impedance increase or decrease?

3. Given that $R_f = 100\text{ k}\Omega$; $R_i = 2\text{ k}\Omega$; $A_{ol} = 120{,}000$; $Z_{in} = 2\text{ M}\Omega$; and $Z_{out} = 60\ \Omega$, what are $Z_{in(I)}$ and $Z_{out(I)}$ for an inverting amplifier configuration?

SUMMARY OF OP-AMP CONFIGURATIONS

BASIC OP-AMP

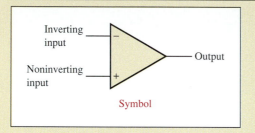

- Very high open loop voltage gain
- Very high input impedance
- Very low output impedance

NONINVERTING AMPLIFIER

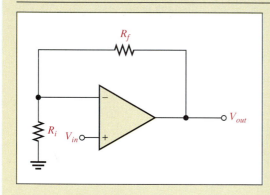

- Voltage gain:

$$A_{cl(\text{NI})} = 1 + \frac{R_f}{R_i}$$

- Input impedance:

$$Z_{in(\text{NI})} = (1 + A_{ol}B)Z_{in}$$

- Output impedance:

$$Z_{out(\text{NI})} = \frac{Z_{out}}{1 + A_{ol}B}$$

VOLTAGE-FOLLOWER

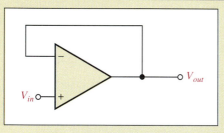

- Voltage gain:

$$A_{cl(\text{VF})} = 1$$

- Input impedance:

$$Z_{in(\text{VF})} = (1 + A_{ol})Z_{in}$$

- Output impedance:

$$Z_{out(\text{VF})} = \frac{Z_{out}}{1 + A_{ol}}$$

SUMMARY OF OP-AMP CONFIGURATIONS, *continued*

INVERTING AMPLIFIER

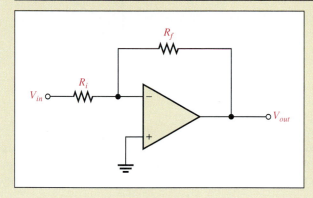

■ Voltage gain:

$$A_{cl(\text{I})} = -\frac{R_f}{R_i}$$

■ Input impedance:

$$Z_{in(\text{I})} \cong R_i$$

■ Output impedance:

$$Z_{out(\text{I})} \cong Z_{out}$$

12–7 ■ BIAS CURRENT AND OFFSET VOLTAGE COMPENSATION

Until now, we have treated the op-amp as an ideal device in many of our discussions. However, certain deviations from the ideal op-amp must be recognized because of their effects on its operation. Transistors within the op-amp must be biased so that they have the correct values of base and collector current and collector-to-emitter voltages. The ideal op-amp has no input current at its terminals, but in fact, the practical op-amp has small input bias currents typically in the nA range. Also, small internal imbalances in the transistors effectively produce a small offset voltage between the inputs. These non-ideal parameters were described in Section 12–3.

After completing this section, you should be able to

■ **Discuss op-amp compensation**
   □ Describe the effect of input bias current
   □ Explain bias current compensation
   □ Describe the effect of input offset voltage
   □ Explain input offset voltage compensation

Effect of an Input Bias Current

Figure 12–33 is an inverting amplifier with zero input voltage. Ideally, the current through R_i is zero because the input voltage is zero and the voltage at the inverting (−) terminal is zero. The small input current I_1 is furnished through R_f. I_1 creates a voltage drop across R_f, as indicated. The positive side of R_f is the output terminal, and therefore, the output error voltage is $I_1 R_f$ when it should be zero.

Figure 12–34 is a voltage-follower with zero input voltage and a source resistance R_s. In this case, an input current I_1 produces a drop across R_s and creates an output volt-

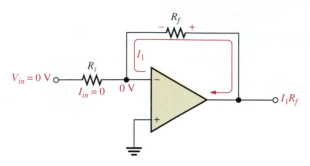

FIGURE 12–33

Input bias current creates output error voltage (I_1R_f) in inverting amplifier.

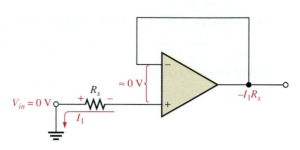

FIGURE 12–34

Input bias current creates output error voltage in a voltage-follower.

age error as shown. The voltage at the inverting input terminal decreases to $-I_1R_s$ because the negative feedback tends to maintain a differential voltage of zero, as indicated. Since the inverting terminal is connected directly to the output terminal, the output error voltage is $-I_1R_s$.

Figure 12–35 is a noninverting amplifier with zero input voltage. Ideally, the voltage at the inverting terminal is also zero, as indicated. The input current I_1 produces a voltage drop across R_f and thus creates an output error voltage of I_1R_f, just as with the inverting amplifier.

FIGURE 12–35

Input bias current creates output error voltage in noninverting amplifier.

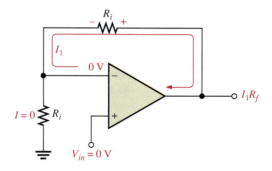

Bias Current Compensation in a Voltage-Follower

The output error voltage due to bias currents in a voltage-follower can be sufficiently reduced by adding a resistor equal to R_s in the feedback path, as shown in Figure 12–36. The voltage drop created by I_1 across the added resistor subtracts from the $-I_2R_s$ output error voltage. If $I_1 = I_2$, then the output voltage is zero. Usually I_1 does not quite equal I_2; but even in this case, the output error voltage is reduced as follows, because I_{OS} is less than I_2.

$$V_{OUT(error)} = |I_1 - I_2|R_s$$

$$V_{OUT(error)} = I_{OS}R_s \qquad \textbf{(12–23)}$$

where I_{OS} is the input offset current.

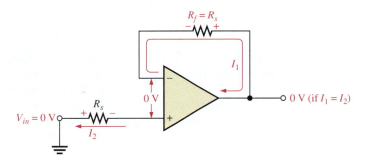

FIGURE 12–36
Bias current compensation in a voltage-follower.

Bias Current Compensation in Other Op-Amp Configurations

To compensate for the effect of bias current in the noninverting amplifier, a resistor R_c is added, as shown in Figure 12–37(a). The compensating resistor value equals the parallel combination of R_i and R_f. The input current creates a voltage drop across R_c that offsets the voltage across the combination of R_i and R_f, thus sufficiently reducing the output error voltage. The inverting amplifier is similarly compensated, as shown in Figure 12–37(b).

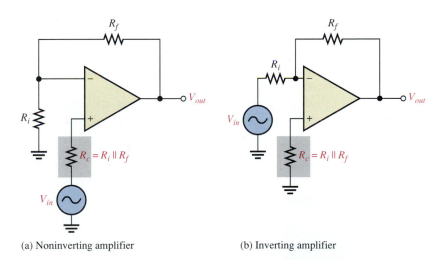

(a) Noninverting amplifier   (b) Inverting amplifier

FIGURE 12–37
Bias current compensation in the noninverting and inverting configurations.

Use of a BIFET Op-Amp to Eliminate the Need for Bias Current Compensation

The BIFET op-amp uses both bipolar junction transistors and JFETs in its internal circuitry. The JFETs are used as the input devices to achieve a higher input impedance than is possible with standard bipolar amplifiers. Because of their very high input impedance, BIFETs typically have input bias currents that are much smaller than in bipolar op-amps, thus reducing or eliminating the need for bias current compensation.

Effect of Input Offset Voltage

The output voltage of an op-amp should be zero when the differential input is zero. However, there is always a small output error voltage present whose value typically ranges from microvolts to millivolts. This is due to unavoidable imbalances within the internal op-amp transistors aside from the bias currents previously discussed. In a negative feedback configuration, the input offset voltage V_{IO} can be visualized as an equivalent small dc voltage source, as illustrated in Figure 12–38 for a voltage-follower. The output error voltage due to the input offset voltage in this case is

$$V_{OUT(error)} = V_{IO}$$

(12–24)

FIGURE 12–38
Input offset voltage equivalent.

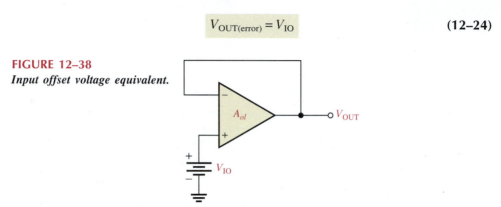

Input Offset Voltage Compensation

Most integrated circuit op-amps provide a means of compensating for offset voltage. This is usually done by connecting an external potentiometer to designated pins on the IC package, as illustrated in Figure 12–39(a) and (b) for a 741 op-amp. The two terminals are labelled *offset null*. With no input, the potentiometer is simply adjusted until the output voltage reads 0, as shown in Figure 12–39(c).

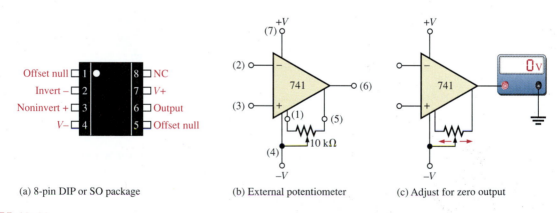

(a) 8-pin DIP or SO package   (b) External potentiometer   (c) Adjust for zero output

FIGURE 12–39
Input offset voltage compensation for a 741.

SECTION 12–7 REVIEW	
	1. What are two sources of dc output error voltages?
	2. How do you compensate for bias current in a voltage-follower?

12–8 ■ TROUBLESHOOTING

As a technician, you will encounter situations in which an op-amp or its associated circuitry has malfunctioned. The op-amp is a complex integrated circuit with many types of internal failures possible. However, since you cannot troubleshoot the op-amp internally, you treat it as a single device with only a few connections to it. If it fails, you replace it just as you would a resistor, capacitor, or transistor.

After completing this section, you should be able to

■ **Troubleshoot op-amp circuits**
   ☐ Analyze faults in a noninverting amplifier
   ☐ Analyze faults in a voltage-follower
   ☐ Analyze faults in an inverting amplifier

In the basic op-amp configurations, there are only a few external components that can fail. These are the feedback resistor, the input resistor, and the potentiometer used for offset voltage compensation. Also, of course, the op-amp itself can fail or there can be faulty contacts in the circuit. We will now examine the three basic configurations for possible faults and the associated symptoms.

Faults in the Noninverting Amplifier

The first thing to do when you suspect a faulty circuit is to check for the proper supply voltage and ground. Having done that, several other possible faults are as follows.

Open Feedback Resistor   If the feedback resistor, R_f, in Figure 12–40 opens, the op-amp is operating with its very high open-loop gain, which causes the input signal to drive the device into nonlinear operation and results in a severely clipped output signal as shown in part (a).

Open Input Resistor   In this case, you still have a closed-loop configuration. But, since R_i is open and effectively equal to infinity, the closed-loop gain from Equation (12–12) is

$$A_{cl(\text{NI})} = 1 + \frac{R_f}{R_i} = 1 + \frac{R_f}{\infty} = 1 + 0 = 1$$

This shows that the amplifier acts like a voltage-follower. You would observe an output signal that is the same as the input, as indicated in Figure 12–40(b).

Open or Incorrectly Adjusted Offset Null Potentiometer   In this situation, the output offset voltage will cause the output signal to begin clipping on only one peak as the input signal is increased to a sufficient amplitude. This is indicated in Figure 12–40(c).

Faulty Op-Amp   As mentioned, many things can happen to an op-amp. In general, an internal failure will result in a loss or distortion of the output signal. The best approach is to first make sure that there are no external failures or faulty conditions. If everything else is good, then the op-amp must be bad.

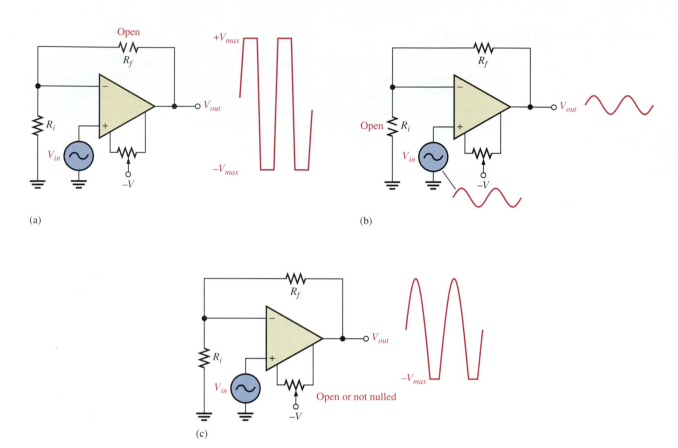

FIGURE 12–40
Faults in the noninverting amplifier.

Faults in the Voltage-Follower

The voltage-follower is a special case of the noninverting amplifier. Except for a bad op-amp, a bad external connection, or a problem with the offset null potentiometer, about the only thing that can happen in a voltage-follower circuit is an open feedback loop. This would have the same effect as an open feedback resistor as previously discussed.

Faults in the Inverting Amplifier

Open Feedback Resistor If R_f opens as indicated in Figure 12–41(a), the input signal still feeds through the input resistor and is amplified by the high open-loop gain of the op-amp. This forces the device to be driven into nonlinear operation, and you will see an output something like that shown. This is the same result as in the noninverting configuration.

Open Input Resistor This prevents the input signal from getting to the op-amp input, so there will be no output signal, as indicated in Figure 12–41(b).

Failures in the op-amp itself or the offset null potentiometer have the same effects as previously discussed for the noninverting amplifier.

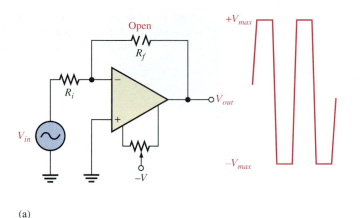

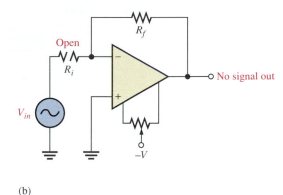

(a)

(b)

FIGURE 12–41
Faults in the inverting amplifier.

SECTION 12–8 REVIEW

1. If you notice that the op-amp output signal is beginning to clip on one peak as you increase the input signal, what should you check?

2. If there is no op-amp output signal when there is a verified input signal, what would you suspect as being faulty?

12–9 ■ SYSTEM APPLICATION

This application involves a spectrophotometer system that uses light optics and electronic circuits to analyze the chemical makeup of various solutions. Your company plans to market this system to the chemical industry and to medical laboratories. Because the system includes electronic, optical, and mechanical components, it is classified as a mixed system. Mixed systems of various types are common in industrial applications. The knowledge you have gained in this chapter will be applied in completing your assignment.

Basic Operation of the System

The light source shown in Figure 12–42 produces a visible light containing a wide spectrum of wavelengths. Each component wavelength in the beam of light is refracted at a different angle by the prism, as indicated. Depending on the angle of the platform, which is set by the pivot angle controller, a certain wavelength of light passes through the narrow slit and is transmitted through the solution under analysis. By precisely pivoting the light source and prism assembly, a selected wavelength of light can be transmitted through the solution.

Since every chemical and compound absorbs different wavelengths of light in different ways, the light coming out of the solution has a unique "signature" that can be used to define the chemicals in the solution.

FIGURE 12–42
Diagram of the spectrophotometer system.

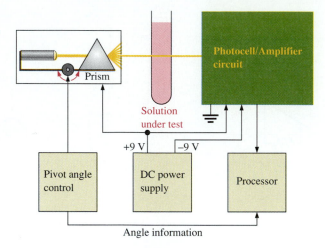

The photocell on the circuit board in Figure 12–43 produces a voltage that is proportional to the amount and wavelength of the light. The op-amp circuit amplifies the photocell output and sends the resulting signal to the processor where the type of chemical(s) in the solution is identified. The focus of this system application is the photocell/amplifier circuit.

The Photocell/Amplifier Circuit Board

☐ Make sure that the circuit board shown in Figure 12–43 is correctly assembled by checking it against the schematic in Figure 12–44. The op-amp is a 741. There are four interconnections on the back side of the board between each pair of horizontally oriented feedthrough pads.

☐ Label a copy of the board with component and input/output designations in agreement with the schematic.

Analysis of the Photocell/Amplifier Circuit

☐ Determine the resistance to which the variable feedback resistor must be adjusted for a voltage gain of 10.

☐ Assume the maximum linear output of the op-amp is 1 V less than the dc supply voltage. Determine the voltage gain required and the value to which the feedback resistor must be set to achieve the maximum linear output voltage. The light source produces wavelengths ranging from 400 nm to 700 nm, which is approximately the full range of visible light from violet to red. The voltage from the photocell is 0.5 V at 825 nm.

☐ Using the gain found in the previous step, determine the op-amp output voltage over the range of wavelengths from 400 nm to 700 nm in 50 nm intervals and plot a graph of the results. Refer to the photocell response curve in Figure 12–45.

Test Procedure

☐ Develop a step-by-step set of instructions on how to check the photocell/amplifier circuit board for proper operation.

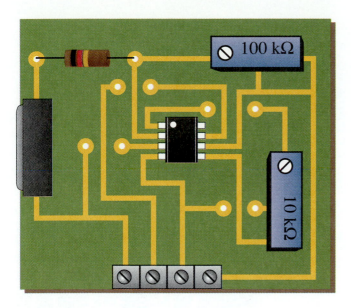

FIGURE 12–43
Photocell/Amplifier circuit board.

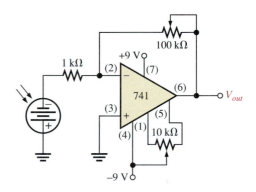

FIGURE 12–44
Photocell/Amplifier schematic.

FIGURE 12–45
Photocell response curve.

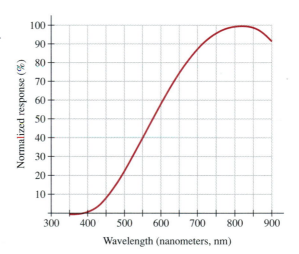

☐ Specify voltage values for all the measurements to be made.

☐ Provide a fault analysis for all possible component failures.

Troubleshooting

Problems have developed in three boards. Based on the test bench measurements for each board indicated in Figure 12–46, determine the most likely fault in each case. The circled numbers indicate test point connections to the circuit board. Assume that each board has the proper dc supply voltage.

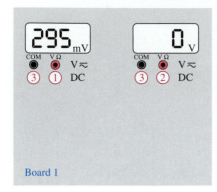

Board 1

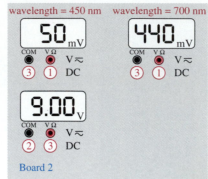

wavelength = 450 nm   wavelength = 700 nm

Board 2

No light

Board 3

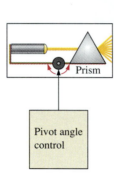

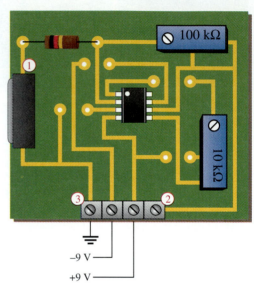

FIGURE 12–46
Test results for three faulty circuit boards.

Final Report

Submit a final written report on the photocell/amplifier circuit board using an organized format that includes the following:

1. A physical description of the circuit.
2. A discussion of the operation of the circuit.
3. A list of the specifications.
4. A list of parts with part numbers if available.
5. A list of the types of problems on the three faulty circuit boards.
6. A description of how you determined the problem on each of the faulty circuit boards.

■ CHAPTER SUMMARY

- The basic op-amp has three terminals not including power and ground: inverting input (−), non-inverting input (+), and output.
- Most op-amps require both a positive and a negative dc supply voltage.
- The ideal op-amp has infinite input impedance, zero output impedance, infinite open-loop voltage gain, infinite bandwidth, and infinite CMRR.
- A practical op-amp has very high input impedance, very low output impedance, and very high open-loop voltage gain.
- A differential amplifier is normally used for the input stage of an op-amp.
- A differential input voltage appears between the inverting and noninverting inputs of a differential amplifier.
- A single-ended input voltage appears between one output and ground (with the other input grounded).
- A differential output voltage appears between two output terminals of a diff-amp.
- A single-ended output voltage appears between the output and ground of a diff-amp.
- Common-mode occurs when equal in-phase voltages are applied to both input terminals.
- Input offset voltage produces an output error voltage (with no input voltage).
- Input bias current also produces an output error voltage (with no input voltage).
- Input offset current is the difference between the two bias currents.
- Open-loop voltage gain is the gain of the op-amp with no external feedback connections.
- The common-mode rejection ratio (CMRR) is a measure of an op-amp's ability to reject common-mode inputs.
- Slew rate is the rate in volts per microsecond at which the output voltage of an op-amp can change in response to a step input.
- There are three basic op-amp configurations: inverting, noninverting, and voltage-follower.
- The three basic op-amp configurations employ negative feedback. Negative feedback occurs when a portion of the output voltage is connected back to the inverting input such that it subtracts from the input voltage, thus reducing the voltage gain but increasing the stability and bandwidth.
- A noninverting amplifier configuration has a higher input impedance and a lower output impedance than the op-amp itself (without feedback).
- An inverting amplifier configuration has an input impedance approximately equal to the input resistor R_i and an output impedance approximately equal to the output impedance of the op-amp itself.
- The voltage-follower has the highest input impedance and the lowest output impedance of the three configurations.
- All practical op-amps have small input bias currents and input offset voltages that produce small output error voltages.
- The input bias current effect can be compensated for with external resistors.
- The input offset voltage can be compensated for with an external potentiometer between the two offset null pins provided on the IC op-amp package and as recommended by the manufacturer.

■ GLOSSARY

Closed-loop   An op-amp configuration in which the output is connected back to the input through a feedback circuit.

Closed-loop gain (A_{cl})   The voltage gain of an op-amp with feedback.

Common mode   A condition characterized by the presence of the same signal on both op-amp inputs.

Common-mode rejection ratio (CMRR) The ratio of open-loop gain to common-mode gain; a measure of an op-amp's ability to reject common-mode signals.

Differential amplifier (diff-amp) An amplifier that produces an output voltage proportional to the difference of the two input voltages.

Inverting amplifier An op-amp closed-loop configuration in which the input signal is applied to the inverting input.

Negative feedback The process of returning a portion of the output signal to the input of an amplifier such that it is out of phase with the input signal.

Noninverting amplifier An op-amp closed-loop configuration in which the input signal is applied to the noninverting input.

Open-loop gain (A_{ol}) The voltage gain of an op-amp without feedback.

Operational amplifier (op-amp) A type of amplifier that has very high voltage gain, very high input impedance, very low output impedance, and good rejection of common-mode signals.

Slew rate The rate of change of the output voltage of an op-amp in response to a step input.

Voltage-follower A closed-loop, noninverting op-amp with a voltage gain of one.

■ **FORMULAS**

Differential Amplifiers

(12–1) $$\text{CMRR} = \frac{A_{v(d)}}{A_{cm}}$$ Common-mode rejection ratio

(12–2) $$\text{CMRR} = 20 \log\left(\frac{A_{v(d)}}{A_{cm}}\right)$$ Common-mode rejection ratio (dB)

Op-Amp Parameters

(12–3) $V_{\text{OUT}} = I_{C2}R_C - I_{C1}R_C$ Differential output

(12–4) $$I_{\text{BIAS}} = \frac{I_1 + I_2}{2}$$ Input bias current

(12–5) $I_{\text{OS}} = |I_1 - I_2|$ Input offset current

(12–6) $V_{\text{OS}} = I_{\text{OS}}R_{in}$ Offset voltage

(12–7) $V_{\text{OUT(error)}} = A_v I_{\text{OS}}R_{in}$ Output error voltage

(12–8) $$\text{CMRR} = \frac{A_{ol}}{A_{cm}}$$ Common-mode rejection ratio

(12–9) $$\text{CMRR} = 20 \log\left(\frac{A_{ol}}{A_{cm}}\right)$$ Common-mode rejection ratio (dB)

(12–10) $$\text{Slew rate} = \frac{\Delta V_{out}}{\Delta t}$$ Slew rate

Op-Amp Configurations

(12–11) $$\frac{V_{out}}{V_{in}} = \frac{A_{ol}}{1 + A_{ol}B}$$ Voltage gain (noninverting)

(12–12) $$A_{cl(\text{NI})} = 1 + \frac{R_f}{R_i}$$ Voltage gain (noninverting)

(12–13) $A_{cl(\text{VF})} = 1$ Voltage gain (voltage-follower)

(12–14) $$A_{cl(\text{I})} = -\frac{R_f}{R_i}$$ Voltage gain (inverting)

Op-Amp Impedances

(12–15)　　$Z_{in(\text{NI})} = (1 + A_{ol}B)Z_{in}$　　Input impedance (noninverting)

(12–16)　　$Z_{out(\text{NI})} = \dfrac{Z_{out}}{1 + A_{ol}B}$　　Output impedance (noninverting)

(12–17)　　$Z_{in(\text{VF})} = (1 + A_{ol})Z_{in}$　　Input impedance (voltage-follower)

(12–18)　　$Z_{out(\text{VF})} = \dfrac{Z_{out}}{1 + A_{ol}}$　　Output impedance (voltage-follower)

(12–19)　　$Z_{in(Miller)} = \dfrac{R_f}{A_{ol} + 1}$　　Miller input impedance (inverting)

(12–20)　　$Z_{out(Miller)} = \left(\dfrac{A_{ol}}{A_{ol} + 1} \right) R_f$　　Miller output impedance (inverting)

(12–21)　　$Z_{in(\text{I})} \cong R_i$　　Input impedance (inverting)

(12–22)　　$Z_{out(\text{I})} \cong Z_{out}$　　Output impedance (inverting)

Error Voltage

(12–23)　　$V_{\text{OUT(error)}} = I_{OS}R_s$　　Output error voltage due to input offset current

(12–24)　　$V_{\text{OUT(error)}} = V_{\text{IO}}$　　Output error voltage due to input offset voltage

---

■ **SELF-TEST**

1. An integrated circuit (IC) op-amp has
   (a) two inputs and two outputs　　(b) one input and one output
   (c) two inputs and one output

2. Which of the following characteristics does not necessarily apply to an op-amp?
   (a) High gain　　　　　　　　　(b) Low power
   (c) High input impedance　　　　(d) Low output impedance

3. A differential amplifier
   (a) is part of an op-amp　　　(b) has one input and one output
   (c) has two outputs　　　　　(d) answers (a) and (c)

4. When a differential amplifier is operated single-ended,
   (a) the output is grounded
   (b) one input is grounded and a signal is applied to the other
   (c) both inputs are connected together
   (d) the output is not inverted

5. In the differential mode,
   (a) opposite polarity signals are applied to the inputs
   (b) the gain is one
   (c) the outputs are different amplitudes
   (d) only one supply voltage is used

6. In the common mode,
   (a) both inputs are grounded　　　　　　　(b) the outputs are connected together
   (c) an identical signal appears on both inputs　(d) the output signals are in-phase

7. Common-mode gain is
   (a) very high　　(b) very low　　(c) always unity　　(d) unpredictable

8. Differential gain is
   (a) very high　　(b) very low　　(c) dependent on the input voltage　　(d) about 100

9. If $A_{v(d)} = 3500$ and $A_{cm} = 0.35$, the CMRR is
   (a) 1225　　(b) 10,000　　(c) 80 dB　　(d) answers (b) and (c)

10. With zero volts on both inputs, an op-amp ideally should have an output
    (a) equal to the positive supply voltage   (b) equal to the negative supply voltage
    (c) equal to zero   (d) equal to the CMRR

11. Of the values listed, the most realistic value for open-loop gain of an op-amp is
    (a) 1   (b) 2000   (c) 80 dB   (d) 100,000

12. A certain op-amp has bias currents of 50 μA and 49.3 μA. The input offset current is
    (a) 700 nA   (b) 99.3 μA   (c) 49.7 μA   (d) none of these

13. The output of a particular op-amp increases 8 V in 12 μs. The slew rate is
    (a) 96 V/μs   (b) 0.67 V/μs   (c) 1.5 V/μs   (d) none of these

14. The purpose of offset nulling is to
    (a) reduce the gain   (b) equalize the input signals
    (c) zero the output error voltage (d) answers (b) and (c)

15. For an op-amp with negative feedback, the output is
    (a) equal to the input   (b) increased
    (c) fed back to the inverting input (d) fed back to the noninverting input

16. The use of negative feedback
    (a) reduces the voltage gain of an op-amp
    (b) makes the op-amp oscillate
    (c) makes linear operation possible
    (d) answers (a) and (c)

17. Negative feedback
    (a) increases the input and output impedances
    (b) increases the input impedance and the bandwidth
    (c) decreases the output impedance and the bandwidth
    (d) does not affect impedances or bandwidth

18. A certain noninverting amplifier has an R_i of 1 kΩ and an R_f of 100 kΩ. The closed-loop gain is
    (a) 100,000   (b) 1000   (c) 101   (d) 100

19. If the feedback resistor in Question 18 is open, the voltage gain
    (a) increases   (b) decreases   (c) is not affected   (d) depends on R_i

20. A certain inverting amplifier has a closed-loop gain of 25. The op-amp has an open-loop gain of 100,000. If another op-amp with an open-loop gain of 200,000 is substituted in the configuration, the closed-loop gain
    (a) doubles   (b) drops to 12.5   (c) remains at 25   (d) increases slightly

21. A voltage-follower
    (a) has a gain of one   (b) is noninverting
    (c) has no feedback resistor   (d) has all of these

■ **BASIC PROBLEMS**

SECTION 12–1   Introduction to Operational Amplifiers

1. Compare a practical op-amp to the ideal.

2. Two IC op-amps are available to you. Their characteristics are listed below. Choose the one you think is more desirable.
   Op-amp 1: $Z_{in} = 5$ MΩ, $Z_{out} = 100$ Ω, $A_{ol} = 50,000$
   Op-amp 2: $Z_{in} = 10$ MΩ, $Z_{out} = 75$ Ω, $A_{ol} = 150,000$

SECTION 12–2   The Differential Amplifier

3. Identify the type of input and output configuration for each basic differential amplifier in Figure 12–47.

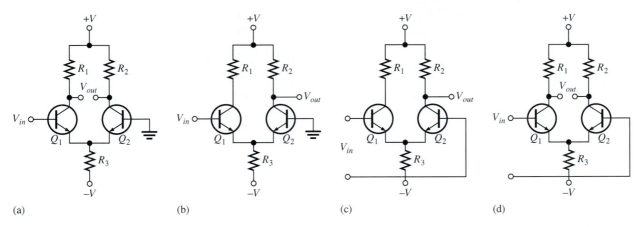

FIGURE 12–47

4. The dc base voltages in Figure 12–48 are zero. Using your knowledge of transistor analysis, determine the dc differential output voltage. Assume that Q_1 has an $\alpha = 0.980$ and Q_2 has an $\alpha = 0.975$.

5. Identify the quantity being measured by each meter in Figure 12–49.

6. A differential amplifier stage has collector resistors of 5.1 kΩ each. If $I_{C1} = 1.35$ mA and $I_{C2} = 1.29$ mA, what is the differential output voltage?

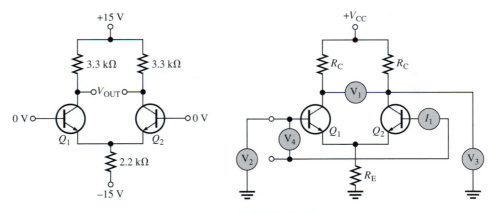

FIGURE 12–48   **FIGURE 12–49**

SECTION 12–3 Op-Amp Parameters

7. Determine the bias current, I_{BIAS}, given that the input currents to an op-amp are 8.3 μA and 7.9 μA.

8. Distinguish between input bias current and input offset current, and then calculate the input offset current in Problem 7.

9. A certain op-amp has a CMRR of 250,000. Convert this to dB.

10. The open-loop gain of a certain op-amp is 175,000. Its common-mode gain is 0.18. Determine the CMRR in dB.

11. An op-amp data sheet specifies a CMRR of 300,000 and an A_{ol} of 90,000. What is the common-mode gain?

FIGURE 12–50

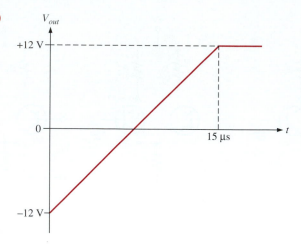

12. Figure 12–50 shows the output voltage of an op-amp in response to a step input. What is the slew rate?

13. How long does it take the output voltage of an op-amp to go from −10 V to +10 V, if the slew rate is 0.5 V/μs?

SECTION 12–5   Op-Amp Configurations with Negative Feedback

14. Identify each of the op-amp configurations in Figure 12–51.

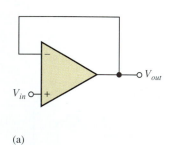

(a)

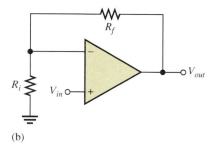

(b)

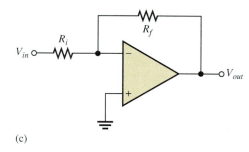

(c)

FIGURE 12–51

15. A noninverting amplifier has an R_i of 1 kΩ and an R_f of 100 kΩ. Determine V_f and B, if $V_{out} =$ 5 V.

16. For the amplifier in Figure 12–52, determine the following:
    (a) $A_{cl(NI)}$ (b) V_{out} (c) V_f

FIGURE 12–52

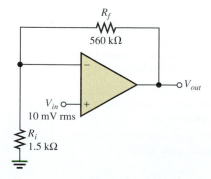

17. Determine the closed-loop gain of each amplifier in Figure 12–53.

18. Find the value of R_f that will produce the indicated closed-loop gain in each amplifier in Figure 12–54.

FIGURE 12–53

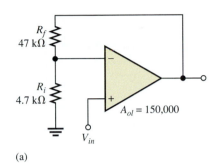

(a)

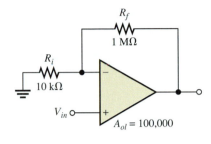

(b)

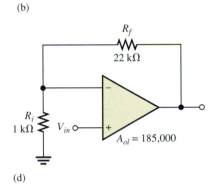

(c)

(d)

FIGURE 12–54

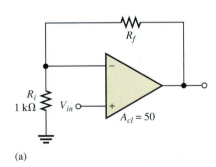

(a)

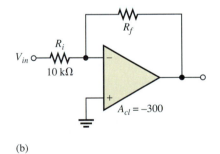

(b)

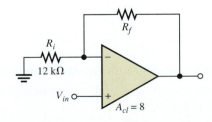

(c)

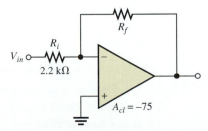

(d)

19. Find the gain of each amplifier in Figure 12–55.

20. If a signal voltage of 10 mV rms is applied to each amplifier in Figure 12–55, what are the output voltages and what is their phase relationship with inputs?

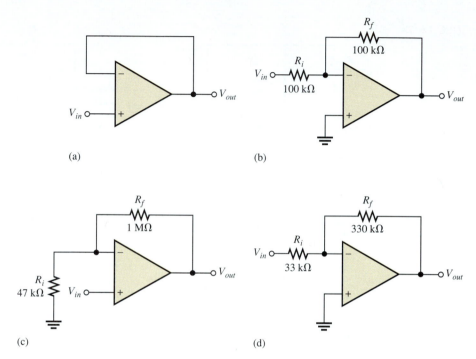

(a)   (b)

(c)   (d)

FIGURE 12–55

21. Determine the approximate values for each of the following quantities in Figure 12–56.
   (a) I_{in} **(b)** I_f **(c)** V_{out} **(d)** closed-loop gain

FIGURE 12–56

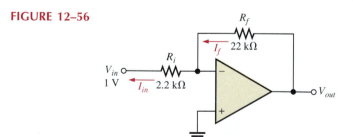

SECTION 12–6   Effects of Negative Feedback on Op-Amp Impedances

22. Determine the input and output impedances for each amplifier configuration in Figure 12–57.

23. Repeat Problem 22 for each circuit in Figure 12–58.

24. Repeat Problem 22 for each circuit in Figure 12–59.

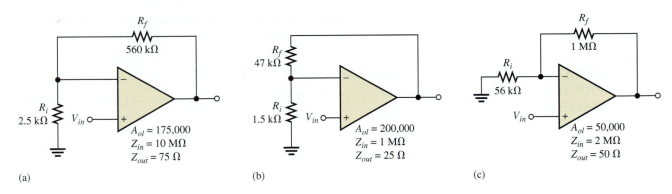

FIGURE 12–57

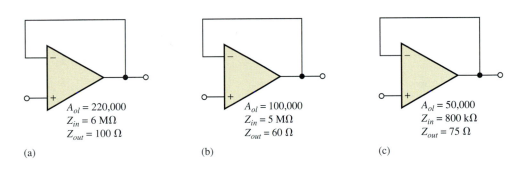

FIGURE 12–58

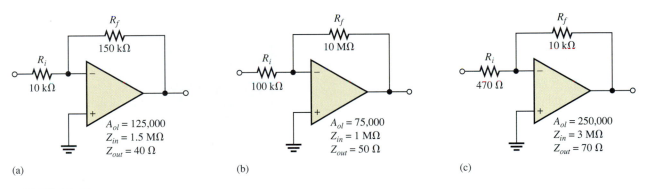

FIGURE 12–59

SECTION 12–7 Bias Current and Offset Voltage Compensation

25. A voltage-follower is driven by a voltage source with a source resistance of 75 Ω.
   (a) What value of compensating resistor is required for bias current, and where should the resistor be placed?
   (b) If the two input currents after compensation are 42 μA and 40 μA, what is the output error voltage?

26. Determine the compensating resistor value for each amplifier configuration in Figure 12–57, and indicate the placement of the resistor.

27. A particular op-amp has an input offset voltage of 2 nV and an open-loop gain of 100,000. What is the output error voltage?

28. What is the input offset voltage of an op-amp if a dc output voltage of 35 mV is measured when the input voltage is zero? The op-amp's open-loop gain is specified to be 200,000.

■ TROUBLE-SHOOTING PROBLEMS

SECTION 12–8 Troubleshooting

29. Determine the most likely fault(s) for each of the following symptoms in Figure 12–60 with a 100 mV signal applied.

   (a) No output signal.

   (b) Output severely clipped on both positive and negative swings.

   (c) Clipping on only positive peaks when input signal is increased to a certain point.

30. Identify the most likely fault(s) for each of the circuits and output waveforms in Figure 12–61.

FIGURE 12–60

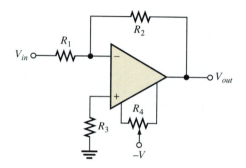

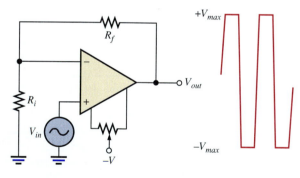

(a)

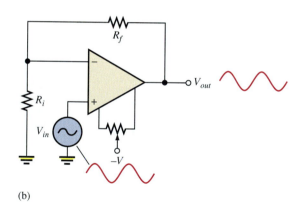

(b)

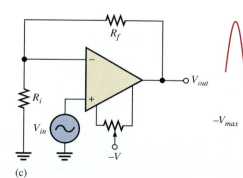

(c)

FIGURE 12–61

SECTION 12–9   System Application

31. On the circuit board in Figure 12–43, what happens if the middle lead (wiper) of the 100 kΩ potentiometer is broken?

32. What is the symptom that indicates pin 1 of the 741 op-amp is open in Figure 12–44?

■ **DATA SHEET PROBLEMS**

33. Refer to the partial 741 data sheet (MC1741) in Figure 12–62. Determine the input and output resistance of a noninverting amplifier which uses a 741 op-amp with R_f = 47 kΩ and R_i = 470 Ω. Use typical values.

34. Refer to the partial data sheet in Figure 12–62. Determine the input and output impedances of a 741 op-amp connected as an inverting amplifier with a closed-loop voltage gain of 100 and R_f = 100 kΩ.

35. Refer to Figure 12–62 and determine the minimum open-loop voltage gain for an MC1741 expressed as a ratio of output volts to input volts.

36. Refer to Figure 12–62. How long does it typically take the output voltage of a 741 to make a transition from −8 V to +8 V in response to a step input?

Electrical Characteristics (V_{CC} = +15 V, V_{EE} = −15 V, T_A = 25°C unless otherwise noted).

Characteristic	Symbol	MC1741 Min	MC1741 Typ	MC1741 Max	MC1741C Min	MC1741C Typ	MC1741C Max	Unit
Input offset voltage (R_S ≤ 10 k)	V_{IO}	–	1.0	5.0	–	2.0	6.0	mV
Input offset current	I_{IO}	–	20	200	–	20	200	nA
Input bias current	I_{IB}	–	80	500	–	80	500	nA
Input resistance	r_i	0.3	2.0	–	0.3	2.0	–	MΩ
Input capacitance	C_i	–	1.4	–	–	1.4	–	pF
Offset voltage adjustment range	V_{IOR}	–	±15	–	–	±15	–	mV
Common mode input voltage range	V_{ICR}	±12	±13	–	±12	±13	–	V
Large-signal voltage gain (V_O = ±10 V, R_L ≥ 2.0 k)	A_v	50	200	–	20	200	–	V/mV
Output resistance	r_o	–	75	–	–	75	–	Ω
Common-mode rejection ratio (R_S ≤ 10 k)	CMRR	70	90	–	70	90	–	dB
Supply voltage rejection ratio (R_S ≤ 10 k)	PSRR	–	30	150	–	30	150	μV/V
Output voltage swing	V_O							V
(R_L ≥ 10 k)		±12	±14	–	±12	±14	–	
(R_L ≥ 2 k)		±10	±13	–	±10	±13	–	
Output short-circuit current	I_{os}	–	20	–	–	20	–	mA
Supply current	I_D	–	1.7	2.8	–	1.7	2.8	mA
Power consumption	P_C	–	50	85	–	50	85	mW
Transient response (unity gain – non-inverting) (V_I = 20 mV, R_L ≥ 2 k, C_L ≤ 100 pF) Rise Time	t_{TLH}	–	0.3	–	–	0.3	–	μs
(V_I = 20 mV, R_L ≥ 2 k, C_L ≤ 100 pF) Overshoot	os	–	15	–	–	15	–	%
(V_I = 10 V, R_L ≥ 2 k, C_L ≤ 100 pF) Slew Rate	SR	–	0.5	–	–	0.5	–	V/μs

FIGURE 12–62

■ **ADVANCED PROBLEMS**

37. Design a noninverting amplifier with an appropriate closed-loop voltage gain of 150 and a minimum input impedance of 100 MΩ using a 741 op-amp.

38. Design an inverting amplifier using a 741 op-amp. The voltage gain must be 68 ± 5% and the input impedance must be approximately 10 kΩ.

39. Refer to Figure 12–62. Calculate the typical common-mode gain for the MC1741 op-amp.

▪ **ANSWERS TO SECTION REVIEWS**

Section 12–1

1. Inverting input, noninverting input, output, positive and negative supply voltages
2. A practical op-amp has very high input impedance, very low output impedance, and very high voltage gain.

Section 12–2

1. Differential input is between two input terminals. Single-ended input is from one input terminal to ground (with other input grounded).
2. Common-mode rejection is the ability of an op-amp to produce very little output when the same signal is applied to both inputs.
3. A higher CMRR results in a lower common-mode gain.

Section 12–3

1. Input bias current, input offset voltage, drift, input offset current, input impedance, output impedance, common-mode input voltage range, CMRR, open-loop voltage gain, slew rate, frequency response.
2. Slew rate and voltage gain are both frequency dependent.

Section 12–4

1. Negative feedback provides a stable controlled gain, control of impedances, and wider bandwidth.
2. The open-loop gain is so high that a very small signal on the input will drive the op-amp into saturation.

Section 12–5

1. The main purpose of negative feedback is to stabilize the gain.
2. False
3. $A_{cl} = 1/0.02 = 50$

Section 12–6

1. The noninverting configuration has a higher Z_{in} than the op-amp alone.
2. Z_{in} increases in a voltage-follower.
3. $Z_{in(I)} \cong R_i = 2 \text{ k}\Omega$, $Z_{out(I)} \cong Z_{out} = 60 \text{ }\Omega$.

Section 12–7

1. Input bias current and input offset voltage are sources of output error.
2. Add a resistor in the feedback path equal to the input source resistance.

Section 12–8

1. Check the output null adjustment.
2. The op-amp is probably bad.

■ **ANSWERS TO RELATED EXERCISES FOR EXAMPLES**

12–1 34,000; 90.6 dB

12–2 **(a)** 0.168   **(b)** 88 dB   **(c)** 2.1 V rms, 4.2 V rms   **(d)** 0.168 V

12–3 12,649

12–4 20 V/μs

12–5 32.9

12–6 67.5 kΩ

12–7 **(a)** 20.5 GΩ, 0.014 Ω   **(b)** 23

12–8 Z_{in} increases, Z_{out} decreases.

12–9 $Z_{in(I)} = 560 \ \Omega$; $Z_{out(I)} = 75 \ \Omega$; $A_{cl} = -146$

13

OP-AMP FREQUENCY RESPONSE, STABILITY, AND COMPENSATION

■ **CHAPTER OBJECTIVES**

☐ Discuss the basic area of op-amp responses
☐ Analyze the open-loop response of an op-amp
☐ Analyze the closed-loop response of an op-amp
☐ Discuss positive feedback and stability in op-amp circuits
☐ Explain op-amp phase compensation

In this chapter, you will learn more about frequency response, bandwidth, phase shift, and other frequency-related parameters. The effects of negative feedback will be further examined. You will learn about stability requirements and how to compensate op-amp circuits to ensure stable operation.

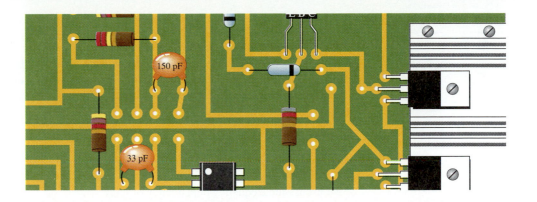

■ SYSTEM APPLICATION

For the system application in Section 13–6, the op-amp is used as an audio preamplifier in an AM receiver. The AM receiver receives amplitude-modulated frequencies from 535 kHz to 1605 kHz, extracts the audio signal from the modulated carrier frequency, and amplifies the audio signal to drive a speaker. AM is a process by which the amplitude of a higher frequency signal (carrier) is varied (modulated) by a lower frequency signal (audio in this case). The focus of the system application is on the audio amplifier circuit, which includes an op-amp as well as a push-pull power amplifier.

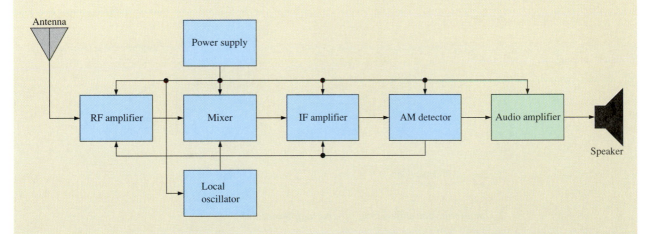

13–1 ▪ BASIC CONCEPTS

The last chapter demonstrated how closed-loop voltage gains of the basic op-amp configurations are determined, and the distinction between open-loop gain and closed-loop gain was established. Because of the importance of these two different types of gain, the definitions are restated in this section.

After completing this section, you should be able to

▪ **Discuss the basic areas of op-amp responses**
  ☐ Define *open-loop gain*
  ☐ Define *closed-loop gain*
  ☐ Discuss the frequency dependency of gain
  ☐ Explain the open-loop bandwidth
  ☐ Explain the unity-gain bandwidth
  ☐ Determine phase shift

Open-Loop Gain

The **open-loop gain,** A_{ol}, of an op-amp is the internal voltage gain of the device and represents the ratio of output voltage to input voltage, as indicated in Figure 13–1(a). Notice that there are no external components, so the open-loop gain is set entirely by the internal design. Open-loop gain can range up to 200,000 and is not a well-controlled parameter. Data sheets often refer to the open-loop gain as the *large-signal voltage gain.*

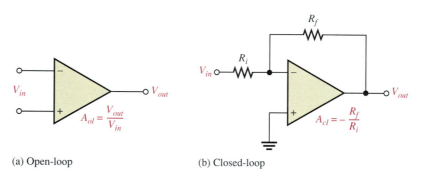

(a) Open-loop   (b) Closed-loop

FIGURE 13–1
Open-loop and closed-loop op-amp configurations.

Closed-Loop Gain

The **closed-loop gain,** A_{cl}, is the voltage gain of an op-amp with external feedback. The amplifier configuration consists of the op-amp and an external negative feedback circuit that connects the output to the inverting input. The closed-loop gain is determined by the external component values, as illustrated in Figure 13–1(b) for an inverting amplifier configuration. The closed-loop gain can be precisely controlled by external component values.

The Gain Is Frequency Dependent

In the last chapter, all of the gain expressions applied to the midrange gain and were considered independent of the frequency. The midrange open-loop gain of an op-amp extends from zero frequency (dc) up to a critical frequency at which the gain is 3 dB less than the midrange value. This concept should be familiar from your study of Chapter 10. Op-amps are dc amplifiers (no capacitive coupling between stages), and therefore, there is no lower critical frequency. This means that the midrange gain extends down to zero frequency, and dc voltages are amplified the same as midrange signal frequencies.

An open-loop response curve (Bode plot) for a certain op-amp is shown in Figure 13–2. Most op-amp data sheets show this type of curve or specify the midrange open-loop gain. Notice that the curve rolls off at −20 dB per decade (−6 dB per octave). The midrange gain is 200,000, which is 106 dB, and the critical (cutoff) frequency is approximately 10 Hz.

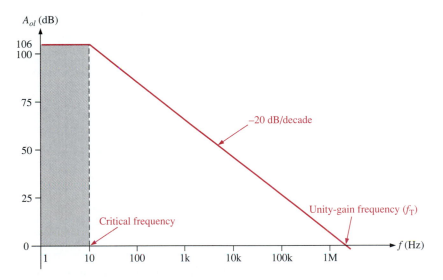

FIGURE 13–2

Ideal plot of open-loop voltage gain versus frequency for a typical op-amp. The frequency scale is logarithmic.

3 dB Open-Loop Bandwidth

Recall from Chapter 10 that the bandwidth of an ac amplifier is the frequency range between the points where the gain is 3 dB less than midrange, equalling the upper critical frequency (f_{cu}) minus the lower critical frequency (f_{cl}).

$$BW = f_{cu} - f_{cl}$$

Since f_{cl} for an op-amp is zero, the bandwidth is simply equal to the upper critical frequency.

$$BW = f_{cu} \qquad \text{(13–1)}$$

From now on, we will refer to f_{cu} as f_c; and we will use open-loop (*ol*) or closed-loop (*cl*) subscript designators, for example, $f_{c(ol)}$.

Unity-Gain Bandwidth

Notice in Figure 13–2 that the gain steadily decreases to a point where it is equal to one (0 dB). The value of the frequency at which this unity gain occurs is the *unity-gain bandwidth*.

Gain-Versus-Frequency Analysis

The RC lag (low-pass) circuits within an op-amp are responsible for the roll-off in gain as the frequency increases, just as was discussed for the discrete amplifiers in Chapter 10. From basic ac circuit theory, the attenuation of an RC lag circuit, such as in Figure 13–3, is expressed as

$$\frac{V_{out}}{V_{in}} = \frac{X_C}{\sqrt{R^2 + X_C^2}}$$

Dividing both the numerator and denominator to the right of the equal sign by X_C, you get

$$\frac{V_{out}}{V_{in}} = \frac{1}{\sqrt{1 + R^2/X_C^2}} \tag{13-2}$$

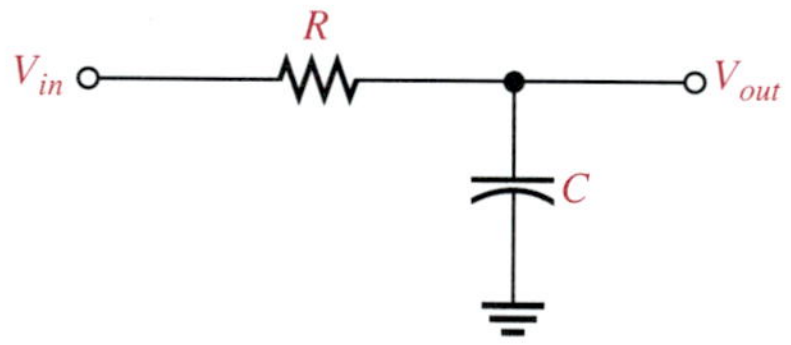

FIGURE 13–3
RC lag circuit.

The critical frequency of an RC circuit is

$$f_c = \frac{1}{2\pi RC}$$

Dividing both sides by f gives

$$\frac{f_c}{f} = \frac{1}{2\pi RCf} = \frac{1}{(2\pi fC)R}$$

Since $X_C = 1/(2\pi fC)$, the above expression can be written as

$$\frac{f_c}{f} = \frac{X_C}{R}$$

Substituting this result in Equation (13–2) produces the following expression for the attenuation of an RC lag circuit:

$$\frac{V_{out}}{V_{in}} = \frac{1}{\sqrt{1 + f^2/f_c^2}} \tag{13-3}$$

FIGURE 13–4

Op-amp represented by gain element and internal RC circuit.

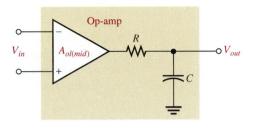

If an op-amp is represented by a voltage gain element and a single RC lag circuit, as shown in Figure 13–4, then the total open-loop gain is the product of the midrange open-loop gain $A_{ol(mid)}$ and the attenuation of the RC circuit:

$$A_{ol} = \frac{A_{ol(mid)}}{\sqrt{1 + f^2/f_c^2}} \qquad (13\text{–}4)$$

As you can see from Equation (13–4), the open-loop gain equals the midrange value when the signal frequency f is much less than the critical frequency f_c and drops off as the frequency increases. Since f_c is part of the open-loop response of an op-amp, we will refer to it as $f_{c(ol)}$.

The following example demonstrates how the open-loop gain decreases as the frequency increases above $f_{c(ol)}$.

EXAMPLE 13–1

Determine A_{ol} for the following values of f. Assume $f_{c(ol)} = 100$ Hz and $A_{ol(mid)} = 100,000$.

(a) $f = 0$ Hz   (b) $f = 10$ Hz   (c) $f = 100$ Hz   (d) $f = 1000$ Hz

Solution

(a) $A_{ol} = \dfrac{A_{ol(mid)}}{\sqrt{1 + f^2/f_{c(ol)}^2}} = \dfrac{100,000}{\sqrt{1 + 0}} = 100,000$

(b) $A_{ol} = \dfrac{100,000}{\sqrt{1 + (0.1)^2}} = 99,503$

(c) $A_{ol} = \dfrac{100,000}{\sqrt{1 + (1)^2}} = \dfrac{100,000}{\sqrt{2}} = 70,710$

(d) $A_{ol} = \dfrac{100,000}{\sqrt{1 + (10)^2}} = 9950$

Related Exercise Find A_{ol} for the following frequencies. Assume $f_{c(ol)} = 200$ Hz, $A_{ol(mid)} = 80,000$.

(a) $f = 2$ Hz   (b) $f = 10$ Hz   (c) $f = 2500$ Hz

Phase Shift

As you know from Chapter 10, an *RC* circuit causes a propagation delay from input to output, thus creating a **phase** difference between the input signal and the output signal. An *RC* lag circuit such as found in an op-amp stage causes the output signal voltage to lag the input, as shown in Figure 13–5. From basic ac circuit theory, the phase shift, θ, is

$$\theta = -\tan^{-1}\left(\frac{R}{X_C}\right) \qquad (13\text{–}5)$$

Since $R/X_C = f/f_c$,

$$\theta = -\tan^{-1}\left(\frac{f}{f_c}\right) \qquad (13\text{–}6)$$

The negative sign indicates that the output lags the input. This equation shows that the phase shift increases with frequency and approaches $-90°$ as f becomes much greater than f_c.

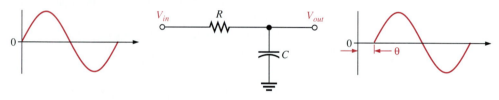

FIGURE 13–5
Output voltage lags input voltage.

EXAMPLE 13–2

Calculate the phase shift for an *RC* lag circuit for each of the following frequencies, and then plot the curve of phase shift versus frequency. Assume $f_c = 100$ Hz.

(a) $f = 1$ Hz   (b) $f = 10$ Hz   (c) $f = 100$ Hz
(d) $f = 1000$ Hz   (e) $f = 10,000$ Hz

Solution

(a) $\theta = -\tan^{-1}\left(\dfrac{f}{f_c}\right) = -\tan^{-1}\left(\dfrac{1\ \text{Hz}}{100\ \text{Hz}}\right) = -0.573°$

(b) $\theta = -\tan^{-1}\left(\dfrac{10\ \text{Hz}}{100\ \text{Hz}}\right) = -5.71°$

(c) $\theta = -\tan^{-1}\left(\dfrac{100\ \text{Hz}}{100\ \text{Hz}}\right) = -45°$

(d) $\theta = -\tan^{-1}\left(\dfrac{1000\ \text{Hz}}{100\ \text{Hz}}\right) = -84.3°$

(e) $\theta = -\tan^{-1}\left(\dfrac{10,000\ \text{Hz}}{100\ \text{Hz}}\right) = -89.4°$

The phase shift-versus-frequency curve is plotted in Figure 13–6. Note that the frequency axis is logarithmic.

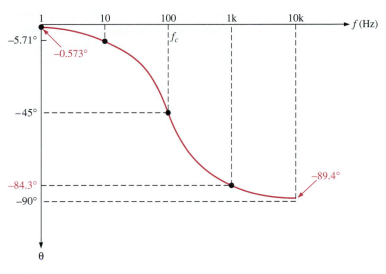

FIGURE 13–6

Related Exercise   At what frequency, in this example, is the phase shift 60°?

SECTION 13–1 REVIEW

1. How do the open-loop gain and the closed-loop gain of an op-amp differ?

2. The upper critical frequency of a particular op-amp is 100 Hz. What is its open-loop 3 dB bandwidth?

3. Does the open-loop gain increase or decrease with frequency above the critical frequency?

13–2 ■ OPEN-LOOP RESPONSE

In this section, you will learn about the open-loop frequency response and the open-loop phase response of an op-amp. Open-loop responses relate to an op-amp with no external feedback. The frequency response indicates how the voltage gain changes with frequency, and the phase response indicates how the phase shift between the input and output signal changes with frequency. The open-loop gain, like the β of a transistor, varies greatly from one device to the next of the same type and cannot be depended upon to have a constant value.

After completing this section, you should be able to

■ Analyze the open-loop response of an op-amp
   ☐ Discuss how internal stages affect the overall response
   ☐ Discuss critical frequencies and roll-off rates
   ☐ Determine overall phase response

Frequency Response

In the previous section, an op-amp was assumed to have a constant roll-off of −20 dB/decade above its critical frequency. For most op-amps this is the case; for some, however, the situation is more complex. The more complex IC operational amplifier may consist of two or more cascaded amplifier stages. The gain of each stage is frequency dependent and rolls off at −20 dB/decade above its critical frequency. Therefore, the total response of an op-amp is a composite of the individual responses of the internal stages. As an example, a three-stage op-amp is represented in Figure 13–7(a), and the frequency response of each stage is shown in Figure 13–7(b). As you know, dB gains are added so that the total op-amp frequency response is as shown in Figure 13–7(c). Since the roll-off rates are additive, the total roll-off rate increases by −20 dB/decade (−6 dB/octave) as each critical frequency is reached.

Phase Response

In a multistage amplifier, each stage contributes to the total phase lag. As you have seen, each RC lag circuit can produce up to a −90° phase shift. Since each stage in an op-amp includes an RC lag circuit, a three-stage op-amp, for example, can have a maximum phase lag of −270°. Also, the phase lag of each stage is less than −45° below the critical

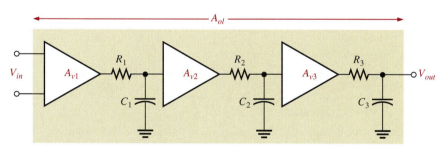

(a) Representation of an op-amp with three internal stages

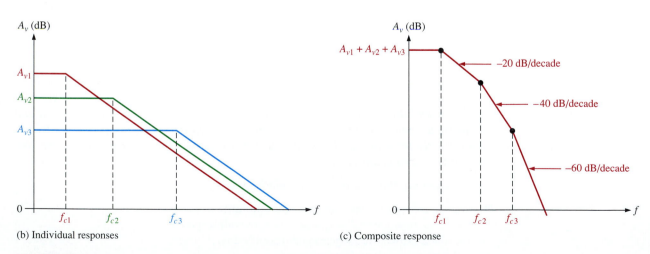

(b) Individual responses

(c) Composite response

FIGURE 13–7
Op-amp open-loop frequency response.

frequency, equal to $-45°$ at the critical frequency, and greater than $-45°$ above the critical frequency. The phase lags of the stages of an op-amp are added to produce a total phase lag, according to the following formula for three stages:

$$\theta_{tot} = -\tan^{-1}\left(\frac{f}{f_{c1}}\right) - \tan^{-1}\left(\frac{f}{f_{c2}}\right) - \tan^{-1}\left(\frac{f}{f_{c3}}\right) \qquad \textbf{(13–7)}$$

EXAMPLE 13–3

A certain op-amp has three internal amplifier stages with the following gains and critical frequencies:

Stage 1: $A_{v1} = 40$ dB, $f_{c1} = 2000$ Hz
Stage 2: $A_{v2} = 32$ dB, $f_{c2} = 40$ kHz
Stage 3: $A_{v3} = 20$ dB, $f_{c3} = 150$ kHz

Determine the open-loop midrange gain in decibels and the total phase lag when $f = f_{c1}$.

Solution

$$A_{ol(mid)} = A_{v1} + A_{v2} + A_{v3} = 40 \text{ dB} + 32 \text{ dB} + 20 \text{ dB} = 92 \text{ dB}$$

$$\theta_{tot} = -\tan^{-1}\left(\frac{f}{f_{c1}}\right) - \tan^{-1}\left(\frac{f}{f_{c2}}\right) - \tan^{-1}\left(\frac{f}{f_{c3}}\right)$$

$$= -\tan^{-1}(1) - \tan^{-1}\left(\frac{2}{40}\right) - \tan^{-1}\left(\frac{2}{150}\right) = -45° - 2.86° - 0.76° = -48.6°$$

Related Exercise   The internal stages of a two-stage amplifier have the following characteristics: $A_{v1} = 50$ dB, $A_{v2} = 25$ dB, $f_{c1} = 1500$ Hz, and $f_{c2} = 3000$ Hz. Determine the open-loop midrange gain in dB and the total phase lag when $f = f_{c1}$.

SECTION 13–2 REVIEW

1. If the individual stage gains of an op-amp are 20 dB and 30 dB, what is the total gain in dB?

2. If the individual phase lags are $-49°$ and $-5.2°$, what is the total phase lag?

13–3 ■ CLOSED-LOOP RESPONSE

Op-amps are normally used in a closed-loop configuration with negative feedback in order to achieve precise control of the gain and bandwidth. In this section, you will see how feedback affects the gain and frequency response of an op-amp.

After completing this section, you should be able to

■ **Analyze the closed-loop response of an op-amp**
  □ Determine the closed-loop gain
  □ Explain the effect of negative feedback on bandwidth
  □ Explain gain-bandwidth product

Recall from Chapter 12 that midrange gain is reduced by negative feedback, as indicated by the following closed-loop gain expressions for the three configurations previously covered, where B is the feedback attenuation. For a noninverting amplifier,

$$A_{cl(NI)} = \frac{A_{ol}}{1 + A_{ol}B} \cong \frac{1}{B}$$

For an inverting amplifier,

$$A_{cl(I)} \cong -\frac{R_f}{R_i}$$

For a voltage-follower,

$$A_{cl(VF)} = 1$$

Effect of Negative Feedback on Bandwidth

You know how negative feedback affects the gain; now you will learn how it affects the amplifier's bandwidth. The closed-loop critical frequency is

$$f_{c(cl)} = f_{c(ol)}(1 + BA_{ol(mid)}) \tag{13–8}$$

This expression shows that the closed-loop critical frequency, $f_{c(cl)}$, is higher than the open-loop critical frequency $f_{c(ol)}$ by the factor $1 + BA_{ol(mid)}$. A derivation of Equation (13–8) can be found in Appendix B.

Since $f_{c(cl)}$ equals the bandwidth for the closed-loop amplifier, the closed-loop bandwidth (BW_{cl}) is also increased by the same factor.

$$BW_{cl} = BW_{ol}(1 + BA_{ol(mid)}) \tag{13–9}$$

EXAMPLE 13–4   A certain amplifier has an open-loop midrange gain of 150,000 and an open-loop 3 dB bandwidth of 200 Hz. The attenuation of the feedback loop is 0.002. What is the closed-loop bandwidth?

Solution $BW_{cl} = BW_{ol}(1 + BA_{ol(mid)}) = 200 \text{ Hz}[1 + (0.002)(150,000)] = 60.2 \text{ kHz}$

Related Exercise   If $A_{ol(mid)} = 200,000$ and $B = 0.05$, what is the closed-loop bandwidth?

Figure 13–8 graphically illustrates the concept of closed-loop response. When the open-loop gain of an op-amp is reduced by negative feedback, the bandwidth is increased. The closed-loop gain is independent of the open-loop gain up to the point of intersection of the two gain curves. This point of intersection is the critical frequency for the closed-loop response. Notice that the closed-loop gain has the same roll-off rate as the open-loop gain, beyond the closed-loop critical frequency.

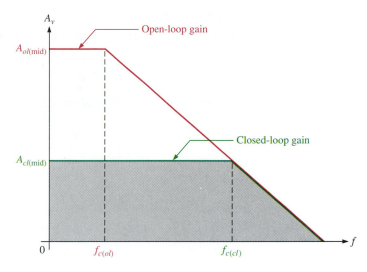

FIGURE 13–8
Closed-loop gain compared to open-loop gain.

Gain-Bandwidth Product

An increase in closed-loop gain causes a decrease in the bandwidth and vice versa, such that the product of gain and bandwidth is a constant. This is true as long as the roll-off rate is fixed. If you let A_{cl} represent the gain of any of the closed-loop configurations and $f_{c(cl)}$ represent the closed-loop critical frequency (same as the bandwidth), then

$$A_{cl}f_{c(cl)} = A_{ol}f_{c(ol)}$$

The gain-bandwidth product is always equal to the frequency at which the op-amp's open-loop gain is unity or 0 dB (unity-gain bandwidth).

$$A_{cl}f_{c(cl)} = \text{unity-gain bandwidth} \qquad \textbf{(13–10)}$$

EXAMPLE 13–5 Determine the bandwidth of each of the amplifiers in Figure 13–9. Both op-amps have an open-loop gain of 100 dB and a unity-gain bandwidth of 3 MHz.

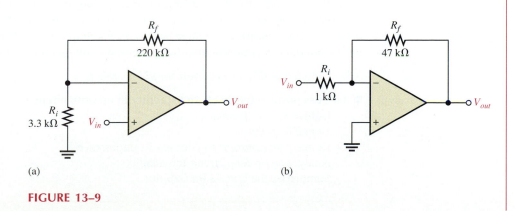

FIGURE 13–9

Solution

(a) For the noninverting amplifier in Figure 13–9(a), the closed-loop gain is

$$A_{cl} \cong \frac{1}{B} = 1 + \frac{R_f}{R_i} = 1 + \frac{220 \text{ k}\Omega}{3.3 \text{ k}\Omega} = 67.7$$

Use Equation (13–10) and solve for $f_{c(cl)}$ (where $f_{c(cl)} = BW_{cl}$):

$$f_{c(cl)} = BW_{cl} = \frac{\text{unity-gain } BW}{A_{cl}}$$

$$BW_{cl} = \frac{3 \text{ MHz}}{67.7} = 44.3 \text{ kHz}$$

(b) For the inverting amplifier in Figure 13–9(b), the closed-loop gain is

$$A_{cl} = -\frac{R_f}{R_i} = -\frac{47 \text{ k}\Omega}{1 \text{ k}\Omega} = -47$$

Using the absolute value of A_{cl}, the closed-loop bandwidth is

$$BW_{cl} = \frac{3 \text{ MHz}}{47} = 63.8 \text{ kHz}$$

Related Exercise Determine the bandwidth of each of the amplifiers in Figure 13–9. Both op-amps have an A_{ol} of 90 dB and a unity-gain bandwidth of 2 MHz.

SECTION 13–3 REVIEW

1. Is the closed-loop gain always less than the open-loop gain?
2. A certain op-amp is used in a feedback configuration having a gain of 30 and a bandwidth of 100 kHz. If the external resistor values are changed to increase the gain to 60, what is the new bandwidth?
3. What is the unity-gain bandwidth of the op-amp in Question 2?

13–4 ■ POSITIVE FEEDBACK AND STABILITY

Stability is an important consideration when using op-amps. Stable operation means that the op-amp does not oscillate under any condition. Instability produces oscillations, which are unwanted voltage swings on the output when there is no signal present on the input, or in response to noise or transient voltages on the input.

After completing this section, you should be able to

■ Discuss positive feedback and stability in op-amp circuits
  □ Define *positive feedback*
  □ Define *loop gain*
  □ Define *phase margin* and discuss its importance
  □ Analyze an op-amp circuit for stability
  □ Summarize the criteria for stability

Positive Feedback

To understand stability, we must first examine instability and its causes. As you know, with negative feedback, the signal fed back to the input is out of phase with the input signal, thus subtracting from it and effectively reducing the voltage gain. As long as the feedback is negative, the amplifier is stable.

When the signal fed back from output to input is in phase with the input signal, a **positive feedback** condition exists. That is, positive feedback occurs when the total phase shift through the op-amp and feedback circuit is 360°, which is equivalent to no phase shift (0°).

Loop Gain

For instability to occur, (a) there must be positive feedback, and (b) the loop gain of the closed-loop amplifier must be greater than 1. The **loop gain** of a closed-loop amplifier is the op-amp's open-loop gain (A_{ol}) times the attenuation, (B), of the feedback circuit:

$$\text{Loop gain} = A_{ol}B \qquad \textbf{(13–11)}$$

Phase Margin

Notice that for each amplifier configuration in Figure 13–10, the feedback loop is connected to the inverting input. There is an inherent phase shift of 180° because of the *inversion* between input and output. Additional phase shift (θ_{tot}) is produced by the *RC* lag circuits within the amplifier. So, the total phase shift around the loop is $180° + \theta_{tot}$.

The **phase margin**, θ_{pm}, is the amount of additional phase shift required to make the total phase shift around the loop 360° (360° is equivalent to 0°.).

$$180° + \theta_{tot} + \theta_{pm} = 360°$$

$$\theta_{pm} = 180° - |\theta_{tot}| \qquad \textbf{(13–12)}$$

If the phase margin is positive, the total phase shift is less than 360° and the amplifier is stable. If the phase margin is zero or negative, then the amplifier is potentially

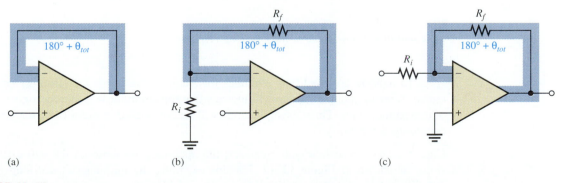

(a) (b) (c)

FIGURE 13–10
Feedback-loop phase shift.

unstable because the signal fed back can be in phase with the input. As you can see from Equation (13–12), when the total lag circuit phase shift θ_{tot} equals or exceeds 180°, then the phase margin is 0° or negative and an unstable condition exists.

Stability Analysis

Since most op-amp configurations use a loop gain, $A_{ol}B$, greater than 1, the criteria for stability are based on the phase angle of the internal lag circuits. As previously mentioned, operational amplifiers are composed of multiple stages, each of which has a critical frequency. For purposes of illustrating the concept of **stability,** we will use a three-stage op-amp with an open-loop response as shown in the Bode plot of Figure 13–11. Notice that there are three different critical frequencies, which indicates three internal RC lag circuits. At the first critical frequency, f_{c1}, the gain begins rolling off at −20 dB/decade; when the second critical frequency, f_{c2}, is reached, the gain decreases at −40 dB/decade; and when the third critical frequency, f_{c3}, is reached, the gain drops at −60 dB/decade.

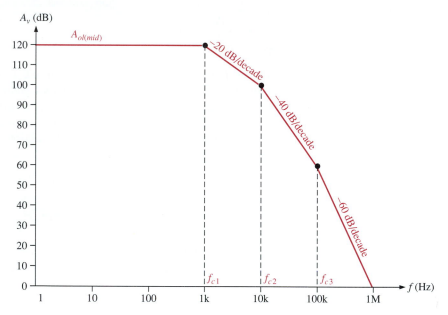

FIGURE 13–11
Bode plot of example of three-stage op-amp response.

To analyze a closed-loop amplifier for stability, the phase margin must be determined. A positive phase margin will indicate that the amplifier is stable for a given value of closed-loop gain. Three example cases will be considered in order to demonstrate the conditions for stability.

Case 1   The closed-loop gain intersects the open-loop response on the −20 dB/decade slope, as shown in Figure 13–12. For this example, the midrange closed-loop gain is 106 dB, and the closed-loop critical frequency is 5 kHz. If we assume that the amplifier is not operated out of its midrange, the maximum phase shift for the 106 dB amplifier occurs at the highest midrange frequency (in this case, 5 kHz). The total phase shift at this frequency due to the three lag circuits is calculated as follows:

$$\theta_{tot} = -\tan^{-1}\left(\frac{f}{f_{c1}}\right) - \tan^{-1}\left(\frac{f}{f_{c2}}\right) - \tan^{-1}\left(\frac{f}{f_{c3}}\right)$$

where $f = 5$ kHz, $f_{c1} = 1$ kHz, $f_{c2} = 10$ kHz, and $f_{c3} = 100$ kHz. Therefore,

$$\theta_{tot} = -\tan^{-1}\left(\frac{5\text{ kHz}}{1\text{ kHz}}\right) - \tan^{-1}\left(\frac{5\text{ kHz}}{10\text{ kHz}}\right) - \tan^{-1}\left(\frac{5\text{ kHz}}{100\text{ kHz}}\right)$$

$$= -78.7° - 26.6° - 2.86° = -108°$$

The phase margin, θ_{pm}, is

$$\theta_{pm} = 180° - |\theta_{tot}| = 180° - 108° = +72°$$

The phase margin is positive, so the amplifier is stable for all frequencies in its midrange. In general, an amplifier is stable for all midrange frequencies if its closed-loop gain intersects the open-loop response curve on a −20 dB/decade slope.

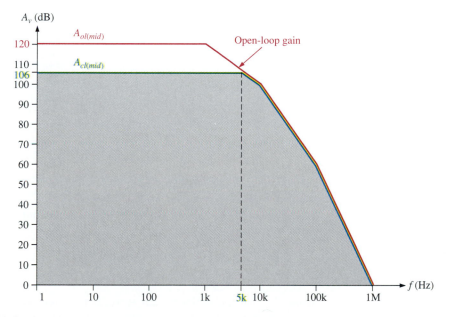

FIGURE 13–12

Case where closed-loop gain intersects open-loop gain on −20 dB/decade slope (stable operation).

Case 2 The closed-loop gain is lowered to where it intersects the open-loop response on the −40 dB/decade slope, as shown in Figure 13–13. The midrange closed-loop gain in this case is 72 dB, and the closed-loop critical frequency is approximately 30 kHz. The total phase shift at $f = 30$ kHz due to the three lag circuits is

$$\theta_{tot} = -\tan^{-1}\left(\frac{30\text{ kHz}}{1\text{ kHz}}\right) - \tan^{-1}\left(\frac{30\text{ kHz}}{10\text{ kHz}}\right) - \tan^{-1}\left(\frac{30\text{ kHz}}{100\text{ kHz}}\right)$$

$$= -88.1° - 71.6° - 16.7° = -176°$$

The phase margin is

$$\theta_{pm} = 180° - 176° = +4°$$

The phase margin is positive, so the amplifier is still stable for all frequencies in its midrange, but a very slight increase in frequency above f_c would cause it to oscillate. Therefore, it is marginally stable and very close to instability because instability occurs where $\theta_{pm} = 0°$ and, as a general rule, a minimum 45° phase margin is recommended to avoid marginal conditions.

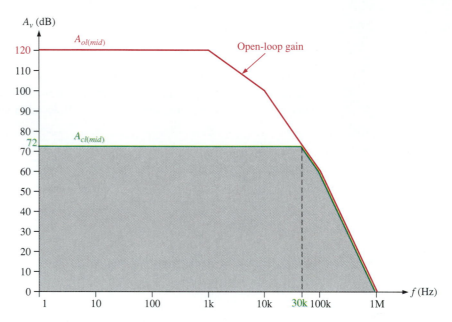

FIGURE 13–13

Case where closed-loop gain intersects open-loop gain on −40 dB/decade slope (marginally stable operation).

Case 3 The closed-loop gain is further decreased until it intersects the open-loop response on the −60 dB/decade slope, as shown in Figure 13–14. The midrange closed-loop gain in this case is 18 dB, and the closed-loop critical frequency is 500 kHz. The total phase shift at $f = 500$ kHz due to the three lag circuits is

$$\theta_{tot} = -\tan^{-1}\left(\frac{500\text{ kHz}}{1\text{ kHz}}\right) - \tan^{-1}\left(\frac{500\text{ kHz}}{10\text{ kHz}}\right) - \tan^{-1}\left(\frac{500\text{ kHz}}{100\text{ kHz}}\right)$$

$$= -89.9° - 88.9° - 78.7° = -258°$$

The phase margin is

$$\theta_{pm} = 180° - 258° = -78°$$

Here the phase margin is negative and the amplifier is unstable at the upper end of its midrange.

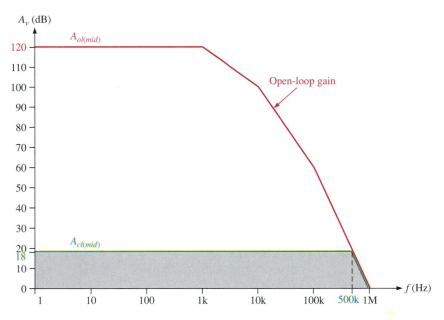

FIGURE 13–14

Case where closed-loop gain intersects open-loop gain on −60 dB/decade slope (unstable operation).

Summary of Stability Criteria

The stability analysis of three example cases has demonstrated that an amplifier's closed-loop gain must intersect the open-loop gain curve on a −20 dB/decade slope to ensure stability for all of its midrange frequencies. If the closed-loop gain is lowered to a value that intersects on a −40 dB/decade slope, then marginal stability or complete instability can occur. In the previous analyses (Cases 1, 2, and 3), the closed-loop gain should be greater than 72 dB.

If the closed-loop gain intersects the open-loop response on a −60 dB/decade slope, definite instability will occur at some frequency within the amplifier's midrange. Therefore, to ensure stability for all of the midrange frequencies, an op-amp must be operated at a closed-loop gain such that the roll-off rate beginning at its dominant critical frequency does not exceed −20 dB/decade.

Using the following BASIC program, you can determine the approximate midrange frequency at which instability occurs. The program computes the phase margin for each of a specified number of frequency values and determines whether the op-amp is stable or unstable. The display shows a list of frequencies with corresponding phase shifts, phase margins, and stability conditions. Instability occurs between the last two frequencies printed. Inputs required for this program are all of the critical frequencies in the open-loop response of the given op-amp, the highest frequency of operation, and the increments of frequency for which the parameters are to be computed.

```
10  CLS
20  PRINT"THIS PROGRAM COMPUTES THE PHASE MARGINS FOR EACH"
30  PRINT"FREQUENCY IN A SPECIFIED RANGE AND DETERMINES THE"
40  PRINT"MAXIMUM FREQUENCY FOR STABLE OPERATION OF THE OP-AMP"
50  PRINT:PRINT:PRINT
60  INPUT "TO CONTINUE PRESS 'ENTER'";X
70  CLS
80  INPUT "NUMBER OF OPEN-LOOP CRITICAL FREQUENCIES";N
90  CLS
100 FOR Y=1 TO N
110 INPUT "CRITICAL FREQUENCY IN HERTZ";FC(Y)
120 NEXT
130 INPUT "THE HIGHEST INPUT FREQUENCY";FH
140 INPUT "INCREMENTS OF FREQUENCY FROM ZERO TO THE HIGHEST";FI
150 CLS
160 PRINT"FREQUENCY","PHASE SHIFT","PHASE MARGIN","STABILITY"
170 FOR F=0 TO FH STEP FI
180 PH=0
190 FOR Y=1 TO N
200 PH=PH-ATN(F/FC(Y))*57.29578
210 NEXT Y
220 PM=180+PH
230 IF PM<=0 THEN S$="UNSTABLE" ELSE S$="STABLE"
240 PRINT F,PH,PM,S$
250 IF PM<=0 THEN END
260 NEXT F
```

SECTION 13–4 REVIEW

1. Under what feedback condition can an amplifier oscillate?

2. How much can the phase shift of an amplifier's internal *RC* circuit be before instability occurs? What is the phase margin at the point where instability begins?

3. What is the maximum roll-off rate of the open-loop gain of an op-amp for which the device will still be stable?

13–5 ■ COMPENSATION

The last section demonstrated that instability can occur when an op-amp's response has roll-off rates exceeding −20 dB/decade and the op-amp is operated in a closed-loop configuration having a gain curve that intersects a higher roll-off rate portion of the open-loop response. In situations like those examined in the last section, the closed-loop voltage gain is restricted to very high values. In many applications, lower values of closed-loop gain are necessary or desirable. To allow op-amps to be operated at low closed-loop gain, phase lag compensation is required.

After completing this section, you should be able to

■ **Explain op-amp phase compensation**
   □ Describe phase-lag compensation
   □ Analyze a compensating circuit
   □ Apply single-capacitor compensation
   □ Apply feedforward compensation

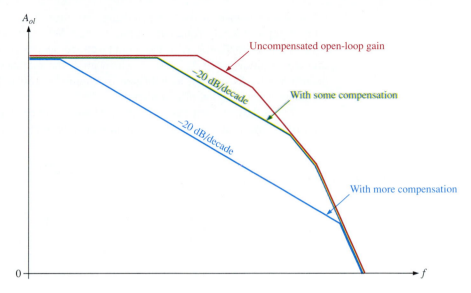

FIGURE 13–15
Bode plot illustrating the effect of phase compensation on open-loop gain of a typical op-amp.

Phase Lag Compensation

As you have seen, the cause of instability is excessive phase shift through an op-amp's internal lag circuits. When these phase shifts equal or exceed 180°, the amplifier can oscillate. **Compensation** is used to either eliminate open-loop roll-off rates greater than −20 dB/decade or extend the −20 dB/decade rate to a lower gain. These concepts are illustrated in Figure 13–15.

Compensating Circuit

There are two basic methods of compensation for integrated circuit op-amps: internal and external. In either case an *RC* circuit is added. The basic compensating action is as follows. Consider first the *RC* circuit shown in Figure 13–16(a). At low frequencies

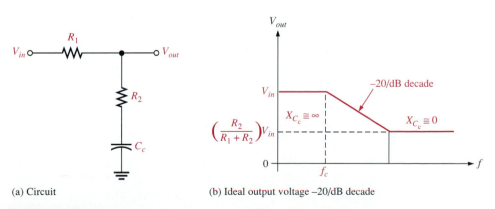

(a) Circuit

(b) Ideal output voltage −20/dB decade

FIGURE 13–16
Basic compensating circuit action.

where the reactance of the compensating capacitor, X_{C_c}, is extremely large, the output voltage approximately equals the input voltage. When the frequency reaches its critical value, $f_c = 1/[2\pi(R_1 + R_2)C_c]$, the output voltage decreases at -20 dB/decade. This roll-off rate continues until $X_{C_c} \cong 0$, at which point the output voltage levels off to a value determined by R_1 and R_2, as indicated in Figure 13–16(b). This is the principle used in the phase compensation of an op-amp.

To see how a compensating circuit changes the open-loop response of an op-amp, refer to Figure 13–17. This diagram represents a two-stage op-amp. The individual stages

FIGURE 13–17

Representation of op-amp with compensation.

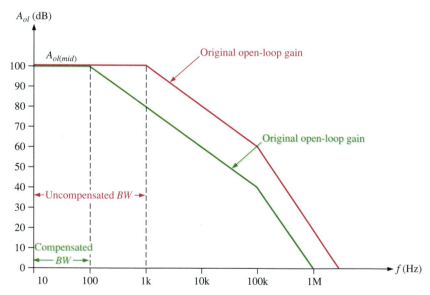

FIGURE 13–18

Example of compensated op-amp frequency response.

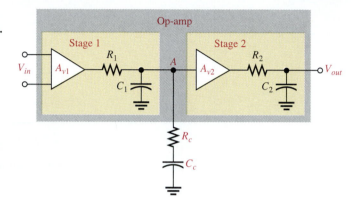

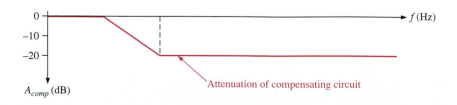

are within the colored blocks along with the associated lag circuit. A compensating circuit is shown connected at point A on the output of stage 1.

The critical frequency of the compensating circuit is set to a value less than the dominant (lowest) critical frequency of the internal lag circuits. This causes the -20 dB/decade roll-off to begin at the compensating circuit's critical frequency. The roll-off of the compensating circuit continues up to the critical frequency of the dominant lag circuit. At this point, the response of the compensating circuit levels off, and the -20 dB/decade roll-off of the dominant lag circuit takes over. The net result is a shift of the open-loop response to the left, thus reducing the bandwidth, as shown in Figure 13–18. The response curve of the compensating circuit is shown in proper relation to the overall open-loop response.

EXAMPLE 13–6

A certain op-amp has the open-loop response in Figure 13–19. As you can see, the lowest closed-loop gain for which stability is assured is approximately 40 dB (where the closed-loop gain line still intersects the -20 dB/decade slope). In a particular application, a 20 dB closed-loop gain is required.

(a) Determine the critical frequency for the compensating circuit.
(b) Sketch the ideal response curve for the compensating circuit.
(c) Sketch the total ideal compensated open-loop response.

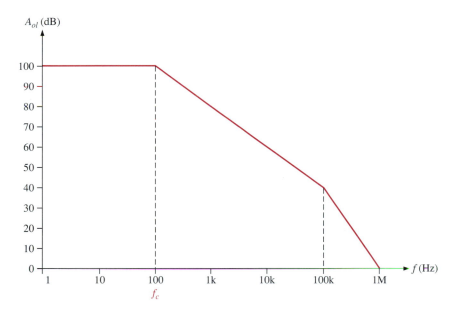

FIGURE 13–19
Original open-loop response.

Solution

(a) The gain must be dropped so that the -20 dB/decade roll-off extends down to 20 dB rather than to 40 dB. To achieve this, the midrange open-loop gain must be made to roll off a decade sooner. Therefore, the critical frequency of the compensating circuit must be 10 Hz.

(b) The roll-off of the compensating circuit must end at 100 Hz, as shown in Figure 13–20(a).

(c) The total open-loop response resulting from compensation is shown in Figure 13–20(b).

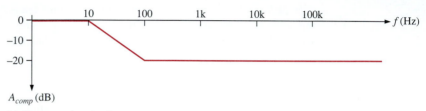

(a) Compensating circuit response

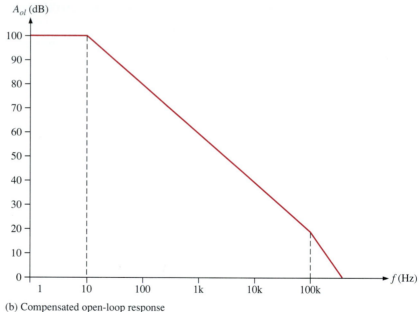

(b) Compensated open-loop response

FIGURE 13–20

Related Exercise   In this example, what is the uncompensated bandwidth? What is the compensated bandwidth?

Extent of Compensation

A larger compensating capacitor will cause the open-loop roll-off to begin at a lower frequency and thus extend the −20 dB/decade roll-off to lower gain levels, as shown in Figure 13–21(a). With a sufficiently large compensating capacitor, an op-amp can be made unconditionally stable, as illustrated in Figure 13–21(b), where the −20 dB/decade slope

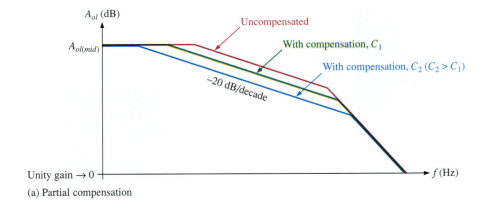

(a) Partial compensation

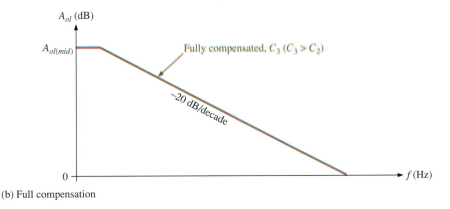

(b) Full compensation

FIGURE 13–21
Extent of compensation.

is extended all the way down to unity gain. This is normally the case when internal compensation is provided by the manufacturer. An internally, fully compensated op-amp can be used for any value of closed-loop gain and remain stable. The 741 is an example of an internally compensated device.

A disadvantage of fully compensated op-amps is that bandwidth is sacrificed; thus the slew rate is decreased. Therefore, many IC op-amps have provisions for external compensation. Figure 13–22 shows typical package layouts of an LM101A op-amp with pins available for external compensation with a small capacitor. With provisions for external connections, just enough compensation can be used for a given application without sacrificing more performance than necessary.

Single-Capacitor Compensation

As an example of compensating an IC op-amp, a capacitor C_1 is connected to pins 1 and 8 of an LM101A in an inverting amplifier configuration, as shown in Figure 13–23(a). Part (b) of the figure shows the open-loop frequency response curves for two values of C_1. The 3 pF compensating capacitor produces a unity-gain bandwidth approaching 10 MHz. Notice that the −20 dB/decade slope extends to a very low gain value. When C_1 is increased ten times to 30 pF, the bandwidth is reduced by a factor of ten. Notice that the −20 dB/decade slope now extends through unity gain.

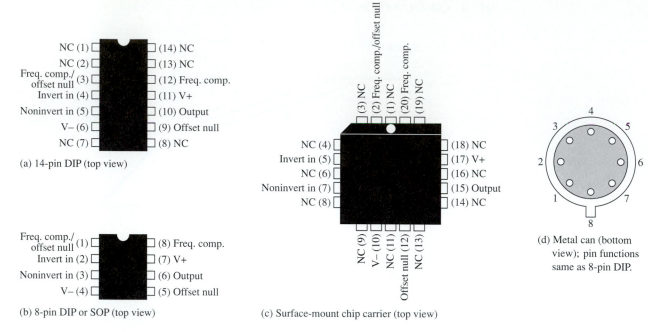

FIGURE 13–22
Typical op-amp packages.

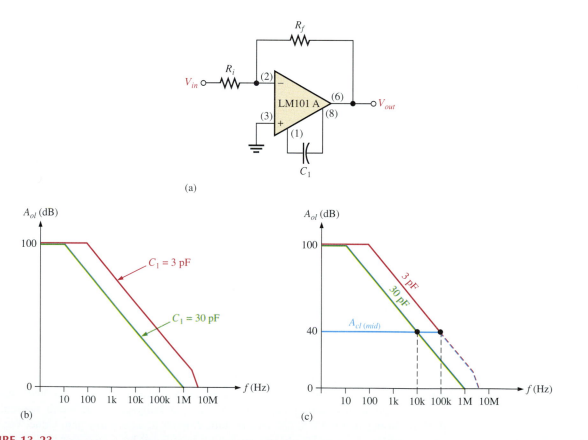

FIGURE 13–23
Example of single-capacitor compensation of an LM101A op-amp.

When the op-amp is used in a closed-loop configuration, as in Figure 13–23(c), the useful frequency range depends on the compensating capacitor. For example, with a closed-loop gain of 40 dB as shown in part (c), the bandwidth is approximately 10 kHz for $C_1 = 30$ pF and increases to approximately 100 kHz when C_1 is decreased to 3 pF.

Feedforward Compensation

Another method of phase compensation is called **feedforward.** This type of compensation results in less bandwidth reduction than the method previously discussed. The basic concept is to bypass the internal input stage of the op-amp at high frequencies and drive the higher-frequency second stage, as shown in Figure 13–24.

FIGURE 13–24
Feedforward compensation showing high-frequency bypassing of first stage.

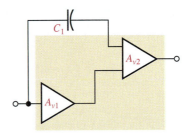

Feedforward compensation of an LM101A is shown in Figure 13–25(a). The feedforward capacitor C_1 is connected from the inverting input to the compensating terminal. A small capacitor is needed across R_f to ensure stability. The Bode plot in Figure 13–25(b) shows the feedforward compensated response and the standard compensated response that was discussed previously. The use of feedforward compensation is restricted to the inverting amplifier configuration. Other compensation methods are also used. Often, recommendations are provided by the manufacturer on the data sheet.

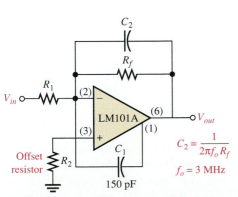

(a) Manufacturers' recommended configuration

$$C_2 = \frac{1}{2\pi f_o R_f}$$

$$f_o = 3 \text{ MHz}$$

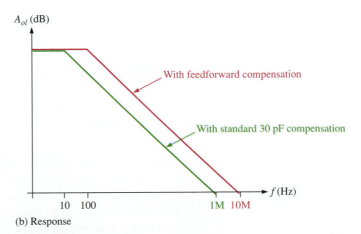

(b) Response

FIGURE 13–25
Feedforward compensation of an LM101A op-amp and the response curves.

1. What is the purpose of phase compensation?
2. What is the main difference between internal and external compensation?
3. When you compensate an amplifier, does the bandwidth increase or decrease?

13–6 ■ SYSTEM APPLICATION

You have been assigned to work on an AM receiver that your company is developing for production. In particular, you will concentrate on the audio amplifier board. The circuit is designed with both discrete components and an integrated circuit (LM101A op-amp). You will apply the knowledge of op-amps acquired in this chapter and knowledge from previous chapters to this assignment.

Basic Operation of the System

Figure 13–26 shows the block diagram of a superheterodyne AM (amplitude modulation) receiver. The antenna picks up all signals across the AM broadcast band from 535 kHz to 1605 kHz and feeds them to the RF amplifier. A signal in the AM band is called a carrier and it is modulated with the audio signal at the transmitting station. Amplitude modulation is accomplished by varying the amplitude of the carrier proportional to the audio signal.

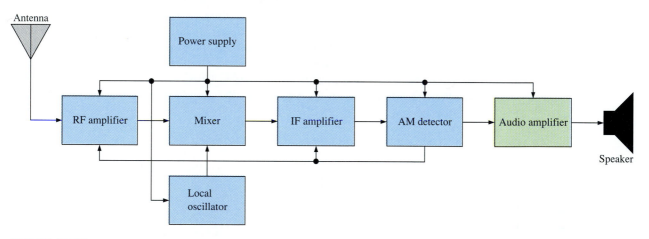

FIGURE 13–26
Block diagram of basic superheterodyne AM receiver.

The radio frequency (RF) amplifier selects a desired frequency from all of those received and amplifies the extremely small signal coming from the antenna. Since it is a tuned amplifier, it is highly frequency selective and eliminates essentially all signals but the one to which it is tuned. The amplified AM signal from the RF amplifier goes to the mixer where it is combined with the output of the local oscillator which has a frequency of 455 kHz above the frequency of the selected carrier frequency. The mixer, through a process called *heterodyning*, produces output frequencies equal to both the sum and difference of the selected carrier frequency and the local oscillator frequency. The sum frequency is filtered out and only the difference frequency of 455 kHz is used. The 455 kHz is called the intermediate frequency (IF) and still carries the same audio modulation that was on the carrier frequency.

The IF amplifier is tuned to 455 kHz and amplifies the signal. The detector takes the AM signal from the IF amplifier and recovers the audio signal while eliminating the intermediate frequency. The small audio signal from the output of the detector is amplified by the audio amplifier circuits, which includes both a preamplifier and a power amplifier. The power amplifier drives the speaker which converts the audio signal to sound.

The Photocell/Amplifier Circuit Board

☐ Make sure that the circuit board shown in Figure 13–27 is correctly assembled by checking it against the schematic in Figure 13–28. A pin diagram for the power transistors is shown for reference. There are a number of interconnections on the back side of the board. Corresponding feedthrough pads are aligned either vertically or horizontally.

☐ Label a copy of the board with component and input/output designations in agreement with the schematic.

Analysis of the Audio Amplifier Circuit

☐ Determine the midrange voltage gain of the amplifier.

☐ Determine the lower critical frequency. Given that the upper critical frequency is 15 kHz, what is the bandwidth?

FIGURE 13–27
Audio amplifier board.

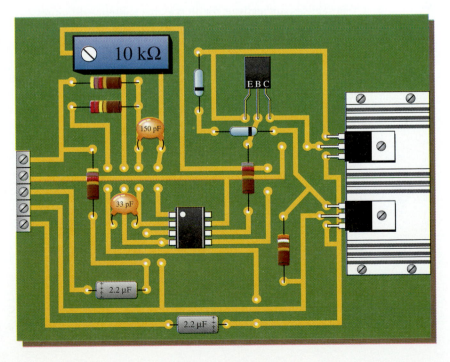

FIGURE 13–28
Audio amplifier schematic.

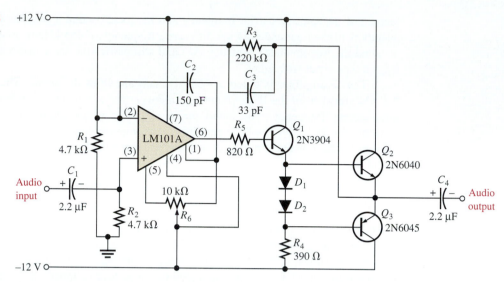

□ Determine the maximum peak-to-peak input voltage that can be applied to the audio amplifier without producing a distorted output signal. Assume that the maximum output peaks are 1 V less than the dc supply voltages.

□ Calculate the power to the speaker for the maximum voltage output.

Test Procedure

□ Develop a step-by-step set of instructions on how to check the audio amplifier circuit board for proper operation.

□ Specify voltage values for all the measurements to be made.

□ Provide a fault analysis for all possible component failures.

Troubleshooting

Problems have developed in three boards. Based on the test bench measurements for each board indicated in Figure 13–29, determine the most likely fault in each case. The circled numbers indicate test point connections to the circuit board. Assume that each board has the proper dc supply voltage.

Final Report

Submit a final written report on the photocell/amplifier circuit board using an organized format that includes the following:

1. A physical description of the circuit.

2. A discussion of the operation of the circuit.

3. A list of the specifications.

4. A list of parts with part numbers if available.

5. A list of the types of problems on the three faulty circuit boards.

6. A description of how you determined the problem on each of the faulty circuit boards.

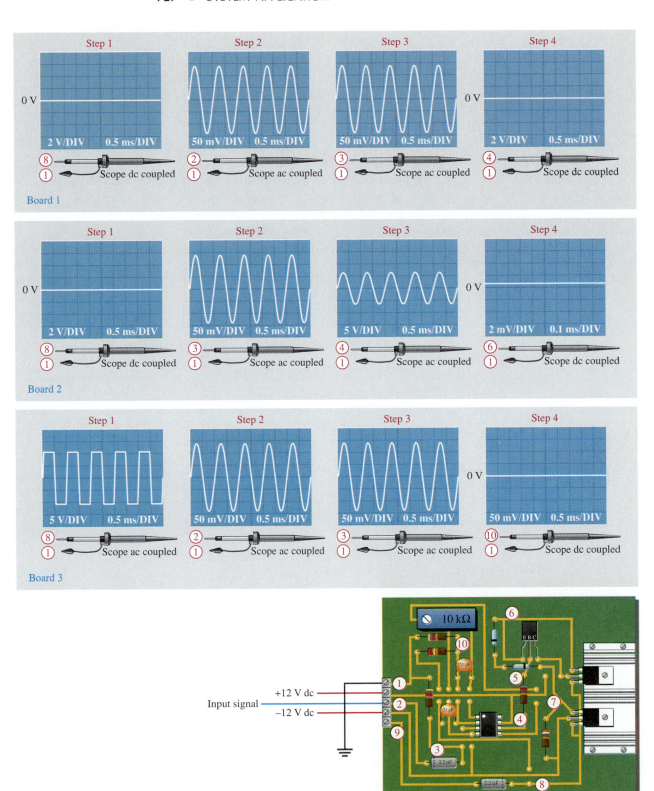

FIGURE 13–29

Test results for three faulty circuit boards.

■ CHAPTER SUMMARY

- The closed-loop gain is always less than the open-loop gain.
- The midrange gain of an op-amp extends down to dc.
- The gain of an op-amp decreases as frequency increases above the critical frequency.
- The bandwidth of an op-amp equals the upper critical frequency.
- The internal *RC* lag circuits that are inherently part of the amplifier stages cause the gain to roll off as frequency goes up.
- The internal *RC* lag circuits also cause a phase shift between input and output signals.
- Negative feedback lowers the gain and increases the bandwidth.
- The product of gain and bandwidth is constant for a given op-amp.
- The gain-bandwidth product equals the frequency at which unity voltage gain occurs.
- Positive feedback occurs when the total phase shift through the op-amp (including 180° inversion) and feedback circuit is 0° (equivalent to 360°) or more.
- The phase margin is the amount of additional phase shift required to make the total phase shift around the loop 360°.
- When the closed-loop gain of an op-amp intersects the open-loop response curve on a −20 dB/decade (−6 dB/octave) slope, the amplifier is stable.
- When the closed-loop gain intersects the open-loop response curve on a slope greater than −20 db/decade, the amplifier can be either marginally stable or unstable.
- A minimum phase margin of 45° is recommended to provide a sufficient safety factor for stable operation.
- A fully compensated op-amp has a −20 dB/decade roll-off all the way down to unity gain.
- Compensation reduces bandwidth and increases slew rate.
- Internally compensated op-amps such as the 741 are available. These are usually fully compensated with a large sacrifice in bandwidth.
- Externally compensated op-amps such as LM101A are available. External compensating circuits can be connected to specified pins, and the compensation can be tailored to a specific application. In this way, bandwidth and slew rate are not degraded more than necessary.

■ GLOSSARY

Closed-loop gain   The voltage gain of an op-amp with external feedback.

Compensation   The process of modifying the roll-off rate of an amplifier to ensure stability.

Feedforward   A method of frequency compensation in op-amp circuits.

Open-loop gain   The voltage gain of an op-amp without feedback.

Phase   The relative angular displacement of a time-varying function relative to a reference.

Phase margin   The difference between the total phase shift through an amplifier and 180°. The additional amount of phase shift that can be allowed before instability occurs.

Stability   A condition in which an amplifier circuit does not oscillate.

■ FORMULAS

(13–1)	$BW = f_{cu}$		Op-amp bandwidth
(13–2)	$\dfrac{V_{out}}{V_{in}} = \dfrac{1}{\sqrt{1 + R^2/X_C^2}}$		*RC* attenuation, lag circuit
(13–3)	$\dfrac{V_{out}}{V_{in}} = \dfrac{1}{\sqrt{1 + f^2/f_c^2}}$		*RC* attenuation
(13–4)	$A_{ol} = \dfrac{A_{ol(mid)}}{\sqrt{1 + f^2/f_c^2}}$		Open-loop gain

$(13-5)$ $\quad \theta = -\tan^{-1}\left(\dfrac{R}{X_C}\right)$ $\qquad\qquad\qquad$ *RC* phase shift

$(13-6)$ $\quad \theta = -\tan^{-1}\left(\dfrac{f}{f_c}\right)$ $\qquad\qquad\qquad$ *RC* phase shift

$(13-7)$ $\quad \theta_{tot} = -\tan^{-1}\left(\dfrac{f}{f_{c1}}\right) - \tan^{-1}\left(\dfrac{f}{f_{c2}}\right) - \tan^{-1}\left(\dfrac{f}{f_{c3}}\right)$ $\qquad$ Total phase shift

$(13-8)$ $\quad f_{c(cl)} = f_{c(ol)}(1 + BA_{ol(mid)})$ $\qquad\qquad\qquad$ Closed-loop critical frequency

$(13-9)$ $\quad BW_{cl} = BW_{ol}(1 + BA_{ol(mid)})$ $\qquad\qquad\qquad$ Closed-loop bandwidth

$(13-10)$ $\quad A_{cl}f_{c(cl)} =$ **unity-gain bandwidth**

$(13-11)$ $\quad$ **Loop gain $= A_{ol}B$** $\qquad\qquad\qquad\qquad\qquad$ Loop gain

$(13-12)$ $\quad \theta_{pm} = 180° - |\theta_{tot}|$ $\qquad\qquad\qquad\qquad$ Phase margin

---

■ **SELF-TEST**

1. The open-loop gain of an op-amp is always
   (a) less than the closed-loop gain
   (b) equal to the closed-loop gain
   (c) greater than the closed-loop gain
   (d) a very stable and constant quantity for a given type of op-amp

2. The bandwidth of an ac amplifier having a lower critical frequency of 1 kHz and an upper critical frequency of 10 kHz is
   (a) 1 kHz   (b) 9 kHz   (c) 10 kHz   (d) 11 kHz

3. The bandwidth of a dc amplifier having an upper critical frequency of 100 kHz is
   (a) 100 kHz   (b) unknown   (c) infinity   (d) 0 kHz

4. The midrange open-loop gain of an op-amp
   (a) extends from the lower critical frequency to the upper critical frequency
   (b) extends from 0 Hz to the upper critical frequency
   (c) rolls off at 20 dB/decade beginning at 0 Hz
   (d) answers (b) and (c)

5. The frequency at which the open-loop gain is equal to one is called
   (a) the upper critical frequency   (b) the cutoff frequency
   (c) the notch frequency   (d) the unity-gain frequency

6. Phase shift through an op-amp is caused by
   (a) the internal *RC* circuits   (b) the external *RC* circuits
   (c) the gain roll off   (d) negative feedback

7. Each *RC* circuit in an op-amp
   (a) causes the gain to roll off at −6 dB/octave
   (b) causes the gain to roll off at −20 dB/decade
   (c) reduces the midrange gain by 3 dB
   (d) answers (a) and (b)

8. When negative feedback is used, the gain-bandwidth product of an op-amp
   (a) increases   (b) decreases   (c) stays the same   (d) fluctuates

9. If a certain op-amp has a midrange open-loop gain of 200,000 and a unity-gain frequency of 5 MHz, the gain-bandwidth product is
   (a) 200,000 Hz   (b) 5,000,000 Hz
   (c) 1×10^{12} Hz   (d) not determinable from the information

10. If a certain op-amp has a closed-loop gain of 20 and an upper critical frequency of 10 MHz, the gain-bandwidth product is
    (a) 200 MHz   (b) 10 MHz   (c) the unity-gain frequency

11. Positive feedback occurs when
   (a) the output signal is fed back to the input in-phase with the input signal
   (b) the output signal is fed back to the input out-of-phase with the input signal
   (c) the total phase shift through the op-amp and feedback circuit is 360°
   (d) answers (a) and (c)

12. For a closed-loop op-amp circuit to be unstable,
   (a) there must be positive feedback
   (b) the loop gain must be greater than one
   (c) the loop gain must be less than one
   (d) answers (a) and (b)

13. The amount of additional phase shift required to make the total phase shift around a closed loop equal to zero is called
   (a) the unity-gain phase shift   (b) phase margin
   (c) phase lag   (d) phase bandwidth

14. For a given value of closed-loop gain, a positive phase margin indicates
   (a) an unstable condition   (b) too much phase shift
   (c) a stable condition   (d) nothing

15. The purpose of phase-lag compensation is to
   (a) make the op-amp stable at very high values of gain
   (b) make the op-amp stable at low values of gain
   (c) reduce the unity-gain frequency
   (d) increase the bandwidth

■ BASIC PROBLEMS

SECTION 13–1   Basic Concepts

1. The midrange open-loop gain of a certain op-amp is 120 dB. Negative feedback reduces this gain by 50 dB. What is the closed-loop gain?

2. The upper critical frequency of an op-amp's open-loop response is 200 Hz. If the midrange gain is 175,000, what is the ideal gain at 200 Hz? What is the actual gain? What is the op-amp's open-loop bandwidth?

3. An RC lag circuit has a critical frequency of 5 kHz. If the resistance value is 1 kΩ, what is X_C when $f = 3$ kHz?

4. Determine the attenuation of an RC lag circuit with $f_c = 12$ kHz for each of the following frequencies.
   (a) 1 kHz   (b) 5 kHz   (c) 12 kHz   (d) 20 kHz   (e) 100 kHz

5. The midrange open-loop gain of a certain op-amp is 80,000. If the open-loop critical frequency is 1 kHz, what is the open-loop gain at each of the following frequencies?
   (a) 100 Hz   (b) 1 kHz   (c) 10 kHz   (d) 1 MHz

6. Determine the phase shift through each circuit in Figure 13–30 at a frequency of 2 kHz.

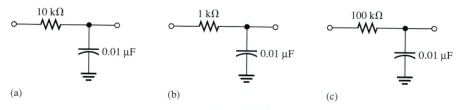

(a)   (b)   (c)

FIGURE 13–30

7. An *RC* lag circuit has a critical frequency of 8.5 kHz. Determine the phase for each frequency and plot a graph of its phase angle versus frequency.
(a) 100 Hz (b) 400 Hz (c) 850 Hz
(d) 8.5 kHz (e) 25 kHz (f) 85 kHz

SECTION 13–2 Open-Loop Response

8. A certain op-amp has three internal amplifier stages with midrange gains of 30 dB, 40 dB, and 20 dB. Each stage also has a critical frequency associated with it as follows: $f_{c1} = 600$ Hz, $f_{c2} = 50$ kHz, and $f_{c3} = 200$ kHz.
(a) What is the midrange open-loop gain of the op-amp, expressed in dB?
(b) What is the total phase shift through the amplifier, including inversion, when the signal frequency is 10 kHz?

9. What is the gain roll-off rate in Problem 8 between the following frequencies?
(a) 0 Hz and 600 Hz (b) 600 Hz and 50 kHz
(c) 50 kHz and 200 kHz (d) 200 kHz and 1 MHz

SECTION 13–3 Closed-Loop Response

10. Determine the midrange gain in dB of each amplifier in Figure 13–31. Are these open-loop or closed-loop gains?

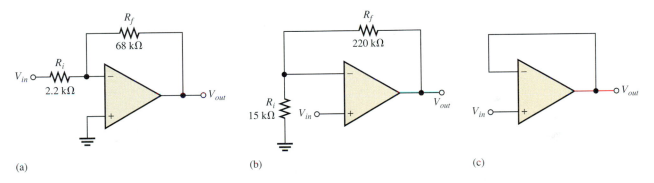

(a)   (b)   (c)

FIGURE 13–31

11. A certain amplifier has an open-loop gain in midrange of 180,000 and an open-loop critical frequency of 1500 Hz. If the attenuation of the feedback path is 0.015, what is the closed-loop bandwidth?

12. Given that $f_{c(ol)} = 750$ Hz, $A_{ol} = 89$ dB, and $f_{c(cl)} = 5.5$ kHz, determine the closed-loop gain in dB.

13. What is the unity-gain bandwidth in Problem 12?

14. For each amplifier in Figure 13–32, determine the closed-loop gain and bandwidth. The op-amps in each circuit exhibit an open-loop gain of 125 dB and a unity-gain bandwidth of 2.8 MHz.

15. Which of the amplifiers in Figure 13–33 has the smaller bandwidth?

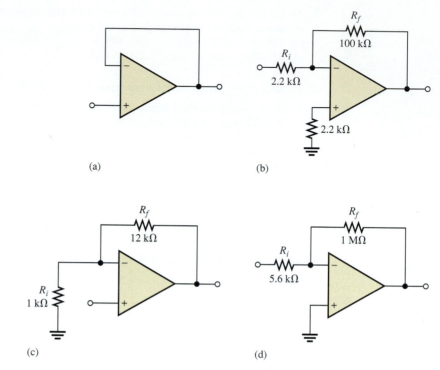

FIGURE 13–32

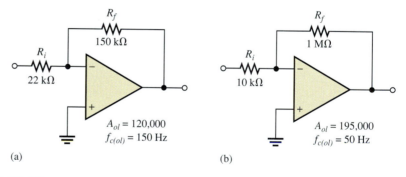

FIGURE 13–33

SECTION 13–4 Positive Feedback and Stability

16. It has been determined that the op-amp circuit in Figure 13–34 has three internal critical frequencies as follows: 1.2 kHz, 50 kHz, 250 kHz. If the midrange open-loop gain is 100 dB, is the amplifier configuration stable, marginally stable, or unstable?

17. Determine the phase margin for each value of phase lag.
    (a) 30°   **(b)** 60°   **(c)** 120°   **(d)** 180°   **(e)** 210°

18. A certain op-amp has the following internal critical frequencies in its open-loop response: 125 Hz, 25 kHz, and 180 kHz. What is the total phase shift through the amplifier when the signal frequency is 50 kHz?

19. Each graph in Figure 13–35 shows both the open-loop and the closed-loop response of a particular op-amp configuration. Analyze each case for stability.

FIGURE 13–34

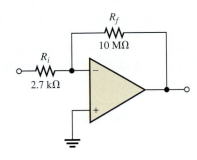

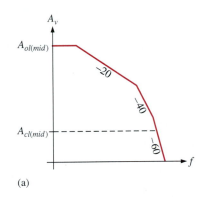

(a)

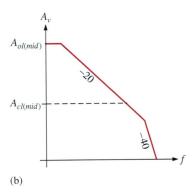

(b)

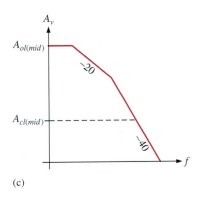

(c)

FIGURE 13–35

SECTION 13–5   Compensation

20. A certain operational amplifier has an open-loop response curve as shown in Figure 13–36. A particular application requires a 30 dB closed-loop midrange gain. In order to achieve a 30 dB gain, compensation must be added because the 30 dB line intersects the uncompensated open-loop gain on the −40 dB/decade slope and, therefore, stability is not assured.

   (a) Find the critical frequency of the compensating circuit such that the −20 dB/decade slope is lowered to a point where it intersects the 30 dB gain line.

   (b) Sketch the ideal response curve for the compensating circuit.

   (c) Sketch the total ideal compensated open-loop response.

FIGURE 13–36

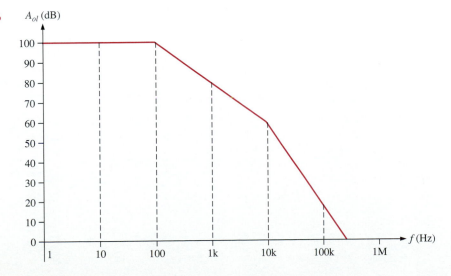

21. The open-loop gain of a certain op-amp rolls off at -20 dB/decade, beginning at $f = 250$ Hz. This roll-off rate extends down to a gain of 60 dB. If a 40 dB closed-loop gain is required, what is the critical frequency for the compensating circuit?

22. Repeat Problem 21 for a closed-loop gain of 20 dB.

■ **TROUBLE-SHOOTING PROBLEMS**

SECTION 13–6 System Application

23. In the amplifier circuit of Figure 13–28, list the possible faults that will cause the push-pull stage to operate nonlinearly.

24. What indication would you observe if a 2.2 MΩ resistor is incorrectly installed for R_3 in Figure 13–28?

25. What voltage will you measure on the output of the amplifier in Figure 13–28 if diode D_1 opens?

■ **DATA SHEET PROBLEMS**

26. Refer to the partial data sheet in Figure 13–37 to determine the following:
    (a) the highest dc supply voltages that can be used for an LM201A
    (b) the maximum power that an LM201A in a plastic dual-in-line package can dissipate at 65°C
    (c) the maximum peak output voltages for an LM101A with dc supply voltages of ±15 V and a load of 10 kΩ
    (d) the minimum open-loop (large signal) voltage gain for an LM101A

27. What voltage ratio is represented by the typical CMRR specification for the LM101A?

Maximum Ratings

Rating	Symbol	LM101A	LM201A	LM301A	Unit
		Value			
Power supply voltage	V_{CC}, V_{EE}	±22	±22	±18	V dc
Input differential voltage	V_{ID}	◄——————— ±30 ———————►			Volts
Input common-mode range*	V_{ICR}	◄——————— ±15 ———————►			Volts
Output short-circuit duration	t_S	◄——————— Continuous ———————►			
Power dissipation (package limitation) Metal can	P_D	◄——————— 500 ———————►			mW
Derate above T_A = +75°C		◄——————— 6.8 ———————►			mW/C°
Plastic dual in-line package (LM201A/		—	625	625	mW
Derate above T_A = +25°C 301A)		—	5.0	5.0	mW/C°
Ceramic package		◄——————— 750 ———————►			mW
Derate above 25°C		◄——————— 6.6 ———————►			mW/C°
Operating ambient temperature range	T_A	−55 to +125	−25 to +85	0 to +70	°C
Storage channel temperature range	T_{stg}	◄——————— −65 to +150 ———————►			°C

* For supply voltages less than ±15 V, the absolute maximum input voltage is equal to the supply voltage.

Electrical Characteristics (T_A = +25°C unless otherwise noted.) Unless otherwise specified, these specifications apply for supply voltages from ±5.0 V to ±20 V for the LM101A and LM201A, and from ±5.0 V to ±15 V for the LM301A.

Characteristics	Symbol	LM101A LM201A			LM301A			Unit
		Min	Typ	Max	Min	Typ	Max	
Input offset voltage ($R_S \leq 50$ kΩ)	V_{IO}	—	0.7	2.0	—	2.0	7.5	mV
Input offset current	I_{IO}	—	1.5	10	—	3.0	50	nA
Input bias current	I_{IB}	—	30	75	—	70	250	nA
Input resistance	r_i	1.5	4.0	—	0.5	2.0	—	Megohms
Supply current V_{CC}/V_{EE} = ±20 V V_{CC}/V_{EE} = ±15 V	I_{CC}, I_{EE}	— —	1.8 —	3.0 —	— —	— 1.8	— 3.0	mA
Large-signal voltage gain (V_{CC}/V_{EE} = ±15 V, V_O = ±10 V, R_L > 2.0 kΩ)	A_v	50	160	—	25	160	—	V/mV

The following specifications apply over the operating temperature range.

Characteristics	Symbol	Min	Typ	Max	Min	Typ	Max	Unit
Input offset voltage ($R_S \leq 50$ kΩ)	V_{IO}	—	—	3.0	—	—	10	mV
Input offset current	I_{IO}	—	—	20	—	—	70	nA
Average temperature coefficient of input offset voltage $T_{A(min)} \leq T_A \leq T_{A(max)}$	$\Delta V_{IO}/\Delta T$	—	3.0	15	—	6.0	30	μV/C°
Average temperature coefficient of input offset current +25°C $\leq T_A \leq T_{A(max)}$ $T_{A(min)} \leq T_A \leq$ 25°C	$\Delta I_{IO}/\Delta T$	— —	0.01 0.02	0.1 0.2	— —	0.01 0.02	0.3 0.6	nA/C°
Input bias current	I_{IB}	—	—	100	—	—	300	nA
Large-signal voltage gain (V_{CC}/V_{EE} = ±15 V, V_O = ±10 V, R_L > 2.0 kΩ)	A_v	25	—	—	15	—	—	V/mV
Input voltage range (V_{CC}/V_{EE} = ±20 V) (V_{CC}/V_{EE} = ±15 V)	V_I	±15 —	— —	— —	— ±12	— —	— —	V
Common-mode rejection ratio $R_S \leq 50$ kΩ	CMRR	80	96	—	70	90	—	dB
Supply voltage rejection ratio $R_S \leq 50$ kΩ	PSRR	80	96	—	70	96	—	dB
Output voltage swing (V_{CC}/V_{EE} = ±15 V, R_L = 10 kΩ, R_L = 2.0 kΩ	V_O	±12 ±10	±14 ±13	— —	±12 ±10	±14 ±13	— —	V
Supply currents ($T_A = T_{A(max)}$, V_{CC}/V_{EE} = ±20 V)	I_{CC}, I_{EE}	—	1.2	2.5	—	—	—	mA

FIGURE 13–37

FIGURE 13–38

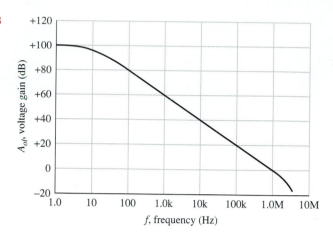

28. Refer to the data sheet graph in Figure 13–38 for the 741 op-amp to determine the following:
    (a) the maximum open-loop voltage gain
    (b) the approximate open-loop −3 dB bandwidth
    (c) the approximate unity-gain bandwidth
    (d) the open-loop roll-off rate

■ **ADVANCED PROBLEMS**

29. Design a noninverting amplifier with an upper critical frequency, $f_{c(cl)}$, of 10 kHz using a 741 op-amp. The dc supply voltages are ±15 V.

30. For the circuit you designed in Problem 29, determine the minimum load resistance if the minimum output voltage swing is to be ±10 V. Refer to the data sheet graphs in Figure 13–39.

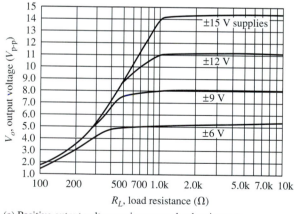

(a) Positive output voltage swing versus load resistance

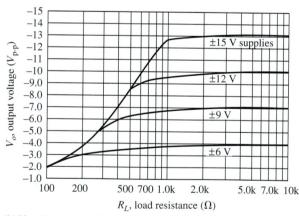

(b) Negative output voltage swing versus load resistance

FIGURE 13–39

31. Design an inverting amplifier using a 741 op-amp if a midrange voltage gain of 50 and a bandwidth of 20 kHz is required.

32. What is the maximum closed-loop voltage gain that can be achieved with a 741 op-amp if the bandwidth must be no less than 5 kHz?

■ **ANSWERS TO SECTION REVIEWS**

Section 13–1

1. Open-loop gain is without feedback, and closed-loop gain is with negative feedback. Open-loop gain is larger.
2. $BW = 100$ Hz
3. A_{ol} decreases.

Section 13–2

1. $A_{v(tot)} = 20$ dB $+ 30$ dB $= 50$ dB
2. $\theta_{tot} = -49° + (-5.2°) = -54.2°$

Section 13–3

1. Yes, A_{cl} is always less than A_{ol}.
2. $BW = 3,000$ kHz/60 = 50 kHz
3. unity-gain $BW = 3,000$ kHz/1 = 3 MHz

Section 13–4

1. Positive feedback can cause oscillation (instability).
2. 180°, 0°
3. −20 dB/decade (−6 dB/octave)

Section 13–5

1. Phase compensation increases the phase margin at a given frequency.
2. Internal compensation is full compensation; external compensation can be tailored to maximize bandwidth.
3. Bandwidth decreases with compensation.

■ **ANSWERS TO RELATED EXERCISES FOR EXAMPLES**

13–1 (a) 79,996   (b) 79,900   (c) 6380

13–2 173 Hz

13–3 75 dB; −71.6°

13–4 2 MHz

13–5 (a) 29.6 kHz   (b) 42.6 kHz

13–6 100 Hz; 10 Hz

14

BASIC OP-AMP CIRCUITS

In the last two chapters, you learned about the principles, operation, and characteristics of the operational amplifier. Op-amps are used in such a wide variety of applications that it is impossible to cover all of them in one chapter, or even in one book. Therefore, in this chapter, three fundamental applications are covered to give you a foundation in op-amp circuits.

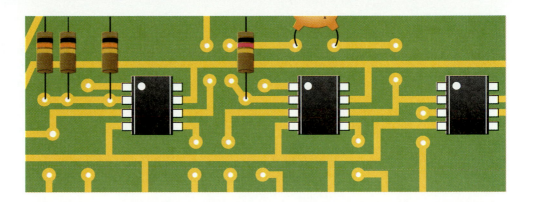

■ SYSTEM APPLICATION

The system application in Section 14–5 is an example of the use of three types of op-amp circuits that you will learn about in this chapter: the summing amplifier, the integrator, and the comparator. The system includes both analog and digital circuits; however, you will focus on the analog-to-digital converter board which incorporates the op-amps. The analog-to-digital converter board converts an audio signal to a digital code in order to record the sound in a digital format. After studying this chapter, you should be able to complete the system application assignment.

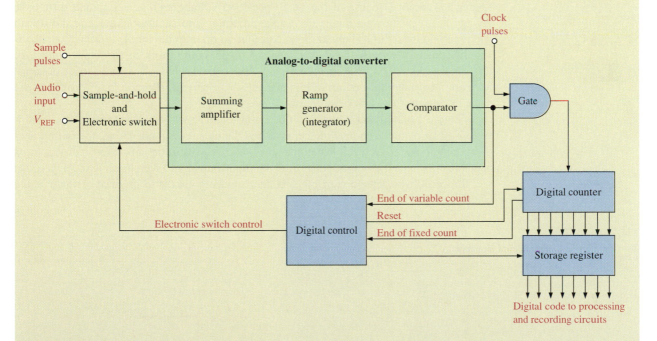

14–1 ■ COMPARATORS

Operational amplifiers are often used to compare the amplitude of one voltage with another. In this application, the op-amp is used in the open-loop configuration, with the input voltage on one input and a reference voltage on the other.

After completing this section, you should be able to

■ **Analyze the operation of several basic comparator circuits**
   ☐ Describe the operation of a zero-level detector
   ☐ Describe the operation of a nonzero-level detector
   ☐ Discuss how input noise affects comparator operation
   ☐ Define *hysteresis*
   ☐ Explain how hysteresis reduces noise effects
   ☐ Describe a Schmitt trigger circuit
   ☐ Describe the operation of bounded comparators
   ☐ Describe the operation of a window comparator
   ☐ Discuss two comparator applications including analog-to-digital conversion

Zero-Level Detection

A basic application of the op-amp as a **comparator** is in determining when an input voltage exceeds a certain level. Figure 14–1(a) shows a zero-level detector. Notice that the inverting (−) input is grounded and that the input signal voltage is applied to the noninverting (+) input. Because of the high open-loop voltage gain, a very small difference voltage between the two inputs drives the amplifier into saturation, causing the output voltage to go to its limit. For example, consider an op-amp having $A_{ol} = 100,000$. A voltage difference of only 0.25 mV between the inputs could produce an output voltage of $(0.25 \text{ mV})(100,000) = 25$ V if the op-amp were capable. However, since most op-amps have output voltage limitations of less than ±15 V because of their dc supply voltages, the device would be driven into saturation.

Figure 14–1(b) shows the result of a sinusoidal input voltage applied to the noninverting input of the zero-level detector. When the sine wave is negative, the output is at its maximum negative level. When the sine wave crosses 0, the amplifier is driven to its opposite state and the output goes to its maximum positive level, as shown. As you can see, the zero-level detector can be used as a squaring circuit to produce a square wave from a sine wave.

FIGURE 14–1

The op-amp as a zero-level detector.

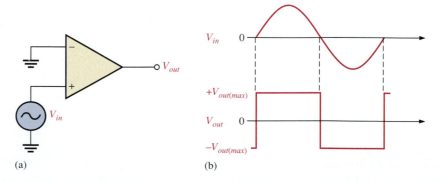

(a)   (b)

Nonzero-Level Detection

The zero-level detector in Figure 14–1 can be modified to detect voltages other than zero by connecting a fixed reference voltage source, as shown in Figure 14–2(a). A more practical arrangement is shown in Figure 14–2(b) using a voltage divider to set the reference voltage as follows:

$$V_{REF} = \frac{R_2}{R_1 + R_2}(+V) \tag{14–1}$$

where $+V$ is the positive op-amp dc supply voltage. The circuit in Figure 14–2(c) uses a zener diode to set the reference voltage ($V_{REF} = V_Z$). As long as the input voltage V_{in} is less than V_{REF}, the output remains at the maximum negative level. When the input voltage exceeds the reference voltage, the output goes to its maximum positive state, as shown in Figure 14–2(d) with a sinusoidal input voltage.

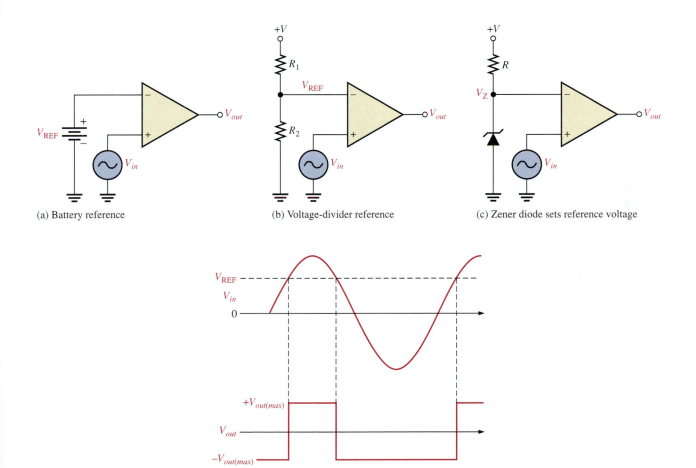

(a) Battery reference   (b) Voltage-divider reference   (c) Zener diode sets reference voltage

(d) Waveforms

FIGURE 14–2
Nonzero-level detectors.

EXAMPLE 14–1   The input signal in Figure 14–3(a) is applied to the comparator circuit in Figure 14–3(b). Make a sketch of the output showing its proper relationship to the input signal. Assume the maximum output levels are ±12 V.

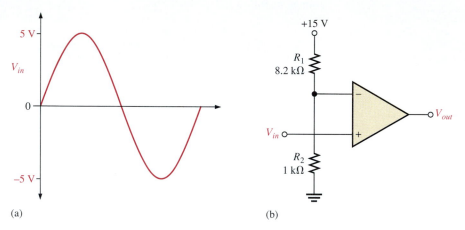

(a)   (b)

FIGURE 14–3

Solution   The reference voltage is set by R_1 and R_2 as follows:

$$V_{\text{REF}} = \frac{R_2}{R_1 + R_2}(+V) = \frac{1\ \text{k}\Omega}{8.2\ \text{k}\Omega + 1\ \text{k}\Omega}(+15\ \text{V}) = 1.63\ \text{V}$$

Each time the input exceeds +1.63 V, the output voltage switches to its +12 V level. Each time the input goes below +1.63 V, the output switches back to its −12 V level, as shown in Figure 14–4.

FIGURE 14–4

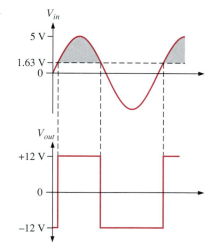

Related Exercise   Determine the reference voltage in Figure 14–3 if $R_1 = 22$ kΩ and $R_2 = 3.3$ kΩ.

Effects of Input Noise on Comparator Operation

In many practical situations, unwanted voltage fluctuations (**noise**) appear on the input line. This noise voltage becomes superimposed on the input voltage, as shown in Figure 14–5, and can cause a comparator to erratically switch output states.

FIGURE 14–5

Sine wave with superimposed noise.

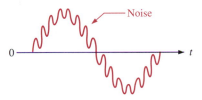

In order to understand the potential effects of noise voltage, consider a low-frequency sinusoidal voltage applied to the input of an op-amp comparator used as a zero-level detector, as shown in Figure 14–6(a). Part (b) of the figure shows the input sine wave plus noise and the resulting output. As you can see, when the sine wave approaches 0, the fluctuations due to noise cause the total input to vary above and below 0 several times, thus producing an erratic output.

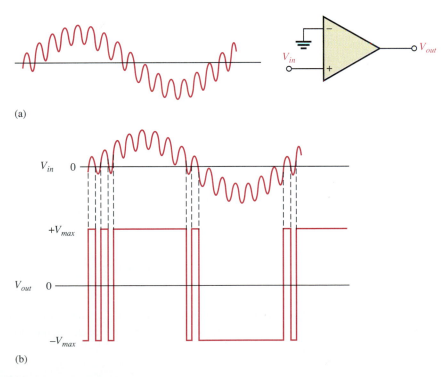

FIGURE 14–6

Effects of noise on comparator circuit.

Reducing Noise Effects with Hysteresis

An erratic output voltage caused by noise on the input occurs because the op-amp comparator switches from its negative output state to its positive output state at the same input voltage level that causes it to switch in the opposite direction, from positive to negative. This unstable condition occurs when the input voltage hovers around the reference voltage, and any small noise fluctuations cause the comparator to switch first one way and then the other.

In order to make the comparator less sensitive to noise, a technique incorporating positive feedback, called **hysteresis,** can be used. Basically, hysteresis means that there is a higher reference level when the input voltage goes from a lower to higher value than when it goes from a higher to a lower value. A good example of hysteresis is a common household thermostat that turns the furnace on at one temperature and off at another.

The two reference levels are referred to as the upper trigger point (UTP) and the lower trigger point (LTP). This two-level hysteresis is established with a positive feedback arrangement, as shown in Figure 14–7. Notice that the noninverting (+) input is connected to a resistive voltage divider such that a portion of the output voltage is fed back to the input. The input signal is applied to the inverting input (−) in this case.

FIGURE 14–7

Comparator with positive feedback for hysteresis.

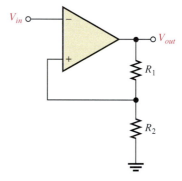

The basic operation of the comparator with hysteresis is as follows and is illustrated in Figure 14–8. Assume that the output voltage is at its positive maximum, $+V_{out(max)}$. The voltage fed back to the noninverting input is V_{UTP} and is expressed as

$$V_{\text{UTP}} = \frac{R_2}{R_1 + R_2}(+V_{out(max)}) \qquad \textbf{(14–2)}$$

When the input voltage V_{in} exceeds V_{UTP}, the output voltage drops to its negative maximum, $-V_{out(max)}$. Now the voltage fed back to the noninverting input is V_{LTP} and is expressed as

$$V_{\text{LTP}} = \frac{R_2}{R_1 + R_2}(-V_{out(max)}) \qquad \textbf{(14–3)}$$

FIGURE 14–8

Operation of a comparator with hysteresis.

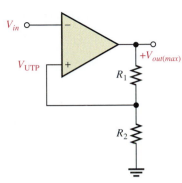

(a) Output in maximum positive state

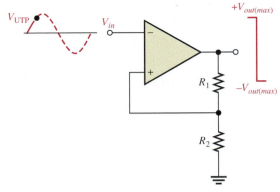

(b) Input exceeds UTP; output switches to maximum negative state.

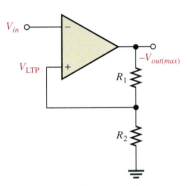

(c) Output in maximum negative state

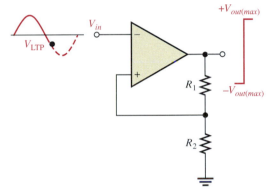

(d) Input goes below LTP; output switches back to maximum positive state.

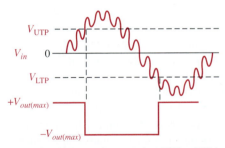

(e) Device triggers only once when UTP or LTP is reached; thus, there is immunity to noise that is riding on the input signal.

The input voltage must now fall below V_{LTP} before the device will switch back to its other state. This means that a small amount of noise voltage has no effect on the output, as illustrated by Figure 14–8(e).

A comparator with hysteresis is sometimes known as a **Schmitt trigger.** The amount of hysteresis is defined by the difference of the two trigger levels.

$$V_{\text{HYS}} = V_{\text{UTP}} - V_{\text{LTP}} \qquad \textbf{(14–4)}$$

EXAMPLE 14–2   Determine the upper and lower trigger points for the comparator circuit in Figure 14–9. Assume that $+V_{out(max)} = +5$ V and $-V_{out(max)} = -5$ V.

FIGURE 14–9

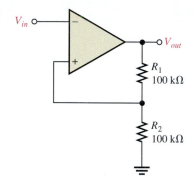

Solution

$$V_{UTP} = \frac{R_2}{R_1 + R_2}(+V_{out(max)}) = 0.5(5 \text{ V}) = +2.5 \text{ V}$$

$$V_{LTP} = \frac{R_2}{R_1 + R_2}(-V_{out(max)}) = 0.5(-5 \text{ V}) = -2.5 \text{ V}$$

Related Exercise   Determine the upper and lower trigger points in Figure 14–9 for $R_1 = 68$ kΩ and $R_2 = 82$ kΩ. The maximum output levels are ±7 V.

Output Bounding

In some applications, it is necessary to limit the output voltage levels of a comparator to a value less than that provided by the saturated op-amp. A single zener diode can be used as shown in Figure 14–10 to limit the output voltage swing to the zener voltage in one direction and to the forward diode drop in the other. This process of limiting the output range is called **bounding.**

The operation is as follows. Since the anode of the zener is connected to the inverting input, it is at virtual ground ($\cong 0$ V). Therefore, when the output reaches a positive value equal to the zener voltage, it limits at that value, as illustrated in Figure 14–11. When the output switches negative, the zener acts as a regular diode and becomes forward-biased at 0.7 V, limiting the negative output swing to this value, as shown. Turning the zener around limits the output in the opposite direction.

FIGURE 14–10
Comparator with output bounding.

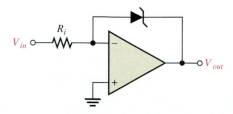

FIGURE 14–11

Operation of a bounded comparator.

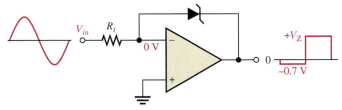

(a) Bounded at a positive value

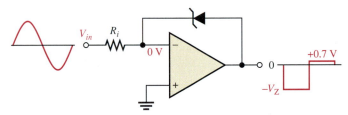

(b) Bounded at a negative value

Two zener diodes arranged as in Figure 14–12 limit the output voltage to the zener voltage plus the voltage drop (0.7 V) of the forward-biased zener, both positively and negatively, as shown in Figure 14–12. Op-amp comparators are available in IC form and are optimized for the comparison operation. Typical of these is the LM307.

FIGURE 14–12

Double-bounded comparator.

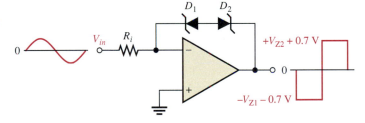

EXAMPLE 14–3   Determine the output voltage waveform for Figure 14–13.

FIGURE 14–13

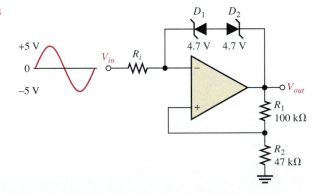

Solution   This comparator has both hysteresis and zener bounding.

The voltage across D_1 and D_2 in either direction is 4.7 V + 0.7 V = 5.4 V. This is because one zener is always forward-biased with a drop of 0.7 V when the other one is in breakdown.

The voltage at the inverting (−) op-amp input is $V_{out} \pm 5.4$ V. Since the differential voltage is negligible, the voltage at the noninverting (+) op-amp input is also $V_{out} \pm$ 5.4 V. Thus,

$$V_{R1} = V_{out} - (V_{out} \pm 5.4 \text{ V}) = \pm 5.4 \text{ V}$$

$$I_{R1} = \frac{V_{R1}}{R_1} = \frac{\pm 5.4 \text{ V}}{100 \text{ k}\Omega} = \pm 54 \text{ }\mu\text{A}$$

Since the current into the noninverting input is negligible,

$$I_{R2} = I_{R1} = \pm 54 \text{ }\mu\text{A}$$

$$V_{R2} = (47 \text{ k}\Omega)(\pm 54 \text{ }\mu\text{A}) = \pm 2.54 \text{ V}$$

$$V_{out} = V_{R1} + V_{R2} = \pm 5.4 \text{ V} \pm 2.54 \text{ V} = \pm 7.94 \text{ V}$$

The upper trigger point (UTP) and the lower trigger point (LTP) are as follows:

$$V_{\text{UTP}} = \left(\frac{R_2}{R_1 + R_2} \right)(+V_{out}) = \left(\frac{47 \text{ k}\Omega}{147 \text{ k}\Omega} \right)(+7.94 \text{ V}) = +2.54 \text{ V}$$

$$V_{\text{LTP}} = \left(\frac{R_2}{R_1 + R_2} \right)(-V_{out}) = \left(\frac{47 \text{ k}\Omega}{147 \text{ k}\Omega} \right)(-7.94 \text{ V}) = -2.54 \text{ V}$$

The output waveform for the given input voltage is shown in Figure 14–14.

FIGURE 14–14

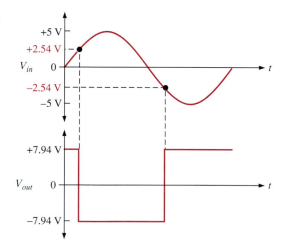

Related Exercise   Determine the upper and lower trigger points for Figure 14–13 if $R_1 = 150$ kΩ, $R_2 = 68$ kΩ, and the zener diodes are 3.3 V devices.

Window Comparator

Two individual op-amp comparators arranged as in Figure 14–15 form what is known as a *window comparator*. This circuit detects when an input voltage is between two limits, an upper and a lower, called the "window."

FIGURE 14–15
A basic window comparator.

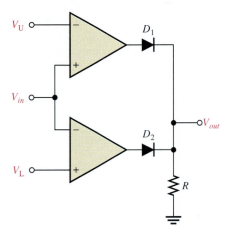

FIGURE 14–15
A basic window comparator.

The upper and lower limits are set by reference voltages designated V_U and V_L. These voltages can be established with voltage dividers, zener diodes, or any type of voltage source. As long as V_{in} is within the window (less than V_U and greater than V_L), the output of each comparator is at its low saturated level. Under this condition, both diodes are reverse-biased and V_{out} is held at zero by the resistor to ground. When V_{in} goes above V_U or below V_L, the output of the associated comparator goes to its high saturated level. This action forward-biases the diode and produces a high-level V_{out}. This is illustrated in Figure 14–16 with V_{in} changing arbitrarily.

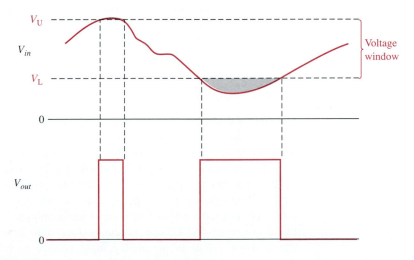

FIGURE 14–16
Example of window comparator operation.

A Comparator Application: Over-Temperature Sensing Circuit

Figure 14–17 shows an op-amp comparator used in a precision over-temperature sensing circuit. The circuit consists of a Wheatstone bridge with the op-amp used to detect when the bridge is balanced. One leg of the bridge contains a thermistor (R_1), which is a temperature-sensing resistor with a negative temperature coefficient (its resistance decreases as temperature increases). The potentiometer (R_2) is set at a value equal to the resistance of the thermistor at the critical temperature. At normal temperatures (below critical), R_1 is greater than R_2, thus creating an unbalanced condition that drives the op-amp to its low saturated output level and keeps transistor Q_1 off.

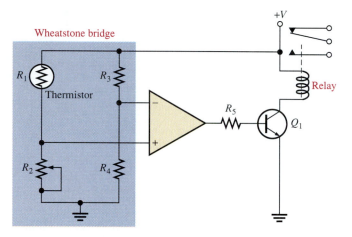

FIGURE 14–17
An over-temperature sensing circuit.

As the temperature increases, the resistance of the thermistor decreases. When the temperature reaches the critical value, R_1 becomes equal to R_2, and the bridge becomes balanced (since $R_3 = R_4$). At this point the op-amp switches to its high saturated output level, turning Q_1 on. This energizes the relay, which can be used to activate an alarm or initiate an appropriate response to the over-temperature condition.

A Comparator Application: Analog-to-Digital (A/D) Conversion

A/D conversion is a common interfacing process often used when a linear **analog** system must provide inputs to a **digital** system. Many methods for A/D conversion are available. However, in this discussion, only one type is used to demonstrate the concept.

The *simultaneous,* or *flash,* method of A/D conversion uses parallel comparators to compare the linear input signal with various reference voltages developed by a voltage divider. When the input voltage exceeds the reference voltage for a given comparator, a high level is produced on that comparator's output. Figure 14–18 shows an analog-to-digital converter (ADC) that produces three-digit binary numbers on its output, which represent the values of the analog input voltage as it changes. This converter requires seven comparators. In general, $2^n - 1$ comparators are required for conversion to an n-digit binary number. The large number of comparators necessary for a reasonably sized

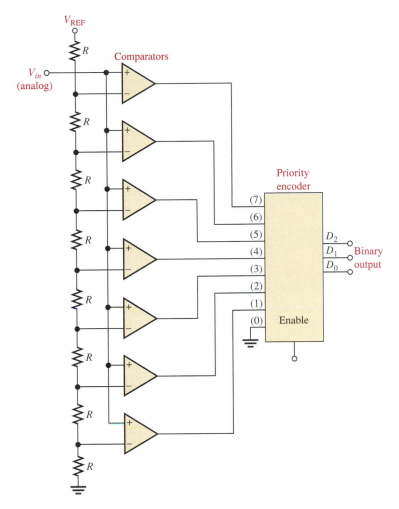

FIGURE 14–18

A simultaneous (flash) analog-to-digital converter (ADC) using op-amps as comparators.

binary number is one of the drawbacks of the simultaneous ADC. Its chief advantage is that it provides a fast conversion time.

The reference voltage for each comparator is set by the resistive voltage-divider circuit and V_{REF}. The output of each comparator is connected to an input of the priority encoder. The *priority encoder* is a digital device that produces a binary number representing the highest value input.

The encoder samples its input when a pulse occurs on the enable line (sampling pulse), and a three-digit binary number proportional to the value of the analog input signal appears on the encoder's outputs.

The sampling rate determines the accuracy with which the sequence of binary numbers represents the changing input signal. The more samples taken in a given unit of time, the more accurately the analog signal is represented in digital form.

The following example illustrates the basic operation of the simultaneous ADC in Figure 14–18.

EXAMPLE 14–4   Determine the binary number sequence of the three-digit simultaneous ADC in Figure 14–18 for the input signal in Figure 14–19 and the sampling pulses (encoder enable) shown. Sketch the resulting digital output waveforms.

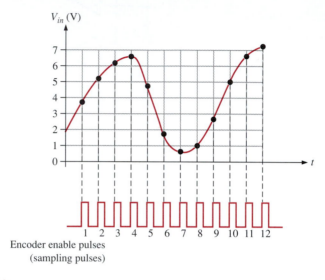

FIGURE 14–19
Sampling of values on analog waveform for conversion to digital.

Solution   The resulting binary output sequence is listed as follows and is shown in the waveform diagram of Figure 14–20 in relation to the sampling pulses.

$$011, 101, 110, 110, 100, 001, 000, 001, 010, 101, 110, 111$$

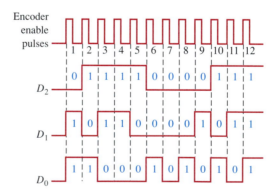

FIGURE 14–20
Resulting digital outputs for sampled values in Figure 14–19. D_0 is the least significant digit.

Related Exercise   If the frequency of the enable pulses in Figure 14–19 is doubled, does the resulting binary output sequence represent the analog waveform more or less accurately?

1. What is the reference voltage for each comparator in Figure 14–21?
2. What is the purpose of hysteresis in a comparator?
3. Define the term *bounding* in relation to a comparator's output.

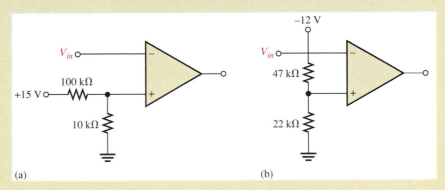

(a)   (b)

FIGURE 14–21

14–2 ■ SUMMING AMPLIFIERS

The summing amplifier is a variation of the inverting op-amp configuration covered in Chapter 12. The summing amplifier has two or more inputs, and its output voltage is proportional to the negative of the algebraic sum of its input voltages. In this section, you will see how a summing amplifier works, and you will learn about the averaging amplifier and the scaling amplifier, which are variations of the basic summing amplifier.

After completing this section, you should be able to

■ **Analyze the operation of several types of summing amplifiers**
  ☐ Describe the operation of a unity-gain summing amplifier
  ☐ Discuss how to achieve any specified gain greater than unity
  ☐ Describe the operation of an averaging amplifier
  ☐ Describe the operation of a scaling adder
  ☐ Discuss a scaling adder used as a digital-to-analog converter

Summing Amplifier with Unity Gain

A two-input summing amplifier is shown in Figure 14–22, but any number of inputs can be used. The operation of the circuit and derivation of the output expression are as follows. Two voltages, V_{IN1} and V_{IN2}, are applied to the inputs and produce currents I_1 and I_2, as shown.

Using the concepts of infinite input impedance and virtual ground, you can see that the inverting input of the op-amp is approximately 0 V, and there is no current out of the input. This means that both input currents I_1 and I_2 combine at this summing point and form the total current (I_T), which goes through R_f, as indicated in Figure 14–22.

$$I_T = I_1 + I_2$$

FIGURE 14–22

Two-input inverting summing amplifier.

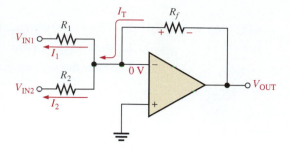

Since $V_{OUT} = -I_T R_f$, the following steps apply:

$$V_{OUT} = -(I_1 + I_2)R_f = -\left(\frac{V_{IN1}}{R_1} + \frac{V_{IN2}}{R_2}\right)R_f$$

If all three of the resistors are equal to the same value R ($R_1 = R_2 = R_f = R$), then

$$V_{OUT} = -\left(\frac{V_{IN1}}{R} + \frac{V_{IN2}}{R}\right)R$$

$$\boxed{V_{OUT} = -(V_{IN1} + V_{IN2})} \tag{14–5}$$

Equation (14–5) shows that the output voltage is the sum of the two input voltages. A general expression is given in Equation (14–6) for a unity-gain summing amplifier with n inputs, as shown in Figure 14–23 where all resistors are equal in value.

$$\boxed{V_{OUT} = -(V_{IN1} + V_{IN2} + V_{IN3} + \cdots + V_{INn})} \tag{14–6}$$

FIGURE 14–23

Summing amplifier with n inputs.

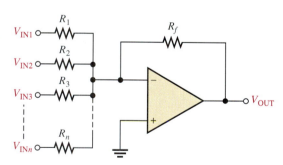

EXAMPLE 14–5

Determine the output voltage in Figure 14–24.

FIGURE 14–24

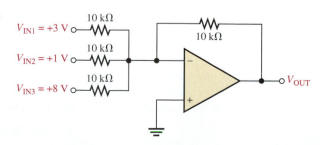

Solution $V_{OUT} = -(V_{IN1} + V_{IN2} + V_{IN3}) = -(3\ V + 1\ V + 8\ V) = -12\ V$

Related Exercise   If a fourth input of +0.5 V is added to Figure 14–24 with a 10 kΩ resistor, what is the output voltage?

Summing Amplifier with Gain Greater Than Unity

When R_f is larger than the input resistors, the amplifier has a gain of R_f/R, where R is the value of each input resistor. The general expression for the output is

$$V_{OUT} = -\frac{R_f}{R}(V_{IN1} + V_{IN2} + \cdot\ \cdot\ \cdot + V_{INn})$$ **(14–7)**

As you can see, the output is the sum of all the input voltages multiplied by a constant determined by the ratio R_f/R.

EXAMPLE 14–6   Determine the output voltage for the summing amplifier in Figure 14–25.

FIGURE 14–25

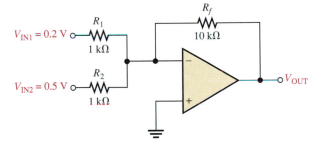

Solution $R_f = 10\ \text{k}\Omega$ and $R = R_1 = R_2 = 1\ \text{k}\Omega$. Therefore,

$$V_{OUT} = -\frac{R_f}{R}(V_{IN1} + V_{IN2}) = -\frac{10\ \text{k}\Omega}{1\ \text{k}\Omega}(0.2\ V + 0.5\ V) = -10(0.7\ V) = -7\ V$$

Related Exercise   Determine the output voltage in Figure 14–25 if the two input resistors are 2.2 kΩ and the feedback resistor is 18 kΩ.

Averaging Amplifier

A summing amplifier can be made to produce the mathematical average of the input voltages. This is done by setting the ratio R_f/R equal to the reciprocal of the number of inputs (n).

$$\frac{R_f}{R} = \frac{1}{n}$$

You obtain the average of several numbers by first adding the numbers and then dividing by the quantity of numbers you have. Examination of Equation (14–7) and a little thought will convince you that a summing amplifier will do this. The next example will illustrate.

EXAMPLE 14–7

Show that the amplifier in Figure 14–26 produces an output whose magnitude is the mathematical average of the input voltages.

FIGURE 14–26

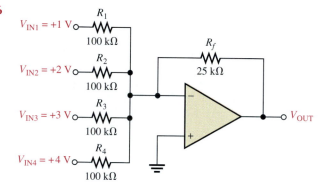

Solution   Since the input resistors are equal, $R = 100$ kΩ. The output voltage is

$$V_{OUT} = -\frac{R_f}{R}(V_{IN1} + V_{IN2} + V_{IN3} + V_{IN4})$$

$$= -\frac{25 \text{ k}\Omega}{100 \text{ k}\Omega}(1 \text{ V} + 2 \text{ V} + 3 \text{ V} + 4 \text{ V}) = -\frac{1}{4}(10 \text{ V}) = -2.5 \text{ V}$$

A simple calculation shows that the average of the input values is the same magnitude as V_{OUT} but of opposite sign.

$$V_{IN(avg)} = \frac{1 \text{ V} + 2 \text{ V} + 3 \text{ V} + 4 \text{ V}}{4} = \frac{10 \text{ V}}{4} = 2.5 \text{ V}$$

Related Exercise   Specify the changes required in the averaging amplifier in Figure 14–26 in order to handle five inputs.

Scaling Adder

A different weight can be assigned to each input of a summing amplifier by simply adjusting the values of the input resistors. As you have seen, the output voltage can be expressed as

$$V_{OUT} = -\left(\frac{R_f}{R_1}V_{IN1} + \frac{R_f}{R_2}V_{IN2} + \cdots + \frac{R_f}{R_n}V_{INn}\right) \qquad \text{(14–8)}$$

The weight of a particular input is set by the ratio of R_f to the input resistance for that input. For example, if an input voltage is to have a weight of 1, then $R = R_f$. Or, if a weight of 0.5 is required, $R = 2R_f$. The smaller the value of input resistance R, the greater the weight, and vice versa.

EXAMPLE 14–8

Determine the weight of each input voltage for the scaling adder in Figure 14–27 and find the output voltage.

FIGURE 14–27

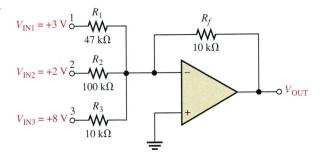

Solution

Weight of input 1: $\dfrac{R_f}{R_1} = \dfrac{10\text{ k}\Omega}{47\text{ k}\Omega} = 0.213$

Weight of input 2: $\dfrac{R_f}{R_2} = \dfrac{10\text{ k}\Omega}{100\text{ k}\Omega} = 0.100$

Weight of input 3: $\dfrac{R_f}{R_3} = \dfrac{10\text{ k}\Omega}{10\text{ k}\Omega} = 1.00$

The output voltage is

$$V_{OUT} = -\left(\frac{R_f}{R_1}V_{IN1} + \frac{R_f}{R_2}V_{IN2} + \frac{R_f}{R_3}V_{IN3}\right) = -[0.213(3\text{ V}) + 0.100(2\text{ V}) + 1.00(8\text{ V})]$$

$$= -(0.639\text{ V} + 0.2\text{ V} + 8\text{ V}) = -8.84\text{ V}$$

Related Exercise Determine the weight of each input voltage in Figure 14–27 if $R_1 = 22\text{ k}\Omega$, $R_2 = 82\text{ k}\Omega$, $R_3 = 56\text{ k}\Omega$, and $R_f = 10\text{ k}\Omega$. Also find V_{OUT}.

A Scaling Adder Application: Digital-to-Analog (D/A) Conversion

D/A conversion is an important interface process for converting digital signals to analog (linear) signals. An example is a voice signal that is digitized for storage, processing, or transmission and must be changed back into an approximation of the original audio signal in order to drive a speaker.

One method of D/A conversion uses a scaling adder with input resistor values that represent the binary weights of the input code. Figure 14–28 shows a four-digit digital-to-

FIGURE 14–28

A scaling adder as a four-digit digital-to-analog converter (DAC).

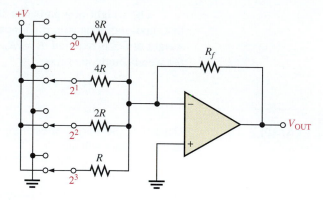

analog converter (DAC) of this type (called a *binary-weighted resistor DAC*). The switch symbols represent transistor switches for applying each of the four binary digits to the inputs.

The inverting input is at virtual ground, so that the output is proportional to the current through the feedback resistor R_f (sum of input currents).

The lowest value resistor R corresponds to the highest weighted binary input (2^3). All of the other resistors are multiples of R and correspond to the binary weights 2^2, 2^1, and 2^0.

EXAMPLE 14–9

Determine the output of the DAC in Figure 14–29(a). The sequence of four-digit binary numbers represented by the waveforms in Figure 14–29(b) are applied to the inputs. A high level is a binary 1, and low level is a binary 0. The least significant binary digit is D_0.

FIGURE 14–29

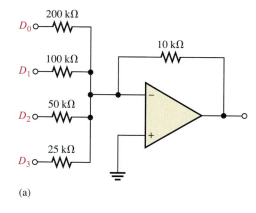

(a)

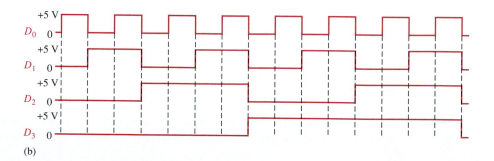

(b)

Solution First, determine the output voltage for each of the weighted inputs. Since the inverting input of the op-amp is at 0 V (virtual ground), and a binary 1 corresponds to a high level (+5 V), the current through any of the input resistors equals 5 V divided by the resistance value.

$$I_0 = \frac{5\ V}{200\ k\Omega} = 0.025\ mA$$

$$I_1 = \frac{5\ V}{100\ k\Omega} = 0.05\ mA$$

$$I_2 = \frac{5\ V}{50\ k\Omega} = 0.1\ mA$$

$$I_3 = \frac{5\ V}{25\ k\Omega} = 0.2\ mA$$

Practically none of the input current is from the inverting op-amp input because of its extremely high impedance. Therefore, practically all of the input current goes through R_f. Since one end of R_f is at 0 V (virtual ground), the drop across R_f equals the output voltage.

$$V_{OUT(D0)} = (10\ k\Omega)(-0.025\ mA) = -0.25\ V$$
$$V_{OUT(D1)} = (10\ k\Omega)(-0.05\ mA) = -0.5\ V$$
$$V_{OUT(D2)} = (10\ k\Omega)(-0.1\ mA) = -1\ V$$
$$V_{OUT(D3)} = (10\ k\Omega)(-0.2\ mA) = -2\ V$$

From Figure 14–29(b), the first binary number on the inputs is 0001 (it stands for decimal 1); for this, the output voltage is −0.25 V. The next binary number is 0010, which produces an output voltage of −0.5 V. The next binary number is 0011, which produces an output voltage of −0.25 V + (−0.5 V) = −0.75 V. Each successive binary number increases the output voltage by −0.25 V. So, for this particular sequence of binary numbers, the output is a stairstep waveform going from 0 V to −3.75 V in −0.25 V steps, as shown in Figure 14–30. If the steps are very small, the output approximates a straight line (linear).

FIGURE 14–30

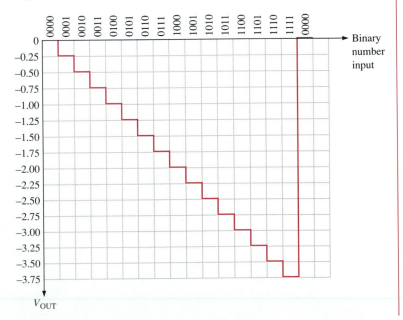

Related Exercise If the 200 kΩ resistor in Figure 14–29(a) is changed to 400 kΩ, would the other resistor values have to be changed? If so, specify the values.

SECTION 14–2
REVIEW

1. Define *summing point*.
2. What is the value of R_f/R for a five-input averaging amplifier?
3. A certain scaling adder has two inputs, one having twice the weight of the other. If the resistor value for the lower weighted input is 10 kΩ, what is the value of the other input resistor?

14–3 ■ THE INTEGRATOR AND DIFFERENTIATOR

An op-amp integrator simulates mathematical integration, which is basically a summing process that determines the total area under the curve of a function. An op-amp differentiator simulates mathematical differentiation, which is a process of determining the instantaneous rate of change of a function. It is not necessary for you to understand mathematical integration or differentiation, at this point, in order to learn how an integrator and differentiator work.

After completing this section, you should be able to

■ **Analyze the operation of integrators and differentiators**
  □ Identify an integrator circuit
  □ Discuss how a capacitor charges
  □ Determine the rate of change of an integrator's output
  □ Identify a differentiator circuit
  □ Determine the output voltage of a differentiator.

The Op-Amp Integrator

A basic **integrator** circuit is shown in Figure 14–31. Notice that the feedback element is a capacitor that forms an *RC* circuit with the input resistor.

FIGURE 14–31
An op-amp integrator.

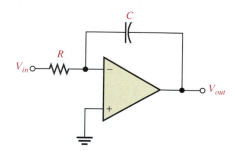

How a Capacitor Charges   To understand how the integrator works, it is important to review how a capacitor charges. Recall that the charge Q on a capacitor is proportional to the charging current (I_C) and the time (t).

$$Q = I_C t$$

Also, in terms of the voltage, the charge on a capacitor is

$$Q = CV_C$$

From these two relationships, the capacitor voltage can be expressed as

$$V_C = \left(\frac{I_C}{C}\right)t$$

This expression is an equation for a straight line which begins at zero with a constant slope of I_C/C. (Remember from algebra that the general formula for a straight line is $y = mx + b$. In this case, $y = V_C$, $m = I_C/C$, $x = t$, and $b = 0$).

Recall that the capacitor voltage in a simple *RC* circuit is not linear but is exponential. This is because the charging current continuously decreases as the capacitor charges and causes the rate of change of the voltage to continuously decrease. The key thing about using an op-amp with an *RC* circuit to form an integrator is that the capacitor's charging current is made constant, thus producing a straight-line (linear) voltage rather than an exponential voltage. Now let's see why this is true.

In Figure 14–32, the inverting input of the op-amp is at virtual ground (0 V), so the voltage across R_i equals V_{in}. Therefore, the input current is

$$I_{in} = \frac{V_{in}}{R_i}$$

FIGURE 14–32

Currents in an integrator.

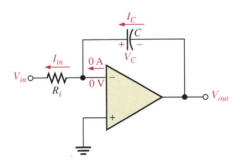

If V_{in} is a constant voltage, then I_{in} is also a constant because the inverting input always remains at 0 V, keeping a constant voltage across R_i. Because of the very high input impedance of the op-amp, there is negligible current from the inverting input. This makes all of the input current flow through the capacitor, as indicated in the figure, so

$$I_C = I_{in}$$

The Capacitor Voltage   Since I_{in} is constant, so is I_C. The constant I_C charges the capacitor linearly and produces a linear voltage across C. The positive side of the capacitor is held at 0 V by the virtual ground of the op-amp. The voltage on the negative side of the capacitor decreases linearly from zero as the capacitor charges, as shown in Figure 14–33. This voltage is called a *negative ramp*.

FIGURE 14–33

A linear ramp voltage is produced across C by the constant charging current.

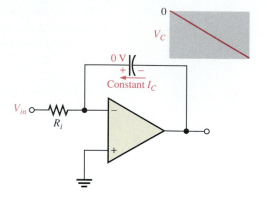

The Output Voltage V_{out} is the same as the voltage on the negative side of the capacitor. When a constant input voltage in the form of a step or pulse (a pulse has a constant amplitude when high) is applied, the output ramp decreases negatively until the op-amp saturates at its maximum negative level. This is indicated in Figure 14–34.

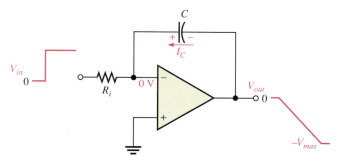

FIGURE 14–34

A constant input voltage produces a ramp on the output of the integrator.

Rate of Change of the Output The rate at which the capacitor charges, and therefore the slope of the output ramp, is set by the ratio I_C/C, as you have seen. Since $I_C = V_{in}/R_i$, the rate of change or slope of the integrator's output voltage is

$$\frac{\Delta V_{out}}{\Delta t} = -\frac{V_{in}}{R_i C} \qquad \textbf{(14–9)}$$

Integrators are especially useful in triangular-wave generators as you will see in Chapter 17.

EXAMPLE 14–10   (a) Determine the rate of change of the output voltage in response to a single pulse input, as shown for the integrator in Figure 14–35(a). The output voltage is initially zero.

(b) Draw the output waveform.

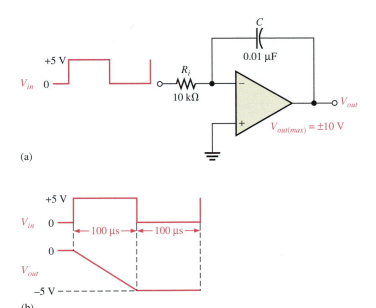

(a)

(b)

FIGURE 14–35

Solution

(a) The rate of change of the output voltage is

$$\frac{\Delta V_{out}}{\Delta t} = -\frac{V_{in}}{R_i C} = -\frac{5\ \text{V}}{(10\ \text{k}\Omega)(0.01\ \mu\text{F})} = -50\ \text{kV/s} = -50\ \text{mV/}\mu\text{s}$$

(b) The rate of change was found to be −50 mV/μs in part (a). When the input is at +5 V, the output is a negative-going ramp. When the input is at 0 V, the output is a constant level. In 100 μs, the voltage decreases.

$$\Delta V_{out} = (-50\ \text{mV/}\mu\text{s})(100\ \mu\text{s}) = -5\ \text{V}$$

Therefore, the negative-going ramp reaches −5 V at the end of the pulse. The output voltage then remains constant at −5 V for the time that the input is zero. The waveforms are shown in Figure 14–35(b).

Related Exercise Modify the integrator in Figure 14–35 to make the output change from 0 to −5 V in 50 μs with the same input.

The Op-Amp Differentiator

A basic **differentiator** is shown in Figure 14–36. Notice how the placement of the capacitor and resistor differ from the integrator. The capacitor is now the input element. A differentiator produces an output that is proportional to the rate of change of the input voltage.

To see how the differentiator works, we will apply a positive-going ramp voltage to the input as indicated in Figure 14–37. In this case, $I_C = I_{in}$ and the voltage across the

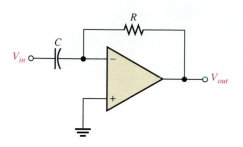

FIGURE 14–36

An op-amp differentiator.

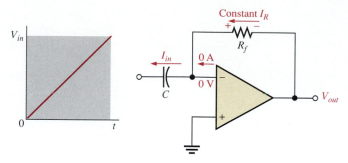

FIGURE 14–37

A differentiator with a ramp input.

capacitor is equal to V_{in} at all times ($V_C = V_{in}$) because of virtual ground on the inverting input.

From the basic formula, $V_C = (I_C/C)t$, we get

$$I_C = \left(\frac{V_C}{t}\right)C$$

Since the current from the inverting input is negligible, $I_R = I_C$. Both currents are constant because the slope of the capacitor voltage (V_C/t) is constant. The output voltage is also constant and equal to the voltage across R_f because one side of the feedback resistor is always 0 V (virtual ground).

$$V_{out} = I_R R_f = I_C R_f$$

$$V_{out} = \left(\frac{V_C}{t}\right)R_f C \qquad\qquad (14\text{–}10)$$

The output is negative when the input is a positive-going ramp and positive when the input is a negative-going ramp as illustrated in Figure 14–38. During the positive slope of

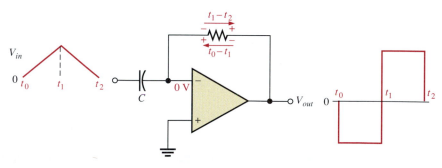

FIGURE 14–38

Output of a differentiator with a series of positive and negative ramps (triangle wave) on the input.

the input, the capacitor is charging from the input source and the constant current through the feedback resistor is in the direction shown. During the negative slope of the input, the current is in the opposite direction because the capacitor is discharging.

Notice in Equation (14–10) that the term V_C/t is the slope of the input. If the slope increases, V_{out} increases. If the slope decreases, V_{out} decreases. So, the output voltage is proportional to the slope (rate of change) of the input. The constant of proportionality is the time constant, R_fC.

EXAMPLE 14–11   Determine the output voltage of the op-amp differentiator in Figure 14–39 for the triangular-wave input shown.

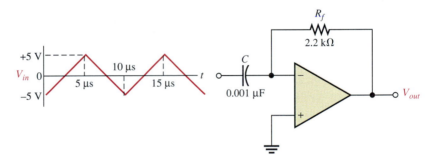

FIGURE 14–39

Solution   Starting at $t = 0$, the input voltage is a positive-going ramp ranging from −5 V to +5 V (a +10 V change) in 5 μs. Then it changes to a negative-going ramp ranging from +5 V to −5 V (a −10 V change) in 5 μs.

The time constant is

$$R_fC = (2.2\ \text{k}\Omega)(0.001\ \mu\text{F}) = 2.2\ \mu\text{s}$$

The slope or rate of change (V_C/t) of the positive-going ramp is determined, and the output voltage is calculated as follows:

$$\frac{V_C}{t} = \frac{10\ \text{V}}{5\ \mu\text{s}} = 2\ \text{V/}\mu\text{s}$$

$$V_{out} = (2\ \text{V/}\mu\text{s})2.2\ \mu\text{s} = +4.4\ \text{V}$$

Likewise, the slope of the negative-going ramp is −2 V/μs, and the output voltage is calculated as follows:

$$V_{out} = (-2\ \text{V/}\mu\text{s})2.2\ \mu\text{s} = -4.4\ \text{V}$$

Finally, the output voltage waveform is graphed relative to the input as shown in Figure 14–40.

FIGURE 14–40

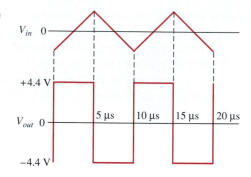

Related Exercise What would the output voltage be if the feedback resistor in Figure 14–39 is changed to 3.3 kΩ?

SECTION 14–3 REVIEW

1. What is the feedback element in an op-amp integrator?
2. For a constant input voltage to an integrator, why is the voltage across the capacitor linear?
3. What is the feedback element in an op-amp differentiator?
4. How is the output of a differentiator related to the input?

SUMMARY OF OP-AMP CIRCUITS

COMPARATORS

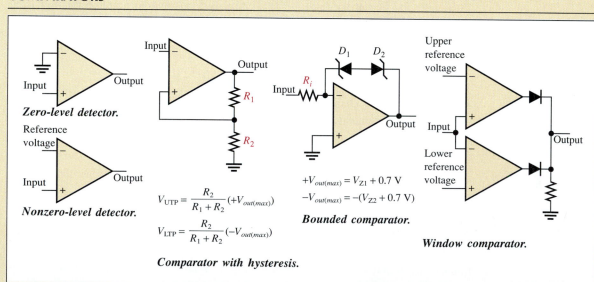

Zero-level detector.

Nonzero-level detector.

$$V_{UTP} = \frac{R_2}{R_1 + R_2}(+V_{out(max)})$$

$$V_{LTP} = \frac{R_2}{R_1 + R_2}(-V_{out(max)})$$

Comparator with hysteresis.

$+V_{out(max)} = V_{Z1} + 0.7 \text{ V}$

$-V_{out(max)} = -(V_{Z2} + 0.7 \text{ V})$

Bounded comparator.

Window comparator.

SUMMARY OF OP-AMP CIRCUITS, *continued*

SUMMING AMPLIFIERS

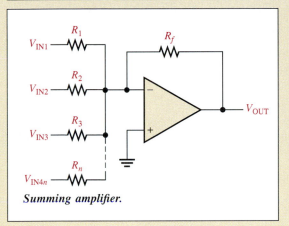

Summing amplifier.

■ Unity gain amplifier:

$$R_f = R_1 = R_2 = R_2 = \cdots = R_n$$

$$V_{OUT} = -(V_{IN1} + V_{IN2} + V_{IN3} + \cdots + V_{INn})$$

■ Averaging amplifier:

$$\frac{R_f}{R} = \frac{1}{n}$$

$$R = R_1 = R_2 = R_3 = \cdots = R_n$$

$$V_{OUT} = -\frac{R_f}{R}(V_{IN1} + V_{IN2} + V_{IN3} + \cdots + V_{INn})$$

■ Greater than unity gain amplifier:

$$R_f > R$$

$$R = R_1 = R_2 = R_3 = \cdots = R_n$$

$$V_{OUT} = -\frac{R_f}{R}(V_{IN1} + V_{IN2} + V_{IN3} + \cdots + V_{INn})$$

■ Scaling adder:

$$V_{OUT} = -\left(\frac{R_f}{R_1}V_{IN1} + \frac{R_f}{R_2}V_{IN2} + \frac{R_f}{R_3}V_{IN3} + \cdots + \frac{R_f}{R_n}V_{INn}\right)$$

INTEGRATOR AND DIFFERENTIATOR

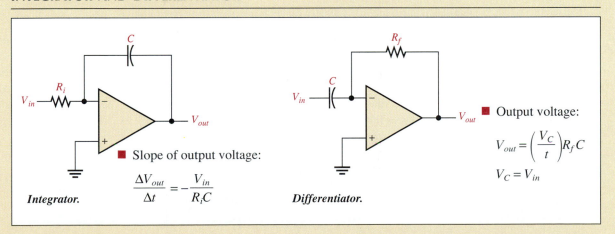

Integrator.

■ Slope of output voltage:

$$\frac{\Delta V_{out}}{\Delta t} = -\frac{V_{in}}{R_i C}$$

Differentiator.

■ Output voltage:

$$V_{out} = \left(\frac{V_C}{t}\right)R_f C$$

$$V_C = V_{in}$$

14–4 ■ TROUBLESHOOTING

Although integrated circuit op-amps are extremely reliable and trouble-free, failures do occur from time to time. One type of internal failure mode is a condition where the op-amp output is in a saturated state resulting in a constant high or constant low level, regardless of the input. Also, external component failures will produce various types of failure modes in op-amp circuits. Some examples are presented in this section.

After completing this section, you should be able to

■ **Troubleshoot basic op-amp circuits**
  □ Identify failures in comparator circuits
  □ Identify failures in summing amplifiers

Figure 14–41 illustrates an internal failure of a comparator circuit that results in a failed output.

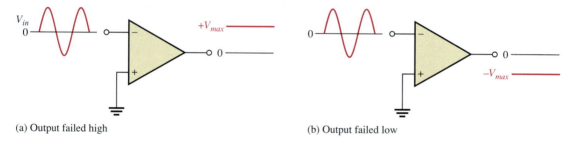

(a) Output failed high   (b) Output failed low

FIGURE 14–41
Typical internal comparator failures.

External Component Failures in Comparator Circuits

Comparators with zener-bounding and hysteresis are shown in Figure 14–42. In addition to a failure of the op-amp itself, a zener diode or one of the resistors could go bad. For example, suppose one of the zener diodes opens. This effectively eliminates both zeners, and the circuit operates as an unbounded comparator, as indicated in Figure 14–43(a). With a shorted diode, the output is limited to the zener voltage (bounded) only in one direction depending on which diode remains operational, as illustrated in Figure 14–43(b). In the other direction, the output is held at the forward diode voltage.

FIGURE 14–42
A bounded comparator.

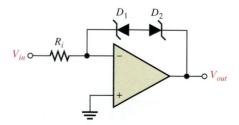

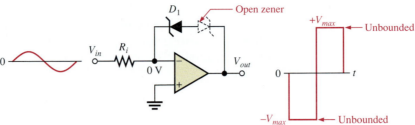

(a) The effect of an open zener

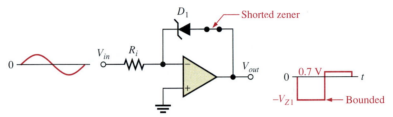

(b) The effect of a shorted zener

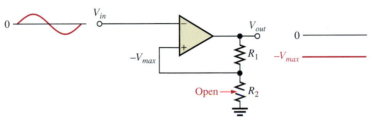

(c) Open R_2 causes output to "stick" in one state

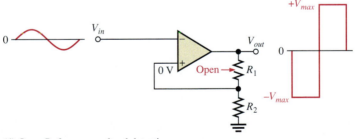

(d) Open R_1 forces zero-level detection

FIGURE 14–43

Examples of comparator circuit failures and their effects.

Recall that R_1 and R_2 set the UTP and LTP for the hysteresis comparator. Now, suppose that R_2 opens. Essentially all of the output voltage is fed back to the noninverting input, and, since the input voltage will never exceed the output, the device will remain in one of its saturated states. This symptom can also indicate a faulty op-amp, as mentioned before. Now, assume that R_1 opens. This leaves the noninverting input near ground potential and causes the circuit to operate as a zero-level detector. These conditions are shown in parts (c) and (d) of Figure 14–43.

EXAMPLE 14–12   One channel of a dual-trace oscilloscope is connected to the comparator output and the other channel is connected to the input, as shown in Figure 14–44. From the observed waveforms, determine if the circuit is operating properly, and if not, what the most likely failure is.

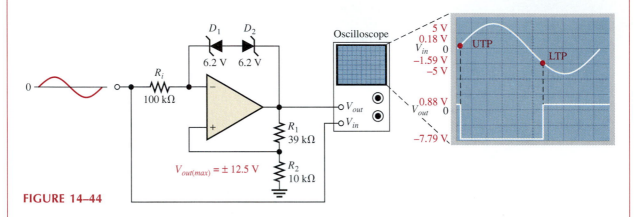

FIGURE 14–44

Solution   The output should be limited to ±8.67 V. However, the positive maximum is +0.88 V and the negative maximum is −7.79 V. This indicates that D_2 is shorted. Refer to Example 14–3 for analysis of the bounded comparator.

Related Exercise   What would the output voltage look like if D_1 shorted rather than D_2?

Component Failures in Summing Amplifiers

If one of the input resistors in a unity-gain summing amplifier opens, the output will be less than the normal value by the amount of the voltage applied to the open input. Stated another way, the output will be the sum of the remaining input voltages.

If the summing amplifier has a nonunity gain, an open input resistor causes the output to be less than normal by an amount equal to the gain times the voltage at the open input.

EXAMPLE 14–13   **(a)** What is the normal output voltage in Figure 14–45?
(b) What is the output voltage if R_2 opens?
(c) What happens if R_5 opens?

FIGURE 14–45

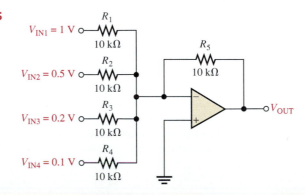

Solution
(a) $V_{OUT} = -(V_{IN1} + V_{IN2} + \cdots + V_{INn}) = -(1\ V + 0.5\ V + 0.2\ V + 0.1\ V) = -1.8\ V$
(b) $V_{OUT} = -(1\ V + 0.2\ V + 0.1\ V) = -1.3\ V$
(c) If the feedback resistor opens, the circuit becomes a comparator and the output goes to $-V_{max}$.

Related Exercise   In Figure 14–45, $R_5 = 47\ k\Omega$. What is the output voltage if R_1 opens?

As another example, let's look at an averaging amplifier. An open input resistor will result in an output voltage that is the average of all the inputs with the open input averaged in as a zero.

EXAMPLE 14–14   (a) What is the normal output voltage for the averaging amplifier in Figure 14–46?
(b) If R_4 opens, what is the output voltage? What does the output voltage represent?

FIGURE 14–46

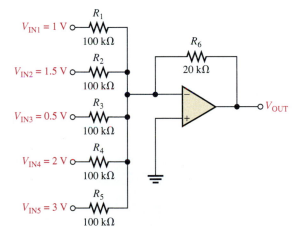

Solution   Since the input resistors are equal, $R = 100\ k\Omega$. $R_f = R_6$.

(a) $V_{OUT} = -\dfrac{R_f}{R}(V_{IN1} + V_{IN2} + \cdots + V_{INn})$

$= -\dfrac{20\ k\Omega}{100\ k\Omega}(1\ V + 1.5\ V + 0.5\ V + 2\ V + 3\ V) = -\dfrac{1}{5}(8\ V) = -1.6\ V$

(b) $V_{OUT} = -\dfrac{20\ k\Omega}{100\ k\Omega}(1\ V + 1.5\ V + 0.5 + 3\ V) = -\dfrac{1}{5}(6\ V) = -1.2\ V$

1.2 V is the average of five voltages with the 2 V input replaced by 0 V. Notice that the output is not the average of the four remaining input voltages.

Related Exercise   If R_4 is open, as was the case in this example, what would you have to do to make the output equal to the average of the remaining four input voltages?

14–5 ■ SYSTEM APPLICATION

You have been assigned to work on an analog-to-digital converter (ADC) that is used in a recording system for changing the audio signal into digital form for recording. You were introduced to one method of A/D conversion in this chapter, called the simultaneous, or flash, method. There are several other methods of A/D conversion, and this system application uses a method called dual-slope. Although many parts of this system are digital, you will focus on the ADC circuit board which incorporates types of op-amp circuits with which you are familiar.

Basic Operation of the System

The dual-slope ADC in Figure 14–47 accepts an audio signal voltage and converts it to a series of digital codes for the purpose of recording. The audio signal voltage is applied to the sample-and-hold circuit. At fixed intervals, sample pulses cause the instantaneous amplitude of the audio waveform to be converted to proportional dc levels that are processed by the rest of the circuits and represented by a series of digital codes.

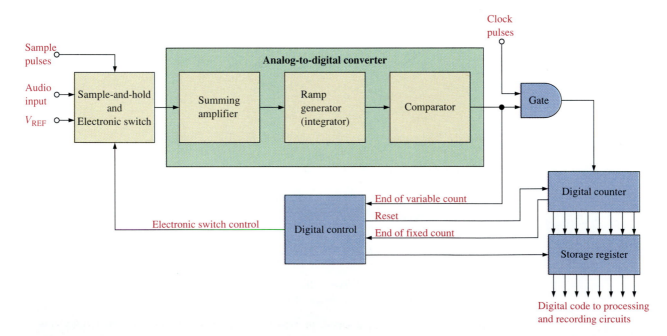

FIGURE 14–47
Block diagram of dual-slope analog-to-digital converter.

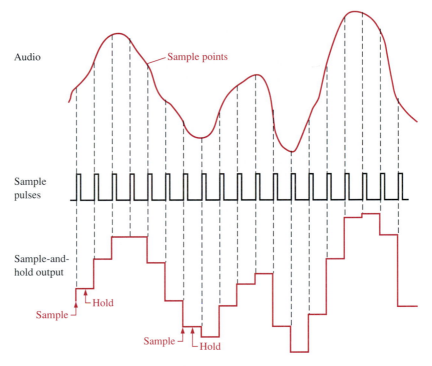

FIGURE 14–48

Illustration of the sample-and-hold process. The output of the sample-and-hold circuit is shown as a rough approximation of the audio voltage for simplification.

The sample pulses occur at a much higher frequency than the audio signal so that a sufficient number of points on the audio waveform are sampled and converted to obtain an accurate digital representation of the audio signal. A rough approximation of the sampling process is illustrated in Figure 14–48. As the frequency of the sample pulses increases relative to the audio frequency, an increasingly accurate representation is achieved.

The summing amplifier has only one input active at a time. For example, when the audio input is switched in, the reference voltage input is zero and vice versa.

During the time between each sample pulse, the dc level from the sample-and-hold circuit is switched electronically into the summing amplifier on the ADC board. The output of the summing amplifier goes to the ramp generator which is an integrator circuit. At the same time, the digital counter starts counting up from zero. During the fixed time interval of the counting sequence, the integrator (ramp generator) produces a positive-going ramp voltage with a slope that depends on the level of the sampled audio voltage. At the end of the fixed time interval, the ramp voltage at the output of the integrator has reached a value that is proportional to the sampled audio voltage. At this time, the digital control logic switches from the sample-and-hold input to the negative dc reference voltage input (V_{REF}) and resets the digital counter to zero.

The summing amplifier inverts the negative reference voltage and applies it to the integrator input, which starts a negative-going ramp on the integrator output. This ramp voltage has a slope that is fixed by the value of V_{REF}. When the negative-ramp starts, the digital counter begins to count up again from zero and will continue to count up until the integrator output reaches zero volts.

At the time when the positive ramp reaches zero, the comparator switches to its negative saturated output level and disables the gate so that there are no additional clock pulses to the counter. At this time, the digital code in the counter is proportional to the time that was required for the positive ramp at the integrator output to reach zero from its maximum negative value. The code in the counter will vary for each different sampled value of the audio.

Recall that the positive-going ramp started at a negative voltage that was dependent on the sampled value of the audio signal. Therefore, the digital code in the counter is also proportional to and represents the amplitude of the sampled audio voltage. The code is shifted out of the counter for temporary storage in the register from which it is processed and recorded.

The conversion process for each sampled value is repeated many times during the period of the highest audio harmonic frequency that is to be recorded accurately. The result is a sequence of digital codes that represents the audio voltage as it varies with time.

Figure 14–49 illustrates this process for several sampled values. In this application, you will focus on the ADC circuit which contains the summing amplifier, integrator, and comparator.

FIGURE 14–49

During the fixed-time interval of the negative-going ramp, the sampled audio input is applied to the integrator. During the variable time interval of the fixed-slope positive-going ramp, the reference voltage is applied to the integrator. The counter controls the fixed-time interval and is reset. Another count begins during the variable interval and the digital code in the counter at the end of this interval represents the sampled audio value.

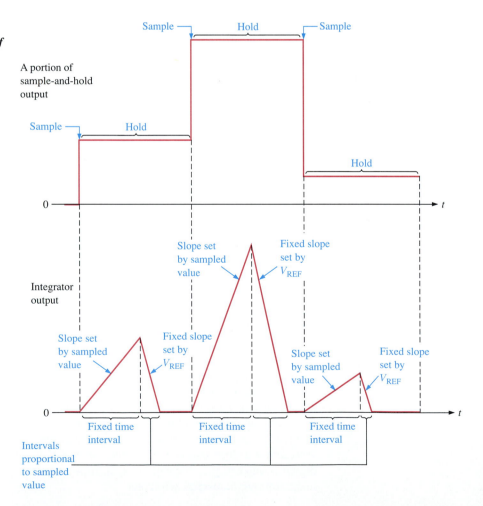

The Analog-to-Digital Converter Board

☐ Make sure that the circuit board shown in Figure 14–50 is correctly assembled by checking it against the schematic in Figure 14–51. A pin diagram for the power transistors is shown for reference. There are several backside interconnections. Corresponding feedthrough pads are vertically aligned.

☐ Label a copy of the board with component and input/output designations in agreement with the schematic.

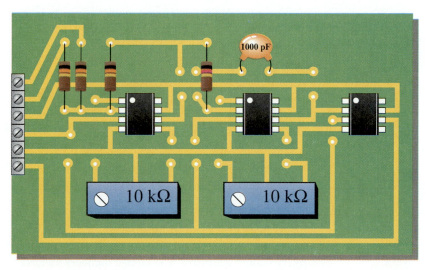

FIGURE 14–50
Analog-to-digital converter (ADC) board.

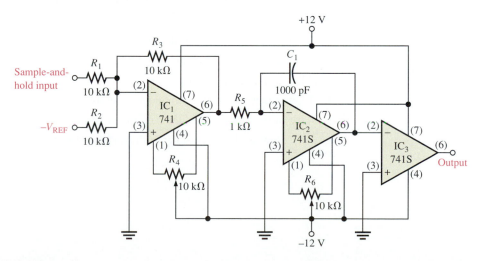

FIGURE 14–51
Schematic of the ADC.

Analysis of the ADC Circuit

☐ Determine the gain of the summing amplifier.

☐ Determine the slope of the integrator ramp in volts per microsecond when a sampled audio voltage of + 2 V is applied.

☐ Determine the slope of the integrator ramp in volts per microsecond when a dc reference voltage of −8 V is applied.

☐ Given that the reference voltage is −8 V and the fixed time interval of the positive-going slope is 1 μs, sketch the dual-slope output of the integrator when an instantaneous audio voltage of +3 V is applied.

☐ Assuming that the maximum audio voltage to be sampled is +6 V, determine the maximum audio frequency that can be sampled by this particular system if there are to be 100 samples per cycle. What is the sample pulse rate in this case?

Test Procedure

☐ Develop a step-by-step set of instructions on how to check the ADC circuit board for proper operation independent of the rest of the system.

☐ Specify voltage values for all the measurements to be made.

☐ Provide a fault analysis for all possible component failures.

Troubleshooting

Faults have developed in three boards. Based on the test bench measurements for each board with specified test signals applied as indicated in Figure 14–52, determine the most likely fault in each case. The circled numbers indicate test point connections to the circuit board. Assume that each board has the proper dc supply voltage.

Final Report

Submit a final written report on the ADC board using an organized format that includes the following:

1. A physical description of the circuit.
2. A discussion of the operation of the circuit.
3. A list of the specifications.
4. A list of parts with part numbers if available.
5. A list of the types of problems on the three faulty circuit boards.
6. A description of how you determined the problem on each of the faulty circuit boards.

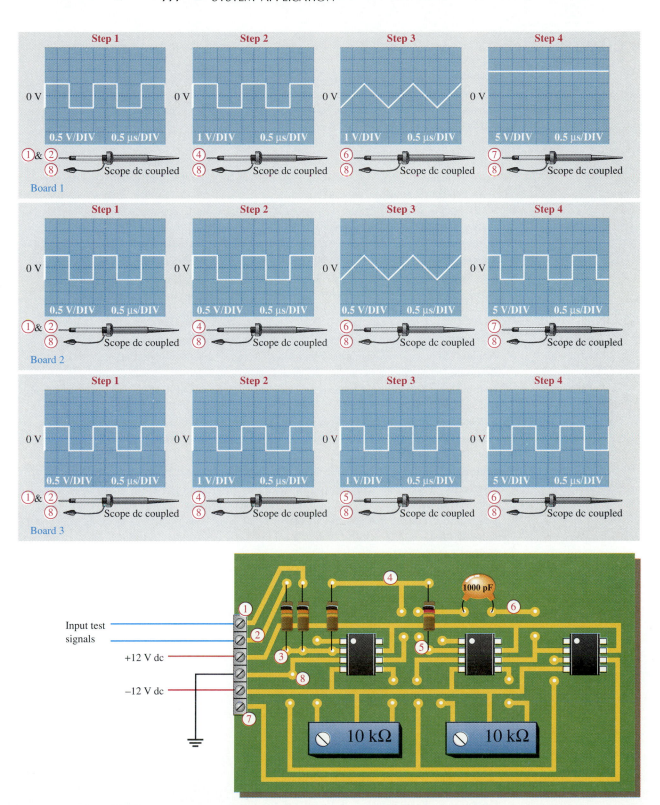

FIGURE 14–52

Test results for three faulty ADC boards.

■ CHAPTER SUMMARY

- In an op-amp comparator, when the input voltage exceeds a specified reference voltage, the output changes state.
- Hysteresis gives an op-amp noise immunity.
- A comparator switches to one state when the input reaches the upper trigger point (UTP) and back to the other state when the input drops below the lower trigger point (LTP).
- The difference between the UTP and the LTP is the hysteresis voltage.
- Bounding limits the output amplitude of a comparator.
- The output voltage of a summing amplifier is proportional to the sum of the input voltages.
- An averaging amplifier is a summing amplifier with a closed-loop gain equal to the reciprocal of the number of inputs.
- In a scaling adder, a different weight can be assigned to each input, thus making the input contribute more or contribute less to the output.
- Integration is a mathematical process for determining the area under a curve.
- Integration of a step produces a ramp with a slope proportional to the amplitude.
- Differentiation is a mathematical process for determining the rate of change of a function.
- Differentiation of a ramp produces a step with an amplitude proportional to the slope.

■ GLOSSARY

A/D conversion A process whereby information in analog form is converted into digital form.

Analog Characterized by a linear process in which a variable takes on a continuous set of values.

Bounding The process of limiting the output range of an amplifier or other circuit.

Comparator A circuit which compares two input voltages and produces an output in either of two states indicating the greater than or less than relationship of the inputs.

D/A conversion The process of converting a sequence of digital codes to an analog form.

Differentiator A circuit that produces an output which approximates the instantaneous rate of change of the input function.

Digital Characterized by a process in which a variable takes on either of two values.

Hysteresis Characteristic of a circuit in which two different trigger levels create an offset or lag in the switching action.

Integrator A circuit that produces an output which approximates the area under the curve of the input function.

Noise An unwanted signal.

Schmitt trigger A comparator with hysteresis.

■ FORMULAS

Comparators

(14–1)	$V_{\text{REF}} = \dfrac{R_2}{R_1 + R_2}(+V)$	Comparator reference
(14–2)	$V_{\text{UTP}} = \dfrac{R_2}{R_1 + R_2}(+V_{out(max)})$	Upper trigger point
(14–3)	$V_{\text{LTP}} = \dfrac{R_2}{R_1 + R_2}(-V_{out(max)})$	Lower trigger point
(14–4)	$V_{\text{HYS}} = V_{\text{UTP}} - V_{\text{LTP}}$	Hysteresis voltage

Summing Amplifier

(14–5)	$V_{\text{OUT}} = -(V_{\text{IN1}} + V_{\text{IN2}})$	Two-input adder
(14–6)	$V_{\text{OUT}} = -(V_{\text{IN1}} + V_{\text{IN2}} + \cdots + V_{\text{IN}n})$	n-input adder

(14–7) $V_{\text{OUT}} = -\dfrac{R_f}{R}(V_{\text{IN1}} + V_{\text{IN2}} + \cdot \cdot \cdot + V_{\text{IN}n})$ Adder with gain

(14–8) $V_{\text{OUT}} = -\left(\dfrac{R_f}{R_1}V_{\text{IN1}} + \dfrac{R_f}{R_2}V_{\text{IN2}} + \cdot \cdot \cdot + \dfrac{R_f}{R_n}V_{\text{IN}n}\right)$ Adder with gain

Integrator and Differentiator

(14–9) $\dfrac{\Delta V_{out}}{\Delta t} = -\dfrac{V_{in}}{R_iC}$ Integrator, rate of change

(14–10) $V_{out} = \left(\dfrac{V_C}{t}\right)R_fC$ Differentiator output with ramp input

---

■ **SELF-TEST**

1. In a zero-level detector, the output changes state when the input
   (a) is positive   (b) is negative
   (c) crosses zero   (d) has a zero rate of change

2. The zero-level detector is one application of a
   (a) comparator   (b) differentiator
   (c) summing amplifier   (d) diode

3. Noise on the input of a comparator can cause the output to
   (a) hang up in one state
   (b) go to zero
   (c) change back and forth erratically between two states
   (d) produce the amplified noise signal

4. The effects of noise can be reduced by
   (a) lowering the supply voltage   (b) using positive feedback
   (c) using negative feedback   (d) using hysteresis
   (e) answers (b) and (d)

5. A comparator with hysteresis
   (a) has one trigger point   (b) has two trigger points
   (c) has a variable trigger point   (d) is like a magnetic circuit

6. In a comparator with hysteresis,
   (a) a bias voltage is applied between the two inputs
   (b) only one supply voltage is used
   (c) a portion of the output is fed back to the inverting input
   (d) a portion of the output is fed back to the noninverting input

7. Using output bounding in a comparator
   (a) makes it faster   (b) keeps the output positive
   (c) limits the output levels   (d) stabilizes the output

8. A window comparator detects when
   (a) the input is between two specified limits
   (b) the input is not changing
   (c) the input is changing too fast
   (d) the amount of light exceeds a certain value

9. A summing amplifier can have
   (a) only one input   (b) only two inputs   (c) any number of inputs

10. If the voltage gain for each input of a summing amplifier with a 4.7 kΩ feedback resistor is unity, the input resistors must have a value of
   (a) 4.7 kΩ
   (b) 4.7 kΩ divided by the number of inputs
   (c) 4.7 kΩ times the number of inputs

11. An averaging amplifier has five inputs. The ratio R_f/R_i must be
   (a) 5   (b) 0.2   (c) 1

12. In a scaling adder, the input resistors are
   (a) all the same value
   (b) all of different values
   (c) each proportional to the weight of its input
   (d) related by a factor of two

13. In an integrator, the feedback element is a
   (a) resistor   (b) capacitor
   (c) zener diode   (d) voltage divider

14. For a step input, the output of an integrator is
   (a) a pulse   (b) a triangular waveform
   (c) a spike   (d) a ramp

15. The rate of change of an integrator's output voltage in response to a step input is set by
   (a) the RC time constant   (b) the amplitude of the step input
   (c) the current through the capacitor   (d) all of these

16. In a differentiator, the feedback element is a
   (a) resistor   (b) capacitor
   (c) zener diode   (d) voltage divider

17. The output of a differentiator is proportional to
   (a) the RC time constant   (b) the rate at which the input is changing
   (c) the amplitude of the input   (d) answers (a) and (b)

18. When you apply a triangular waveform to the input of a differentiator, the output is
   (a) a dc level   (b) an inverted triangular waveform
   (c) a square waveform   (d) the first harmonic of the triangular waveform

---

▪ BASIC PROBLEMS

SECTION 14–1 Comparators

1. A certain op-amp has an open-loop gain of 80,000. The maximum saturated output levels of this particular device are ±12 V when the dc supply voltages are ±15 V. If a differential voltage of 0.15 mV rms is applied between the inputs, what is the peak-to-peak value of the output?

2. Determine the output level (maximum positive or maximum negative) for each comparator in Figure 14–53.

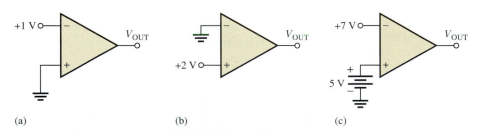

(a)   (b)   (c)

FIGURE 14–53

3. Calculate the V_{UTP} and V_{LTP} in Figure 14–54. $V_{out(max)} = \pm10$ V.

4. What is the hysteresis voltage in Figure 14–54?

FIGURE 14–54

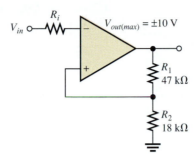

5. Sketch the output voltage waveform for each circuit in Figure 14–55 with respect to the input. Show voltage levels.

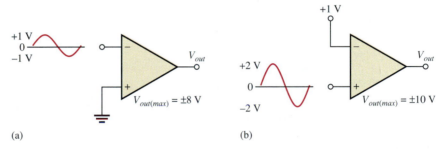

(a)   (b)

FIGURE 14–55

6. Determine the hysteresis voltage for each comparator in Figure 14–56. The maximum output levels are ±11 V.

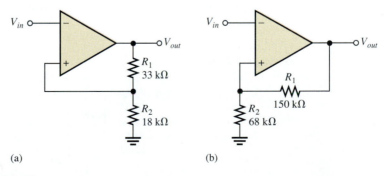

(a)   (b)

FIGURE 14–56

7. A 6.2 V zener diode is connected from the output to the inverting input in Figure 14–54 with the cathode at the output. What are the positive and negative output levels?

FIGURE 14–57

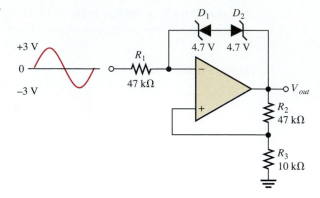

8. Determine the output voltage waveform in Figure 14–57.

SECTION 14–2 Summing Amplifiers

9. Determine the output voltage for each circuit in Figure 14–58.

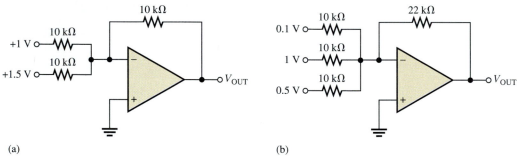

(a)   (b)

FIGURE 14–58

10. Refer to Figure 14–59. Determine the following:
  (a) V_{R1} and V_{R2} **(b)** Current through R_f **(c)** V_{OUT}
11. Find the value of R_f necessary to produce an output that is five times the sum of the inputs in Figure 14–59.

FIGURE 14–59

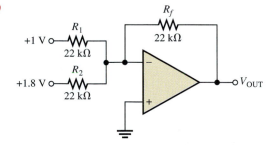

12. Show a summing amplifier that will average eight input voltages. Use input resistances of 10 kΩ each.

FIGURE 14–60

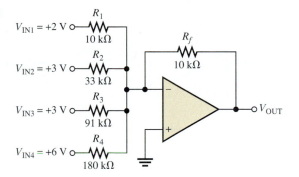

13. Find the output voltage when the input voltages shown in Figure 14–60 are applied to the scaling adder. What is the current through R_f?

14. Determine the values of the input resistors required in a six-input scaling adder so that the lowest weighted input is 1 and each successive input has a weight twice the previous one. Use $R_f = 100\ \text{k}\Omega$.

SECTION 14–3 The Integrator and Differentiator

15. Determine the rate of change of the output voltage in response to the step input to the integrator in Figure 14–61.

FIGURE 14–61

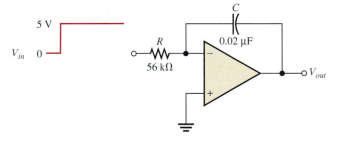

16. A triangular waveform is applied to the input of the circuit in Figure 14–62 as shown. Determine what the output should be and sketch its waveform in relation to the input.

17. What is the magnitude of the capacitor current in Problem 16?

FIGURE 14–62

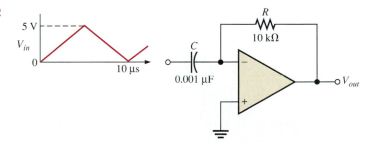

18. A triangular waveform with a peak-to-peak voltage of 2 V and a period of 1 ms is applied to the differentiator in Figure 14–63(a). What is the output voltage?

19. Beginning in position 1 in Figure 14–63(b), the switch is thrown into position 2 and held there for 10 ms, then back to position 1 for 10 ms, and so forth. Sketch the resulting output waveform if its initial value is 0 V. The saturated output levels of the op-amp are ±12 V.

FIGURE 14–63

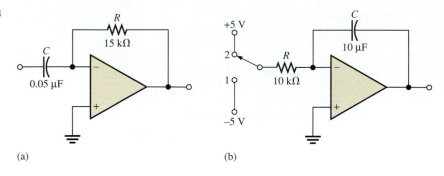

(a)   (b)

■ **TROUBLE-SHOOTING PROBLEMS**

SECTION 14–4   Troubleshooting

20. The waveforms given in Figure 14–64(a) are observed at the indicated points in Figure 14–64(b). Is the circuit operating properly? If not, what is a likely fault?

21. The waveforms shown for the window comparator in Figure 14–65 are measured. Determine if the output waveform is correct and, if not, specify the possible fault(s).

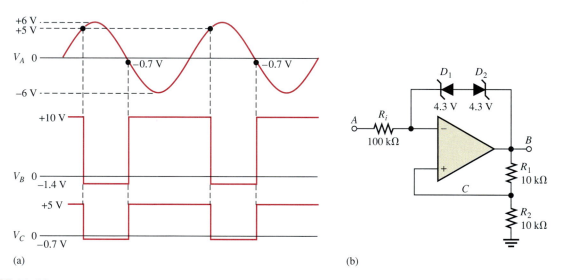

(a)   (b)

FIGURE 14–64

FIGURE 14–65

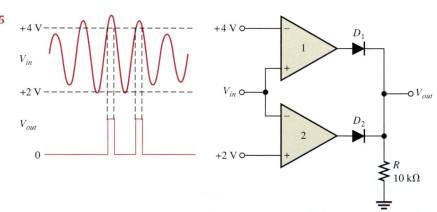

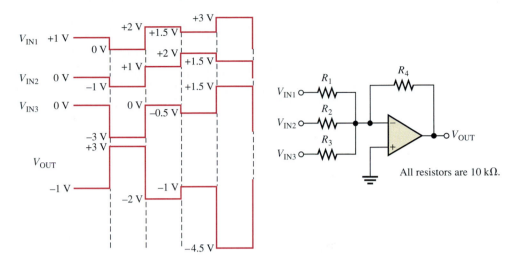

FIGURE 14–66

22. The sequences of voltage levels shown in Figure 14–66 are applied to the summing amplifier and the indicated output is observed. First, determine if this output is correct. If it is not correct, determine the fault.

23. The given ramp voltages are applied to the op-amp circuit in Figure 14–67. Is the given output correct? If it isn't, what is the problem?

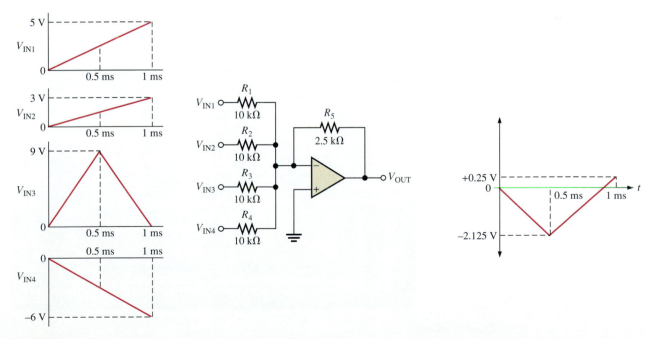

FIGURE 14–67

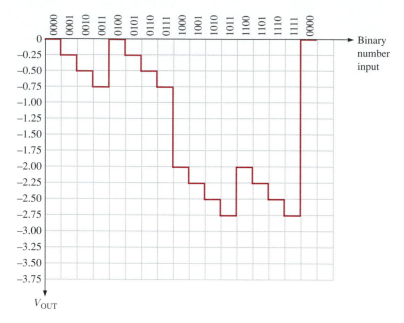

FIGURE 14–68

24. The DAC with inputs as shown in Figure 14–29 produces the output shown in Figure 14–68. Determine the fault in the circuit.

SECTION 14–5   System Application

25. The ADC board, shown in Figure 14–69, for the system application has just come off the assembly line and a pass/fail test indicates that it doesn't work. The board now comes to you for troubleshooting. What is the very first thing you should do? Can you isolate the problem(s) by this first step in this case?

FIGURE 14–69

26. Describe the effect of an open integrator capacitor on the ADC in Figure 14–51.

27. Assume that a 1 kΩ resistor is inadvertently used for R_1 in Figure 14–51. What effect does this have on the circuit operation?

■ **ADVANCED PROBLEMS**

28. The schematic in Figure 14–51 for the analog-to-digital converter shows 741S op-amps used for the integrator and comparator functions. Data sheets indicate that a 741 has a typical slew rate of 0.5 V/μs and a 741S has a typical slew rate of 12 V/μs. If the sample pulse rate is 500 kHz and the reference voltage is −8 V, determine the maximum input voltage from the sample-and-hold circuit that the ADC can handle.

29. Repeat Problem 28 if the 741S op-amps are replaced by 741 op-amps.

30. Design an integrator that will produce an output voltage with a slope of 100 mV/μs when the input voltage is a constant 5 V. Specify the input frequency of a square wave with an amplitude of 5 V that will result in a 5 V peak-to-peak triangular wave output.

■ **ANSWERS TO SECTION REVIEWS**

Section 14–1

1. (a) $V = (10 \text{ k}\Omega/110 \text{ k}\Omega)15 \text{ V} = 1.36 \text{ V}$
   (b) $V = (22 \text{ k}\Omega/69 \text{ k}\Omega)(-12 \text{ V}) = -3.83 \text{ V}$

2. Hysteresis makes the comparator noise free.

3. Bounding limits the output amplitude to a specified level.

Section 14–2

1. The summing point is the point where the input resistors are commonly connected.

2. $R_f/R = 1/5 = 0.2$

3. 5 kΩ

Section 14–3

1. The feedback element in an integrator is a capacitor.

2. The capacitor voltage is linear because the capacitor current is constant.

3. The feedback element in a differentiator is a resistor.

4. The output of a differentiator is proportional to the rate of change of the input.

Section 14–4

1. An op-amp can fail with a shorted output.

2. Replace suspected components one by one.

■ **ANSWERS TO**
RELATED
EXERCISES FOR
EXAMPLES

14–1 1.96 V

14–2 +3.83 V; −3.83 V

14–3 +1.81 V; −1.81 V

14–4 More accurately

14–5 −12.5 V

14–6 −5.73 V

14–7 Changes require an additional 100 kΩ input resistor and a change of R_f to 20 kΩ.

14–8 0.45, 0.12, 0.18; $V_{OUT} = -3.03$ V

14–9 Yes. All should be doubled.

14–10 Change C to 5000 pF.

14–11 Same waveform with an amplitude of 6.6 V

14–12 A pulse from −0.88 V to +7.79 V

14–13 −3.76 V

14–14 Change R_6 to 25 kΩ.

15

MORE OP-AMP CIRCUITS

■ **CHAPTER OBJECTIVES**

☐ Analyze and explain the operation of an instrumentation amplifier

☐ Analyze and explain the operation of an isolation amplifier

☐ Analyze and explain the operation of the OTA

☐ Analyze and explain the operation of log and antilog amplifiers

☐ Analyze and explain several special types of op-amp circuits

A general-purpose op-amp, such as the 741, is a versatile and widely used device. However, some specialized IC amplifiers are available that have certain features or characteristics oriented to special applications. Most of these devices are actually derived from the basic op-amp. These special circuits include the instrumentation amplifier that is used in high-noise environments, the isolation amplifier that is used in high-voltage and medical applications, the operational transconductance amplifier (OTA) that is used as a voltage-to-current amplifier, and the logarithmic amplifiers that are used for linearizing certain types of inputs and for mathematical operations.

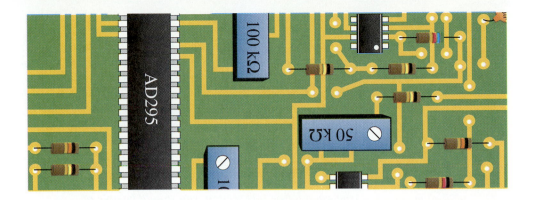

■ SYSTEM APPLICATION

Medical electronics is an important area of application for electronic devices and one of the most beneficial. The electrocardiograph (ECG) is a common instrument used to monitor the heart function of patients to detect irregularities or abnormalities in the heartbeat. Electrode sensors are placed at points on the body to pick up the small electrical signal produced by the heart. This signal is amplified and fed to a video monitor or chart recorder for analysis. Due to the safety hazards related to electrical equipment, it is important that the patient be protected from the possibility of severe electrical shock. For this reason, the isolation amplifier is used in medical equipment that comes in contact with the human body. The block diagram shows a basic ECG system that is the topic of the system application in Section 15–6.

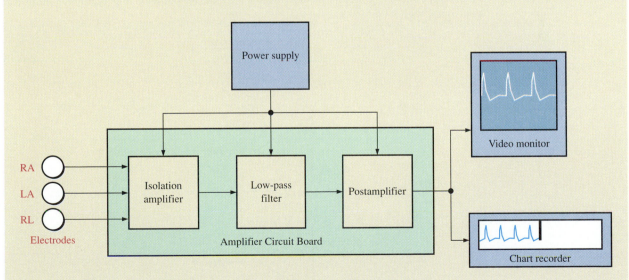

15–1 ■ INSTRUMENTATION AMPLIFIERS

An instrumentation amplifier is a differential voltage-gain device that amplifies the dif-ference between the voltages existing at its two input terminals. The main purpose of an instrumentation amplifier is to amplify small signals riding on large common-mode voltages. The key characteristics are high input impedance, high common-mode rejec-tion, low output offset, and low output impedance. A basic instrumentation amplifier is made up of three operational amplifiers and several resistors. The voltage gain is set with an external resistor. Instrumentation amplifiers are commonly used in environ-ments with high common-mode noise such as in data acquisition systems where remote sensing of input variables is required.

After completing this section, you should be able to

■ **Analyze and explain the operation of an instrumentation amplifier**
  □ Explain how op-amps are connected to form an instrumentation amplifier
  □ Describe how the voltage gain is set
  □ Discuss an application
  □ Describe the features of the AD521 instrumentation amplifier

The Basic Instrumentation Amplifier

A basic **instrumentation** amplifier is shown in Figure 15–1. Op-amps 1 and 2 are nonin-verting configurations that provide high input impedance and voltage gain. Op-amp 3 is used as a unity-gain differential amplifier.

The gain-setting resistor, R_G, is connected externally as shown in Figure 15–2. Op-amp 1 receives the differential input signal V_{in1} on its noninverting input and amplifies this signal with a voltage gain of

$$A_v = 1 + \frac{R_1}{R_G}$$

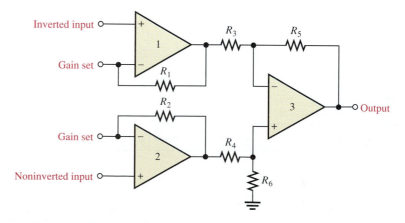

FIGURE 15–1
The basic instrumentation amplifier.

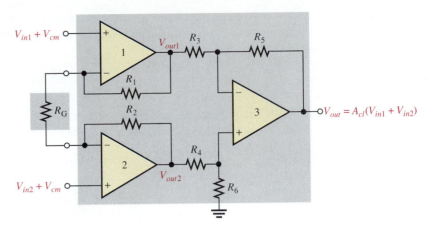

FIGURE 15–2
The instrumentation amplifier with the external gain-setting resistor R_G. Differential and common-mode signals are indicated.

Op-amp 1 also receives the input signal V_{in2} through op-amp 2 and the path formed by R_2, R_G, and R_1. V_{in2} effectively appears on the inverting input of op-amp 1 and is amplified by a voltage gain of

$$A_v = \frac{R_1}{R_G}$$

Also, the common-mode voltage, V_{cm}, on the noninverting input is amplified by the small common-mode gain of op-amp 1. (A_{cm} is typically less than 1). The total output voltage of op-amp 1 is

$$V_{out1} = \left(1 + \frac{R_1}{R_G}\right)V_{in1} - \left(\frac{R_1}{R_G}\right)V_{in2} + V_{cm}$$

A similar analysis can be applied to op-amp 2 and results in the following output expression:

$$V_{out2} = \left(1 + \frac{R_2}{R_G}\right)V_{in2} - \left(\frac{R_2}{R_G}\right)V_{in1} + V_{cm}$$

Op-amp 3 has V_{out1} on one of its inputs and V_{out2} on the other. Therefore, the differential input voltage to op-amp 3 is $V_{out2} - V_{out1}$.

$$V_{out2} - V_{out1} = \left(1 + \frac{R_2}{R_G} + \frac{R_1}{R_G}\right)V_{in2} - \left(1 + \frac{R_2}{R_G} + \frac{R_1}{R_G}\right)V_{in1} + V_{cm} - V_{cm}$$

For $R_1 = R_2 = R$,

$$V_{out2} - V_{out1} = \left(1 + \frac{2R}{R_G}\right)V_{in2} - \left(1 + \frac{2R}{R_G}\right)V_{in1} + V_{cm} - V_{cm}$$

Notice that, since the common-mode voltages (V_{cm}) are equal, they cancel each other. Factoring out the differential gain gives the following expression for the differential input to op-amp 3:

$$V_{out2} - V_{out1} = \left(1 + \frac{2R}{R_G}\right)(V_{in2} - V_{in1})$$

Op-amp 3 has unity gain because $R_3 = R_5 = R_4 = R_6$ and $A_v = R_5/R_3 = R_6/R_4$. Therefore, the final output of the instrumentation amplifier (the output of op-amp 3) is

$$V_{out} = 1(V_{out2} - V_{out1})$$

$$V_{out} = \left(1 + \frac{2R}{R_G}\right)(V_{in2} - V_{in1}) \qquad \text{(15–1)}$$

The closed-loop gain is

$$A_{cl} = \frac{V_{out}}{V_{in2} - V_{in1}}$$

$$A_{cl} = 1 + \frac{2R}{R_G} \qquad \text{(15–2)}$$

where $R_1 = R_2 = R$. Equation (15–2) shows that the gain of the instrumentation amplifier can be set by the value of the external resistor R_G when R_1 and R_2 have known fixed values.

The external gain-setting resistor R_G can be calculated for a desired voltage gain by using the following formula:

$$R_G = \frac{2R}{A_{cl} - 1} \qquad \text{(15–3)}$$

EXAMPLE 15–1

Determine the value of the external gain-setting resistor R_G for a certain IC instrumentation amplifier with $R_1 = R_2 = 25 \text{ k}\Omega$. The closed-loop voltage gain is to be 500.

Solution

$$R_G = \frac{2R}{A_{cl} - 1} = \frac{50 \text{ k}\Omega}{500 - 1} \cong 100 \ \Omega$$

Related Exercise   What value of external gain-setting resistor is required for an instrumentation amplifier with $R_1 = R_2 = 39 \text{ k}\Omega$ to produce a gain of 325?

Applications

As mentioned in the introduction to this section, the instrumentation amplifier is normally used to measure small differential signal voltages that are superimposed on a common-mode voltage often much larger than the signal voltage. Applications include situations where a quantity is sensed by a remote device, such as a temperature- or pressure-sensi-

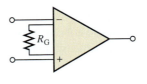
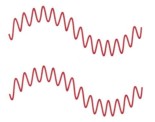

Small differential signals riding on larger common-mode signals

Instrumentation amplifier

Amplified differential signal. No common-mode signal.

FIGURE 15–3
Illustration of the rejection of large common-mode voltages and the amplification of smaller signal voltages by an instrumentation amplifier.

tive transducer, and the resulting small electrical signal is sent over a long line subject to electrical noise that produces common-mode voltages in the line. The instrumentation amplifier at the end of the line must amplify the small signal from the remote sensor and reject the large common-mode voltage. Figure 15–3 illustrates this.

A Specific Instrumentation Amplifier

Now that you have the basic idea of how an instrumentation amplifier works, let's look at a specific device. A representative device, the AD521, is shown in Figure 15–4 where IC pin numbers are given for reference. As you can see, there are some additional inputs and outputs that did not appear on the basic circuit. These provide additional features that are typical of many IC instrumentation amplifiers on the market.

FIGURE 15–4
The AD521 instrumentation amplifier.

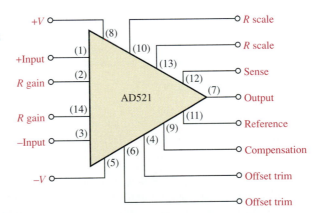

Some of the features of the AD521 are as follows. The voltage gain can be adjusted from 0.1 to 1000 with two external resistors. The input impedance is 3000 MΩ. The common-mode rejection ratio (CMRR) has a minimum value of 110 dB. Recall that a higher CMRR means better rejection of common-mode voltages. The AD521 has a gain-bandwidth product of 40 MHz. There is also an external provision for limiting the bandwidth, and the device is protected against excessive input voltages.

Setting the Voltage Gain For the AD521, two external resistors must be used to set the voltage gain as indicated in Figure 15–5. Resistor R_G is connected between the R-gain terminals. (pins 2 and 14). Resistor R_S is connected between the R-scale terminals (pins 10 and 13). R_S must be within ±15% of 100 kΩ, and R_G is selected for the desired gain based on the following formula:

$$A_v = \frac{R_S}{R_G}$$ **(15–4)**

Don't be concerned about the difference in this gain expression and the closed-loop gain for the basic circuit stated in Equation (15–2). This difference is due to the subtle differences in design. If you were constructing an instrumentation amplifier from separate op-amps and discrete resistors, you would use the formulas discussed earlier.

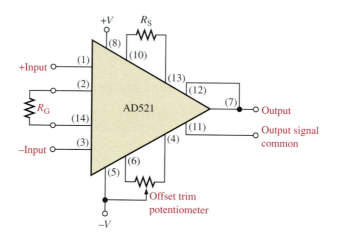

FIGURE 15–5
The AD521 with gain-setting resistors and output offset adjustment.

Offset Trim The offset **trim** adjustment (pins 4 and 6) is used to zero any output offset voltage caused by an input offset voltage multiplied by the gain. A potentiometer connected between pins 4 and 6 as shown in Figure 15–5 can be used to adjust the offset.

Bandwidth Control When you need to set the amplifier's bandwidth to a desired value, the compensation input (pin 9) can be used in conjunction with an external RC circuit. A recommended configuration is shown in Figure 15–6. The values of the two resistors and one of the capacitors in the compensation circuit are set as recommended by the manufacturer and then the value of the capacitor C_x is selected for the desired bandwidth according to the following formula:

$$C_x = \frac{1}{100\pi f_c}$$ **(15–5)**

where $BW = f_c$ is in kilohertz (kHz) and C_x is in microfarads (μF).

FIGURE 15–6
The AD521 with a compensation circuit for controlling the bandwidth.

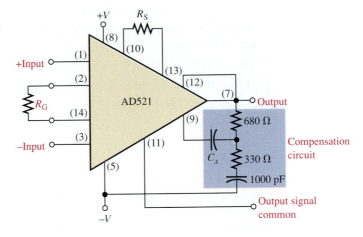

EXAMPLE 15–2

Determine the voltage gain and the bandwidth for the instrumentation amplifier in Figure 15–7.

FIGURE 15–7

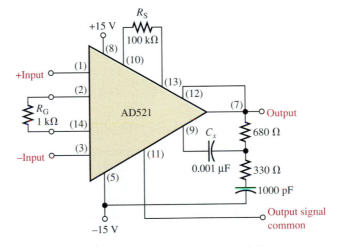

Solution   The voltage gain is determined by R_S and R_G as follows:

$$A_v = \frac{R_S}{R_G} = \frac{100 \text{ k}\Omega}{1 \text{ k}\Omega} = 100$$

The bandwidth is determined as follows:

$$C_x = \frac{1}{100\pi f_c}$$

$$BW = f_c = \frac{1}{100\pi C_x} = \frac{1}{100\pi(0.001\mu\text{F})} = 3.18 \text{ kHz}$$

Notice that the number of microfarads (0.001) is substituted into the formula, not the number of farads (0.001×10^{-6}).

Related Exercise   Modify the circuit in Figure 15–7 for a gain of approximately 45 and a bandwidth of approximately 10 kHz.

15–2 ■ ISOLATION AMPLIFIERS

An isolation amplifier provides dc isolation between input and output for the protection of human life or sensitive equipment in those applications where hazardous power-line leakage or high-voltage transients are possible. The principal areas of application for isolation amplifiers are in medical instrumentation, power plant instrumentation, industrial processing, and automated testing.

After completing this section, you should be able to

■ **Analyze and explain the operation of an isolation amplifier**
  □ Explain the basic configuration of an isolation amplifier
  □ Discuss an application in medical electronics
  □ Describe the features of the AD295 isolation amplifier

The Basic Isolation Amplifier

In some ways, the isolation amplifier can be viewed as an elaborate op-amp or instrumentation amplifier. The isolation amplifier has an input circuit that is electrically isolated from the output and power supply circuits using transformers or optical coupling. Transformer coupling is more common. The circuits are in IC form, but the transformers are not integrated. Although the packages that contain both the circuits and transformers are somewhat larger than standard IC packages, they are generally pin compatible for easy circuit board assembly.

The typical three-port isolation amplifier has three basic isolated sections that are transformer coupled. As shown in the block diagram of Figure 15–8, the sections are the input circuit, the output circuit, and the power section. The input section contains an instrumentation amplifier or an op-amp (in this case, it is an instrumentation amplifier), a power supply, and a modulator. The output section contains an op-amp, a power supply, and a demodulator. The power section contains an oscillator which produces an ac voltage from the dc input.

Operation An external dc supply voltage is applied to the power section where the internal oscillator is energized and converts the dc input power to ac power. The frequency of the oscillator is fairly high in order to keep the size of the transformers small; for example, it is 80 kHz in the AD295 isolation amplifier. The ac power signal from the oscillator is coupled to both the input and output sections through transformer T2. In the input and output sections, the ac power signal is rectified and filtered by the power-supply circuits to provide dc power for the amplifiers.

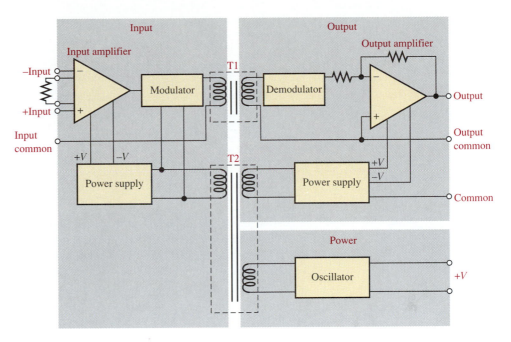

FIGURE 15–8
Basic isolation amplifier block diagram.

The ac power signal is also sent to the modulator in the input section where it is modulated by the output signal from the input amplifier. The modulated signal is coupled through transformer T1 to the output section where it is demodulated. The demodulation process recovers the original signal from the ac power signal. The output amplifier provides further gain before the signal goes to the final output.

Although the isolation amplifier is a fairly complex system in itself, in terms of its overall function, it is still simply an amplifier. You apply a dc voltage, put a signal in, and you get an amplified signal out. The isolation function is an unseen process.

Applications

As previously mentioned, the isolation amplifier is mainly used in medical equipment and for remote sensing in high-noise industrial environments where interfacing to sensitive equipment is required. In medical applications where body functions such as heart and blood pressure are monitored, the very small monitored signals are combined with large common-mode signals, such as 60 Hz power line pickup from the skin. In these situations, without isolation, dc leakage or equipment failure could be fatal. In chemical, nuclear, and metal-processing industries, for example, millivolt signals typically exist in the presence of common-mode voltages that can be in the kilovolt range. In this type of environment, the isolation amplifier can amplify small signals from very noisy equipment and provide a safe output to sensitive equipment such as computers.

Figure 15–9 shows a simplified diagram of an isolation amplifier in a cardiac monitoring application. In this situation, there are heart signals, which are very small, combined with much larger common-mode signals caused by muscle noise, electrochemical noise, residual electrode voltages, and 60 Hz line pickup from the skin. The monitoring of

FIGURE 15–9

Fetal heartbeat monitoring using an isolation amplifier.

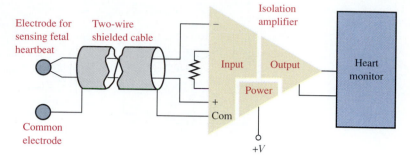

fetal heartbeat, as illustrated, is the most demanding type of cardiac monitoring because in addition to the fetal heartbeat that typically generates 50 μV, there is also the mother's heartbeat that typically generates 1 mV. The common-mode voltages can run from about 1 mV to about 100 mV. The CMR (common-mode rejection) of the isolation amplifier separates the signal of the fetal heartbeat from that of the mother's heartbeat and from those signals of the common-mode voltages. So, the signal from the fetal heartbeat is essentially all that the amplifier sends to the monitoring equipment.

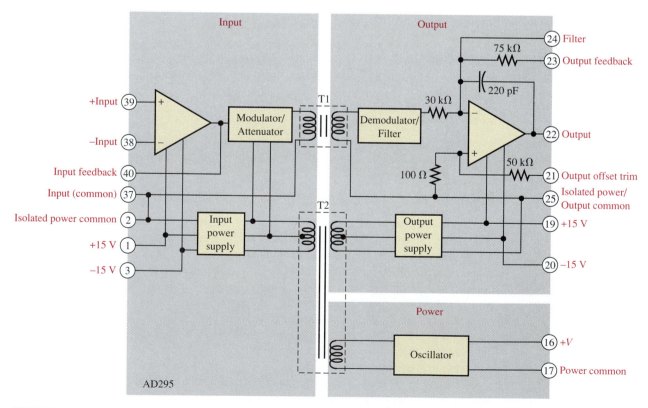

FIGURE 15–10

The AD295 isolation amplifier.

A Specific Isolation Amplifier

Now that you have learned basically what an isolation amplifier is and what it does, let's look at a representative device, the AD295, for a good introduction to practical IC isolation amplifiers. As you can see in Figure 15–10, the AD295 is similar to the basic isolation amplifier in Figure 15–8 except that the input amplifier is an op-amp, and it has a few more inputs and outputs. These additional pins provide for gain adjustments, offset adjustments, isolated dc voltage outputs, and other functions.

Isolated Power Outputs   The AD295 isolation amplifier provides ±15 V from both isolated power supplies. These voltages are available for powering associated external circuits such as preamplifiers, transducers, and the like.

Voltage Gain   The gains of both the input and the output amplifiers can be set with external resistors. The overall voltage gain of the device can be set at any value from 1 to 1000. Figure 15–11 shows the circuit connected for unity gain. In the input circuit, the amplifier output is connected directly back to the inverting input (pin 40 to pin 38) creating a voltage-follower configuration with a gain of 1. The attenuation circuit within the modulator/attenuator block has an inherent fixed attenuation of 0.4. To overcome this

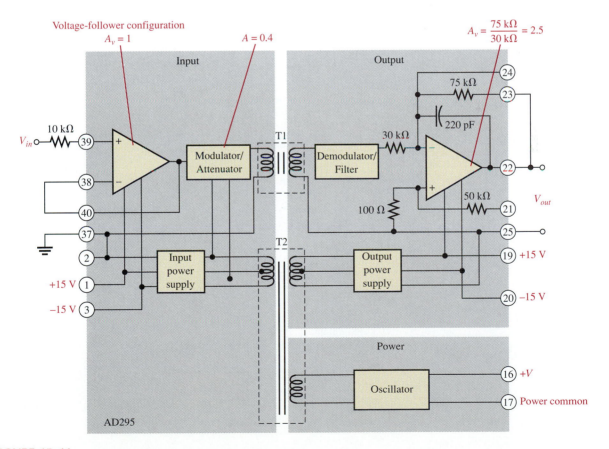

FIGURE 15–11
Unity-gain connections for the AD295 isolation amplifier.

attenuation, the output amplifier has a gain of 2.5 ($A_v = 75$ kΩ/30 k$\Omega = 2.5$) when the output is connected directly to the 75 kΩ feedback resistor (pin 22 to pin 23). This produces a combined gain of 1 ($0.4 \times 2.5 = 1$).

Gains up to 1000 can be achieved by connecting external resistors as shown in Figure 15–12. Although the connection for the input amplifier is for a noninverting configuration, it can also be connected in an inverting configuration. The voltage gain of the input amplifier is

$$A_{v(input)} = 1 + \frac{R_f}{R_1} \qquad \textbf{(15–6)}$$

For the AD295, there must also be a 10 kΩ resistor in series with the input as indicated in the figure. Also, $R_f + R_1$ must be equal to or greater than 10 kΩ. R_f is the feedback resistor and R_1 is the input resistor for the op-amp configuration.

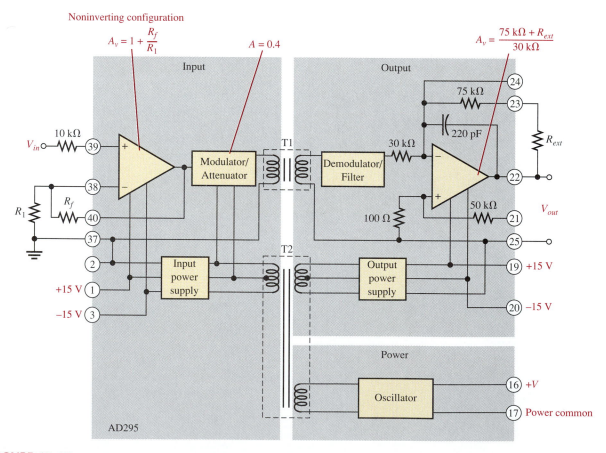

FIGURE 15–12
Nonunity gain connections for the AD295 isolation amplifier.

The voltage gain of the output amplifier can be increased above 2.5 by adding an external resistor in series with the internal 75 kΩ resistor as shown in Figure 15–12. For this case, the voltage gain of the output amplifier is

$$A_{v(output)} = \frac{75 \text{ k}\Omega + R_{ext}}{30 \text{ k}\Omega} \tag{15–7}$$

Offset Adjustments The external connections for adjustment of the input and output offset voltages are shown in Figure 15–13 in conjunction with a unity-gain configuration. The resistor values shown are recommended by the manufacturer for this particular device.

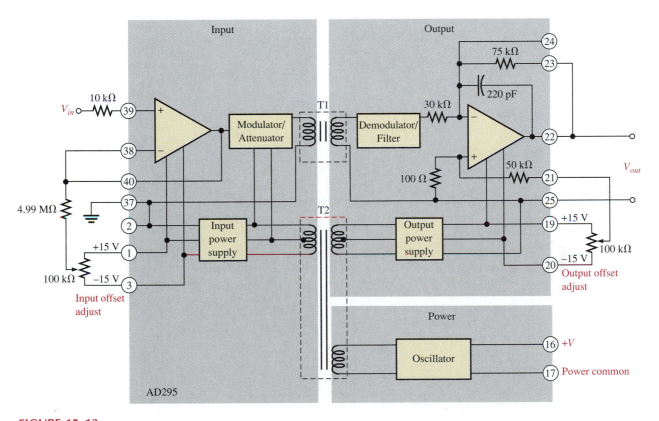

FIGURE 15–13
Offset voltage adjustments for the AD295 isolation amplifier.

EXAMPLE 15–3   Determine the overall voltage gain of the AD295 isolation amplifier in Figure 15–14.

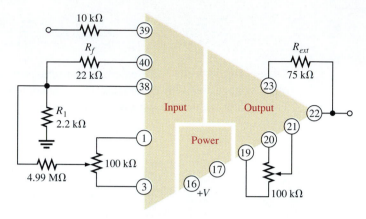

FIGURE 15–14

Solution   The voltage gain of the input amplifier is

$$A_{v(input)} = 1 + \frac{R_f}{R_1} = 1 + \frac{22 \text{ k}\Omega}{2.2 \text{ k}\Omega} = 1 + 10 = 11$$

The voltage gain of the output amplifier is

$$A_{v(output)} = \frac{75 \text{ k}\Omega + R_{ext}}{30 \text{ k}\Omega} = \frac{75 \text{ k}\Omega + 75 \text{ k}\Omega}{30 \text{ k}\Omega} = 5$$

Since the fixed internal attentuation is 0.4 for the AD295, the overall voltage gain of the isolation amplifier is

$$A'_v = A_{v(input)} \times A_{v(output)} \times \text{Atten} = (11)(5)(0.4) = 22$$

Related Exercise   Select resistor values and specify the connections in Figure 15–14 that will produce an overall voltage gain of approximately 10.

SECTION 15–2 REVIEW

1. In what types of applications are isolation amplifiers used?
2. What are the three sections in a typical isolation amplifier?
3. How are the sections in an isolation amplifier connected?
4. What is the purpose of the oscillator in an isolation amplifier?

15–3 ■ OPERATIONAL TRANSCONDUCTANCE AMPLIFIERS (OTAs)

Conventional op-amps are, as you know, primarily voltage amplifiers in which the output voltage equals the gain times the input voltage. The OTA is primarily a voltage-to-current amplifier in which the output current equals the gain times the input voltage.

After completing this section, you should be able to

■ **Analyze and explain the operation of the OTA**
   ☐ Identify the OTA symbol
   ☐ Define *transconductance*
   ☐ Discuss the relationship between transconductance and bias current
   ☐ Describe the features of the CA3080 OTA
   ☐ Discuss OTA applications

Figure 15–15 shows the symbol for an operational transconductance amplifier (OTA). The double circle symbol at the output represents an output current source that is dependent on a bias current. Like the conventional op-amp, the OTA has two differential input terminals, a high input impedance, and a high CMRR. Unlike the conventional op-amp, the OTA has a bias-current input terminal, a high output impedance, and no fixed open-loop voltage gain.

FIGURE 15–15

Symbol for an operational transconductance amplifier (OTA).

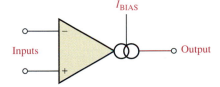

The Transconductance Is the Gain of an OTA

By definition, the **transconductance** of an electronic device is the ratio of the output current to the input voltage. For an OTA, voltage is the input variable and current is the output variable; therefore, the ratio of output current to input voltage is its gain. Consequently, the voltage-to-current gain of an OTA is the transconductance, g_m.

$$g_m = \frac{I_{out}}{V_{in}}$$ **(15–8)**

In an OTA, the transconductance is dependent on a constant (K) times the bias current (I_{BIAS}) as indicated in Equation (15–9). The value of the constant is dependent on the internal circuit design.

$$g_m = KI_{BIAS}$$ **(15–9)**

The output current is controlled by the input voltage and the bias current as shown by the following formula:

$$I_{out} = g_m V_{in} = KI_{BIAS}V_{in}$$

The Transconductance Is a Function of Bias Current

The relationship of the transconductance and the bias current in an OTA is a very important characteristic. The graph in Figure 15–16 illustrates a typical relationship. Notice that the transconductance increases linearly with the bias current. The constant of proportionality, K, is the slope of the line. In this case, K is approximately 16.

FIGURE 15–16

Graph of transconductance versus bias current for a typical OTA.

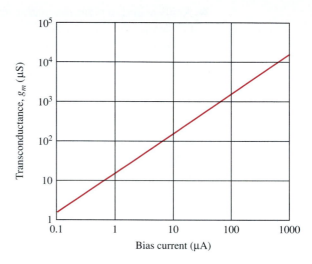

EXAMPLE 15–4   If an OTA has a $g_m = 1000$ μS, what is the output current when the input voltage is 50 mV?

Solution $I_{out} = g_m V_{in} = (1000 \ \mu S)(50 \ mV) = 50 \ \mu A$

Related Exercise   From the graph in Figure 15–16, determine the approximate bias current required to produce $g_m = 1000$ μS.

Basic OTA Circuits

Figure 15–17 shows the OTA used as an inverting amplifier with a fixed voltage gain. The voltage gain is set by the transconductance and the load resistance as follows.

$$V_{out} = I_{out} R_L$$

$$I_{out} = \frac{V_{out}}{R_L}$$

$$g_m = \frac{I_{out}}{V_{in}} = \frac{(V_{out}/R_L)}{V_{in}} = \left(\frac{V_{out}}{V_{in}}\right)\left(\frac{1}{R_L}\right)$$

$$g_m R_L = \frac{V_{out}}{V_{in}}$$

Since V_{out}/V_{in} is the voltage gain,

$$A_v = g_m R_L$$

FIGURE 15–17

An OTA as an inverting amplifier with a fixed voltage gain.

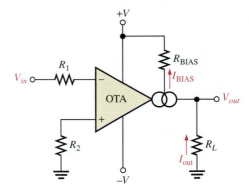

The transconductance of the amplifier in Figure 15–17 is determined by the amount of bias current, which is set by the dc supply voltages and the bias resistor R_{BIAS}.

One of the most useful features of an OTA is that the voltage gain can be controlled by the amount of bias current. This can be done manually, as shown in Figure 15–18(a), by using a variable resistor in series with R_{BIAS} in the circuit of Figure 15–17. By changing the resistance, you can produce a change in I_{BIAS}, which changes the transconductance. A change in the transconductance changes the voltage gain. The voltage gain can also be controlled with an externally applied variable voltage as shown in Figure 15–18(b). A variation in the applied bias voltage causes a change in the bias current.

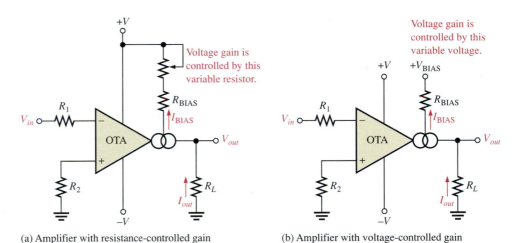

(a) Amplifier with resistance-controlled gain

(b) Amplifier with voltage-controlled gain

FIGURE 15–18

An OTA as an inverting amplifier with a variable-voltage gain.

A Specific OTA

The CA3080 is a typical OTA and serves as a representative device. Figure 15–19 shows its pin configuration for an eight-pin DIP. The maximum dc supply voltage is ±15 V, and its transconductance characteristic happens to be the same as indicated by the graph in Figure 15–16. For a CA3080, the bias current is determined by the following formula:

$$I_{BIAS} = \frac{(+V) - (-V) - 0.7 \text{ V}}{R_{BIAS}}$$

(15–10)

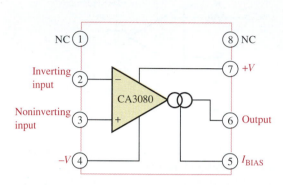

FIGURE 15–19
The CA3080 OTA.

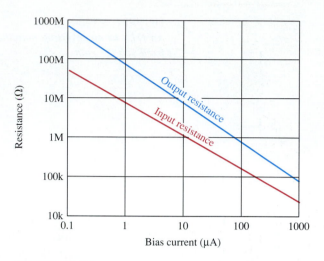

FIGURE 15–20
Input and output resistances versus bias current.

The 0.7 V is due to the internal circuit where a base-emitter junction connects the external R_{BIAS} with the negative supply voltage ($-V$). The positive supply voltage is $+V$.

Not only does the transconductance of an OTA vary with bias current, but so do the input and output resistances. Both the input and output resistances decrease as the bias current increases, as shown in Figure 15–20 for a CA3080.

EXAMPLE 15–5

The OTA in Figure 15–21 is connected as an inverting fixed-gain amplifier. Determine the approximate voltage gain.

FIGURE 15–21

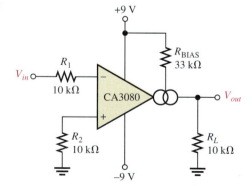

Solution The bias current is calculated as follows:

$$I_{\text{BIAS}} = \frac{(+V) - (-V) - 0.7\ \text{V}}{R_{\text{BIAS}}} = \frac{9\ \text{V} - (-9\ \text{V}) - 0.7\ \text{V}}{33\ \text{k}\Omega} = 524\ \mu\text{A}$$

From the graph in Figure 15–16, $K \cong 16$. Therefore,

$$g_m = KI_{BIAS} = 16(524\ \mu A) = 8.38\ mS$$

Using this value of g_m, the voltage gain is calculated.

$$A_v = g_m R_L = (8.38\ mS)(10\ k\Omega) = 83.8$$

Related Exercise If the OTA in Figure 15–21 is operated with dc supply voltages of
±12 V, will this change the voltage gain and, if so, to what value?

Two OTA Applications

Amplitude Modulator Figure 15–22 illustrates an OTA connected as an amplitude
modulator. The voltage gain is varied by applying a modulation voltage to the bias input.
When a constant-amplitude input signal is applied, the amplitude of the output signal will
vary according to the modulation voltage on the bias input. The gain is dependent on bias
current, and bias current is related to the modulation voltage by the following relationship:

$$I_{BIAS} = \frac{V_{MOD} - (-V) - 0.7\ V}{R_{BIAS}}$$

This modulating action is shown in Figure 15–22 for a higher frequency sinusoidal input
voltage and a lower frequency sinusoidal modulating voltage.

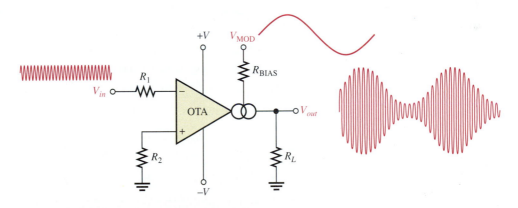

FIGURE 15–22
The OTA as an amplitude modulator.

EXAMPLE 15–6

The input to the OTA amplitude modulator in Figure 15–23 is a 50 mV peak-to-peak,
1 MHz sine wave. Determine the output signal, given the modulation voltage shown is
applied to the bias input.

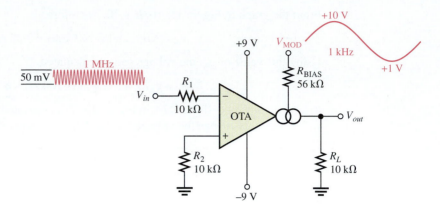

FIGURE 15–23

Solution The maximum voltage gain is when I_{BIAS}, and thus g_m, is maximum. This occurs at the maximum peak of the modulating voltage, V_{MOD}.

$$I_{\text{BIAS(max)}} = \frac{V_{\text{MOD(max)}} - (-V) - 0.7\text{ V}}{R_{\text{BIAS}}} = \frac{10\text{ V} - (-9\text{ V}) - 0.7\text{ V}}{56\text{ k}\Omega} = 327\text{ }\mu\text{A}$$

From the graph in Figure 15–16, the constant K is approximately 16.

$$g_m = KI_{\text{BIAS(max)}} = 16(327\text{ }\mu\text{A}) = 5.23\text{ mS}$$
$$A_{v(max)} = g_m R_L = (5.23\text{ mS})(10\text{ k}\Omega) = 52.3$$
$$V_{out(max)} = A_{v(max)}V_{in} = (52.3)(50\text{ mV}) = 2.62\text{ V}$$

The minimum bias current is

$$I_{\text{BIAS(min)}} = \frac{V_{\text{MOD(min)}} - (-V) - 0.7\text{ V}}{R_{\text{BIAS}}} = \frac{1\text{ V} - (-9\text{ V}) - 0.7\text{ V}}{56\text{ k}\Omega} = 166\text{ }\mu\text{A}$$

$$g_m = KI_{\text{BIAS(min)}} = 16(166\text{ }\mu\text{A}) = 2.66\text{ mS}$$
$$A_{v(min)} = g_m R_L = (2.66\text{ mS})(10\text{ k}\Omega) = 26.6$$
$$V_{out(min)} = A_{v(min)}V_{in} = (26.6)(50\text{ mV}) = 1.33\text{ V}$$

The resulting output voltage is shown in Figure 15–24.

FIGURE 15–24

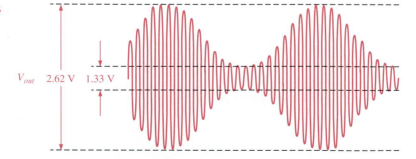

Related Exercise Repeat this example with the sine-wave modulating signal replaced by a square wave with the same maximum and minimum levels and a bias resistor of 39 kΩ.

Schmitt Trigger Figure 15–25 shows an OTA used in a Schmitt-trigger configuration. Basically, a Schmitt trigger is a comparator with hysteresis where the input voltage is large enough to drive the device into its saturated states. When the input voltage exceeds a certain threshold value or trigger point, the device switches to one of its saturated output states. When the input falls back below another threshold value, the device switches back to its other saturated output state.

FIGURE 15–25
The OTA as a Schmitt trigger.

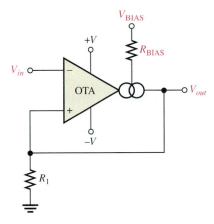

In the case of the OTA Schmitt trigger, the threshold levels are set by the current through resistor R_1. The maximum output current in an OTA equals the bias current. Therefore, in the saturated output states, $I_{out} = I_{BIAS}$. The maximum positive output voltage is $I_{out}R_1$, and this voltage is the positive threshold value or upper trigger point. When the input voltage exceeds this value, the output switches to its maximum negative voltage, which is $-I_{out}R_1$. Since $I_{out} = I_{BIAS}$, the trigger points can be controlled by the bias current. Figure 15–26 illustrates this operation.

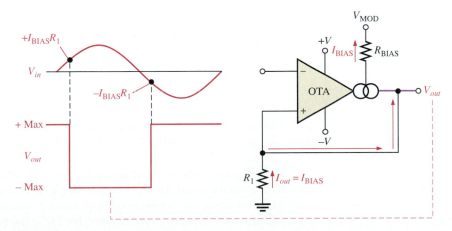

FIGURE 15–26
Basic operation of the OTA Schmitt trigger.

1. What does OTA stand for?
2. If the bias current in an OTA is increased, does the transconductance increase or decrease?
3. What happens to the voltage gain if the OTA is connected as a fixed-voltage amplifier and the supply voltages are increased?
4. What happens to the voltage gain if the OTA is connected as a variable-gain voltage amplifier and the voltage at the bias terminal is decreased?

15–4 ■ LOG AND ANTILOG AMPLIFIERS

A logarithmic (log) amplifier produces an output that is proportional to the logarithm of the input, and antilogarithmic (antilog) amplifiers take the antilog or inverse log of the input. Log amplifiers are used in applications that require compression of analog input data, linearization of transducers that have exponential outputs, and analog multiplication and division. In this section, we will discuss the principles of logarithmic amplifiers.

After completing this section, you should be able to

■ **Analyze and explain the operation of log and antilog amplifiers**
   □ Describe the feedback configurations
   □ Define *logarithm* and *natural logarithm*
   □ Discuss signal compression with logarithmic amplifiers

The Basic Logarithmic Amplifier

The key element in a log amplifier is a device that exhibits a logarithmic characteristic that, when placed in the feedback loop of an op-amp, produces a logarithmic response. This means that the output voltage is a function of the logarithm of the input voltage, as expressed by the following general equation. K is a constant, and ln is the **natural logarithm** to the base e.

$$V_{out} = -K \ln(V_{in}) \tag{15–11}$$

Although we will use natural logarithms in the formulas in this section, each expression can be converted to a logarithm to the base 10 ($\log_{10}$) using the relationship $\ln x = 2.3 \log_{10} x$.

The semiconductor *pn* junction in the form of either a diode or the base-emitter junction of a bipolar transistor provides a logarithmic characteristic. You may recall that a diode has a nonlinear characteristic up to a forward voltage of approximately 0.7 V. Figure 15–27 shows the characteristic curve, where I_F is the forward diode current and V_F is the forward diode voltage.

FIGURE 15–27

A portion of a diode (pn junction) characteristic curve (I_F versus V_F).

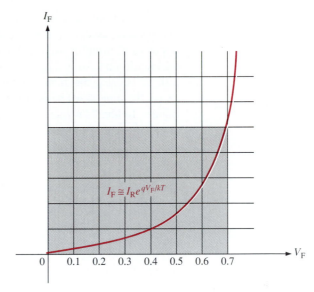

As you can see on the graph, the diode curve is nonlinear. Not only is the characteristic curve nonlinear, it is logarithmic and is specifically defined by the following formula:

$$I_\mathrm{F} \cong I_R e^{qV_\mathrm{F}/kT} \tag{15–12}$$

where I_R is the reverse leakage current, q is the charge on an electron, k is Boltzmann's constant, and T is the absolute temperature in degrees Kelvin. Equation (15–12) can be solved for the diode forward voltage, V_F, as follows. Take the natural logarithm (ln is the logarithm to the base e) of both sides.

$$\ln I_\mathrm{F} = \ln I_R e^{qV_\mathrm{F}/kT}$$

The ln of a product of two terms equals the sum of the ln of each term.

$$\ln I_\mathrm{F} = \ln I_\mathrm{R} + \ln e^{qV_\mathrm{F}/kT} = \ln I_\mathrm{R} + \frac{qV_\mathrm{F}}{kT}$$

$$\ln I_F - \ln I_\mathrm{R} = \frac{qV_\mathrm{F}}{kT}$$

The difference of two ln terms equals the ln of the quotient of the terms.

$$\ln\!\left(\frac{I_\mathrm{F}}{I_\mathrm{R}}\right) = \frac{qV_\mathrm{F}}{kT}$$

Solving for V_F,

$$V_\mathrm{F} = \left(\frac{kT}{q}\right)\ln\!\left(\frac{I_\mathrm{F}}{I_\mathrm{R}}\right) \tag{15–13}$$

Log Amplifier with a Diode   When you place a diode in the feedback loop of an op-amp circuit, as shown in Figure 15–28, you have a basic log amplifier. Since the inverting input is at virtual ground (0 V), the output is at $-V_F$ when the input is positive. Since V_F is logarithmic, so is V_{out}. The output is limited to a maximum value of approximately −0.7 V because the diode's logarithmic characteristic is restricted to voltages below 0.7 V. Also, the input must be positive when the diode is connected in the direction shown in the figure. To handle negative inputs, you must turn the diode around.

FIGURE 15–28

A basic log amplifier using a diode as the feedback element.

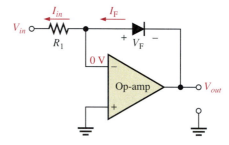

An analysis of the circuit in Figure 15–28 is as follows, beginning with the fact that $V_{out} = -V_F$ and the fact that $I_F = I_{in}$ because there is no current at the inverting input.

$$V_{out} = -V_F$$

$$I_F = I_{in} = \frac{V_{in}}{R_1}$$

Substituting into the formula for V_F,

$$V_{out} = -\left(\frac{kT}{q}\right)\ln\left(\frac{V_{in}}{I_R R_1}\right) \qquad \text{(15–14)}$$

The term kT/q is a constant equal to approximately 25 mV at 25°C. Therefore, the output voltage can be expressed as

$$V_{out} \cong -(0.025\ \text{V})\ln\left(\frac{V_{in}}{I_R R_1}\right) \qquad \text{(15–15)}$$

From Equation (15–15), you can see that the output voltage is the negative of a logarithmic function of the input voltage. The value of the output is controlled by the value of the input voltage and the value of the resistor R_1. The other factor, I_R, is a constant for a given diode.

EXAMPLE 15–7   Determine the output voltage for the log amplifier in Figure 15–29. Assume $I_R = 50$ nA.

FIGURE 15–29

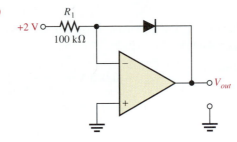

Solution   The input voltage and the resistor value are given in Figure 15–29.

$$V_{out} = -(0.025 \text{ V})\ln\left(\frac{V_{in}}{I_R R_1}\right) = -(0.025 \text{ V})\ln\left(\frac{2 \text{ V}}{50 \text{ nA} \times 100 \text{ k}\Omega}\right)$$

$$= -(0.025 \text{ V})\ln(400) = -(0.025 \text{ V})(5.99) = -0.150 \text{ V}$$

Related Exercise   Calculate the output voltage of the log amplifier with a +4 V input.

Log Amplifier with a BJT   The base-emitter junction of a bipolar transistor exhibits the same type of logarithmic characteristic as a diode because it is also a *pn* junction. A log amplifier with a BJT connected in a common-base form in the feedback loop is shown in Figure 15–30. Notice that V_{out} with respect to ground is equal to V_{BE}.

FIGURE 15–30
A basic log amplifier using a transistor as the feedback element.

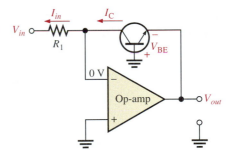

The analysis for this circuit is the same as for the diode log amplifier except that V_{BE} replaces V_F, I_C replaces I_F, and I_{EBO} replaces I_R. The expression for the I_C versus V_{BE} characteristic curve is

$$I_C = I_{EBO} e^{qV_{BE}/kT} \qquad \text{(15–16)}$$

where I_{EBO} is the emitter-to-base leakage current. The expression for the output voltage is

$$V_{out} = -(0.025 \text{ V})\ln\left(\frac{V_{in}}{I_{EBO} R_1}\right) \qquad \text{(15–17)}$$

EXAMPLE 15–8

What is V_{in} for a transistor log amplifier with $V_{in} = 3$ V and $R_1 = 68$ kΩ? Assume $I_{EBO} = 40$ nA.

Solution

$$V_{out} = -(0.025 \text{ V})\ln\left(\frac{V_{in}}{I_{EBO}R_1}\right) = -(0.025 \text{ V})\ln\left(\frac{3 \text{ V}}{40 \text{ nA} \times 68 \text{ kΩ}}\right)$$

$$= -(0.025 \text{ V})\ln(1102.94) = -0.1751 \text{ V} = -175.1 \text{ mV}$$

Related Exercise   Calculate V_{out} if R_1 is changed to 33 kΩ.

The Basic Antilog Amplifier

The **logarithm** of a number is the power to which the base must be raised to get that number. The antilogarithm of a number is the result obtained when the base is raised to a power equal to the logarithm of that number. To get the antilogarithm, you must take the exponential of the logarithm (antilogarithm of $x = e^{\ln x}$).

An antilog amplifier is formed by connecting a transistor (or diode) as the input element as shown in Figure 15–31. The exponential formula in Equation (15–16) still applies to the base-emitter *pn* junction. The output voltage is determined by the current (equal to the collector current) through the feedback resistor.

$$V_{out} = -R_f I_C$$

The characteristic equation of the *pn* junction is

$$I_C = I_{EBO}e^{qV_{BE}/kT}$$

Substituting into the equation for V_{out}, you get

$$V_{out} = -R_f I_{EBO}e^{qV_{BE}/kT}$$

As you can see in Figure 15–31, $V_{in} = V_{BE}$.

$$V_{out} = -R_f I_{EBO}e^{qV_{in}/kT}$$

The exponential term can be expressed as an antilogarithm as follows:

$$V_{out} = -R_f I_{EBO}\text{antilog}\left(\frac{V_{in}q}{kT}\right)$$

Since kT/q is approximately 25 mV,

$$V_{out} = -R_f I_{EBO}\text{antilog}\left(\frac{V_{in}}{25 \text{ mV}}\right) \qquad (15\text{--}18)$$

FIGURE 15–31
A basic antilog amplifier.

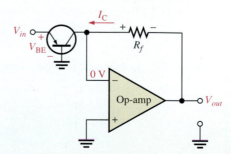

EXAMPLE 15–9

For the antilog amplifier in Figure 15–32, find the output voltage. Assume $I_{\text{EBO}} = 40$ nA.

FIGURE 15–32

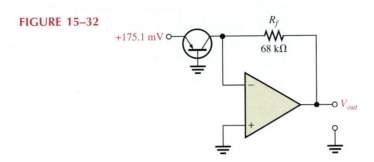

Solution   First of all, notice that the input voltage in Figure 15–32 is the same as the output voltage of the log amplifier in Example 15–8, where the output voltage is proportional to the logarithm of the input voltage. In this case, the antilog amplifier reverses the process and produces an output that is proportional to the antilog of the input. Stated another way, the input of an antilog amplifier is proportional to the logarithm of the output. So, the output voltage of the antilog amplifier in Figure 15–32 should have the same magnitude as the input voltage of the log amplifier in Example 15–8 because all the constants are the same. Let's see if it does.

$$V_{out} = -R_f I_{\text{EBO}} \text{antilog}\left(\frac{V_{in}}{25 \text{ mV}}\right) = -(68 \text{ k}\Omega)(40 \text{ nA})\text{antilog}\left(\frac{175.1 \text{ mV}}{25 \text{ mV}}\right)$$

$$= -(68 \text{ k}\Omega)(40 \text{ nA})(1101) = -3 \text{ V}$$

Related Exercise   Determine V_{out} for the amplifier in Figure 15–32 if the feedback resistor is changed to 100 kΩ.

Signal Compression with Logarithmic Amplifiers

In certain applications, a signal may be too large in magnitude for a particular system to handle. The term *dynamic range* is often used to describe the range of voltages contained in a signal. In these cases, the signal voltage must be scaled down by a process called **signal compression** so that it can be properly handled by the system. If a linear circuit is used to scale a signal down in amplitude, the lower voltages are reduced by the same percentage as the higher voltages. Linear signal compression often results in the lower voltages becoming obscured by noise and difficult to accurately distinguish, as illustrated in Figure 15–33(a). To overcome this problem, a signal with a large dynamic range can be compressed using a logarithmic response, as shown in Figure 15–33(b). In logarithmic signal compression, the higher voltages are reduced by a greater percentage than the lower voltages, thus keeping the lower voltage signals from being lost in noise.

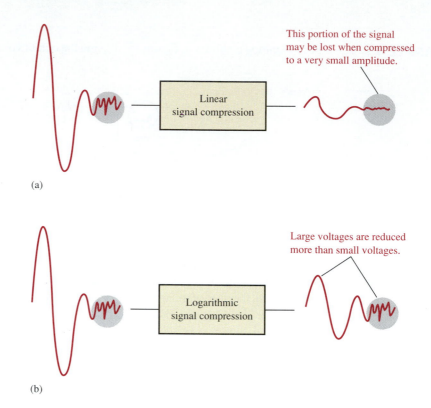

FIGURE 15–33

The basic concept of signal compression with a logarithmic amplifier.

1. What purpose does the diode or transistor perform in the feedback loop of a log amplifier?

2. Why is the output of a log amplifier limited to about 0.7 V?

3. What are the factors that determine the output voltage of a basic log amplifier?

4. In terms of implementation, how does a basic antilog amplifier differ from a basic log amplifier?

15–5 ■ CONVERTERS AND OTHER OP-AMP CIRCUITS

This section introduces a few more op-amp circuits that represent basic applications of the op-amp. You will learn about the constant-current source, the current-to-voltage converter, the voltage-to-current converter, and the peak detector. This is, of course, not a comprehensive coverage of all possible op-amp circuits but is intended only to introduce you to some common and basic uses.

After completing this section, you should be able to

■ **Analyze and explain several special types of op-amp circuits**
  ☐ Describe how the op-amp is used as a constant-current source
  ☐ Explain the operation of a current-to-voltage converter
  ☐ Explain the operation of a voltage-to-current converter
  ☐ Describe the operation of a peak detector

Constant-Current Source

A constant-current source delivers a load current that remains constant when the load resistance changes. Figure 15–34 shows a basic circuit in which a stable voltage source (V_{IN}) provides a constant current (I_i) through the input resistor (R_i). Since the inverting (−) input of the op-amp is at virtual ground (0 V), the value of I_i is determined by V_{IN} and R_i as

$$I_i = \frac{V_{IN}}{R_i}$$

FIGURE 15–34

A basic constant-current source.

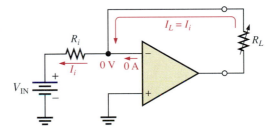

Now, since the internal input impedance of the op-amp is extremely high (ideally infinite), practically all of I_i flows through R_L, which is connected in the feedback path. Since $I_i = I_L$,

$$I_L = \frac{V_{IN}}{R_i} \qquad\qquad (15\text{–}19)$$

If R_L changes, I_L remains constant as long as V_{IN} and R_i are held constant.

Current-to-Voltage Converter

A current-to-voltage converter converts a variable input current to a proportional output voltage. A basic circuit that accomplishes this is shown in Figure 15–35(a). Since practically all of I_i flows through the feedback path, the voltage dropped across R_f is I_iR_f. Because the left side of R_f is at virtual ground (0 V), the output voltage equals the voltage across R_f, which is proportional to I_i.

$$V_{out} = I_iR_f \qquad\qquad (15\text{–}20)$$

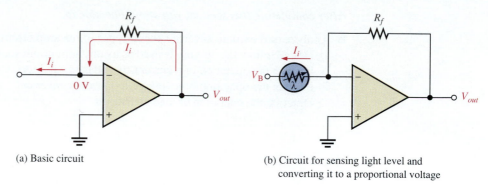

(a) Basic circuit

(b) Circuit for sensing light level and converting it to a proportional voltage

FIGURE 15–35
Current-to-voltage converter.

A specific application of this circuit is illustrated in Figure 15–35(b), where a photoconductive cell is used to sense changes in light level. As the amount of light changes, the current through the photoconductive cell varies because of the cell's change in resistance. This change in resistance produces a proportional change in the output voltage ($\Delta V_{out} = \Delta I_i R_f$).

Voltage-to-Current Converter

A basic voltage-to-current converter is shown in Figure 15–36. This circuit is used in applications where it is necessary to have an output (load) current that is controlled by an input voltage.

FIGURE 15–36
Voltage-to-current converter.

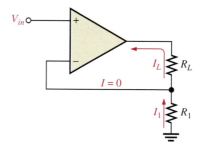

Neglecting the input offset voltage, both inverting and noninverting input terminals of the op-amp are at the same voltage, V_{in}. Therefore, the voltage across R_1 equals V_{in}. Since negligible current flows from the inverting input, the same current that flows through R_1 also flows through R_L; thus

$$I_L = \frac{V_{in}}{R_1}$$

(15–21)

Peak Detector

An interesting application of the op-amp is in a peak detector circuit such as the one shown in Figure 15–37. In this case the op-amp is used as a comparator. This circuit is used to detect the peak of the input voltage and store that peak voltage on a capacitor. For example, this circuit can be used to detect and store the maximum value of a voltage surge; this value can then be measured at the output with a voltmeter or recording device. The basic operation is as follows. When a positive voltage is applied to the noninverting input of the op-amp through R_i, the high-level output voltage of the op-amp forward-biases the diode and charges the capacitor. The capacitor continues to charge until its voltage reaches a value equal to the input voltage and thus both op-amp inputs are at the same voltage. At this point, the op-amp comparator switches, and its output goes to the low level. The diode is now reverse-biased, and the capacitor stops charging. It has reached a voltage equal to the peak of V_{in} and will hold this voltage until the charge eventually leaks off. If a greater input peak occurs, the capacitor charges to the new peak.

FIGURE 15–37

A basic peak detector.

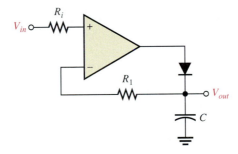

SECTION 15–5 REVIEW

1. For the constant-current source in Figure 15–34, the input reference voltage is 6.8 V and R_i is 10 kΩ. What value of constant current does the circuit supply to a 1 kΩ load? To a 5 kΩ load?

2. What element determines the constant of proportionality that relates input current to output voltage in the current-to-voltage converter?

15–6 ■ SYSTEM APPLICATION

Your company is trying to get into the medical instrumentation market with an ECG system and you have been assigned to work on the amplifier circuits which include an isolation amplifier, a filter, and a postamplifier with summing capability. The ECG, or electrocardiograph, system is used for monitoring heart signals. From the output waveform, the doctor can detect abnormalities in the heartbeat.

Basic Operation of the System

The human heart produces an electrical signal that can be picked up by electrodes in contact with the skin. When the heart signal is displayed on a chart recorder or on a video monitor, it is called an electrocardiograph (ECG). Typically, the heart signal picked up by

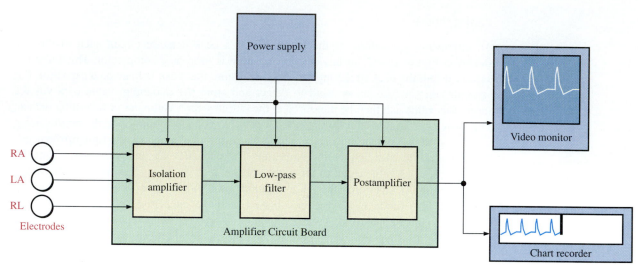

FIGURE 15–38
Block diagram of the ECG system.

the electrode is about 1 mV and has significant frequency components from less than 1 Hz to about 100 Hz.

As indicated in the block diagram in Figure 15–38, an ECG system has at least three electrodes. There is a right-arm (RA) electrode, a left-arm (LA) electrode, and a right-leg (RL) electrode that is the common terminal. The isolation amplifier provides for differential inputs from the electrodes, provides a high CMR (common-mode rejection) to eliminate the relatively high common-mode noise voltages associated with heart signals, and provides electrical isolation for protection of the patient. The low-pass filter rejects frequencies above those contained in the heart signal. The postamplifier provides most of the amplification in the system and drives a video monitor and/or a chart recorder. The three op-amp circuits are contained on the amplifier circuit board shown in Figure 15–39.

The inputs from the electrode sensors are connected to the amplifier board with shielded cable to prevent noise pickup. The schematic for the amplifier board is shown in Figure 15–40. The shielded cable is basically a twisted pair of wires surrounded by a braided metal sheathing that is covered by an insulated sleeve. The braided metal shield serves as the conduit for the common connection. The incoming differential signal is amplified by the fixed gain of the AD295 isolation amplifier. The AD295 is housed in a package that is larger than standard IC DIPs or SOPs. The larger configuration is required to accommodate the coupling transformers used in the isolation circuitry. The particular package used in this system has 40 pins. As shown in Figure 15–39, the dot indicates pin 1 and pin 40 is directly opposite pin 1.

The low-pass filter is an active filter of a type to be studied in Chapter 16. All you need to know for this assignment is how to calculate its critical frequency and gain, and those formulas will be provided.

The postamplifier is an inverting amplifier with an adjustable voltage gain. The inverting input also serves as a summing point for the signal voltage, and a dc voltage used for adding an adjustable dc level to the output for adjusting the vertical position of the display.

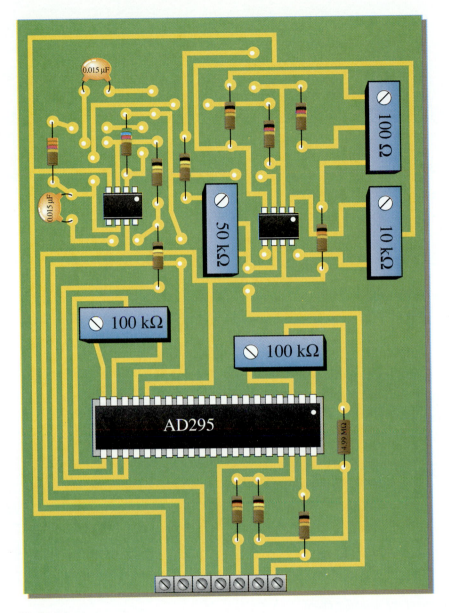

FIGURE 15–39
Amplifier board.

The Amplifier Board

☐ Make sure that the circuit board shown in Figure 15–39 is correctly assembled by checking it against the schematic in Figure 15–40. There are several backside connections. Corresponding feedthrough pads are aligned either vertically or horizontally.

☐ Label a copy of the board with component and input/output designations in agreement with the schematic.

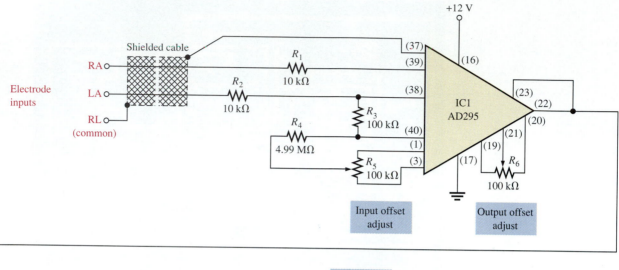

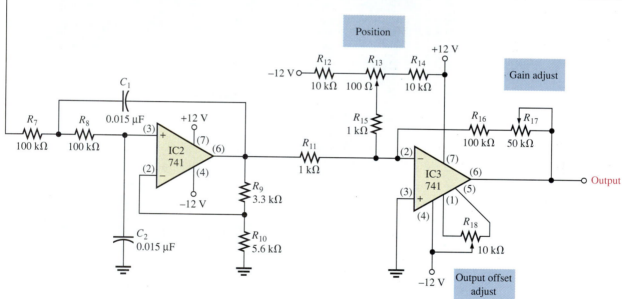

FIGURE 15–40
Schematic of the amplifier board.

The Circuits

☐ Determine the voltage gain of the isolation amplifier.

☐ Determine the bandwidth and voltage gain of the filter using the following formulas, given that the response is from dc to the critical frequency.

$$f_c = \frac{1}{2\pi\sqrt{R_7 R_8 C_1 C_2}}$$

$$A_v = 1 + \frac{R_9}{R_{10}}$$

☐ Determine the minimum and maximum voltage gains of the postamplifier.

☐ Determine the overall gain range of the amplifier board.

☐ Determine the dc voltage range at the wiper of the position adjustment potentiometer.

Test Procedure

☐ Develop a step-by-step set of instructions on how to check the amplifier board for proper operation independent of the rest of the system.

☐ Specify voltage values for all the measurements to be made.

Troubleshooting

Faults have developed in four boards. Based on the indication for each board listed below, determine the most likely fault in each case. The circled numbers indicate the referenced test points on the circuit board in Figure 15–41. Assume that there is a verified sinusoidal input signal in each case and that each board has the proper dc supply voltages.

Board 1:   There is no voltage at test point 1.

Board 2:   With an input signal of 1 mV, there is an 11 mV signal at test point 2 but no signal at test point 3.

Board 3:   With an input signal of 1 mV, there is a 17.5 mV signal at test point 4, but no signal at test point 5.

Board 4:   There is an approximate square wave voltage with peaks at +11 V and −11 V at test point 1.

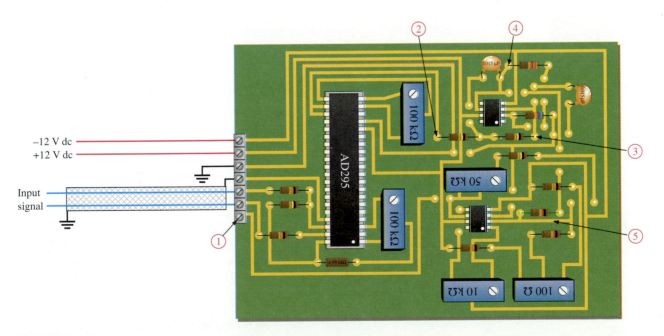

FIGURE 15–41
Amplifier board with designated test points.

Final Report

Submit a final written report on the amplifier board using an organized format that includes the following:

1. A physical description of the circuit.
2. A discussion of the operation of the circuit.
3. A list of the specifications.
4. A list of parts with part numbers if available.
5. A list of the types of problems on the four faulty circuit boards.
6. A description of how you determined the problem on each of the faulty circuit boards.

■ CHAPTER SUMMARY

- A basic instrumentation amplifier is formed by three op-amps and seven resistors, including the gain-setting resistor R_G.
- An instrumentation amplifier has high input impedance, high CMRR, low output offset, and low output impedance.
- The voltage gain of a basic instrumentation amplifier is set by a single external resistor.
- An instrumentation amplifier is useful in applications where small signals are embedded in large common-mode noise.
- A basic isolation amplifier has three electrically isolated sections: input, output, and power.
- Most isolation amplifiers use transformer coupling for isolation.
- Isolation amplifiers are used to interface sensitive equipment with high-voltage environments and to provide protection from electrical shock in certain medical applications.
- The operational transconductance amplifier (OTA) is a voltage-to-current amplifier.
- The output current of an OTA is the input voltage times the transconductance.
- In an OTA, transconductance varies with bias current; therefore, the gain of an OTA can be varied with a bias voltage.
- The operation of log and antilog amplifiers is based on the nonlinear (logarithmic) characteristics of a *pn* junction.
- A log amplifier has a BJT in the feedback loop.
- An antilog amplifier has a BJT in series with the input.
- Logarithmic amplifiers are used for analog multiplication and division.

■ GLOSSARY

Instrumentation   Related to an arrangement of instruments for the purpose of monitoring or measuring certain quantities.

Logarithm   An exponent; the logarithm of a quantity is the exponent or power to which a given number called the base must be raised in order to equal the quantity.

Natural logarithm   The exponent to which the base e ($e = 2.71828$) must be raised in order to equal a given quantity.

Signal compression   The process of scaling down the amplitude of a signal voltage.

Transconductance   In an electronic device, the ratio of the output current to the input voltage.

Trim   To precisely adjust or fine tune a value.

■ FORMULAS

Instrumentation Amplifier

$$(15\text{--}1) \qquad V_{out} = \left(1 + \frac{2R}{R_\text{G}}\right)(V_{in2} - V_{in1})$$

$$(15\text{--}2) \qquad A_{cl} = 1 + \frac{2R}{R_\text{G}}$$

$$(15\text{--}3) \qquad R_\text{G} = \frac{2R}{A_{cl} - 1}$$

$$(15\text{--}4) \qquad A_v = \frac{R_\text{S}}{R_\text{G}}$$

$$(15\text{--}5) \qquad C_x = \frac{1}{100\pi f_c}$$

Isolation Amplifier

$$(15\text{--}6) \qquad A_{v(input)} = 1 + \frac{R_f}{R_1}$$

$$(15\text{--}7) \qquad A_{v(output)} = \frac{75\ \text{k}\Omega + R_{ext}}{30\ \text{k}\Omega}$$

Operational Transconductance Amplifier (OTA)

$$(15\text{--}8) \qquad g_m = \frac{I_{out}}{V_{in}}$$

$$(15\text{--}9) \qquad g_m = KI_\text{BIAS}$$

$$(15\text{--}10) \qquad I_\text{BIAS} = \frac{(+V) - (-V) - 0.7\ \text{V}}{R_\text{BIAS}}$$

Logarithmic Amplifier

$$(15\text{--}11) \qquad V_{out} = -K\ln(V_{in})$$

$$(15\text{--}12) \qquad I_\text{F} \cong I_\text{R}e^{qV_\text{F}/kT}$$

$$(15\text{--}13) \qquad V_\text{F} = \left(\frac{kT}{q}\right)\ln\left(\frac{I_\text{F}}{I_\text{R}}\right)$$

$$(15\text{--}14) \qquad V_{out} = -\left(\frac{kT}{q}\right)\ln\left(\frac{V_{in}}{I_\text{R}R_1}\right)$$

$$(15\text{--}15) \qquad V_{out} \cong -(0.025\ \text{V})\ln\left(\frac{V_{in}}{I_\text{R}R_1}\right)$$

$$(15\text{--}16) \qquad I_\text{C} = I_\text{EBO}e^{qV_\text{BE}/kT}$$

$$(15\text{--}17) \qquad V_{out} = -(0.025\ \text{V})\ln\left(\frac{V_{in}}{I_\text{EBO}R_1}\right)$$

$$(15\text{--}18) \qquad V_{out} = -R_f I_\text{EBO}\text{antilog}\left(\frac{V_{in}}{25\ \text{mV}}\right)$$

Converters and Other Op-Amp Circuits

$$(15\text{--}19) \qquad I_L = \frac{V_\text{IN}}{R_i} \qquad\qquad \text{Constant-current source}$$

$$(15\text{--}20) \qquad V_{out} = I_i R_f \qquad\qquad \text{Current-to-voltage converter}$$

$$(15\text{--}21) \qquad I_L = \frac{V_{in}}{R_1} \qquad\qquad \text{Voltage-to-current converter}$$

■ **SELF-TEST**

1. To make a basic instrumentation amplifier, it takes
   (a) one op-amp with a certain feedback arrangement
   (b) two op-amps and seven resistors
   (c) three op-amps and seven capacitors
   (d) three op-amps and seven resistors

2. Typically, an instrumentation amplifier has an external resistor used for
   (a) establishing the input impedance   (b) setting the voltage gain
   (c) setting the current gain   (d) interfacing with an instrument

3. Instrumentation amplifiers are used primarily in
   (a) high-noise environments   (b) medical equipment
   (c) test instruments   (d) filter circuits

4. Isolation amplifiers are used primarily in
   (a) remote, isolated locations
   (b) systems that isolate a single signal from many different signals
   (c) applications where there are high voltages and sensitive equipment
   (d) applications where human safety is a concern
   (e) answers (c) and (d)

5. The three sections of a basic isolation amplifier are
   (a) amplifier, filter, and power   (b) input, output, and coupling
   (c) input, output, and power   (d) gain, attenuation, and offset

6. The sections of most isolation amplifiers are connected by
   (a) copper strips   (b) transformers
   (c) microwave links   (d) current loops

7. The characteristic that allows an isolation amplifier to amplify small signal voltages in the presence of much greater noise voltages is its
   (a) CMRR   (b) high gain
   (c) high input impedance   (d) magnetic coupling between input and output

8. The term *OTA* means
   (a) operational transistor amplifier
   (b) operational transformer amplifier
   (c) operational transconductance amplifier
   (d) output transducer amplifier

9. In an OTA, the transconductance is controlled by
   (a) the dc supply voltage   (b) the input signal voltage
   (c) the manufacturing process   (d) a bias current

10. The voltage gain of an OTA circuit is set by
    (a) a feedback resistor
    (b) the transconductance only
    (c) the transconductance and the load resistor
    (d) the bias current and supply voltage

11. An OTA is basically a
    (a) voltage-to-current amplifier   (b) current-to-voltage amplifier
    (c) current-to-current amplifier   (d) voltage-to-voltage amplifier

12. The operation of a logarithmic amplifier is based on
    (a) the nonlinear operation of an op-amp
    (b) the logarithmic characteristic of a *pn* junction
    (c) the reverse breakdown characteristic of a *pn* junction
    (d) the logarithmic charge and discharge of an *RC* circuit

13. If the input to a log amplifier is *x*, the output is proportional to
    (a) e^x (b) $\ln x$ (c) $\log_{10} x$
    (d) $2.3 \log_{10} x$ (e) answers (a) and (c)   (f) answers (b) and (d)

14. If the input to an antilog amplifier is x, the output is proportional to
(a) $e^{\ln x}$ (b) e^x (c) $\ln x$ (d) e^{-x}

15. The logarithm of the product of two numbers is equal to the
(a) sum of the two numbers
(b) sum of the logarithms of each of the numbers
(c) difference of the logarithms of each of the numbers
(d) ratio of the logarithms of the numbers

16. If you subtract $\ln y$ from $\ln x$, you get
(a) $\ln x/\ln y$ (b) $(\ln x)(\ln y)$ (c) $\ln(x/y)$ (d) $\ln(y/x)$

■ **BASIC PROBLEMS**

SECTION 15–1   Instrumentation Amplifiers

1. Determine the voltage gains of op-amps 1 and 2 for the instrumentation amplifier configuration in Figure 15–42.

2. Find the overall voltage gain of the instrumentation amplifier in Figure 15–42.

3. The following voltages are applied to the instrumentation amplifier in Figure 15–42: $V_{in1} = 5$ mV, $V_{in2} = 10$ mV, and $V_{cm} = 225$ mV. Determine the final output voltage.

4. What value of R_G must be used to change the gain of the instrumentation amplifier in Figure 15–42 to 1000?

5. What is the voltage gain of the AD521 instrumentation amplifier in Figure 15–43?

FIGURE 15–42

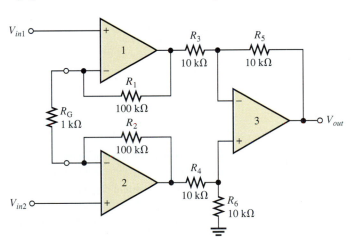

FIGURE 15–43

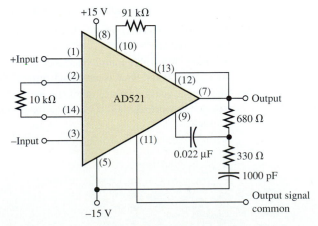

6. Determine the bandwidth of the amplifier in Figure 15–43.

7. Specify what you must do to change the gain of the amplifier in Figure 15–43 to approximately 50.

8. Specify what you must do to change the bandwidth of the amplifier in Figure 15–43 to approximately 15 kHz.

SECTION 15–2   Isolation Amplifiers

9. The op-amp in the input section of a certain isolation amplifier has a voltage gain of 30 and the attenuation circuit has an attenuation of 0.75. The output section is set for a gain of 10. What is the overall voltage gain of this device?

10. Determine the overall voltage gain of each AD295 in Figure 15–44.

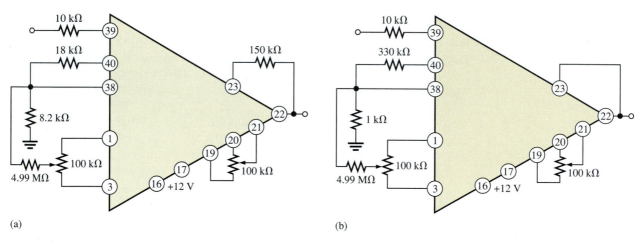

(a)   (b)

FIGURE 15–44

11. Specify how you would change the overall gain of the amplifier in Figure 15–44(a) to approximately 100 by changing only the gain of the input section.

12. Specify how you would change the overall gain in Figure 15–44(b) to approximately 100 by changing only the gain of the output section.

13. Specify how you would connect each amplifier in Figure 15–44 for unity gain.

SECTION 15–3   Operational Transconductance Amplifiers (OTAs)

14. A certain OTA has an input voltage of 10 mV and an output current of 10 μA. What is the transconductance?

15. A certain OTA with a transconductance of 5000 μS has a load resistance of 10 kΩ. If the input voltage is 100 mV, what is the output current? What is the output voltage?

16. The output voltage of a certain OTA with a load resistance is determined to be 3.5 V. If its transconductance is 4000 μS and the input voltage is 100 mV, what is the value of the load resistance?

17. Determine the voltage gain of the OTA in Figure 15–45. Assume $K = 16$ for the graph in Figure 15–46.

18. If a 10 kΩ rheostat is added in series with the bias resistor in Figure 15–45, what are the minimum and maximum voltage gains?

19. The OTA in Figure 15–47 functions as an amplitude modulation circuit. Determine the output voltage waveform for the given input waveforms assuming $K = 16$.

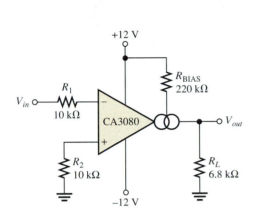

FIGURE 15–45

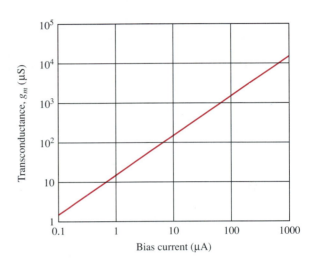

FIGURE 15–46

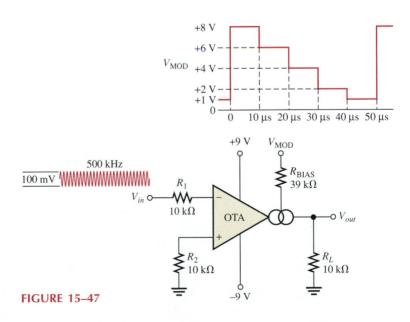

FIGURE 15–47

20. Determine the trigger points for the Schmitt-trigger circuit in Figure 15–48.

21. Determine the output voltage waveform for the Schmitt trigger in Figure 15–48 in relation to a 1 kHz sine wave with peak values of ±10 V.

FIGURE 15–48

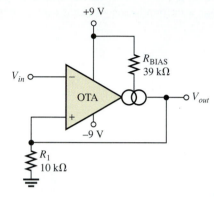

SECTION 15–4   Log and Antilog Amplifiers

22. Using your calculator, find the natural logarithm (ln) of each of the following numbers:

    (a) 0.5   **(b)** 2   **(c)** 50   **(d)** 130

23. Repeat Problem 22 for $\log_{10}$.

24. What is the antilog of 1.6?

25. Explain why the output of a log amplifier is limited to approximately 0.7 V.

26. What is the output voltage of a certain log amplifier with a diode in the feedback path when the input voltage is 3 V? The input resistor is 82 kΩ and the reverse leakage current is 100 nA.

27. Determine the output voltage for the amplifier in Figure 15–49. Assume $I_{EBO} = 60$ nA.

28. Determine the output voltage for the amplifier in Figure 15–50. Assume $I_{EBO} = 60$ nA.

29. Signal compression is one application of logarithmic amplifiers. Suppose an audio signal with a maximum voltage of 1 V and a minimum voltage of 100 mV is applied to the log amplifier in Figure 15–49. What will be the maximum and minimum output voltages? What conclusion can you draw from this result?

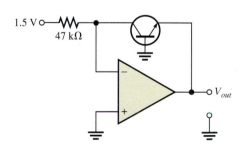

FIGURE 15–49

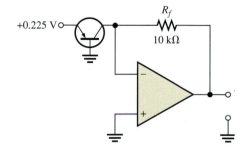

FIGURE 15–50

SECTION 15–5   Converters and Other Op-Amp Circuits

30. Determine the load current in each circuit of Figure 15–51.

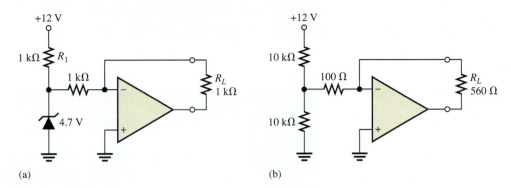

(a)   (b)

FIGURE 15–51

31. Devise a circuit for remotely sensing temperature and producing a proportional voltage that can then be converted to digital form for display. A thermistor can be used as the temperature-sensing element.

■ TROUBLE-SHOOTING PROBLEMS

SECTION 15–6   System Application

32. With a 1 mV, 50 Hz signal applied to the ECG amplifier board in Figure 15–52, what voltage would you expect to see at each of the test points? Assume that all offset voltages are nulled out and the position control is adjusted for zero deflection.

33. Repeat Problem 32 for a 2 mV, 1 kHz input signal.

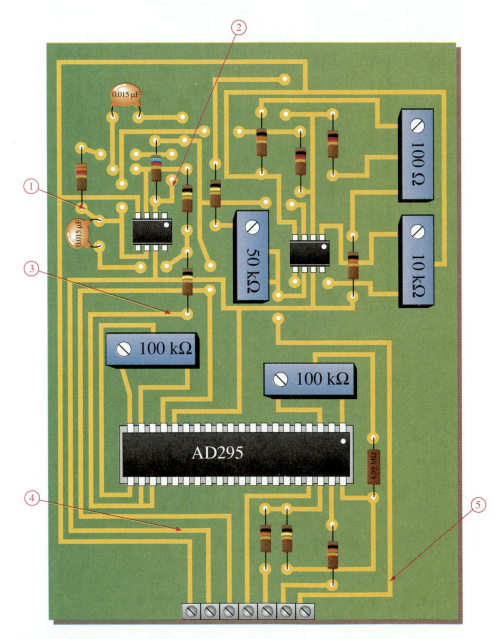

FIGURE 15–52

■ ANSWERS TO SECTION REVIEWS

Section 15–1

1. The main purpose of an instrumentation amplifier is to amplify small signals that occur on large common-mode voltages. The key characteristics are high input impedance, high CMRR, low output impedance, and low output offset.

2. Three op-amps and seven resistors including the gain resistor are required to construct a basic instrumentation amplifier.

3. The gain is set by the internal feedback resistors and an external resistor.

4. The gain is greater than unity.

Section 15–2

1. Isolation amplifiers are used in medical equipment, power plant instrumentation, industrial processing, and automated testing.

2. The three sections of an isolation amplifier are input, output, and power.

3. The sections are connected by transformer coupling and in some devices by optical coupling.

4. The oscillator acts as a dc to ac converter so that the dc power can be ac coupled to the input and output sections.

Section 15–3

1. OTA stands for Operational Transconductance Amplifier.

2. Transconductance increases with bias current.

3. Assuming that the bias input is connected to the supply voltage, the voltage gain increases when the supply voltage is increased because this increases the bias current.

4. The voltage gain decreases as the bias voltage decreases.

Section 15–4

1. A diode or transistor in the feedback loop provides the exponential (nonlinear) characteristic.

2. The output of a log amplifier is limited to the barrier potential of the *pn* junction (about 0.7 V).

3. The output voltage is determined by the input voltage, the input resistor, and the emitter-to-base leakage current.

4. The transistor in an antilog amplifier is in series with the input rather than in the feedback loop.

Section 15–5

1. $I_L = 6.8$ V/10 kΩ = 0.68 mA; same value to 5 kΩ load.

2. The feedback resistor is the constant of proportionality.

■ ANSWERS TO RELATED EXERCISES FOR EXAMPLES

15–1 240 Ω

15–2 Leave $R_S = 100$ kΩ and make $R_G = 2.2$ kΩ, $C_x = 0.00033$ μF.

15–3 Many combinations are possible. Here is one: Connect the input for unity gain (pin 38 to 40, no R_1 or R_F). Connect a 680 kΩ resistor from pin 22 to pin 23.

15–4 Approximately 62 μA

15–5 The gain will change to approximately 113.

15–6 The output is a square-wave modulated signal with a maximum amplitude of 3.75 V and a minimum amplitude of 1.91 V.

15–7 -0.167 V

15–8 -0.193 V

15–9 -4.39 V

16

ACTIVE FILTERS

■ CHAPTER OBJECTIVES

□ Describe the gain-versus-frequency responses of the basic filters
□ Describe the three basic filter response characteristics and other filter parameters
□ Analyze active low-pass filters
□ Analyze active high-pass filters
□ Analyze active band-pass filters
□ Analyze active band-stop filters
□ Discuss two methods for measuring frequency response

P ower supply filters were introduced in Chapter 2. In this chapter, we introduce active filters that are used for signal processing. Filters are circuits that are capable of passing signals with certain selected frequencies while rejecting signals with other frequencies. This property is called *selectivity.*

Active filters use transistors or op-amps combined with passive *RC, RL,* or *RLC* circuits. The active devices provide voltage gain, and the passive circuits provide frequency selectivity. In terms of general response, the four basic categories of active filters are low-pass, high-pass, band-pass, and band-stop. In this chapter, you will study active filters using op-amps and *RC* circuits.

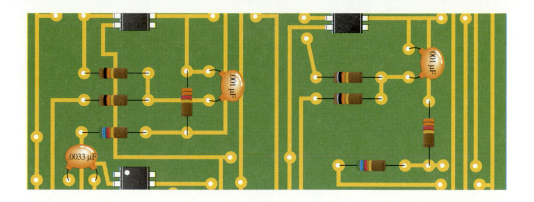

■ SYSTEM APPLICATION

The FM stereo receiver block diagram shown below accepts carrier signals in the frequency range from 88 MHz to 108 MHz. In frequency modulation, the frequency of the carrier is varied in proportion to the amplitude and frequency of the modulating audio signal. The actual FM stereo multiplex signal that is received is quite complex and beyond the scope of our coverage; therefore, the focus of the system application in Section 16–8 is limited to the filter circuits which are part of the channel separation circuits. You will see how the active filters covered in the chapter can be applied in the FM system to separate out the audio signals that go to the left and right channel speakers.

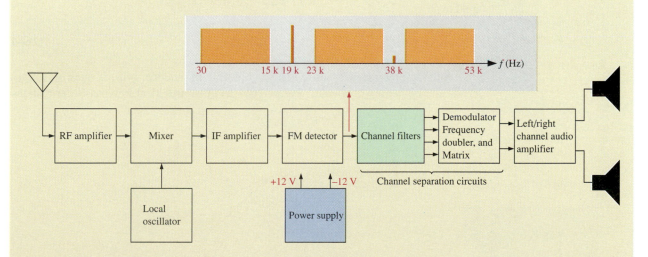

16–1 ■ BASIC FILTER RESPONSES

Filters are usually categorized by the manner in which the output voltage varies with the frequency of the input voltage. The categories of active filters are low-pass, high-pass, band-pass, and band-stop. We will examine each of these general responses in this section.

After completing this section, you should be able to

■ **Describe the gain-versus-frequency responses of the basic filters**
  □ Explain the low-pass response
  □ Determine the critical frequency and bandwidth of a low-pass filter
  □ Explain the high-pass response
  □ Determine the critical frequency of a high-pass filter
  □ Explain the band-pass response
  □ Define *quality factor* and explain its significance
  □ Determine the critical frequency, bandwidth, quality factor, and damping factor of a band-pass filter
  □ Explain the band-stop response

Low-Pass Filter Response

The pass band of the basic **low-pass filter** is defined to be from 0 Hz (dc) up to the critical (cutoff) frequency, f_c, at which the output voltage is 70.7 percent of the pass-band voltage, as indicated in Figure 16–1(a). The ideal pass band, shown by the shaded region within the dashed lines, has an instantaneous roll-off at f_c. The bandwidth of this filter is equal to f_c.

$$BW = f_c \tag{16–1}$$

Although the ideal response is not achievable in practice, roll-off rates of −20 dB/decade and higher are obtainable. Figure 16–1(b) illustrates ideal low-pass filter response

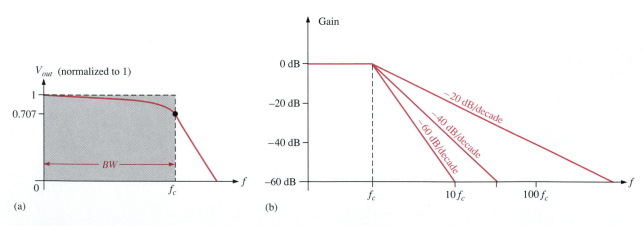

FIGURE 16–1
Low-pass filter responses.

curves with several roll-off rates. The −20 dB/decade rate is obtained with a single RC circuit consisting of one resistor and one capacitor. The higher roll-off rates require additional RC circuits. Each circuit is called a **pole**.

As explained in Chapter 10, the critical frequency of the RC low-pass filter occurs when $X_C = R$, where

$$f_c = \frac{1}{2\pi RC} \tag{16-2}$$

High-Pass Filter Response

A **high-pass filter** response is one that significantly attenuates all frequencies below f_c and passes all frequencies above f_c. The critical frequency is, of course, the frequency at which the output voltage is 70.7 percent of the pass-band voltage, as shown in Figure 16–2(a). The ideal response, shown by the shaded region within the dashed lines, has an instantaneous drop at f_c, which is not achievable. Roll-off rates of 20 dB/decade/pole are realizable. Figure 16–2(b) illustrates high-pass filter responses with several roll-off rates.

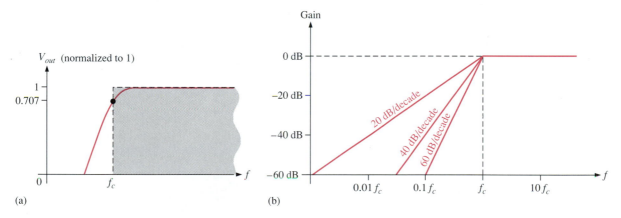

FIGURE 16–2

High-pass filter responses.

As with the RC low-pass filter, the high-pass critical frequency corresponds to the value where $X_C = R$ and is calculated with the following formula:

$$f_c = \frac{1}{2\pi RC}$$

The response of a high-pass filter extends from f_c up to a frequency that is determined by the limitations of the active element (transistor or op-amp) used.

Band-Pass Filter Response

A **band-pass filter** passes all signals lying within a band between a lower-frequency and an upper-frequency limit and essentially rejects all other frequencies that are outside this specified band. A generalized band-pass response curve is shown in Figure 16–3. The

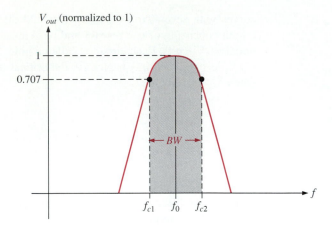

FIGURE 16–3
General band-pass response curve.

bandwidth (*BW*) is defined as the difference between the upper critical frequency (f_{c2}) and the lower critical frequency (f_{c1}).

$$BW = f_{c2} - f_{c1} \qquad \text{(16–3)}$$

The critical frequencies are, of course, the points at which the response curve is 70.7 percent of its maximum. Recall from Chapter 13 that these critical frequencies are also called *3 dB frequencies*. The frequency about which the pass band is centered is called the *center frequency*, f_0, defined as the geometric mean of the critical frequencies.

$$f_0 = \sqrt{f_{c1} f_{c2}} \qquad \text{(16–4)}$$

Quality Factor   The **quality factor** (*Q*) of a band-pass filter is the ratio of the center frequency to the bandwidth.

$$Q = \frac{f_0}{BW} \qquad \text{(16–5)}$$

The value of *Q* is an indication of the selectivity of a band-pass filter. The higher the value of *Q*, the narrower the bandwidth and the better the selectivity for a given value of f_0. Band-pass filters are sometimes classified as narrow-band ($Q > 10$) or wide-band ($Q < 10$). The quality factor (*Q*) can also be expressed in terms of the damping factor (*DF*) of the filter as

$$Q = \frac{1}{DF} \qquad \text{(16–6)}$$

You will study the damping factor in Section 16–2.

EXAMPLE 16–1   A certain band-pass filter has a center frequency of 15 kHz and a bandwidth of 1 kHz. Determine Q and classify the filter as narrow-band or wide-band.

Solution
$$Q = \frac{f_0}{BW} = \frac{15 \text{ kHz}}{1 \text{ kHz}} = 15$$

Because $Q > 10$, this is a narrow-band filter.

Related Exercise   If the quality factor of the filter is doubled, what will the bandwidth be?

Band-Stop Filter Response

Another category of active filter is the **band-stop filter,** also known as *notch, band-reject,* or *band-elimination* filter. You can think of the operation as opposite to that of the band-pass filter because frequencies within a certain bandwidth are rejected, and frequencies outside the bandwidth are passed. A general response curve for a band-stop filter is shown in Figure 16–4. Notice that the bandwidth is the band of frequencies between the 3 dB points, just as in the case of the band-pass filter response.

FIGURE 16–4
General band-stop filter response.

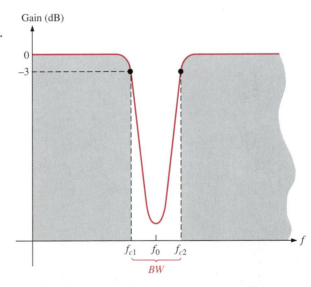

SECTION 16–1 REVIEW

1. What determines the bandwidth of a low-pass filter?
2. What limits the bandwidth of an active high-pass filter?
3. How are the Q and the bandwidth of a band-pass filter related? Explain how the selectivity is affected by the Q of a filter.

16–2 ■ FILTER RESPONSE CHARACTERISTICS

Each type of filter response (low-pass, high-pass, band-pass, or band-stop) can be tailored by circuit component values to have either a Butterworth, Chebyshev, or Bessel characteristic. Each of these characteristics is identified by the shape of the response curve, and each has an advantage in certain applications.

After completing this section, you should be able to

■ **Describe the three basic filter response characteristics and other filter parameters**
  ☐ Describe the Butterworth characteristic
  ☐ Describe the Chebyshev characteristic
  ☐ Describe the Bessel characteristic
  ☐ Define *damping factor* and discuss its significance
  ☐ Calculate the damping factor of a filter
  ☐ Define *pole*
  ☐ Discuss the order of a filter and its affect on the roll-off rate

Butterworth, Chebyshev, or Bessel response characteristics can be realized with most active filter circuit configurations by proper selection of certain component values. A general comparison of the three response characteristics for a low-pass filter response curve is shown in Figure 16–5. High-pass and band-pass filters can also be designed to have any one of the characteristics.

FIGURE 16–5

Comparative plots of three types of filter response characteristics.

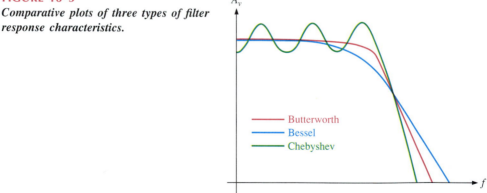

The Butterworth Characteristic   The **Butterworth** characteristic provides a very flat amplitude response in the pass band and a roll-off rate of 20 dB/decade/pole. The phase response is not linear, however, and the phase shift (thus, time delay) of signals passing through the filter varies nonlinearly with frequency. Therefore, a pulse applied to a filter with a Butterworth response will cause overshoots on the output, because each frequency component of the pulse's rising and falling edges experiences a different time delay. Filters with the Butterworth response are normally used when all frequencies in the pass band must have the same gain. The Butterworth response is often referred to as a maximally flat response.

The Chebyshev Characteristic   Filters with the **Chebyshev** response characteristic are useful when a rapid roll-off is required because it provides a roll-off rate greater than 20 dB/decade/pole. This is a greater rate than that of the Butterworth, so filters can be implemented with the Chebyshev response with fewer poles and less complex circuitry for a given roll-off rate. This type of filter response is characterized by overshoot or ripples in the pass band (depending on the number of poles) and an even less linear phase response than the Butterworth.

The Bessel Characteristic   The **Bessel** response exhibits a linear phase characteristic, meaning that the phase shift increases linearly with frequency. The result is almost no overshoot on the output with a pulse input. For this reason, filters with the Bessel response are used for filtering pulse waveforms without distorting the shape of the waveform.

The Damping Factor

As mentioned, an active filter can be designed to have either a Butterworth, Chebyshev, or Bessel response characteristic regardless of whether it is a low-pass, high-pass, band-pass, or band-stop type. The **damping factor** (*DF*) of an active filter circuit determines which response characteristic the filter exhibits. To explain the basic concept, a generalized active filter is shown in Figure 16–6. It includes an amplifier, a negative feedback circuit, and a filter section. The amplifier and feedback are connected in a noninverting configuration. The damping factor is determined by the negative feedback circuit and is defined by the following equation:

$$DF = 2 - \frac{R_1}{R_2}$$

(16–7)

FIGURE 16–6
General diagram of an active filter.

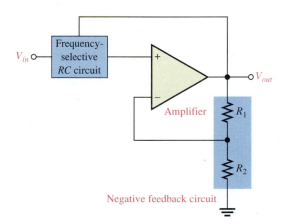

Basically, the damping factor affects the filter response by negative feedback action. Any attempted increase or decrease in the output voltage is offset by the opposing effect of the negative feedback. This tends to make the response curve flat in the pass band of the filter if the value for the damping factor is precisely set. By advanced mathematics, which we will not cover, values for the damping factor have been derived for various orders of filters to achieve the maximally flat response of the Butterworth characteristic.

The value of the damping factor required to produce a desired response characteristic depends on the *order* (number of poles) of the filter. A **pole,** for our purposes, is simply a circuit with one resistor and one capacitor. The more poles a filter has, the faster its roll-off rate is. To achieve a second-order Butterworth response, for example, the damping factor must be 1.414. To implement this damping factor, the feedback resistor ratio must be

$$\frac{R_1}{R_2} = 2 - DF = 2 - 1.414 = 0.586$$

This ratio gives the closed-loop gain of the noninverting filter amplifier, $A_{cl(NI)}$, a value of 1.586, derived as follows:

$$A_{cl(NI)} = \frac{1}{B} = \frac{R_1 + R_2}{R_2} = \frac{R_1}{R_2} + 1 = 0.586 + 1 = 1.586$$

EXAMPLE 16–2

If resistor R_2 in the feedback circuit of an active two-pole filter of the type in Figure 16–6 is 10 kΩ, what value must R_1 be to obtain a maximally flat Butterworth response?

Solution

$$\frac{R_1}{R_2} = 0.586$$

$$R_1 = 0.586R_2 = 0.586(10 \text{ k}\Omega) = 5.86 \text{ k}\Omega$$

Using the nearest standard 5 percent value of 5.6 kΩ will get very close to the ideal Butterworth response.

Related Exercise   What is the damping factor for $R_2 = 10$ kΩ and $R_1 = 5.6$ kΩ?

Critical Frequency and Roll-Off Rate

The critical frequency is determined by the values of the resistor and capacitors in the frequency-selective *RC* circuit, as shown in Figure 16–6. For a single-pole (first-order) filter, as shown in Figure 16–7, the critical frequency is

$$f_c = \frac{1}{2\pi RC}$$

Although we show a low-pass configuration, the same formula is used for the f_c of a single-pole high-pass filter. The number of poles determines the roll-off rate of the filter. A Butterworth response produces 20 dB/decade/pole. So, a first-order (one-pole) filter has a roll-off of 20 dB/decade; a second-order (two-pole) filter has a roll-off rate of 40 dB/decade; a third-order (three-pole) filter has a roll-off rate of 60 dB/decade; and so on.

Generally, to obtain a filter with three poles or more, one-pole or two-pole filters are cascaded, as shown in Figure 16–8. To obtain a third-order filter, for example, cascade a second-order and a first-order filter; to obtain a fourth-order filter, cascade two second-order filters; and so on. Each filter in a cascaded arrangement is called a *stage* or *section.*

Because of its maximally flat response, the Butterworth characteristic is the most widely used. Therefore, we will limit our coverage to the Butterworth response to illustrate basic filter concepts. Table 16–1 lists the roll-off rates, damping factors, and R_1/R_2 ratios for up to sixth-order Butterworth filters.

FIGURE 16–7

First-order (one-pole) low-pass filter.

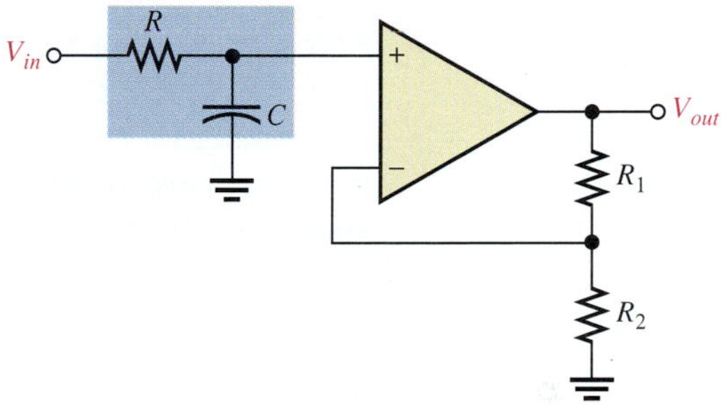

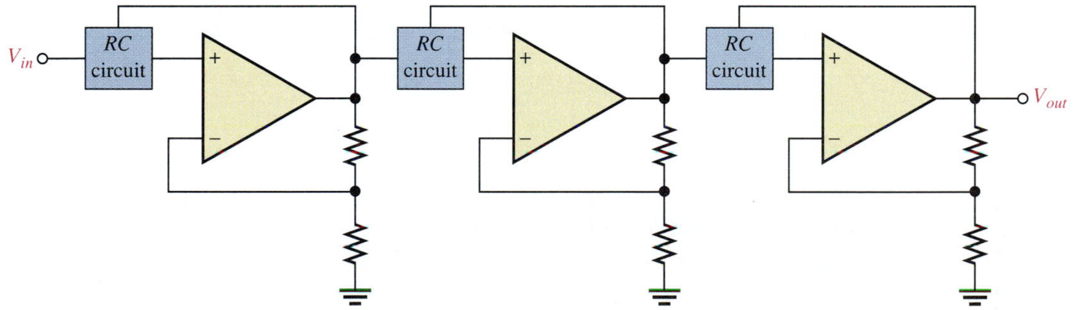

FIGURE 16–8

The number of filter poles can be increased by cascading.

TABLE 16–1

Values for the Butterworth response

Order	Roll-off dB/decade	1st stage			2nd stage			3rd stage		
		Poles	DF	R_1/R_2	Poles	DF	R_1/R_2	Poles	DF	R_1/R_2
1	20	1	Optional							
2	40	2	1.414	0.586						
3	60	2	1.00	1	1	1.00	1			
4	80	2	1.848	0.152	2	0.765	1.235			
5	100	2	1.00	1	2	1.618	0.382	1	1.618	1.382
6	120	2	1.932	0.068	2	1.414	0.586	2	0.518	1.482

16–3 ■ ACTIVE LOW-PASS FILTERS

Filters that use op-amps as the active element provide several advantages over passive filters (R, L, and C elements only). The op-amp provides gain, so that the signal is not attenuated as it passes through the filter. The high input impedance of the op-amp prevents excessive loading of the driving source, and the low output impedance of the op-amp prevents the filter from being affected by the load that it is driving. Active filters are also easy to adjust over a wide frequency range without altering the desired response.

After completing this section, you should be able to

■ **Analyze active low-pass filters**
  □ Identify a single-pole filter and determine its gain and critical frequency
  □ Identify a two-pole Sallen-Key filter and determine its gain and critical frequency
  □ Explain how a higher roll-off rate is achieved by cascading low-pass filters

A Single-Pole Filter

Figure 16–9(a) shows an active filter with a single low-pass *RC* frequency-selective circuit that provides a roll-off of −20 dB/decade above the critical frequency, as indicated by the response curve in Figure 16–9(b). The critical frequency of the single-pole filter is $f_c =$

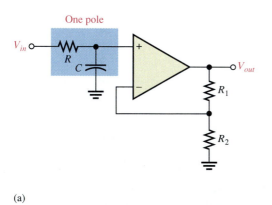

(a)

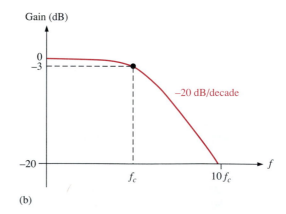

(b)

FIGURE 16–9

Single-pole active low-pass filter and response curve.

$1/2\pi RC$. The op-amp in this filter is connected as a noninverting amplifier with the closed-loop voltage gain in the pass band set by the values of R_1 and R_2.

$$A_{cl} = \frac{R_1}{R_2} + 1 \qquad \textbf{(16–8)}$$

The Sallen-Key Low-Pass Filter

The Sallen-Key is one of the most common configurations for a second-order (two-pole) filter. It is also known as a VCVS (voltage-controlled voltage source) filter. A low-pass version of the Sallen-Key filter is shown in Figure 16–10. Notice that there are two low-pass RC circuits that provide a roll-off of −40 dB/decade above the critical frequency (assuming a Butterworth characteristic). One RC circuit consists of R_A and C_A, and the second circuit consists of R_B and C_B. A unique feature is the capacitor C_A that provides feedback for shaping the response near the edge of the pass band. The critical frequency for the second-order Sallen-Key filter is

$$f_c = \frac{1}{2\pi\sqrt{R_A R_B C_A C_B}} \qquad \textbf{(16–9)}$$

For simplicity, the component values can be made equal so that $R_A = R_B = R$ and $C_A = C_B = C$. In this case, the expression for the critical frequency simplifies to

$$f_c = \frac{1}{2\pi RC}$$

FIGURE 16–10

Basic Sallen-Key second-order low-pass filter.

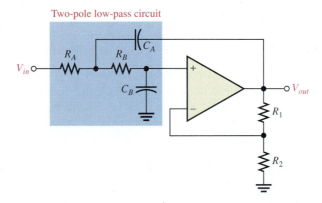

Two-pole low-pass circuit

As in the single-pole filter, the op-amp in the second-order Sallen-Key filter acts as a noninverting amplifier with the negative feedback provided by the R_1/R_2 circuit. As you have learned, the damping factor is set by the values of R_1 and R_2, thus making the filter response either Butterworth, Chebyshev, or Bessel. For example, from Table 16–1, the R_1/R_2 ratio must be 0.586 to produce the damping factor of 1.414 required for a second-order Butterworth response.

EXAMPLE 16–3   Determine the critical frequency of the low-pass filter in Figure 16–11, and set the value of R_1 for an approximate Butterworth response.

FIGURE 16–11

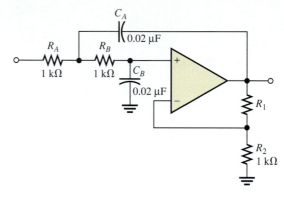

Solution   Since $R_A = R_B = 1$ kΩ and $C_A = C_B = 0.02$ μF,

$$f_c = \frac{1}{2\pi RC} = \frac{1}{2\pi(1 \text{ k}\Omega)(0.02 \text{ μF})} = 7.96 \text{ kHz}$$

For a Butterworth response, $R_1/R_2 = 0.586$.

$$R_1 = 0.586R_2 = 0.586(1 \text{ k}\Omega) = 586 \text{ }\Omega$$

Select a standard value as near as possible to this calculated value.

Related Exercise   Determine f_c for Figure 16–11 if $R_A = R_B = R_2 = 2.2$ kΩ and $C_A = C_B = 0.01$ μF. Also determine the value of R_1 for a Butterworth response.

Cascaded Low-Pass Filters Achieve a Higher Roll-Off Rate

A three-pole filter is required to get a third-order low-pass response (−60 dB/decade). This is done by cascading a two-pole low-pass filter and a single-pole low-pass filter, as shown in Figure 16–12(a). Figure 16–12(b) shows a four-pole configuration obtained by cascading two two-pole filters.

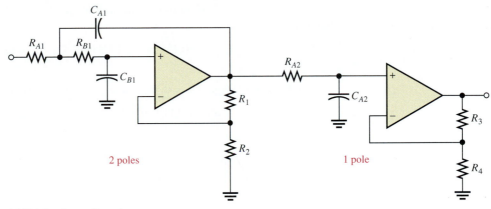

(a) Third-order configuration

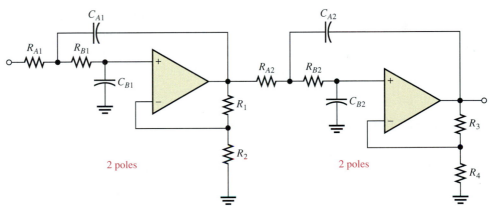

(b) Fourth-order configuration

FIGURE 16–12
Cascaded low-pass filters.

EXAMPLE 16–4

For the four-pole filter in Figure 16–12(b), determine the capacitance values required to produce a critical frequency of 2680 Hz if all the resistors are 1.8 kΩ. Also select values for the feedback resistors to get a Butterworth response.

Solution   Both stages must have the same f_c. Assuming equal-value capacitors,

$$f_c = \frac{1}{2\pi RC}$$

$$C = \frac{1}{2\pi R f_c} = \frac{1}{2\pi (1.8 \text{ k}\Omega)(2680 \text{ Hz})} = 0.033 \text{ μF}$$

$$C_{A1} = C_{B1} = C_{A2} = C_{B2} = 0.033 \text{ μF}$$

Select $R_2 = R_4 = 1.8$ kΩ for simplicity. Refer to Table 16–1.

For a Butterworth response in the first stage, $DF = 1.848$ and $R_1/R_2 = 0.152$. Therefore,

$$R_1 = 0.152R_2 = 0.152(1800 \ \Omega) = 274 \ \Omega$$

Choose $R_1 = 270 \ \Omega$.

In the second stage, $DF = 0.765$ and $R_3/R_4 = 1.235$. Therefore,

$$R_3 = 1.235R_4 = 1.235(1800 \ \Omega) = 2.22 \ k\Omega$$

Choose $R_3 = 2.2 \ k\Omega$.

Related Exercise For the filter in Figure 16–12(b), determine the capacitance values for $f_c = 1$ kHz if all the filter resistors are 680 Ω. Also specify the values for the feedback resistors to produce a Butterworth response.

SECTION 16–3 REVIEW

1. How many poles does a second-order low-pass filter have? How many resistors and how many capacitors are used in the frequency-selective circuit?

2. Why is the damping factor of a filter important?

3. What is the primary purpose of cascading low-pass filters?

16–4 ■ ACTIVE HIGH-PASS FILTERS

In high-pass filters, the roles of the capacitor and resistor are reversed in the RC circuits. Otherwise, the basic considerations are the same as for the low-pass filters.

After completing this section, you should be able to

■ **Analyze active high-pass filters**
  □ Identify a single-pole filter and determine its gain and critical frequency
  □ Identify a two-pole Sallen-Key filter and determine its gain and critical frequency
  □ Explain how a higher roll-off rate is achieved by cascading high-pass filters

A Single-Pole Filter

A high-pass active filter with a 20 dB/decade roll-off is shown in Figure 16–13(a). Notice that the input circuit is a single high-pass *RC* circuit. The negative feedback circuit is the same as for the low-pass filters previously discussed. The high-pass response curve is shown in Figure 16–13(b).

Ideally, a high-pass filter passes all frequencies above f_c without limit, as indicated in Figure 16–14(a), although in practice, this is not the case. As you have learned, all op-amps inherently have internal *RC* circuits that limit the amplifier's response at high frequencies. Therefore, there is an upper-frequency limit on the high-pass filter's response which, in effect, makes it a band-pass filter with a very wide bandwidth. In the majority of applications, the internal high-frequency limitation is so much greater than that of the filter's critical frequency that the limitation can be neglected. In some applications, discrete transistors are used for the gain element to increase the high-frequency limitation beyond that realizable with available op-amps.

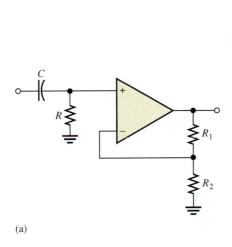

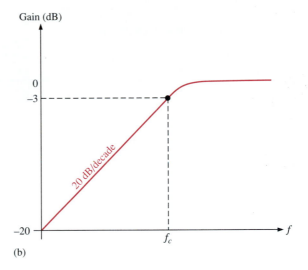

(a)

(b)

FIGURE 16–13

Single-pole active high-pass filter and response curve.

FIGURE 16–14

High-pass filter response.

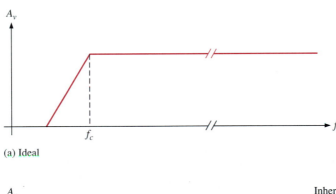

(a) Ideal

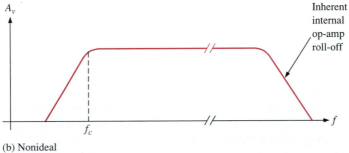

(b) Nonideal

The Sallen-Key High-Pass Filter

A high-pass second-order Sallen-Key configuration is shown in Figure 16–15. The components R_A, C_A, R_B, and C_B form the two-pole frequency-selective circuit. Notice that the positions of the resistors and capacitors in the frequency-selective circuit are opposite to those in the low-pass configuration. As with the other filters, the response characteristic can be optimized by proper selection of the feedback resistors, R_1 and R_2.

FIGURE 16–15

Basic Sallen-Key second-order high-pass filter.

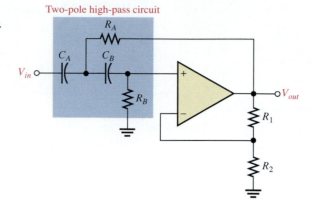

EXAMPLE 16–5

Choose values for the Sallen-Key high-pass filter in Figure 16–15 to implement an equal-value second-order Butterworth response with a critical frequency of approximately 10 kHz.

Solution Start by selecting a value for R_A and R_B (R_1 or R_2 can also be the same value as R_A and R_B for simplicity).

$$R = R_A = R_B = R_2 = 3.3 \text{ k}\Omega \qquad \text{(an arbitrary selection)}$$

Next, calculate the capacitance value from $f_c = 1/2\pi RC$.

$$C = C_A = C_B = \frac{1}{2\pi R f_c} = \frac{1}{2\pi (3.3 \text{ k}\Omega)(10 \text{ kHz})} = 0.0048 \ \mu\text{F}$$

For a Butterworth response, the damping factor must be 1.414 and $R_1/R_2 = 0.586$.

$$R_1 = 0.586 R_2 = 0.586(3.3 \text{ k}\Omega) = 1.93 \text{ k}\Omega$$

If you had let $R_1 = 3.3 \text{ k}\Omega$, then

$$R_2 = \frac{R_1}{0.586} = \frac{3.3 \text{ k}\Omega}{0.586} = 5.63 \text{ k}\Omega$$

Either way, an approximate Butterworth response is realized by choosing the nearest standard value.

Related Exercise Select values for all the components in the high-pass filter of Figure 16–15 to obtain an $f_c = 300$ Hz. Use equal-value components with $R = 10 \text{ k}\Omega$ and optimize for a Butterworth response.

Cascading High-Pass Filters

As with the low-pass configuration, first- and second-order high-pass filters can be cascaded to provide three or more poles and thereby create faster roll-off rates. Figure 16–16 shows a six-pole high-pass filter consisting of three two-pole stages. With this configuration optimized for a Butterworth response, a roll-off of 120 dB/decade is achieved.

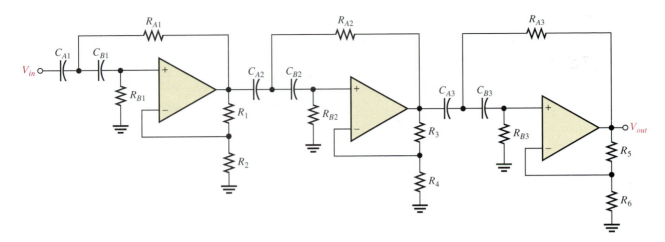

FIGURE 16–16
Sixth-order high-pass filter.

16–5 ■ ACTIVE BAND-PASS FILTERS

As mentioned, band-pass filters pass all frequencies bounded by a lower-frequency limit and an upper-frequency limit and reject all others lying outside this specified band. A band-pass response can be thought of as the overlapping of a low-frequency response curve and a high-frequency response curve.

After completing this section, you should be able to

■ **Analyze active band-pass filters**
  ☐ Describe a band-pass filter composed of a low-pass and a high-pass filter
  ☐ Determine the critical frequencies and center frequency of a cascaded band-pass filter
  ☐ Analyze a multiple-feedback band-pass filter to determine center frequency, bandwidth, and gain
  ☐ Analyze a state-variable band-pass filter

Cascaded Low-Pass and High-Pass Filters Achieve a Band-Pass Response

One way to implement a band-pass filter is a cascaded arrangement of a high-pass filter and a low-pass filter, as shown in Figure 16–17(a), as long as the critical frequencies are sufficiently separated. Each of the filters shown is a two-pole Sallen-Key Butterworth

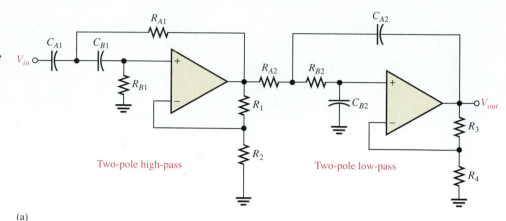

(a)

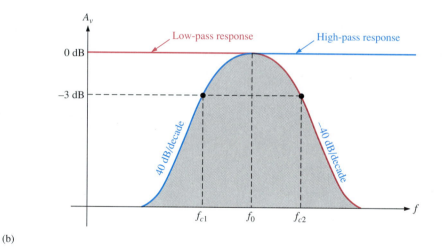

(b)

configuration so that the roll-off rates are ±40 dB/decade, indicated in the composite response curve of Figure 16–17(b). The critical frequency of each filter is chosen so that the response curves overlap sufficiently, as indicated. The critical frequency of the high-pass filter must be sufficiently lower than that of the low-pass stage.

The lower frequency f_{c1} of the pass band is the critical frequency of the high-pass filter. The upper frequency f_{c2} is the critical frequency of the low-pass filter. Ideally, as discussed earlier, the center frequency f_0 of the pass band is the geometric mean of f_{c1} and f_{c2}. The following formulas express the three frequencies of the band-pass filter in Figure 16–17.

$$f_{c1} = \frac{1}{2\pi\sqrt{R_{A1}R_{B1}C_{A1}C_{B1}}}$$

$$f_{c2} = \frac{1}{2\pi\sqrt{R_{A2}R_{B2}C_{A2}C_{B2}}}$$

$$f_0 = \sqrt{f_{c1}f_{c2}}$$

Of course, if equal-value components are used in implementing each filter, the critical frequency equations simplify to the form $f_c = 1/2\pi RC$.

Multiple-Feedback Band-Pass Filter

Another type of filter configuration, shown in Figure 16–18, is a multiple-feedback band-pass filter. The two feedback paths are through R_2 and C_1. Components R_1 and C_1 provide the low-pass response, and R_2 and C_2 provide the high-pass response. The maximum gain, A_0, occurs at the center frequency. Q values of less than 10 are typical in this type of filter. An expression for the center frequency is developed as follows, recognizing that R_1 and R_3 appear in parallel as viewed from the C_1 feedback path (with the V_{in} source replaced by a short).

$$f_0 = \frac{1}{2\pi\sqrt{(R_1 \parallel R_3)R_2C_1C_2}}$$

Making $C_1 = C_2 = C$ yields

$$f_0 = \frac{1}{2\pi\sqrt{(R_1 \parallel R_3)R_2C^2}} = \frac{1}{2\pi C\sqrt{(R_1 \parallel R_3)R_2}}$$

$$= \frac{1}{2\pi C}\sqrt{\frac{1}{R_2(R_1 \parallel R_3)}} = \frac{1}{2\pi C}\sqrt{\left(\frac{1}{R_2}\right)\left(\frac{1}{R_1R_3/(R_1 + R_3)}\right)}$$

$$f_0 = \frac{1}{2\pi C}\sqrt{\frac{R_1 + R_3}{R_1R_2R_3}} \qquad\qquad \textbf{(16–10)}$$

FIGURE 16–18
Multiple-feedback band-pass filter.

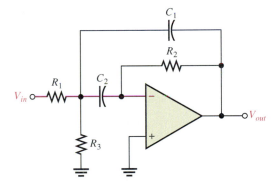

A convenient value for the capacitors is chosen; then the three resistor values are calculated based on the desired values for f_0, BW, and A_0. As you know, the Q can be determined from the relation $Q = f_0/BW$. The resistor values can be found using the following formulas (stated without derivation):

$$R_1 = \frac{Q}{2\pi f_0 C A_0}$$

$$R_2 = \frac{Q}{\pi f_0 C}$$

$$R_3 = \frac{Q}{2\pi f_0 C(2Q^2 - A_0)}$$

To develop a gain expression, solve for Q in the R_1 and R_2 formulas as follows:

$$Q = 2\pi f_0 A_0 C R_1$$
$$Q = \pi f_0 C R_2$$

Then,

$$2\pi f_0 A_0 C R_1 = \pi f_0 C R_2$$

Cancelling, you get

$$2A_0 R_1 = R_2$$

$$A_0 = \frac{R_2}{2R_1} \qquad\qquad (16\text{--}11)$$

In order for the denominator of the equation $R_3 = Q/2\pi f_0 C(2Q^2 - A_0)$ to be positive, $A_0 < 2Q^2$, which imposes a limitation on the gain.

EXAMPLE 16–6

Determine the center frequency, maximum gain, and bandwidth for the filter in Figure 16–19.

FIGURE 16–19

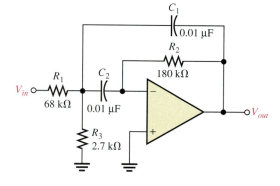

Solution

$$f_0 = \frac{1}{2\pi C}\sqrt{\frac{R_1 + R_3}{R_1 R_2 R_3}} = \frac{1}{2\pi(0.01\ \mu F)}\sqrt{\frac{68\ k\Omega + 2.7\ k\Omega}{(68\ k\Omega)(180\ k\Omega)(2.7\ k\Omega)}} = 736\ Hz$$

$$A_0 = \frac{R_2}{2R_1} = \frac{180\ k\Omega}{2(68\ k\Omega)} = 1.32$$

$$Q = \pi f_0 C R_2 = \pi(736\ Hz)(0.01\ \mu F)(180\ k\Omega) = 4.16$$

$$BW = \frac{f_0}{Q} = \frac{736\ Hz}{4.16} = 177\ Hz$$

Related Exercise If R_2 in Figure 16–19 is increased to 330 kΩ, determine the gain, center frequency, and bandwidth of the filter?

State-Variable Band-Pass Filter

The state-variable or universal active filter is widely used for band-pass applications. As shown in Figure 16–20, it consists of a summing amplifier and two op-amp integrators (which act as single-pole low-pass filters) that are combined in a cascaded arrangement to form a second-order filter. Although used primarily as a band-pass (BP) filter, the state-variable configuration also provides low-pass (LP) and high-pass (HP) outputs. The center frequency is set by the RC circuits in both integrators. When used as a band-pass filter, the critical frequencies of the integrators are usually made equal, thus setting the center frequency of the pass band.

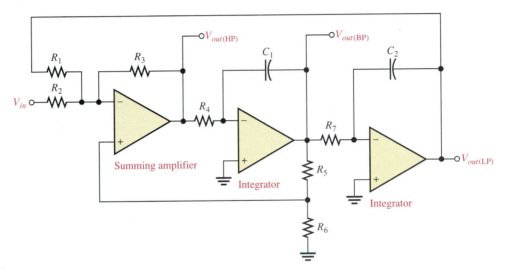

FIGURE 16–20
State-variable band-pass filter.

Basic Operation   At input frequencies below f_c, the input signal passes through the summing amplifier and integrators and is fed back 180° out of phase. Thus, the feedback signal and input signal cancel for all frequencies below approximately f_c. As the low-pass response of the integrators rolls off, the feedback signal diminishes, thus allowing the input to pass through to the band-pass output. Above f_c, the low-pass response disappears, thus preventing the input signal from passing through the integrators. As a result, the band-pass output peaks sharply at f_c, as indicated in Figure 16–21. Stable Qs up to 100 can be obtained with this type of filter. The Q is set by the feedback resistors R_5 and R_6 according to the following equation.

$$Q = \frac{1}{3}\left(\frac{R_5}{R_6} + 1\right)$$

(16–12)

The state-variable filter cannot be optimized for low-pass, high-pass, and band-pass performance simultaneously for this reason: To optimize for a low-pass or a high-pass Butterworth response, DF must equal 1.414. Since $Q = 1/DF$, a Q of 0.707 will result. Such a low Q provides a very poor band-pass response (large BW and poor selectivity). For optimization as a band-pass filter, the Q must be set high.

FIGURE 16–21

General state-variable response curves.

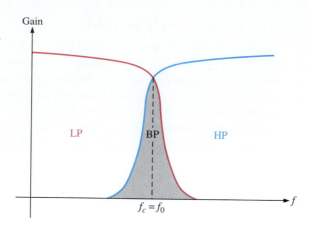

EXAMPLE 16–7   Determine the center frequency, Q, and BW for the band-pass output of the state-variable filter in Figure 16–22.

FIGURE 16–22

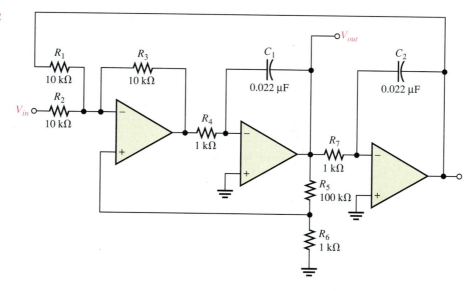

Solution   For each integrator,

$$f_c = \frac{1}{2\pi R_4 C_1} = \frac{1}{2\pi R_7 C_2} = \frac{1}{2\pi(1\ \text{k}\Omega)(0.022\ \mu\text{F})} = 7.23\ \text{kHz}$$

The center frequency is approximately equal to the critical frequencies of the integrators.

$$f_0 = f_c = 7.23\ \text{kHz}$$

$$Q = \frac{1}{3}\left(\frac{R_5}{R_6} + 1\right) = \frac{1}{3}\left(\frac{100\ \text{k}\Omega}{1\ \text{k}\Omega} + 1\right) = 33.7$$

$$BW = \frac{f_0}{Q} = \frac{7.23\ \text{kHz}}{33.7} = 215\ \text{Hz}$$

Related Exercise Determine f_0, Q, and BW for the filter in Figure 16–22 if $R_4 = R_6 = R_7 = 330\ \Omega$ with all other component values the same as shown on the schematic.

SECTION 16–5 REVIEW

1. What determines selectivity in a band-pass filter?
2. One filter has a $Q = 5$ and another has a $Q = 25$. Which has the narrower bandwidth?
3. List the elements that make up a state-variable filter.

16–6 ■ ACTIVE BAND-STOP FILTERS

Band-stop filters reject a specified band of frequencies and pass all others. The response is opposite to that of a band-pass filter.

After completing this section, you should be able to

■ **Analyze active band-stop filters**
  ☐ Identify a multiple-feedback band-stop filter
  ☐ Analyze a state-variable band-stop filter

Multiple-Feedback Band-Stop Filter

Figure 16–23 shows a multiple-feedback band-stop filter. Notice that this configuration is similar to the band-pass version in Figure 16–18 except that R_3 has been moved and R_4 has been added.

FIGURE 16–23
Multiple-feedback band-stop filter.

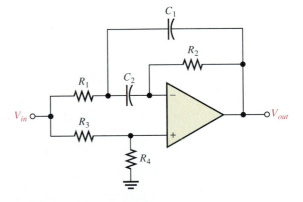

State-Variable Band-Stop Filter

Summing the low-pass and the high-pass responses of the state-variable filter covered in Section 16–5 creates a band-stop response as shown in Figure 16–24. One important application of this filter is minimizing the 60 Hz "hum" in audio systems by setting the center frequency to 60 Hz.

FIGURE 16–24

State-variable band-stop filter.

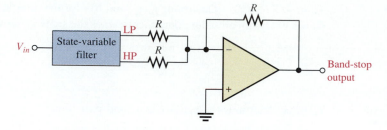

EXAMPLE 16–8

Verify that the band-stop filter in Figure 16–25 has a center frequency of 60 Hz, and optimize it for a Q of 30.

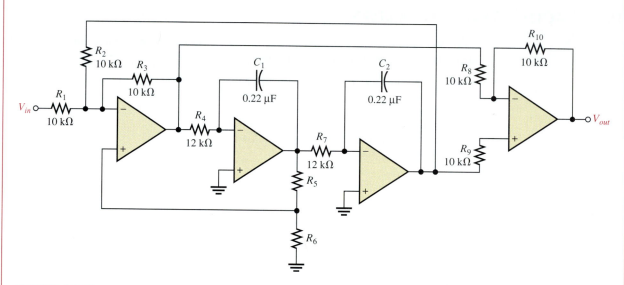

FIGURE 16–25

Solution f_0 equals the f_c of the integrator stages.

$$f_0 = \frac{1}{2\pi R_4 C_1} = \frac{1}{2\pi R_7 C_2} = \frac{1}{2\pi(12 \text{ k}\Omega)(0.22 \text{ }\mu\text{F})} = 60 \text{ Hz}$$

You can obtain a $Q = 30$ by choosing R_6 and then calculating R_5.

$$Q = \frac{1}{3}\left(\frac{R_5}{R_6} + 1\right)$$
$$R_5 = (3Q - 1)R_6$$

Choose $R_6 = 1 \text{ k}\Omega$. Then

$$R_5 = [3(30) - 1]1 \text{ k}\Omega = 89 \text{ k}\Omega$$

Related Exercise How would you change the center frequency to 120 Hz in Figure 16–25?

16–7 ▪ FILTER RESPONSE MEASUREMENTS

In this section, we discuss two methods of determining a filter's response by measurement—discrete point measurement and swept frequency measurement.

After completing this section, you should be able to

▪ **Discuss two methods for measuring frequency response**
- ☐ Explain the discrete-point measurement method
- ☐ Explain the swept frequency measurement method

Discrete Point Measurement

Figure 16–26 shows an arrangement for taking filter output voltage measurements at discrete values of input frequency using common laboratory instruments. The general procedure is as follows:

1. Set the amplitude of the sine wave generator to a desired voltage level.

2. Set the frequency of the sine wave generator to a value well below the expected critical frequency of the filter under test. For a low-pass filter, set the frequency as near as possible to 0 Hz. For a band-pass filter, set the frequency well below the expected lower critical frequency.

3. Increase the frequency in predetermined steps sufficient to allow enough data points for an accurate response curve.

4. Maintain a constant input voltage amplitude while varying the frequency.

5. Record the output voltage at each value of frequency.

6. After recording a sufficient number of points, plot a graph of output voltage versus frequency.

If the frequencies to be measured exceed the response of the DMM, an oscilloscope may have to be used instead.

FIGURE 16–26

Test setup for discrete point measurement of the filter response. (Readings are arbitrary and for display only.)

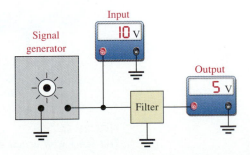

Swept Frequency Measurement

The swept frequency method requires more elaborate test equipment than does the discrete point method, but it is much more efficient and can result in a more accurate response curve. A general test setup is shown in Figure 16–27 using a swept frequency generator and a spectrum analyzer.

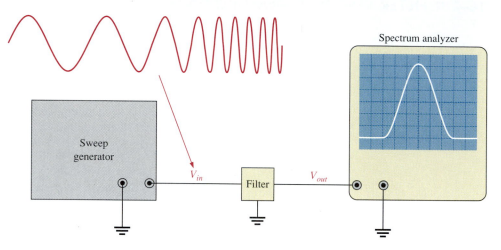

FIGURE 16–27

Test setup for swept frequency measurement of the filter response.

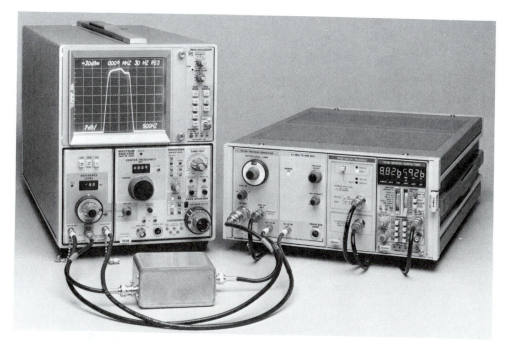

FIGURE 16–28

Test setup with actual equipment (courtesy of Tektronix, Inc.).

The swept frequency generator produces a constant amplitude output signal whose frequency increases linearly between two preset limits, as indicated in Figure 16–27. The spectrum analyzer is essentially an elaborate oscilloscope that can be calibrated for a desired *frequency span/division* rather than for the usual *time/division* setting. Therefore, as the input frequency to the filter sweeps through a preselected range, the response curve is traced out on the screen of the spectrum analyzer. An actual swept frequency test setup using typical equipment is shown in Figure 16–28.

SECTION 16–7 REVIEW	1. What is the purpose of the two tests discussed in this section?
	2. Name one disadvantage and one advantage of each test method.

16–8 ■ SYSTEM APPLICATION

In this system application, the focus is on the filter board which is part of the channel separation circuit in an FM stereo receiver. In addition to the active filters, the left and right channel separation circuit includes a demodulator, a frequency doubler, and a stereo matrix. Except for a brief statement of their purpose, these circuits will not be emphasized. The matrix, however, is an interesting application of summing amplifiers, with which you are familiar, and will be shown in detail on the schematic.

Basic Operation of the System

Stereo FM (**frequency modulation**) signals are transmitted on a **carrier** frequency of 88 MHz to 108 MHz. The standard transmitted stereo signal consists of three modulating signals. These are the sum of the left and right channel audio (L + R), the difference of the left and right channel audio (L − R), and a 19 kHz pilot subcarrier.

The L + R audio extends from 30 Hz to 15 kHz and the L − R audio is contained in two sidebands extending from 23 kHz to 53 kHz, as indicated in Figure 16–29. These frequencies come from the FM detector and go into the filter circuits where they are separated.

The frequency doubler and demodulator are used to extract the audio signal from the 23 kHz to 53 kHz sidebands after which the 30 Hz to 15 kHz L + R signal is passed through a filter.

The L + R and L − R audio signals are then sent to the matrix where they are applied to the summing circuits to produce the left and right channel audio (−2L and − 2R).

The Channel Filter Circuit Board

☐ Make sure that the circuit board shown in Figure 16–30 is correct by checking it against the schematic in Figure 16–31.

☐ Label a copy of the board with component and input/output designations in agreement with the schematic.

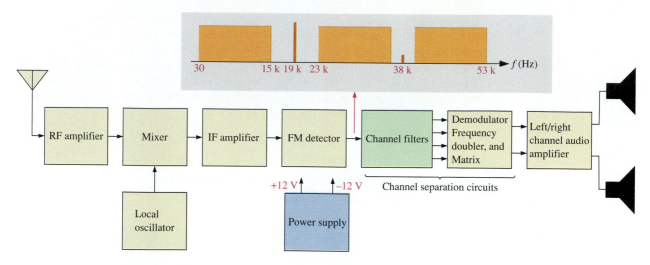

FIGURE 16–29

Basic block diagram of an FM receiver system.

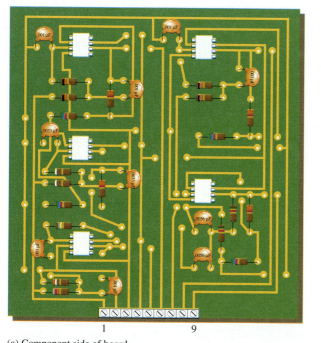

(a) Component side of board

(b) Backside of board

FIGURE 16–30

Channel filter board.

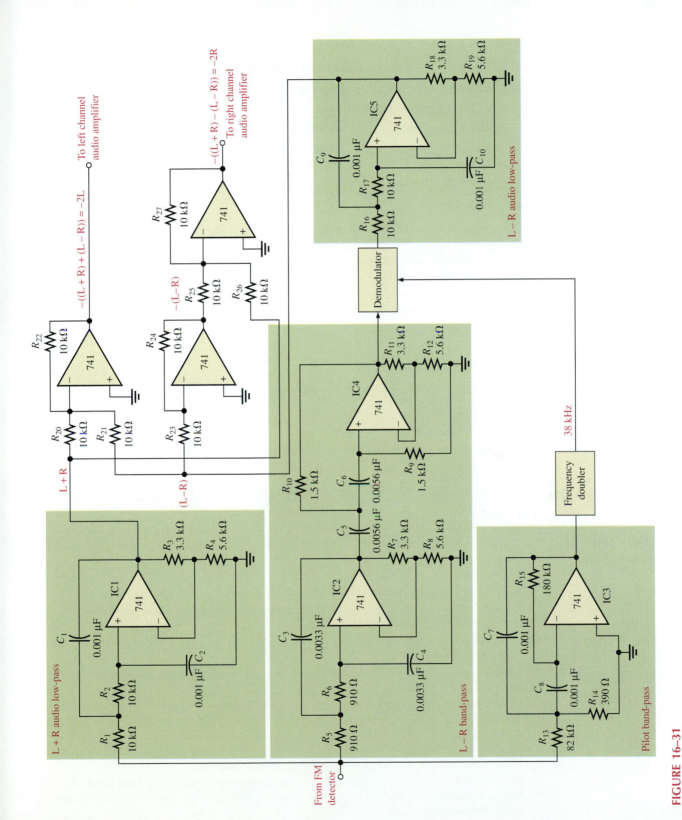

FIGURE 16–31

Channel separation circuits. The circuits on the filter board are shown in the green blocks.

The Filter Circuit

☐ Calculate the critical frequencies of each of the Sallen-Key filters.
☐ Calculate the center frequency of the multiple-feedback filter.
☐ Calculate the bandwidth of each filter.
☐ Determine the voltage gain of each filter.
☐ Verify that the Sallen-Key filters have an approximate Butterworth response characteristic.

Test Procedure

Develop a basic procedure for independently testing the filter board using only the inputs and outputs available on the terminal strip. Use a sweep generator, a spectrum analyzer, and a dual-polarity power supply.

For the sweep generator, a minimum and a maximum frequency are selected, and the instrument produces an output signal that repetitively sweeps through all frequencies between the preset limits. The spectrum analyzer is essentially a type of oscilloscope that displays a frequency response curve.

Final Report

Submit a final written report on the channel filter board using an organized format that includes the following:

1. A physical description of the circuit.
2. A discussion of the operation of the circuit.
3. A list of the specifications.
4. A list of parts with part numbers if available.

■ CHAPTER SUMMARY

- The bandwidth in a low-pass filter equals the critical frequency because the response extends to 0 Hz.
- The bandwidth in a high-pass filter extends above the critical frequency and is limited only by the inherent frequency limitation of the active circuit.
- A band-pass filter passes all frequencies within a band between a lower and an upper critical frequency and rejects all others outside this band.
- The bandwidth of a band-pass filter is the difference between the upper critical frequency and the lower critical frequency.
- A band-stop filter rejects all frequencies within a specified band and passes all those outside this band.
- Filters with the Butterworth response characteristic have a very flat response in the pass band, exhibit a roll-off of 20 dB/decade/pole, and are used when all the frequencies in the pass band must have the same gain.
- Filters with the Chebyshev characteristic have ripples or overshoot in the pass band and exhibit a faster roll-off per pole than filters with the Butterworth characteristic.
- Filters with the Bessel characteristic are used for filtering pulse waveforms. Their linear phase characteristic results in minimal waveshape distortion. The roll-off rate per pole is slower than for the Butterworth.

- In filter terminology, a single *RC* circuit is called a *pole.*
- Each pole in a Butterworth filter causes the output to roll off at a rate of 20 dB/decade.
- The quality factor Q of a band-pass filter determines the filter's selectivity. The higher the Q, the narrower the bandwidth and the better the selectivity.
- The damping factor determines the filter response characteristic (Butterworth, Chebyshev, or Bessel).

■ **GLOSSARY**

Active filter   A frequency-selective circuit consisting of active devices such as transistors or op-amps coupled with reactive components.

Band-pass filter   A type of filter that passes a range of frequencies lying between a certain lower frequency and a certain higher frequency.

Band-stop filter   A type of filter that blocks or rejects a range of frequencies lying between a certain lower frequency and a certain higher frequency.

Bessel   A type of filter response having a linear phase characteristic and less than 20 dB/decade/pole roll-off.

Butterworth   A type of filter response characterized by flatness in the pass band and a 20 dB/decade/pole roll-off.

Carrier   The high frequency (RF) signal that carries modulated information in AM, FM, or other systems.

Chebyshev   A type of filter response characterized by ripples in the pass band and a greater than 20 dB/decade/pole roll-off.

Damping factor   A filter characteristic that determines the type of response.

Frequency modulation (FM)   A communication method in which a lower frequency intelligence-carrying signal modulates (varies) the frequency of a higher frequency signal.

High-pass filter   A type of filter that passes frequencies above a certain frequency while rejecting lower frequencies.

Low-pass filter   A type of filter that passes frequencies below a certain frequency while rejecting higher frequencies.

Pole   A circuit containing one resistor and one capacitor that contributes 20 dB/decade to a filter's roll-off rate.

Quality factor (Q)   The ratio of a band-pass filter's center frequency to its bandwidth.

■ **FORMULAS**

(16–1)	$BW = f_c$	Low-pass bandwidth
(16–2)	$f_c = \dfrac{1}{2\pi RC}$	Filter critical frequency
(16–3)	$BW = f_{c2} - f_{c1}$	Filter bandwidth
(16–4)	$f_0 = \sqrt{f_{c1}f_{c2}}$	Center frequency of a band-pass filter
(16–5)	$Q = \dfrac{f_0}{BW}$	Quality factor of a band-pass filter
(16–6)	$Q = \dfrac{1}{DF}$	Q in terms of damping factor
(16–7)	$DF = 2 - \dfrac{R_1}{R_2}$	Damping factor
(16–8)	$A_{cl} = \dfrac{R_1}{R_2} + 1$	Closed-loop gain

(16–9) $$f_c = \frac{1}{2\pi \sqrt{R_A R_B C_A C_B}}$$ Critical frequency for a second-order Sallen-Key filter

(16–10) $$f_0 = \frac{1}{2\pi C} \sqrt{\frac{R_1 + R_3}{R_1 R_2 R_3}}$$ Center frequency of a multiple-feedback filter

(16–11) $$A_0 = \frac{R_2}{2R_1}$$ Gain of a multiple-feedback filter

(16–12) $$Q = \frac{1}{3}\left(\frac{R_5}{R_6} + 1\right)$$ Q of a state-variable filter

■ **SELF-TEST**

1. The term *pole* in filter terminology refers to
   (a) a high-gain op-amp   (b) one complete active filter
   (c) a single *RC* circuit   (d) the feedback circuit

2. An *RC* circuit produces a roll-off rate of
   (a) −20 dB/decade   (b) −40 dB/decade
   (c) −6 dB/octave   (d) answers (a) and (c)

3. A band-pass response has
   (a) two critical frequencies   (b) one critical frequency
   (c) a flat curve in the pass band   (d) a wide bandwidth

4. The lowest frequency passed by a low-pass filter is
   (a) 1 Hz   (b) 0 Hz
   (c) 10 Hz   (d) dependent on the critical frequency

5. The quality factor (Q) of a band-pass filter depends on
   (a) the critical frequencies
   (b) only the bandwidth
   (c) the center frequency and the bandwidth
   (d) only the center frequency

6. The damping factor of an active filter determines
   (a) the voltage gain   (b) the critical frequency
   (c) the response characteristic   (d) the roll-off rate

7. A maximally flat frequency response is known as
   (a) Chebyshev   (b) Butterworth
   (c) Bessel   (d) Colpitts

8. The damping factor of a filter is set by
   (a) the negative feedback circuit   (b) the positive feedback circuit
   (c) the frequency-selective circuit   (d) the gain of the op-amp

9. The number of poles in a filter affect the
   (a) voltage gain   (b) bandwidth
   (c) center frequency   (d) roll-off rate

10. Sallen-Key filters are
    (a) single-pole filters   (b) second-order filters
    (c) Butterworth filters   (d) band-pass filters

11. When filters are cascaded, the roll-off rate
    (a) increases   (b) decreases   (c) does not change

12. When a low-pass and a high-pass filter are cascaded to get a band-pass filter, the critical frequency of the low-pass filter must be
    (a) equal to the critical frequency of the high-pass filter
    (b) less than the critical frequency of the high-pass filter
    (c) greater than the critical frequency of the high-pass filter

13. A state-variable filter consists of
    (a) one op-amp with multiple-feedback paths
    (b) a summing amplifier and two integrators
    (c) a summing amplifier and two differentiators
    (d) three Butterworth stages

14. When the gain of a filter is minimum at its center frequency, it is
    (a) band-pass filter   **(b)** a band-stop filter
    (c) a notch filter   **(d)** answers (b) and (c)

■ **BASIC PROBLEMS**

SECTION 16–1 Basic Filter Responses

1. Identify each type of filter response (low-pass, high-pass, band-pass, or band-stop) in Figure 16–32.

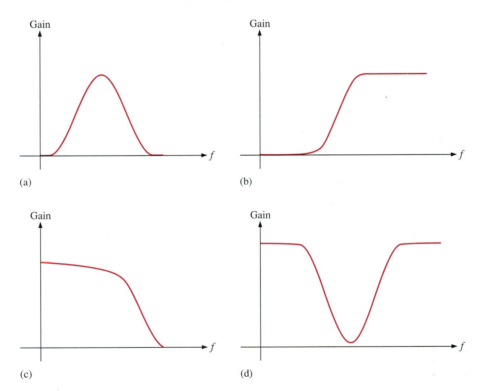

FIGURE 16–32

2. A certain low-pass filter has a critical frequency of 800 Hz. What is its bandwidth?

3. A single-pole high-pass filter has a frequency-selective circuit with $R = 2.2$ kΩ and $C = 0.0015$ μF. What is the critical frequency? Can you determine the bandwidth from the available information?

4. What is the roll-off rate of the filter described in Problem 3?

5. What is the bandwidth of a band-pass filter whose critical frequencies are 3.2 kHz and 3.9 kHz? What is the Q of this filter?

6. What is the center frequency of a filter with a Q of 15 and a bandwidth of 1 kHz?

SECTION 16–2 Filter Response Characteristics

7. What is the damping factor in each active filter shown in Figure 16–33? Which filters are approximately optimized for a Butterworth response characteristic?

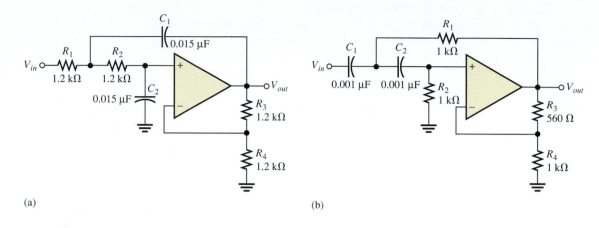

(a)

(b)

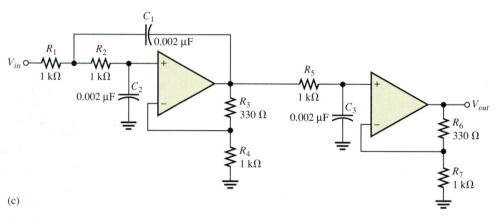

(c)

FIGURE 16–33

8. For the filters in Figure 16–33 that do not have a Butterworth response, specify the changes necessary to convert them to Butterworth responses. (Use nearest standard values.)

9. Response curves for high-pass second-order filters are shown in Figure 16–34. Identify each as Butterworth, Chebyshev, or Bessel.

FIGURE 16–34

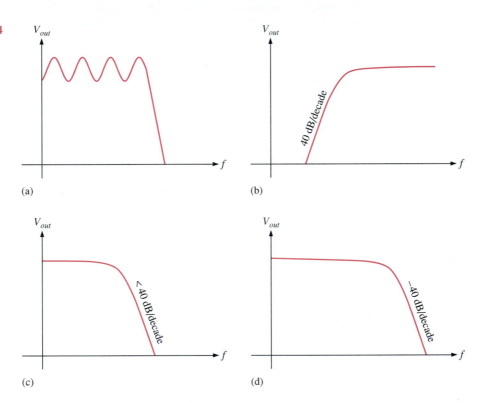

10. Is the four-pole filter in Figure 16–35 a low-pass or a high-pass type? Is it approximately optimized for a Butterworth response? What is the roll-off rate?

SECTION 16–3 Active Low-Pass Filters

10. Is the four-pole filter in Figure 16–35 a low-pass or a high-pass type? Is it approximately optimized for a Butterworth response? What is the roll-off rate?

11. Determine the critical frequency in Figure 16–35.

FIGURE 16–35

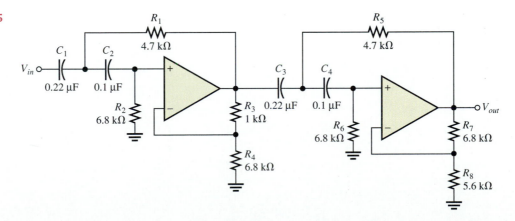

12. Without changing the response curve, adjust the component values in the filter of Figure 16–35 to make it an equal-value filter. Select $C = 0.22 \, \mu F$ for both stages.

13. Modify the filter in Figure 16–35 to increase the roll-off rate to -120 dB/decade while maintaining an approximate Butterworth response.

14. Using a block diagram format, show how to implement the following roll-off rates using single-pole and two-pole low-pass filters with Butterworth responses.

   (a) -40 dB/decade   (b) -20 dB/decade
   (c) -60 dB/decade   (d) -100 dB/decade
   (e) -120 dB/decade

SECTION 16–4   Active High-Pass Filters

15. Convert the filter in Problem 12 to a low-pass with the same critical frequency and response characteristic.

16. Make the necessary circuit modification to reduce by half the critical frequency in Problem 15.

17. For the filter in Figure 16–36, **(a)** how would you increase the critical frequency? **(b)** How would you increase the gain?

FIGURE 16–36

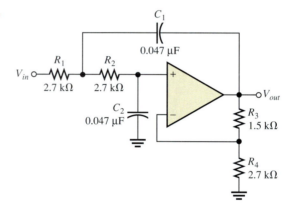

SECTION 16–5   Active Band-Pass Filters

18. Identify each band-pass filter configuration in Figure 16–37.

19. Determine the center frequency and bandwidth for each filter in Figure 16–37.

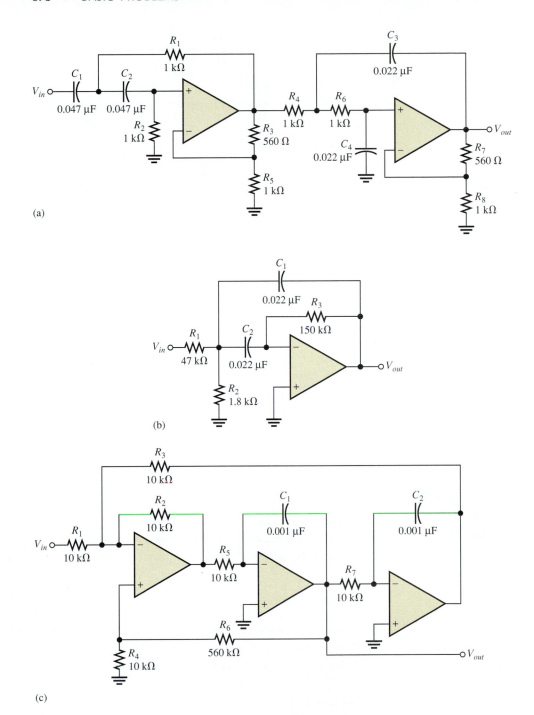

(a)

(b)

(c)

FIGURE 16–37

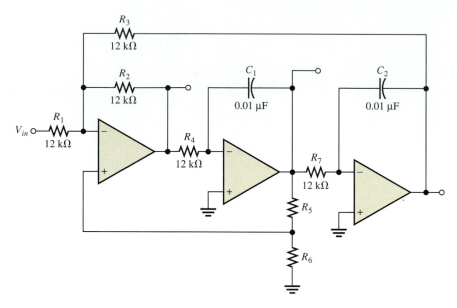

FIGURE 16–38

20. Optimize the state-variable filter in Figure 16–38 for $Q = 50$. What bandwidth is achieved?

SECTION 16–6 Active Band-Stop Filters

21. Show how to make a notch (band-stop) filter using the basic circuit in Figure 16–38.

22. Modify the band-stop filter in Problem 21 for a center frequency of 120 Hz.

■ **ANSWERS TO SECTION REVIEWS**

Section 16–1

1. The critical frequency determines the bandwidth.

2. The inherent frequency limitation of the op-amp limits the bandwidth.

3. Q and BW are inversely related. The higher the Q, the better the selectivity, and vice versa.

Section 16–2

1. Butterworth is very flat in the pass band and has a 20 dB/decade/pole roll-off. Chebyshev has ripples in the pass band and has greater than 20 dB/decade/pole roll-off. Bessel has a linear phase characteristic and less than 20 dB/decade/pole roll-off.

2. The damping factor determines the response characteristic.

3. Frequency-selection circuit, gain element, and negative feedback circuit are the parts of an active filter.

Section 16–3

1. A second-order filter has two poles. Two resistors and two capacitors make up the frequency-selective circuit.

2. The damping factor sets the response characteristic.

3. Cascading increases the roll-off rate.

Section 16–4

1. The positions of the *R*s and *C*s in the frequency-selection circuit are opposite for low-pass and high-pass configurations.

2. Decrease the *R* values to increase f_c.

3. 140 dB/decade

Section 16–5

1. *Q* determines selectivity.

2. $Q = 25$. Higher *Q* gives narrower *BW*.

3. A summing amplifier and two integrators make up a state-variable filter.

Section 16–6

1. A band-stop rejects frequencies within the stop band. A band-pass passes frequencies within the pass band.

2. The low-pass and high-pass outputs are summed.

Section 16–7

1. To check the frequency response of a filter

2. Discrete point measurement: tedious and less complete; simpler equipment.
Swept frequency measurement: uses more expensive equipment; more efficient, can be more accurate and complete.

■ **ANSWERS TO RELATED EXERCISES FOR EXAMPLES**

16–1 500 Hz

16–2 1.44

16–3 7.23 kHz; 1.29 kΩ

16–4 $C_{A1} = C_{A2} = C_{B1} = C_{B2} = 0.234$ μF; $R_2 = R_4 = 680$ Ω; $R_1 = 103$ Ω; $R_3 = 840$ Ω

16–5 $R_A = R_B = R_2 = 10$ kΩ; $C_A = C_B = 0.053$ μF; $R_1 = 586$ kΩ

16–6 Gain increases to 2.43, frequency decreases to 544 Hz, and bandwidth decreases to 96.5 Hz.

16–7 $f_0 = 21.9$ kHz; $Q = 101$; $BW = 217$ Hz

16–8 Decrease the input resistors or the feedback capacitors of the two integrator stages by half.

17

OSCILLATORS AND THE PHASE-LOCKED LOOP

■ CHAPTER OBJECTIVES

☐ Describe the basic concept of an oscillator
☐ Discuss the principles on which the operation of oscillators is based
☐ Describe and analyze the operation of basic *RC* oscillators
☐ Describe and analyze the operation of basic *LC* oscillators
☐ Describe and analyze the operation of basic nonsinusoidal oscillators
☐ Use a 555 timer in an oscillator application
☐ Explain the basic concept of a phase-locked loop (PLL)

O scillators are circuits that generate an output signal without an input signal. They are used as signal sources in all sorts of applications. Different types of oscillators produce various types of outputs including sine waves, square waves, triangular waves, and sawtooth waves. In this chapter, several types of basic oscillator circuits using both discrete transistors and op-amps as the gain element are introduced. Also, a very popular integrated circuit, called the 555 timer, is discussed in relation to its oscillator applications.

Oscillator operation is based on the principle of positive feedback, where a portion of the output signal is fed back to the input in a way that causes it to reinforce itself and thus sustain a continuous output signal. Oscillators are widely used in most communications systems as well as in digital systems, including computers, to generate required frequencies and timing signals. Also, oscillators are found in many types of test instruments like those used in the laboratory.

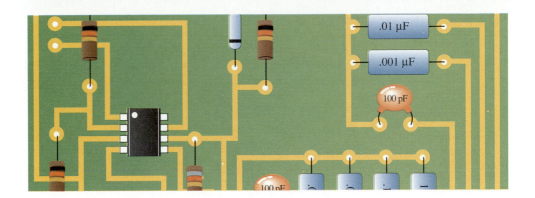

■ SYSTEM APPLICATION

The system application in Section 17–8 is a function generator that uses a sinusoidal oscillator as its signal source. Other circuits in the system are a zero-level detector and an integrator, with which you are already familiar. The function generator is a laboratory test instrument which is, effectively, a multiple-signal source because it can produce not only sinusoidal output voltages but also square waves and triangular waveforms.

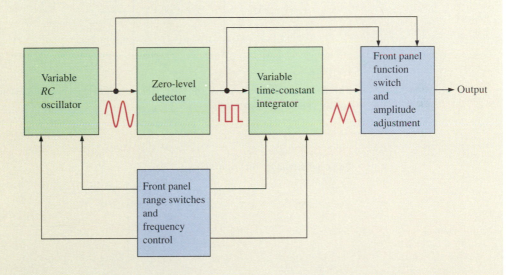

17–1 ■ THE OSCILLATOR

An oscillator is a circuit that produces a repetitive waveform on its output with only the dc supply voltage as an input. A repetitive input signal is not required. The output voltage can be either sinusoidal or nonsinusoidal, depending on the type of oscillator.

After completing this section, you should be able to

■ **Describe the basic concept of an oscillator**
  ☐ Explain the purpose of an oscillator
  ☐ List the basic elements of an oscillator

The basic oscillator concept is illustrated in Figure 17–1. Essentially, an **oscillator** converts electrical energy in the form of dc to electrical energy in the form of ac. A basic oscillator consists of an amplifier for gain (either discrete transistor or op-amp) and a positive feedback circuit that produces phase shift and provides attenuation, as shown in Figure 17–2.

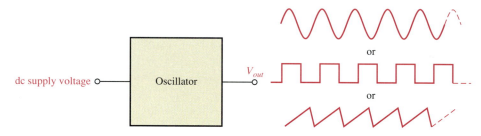

FIGURE 17–1

The basic oscillator concept showing three possible types of output waveforms: sine wave, square wave, and sawtooth.

FIGURE 17–2
Basic elements of an oscillator.

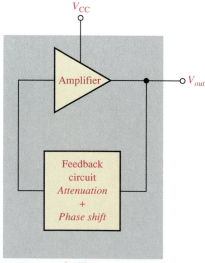

17–2 ■ OSCILLATOR PRINCIPLES

With the exception of the relaxation oscillator, which we will cover in Section 17–5, oscillator operation is based on the principle of positive feedback. In this section, we will examine this concept and look at the general conditions required for oscillation to occur.

After completing this section, you should be able to

■ **Discuss the principles on which the operation of oscillators is based**
  ☐ Explain positive feedback
  ☐ Describe the conditions for oscillation
  ☐ Discuss the start-up conditions

Positive Feedback

Positive feedback is characterized by the condition wherein a portion of the output voltage of an amplifier is fed back to the input with no net phase shift, resulting in a reinforcement of the output signal. This basic idea is illustrated in Figure 17–3. As you can see, the in-phase feedback voltage is amplified to produce the output voltage, which in turn produces the feedback voltage. That is, a loop is created in which the signal sustains itself and a continuous sine wave output is produced. This phenomenon is called *oscillation*.

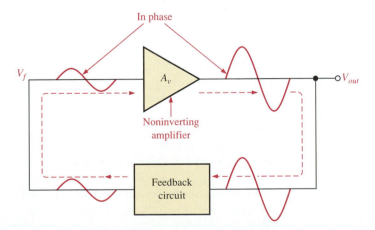

FIGURE 17–3
Positive feedback produces oscillation.

Conditions for Oscillation

Two conditions are required for a sustained state of oscillation:

1. The phase shift around the feedback loop must be 0°.

2. The voltage gain around the closed feedback loop must equal 1 (unity).

The voltage gain around the closed feedback loop (A_{cl}) is the product of the amplifier gain (A_v) and the attenuation (B) of the feedback circuit.

$$A_{cl} = A_v B$$

For example, if the amplifier has a gain of 100, the feedback circuit must have an attenuation of 0.01 to make the loop gain equal to 1 ($A_v B = 100 \times 0.01 = 1$). These conditions for oscillation are illustrated in Figure 17–4.

FIGURE 17–4

Conditions for oscillation.

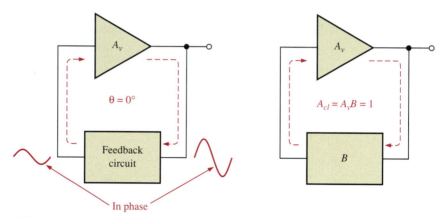

(a) The phase shift around the loop is 0°. (b) The closed loop gain is 1.

Start-Up Conditions

So far, you have seen what it takes for an oscillator to produce a continuous sine wave output. Now let's examine the requirements for the oscillation to start when the dc supply voltage is turned on. As you know, the unity-gain condition must be met for oscillation to be sustained. For oscillation to begin, the voltage gain around the positive feedback loop must be greater than 1 so that the amplitude of the output can build up to a desired level. The gain must then decrease to 1 so that the output stays at the desired level. Ways that certain amplifiers achieve this reduction in gain after start-up are discussed in later sections of this chapter. The voltage gain conditions for both starting and sustaining oscillation are illustrated in Figure 17–5.

A question that normally arises is this: If the oscillator is off (no dc voltage) and there is no output voltage, how does a feedback signal originate to start the positive feedback build-up process? Initially, a small positive feedback voltage develops from thermally produced broad-band noise in the resistors or other components or from turn-on transients. The feedback circuit permits only a voltage with a frequency equal to the selected oscillation frequency to appear in-phase on the amplifier's input. This initial feedback voltage is amplified and continually reinforced, resulting in a buildup of the output voltage as previously discussed.

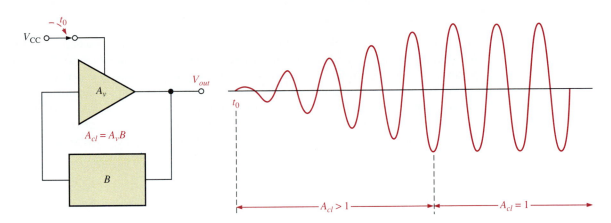

FIGURE 17–5

When oscillation starts at t_0, the condition $A_{cl} > 1$ causes the sinusoidal output voltage amplitude to build up to a desired level, where A_{cl} decreases to 1 and maintains the desired amplitude.

SECTION 17–2 REVIEW	1. What are the conditions required for a circuit to oscillate?
	2. Define positive feedback.
	3. What is the voltage gain condition for oscillator start-up?

17–3 ■ OSCILLATORS WITH *RC* FEEDBACK CIRCUITS

In this section you will learn about three types of RC oscillator circuits that produce sinusoidal outputs: the Wien-bridge oscillator, the phase-shift oscillator, and the twin-T oscillator. Generally, RC oscillators are used for frequencies up to about 1 MHz. The Wien-bridge is by far the most widely used type of RC oscillator for this range of frequencies.

After completing this section, you should be able to

■ **Describe and analyze the operation of basic *RC* oscillators**
  ☐ Identify a Wien-bridge oscillator
  ☐ Calculate the resonant frequency of a Wien-bridge oscillator
  ☐ Analyze oscillator feedback conditions
  ☐ Analyze oscillator start-up conditions
  ☐ Describe a self-starting Wien-bridge oscillator
  ☐ Identify a phase-shift oscillator
  ☐ Calculate the resonant frequency and analyze the feedback conditions for a phase-shift oscillator
  ☐ Identify a twin-T oscillator and describe its operation

The Wien-Bridge Oscillator

One type of sine wave oscillator is the *Wien-bridge oscillator.* A fundamental part of the Wien-bridge oscillator is a lead-lag circuit like that shown in Figure 17–6(a). R_1 and C_1 together form the lag portion of the circuit; R_2 and C_2 form the lead portion. The operation of this circuit is as follows. At lower frequencies, the lead circuit dominates due to the high reactance of C_2. As the frequency increases, X_{C2} decreases, thus allowing the output voltage to increase. At some specified frequency, the response of the lag circuit takes over, and the decreasing value of X_{C1} causes the output voltage to decrease.

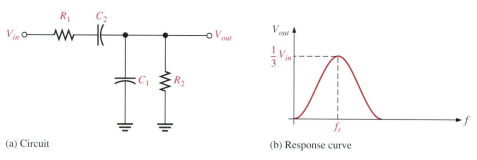

(a) Circuit   (b) Response curve

FIGURE 17–6
A lead-lag circuit.

The response curve for the lead-lag circuit shown in Figure 17–6(b) indicates that the output voltage peaks at a frequency called the resonant frequency, f_r. At this point, the attenuation (V_{out}/V_{in}) of the circuit is 1/3 if $R_1 = R_2$ and $X_{C1} = X_{C2}$ as stated by the following equation (derived in Appendix B):

$$\frac{V_{out}}{V_{in}} = \frac{1}{3} \qquad (17\text{–}1)$$

The formula for the resonant frequency (also derived in Appendix B) is

$$f_r = \frac{1}{2\pi RC} \qquad (17\text{–}2)$$

To summarize, the lead-lag circuit in the Wien-bridge oscillator has a resonant frequency, f_r, at which the phase shift through the circuit is 0° and the attentuation is 1/3. Below f_r, the lead circuit dominates and the output leads the input. Above f_r, the lag circuit dominates and the output lags the input.

The Basic Circuit The lead-lag circuit is used in the positive feedback loop of an op-amp, as shown in Figure 17–7(a). A voltage divider is used in the negative feedback loop. The Wien-bridge oscillator circuit can be viewed as a noninverting amplifier configuration with the input signal fed back from the output through the lead-lag circuit. Recall that the closed-loop gain of the amplifier is determined by the voltage divider.

$$A_{cl} = \frac{1}{B} = \frac{1}{R_2/(R_1 + R_2)} = \frac{R_1 + R_2}{R_2}$$

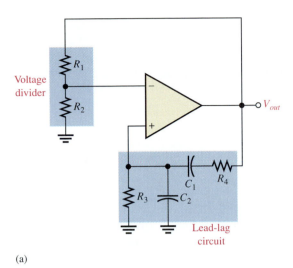

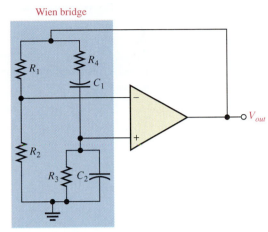

(a)

(b) Wien bridge circuit combines voltage divider and lead-lag circuit.

FIGURE 17–7

The Wien-bridge oscillator schematic shown in two different but equivalent forms.

The circuit is redrawn in Figure 17–7(b) to show that the op-amp is connected across the Wien bridge. One leg of the bridge is the lead-lag circuit, and the other is the voltage divider.

Positive Feedback Conditions for Oscillation As you know, for the circuit to produce a sustained sine wave output (oscillate), the phase shift around the positive feedback loop must be $0°$ and the gain around the loop must be at least unity (1). The $0°$ phase-shift condition is met when the frequency is f_r because the phase shift through the lead-lag circuit is $0°$ and there is no inversion from the noninverting input (+) of the op-amp to the output. This is shown in Figure 17–8(a).

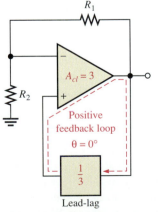

(a) The phase shift around the loop is $0°$.

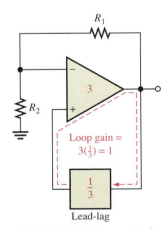

(b) The voltage gain around the loop is 1.

FIGURE 17–8

Conditions for oscillation.

The unity-gain condition in the feedback loop is met when

$$A_{cl} = 3$$

This offsets the 1/3 attenuation of the lead-lag circuit, thus making the gain around the positive feedback loop equal to 1, as depicted in Figure 17–8(b). To achieve a closed-loop gain of 3,

$$R_1 = 2R_2$$

Then

$$A_{cl} = \frac{R_1 + R_2}{R_2} = \frac{2R_2 + R_2}{R_2} = \frac{3R_2}{R_2} = 3$$

Start-Up Conditions Initially, the closed-loop gain of the amplifier must be more than 1 ($A_{cl} > 3$) until the output signal builds up to a desired level. The gain must then decrease to 1 so that the output signal stays at the desired level. This is illustrated in Figure 17–9.

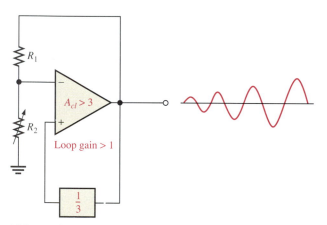

(a) Loop gain greater than 1 causes output to build up.

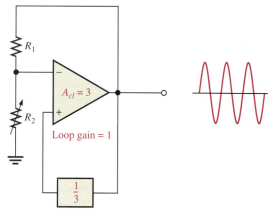

(b) Loop gain of 1 causes a sustained constant output.

FIGURE 17–9

Oscillator start-up conditions.

The circuit in Figure 17–10 illustrates a basic method for achieving the condition just described. Notice that the voltage-divider circuit has been modified to include an additional resistor R_3 in parallel with a back-to-back zener diode arrangement. When dc power is first applied, both zener diodes appear as opens. This places R_3 in series with R_1, thus increasing the closed-loop gain as follows ($R_1 = 2R_2$):

$$A_{cl} = \frac{R_1 + R_2 + R_3}{R_2} = \frac{3R_2 + R_3}{R_2} = 3 + \frac{R_3}{R_2}$$

Initially, a small positive feedback signal develops from noise or turn-on transients. The lead-lag circuit permits only a signal with a frequency equal to f_r to appear in phase on the noninverting input. This feedback signal is amplified and continually reinforced, resulting in a buildup of the output voltage. When the output signal reaches the zener breakdown voltage, the zeners conduct and effectively short out R_3. This lowers the amplifier's closed-loop gain to 3. At this point the output signal levels off and the oscillation is sustained. (Incidentally, the frequency of oscillation can be adjusted by using gang-tuned capacitors in the lead-lag circuit.)

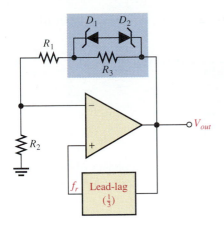

FIGURE 17–10

Self-starting Wien-bridge oscillator using back-to-back zener diodes.

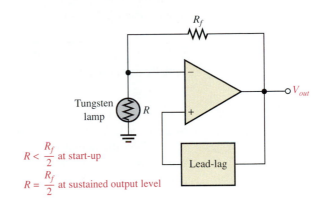

$$R < \frac{R_f}{2} \text{ at start-up}$$

$$R = \frac{R_f}{2} \text{ at sustained output level}$$

FIGURE 17–11

Self-starting Wien-bridge oscillator using a tungsten lamp in the negative feedback loop.

Another method that can be used to ensure self-starting employs a tungsten lamp in the voltage divider, as shown in Figure 17–11. When the power is first turned on, the resistance of the lamp is lower than its nominal value. This keeps the negative feedback small and makes the closed-loop gain of the amplifier greater than 3. As the output voltage builds up, the voltage across the tungsten lamp—and thus its current—increases. As a result, the lamp resistance increases until it reaches a value equal to one-half the feedback resistance. At this point the closed-loop gain is 3, and the output is sustained at a constant level.

EXAMPLE 17–1

Determine the frequency of oscillation for the Wien-bridge oscillator in Figure 17–12. Also, verify that oscillations will start and then continue when the output signal reaches 5.4 V.

FIGURE 17–12

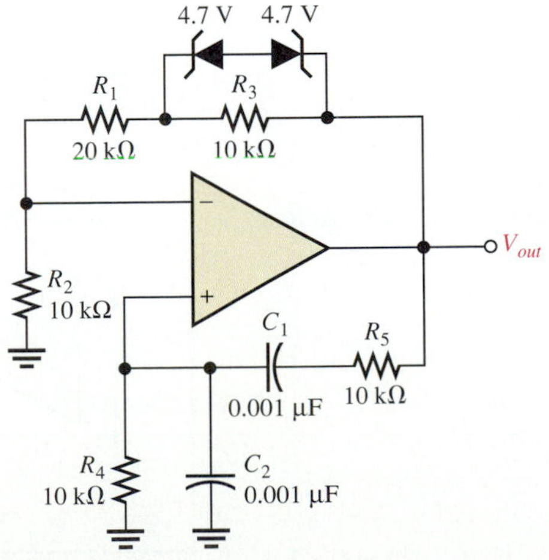

Solution For the lead-lag circuit, $R_4 = R_5 = R = 10\ \text{k}\Omega$ and $C_1 = C_2 = C = 0.001\ \mu\text{F}$. The resonant frequency is

$$f_r = \frac{1}{2\pi RC} = \frac{1}{2\pi(10\ \text{k}\Omega)(0.001\ \mu\text{F})} = 15.9\ \text{kHz}$$

Initially, the closed-loop gain is

$$A_{cl} = \frac{R_1 + R_2 + R_3}{R_2} = \frac{40\ \text{k}\Omega}{10\ \text{k}\Omega} = 4$$

Since $A_{cl} > 3$, the start-up condition is met.

When the output reaches 5.4 V (4.7 V + 0.7 V), the zeners conduct (their forward resistance is assumed small, compared to 10 kΩ), and the closed-loop gain is reached. Thus, oscillation is sustained.

$$A_{cl} = \frac{R_1 + R_2}{R_2} = \frac{30\ \text{k}\Omega}{10\ \text{k}\Omega} = 3$$

Related Exercise What change is required in the oscillator in Figure 17–12 to produce an output with an amplitude of 6.8 V?

The Phase-Shift Oscillator

Figure 17–13 shows a type of sine-wave oscillator called the *phase-shift oscillator.* Each of the three *RC* circuits in the feedback loop can provide a maximum phase shift approaching 90°. Oscillation occurs at the frequency where the total phase shift through the three *RC* circuits is 180°. The inversion of the op-amp, itself, provides the additional 180° to meet the requirement for oscillation.

The attenuation, *B*, of the three-section *RC* feedback circuit is

$$B = \frac{1}{29} \tag{17–3}$$

The derivation of this unusual result is given in Appendix B. To meet the greater-than-unity loop gain requirement, the closed-loop voltage gain of the op-amp must be greater than 29 (set by R_f and R_3). The frequency of oscillation (f_r) is also derived in Appendix B and stated in the following equation, where $R_1 = R_2 = R_3 = R$ and $C_1 = C_2 = C_3 = C$.

$$f_r = \frac{1}{2\pi\sqrt{6}RC} \tag{17–4}$$

FIGURE 17–13
Op-amp phase-shift oscillator.

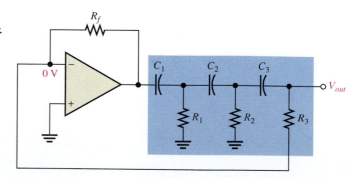

EXAMPLE 17–2

(a) Determine the value of R_f necessary for the circuit in Figure 17–14 to operate as an oscillator.

(b) Determine the frequency of oscillation.

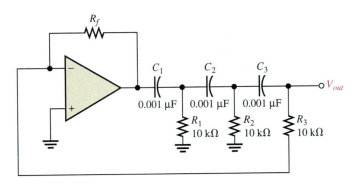

FIGURE 17–14

Solution

(a) $A_{cl} = 29$, and $A_{cl} = \dfrac{R_f}{R_3}$. Therefore, $\dfrac{R_f}{R_3} = 29$.

$$R_f = 29R_3 = 29(10 \text{ k}\Omega) = 290 \text{ k}\Omega$$

(b) $R_1 = R_2 = R_3 = R$ and $C_1 = C_2 = C_3 = C$. Therefore,

$$f_r = \frac{1}{2\pi\sqrt{6}RC} = \frac{1}{2\pi\sqrt{6}(10 \text{ k}\Omega)(0.001 \text{ μF})} \cong 6.5 \text{ kHz}$$

Related Exercise

(a) If R_1, R_2, and R_3 in Figure 17–14 are changed to 8.2 kΩ, what value must R_f be for oscillation?

(b) What is the value of f_r?

Twin-T Oscillator

Another type of *RC* oscillator is called the *twin-T* because of the two T-type *RC* filters used in the negative feedback loop, as shown in Figure 17–15(a). One of the twin-T filters has a low-pass response, and the other has a high-pass response. The combined parallel filters produce a band-stop or notch response with a center frequency equal to the desired frequency of oscillation, f_r, as shown in Figure 17–15(b).

Oscillation cannot occur at frequencies above or below f_r because of the negative feedback through the filters. At f_r, however, there is negligible negative feedback; thus, the positive feedback through the voltage divider (R_1 and R_2) allows the circuit to oscillate. Self-starting can be achieved by using a tungsten lamp in the place of R_2.

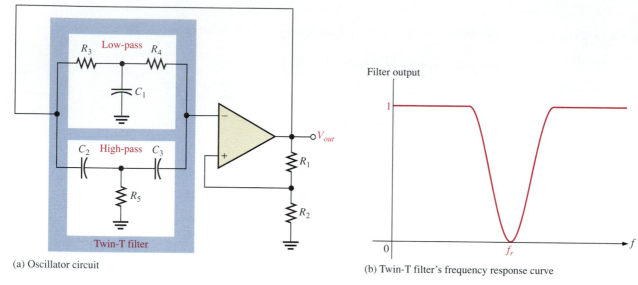

(a) Oscillator circuit

(b) Twin-T filter's frequency response curve

FIGURE 17–15

Twin-T oscillator and twin-T filter response.

17–4 ■ OSCILLATORS WITH *LC* FEEDBACK CIRCUITS

Although the RC oscillators, particularly the Wien bridge, are generally suitable for frequencies up to about 1 MHz, LC feedback elements are normally used in oscillators that require higher frequencies of oscillation. Also, because of the frequency limitation (lower unity-gain frequency) of most op-amps, discrete transistors are often used as the gain element in LC oscillators. This section introduces several types of resonant LC oscillators—the Colpitts, Clapp, Hartley, Armstrong, and crystal-controlled oscillators.

After completing this section, you should be able to

■ **Describe and analyze the operation of basic *LC* oscillators**
  ☐ Identify and analyze a Colpitts oscillator
  ☐ Identify and analyze a Clapp oscillator
  ☐ Identify and analyze a Hartley oscillator
  ☐ Identify and analyze an Armstrong oscillator
  ☐ Discuss the operation of crystal-controlled oscillators

The Colpitts Oscillator

One basic type of resonant circuit oscillator is the Colpitts, named after its inventor—as are most of the others we cover here. As shown in Figure 17–16, this type of oscillator uses an *LC* circuit in the feedback loop to provide the necessary phase shift and to act as a resonant filter that passes only the desired frequency of oscillation.

FIGURE 17–16

A basic Colpitts oscillator with a BJT as the gain element.

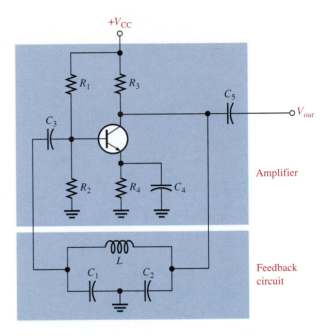

The approximate frequency of oscillation is the resonant frequency of the *LC* circuit and is established by the values of C_1, and C_2 and L according to this familiar formula:

$$f_r \cong \frac{1}{2\pi\sqrt{LC_T}}$$

(17–5)

Because the capacitors effectively appear in series around the tank circuit, the total capacitance (C_T) is

$$C_T = \frac{C_1 C_2}{C_1 + C_2}$$

(17–6)

Conditions for Oscillation and Start-Up The attenuation, *B*, of the resonant feedback circuit in the Colpitts oscillator is basically determined by the values of C_1 and C_2.

Figure 17–17 shows that the circulating tank current flows through both C_1 and C_2 (they are effectively in series). The voltage developed across C_2 is the oscillator's output voltage (V_{out}) and the voltage developed across C_1 is the feedback voltage (V_f), as indicated. The expression for the attenuation (B) is

$$B = \frac{V_f}{V_{out}} \cong \frac{IX_{C1}}{IX_{C2}} = \frac{X_{C1}}{X_{C2}} = \frac{1/2\pi f_r C_1}{1/2\pi f_r C_2}$$

FIGURE 17–17

The attenuation of the tank circuit is the output of the tank (V_f) divided by the input to the tank (V_{out}). $B = V_f/V_{out} = C_2/C_1$. For $A_v B > 1$, A_v must be greater than C_1/C_2.

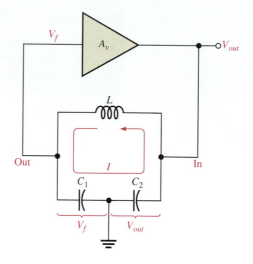

Cancelling the $2\pi f_r$ terms gives

$$B = \frac{C_2}{C_1} \tag{17–7}$$

As you know, a condition for oscillation is $A_v B = 1$. Since $B = C_2/C_1$,

$$A_v = \frac{C_1}{C_2} \tag{17–8}$$

where A_v is the voltage gain of the transistor amplifier. With this condition met, $A_v B = (C_1/C_2)(C_2/C_1) = 1$. Actually, for the oscillator to be self-starting, $A_v B$ must be greater than 1 ($A_v B > 1$). Therefore, the voltage gain must be made slightly greater than C_1/C_2.

$$A_v > \frac{C_1}{C_2} \tag{17–9}$$

Loading of the Feedback Circuit Affects the Frequency of Oscillation As indicated in Figure 17–18, the input impedance of the transistor amplifier acts as a load on the resonant feedback circuit and reduces the Q of the circuit. Recall from your study of resonance that the resonant frequency of a parallel resonant circuit depends on the Q, according to the following formula:

$$f_r = \frac{1}{2\pi\sqrt{LC_T}} \sqrt{\frac{Q^2}{Q^2 + 1}} \tag{17–10}$$

As a rule of thumb, for a Q greater than 10, the frequency is approximately $1/2\pi\sqrt{LC_T}$, as stated in Equation (17–5). When Q is less than 10, however, f_r is reduced significantly.

An FET can be used in place of a bipolar transistor, as shown in Figure 17–19, to minimize the loading effect of the transistor's input impedance. Recall that FETs have much higher input impedances than do bipolar transistors. Also, when an external load is

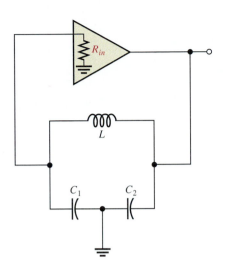

FIGURE 17–18

R_{in} of the transistor amplifier loads the feedback circuit and lowers its Q, thus lowering the resonant frequency.

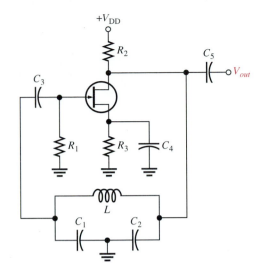

FIGURE 17–19

A basic FET Colpitts oscillator.

connected to the oscillator output, as shown in Figure 17–20(a), f_r may decrease, again because of a reduction in Q. This happens if the load resistance is too small. In some cases, one way to eliminate the effects of a load resistance is by transformer coupling as indicated in Figure 17–20(b).

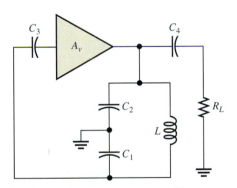

(a) A load capacitively coupled to oscillator output can reduce circuit Q and f_r.

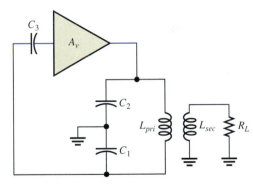

(b) Transformer coupling of load can reduce loading effect by impedance transformation.

FIGURE 17–20

Oscillator loading.

EXAMPLE 17–3

(a) Determine the frequency of oscillation for the oscillator in Figure 17–21. Assume there is negligible loading on the feedback circuit and that its Q is greater than 10.

(b) Find the frequency of oscillation if the oscillator is loaded to a point where the Q drops to 8.

FIGURE 17–21

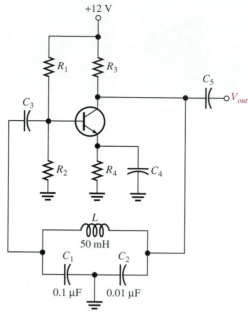

Solution

(a) $C_T = \dfrac{C_1 C_2}{C_1 + C_2} = \dfrac{(0.1\ \mu\text{F})(0.01\ \mu\text{F})}{0.11\ \mu\text{F}} = 0.0091\ \mu\text{F}$

$f_r \cong \dfrac{1}{2\pi\sqrt{LC_T}} = \dfrac{1}{2\pi\sqrt{(50\ \text{mH})(0.0091\ \mu\text{F})}} = 7.46\ \text{kHz}$

(b) $f_r = \dfrac{1}{2\pi\sqrt{LC_T}}\sqrt{\dfrac{Q^2}{Q^2+1}} = (7.46\ \text{kHz})(0.9923) = 7.40\ \text{kHz}$

Related Exercise What frequency does the oscillator in Figure 17–21 produce if it is loaded to a point where $Q = 4$?

The Clapp Oscillator

The Clapp oscillator is a variation of the Colpitts. The basic difference is an additional capacitor, C_3, in series with the inductor in the resonant feedback circuit, as shown in Figure 17–22. Since C_3 is in series with C_1 and C_2 around the tank circuit, the total capacitance is

$$C_T = \dfrac{1}{\dfrac{1}{C_1} + \dfrac{1}{C_2} + \dfrac{1}{C_3}} \qquad (17\text{–}11)$$

and the approximate frequency of oscillation ($Q > 10$) is $f_r \cong 1/2\pi\sqrt{LC_T}$.

If C_3 is much smaller than C_1 and C_2, the resonant frequency is controlled almost entirely by C_3 ($f_r \cong 1/2\pi\sqrt{LC_3}$). Since C_1 and C_2 are both connected to ground at one end,

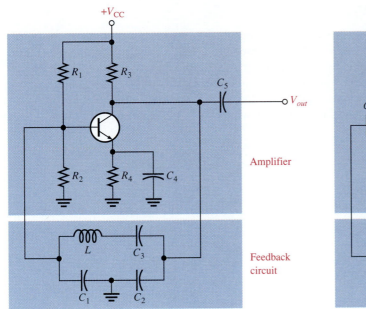

FIGURE 17–22

A basic Clapp oscillator.

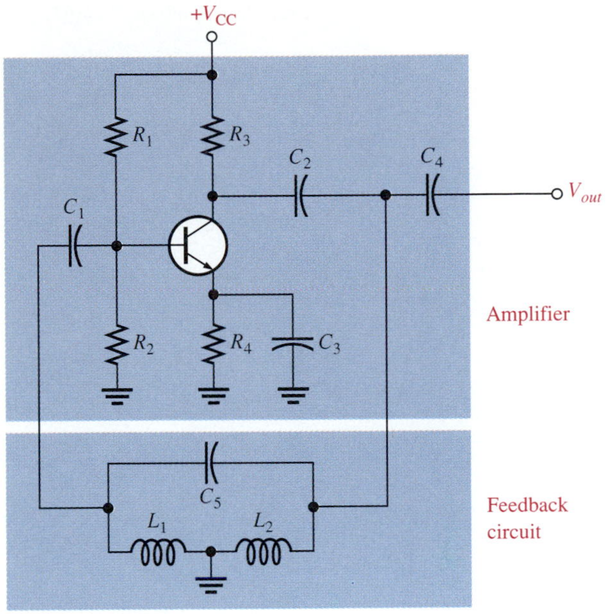

FIGURE 17–23

A basic Hartley oscillator.

the junction capacitance of the transistor and other stray capacitances appear in parallel with C_1 and C_2 to ground, altering their effective values. C_3 is not affected, however, and thus provides a more accurate and stable frequency of oscillation.

The Hartley Oscillator

The Hartley oscillator is similar to the Colpitts except that the feedback circuit consists of two series inductors and a parallel capacitor as shown in Figure 17–23.

In this circuit, the frequency of oscillation for $Q > 10$ is

$$f_r \cong \frac{1}{2\pi\sqrt{L_T C}} \qquad (17\text{–}12)$$

where $L_T = L_1 + L_2$. The inductors act in a role similar to C_1 and C_2 in the Colpitts to determine the attenuation, B, of the feedback circuit.

$$B \cong \frac{L_1}{L_2} \qquad (17\text{–}13)$$

To assure start-up of oscillation, A_v must be greater than $1/B$.

$$A_v > \frac{L_2}{L_1} \qquad (17\text{–}14)$$

Loading of the tank circuit has the same effect in the Hartley as in the Colpitts; that is, the Q is decreased and thus f_r decreases.

The Armstrong Oscillator

This type of *LC* oscillator uses transformer coupling to feed back a portion of the signal voltage, as shown in Figure 17–24. It is sometimes called a "tickler" oscillator in reference to the transformer secondary or "tickler coil" that provides the feedback to keep the oscillation going. The Armstrong is less common than the Colpitts, Clapp, and Hartley, mainly because of the disadvantage of transformer size and cost. The frequency of oscillation is set by the inductance of the primary winding (L_{pri}) in parallel with C_1.

$$f_r = \frac{1}{2\pi\sqrt{L_{pri}C_1}}$$

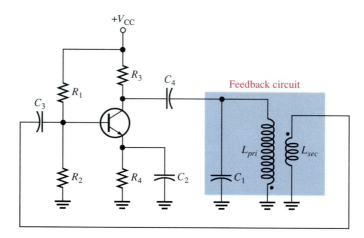

FIGURE 17–24
A basic Armstrong oscillator.

Crystal-Controlled Oscillators

The most stable and accurate type of oscillator uses a piezoelectric **crystal** in the feedback loop to control the frequency.

The Piezoelectric Effect Quartz is one type of crystalline substance found in nature that exhibits a property called the **piezoelectric effect.** When a changing mechanical stress is applied across the crystal to cause it to vibrate, a voltage develops at the frequency of mechanical vibration. Conversely, when an ac voltage is applied across the crystal, it vibrates at the frequency of the applied voltage. The greatest vibration occurs at the crystal's natural resonant frequency, which is determined by the physical dimensions and by the way the crystal is cut.

Crystals used in electronic applications typically consist of a quartz wafer mounted between two electrodes and enclosed in a protective "can" as shown in Figure 17–25(a) and (b). A schematic symbol for a crystal is shown in Figure 17–25(c) and an equivalent *RLC* circuit for the crystal appears in Figure 17–25(d). As you can see, the crystal's equivalent circuit is a series-parallel *RLC* circuit and can operate in either series resonance or parallel resonance. At the series resonant frequency, the inductive reactance is cancelled by the reactance of C_s. The remaining series resistor, R_s, determines the impedance of the crystal. Parallel resonance occurs when the inductive reactance and the reac-

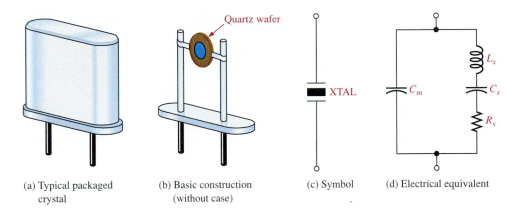

(a) Typical packaged crystal

(b) Basic construction (without case)

(c) Symbol

(d) Electrical equivalent

FIGURE 17–25
A quartz crystal.

tance of the parallel capacitance, C_m, are equal. The parallel resonant frequency is usually at least 1 kHz higher than the series resonant frequency. A great advantage of the crystal is that it exhibits a very high Q (Qs of several thousand are typical).

An oscillator that uses a crystal as a series resonant tank circuit is shown in Figure 17–26(a). The impedance of the crystal is minimum at the series resonant frequency, thus providing maximum feedback. The crystal tuning capacitor, C_C, is used to "fine tune" the oscillator frequency by "pulling" the resonant frequency of the crystal slightly up or down.

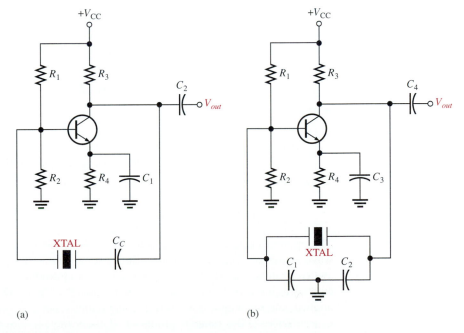

(a)

(b)

FIGURE 17–26
Basic crystal oscillators.

A modified Colpitts configuration is shown in Figure 17–26(b) with a crystal acting as a parallel resonant tank circuit. The impedance of the crystal is maximum at parallel resonance, thus developing the maximum voltage across the capacitors. The voltage across C_1 is fed back to the input.

Modes of Oscillation in the Crystal Piezoelectric crystals can oscillate in either of two modes—fundamental or overtone. The fundamental frequency of a crystal is the lowest frequency at which it is naturally resonant. The fundamental frequency depends on the crystal's mechanical dimensions, type of cut, and other factors, and is inversely proportional to the thickness of the crystal slab. Because a slab of crystal cannot be cut too thin without fracturing, there is an upper limit on the fundamental frequency. For most crystals, this upper limit is less than 20 MHz. For higher frequencies, the crystal must be operated in the overtone mode. Overtones are approximate integer multiples of the fundamental frequency. The overtone frequencies are usually, but not always, odd multiples (3, 5, 7, . . .) of the fundamental.

SECTION 17–4 REVIEW	**1.** What is the basic difference between the Colpitts and the Hartley oscillators? **2.** What is the advantage of an FET amplifier in a Colpitts or Hartley oscillator? **3.** How can you distinguish a Colpitts oscillator from a Clapp oscillator?

17–5 ■ NONSINUSOIDAL OSCILLATORS

In this section, several types of op-amp oscillator circuits that produce triangular, sawtooth, or square waveforms are discussed. Some of these types of oscillators are frequently referred to as signal generators and multivibrators depending on the particular circuit implementation.

After completing this section, you should be able to

■ **Describe and analyze the operation of basic nonsinusoidal oscillators**
  ☐ Discuss the operation of basic triangular-wave oscillators
  ☐ Discuss the operation of a voltage-controlled oscillator (VCO)
  ☐ Discuss the operation of a square-wave relaxation oscillator

A Triangular-Wave Oscillator

The op-amp integrator covered in Chapter 14 can be used as the basis for a triangular wave oscillator. The basic idea is illustrated in Figure 17–27(a) where a dual-polarity, switched input is used. We use the switch only to introduce the concept; it is not a practical way to implement this circuit. When the switch is in position 1, the negative voltage is applied, and the output is a positive-going ramp. When the switch is thrown into position 2, a negative-going ramp is produced. If the switch is thrown back and forth at fixed intervals, the output is a triangular wave consisting of alternating positive-going and negative-going ramps, as shown in Figure 17–27(b).

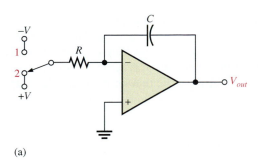

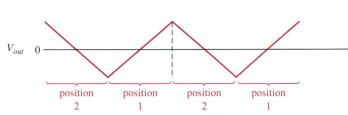

(a)

(b) Output voltage as the switch is thrown back and forth at regular intervals

FIGURE 17–27
Basic triangular-wave oscillator.

A Practical Triangular-Wave Circuit   One practical implementation of a triangular-wave oscillator utilizes an op-amp comparator to perform the switching function, as shown in Figure 17–28. The operation is as follows. To begin, assume that the output voltage of the comparator is at its maximum negative level. This output is connected to the inverting input of the integrator through R_1, producing a positive-going ramp on the output of the integrator. When the ramp voltage reaches the upper trigger point (UTP), the comparator switches to its maximum positive level. This positive level causes the integrator ramp to change to a negative-going direction. The ramp continues in this direction until the lower trigger point (LTP) of the comparator is reached. At this point, the comparator output switches back to the maximum negative level and the cycle repeats. This action is illustrated in Figure 17–29.

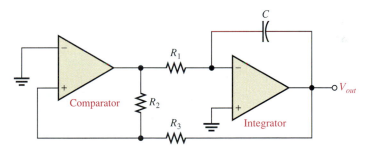

FIGURE 17–28
A triangular-wave oscillator using two op-amps.

FIGURE 17–29
Waveforms for the circuit in Figure 17–28.

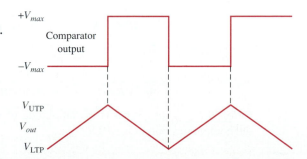

Since the comparator produces a square-wave output, the circuit in Figure 17–28 can be used as both a triangular-wave oscillator and a square-wave oscillator. Devices of this type are commonly known as *function generators* because they produce more than one output function. The output amplitude of the square wave is set by the output swing of the comparator, and the resistors R_2 and R_3 set the amplitude of the triangular output by establishing the UTP and LTP voltages according to the following formulas:

$$V_{\text{UTP}} = +V_{max}\left(\frac{R_3}{R_2}\right)$$

(17–15)

$$V_{\text{LTP}} = -V_{max}\left(\frac{R_3}{R_2}\right)$$

(17–16)

where the comparator output levels, $+V_{max}$ and $-V_{max}$, are equal. The frequency of both waveforms depends on the R_1C time constant as well as the amplitude-setting resistors, R_2 and R_3. By varying R_1, the frequency of oscillation can be adjusted without changing the output amplitude.

$$f_r = \frac{1}{4R_1C}\left(\frac{R_2}{R_3}\right)$$

(17–17)

EXAMPLE 17–4

Determine the frequency of oscillation of the circuit in Figure 17–30. To what value must R_1 be changed to make the frequency 20 kHz?

FIGURE 17–30

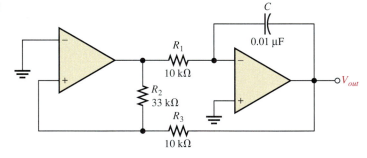

Solution

$$f_r = \frac{1}{4R_1C}\left(\frac{R_2}{R_3}\right) = \left(\frac{1}{4(10\ \text{k}\Omega)(0.01\ \mu\text{F})}\right)\left(\frac{33\ \text{k}\Omega}{10\ \text{k}\Omega}\right) = 8.25\ \text{kHz}$$

To make $f = 20$ kHz,

$$R_1 = \frac{1}{4fC}\left(\frac{R_2}{R_3}\right) = \left(\frac{1}{4(20\ \text{kHz})(0.01\ \mu\text{F})}\right)\left(\frac{33\ \text{k}\Omega}{10\ \text{k}\Omega}\right) = 4.13\ \text{k}\Omega$$

Related Exercise What is the amplitude of the triangular wave in Figure 17–30 if the comparator output is ±10 V?

A Voltage-Controlled Sawtooth Oscillator (VCO)

The voltage-controlled oscillator (VCO) is an oscillator whose frequency can be changed by a variable dc control voltage. VCOs can be either sinusoidal or nonsinusoidal. One way to build a voltage-controlled sawtooth oscillator is with an op-amp integrator that uses a switching device (PUT) in parallel with the feedback capacitor to terminate each ramp at a prescribed level and effectively "reset" the circuit. Figure 17–31(a) shows the implementation.

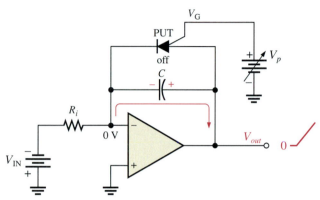

(a) Initially, the capacitor charges, the output ramp begins, and the PUT is off.

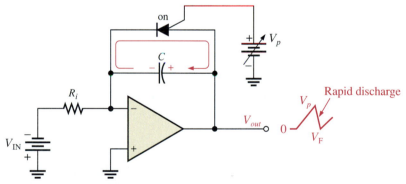

(b) The capacitor rapidly discharges when the PUT momentarily turns on.

FIGURE 17–31
Voltage-controlled sawtooth oscillator operation.

As you learned in Chapter 11, the PUT is a programmable unijunction transistor with an anode, a cathode, and a gate terminal. The gate is always biased positively with respect to the cathode. When the anode voltage exceeds the gate voltage by approximately 0.7 V, the PUT turns on and acts as a forward-biased diode. When the anode voltage falls below this level, the PUT turns off. Also, the current must be above the holding value to maintain conduction.

The operation of the sawtooth oscillator begins when the negative dc input voltage, $-V_{\text{IN}}$, produces a positive-going ramp on the output. During the time that the ramp is increasing, the circuit acts as a regular integrator. The PUT triggers on when the output ramp (at the anode) exceeds the gate voltage by 0.7 V. The gate is set to the approximate desired sawtooth peak voltage. When the PUT turns on, the capacitor rapidly discharges, as shown in Figure 17–31(b). The capacitor does not discharge completely to zero because of the PUT's forward voltage, V_{F}. Discharge continues until the PUT current falls below the holding value. At this point, the PUT turns off and the capacitor begins to charge again, thus generating a new output ramp. The cycle continually repeats, and the resulting output is a repetitive sawtooth waveform, as shown. The sawtooth amplitude and period can be adjusted by varying the PUT gate voltage.

The frequency of oscillation is determined by the $R_i C$ time constant of the integrator and the peak voltage set by the PUT. Recall that the charging rate of the capacitor is $V_{\text{IN}}/R_i C$. The time it takes the capacitor to charge from V_{F} to V_p is the period, T, of the sawtooth (neglecting the rapid discharge time).

$$T = \frac{V_p - V_{\text{F}}}{\dfrac{|V_{\text{IN}}|}{R_i C}} \qquad (17\text{–}18)$$

From $f = 1/T$,

$$f = \frac{|V_{\text{IN}}|}{R_i C}\left(\frac{1}{V_p - V_{\text{F}}}\right) \qquad (17\text{–}19)$$

EXAMPLE 17–5

(a) Find the amplitude and frequency of the sawtooth output in Figure 17–32. Assume that the forward PUT voltage, V_{F}, is approximately 1 V.

(b) Sketch the output waveform.

FIGURE 17–32

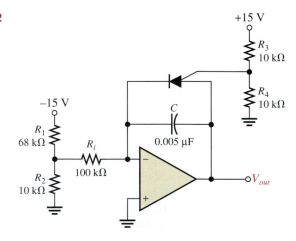

Solution

(a) First, find the gate voltage in order to establish the approximate voltage at which the PUT turns on.

$$V_G = \frac{R_4}{R_3 + R_4}(+V) = \frac{10\ k\Omega}{20\ k\Omega}(15\ V) = 7.5\ V$$

This voltage sets the approximate maximum peak value of the sawtooth output (neglecting the 0.7 V).

$$V_p \cong 7.5\ V$$

The minimum peak value (low point) is

$$V_F \cong 1\ V$$

So the peak-to-peak amplitude is

$$V_{pp} = V_p - V_F = 7.5\ V - 1\ V = 6.5\ V$$

The frequency is determined as follows:

$$V_{IN} = \frac{R_2}{R_1 + R_2}(-V) = \frac{10\ k\Omega}{78\ k\Omega}(-15\ V) = -1.92\ V$$

$$f = \frac{|V_{IN}|}{R_i C}\left(\frac{1}{V_p - V_F}\right) = \left(\frac{1.92\ V}{(100\ k\Omega)(0.005\ \mu F)}\right)\left(\frac{1}{7.5\ V - 1\ V}\right) = 591\ Hz$$

(b) The output waveform is shown in Figure 17–33, where the period is determined as follows:

$$T = \frac{1}{f} = \frac{1}{591\ Hz} = 1.69\ ms$$

FIGURE 17–33
Output of the circuit in Figure 17–32.

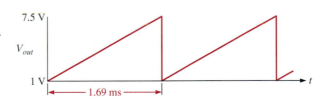

Related Exercise If R_i is changed to 56 kΩ in Figure 17–32, what is the frequency?

A Square-Wave Relaxation Oscillator

The basic square-wave oscillator shown in Figure 17–34 is a type of relaxation oscillator because its operation is based on the charging and discharging of a capacitor. Notice that the op-amp's inverting input is the capacitor voltage and the noninverting input is a portion of the output fed back through resistors R_2 and R_3. When the circuit is first turned on, the capacitor is uncharged, and thus the inverting input is at 0 V. This makes the output a positive maximum, and the capacitor begins to charge toward V_{out} through R_1. When the capacitor voltage (V_C) reaches a value equal to the feedback voltage (V_f) on the

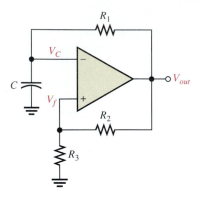

FIGURE 17–34

A square-wave relaxation oscillator.

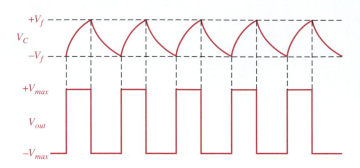

FIGURE 17–35

Waveforms for the square-wave relaxation oscillator.

noninverting input, the op-amp switches to the maximum negative state. At this point, the capacitor begins to discharge from $+V_f$ toward $-V_f$. When the capacitor voltage reaches $-V_f$, the op-amp switches back to the maximum positive state. This action continues to repeat, as shown in Figure 17–35, and a square-wave output voltage is obtained.

SECTION 17–5 REVIEW	**1.** What is a VCO, and basically, what does it do?
	2. Upon what principle does a relaxation oscillator operate?

17–6 ■ THE 555 TIMER AS AN OSCILLATOR

The 555 timer is a versatile integrated circuit with many applications. In this section, you will see how the 555 is configured as an astable or free-running multivibrator, which is essentially a square-wave oscillator. We will also discuss the use of the 555 timer as a voltage-controlled oscillator (VCO).

After completing this section, you should be able to

■ **Use a 555 timer in an oscillator application**
 ☐ Explain what the 555 timer is
 ☐ Discuss astable operation of the 555 timer
 ☐ Explain how to use the 555 timer as a VCO

The 555 timer consists basically of two comparators, a flip-flop, a discharge transistor, and a resistive voltage divider, as shown in Figure 17–36. The flip-flop (bistable multivibrator) is a digital device that may be unfamiliar to you at this point unless you already have taken a digital fundamentals course. Briefly, it is a two-state device whose output can be at either a high voltage level (set, S) or a low voltage level (reset, R). The state of the output can be changed with proper input signals.

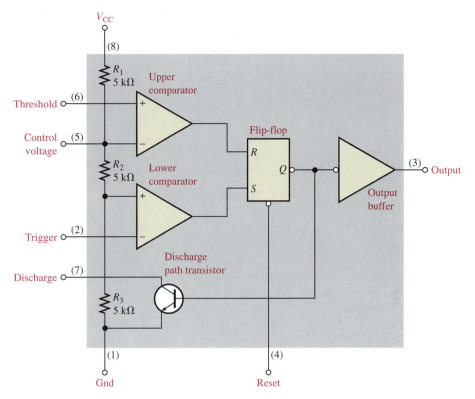

FIGURE 17–36

Internal diagram of a 555 integrated circuit timer. (IC pin numbers are in parentheses.)

The resistive voltage divider is used to set the voltage comparator levels. All three resistors are of equal value; therefore, the upper comparator has a reference of $\frac{2}{3}V_{CC}$, and the lower comparator has a reference of $\frac{1}{3}V_{CC}$. The comparators' outputs control the state of the flip-flop. When the trigger voltage goes below $\frac{1}{3}V_{CC}$, the flip-flop sets and the output jumps to its high level. The threshold input is normally connected to an external RC timing circuit. When the external capacitor voltage exceeds $\frac{2}{3}V_{CC}$, the upper comparator resets the flip-flop, which in turn switches the output back to its low level. When the device output is low, the discharge transistor (Q_d) is turned on and provides a path for rapid discharge of the external timing capacitor. This basic operation allows the timer to be configured with external components as an oscillator, a one-shot, or a time-delay element.

Astable Operation

A 555 timer connected to operate in the astable mode as a free-running nonsinusoidal oscillator (astable multivibrator) is shown in Figure 17–37. Notice that the threshold input (THRESH) is now connected to the trigger input (TRIG). The external components R_1, R_2, and C_{ext} form the timing circuit that sets the frequency of oscillation. The 0.01 μF capacitor connected to the control (CONT) input is strictly for decoupling and has no effect on the operation; in some cases it can be left off.

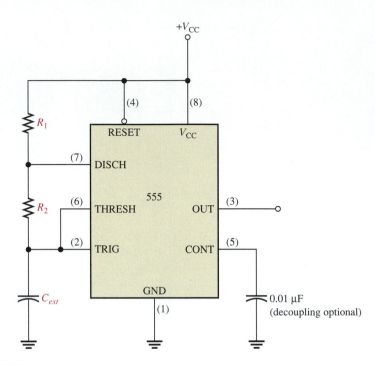

FIGURE 17–37
The 555 timer connected as an astable multivibrator.

Initially, when the power is turned on, the capacitor C_{ext} is uncharged and thus the trigger voltage (2) is at 0 V. This causes the output of the lower comparator to be high and the output of the upper comparator to be low, forcing the output of the flip-flop, and thus the base of Q_d, low and keeping the transistor off. Now, C_{ext} begins charging through R_1 and R_2 as indicated in Figure 17–38. When the capacitor voltage reaches $\frac{1}{3}V_{CC}$, the lower comparator switches to its low output state, and when the capacitor voltage reaches $\frac{2}{3}V_{CC}$, the upper comparator switches to its high output state. This resets the flip-flop, causes the base of Q_d to go high, and turns on the transistor. This sequence creates a discharge path for the capacitor through R_2 and the transistor, as indicated. The capacitor now begins to discharge, causing the upper comparator to go low. At the point where the capacitor discharges down to $\frac{1}{3}V_{CC}$, the lower comparator switches high, setting the flip-flop, which makes the base of Q_d low and turns off the transistor. Another charging cycle begins, and the entire process repeats. The result is a rectangular wave output whose duty cycle depends on the values of R_1 and R_2. The frequency of oscillation is given by the following formula:

$$f_r = \frac{1.44}{(R_1 + 2R_2)C_{ext}} \qquad\qquad \textbf{(17–20)}$$

By selecting R_1 and R_2, the duty cycle of the output can be adjusted. Since C_{ext} charges through $R_1 + R_2$ and discharges only through R_2, duty cycles approaching a minimum of 50 percent can be achieved if $R_2 \gg R_1$ so that the charging and discharging times are approximately equal.

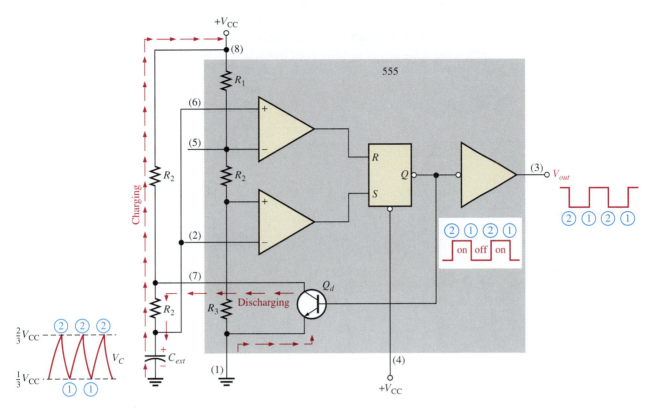

FIGURE 17–38

Operation of the 555 timer in the astable mode.

An expression to calculate the duty cycle is developed as follows. The time that the output is high (t_H) is how long it takes C_{ext} to charge from $\frac{1}{3}V_{CC}$ to $\frac{2}{3}V_{CC}$. It is expressed as

$$t_H = 0.693(R_1 + R_2)C_{ext}$$

The time that the output is low (t_L) is how long it takes C_{ext} to discharge from $\frac{2}{3}V_{CC}$ to $\frac{1}{3}V_{CC}$. It is expressed as

$$t_L = 0.693R_2C_{ext}$$

The period, T, of the output waveform is the sum of t_H and t_L. The following formula for T is the reciprocal of f in Equation (17–20).

$$T = t_H + t_L = 0.693(R_1 + 2R_2)C_{ext}$$

Finally, the duty cycle is

$$\text{Duty cycle} = \frac{t_H}{T} = \frac{t_H}{t_H + t_L}$$

$$\text{Duty cycle} = \frac{R_1 + R_2}{R_1 + 2R_2} \times 100\% \qquad \textbf{(17–21)}$$

To achieve duty cycles of less than 50 percent, the circuit in Figure 17–37 can be modified so that C_{ext} charges through only R_1 and discharges through R_2. This is achieved with a diode D_1 placed as shown in Figure 17–39. The duty cycle can be made less than 50 percent by making R_1 less than R_2. Under this condition, the expression for the duty cycle is

$$\text{Duty cycle} = \frac{R_1}{R_1 + R_2} \times 100\%$$

(17–22)

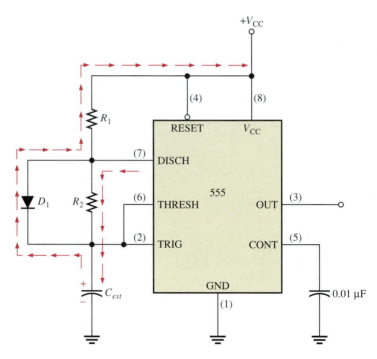

FIGURE 17–39
The addition of diode D_1 allows the duty cycle of the output to be adjusted to less than 50 percent by making $R_1 < R_2$.

EXAMPLE 17–6   A 555 timer configured to run in the astable mode (oscillator) is shown in Figure 17–40. Determine the frequency of the output and the duty cycle.

FIGURE 17–40

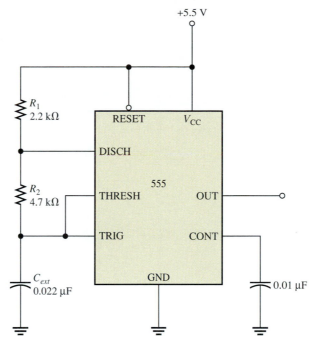

Solution

$$f_r = \frac{1.44}{(R_1 + 2R_2)C_{ext}} = \frac{1.44}{(2.2\ \text{k}\Omega + 9.4\ \text{k}\Omega)0.022\ \mu\text{F}} = 5.64\ \text{kHz}$$

$$\text{Duty cycle} = \frac{R_1 + R_2}{R_1 + 2R_2} \times 100\% = \frac{2.2\ \text{k}\Omega + 4.7\ \text{k}\Omega}{2.2\ \text{k}\Omega + 9.4\ \text{k}\Omega} \times 100\% = 59.5\%$$

Related Exercise Determine the duty cycle in Figure 17–40 if a diode is connected across R_2 as indicated in Figure 17–39.

Operation As a Voltage-Controlled Oscillator (VCO)

A 555 timer can be configured to operate as a VCO by using the same external connections as for astable operation, with the exception that a variable control voltage is applied to the CONT input (pin 5), as indicated in Figure 17–41.

As shown in Figure 17–42, the control voltage (V_{CONT}) changes the threshold values of $\frac{1}{3}V_{CC}$ and $\frac{2}{3}V_{CC}$ for the internal comparators. With the control voltage, the upper value is V_{CONT} and the lower value is $\frac{1}{2}V_{CONT}$, as you can see by examining the internal diagram of the 555 timer. When the control voltage is varied, the output frequency also varies. An increase in V_{CONT} increases the charging and discharging time of the external capacitor and causes the frequency to decrease. A decrease in V_{CONT} decreases the charging and discharging time of the capacitor and causes the frequency to increase.

An interesting application of the VCO is in phase-locked loops, which are used in various types of communication receivers to track variations in the frequency of incoming signals. We will cover the basic operation of a phase-locked loop in the next section.

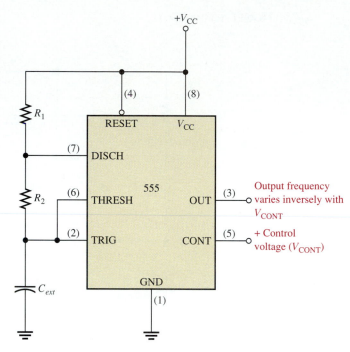

FIGURE 17–41

The 555 timer connected as a voltage-controlled oscillator (VCO). Note the variable control voltage input on pin 5.

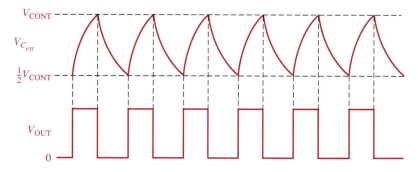

FIGURE 17–42

The VCO output frequency varies inversely with V_{CONT} because the charging and discharging time of C_{ext} is directly dependent on the control voltage.

SECTION 17–6 REVIEW	1. Name the five basic elements in a 555 timer IC.
	2. When the 555 timer is configured as an astable multivibrator, how is the duty cycle determined?

17–7 ■ THE PHASE-LOCKED LOOP

The phase-locked loop (PLL) is an electronic feedback circuit consisting of a phase detector, a low-pass filter, and a voltage-controlled oscillator (VCO). It is capable of locking onto or synchronizing with an incoming signal. When the phase changes, indicating that the incoming frequency is changing, the phase detector's output voltage (error voltage) increases or decreases just enough to keep the oscillator frequency the same as the incoming frequency. Phase-locked loops are used in a wide variety of communication system applications, including TV receivers, FM demodulation, modems, telemetry, and tone decoders.

After completing this section, you should be able to

■ **Explain the basic concept of a phase-locked loop (PLL)**
  □ Discuss basic PLL operation including lock range, capture range, and frequency considerations
  □ Explain when the PLL is in lock or out of lock and discuss the significance

Basic Operation

Using the basic block diagram of a phase-locked loop (PLL) in Figure 17–43 as reference, the general operation is as follows. When there is no input signal, the error voltage is zero and the frequency, f_o, of the voltage-controlled oscillator (VCO) is called the *free-running* or *center frequency*. When an input signal is applied, the phase detector compares the phase and frequency of the input signal with the VCO frequency and produces error voltage, V_e. This error voltage is proportional to the phase and frequency difference of the incoming frequency and the VCO frequency. The error voltage contains components that are the sum and difference of the two compared frequencies. The low-pass filter passes only the difference frequency, V_d, which is the lower of the two components. This signal is amplified and fed back to the VCO as a control voltage, V_{CONT}. The control voltage forces the VCO frequency to change in a direction that reduces the difference

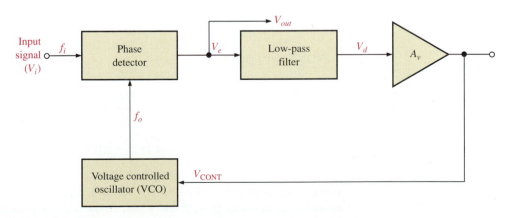

FIGURE 17–43
Basic phase-locked loop block diagram.

between the incoming frequency, f_i, and the VCO frequency, f_o. When f_i and f_o are sufficiently close in value, the feedback action of the PLL causes the VCO to lock onto the incoming signal. Once the VCO locks, its frequency is the same as the input frequency with a slight difference in phase. This phase difference, ϕ, is necessary to keep the PLL in the lock condition.

Lock Range   Once the PLL is locked, it can track frequency changes in the incoming signal. The range of frequencies over which the PLL can maintain lock with the incoming signal is called the *lock* or *tracking range*. The lock range is usually expressed as a percentage of the VCO frequency.

Capture Range   The range of frequencies over which the PLL can acquire lock with an input signal is called the *capture range*. The parameter is normally expressed in percentage of f_o, also.

Sum and Difference Frequencies   The phase detector operates as a multiplier circuit to produce the sum and difference of the input frequency, f_i, and the VCO frequency, f_o. This action can best be described mathematically as follows. Recall from ac circuit theory that a sinusoidal voltage can be expressed as

$$v = V_p \sin 2\pi f t$$

where V_p is the peak value, f is the frequency, and t is the time. Using this basic expression, the input signal voltage, v_i, and the VCO voltage, v_o, can be written as

$$v_i = V_{ip} \sin 2\pi f_i t$$
$$v_o = V_{op} \sin 2\pi f_o t$$

When these two signals are multiplied in the phase detector, we get a product at the output as follows:

$$V_{out} = V_{ip} V_{op} (\sin 2\pi f_i t)(\sin 2\pi f_o t)$$

Applying the trigonometric identity,

$$(\sin A)(\sin B) = \frac{1}{2}[\cos(A - B) - \cos(A + B)]$$

to the preceding equation for V_{out}, we get

$$V_{out} = \frac{V_{ip} V_{op}}{2}[\cos(2\pi f_i t - 2\pi f_o t) - \cos(2\pi f_i t + 2\pi f_o t)]$$

$$V_{out} = \frac{V_{ip} V_{op}}{2} \cos 2\pi (f_i - f_o)t - \frac{V_{ip} V_{op}}{2} \cos 2\pi (f_i + f_o)t \qquad \textbf{(17–23)}$$

You can see in Equation (17–23) that V_{out} of the phase detector consists of a difference frequency component $(f_i - f_o)$ and a sum frequency component $(f_i + f_o)$. This concept is illustrated in Figure 17–44 with frequency spectrum graphs. Each vertical line represents a specific signal frequency, and the height is its amplitude, A.

FIGURE 17–44
Frequency spectrum of the phase detector.

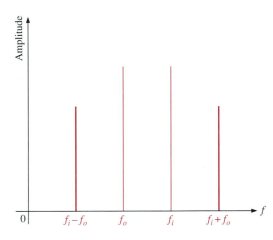

EXAMPLE 17–7   A 10 kHz signal f_i and an 8 kHz signal f_o are applied to a phase detector. Determine the sum and difference frequencies.

Solution
$$f_i + f_o = 10 \text{ kHz} + 8 \text{ kHz} = 18 \text{ kHz}$$
$$f_i - f_o = 10 \text{ kHz} - 8 \text{ kHz} = 2 \text{ kHz}$$

Related Exercise   If the sum and difference frequencies are 30 kHz and 6 kHz respectively, what are f_i and f_o?

When the PLL Is in Lock

When the PLL is in a lock condition, the VCO frequency equals the input frequency ($f_o = f_i$). Thus, the difference frequency is $f_o - f_i = 0$. A zero frequency indicates a dc component. The low-pass filter removes the sum frequency ($f_i + f_o$) and passes the dc (0-frequency) component, which is amplified and fed back to the VCO. When the PLL is in lock, the difference frequency component is always dc and is always passed by the filter, so the lock range is independent of the bandwidth of the low-pass filter. This is illustrated in Figure 17–45.

FIGURE 17–45
When the PLL is in lock, the difference frequency is zero and passes through the filter. The lock range is independent of the filter bandwidth.

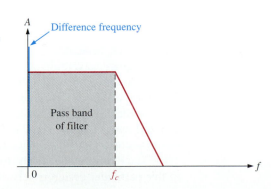

When the PLL Is Not in Lock

In the case where the PLL has not yet locked onto the incoming frequency, the phase detector still produces sum and difference frequencies. In this case, however, the difference frequency may lie outside the pass band of the low-pass filter and will not be fed back to the VCO, as illustrated in Figure 17–46. The VCO will remain at its center frequency (free-running) as long as this condition exists.

As the input frequency approaches the VCO frequency, the difference component produced by the phase detector decreases and eventually falls into the pass band of the filter and drives the VCO toward the incoming frequency. When the VCO reaches the incoming frequency, the PLL locks on it.

FIGURE 17–46

When the PLL is not in lock, the difference frequency can be outside the pass band.

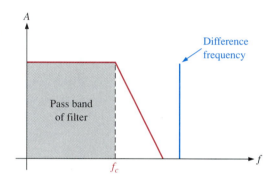

**SECTION 17–7
REVIEW**

1. What are the basic blocks in a phase-locked loop?
2. State the general purpose of a phase-locked loop.

17–8 ■ SYSTEM APPLICATION

The company for which you work has introduced a line of electronic test equipment. You have been given the responsibility for the function generator, an instrument that produces sinusoidal waveforms, square waveforms, and triangular waveforms.

Basic Operation of the System

The function generator is an electronic system that produces either a sinusoidal waveform, a square waveform, or a triangular waveform depending on the function selected by the front panel switches, as shown in Figure 17–47. The frequency of the selected output waveform can be varied from less than 1 Hz to greater than 80 kHz using the frequency range switches and the frequency dial. The amplitude of the output waveform can be adjusted up to approximately +10 V with the front panel amplitude control. Also, any dc offset can be nulled out with the front panel dc offset control.

The block diagram of the function generator is shown in Figure 17–48. The concept of this particular design is very simple. The oscillator produces a sinusoidal output volt-

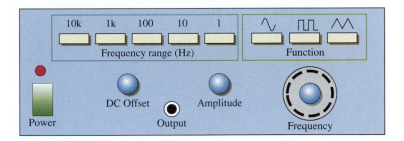

FIGURE 17–47
Front panel of the function generator.

age that drives a zero-level detector (comparator), which produces a square wave of the same frequency as the oscillator output. The output of the level detector goes to an integrator, which produces a triangular output voltage also with a frequency equal to the oscillator output. The type of output waveform is selected by the function switches on the front panel. The frequency is set by the front panel range switches and the frequency dial, and amplitude is set by the front panel control knob.

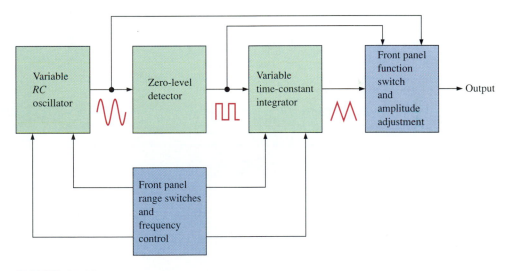

FIGURE 17–48
Block diagram of the function generator.

The schematic of the function generator is shown in Figure 17–49, where the portions in blue are the front panel components. The frequency of the sinusoidal oscillator is controlled by the selection of any two of ten capacitors (C_1 through C_{10}) in the oscillator feedback circuit. These capacitors produce the five frequency ranges indicated on the front panel switches, which are multiplication factors for the setting of the frequency dial. The adjustment of the frequency within each range is accomplished by varying the resistors R_5 and R_6 in the feedback circuit of the oscillator.

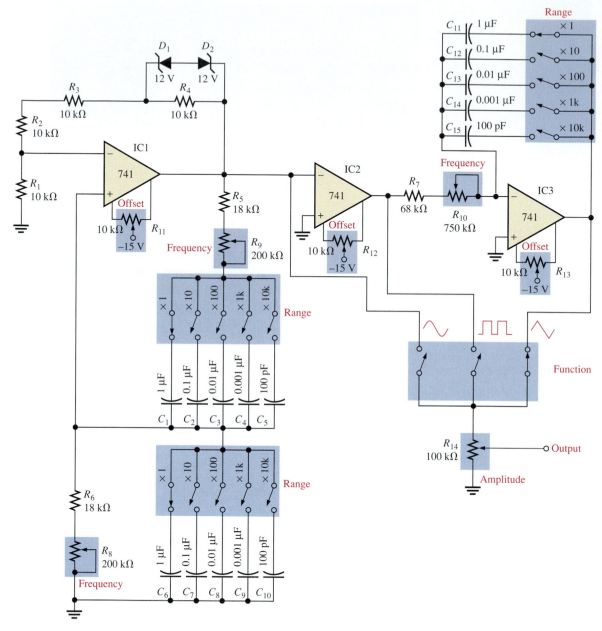

FIGURE 17–49
Schematic of the function generator.

The integrator time constant is adjusted in step with the frequency by selection of the appropriate capacitor (C_{11} through C_{15}) and adjustment of resistor R_{10}. Resistors R_9, R_8, and R_{10} are potentiometers that are ganged together so that they change resistance together as the frequency dial is turned. For example, if the 1k switch is selected and the frequency dial is set at 5, then the resulting output frequency for any of the three types of waveforms is 1 kHz × 5 = 5 kHz.

The Function Generator Circuit Boards

☐ Make sure that the circuit boards shown in Figure 17–50 are correct by checking them against the schematic in Figure 17–49. Both boards have backside interconnections. Corresponding feedthrough pads are aligned horizontally.

☐ Label a copy of the board with component and input/output designations in agreement with the schematic.

☐ Develop a board-to-board wire list specifying which terminals on the two boards connect together and which terminals connect to front panel components.

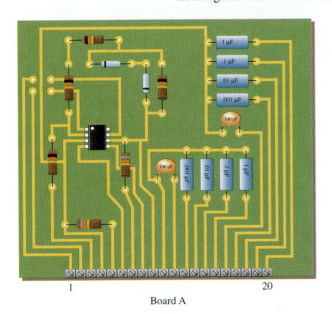

1   20

Board A

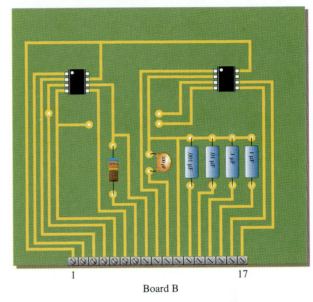

1   17

Board B

FIGURE 17–50
The function generator circuit boards.

The Function Generator Circuit

☐ Determine the maximum frequency of the oscillator for each range switch (×1, ×10, ×100, ×1k, and ×10k). Only one set of three switches corresponding to a given range setting can be closed at a time. There is a set of ganged switches for ×1, a set of ganged switches for ×10, and so on.

☐ Determine the minimum frequency of the oscillator for each range switch.

☐ Determine the approximate maximum peak-to-peak output voltages for each function. The dc supply voltages are +15 V and −15 V.

Test Procedure

Develop a basic procedure for thoroughly testing the function generator.

Troubleshooting

Based on the results indicated in Figure 17–51 for four faulty function generator units, determine the most likely fault or faults in each case.

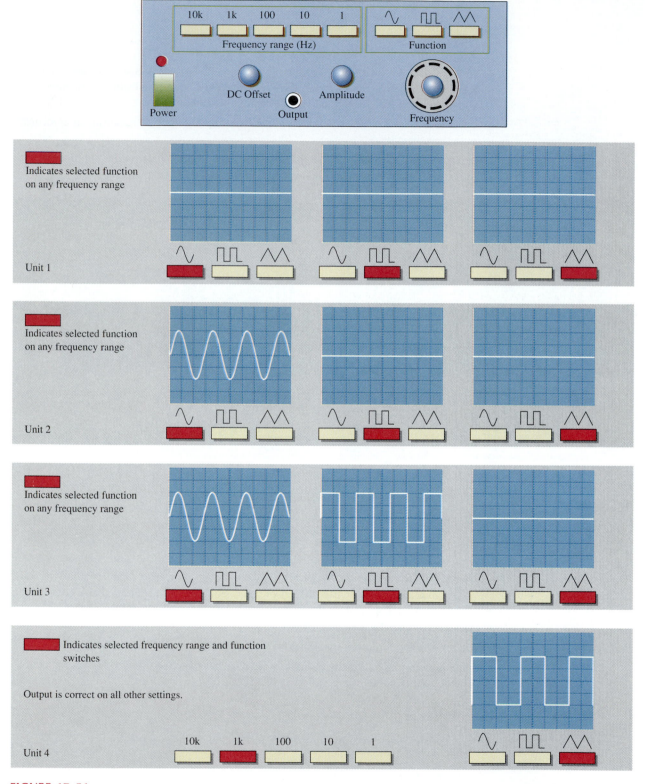

FIGURE 17–51

Results of tests on four faulty units. The scope screen shows the output voltage in each case.

Final Report

Submit a final written report on the channel filter boards using an organized format that includes the following:

1. A physical description of the circuit.
2. A discussion of the operation of the circuit.
3. A list of the specifications.
4. A list of parts with part numbers if available.
5. A list of the types of problems in the four faulty units.

■ **CHAPTER SUMMARY**

- Oscillators operate with positive feedback.
- The two conditions for positive feedback are the phase shift around the feedback loop must be 0° and the voltage gain around the feedback loop must equal 1.
- For initial start-up, the loop gain must be greater than 1.
- Sinusoidal RC oscillators include the Wien-bridge, phase-shift, and twin-T.
- Sinusoidal LC oscillators include the Colpitts, Clapp, Hartley, Armstrong, and crystal-controlled.
- The feedback signal in a Colpitts oscillator is derived from a capacitive voltage divider in the LC circuit.
- The Clapp oscillator is a variation of the Colpitts with a capacitor added in series with the inductor.
- The feedback signal in a Hartley oscillator is derived from an inductive voltage divider in the LC circuit.
- The feedback signal in an Armstrong oscillator is derived by transformer coupling.
- Crystal oscillators are the most stable type.
- The frequency in a voltage-controlled oscillator (VCO) can be varied with a dc control voltage.
- The 555 timer is an integrated circuit that can be used as an oscillator, in addition to many other applications.
- A phase-locked loop can lock onto and track a signal whose frequency is changing.

■ **GLOSSARY**

Crystal   A quartz device that operates on the piezoelectric effect and exhibits very stable resonant properties.

Oscillator   An electronic circuit that operates with positive feedback and produces a time-varying output signal without an external input signal.

Piezoelectric effect   The property of a crystal whereby a changing mechanical stress produces a voltage across the crystal.

Positive feedback   The return of a portion of the output signal to the input such that it sustains the output.

■ **FORMULAS**

(17–1) $\dfrac{V_{out}}{V_{in}} = \dfrac{1}{3}$ Wien-bridge positive feedback attenuation

(17–2) $f_r = \dfrac{1}{2\pi RC}$ Wien-bridge resonant frequency

(17–3) $B = \dfrac{1}{29}$ Phase-shift feedback attenuation

(17–4) $$f_r = \frac{1}{2\pi\sqrt{6}RC}$$ Phase-shift oscillator frequency

(17–5) $$f_r \cong \frac{1}{2\pi\sqrt{LC_T}}$$ Colpitts approximate resonant frequency

(17–6) $$C_T = \frac{C_1 C_2}{C_1 + C_2}$$ Colpitts feedback capacitance

(17–7) $$B = \frac{C_2}{C_1}$$ Colpitts feedback attenuation

(17–8) $$A_v = \frac{C_1}{C_2}$$ Colpitts amplifier gain

(17–9) $$A_v > \frac{C_1}{C_2}$$ Colpitts self-starting gain

(17–10) $$f_r = \frac{1}{2\pi\sqrt{LC_T}}\sqrt{\frac{Q^2}{Q^2 + 1}}$$ Colpitts resonant frequency

(17–11) $$C_T = \frac{1}{1/C_1 + 1/C_2 + 1/C_3}$$ Clapp feedback capacitance

(17–12) $$f_r \cong \frac{1}{2\pi\sqrt{L_T C}}$$ Hartley approximate resonant frequency

(17–13) $$B \cong \frac{L_1}{L_2}$$ Hartley feedback attenuation

(17–14) $$A_v > \frac{L_2}{L_1}$$ Hartley self-starting gain

(17–15) $$V_{\text{UTP}} = +V_{max}\left(\frac{R_3}{R_2}\right)$$ Triangular-wave oscillator upper trigger point

(17–16) $$V_{\text{LTP}} = -V_{max}\left(\frac{R_3}{R_2}\right)$$ Triangular-wave oscillator lower trigger point

(17–17) $$f_r = \frac{1}{4R_1 C}\left(\frac{R_2}{R_3}\right)$$ Triangular-wave oscillator frequency

(17–18) $$T = \frac{V_p - V_{\text{F}}}{|V_{\text{IN}}|/R_i C}$$ Sawtooth VCO period

(17–19) $$f = \frac{|V_{\text{IN}}|}{R_i C}\left(\frac{1}{V_p - V_{\text{F}}}\right)$$ Sawtooth VCO frequency

(17–20) $$f_r = \frac{1.44}{(R_1 + 2R_2)C_{ext}}$$ 555 astable frequency

(17–21) $$\text{Duty cycle} = \frac{R_1 + R_2}{R_1 + 2R_2} \times 100\%$$ 555 astable

(17–22) $$\text{Duty cycle} = \frac{R_1}{R_1 + R_2} \times 100\%$$ 555 astable (duty cycle < 50%)

(17–23) $$V_{out} = \frac{V_{ip}V_{op}}{2}\cos 2\pi(f_i - f_o)t - \frac{V_{ip}V_{op}}{2}\cos 2\pi(f_i + f_o)t$$ PLL output

▪ **SELF-TEST**

1. An oscillator differs from an amplifier because
   (a) it has more gain
   (b) it requires no input signal
   (c) it requires no dc supply
   (d) it always has the same output

2. All oscillators are based on
   (a) positive feedback
   (b) negative feedback
   (c) the piezoelectric effect
   (d) high gain

3. One condition for oscillation is
   (a) a phase shift around the feedback loop of $180°$
   (b) a gain around the feedback loop of one-third
   (c) a phase shift around the feedback loop of $0°$
   (d) a gain around the feedback loop of less than one

4. A second condition for oscillation is
   (a) no gain around the feedback loop
   (b) a gain of one around the feedback loop
   (c) the attenuation of the feedback circuit must be one-third
   (d) the feedback circuit must be capacitive

5. In a certain oscillator, $A_v = 50$. The attenuation of the feedback circuit must be
   (a) 1   (b) 0.01   (c) 10   (d) 0.02

6. For an oscillator to properly start, the gain around the feedback loop must initially be
   (a) 1   (b) less than 1   (c) greater than 1   (d) equal to B

7. In a Wien-bridge oscillator, if the resistances in the positive feedback circuit are decreased, the frequency
   (a) decreases   (b) increases   (c) remains the same

8. The Wien-bridge oscillator's positive feedback circuit is
   (a) an RL circuit
   (b) an LC circuit
   (c) a voltage divider
   (d) a lead-lag circuit

9. A phase-shift oscillator has
   (a) three RC circuits
   (b) three LC circuits
   (c) a T-type circuit
   (d) a π-type circuit

10. Colpitts, Clapp, and Hartley are names that refer to
    (a) types of RC oscillators
    (b) inventors of the transistor
    (c) types of LC oscillators
    (d) types of filters

11. An oscillator whose frequency is changed by a variable dc voltage is known as
    (a) a crystal oscillator
    (b) a VCO
    (c) an Armstrong oscillator
    (d) a piezoelectric device

12. The main feature of a crystal oscillator is
    (a) economy   (b) reliability   (c) stability   (d) high frequency

13. The operation of a relaxation oscillator is based on
    (a) the charging and discharging of a capacitor
    (b) a highly selective resonant circuit
    (c) a very stable supply voltage
    (d) low power consumption

14. Which one of the following is *not* an input or output of the 555 timer?
    (a) Threshold   (b) Control voltage   (c) Clock   (d) Trigger
    (d) Discharge   (f) Reset

15. A type of circuit that is capable of locking onto or synchronizing with an incoming signal is called
    (a) an astable multivibrator
    (b) a monostable multivibrator
    (c) a phase-locked loop
    (d) a phase detector

■ BASIC PROBLEMS

SECTION 17–1   The Oscillator

1. What type of input is required for an oscillator?
2. What are the basic components of an oscillator circuit?

SECTION 17–2   Oscillator Principles

3. If the voltage gain of the amplifier portion of an oscillator is 75, what must be the attenuation of the feedback circuit to sustain the oscillation?
4. Generally describe the change required in the oscillator of Problem 3 in order for oscillation to begin when the power is initially turned on.

SECTION 17–3   Oscillators with *RC* Feedback Circuits

5. A certain lead-lag circuit has a resonant frequency of 3.5 kHz. What is the rms output voltage if an input signal with a frequency equal to f_r and with an rms value of 2.2 V is applied to the input?
6. Calculate the resonant frequency of a lead-lag circuit with the following values: $R_1 = R_2 = 6.2 \text{ k}\Omega$, and $C_1 = C_2 = 0.02 \text{ μF}$.
7. Determine the necessary value of R_2 in Figure 17–52 so that the circuit will oscillate. Neglect the forward resistance of the zener diodes.
8. Explain the purpose of R_3 in Figure 17–52.
9. What is the initial closed-loop gain in Figure 17–52? At what value of output voltage does A_{cl} change and to what value does it change? (The value of R_2 was found in Problem 7).
10. Find the frequency of oscillation for the Wien-bridge oscillator in Figure 17–52.
11. What value of R_f is required in Figure 17–53? What is f_r?

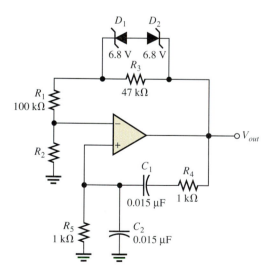

FIGURE 17–52

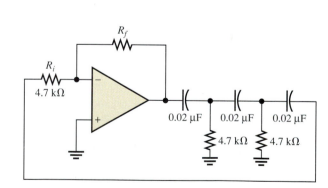

FIGURE 17–53

SECTION 17–4   Oscillators with *LC* Feedback Circuits

12. Calculate the frequency of oscillation for each circuit in Figure 17–54 and identify the type of oscillator. Assume $Q > 10$ in each case.

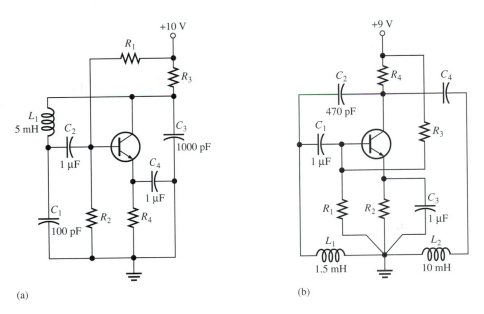

(a)   (b)

FIGURE 17–54

13. Determine what the gain of the amplifier stage must be in Figure 17–55 in order to have sustained oscillation.

FIGURE 17–55

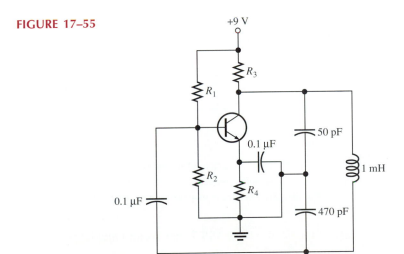

SECTION 17–5   Nonsinusoidal Oscillators

14. What type of signal does the circuit in Figure 17–56 produce? Determine the frequency of the output.

FIGURE 17–56

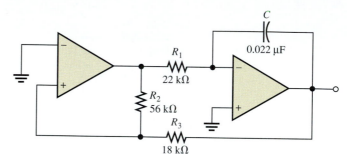

15. Show how to change the frequency of oscillation in Figure 17–56 to 10 kHz.

16. Determine the amplitude and frequency of the output voltage in Figure 17–57. Use 1 V as the forward PUT voltage.

FIGURE 17–57

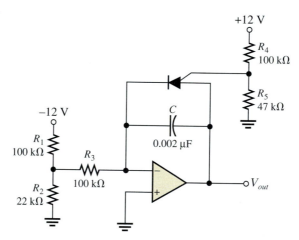

17. Modify the sawtooth generator in Figure 17–57 so that its peak-to-peak output is 4 V.

18. A certain sawtooth generator has the following parameter values: $V_{IN} = 3$ V, $R = 4.7$ kΩ, $C = 0.001$ μF. Determine its peak-to-peak output voltage if the period is 10 μs.

SECTION 17–6   The 555 Timer As an Oscillator

19. What are the two comparator reference voltages in a 555 timer when $V_{CC} = 10$ V?

20. Determine the frequency of oscillation for the 555 astable oscillator in Figure 17–58.

21. To what value must C_{ext} be changed in Figure 17–58 to achieve a frequency of 25 kHz?

22. In an astable 555 configuration, the external resistor $R_1 = 3.3$ kΩ. What must R_2 equal to produce a duty cycle of 75 percent?

FIGURE 17–58

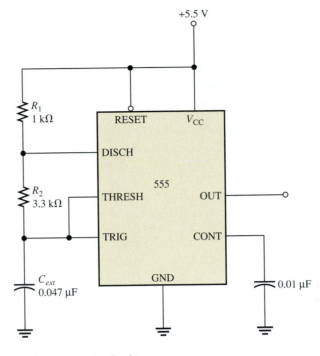

SECTION 17–7 The Phase-Locked Loop

23. The lock range of a certain PLL is specified to be ±15 percent of the center frequency. Determine the minimum and maximum frequencies for which the PLL will maintain lock if $f_o = 50$ kHz.

24. A 15 kHz signal f_o and a 7.5 kHz signal f_i are applied to a PLL phase detector. Determine the sum and difference frequencies.

25. A 25 kHz sine wave with a peak value of 50 mV is applied to a PLL. When the PLL is in lock, what is the VCO frequency?

■ **ANSWERS TO SECTION REVIEWS**

Section 17–1

1. An oscillator is a circuit that produces a repetitive output waveform with only the dc supply voltage as an input.

2. Positive feedback

3. The feedback circuit provides attenuation and phase shift.

Section 17–2

1. Zero phase shift and unity voltage gain around the closed feedback

2. Positive feedback is when a portion of the output signal is fed back to the input of the amplifier such that it reinforces itself.

3. Loop gain greater than 1

Section 17–3

1. The negative feedback loop sets the closed-loop gain; the positive feedback loop sets the frequency of oscillation.

2. 1.67 V

3. The three RC circuits each contribute 60°.

Section 17–4

1. Colpitts uses a capacitive voltage divider in the feedback circuit; Hartley uses an inductive voltage divider.
2. The higher FET input impedance has less loading effect on the resonant feedback circuit.
3. A Clapp has an additional capacitor in series with the inductor in the feedback circuit.

Section 17–5

1. A voltage-controlled oscillator exhibits a frequency that can be varied with a dc control voltage.
2. The basis of a relaxation oscillator is the charging and discharging of a capacitor.

Section 17–6

1. Two comparators, a flip-flop, a discharge transistor, and a resistive voltage divider
2. The duty cycle is set by the external resistors.

Section 17–7

1. The PLL blocks are phase detector, VCO, low-pass filter, and amplifier.
2. A PLL maintains synchronization (lock) with an incoming signal.

■ **ANSWERS TO RELATED EXERCISES FOR EXAMPLES**

17–1 Change the zener diodes to 6.1 V devices.

17–2 (a) 238 kΩ   (b) 7.92 kHz

17–3 7.24 kHz

17–4 6.06 V peak-to-peak

17–5 1.06 kHz

17–6 31.9%

17–7 $f_i = 18$ kHz and $f_o = 12$ kHz

18

VOLTAGE REGULATORS

■ **CHAPTER OBJECTIVES**

☐ Describe the basic concept of voltage regulation
☐ Discuss the principles of series voltage regulators
☐ Discuss the principles of shunt voltage regulators
☐ Discuss the principles of switching regulators
☐ Discuss integrated circuit voltage regulators
☐ Discuss applications of IC voltage regulators

A voltage regulator provides a constant dc output voltage that is essentially independent of the input voltage, output load current, and temperature. The voltage regulator is one part of a power supply. Its input voltage comes from the filtered output of a rectifier derived from an ac voltage or from a battery in the case of portable systems.

Most voltage regulators fall into two broad categories: linear regulators and switching regulators. In the linear regulator category, two general types are the series regulator and the shunt regulator. These are normally available for either positive or negative output voltages. A dual regulator provides both positive and negative outputs. In the switching regulator category, three general configurations are step-down, step-up, and inverting.

Many types of integrated circuit (IC) regulators are available. The most popular types of linear regulator are the three-terminal fixed voltage regulator and the three-terminal adjustable voltage regulator. Switching regulators are also widely used. In this chapter, specific IC devices are introduced as representative of the wide range of available devices.

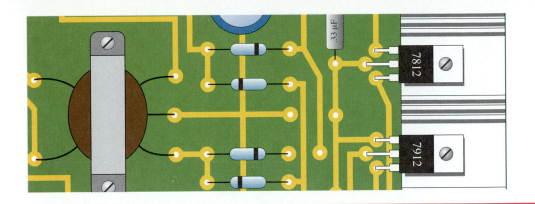

A dual-polarity power supply is to be used for the FM receiver from Chapter 16. Two regulators, one positive and the other negative, provide the positive voltage required for the receiver circuits and the dual-polarity voltages for the op-amp circuits.

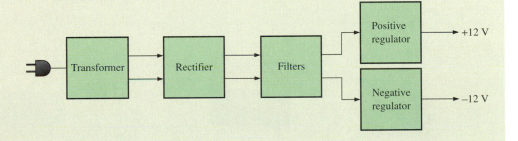

18–1 ■ VOLTAGE REGULATION

Two basic categories of voltage regulation are line regulation and load regulation. Line regulation maintains a nearly constant output voltage when the input voltage varies. Load regulation maintains a nearly constant output voltage when the load varies.

After completing this section, you should be able to

■ **Describe the basic concept of voltage regulation**
 □ Explain line regulation
 □ Calculate line regulation
 □ Explain load regulation
 □ Calculate load regulation

Line Regulation

When the dc input (line) voltage changes, the voltage **regulator** must maintain a nearly constant output voltage, as illustrated in Figure 18–1.

 Line regulation can be defined as the percentage change in the output voltage for a given change in the input (line) voltage. It is usually expressed in units of %/V. For example, a line regulation of 0.05%/V means that the output voltage changes 0.05 percent

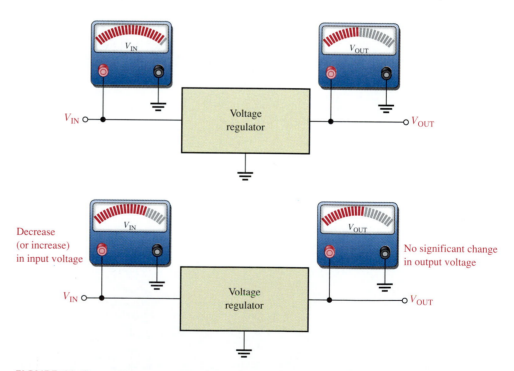

FIGURE 18–1

Line regulation. A change in input (line) voltage does not significantly affect the output voltage of a regulator (within certain limits).

when the input voltage increases or decreases by one volt. Line regulation can be calculated using the following formula (Δ means "a change in"):

$$\text{Line regulation} = \frac{\left(\dfrac{\Delta V_{\text{OUT}}}{V_{\text{OUT}}}\right)100\%}{\Delta V_{\text{IN}}} \qquad \text{(18–1)}$$

EXAMPLE 18–1

When the input to a particular voltage regulator decreases by 5 V, the output decreases by 0.25 V. The nominal output is 15 V. Determine the line regulation in %/V.

Solution   The line regulation is

$$\text{line regulation} = \frac{\left(\dfrac{\Delta V_{\text{OUT}}}{V_{\text{OUT}}}\right)100\%}{\Delta V_{\text{IN}}} = \frac{\left(\dfrac{0.25\ \text{V}}{15\ \text{V}}\right)100\%}{5\ \text{V}} = 0.333\%/\text{V}$$

Related Exercise   The input of a certain regulator increases by 3.5 V. As a result, the output voltage increases by 0.42 V. The nominal output is 20 V. Determine the regulation in %/V.

Load Regulation

When the amount of current through a load changes due to a varying load resistance, the voltage regulator must maintain a nearly constant output voltage across the load, as illustrated in Figure 18–2.

FIGURE 18–2

Load regulation. A change in load current has practically no effect on the output voltage of a regulator (within certain limits).

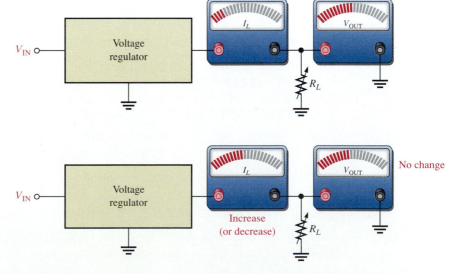

Load regulation can be defined as the percentage change in output voltage for a given change in load current. One way to express load regulation is as a percentage change in output voltage from no-load (NL) to full-load (FL) as follows.

$$\text{Load regulation} = \left(\frac{V_{\text{NL}} - V_{\text{FL}}}{V_{\text{FL}}}\right)100\%$$

(18–2)

Alternately, the load regulation can be expressed as a percentage change in output voltage for each mA change in load current. For example, a load regulation of 0.01%/mA means that the output voltage changes 0.01 percent when the load current increases or decreases 1 mA.

EXAMPLE 18–2   A certain voltage regulator has a 12 V output when there is no load ($I_L = 0$). When there is a full-load current of 10 mA, the output voltage is 11.9 V. Express the voltage regulation as a percentage change from no-load to full-load and also as a percentage change for each mA change in load current.

Solution   The no-load output voltage is

$$V_{\text{NL}} = 12 \text{ V}$$

The full-load output voltage is

$$V_{\text{FL}} = 11.9 \text{ V}$$

The load regulation is

$$\text{load regulation} = \left(\frac{V_{\text{NL}} - V_{\text{FL}}}{V_{\text{FL}}}\right)100\% = \left(\frac{12 \text{ V} - 11.9 \text{ V}}{11.9 \text{ V}}\right)100\% = 0.840\%$$

The load regulation can also be expressed as

$$\text{load regulation} = \frac{0.840\%}{10 \text{ mA}} = 0.084\%/\text{mA}$$

where the change in load current from no-load to full-load is 10 mA.

Related Exercise   A regulator has a no-load output voltage of 18 V and a full-load output of 17.8 V at a load current of 50 mA. Determine the voltage regulation as a percentage change from no-load to full-load and also as a percentage change for each mA change in load current.

SECTION 18–1 REVIEW
1. Define *line regulation*.
2. Define *load regulation*.

18–2 ■ BASIC SERIES REGULATORS

The two fundamental classes of voltage regulators are linear regulators and switching regulators. Both of these are available in integrated circuit form. There are two basic types of linear regulator. One is the series regulator and the other is the shunt regulator. In this section, we will look at the series regulator. The shunt and switching regulators are covered in the next two sections.

After completing this section, you should be able to

■ **Discuss the principles of series voltage regulators**
  □ Explain regulating action
  □ Calculate output voltage of an op-amp series regulator
  □ Discuss overload protection and explain how to use current limiting
  □ Describe a regulator with fold-back current limiting

A simple representation of a series type of linear regulator is shown in Figure 18–3(a), and the basic components are shown in the block diagram in Figure 18–3(b). The control element is in series with the load between input and output. The output sample circuit senses a change in the output voltage. The error detector compares the sample voltage with a reference voltage and causes the control element to compensate in order to maintain a constant output voltage.

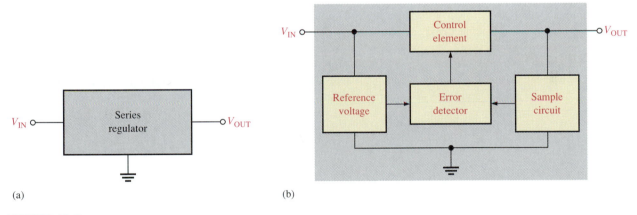

(a)   (b)

FIGURE 18–3
Simple series voltage regulator block diagram.

Regulating Action

A basic op-amp series regulator circuit is shown in Figure 18–4. The operation of the series regulator is illustrated in Figure 18–5 and is as follows. The resistive voltage divider formed by R_2 and R_3 senses any change in the output voltage. When the output tries to decrease, as indicated in part (a), because of a decrease in V_{IN} or because of an increase in I_L, a proportional voltage decrease is applied to the op-amp's inverting input by the voltage divider. Since the zener diode (D_1) holds the other op-amp input at a nearly constant reference voltage V_{REF}, a small difference voltage (error voltage) is developed

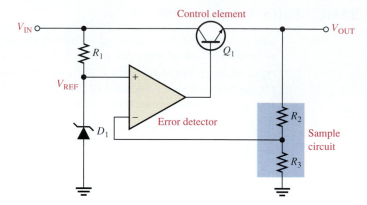

FIGURE 18–4

Basic op-amp series regulator.

across the op-amp's inputs. This difference voltage is amplified, and the op-amp's output voltage, V_B, increases. This increase is applied to the base of Q_1, causing the emitter voltage V_{OUT} to increase until the voltage to the inverting input again equals the reference (zener) voltage. This action offsets the attempted decrease in output voltage, thus keeping it nearly constant, as shown in part (b). The power transistor, Q_1 is usually used with a heat sink because it must handle all of the load current.

The opposite action occurs when the output tries to increase, as indicated in Figure 18–5(c) and (d). The op-amp in the series regulator is actually connected as a noninverting amplifier where the reference voltage V_{REF} is the input at the noninverting terminal, and the R_2/R_3 voltage divider forms the negative feedback circuit. The closed-loop voltage gain is

$$A_{cl} = 1 + \frac{R_2}{R_3} \tag{18–3}$$

Therefore, the regulated output voltage (neglecting the base-emitter voltage of Q_1) is

$$V_{OUT} \cong \left(1 + \frac{R_2}{R_3}\right)V_{REF} \tag{18–4}$$

From this analysis, you can see that the output voltage is determined by the zener voltage and the resistors R_2 and R_3. It is relatively independent of the input voltage, and therefore, regulation is achieved (as long as the input voltage and load current are within specified limits).

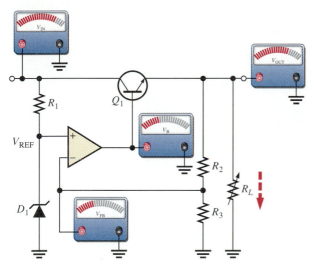

(a) When V_{IN} or R_L decreases, V_{OUT} attempts to decrease. The feedback voltage, V_{FB}, also attempts to decrease, and as a result, the op-amp's output voltage V_B attempts to increase, thus compensating for the attempted decrease in V_{OUT} by increasing the Q_1 emitter voltage. Changes in V_{OUT} are exaggerated for illustration.

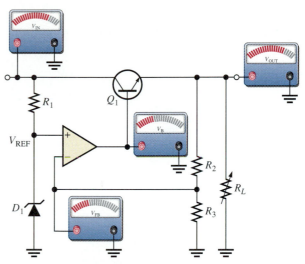

(b) When V_{IN} (or R_L) stabilizes at its new lower value, the voltages return to their original values, thus keeping V_{OUT} constant as a result of the negative feedback.

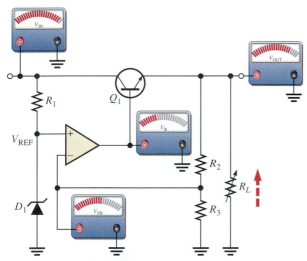

(c) When V_{IN} or R_L increases, V_{OUT} attempts to increase. The feedback voltage, V_{FB}, also attempts to increase, and as a result, V_B applied to the base of the control transistor, attempts to decrease, thus compensating for the attempted increase in V_{OUT} by decreasing the Q_1 emitter voltage.

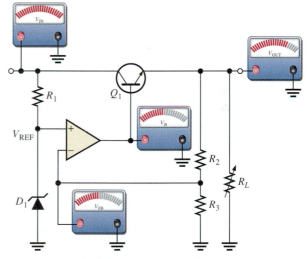

(d) When V_{IN} (or R_L) stabilizes at its new higher value, the voltages return to their original values, thus keeping V_{OUT} constant as a result of the negative feedback.

FIGURE 18–5

Illustration of series regulator action that keeps V_{OUT} constant when V_{IN} or R_L changes.

EXAMPLE 18–3   Determine the output voltage for the regulator in Figure 18–6.

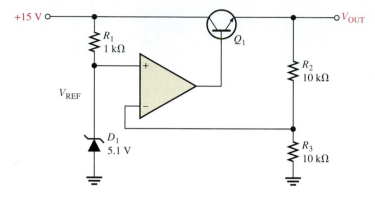

FIGURE 18–6

Solution $V_{REF} = 5.1$ V, the zener voltage. The regulated output voltage is therefore

$$V_{OUT} = \left(1 + \frac{R_2}{R_3}\right)V_{REF} = \left(1 + \frac{10 \text{ k}\Omega}{10 \text{ k}\Omega}\right)5.1 \text{ V} = (2)5.1 \text{ V} = 10.2 \text{ V}$$

Related Exercise   The following changes are made in the circuit in Figure 18–6: A 3.3 V zener replaces the 5.1 V zener, $R_1 = 1.8$ kΩ, $R_2 = 22$ kΩ, and $R_3 = 18$ kΩ. What is the output voltage?

Short-Circuit or Overload Protection

If an excessive amount of load current is drawn, the series-pass transistor can be quickly damaged or destroyed. Most regulators employ some type of excess current protection in the form of a current-limiting mechanism. Figure 18–7 shows one method of current limiting to prevent overloads called *constant-current limiting*. The current-limiting circuit consists of transistor Q_2 and resistor R_4.

FIGURE 18–7
Series regulator with constant-current limiting.

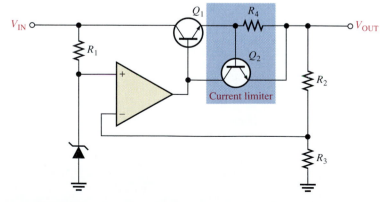

The load current through R_4 creates a voltage from base to emitter of Q_2. When I_L reaches a predetermined maximum value, the voltage drop across R_4 is sufficient to forward-bias the base-emitter junction of Q_2, thus causing it to conduct. Enough Q_1 base current is diverted from the collector of Q_2 so that I_L is limited to its maximum value $I_{L(max)}$. Since the base-to-emitter voltage of Q_2 cannot exceed about 0.7 V for a silicon transistor, the voltage across R_4 is held to this value, and the load current is limited to

$$I_{L(\text{max})} = \frac{0.7\text{ V}}{R_4} \qquad\qquad (18\text{–}5)$$

EXAMPLE 18–4

Determine the maximum current that the regulator in Figure 18–8 can provide to a load.

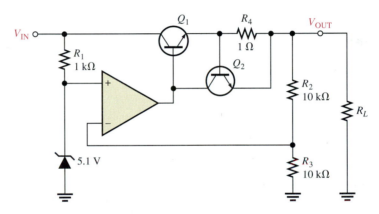

FIGURE 18–8

Solution

$$I_{L(\text{max})} = \frac{0.7\text{ V}}{R_4} = \frac{0.7\text{ V}}{1\ \Omega} = 0.7\text{ A}$$

Related Exercise   If the output of the regulator in Figure 18–8 is shorted, what is the current?

Regulator with Fold-Back Current Limiting

In the previous current-limiting technique, the current is restricted to a maximum constant value.

Fold-back current limiting is a method used particularly in high-current regulators whereby the output current under overload conditions drops to a value well below the peak load current capability to prevent excessive power dissipation.

Basic Idea   The basic concept of fold-back current limiting is as follows, with reference to Figure 18–9. The circuit is similar to the constant current-limiting arrangement in Figure 18–7, with the exception of resistors R_5 and R_6. The voltage drop developed across R_4

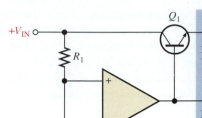
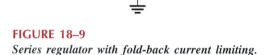

FIGURE 18–9
Series regulator with fold-back current limiting.

by the load current must not only overcome the base-emitter voltage required to turn on Q_2, but it must overcome the voltage across R_5. That is, the voltage across R_4 must be

$$V_{R4} = V_{R5} + V_{BE}$$

In an overload or short-circuit condition the load current increases to a value $I_{L(max)}$ that is sufficient to cause Q_2 to conduct. At this point the current can increase no further. The decrease in output voltage results in a proportional decrease in the voltage across R_5; thus, less current through R_4 is required to maintain the forward-biased condition of Q_1. So, as V_{OUT} decreases, I_L decreases, as shown in the graph of Figure 18–10.

The advantage of this technique is that the regulator is allowed to operate with peak load current up to $I_{L(max)}$; but when the output becomes shorted, the current drops to a lower value to prevent overheating of the device.

FIGURE 18–10
Fold-back current limiting (output voltage versus load current).

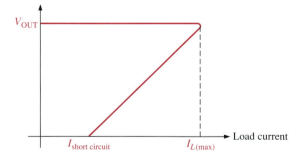

SECTION 18–2 REVIEW

1. What are the basic components in a series regulator?
2. A certain series regulator has an output voltage of 8 V. If the op-amp's closed loop gain is 4, what is the value of the reference voltage?

18–3 ■ BASIC SHUNT REGULATORS

The second basic type of linear voltage regulator is the shunt regulator. As you have learned, the control element in the series regulator is the series-pass transistor. In the shunt regulator, the control element is a transistor in parallel (shunt) with the load.

After completing this section, you should be able to

- **Discuss the principles of shunt voltage regulators**
  - ☐ Describe the operation of a basic op-amp shunt regulator
  - ☐ Compare series and shunt regulators

A simple representation of a shunt type of linear regulator is shown in Figure 18–11(a), and the basic components are shown in the block diagram in part (b).

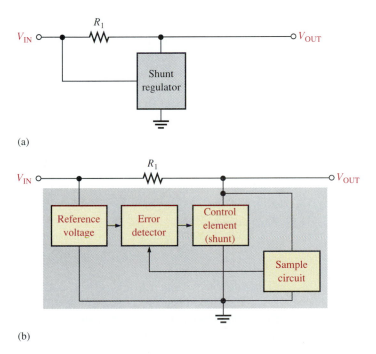

(a)

(b)

FIGURE 18–11
Simple shunt regulator block diagrams.

In the basic shunt regulator, the control element is a transistor Q_1 in parallel with the load, as shown in Figure 18–12. A resistor, R_1 is in series with the load. The operation of the circuit is similar to that of the series regulator, except that regulation is achieved by controlling the current through the parallel transistor Q_1.

When the output voltage tries to decrease due to a change in input voltage or load current, as shown in Figure 18–13(a), the attempted decrease is sensed by R_3 and R_4 and applied to the op-amp's noninverting input. The resulting difference voltage reduces the

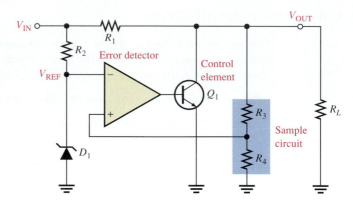

FIGURE 18–12
Basic op-amp shunt regulator with load resistor.

op-amp's output (V_B), driving Q_1 less, thus reducing its collector current (shunt current) and increasing its effective collector-to-emitter resistance r'_{CE}. Since r'_{CE} acts as a voltage divider with R_1, this action offsets the attempted decrease in V_{OUT} and maintains it at an almost constant level.

The opposite action occurs when the output tries to increase, as indicated in Figure 18–13(b). With I_L and V_{OUT} constant, a change in the input voltage produces a change in shunt current (I_S) as follows (Δ means "a change in").

$$\Delta I_S = \frac{\Delta V_{IN}}{R_1} \tag{18–6}$$

With a constant V_{IN} and V_{OUT}, a change in load current causes an opposite change in shunt current.

$$\Delta I_S = -\Delta I_L \tag{18–7}$$

This formula says that if I_L increases, I_S decreases, and vice versa.

The shunt regulator is less efficient than the series type but offers inherent short-circuit protection. If the output is shorted ($V_{OUT} = 0$), the load current is limited by the series resistor R_1 to a maximum value as follows ($I_S = 0$).

$$I_{L(max)} = \frac{V_{IN}}{R_1} \tag{18–8}$$

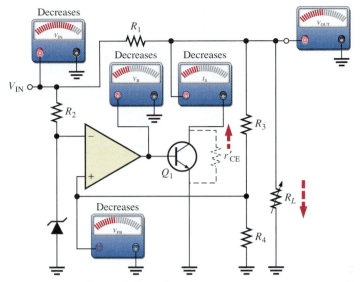

(a) Response to a decrease in V_{IN} or R_L

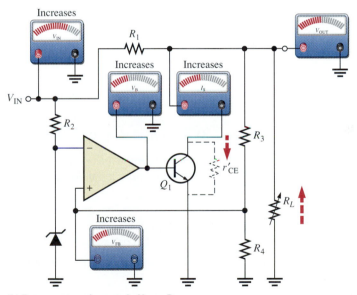

(b) Response to an increase in V_{IN} or R_L

FIGURE 18–13

Sequence of responses when V_{OUT} tries to decrease as a result of a decrease in R_L or V_{IN} (opposite responses for an attempted increase).

EXAMPLE 18–5   In Figure 18–14, what power rating must R_1 have if the maximum input voltage is 12.5 V?

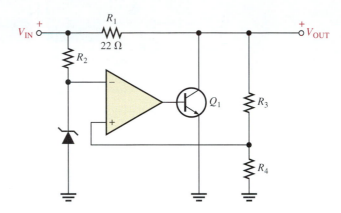

FIGURE 18–14

Solution   The worst-case power dissipation in R_1 occurs when the output is short-circuited and $V_{OUT} = 0$. When $V_{IN} = 12.5$ V, the voltage dropped across R_1 is

$$V_{R1} = V_{IN} - V_{OUT} = 12.5 \text{ V}$$

The power dissipation in R_1 is

$$P_{R1} = \frac{V_{R1}^2}{R_1} = \frac{(12.5 \text{ V})^2}{22 \text{ } \Omega} = 7.10 \text{ W}$$

Therefore, a resistor of at least 10 W should be used.

Related Exercise   In Figure 18–14, R_1 is changed to 33 Ω. What must be the power rating of R_1 if the maximum input voltage is 24 V?

SECTION 18–3 REVIEW

1. How does the control element in a shunt regulator differ from that in a series regulator?
2. What is one advantage of a shunt regulator over a series type? What is a disadvantage?

18–4 ■ BASIC SWITCHING REGULATORS

The two types of linear regulators, series and shunt, have control elements (transistors) that are conducting all the time, with the amount of conduction varied as demanded by changes in the output voltage or current. The switching regulator is different; the control element operates as a switch. A greater efficiency can be realized with this type of voltage regulator than with the linear types because the transistor is not always conducting. Therefore, switching regulators can provide greater load currents at low voltage than linear regulators because the control transistor doesn't dissipate as much power. Three basic configurations of switching regulators are step-down, step-up, and inverting.

After completing this section, you should be able to

■ **Discuss the principles of switching regulators**
  ☐ Describe the step-down configuration of switching regulator
  ☐ Determine output voltage of the step-down configuration
  ☐ Describe the step-up configuration of switching regulator
  ☐ Describe the voltage-inverter configuration

Step-Down Configuration

In the step-down configuration, the output voltage is always less than the input voltage. A basic step-down switching regulator is shown in Figure 18–15(a), and its simplified equivalent is shown in Figure 18–15(b). Transistor Q_1 is used to switch the input voltage at a

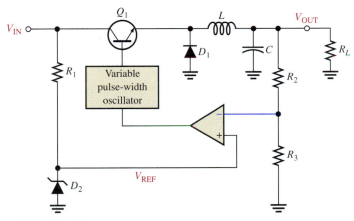

(a) Typical circuit

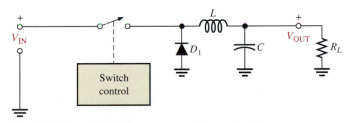

(b) Simplified equivalent circuit

FIGURE 18–15
Basic step-down switching regulator.

duty cycle that is based on the regulator's load requirement. The *LC* filter is then used to average the switched voltage. Since Q_1 is either *on* (saturated) or *off*, the power lost in the control element is relatively small. Therefore, the switching regulator is useful primarily in higher power applications or in applications where efficiency is of utmost concern.

The on and off intervals of Q_1 are shown in the waveform of Figure 18–16(a). The capacitor charges during the on-time (t_{on}) and discharges during the off-time (t_{off}). When the on-time is increased relative to the off-time, the capacitor charges more, thus increasing the output voltage, as indicated in Figure 18–16(b). When the on-time is decreased relative to the off-time, the capacitor discharges more, thus decreasing the output voltage, as in Figure 18–16(c). Therefore, by adjusting the duty cycle $t_{on}/(t_{on} + t_{off})$ of Q_1, the output voltage can be varied. The inductor further smooths the fluctuations of the output voltage caused by the charging and discharging action.

The output voltage is expressed as

$$V_{OUT} = \left(\frac{t_{on}}{T}\right)V_{IN}$$

(18–9)

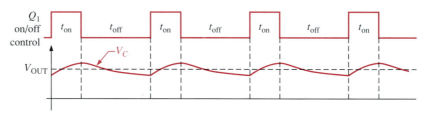

(a) V_{OUT} depends on the duty cycle.

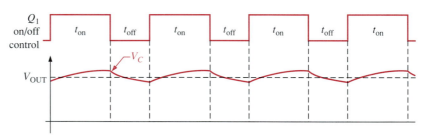

(b) Increase the duty cycle and V_{OUT} increases.

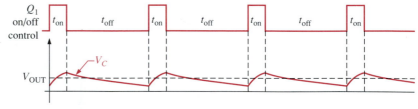

(c) Decrease the duty cycle and V_{OUT} decreases.

FIGURE 18–16

Switching regulator waveforms. The V_C waveform is for no inductive filtering to illustrate the charge and discharge action. L and C smooth V_C to a nearly constant level, as indicated by the dashed line for V_{OUT}.

T is the period of the on-off cycle of Q_1 and is related to the frequency by $T = 1/f$. The period is the sum of the on-time and the off-time.

$$T = t_{on} + t_{off} \qquad (18\text{--}10)$$

The ratio t_{on}/T is called the *duty cycle*.

The regulating action is as follows and is illustrated in Figure 18–17. When V_{OUT} tries to decrease, the on-time of Q_1 is increased, causing an additional charge on C to offset the attempted decrease. When V_{OUT} tries to increase, the on-time of Q_1 is decreased, causing the capacitor to discharge enough to offset the attempted increase.

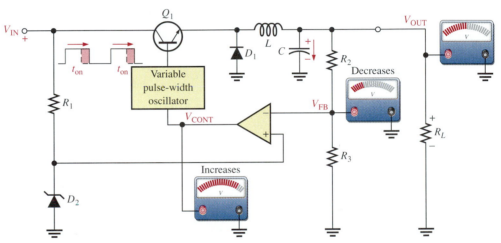

(a) When V_{OUT} attempts to decrease, the on-time of Q_1 increases.

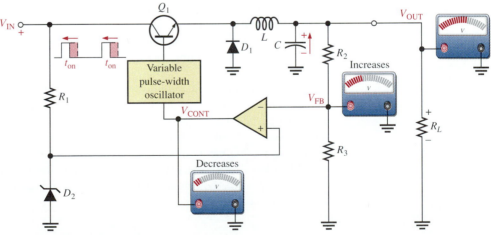

(b) When V_{OUT} attempts to increase, the on-time of Q_1 decreases.

FIGURE 18–17

Regulating action of the basic step-down switching regulator.

Step-Up Configuration

A basic step-up type of switching regulator is shown in Figure 18–18.

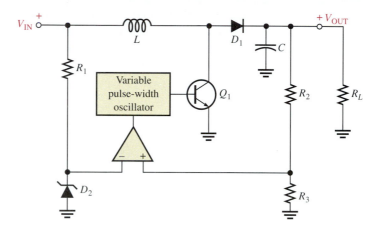

FIGURE 18–18
Basic step-up switching regulator.

The switching action is illustrated in Figure 18–19. When Q_1 turns on, voltage across L increases instantaneously to $V_{IN} - V_{CE(sat)}$, and the inductor's magnetic field expands quickly, as indicated in Figure 18–19(a). During the on-time (t_{on}) of Q_1, V_L decreases from its initial maximum, as shown. The longer Q_1 is on, the smaller V_L becomes. When Q_1 turns off, the inductor's magnetic field collapses; and its polarity reverses so that its voltage adds to V_{IN}, thus producing an output voltage greater than the input, as indicated in Figure 18–19(b). During the off-time (t_{off}) of Q_1, the diode is forward-biased, allowing the capacitor to charge. The variations in the output voltage due to the charging and discharging action are sufficiently smoothed by the filtering action of L and C.

The regulating action is illustrated in Figure 18–20. The shorter the on-time of Q_1, the greater the inductor voltage is, and thus the greater the output voltage is (greater V_L adds to V_{IN}). The longer the on-time of Q_1, the smaller are the inductor voltage and the output voltage (small V_L adds to V_{IN}). When V_{OUT} tries to decrease because of increasing load or decreasing input voltage, t_{on} decreases and the attempted decrease in V_{OUT} is offset. When V_{OUT} tries to increase, t_{on} increases and the attempted increase in V_{OUT} is offset. As you can see, the output voltage is inversely related to the duty cycle of Q_1 and can be expressed as follows.

$$V_{OUT} = \left(\frac{T}{t_{on}}\right)V_{IN}$$

(18–11)

where $T = t_{on} + t_{off}$.

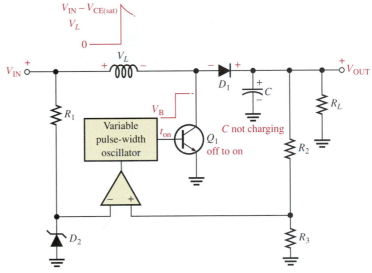

(a) When Q_1 is on

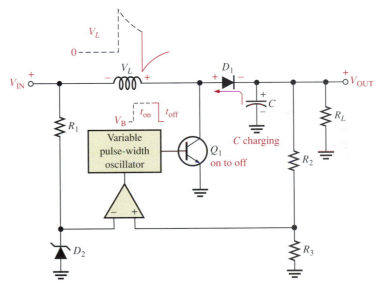

(b) When Q_1 turns off

FIGURE 18–19

Switching action of the basic step-up regulator.

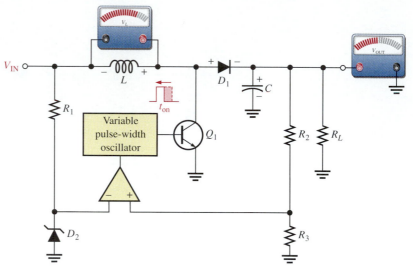

(a) When V_{OUT} tries to decrease, t_{on} decreases, causing V_L to
    increase. This compensates for the attempted decrease in V_{OUT}.

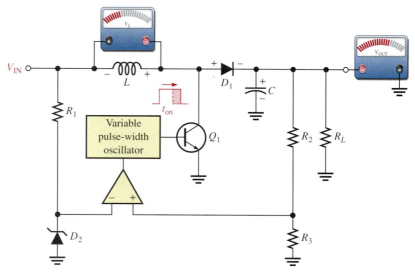

(b) When V_{OUT} tries to increase, t_{on} increases, causing V_L to
    decrease. This compensates for the attempted increase in V_{OUT}.

FIGURE 18–20

Regulating action of the basic step-up switching regulator.

Voltage-Inverter Configuration

A third type of switching regulator produces an output voltage that is opposite in polarity
to the input. A basic diagram is shown in Figure 18–21.

    When Q_1 turns on, the inductor voltage jumps to $V_{IN} - V_{CE(sat)}$ and the magnetic
field rapidly expands, as shown in Figure 18–22(a). While Q_1 is on, the diode is reverse-

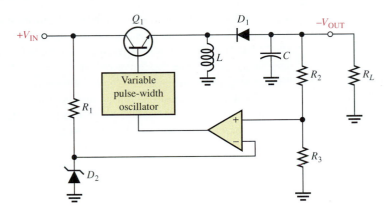

FIGURE 18–21

Basic inverting switching regulator.

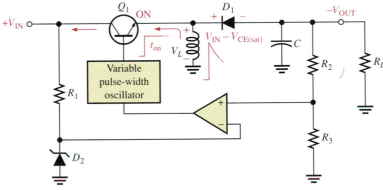

(a) When Q_1 is on, D_1 is reverse-biased.

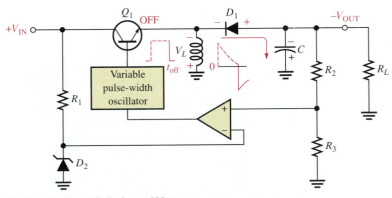

(b) When Q_1 turns off, D_1 forward biases.

FIGURE 18–22

Inverting action of the basic inverting switching regulator.

biased and the inductor voltage decreases from its initial maximum. When Q_1 turns off, the magnetic field collapses and the inductor's polarity reverses, as shown in Figure 18–22(b). This forward-biases the diode, charges C, and produces a negative output voltage, as indicated. The repetitive on-off action of Q_1 produces a repetitive charging and discharging that is smoothed by the LC filter action.

As with the step-up regulator, the less time Q_1 is on, the greater the output voltage is, and vice versa. This regulating action is illustrated in Figure 18–23. Switching regulator efficiencies can be greater than 90 percent.

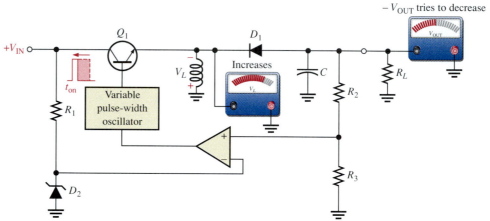

(a) When $-V_{OUT}$ tries to decrease, t_{on} decreases, causing V_L to increase. This compensates for the attempted decrease in $-V_{OUT}$.

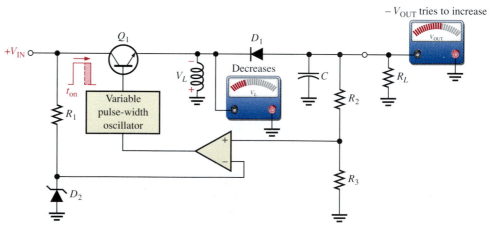

(b) When $-V_{OUT}$ tries to increase, t_{on} increases, causing V_L to decrease. This compensates for the attempted increase in $-V_{OUT}$.

FIGURE 18–23

Regulating action of the basic inverting switching regulator.

1. What are three types of switching regulators?
2. What is the primary advantage of switching regulators over linear regulators?
3. How are changes in output voltage compensated for in the switching regulator?

18–5 ■ INTEGRATED CIRCUIT VOLTAGE REGULATORS

In the previous sections, we presented the basic voltage regulator configurations. Several types of both linear and switching regulators are available in integrated circuit (IC) form. Generally, the linear regulators are three-terminal devices that provide either positive or negative output voltages that can be either fixed or adjustable. In this section, typical linear and switching IC regulators are introduced.

After completing this section, you should be able to

■ Discuss integrated circuit voltage regulators
  □ Describe the 7800 series of positive regulators
  □ Describe the 7900 series of negative regulators
  □ Describe the LM317 adjustable positive regulator
  □ Describe the LM337 adjustable negative regulator
  □ Describe IC switching regulators

Fixed Positive Linear Voltage Regulators

Although many types of IC regulators are available, the 7800 series of IC regulators is representative of three-terminal devices that provide a fixed positive output voltage. The three terminals are input, output, and ground as indicated in the standard fixed voltage configuration in Figure 18–24(a). The last two digits in the part number designate the output voltage. For example, the 7805 is a +5.0 V regulator. Other available output voltages are given in Figure 18–24(b) and common packages are shown in part (c).

Capacitors, although not always necessary, are sometimes used on the input and output as indicated in Figure 18–24(a). The output capacitor acts basically as a line filter to improve transient response. The input capacitor is used to prevent unwanted oscillations when the regulator is some distance from the power supply filter such that the line has a significant inductance.

The 7800 series can produce output current in excess of 1 A when used with an adequate heat sink. The 78L00 series can provide up to 100 mA, the 78M00 series can provide up to 500 mA, and the 78T00 series can provide in excess of 3 A.

The input voltage must be at least 2 V above the output voltage in order to maintain regulation. The circuits have internal thermal overload protection and short-circuit current-limiting features. **Thermal overload** occurs when the internal power dissipation becomes excessive and the temperature of the device exceeds a certain value.

Fixed Negative Linear Voltage Regulators

The 7900 series is typical of three-terminal IC regulators that provide a fixed negative output voltage. This series is the negative-voltage counterpart of the 7800 series and

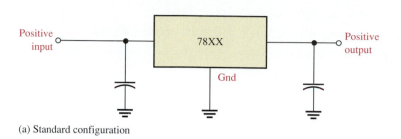

(a) Standard configuration

(b) The 7800 series

Type number	Output voltage
7805	+5.0 V
7806	+6.0 V
7808	+8.0 V
7809	+9.0 V
7812	+12.0 V
7815	+15.0 V
7818	+18.0 V
7824	+24.0 V

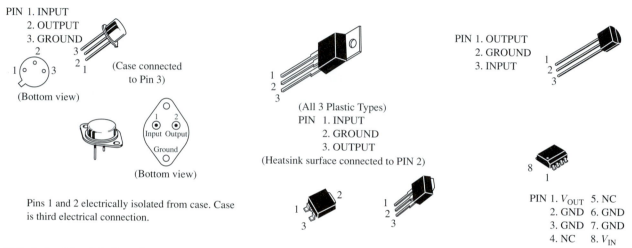

PIN 1. INPUT
    2. OUTPUT
    3. GROUND

(Case connected to Pin 3)

(Bottom view)

(Bottom view)

Pins 1 and 2 electrically isolated from case. Case is third electrical connection.

(All 3 Plastic Types)
PIN 1. INPUT
    2. GROUND
    3. OUTPUT
(Heatsink surface connected to PIN 2)

PIN 1. OUTPUT
    2. GROUND
    3. INPUT

PIN 1. V_{OUT} 5. NC
    2. GND   6. GND
    3. GND   7. GND
    4. NC   8. V_{IN}

(c) Typical metal and plastic packages

FIGURE 18–24

The 7800 series three-terminal fixed positive voltage regulators.

shares most of the same features and characteristics. Figure 18–25 indicates the standard configuration and part numbers with corresponding output voltages that are available.

Adjustable Positive Linear Voltage Regulators

The LM317 is an excellent example of a three-terminal positive regulator with an adjustable output voltage. A data sheet for this device is given in Appendix C. The standard configuration is shown in Figure 18–26. Input and output capacitors, although not shown, are often used for the reasons discussed previously. Notice that there is an input, an output, and an adjustment terminal. The external fixed resistor R_1 and the external variable resistor R_2 provide the output voltage adjustment. V_{OUT} can be varied from 1.2 V to 37 V depending on the resistor values. The LM317 can provide over 1.5 A of output current to a load.

The LM317 is operated as a "floating" regulator because the adjustment terminal is not connected to ground, but floats to whatever voltage is across R_2. This allows the output voltage to be much higher than that of a fixed-voltage regulator.

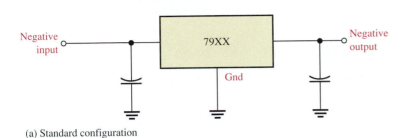

Type number	Output voltage
7905	−5.0 V
7905.2	−5.2 V
7906	−6.0 V
7908	−8.0 V
7912	−12.0 V
7915	−15.0 V
7918	−18.0 V
7924	−24.0 V

(a) Standard configuration

(b) The 7900 series

FIGURE 18–25
The 7900 series three-terminal fixed negative voltage regulators.

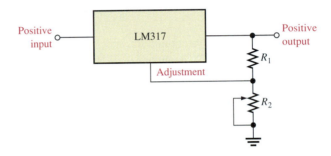

FIGURE 18–26
The LM317 three-terminal adjustable positive voltage regulator.

Basic Operation As indicated in Figure 18–27, a constant 1.2 V reference voltage (V_{REF}) is maintained by the regulator between the output terminal and the adjustment terminal. This constant reference voltage produces a constant current (I_{REF}) through R_1, regardless of the value of R_2. I_{REF} also flows through R_2.

$$I_{REF} = \frac{V_{REF}}{R_1} = \frac{1.25\ V}{R_1}$$

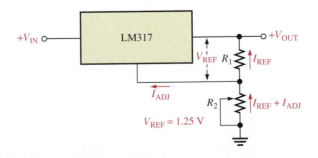

FIGURE 18–27
Operation of the LM317 adjustable voltage regulator.

Also, there is a very small constant current into the adjustment terminal of approximately 50 µA called I_{ADJ}, which flows through R_2. An expression for the output voltage is developed as follows.

$$V_{OUT} = V_{R1} + V_{R2} = I_{REF}R_1 + I_{REF}R_2 + I_{ADJ}R_2$$

$$= I_{REF}(R_1 + R_2) + I_{ADJ}R_2 = \frac{V_{REF}}{R_1}(R_1 + R_2) + I_{ADJ}R_2$$

$$V_{OUT} = V_{REF}\left(1 + \frac{R_2}{R_1}\right) + I_{ADJ}R_2 \qquad \textbf{(18–12)}$$

As you can see, the output voltage is a function of both R_1 and R_2. Once the value of R_1 is set, the output voltage is adjusted by varying R_2.

EXAMPLE 18–6   Determine the minimum and maximum output voltages for the voltage regulator in Figure 18–28. Assume $I_{ADJ} = 50$ µA.

FIGURE 18–28

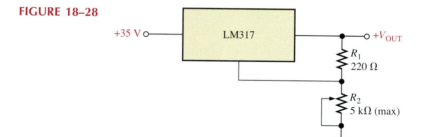

Solution $V_{R1} = V_{REF} = 1.25$ V

When R_2 is set at its maximum of 5 kΩ,

$$V_{OUT} = V_{REF}\left(1 + \frac{R_2}{R_1}\right) + I_{ADJ}R_2 = 1.25 \text{ V}\left(1 + \frac{5 \text{ k}\Omega}{220 \text{ }\Omega}\right) + (50 \text{ µA})5 \text{ k}\Omega$$

$$= 29.66 \text{ V} + 0.25 \text{ V} = 29.9 \text{ V}$$

When R_2 is set at its minimum of 0 Ω,

$$V_{OUT} = V_{REF}(1 + \frac{R_2}{R_1}) + I_{ADJ}R_2 = 1.25 \text{ V}(1) = 1.25 \text{ V}$$

Related Exercise   What is the output voltage of the regulator if R_2 is set at 2 kΩ?

Adjustable Negative Linear Voltage Regulators

The LM337 is the negative output counterpart of the LM317 and is a good example of this type of IC regulator. Like the LM317, the LM337 requires two external resistors for output voltage adjustment as shown in Figure 18–29. The output voltage can be adjusted from −1.2 V to −37 V, depending on the external resistor values.

FIGURE 18–29

The LM337 three-terminal adjustable negative voltage regulator.

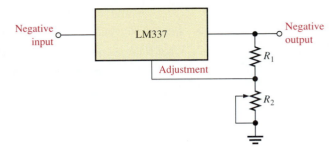

Switching Voltage Regulators

As an example of an IC switching voltage regulator, let's look at the 78S40. This is a universal device that can be used with external components to provide step-up, step-down, and inverting operation.

The internal circuitry of the 78S40 is shown in Figure 18–30. This circuit can be compared to the basic switching regulators that were covered in Section 18–4. For example, look back at Figure 18–15(a). The oscillator and comparator functions are directly comparable. The gate and flip-flop in the 78S40 were not included in the basic circuit of Figure 18–15(a), but they provide additional regulating action. Transistors Q_1 and Q_2 effectively perform the same function as Q_1 in the basic circuit. The 1.25 V reference block in the 78S40 has the same purpose as the zener diode in the basic circuit, and diode D_1 in the 78S40 corresponds to D_1 in the basic circuit.

The 78S40 also has an "uncommited" op-amp thrown in for good measure. It is not used in any of the regulator configurations. External circuitry is required to make this device operate as a regulator, as you will see in Section 18–6.

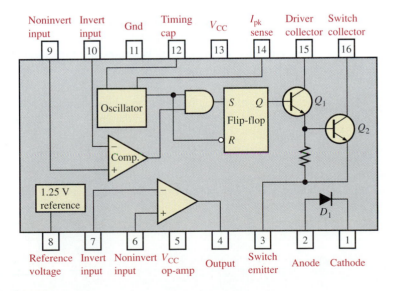

FIGURE 18–30

The 78S40 switching regulator.

18–6 ▪ APPLICATIONS OF IC VOLTAGE REGULATORS

In the last section, you saw several devices that are representative of the general types of IC voltage regulators. Now, we will examine several different ways these devices can be modified with external circuitry to improve or alter their performance.

After completing this section, you should be able to

▪ **Discuss applications of IC voltage regulators**
- ☐ Explain the use of an external pass transistor
- ☐ Explain the use of current limiting
- ☐ Explain how to use a voltage regulator as a constant-current source
- ☐ Discuss some application considerations for switching regulators

The External Pass Transistor

As you know, an IC voltage regulator is capable of delivering only a certain amount of output current for a load. For example, the 7800 series regulators can handle a maximum output current of at least 1.3 A and typically 2.5 A. If the load current exceeds the maximum allowable value, there will be thermal overload and the regulator will shut down. A thermal overload condition means that there is excessive power dissipation inside the device.

If an application requires more than the maximum current that the regulator can deliver, an external pass transistor can be used. Figure 18–31 illustrates a three-terminal regulator with an external pass transistor for handling currents in excess of the output current capability of the basic regulator.

The value of the external current-sensing resistor, R_{ext}, determines the value of current at which Q_{ext} begins to conduct because it sets the V_{BE} voltage of the transistor. As long as the current is less than the value set by R_{ext}, the transistor Q_{ext} is off, and the regu-

FIGURE 18–31

A 7800-series three-terminal regulator with an external pass transistor to increase power dissipation.

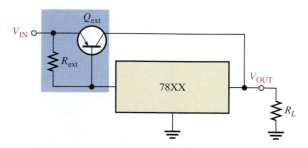

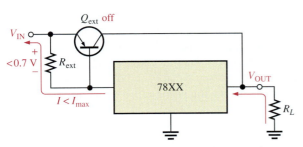

(a) When the regulator current is less than I_{max}, the external pass transistor is off and the regulator is handling all of the current.

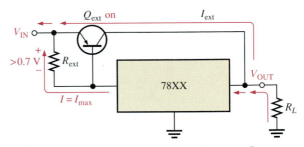

(b) When the load current exceeds I_{max}, the drop across R_{ext} turns Q_{ext} on and it conducts the excess current.

FIGURE 18–32

Operation of the regulator with an external pass transistor.

lator operates normally as shown in Figure 18–32(a). This is because the voltage drop across R_{ext} is less than the 0.7 V base-to-emitter voltage required to turn Q_{ext} on. R_{ext} is determined by the following formula, where I_{max} is the highest current that the voltage regulator is to handle internally.

$$R_{ext} = \frac{0.7 \text{ V}}{I_{max}} \qquad (18\text{–}13)$$

When the current is sufficient to produce at least a 0.7 V drop across R_{ext}, the external pass transistor Q_{ext} turns on and conducts any current in excess of I_{max}, as indicated in Figure 18–32(b). Q_{ext} will conduct more or less, depending on the load requirements. For example, if the total load current is 3 A and I_{max} was selected to be 1 A, the external pass transistor will conduct 2 A, which is the excess over the current I_{max}, flowing internally through the voltage regulator.

EXAMPLE 18–7   What value is R_{ext} if the maximum current to be handled internally by the voltage regulator in Figure 18–31 is set at 700 mA?

Solution
$$R_{ext} = \frac{0.7 \text{ V}}{I_{max}} = \frac{0.7 \text{ V}}{0.7 \text{ A}} = 1 \text{ } \Omega$$

Related Exercise   If R_{ext} is changed to 1.5 Ω, at what current value will Q_{ext} turn on?

The external pass transistor is typically a power transistor with heat sink that must be capable of handling a maximum power of

$$P_{ext} = I_{ext}(V_{IN} - V_{OUT}) \qquad (18\text{–}14)$$

EXAMPLE 18–8

What must be the minimum power rating for the external pass transistor used with a 7824 regulator in a circuit such as that shown in Figure 18–31? The input voltage is 30 V and the load resistance is 10 Ω. The maximum internal current is to be 700 mA. Assume that there is no heat sink for this calculation. Keep in mind that the use of a heat sink increases the effective power rating of the transistor and you can use a lower rated transistor.

Solution   The load current is

$$I_L = \frac{V_{OUT}}{R_L} = \frac{24\ V}{10\ \Omega} = 2.4\ A$$

The current through Q_{ext} is

$$I_{ext} = I_L - I_{max} = 2.4\ A - 0.7\ A = 1.7\ A$$

The power dissipated by Q_{ext} is

$$P_{ext(min)} = I_{ext}(V_{IN} - V_{OUT}) = (1.7\ A)(30\ V - 24\ V) = (1.7\ A)(6\ V) = 10.2\ W$$

For a safety margin, choose a power transistor with a rating greater than 10.2 W, say at least 15 W.

Related Exercise   Rework this example using a 7815 regulator.

Current Limiting

A drawback of the circuit in Figure 18–31 is that the external transistor is not protected from excessive current, such as would result from a shorted output. An additional current-limiting circuit (Q_{lim} and R_{lim}) can be added as shown in Figure 18–33 to protect Q_{ext} from excessive current and possible burn out.

The following describes the way the current-limiting circuit works. The current-sensing resistor R_{lim} sets the V_{BE} of transistor Q_{lim}. The base-to-emitter voltage of Q_{ext} is now determined by $V_{R_{ext}} - V_{R_{lim}}$ because they have opposite polarities. So, for normal operation, the drop across R_{ext} must be sufficient to overcome the opposing drop across

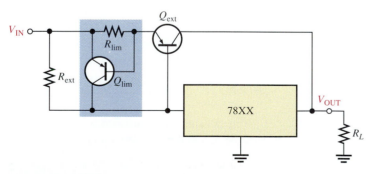

FIGURE 18–33
Regulator with current limiting.

R_{lim}. If the current through Q_{ext} exceeds a certain maximum ($I_{ext(max)}$) because of a shorted output or a faulty load, the voltage across R_{lim} reaches 0.7 V and turns Q_{lim} on. Q_{lim} now conducts current away from Q_{ext} and through the regulator, forcing a thermal overload to occur and shut down the regulator. Remember, the IC regulator is internally protected from thermal overload as part of its design.

This action is illustrated in Figure 18–34. In part (a), the circuit is operating normally with Q_{ext} conducting less than the maximum current that it can handle with Q_{lim} off. Part (b) shows what happens when there is a short across the load. The current through Q_{ext} suddenly increases and causes the voltage drop across R_{lim} to increase, which turns Q_{lim} on. The current is now diverted into the regulator, which causes it to shut down due to thermal overload.

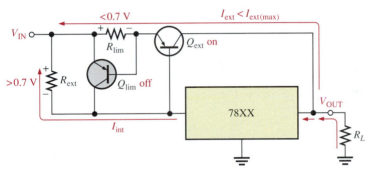

(a) During normal operation, when the load current is not excessive, Q_{lim} is off.

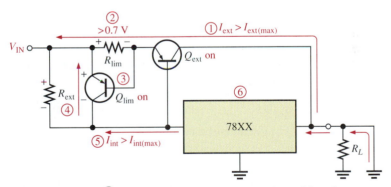

(b) When short occurs ①, the external current becomes excessive and the voltage across R_{lim} increases ② and turns on Q_{lim} ③, which then conducts current away from Q_{ext} and routes it through the regulator ④, causing the internal regulator current to become excessive ⑤ and to force the regulator into thermal shut down ⑥.

FIGURE 18–34

The current-limiting action of the regulator circuit.

A Current Regulator

The three-terminal regulator can be used as a current source when an application requires that a constant current be supplied to a variable load. The basic circuit is shown in Figure 18–35 where R_1 is the current-setting resistor. The regulator provides a fixed constant voltage, V_{OUT}, between the ground terminal (not connected to ground in this case) and the output terminal. This determines the constant load current.

$$I_L = \frac{V_{OUT}}{R_1} + I_G \qquad\qquad (18\text{–}15)$$

The current to the ground terminal I_G is very small compared to the output current and can often be neglected.

FIGURE 18–35

The three-terminal regulator as a current source.

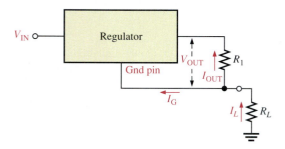

EXAMPLE 18–9   Use a 7805 regulator to provide a constant current of 1 A to a variable load. The input must be at least 2 V greater than the output and $I_G = 1.5$ mA.

Solution   First, 1 A is within the limits of the 7805's capability (remember, it can handle at least 1.3 A without an external pass transistor).

The 7805 produces 5 V between its ground terminal and its output terminal. Therefore, if you want 1 A of current, the current-setting resistor must be (neglecting I_G)

$$R_1 = \frac{V_{OUT}}{I_L} = \frac{5\ \text{V}}{1\ \text{A}} = 5\ \Omega$$

The circuit is shown in Figure 18–36.

FIGURE 18–36

A 1 A constant-current source.

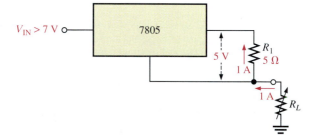

Related Exercise   If a 7812 regulator is used instead of the 7805, to what value would you change R_1 to maintain a constant current of 1 A?

Switching Regulator Configurations

In Section 18–5, we introduced the 78S40 as an example of an IC switching voltage regulator. Figure 18–37 shows the external connections for a step-down configuration where the output voltage is less than the input voltage, and Figure 18–38 shows a step-up configuration in which the output voltage is greater than the input voltage. An inverting configuration is also possible, but it is not shown here.

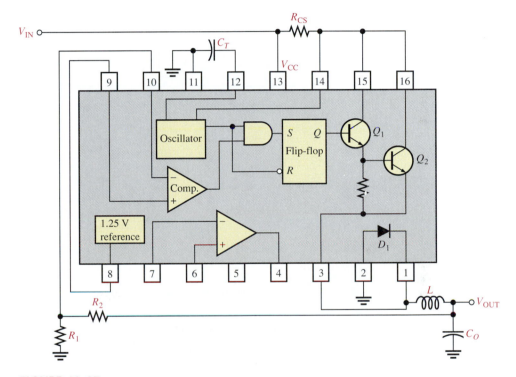

FIGURE 18–37

The step-down configuration of the 78S40 switching regulator.

The timing capacitor, C_T, controls the pulse width and frequency of the oscillator and thus establishes the on-time of transistor Q_2. The voltage across the current-sensing resistor, R_{CS} is used internally by the oscillator to vary the duty cycle based on the desired peak load current. The voltage divider, made up of R_1 and R_2, reduces the output voltage to a nominal value equal to the reference voltage. If V_{OUT} exceeds its set value, the output of the comparator switches to its low state, disabling the gate to turn Q_2 off until the output decreases. This regulating action is in addition to that produced by the duty cycle variation of the oscillator as described in Section 18–4 in relation to the basic switching regulator.

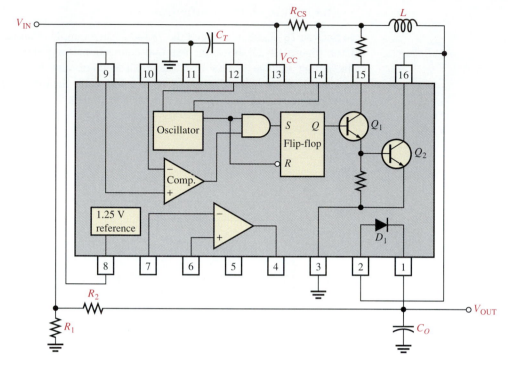

FIGURE 18–38

The step-up configuration of the 78S40 switching regulator.

SECTION 18–6 REVIEW

1. What is the purpose of using an external pass transistor with an IC voltage regulator?

2. What is the advantage of current limiting in a voltage regulator?

3. What does *thermal overload* mean?

18–7 ■ SYSTEM APPLICATION

The regulated power supply in this application provides dual-polarity dc voltages of ±12 V to the FM receiver system which you worked with in Chapter 16. The power supply consists of a transformer-coupled full-wave bridge rectifier and filter with positive and negative three-terminal voltage regulators.

The block diagram of the dual-polarity power supply is shown in Figure 18–39, and the circuit board is shown in Figure 18–40. The large vertically mounted capacitors are the 100 μF filter capacitors. The 0.33 μF and 1 μF capacitors, although not necessary in all applications, are recommended by the manufacturer for stability and improved transient response.

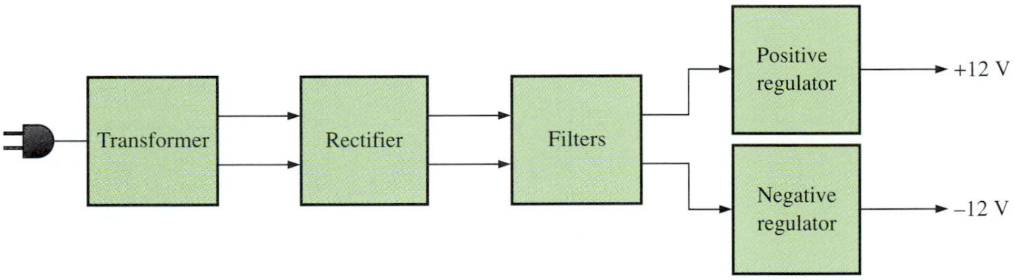

FIGURE 18–39
Block diagram of the dual-polarity power supply.

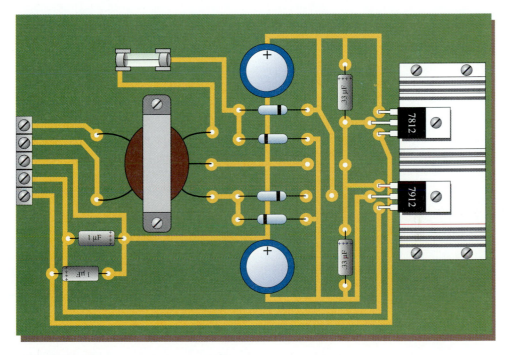

FIGURE 18–40
The dual-polarity power supply circuit board.

The Dual-Polarity Power Supply Board

☐ Make sure that the circuit board shown in Figure 18–40 is correct by checking it against the schematic in Figure 18–41 and by referring to the the 7812 and 7912 data sheets in Appendix C. There are back-side conections. Corresponding feedthrough pads are aligned horizontally.

☐ Label a copy of the board with component and input/output designations in agreement with the schematic.

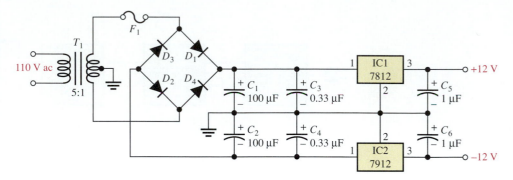

FIGURE 18–41

The dual-polarity power supply schematic.

The Power Supply Circuit

☐ Determine the approximate voltage at each of the four "corners" of the bridge. The transformer is rated at 24 V rms.

☐ Calculate the peak inverse voltage of the rectifier diodes.

☐ Determine the voltage at the input of each voltage regulator.

☐ In the FM receiver from Chapter 16, assume that op-amps are used only in the channel separation circuits. If all the other circuits in the receiver, excluding the channel separation circuits, used +12 V only and draw an average dc current of 100 mA, determine how much total current each regulator must supply.

☐ Based on the amount of dc current required by the receiver, do the regulators need to be attached to the heat sink?

Test Procedure

Develop a basic procedure for thoroughly testing the dual-polarity power supply.

Troubleshooting

Based on the results indicated in Figure 18–42 for four faulty power supply boards, determine the most likely fault or faults in each case.

Final Report

Submit a final written report on the dual-polarity power supply using an organized format that includes the following:

1. A physical description of the circuit.
2. A discussion of the operation of the circuit.
3. A list of the specifications.
4. A list of parts with part numbers if available.
5. A list of the types of problems in the four faulty boards.

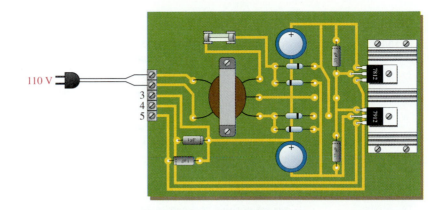

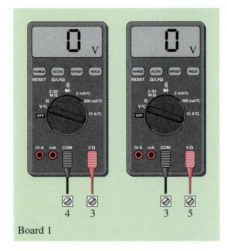

Board 1

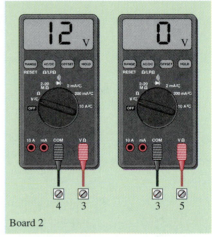

Board 2

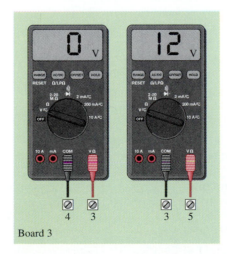

Board 3

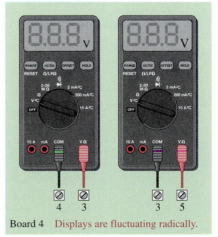

Board 4   Displays are fluctuating radically.

FIGURE 18–42

Results of tests on four faulty power supply boards.

■ CHAPTER SUMMARY

- Voltage regulators keep a constant dc output voltage when the input or load varies within limits.
- A basic voltage regulator consists of a reference voltage source, an error detector, a sampling element, and a control device. Protection circuitry is also found in most regulators.
- Two basic categories of voltage regulators are linear and switching.
- Two basic types of linear regulators are series and shunt.
- In a series linear regulator, the control element is a transistor in series with the load.
- In a shunt linear regulator, the control element is a transistor in parallel with the load.
- Three configurations for switching regulators are step-down, step-up, and inverting.
- Switching regulators are more efficient than linear regulators and are particularly useful in low-voltage, high-current applications.
- Three-terminal linear IC regulators are available for either fixed output or variable output voltages of positive or negative polarities.
- An external pass transistor increases the current capability of a regulator.
- The 7800 series are three-terminal IC regulators with fixed positive output voltage.
- The 7900 series are three-terminal IC regulators with fixed negative output voltage.
- The LM317 is a three-terminal IC regulator with a positive variable output voltage.
- The LM337 is a three-terminal IC regulator with a negative variable output voltage.
- The 78S40 is a switching voltage regulator.

■ GLOSSARY

Fold-back current limiting A method of current limiting in voltage regulators.

Line regulation The percentage change in output voltage for a given change in input (line) voltage.

Load regulation The percentage change in output voltage for a given change in load current.

Regulator An electronic circuit that maintains an essentially constant output voltage with a changing input voltage or load current.

Thermal overload A condition in a rectifier where the internal power dissipation of the circuit exceeds a certain maximum due to excessive current.

■ FORMULAS

Voltage Regulation

$$(18\text{–}1) \qquad \textbf{Line regulation} = \frac{\left(\dfrac{\Delta V_{\text{OUT}}}{V_{\text{OUT}}}\right)100\%}{\Delta V_{\text{IN}}} \qquad \text{Percent line regulation}$$

$$(18\text{–}2) \qquad \textbf{Load regulation} = \left(\frac{V_{\text{NL}} - V_{\text{FL}}}{V_{\text{FL}}}\right)100\% \qquad \text{Percent load regulation}$$

Basic Series Regulator

$$(18\text{–}3) \qquad A_{cl} = 1 + \frac{R_2}{R_3} \qquad \text{Closed-loop voltage gain}$$

$$(18\text{–}4) \qquad V_{\text{OUT}} \cong \left(1 + \frac{R_2}{R_3}\right)V_{\text{REF}} \qquad \text{Regulator output}$$

$$(18\text{–}5) \qquad I_{L(\text{max})} = \frac{0.7\text{ V}}{R_4} \qquad \text{For constant-current limiting}$$

Basic Shunt Regulator

(18–6) $\Delta I_S = \dfrac{\Delta V_{IN}}{R_1}$ Change in shunt current

(18–7) $\Delta I_S = -\Delta I_L$ Change in shunt current

(18–8) $I_{L(max)} = \dfrac{V_{IN}}{R_1}$ Maximum load current

Basic Switching Regulators

(18–9) $V_{OUT} = \left(\dfrac{t_{on}}{T}\right)V_{IN}$ For step-down switching regulator

(18–10) $T = t_{on} + t_{off}$ Switching period

(18–11) $V_{OUT} = \left(\dfrac{T}{t_{on}}\right)V_{IN}$ For step-up switching regulator

IC Voltage Regulators

(18–12) $V_{OUT} = V_{REF}\left(1 + \dfrac{R_2}{R_1}\right) + I_{ADJ}R_2$ IC regulator

(18–13) $R_{ext} = \dfrac{0.7\text{ V}}{I_{max}}$ For external pass circuit

(18–14) $P_{ext} = I_{ext}(V_{IN} - V_{OUT})$ For external pass transistor

(18–15) $I_L = \dfrac{V_{OUT}}{R_1} + I_G$ Regulator as a current source

---

■ **SELF-TEST**

1. In the case of line regulation,
   (a) when the temperature varies, the output voltage stays constant
   (b) when the output voltage changes, the load current stays constant
   (c) when the input voltage changes, the output voltage stays constant
   (d) when the load changes, the output voltage stays constant

2. In the case of load regulation,
   (a) when the temperature varies, the output voltage stays constant
   (b) when the input voltage changes, the load current stays constant
   (c) when the load changes, the load current stays constant
   (d) when the load changes, the output voltage stays constant

3. All of the following are parts of a basic voltage regulator *except*
   (a) control element   (b) sampling circuit
   (c) voltage-follower   (d) error detector
   (e) reference voltage

4. The basic difference between a series regulator and a shunt regulator is
   (a) the amount of current that can be handled
   (b) the position of the control element
   (c) the type of sample circuit
   (d) the type of error detector

5. In a basic series regulator, V_{OUT} is determined by
   (a) the control element   (b) the sample circuit
   (c) the reference voltage   (d) answers (b) and (c)

6. The main purpose of current limiting in a regulator is
    (a) protection of the regulator from excessive current
    (b) protection of the load from excessive current
    (c) to keep the power supply transformer from burning up
    (d) to maintain a constant output voltage

7. In a linear regulator, the control transistor is conducting
    (a) a small part of the time   (b) half the time
    (c) all of the time   (d) only when the load current is excessive

8. In a switching regulator, the control transistor is conducting
    (a) part of the time
    (b) all of the time
    (c) only when the input voltage exceeds a set limit
    (d) only when there is an overload

9. The LM317 is an example of an IC
    (a) three-terminal negative voltage regulator
    (b) fixed positive voltage regulator
    (c) switching regulator
    (d) linear regulator
    (e) variable positive voltage regulator
    (f) answers (b) and (d) only
    (g) answers (d) and (e) only

10. An external pass transistor is used for
    (a) increasing the output voltage
    (b) improving the regulation
    (c) increasing the current that the regulator can handle
    (d) short-circuit protection

■ BASIC PROBLEMS

SECTION 18–1   Voltage Regulation

1. The nominal output voltage of a certain regulator is 8 V. The output changes 2 mV when the input voltage goes from 12 V to 18 V. Determine the line regulation and express it as a percentage change over the entire range of V_{IN}.

2. Express the line regulation found in Problem 1 in units of %/V.

3. A certain regulator has a no-load output voltage of 10 V and a full-load output voltage of 9.90 V. What is the percent load regulation?

4. In Problem 3, if the full-load current is 250 mA, express the load regulation in %/mA.

SECTION 18–2   Basic Series Regulators

5. Label the functional blocks for the voltage regulator in Figure 18–43.

FIGURE 18–43

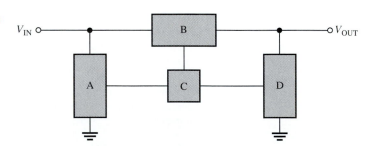

6. Determine the output voltage for the regulator in Figure 18–44.

7. Determine the output voltage for the series regulator in Figure 18–45.

8. If R_3 in Figure 18–45 is increased to 4.7 kΩ, what happens to the output voltage?

9. If the zener voltage is 2.7 V instead of 2.4 V in Figure 18–45, what is the output voltage?

FIGURE 18–44

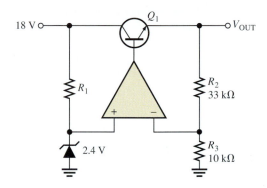

FIGURE 18–45

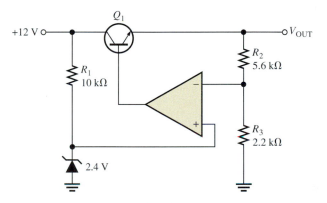

10. A series voltage regulator with constant current limiting is shown in Figure 18–46. Determine the value of R_4 if the load current is to be limited to a maximum value of 250 mA. What power rating must R_4 have?

11. If the R_4 determined in Problem 10 is halved, what is the maximum load current?

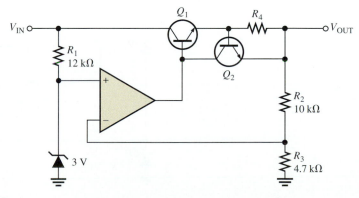

FIGURE 18–46

SECTION 18–3 Basic Shunt Regulators

12. In the shunt regulator of Figure 18–47, when the load current increases, does Q_1 conduct more or less? Why?

13. Assume I_L remains constant and V_{IN} changes by 1 V in Figure 18–47. What is the change in the collector current of Q_1?

14. With a constant input voltage of 17 V, the load resistance in Figure 18–47 is varied from 1 kΩ to 1.2 kΩ. Neglecting any change in output voltage, how much does the shunt current through Q_1 change?

15. If the maximum allowable input voltage in Figure 18–47 is 25 V, what is the maximum possible output current when the output is short-circuited? What power rating should R_1 have?

FIGURE 18–47

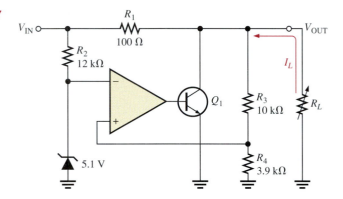

SECTION 18–4 Basic Switching Regulators

16. A basic switching regulator is shown in Figure 18–48. If the switching frequency of the transistor is 100 Hz with an off-time of 6 ms, what is the output voltage?

17. What is the duty cycle of the transistor in Problem 16?

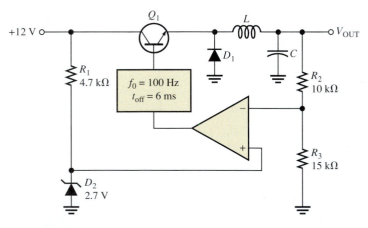

FIGURE 18–48

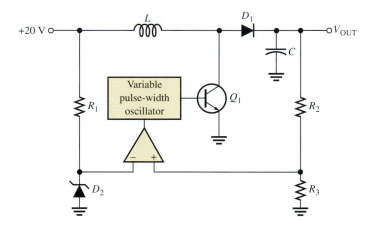

FIGURE 18–49

18. Determine the output voltage for the switching regulator in Figure 18–49 when the duty cycle is 40 percent.

19. If the on-time of Q_1 in Figure 18–49 is decreased, does the output voltage increase or decrease?

SECTION 18–5   Integrated Circuit Voltage Regulators

20. What is the output voltage of each of the following IC regulators?
    (a) 7806   (b) 7905.2   (c) 7818   (d) 7924

21. Determine the output voltage of the regulator in Figure 18–50. $I_{ADJ} = 50 \, \mu A$.

22. Determine the minimum and maximum output voltages for the circuit in Figure 18–51. $I_{ADJ} = 50 \, \mu A$.

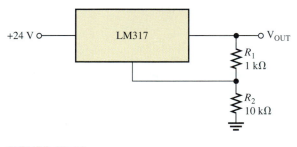

FIGURE 18–50

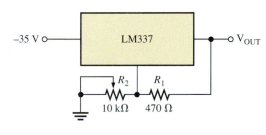

FIGURE 18–51

23. With no load connected, how much current is there through the regulator in Figure 18–50? Neglect the adjustment terminal current.

24. Select the values for the external resistors to be used in an LM317 circuit that is required to produce an output voltage of 12 V with an input of 18 V. The maximum regulator current with no load is to be 2 mA. There is no external pass transistor.

SECTION 18–6 Applications of IC Voltage Regulators

25. In the regulator circuit of Figure 18–52, determine R_{ext} if the maximum internal regulator current is to be 250 mA.

FIGURE 18–52

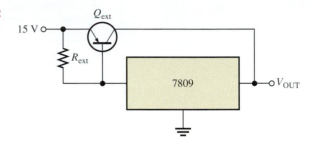

26. Using a 7812 voltage regulator and a 10 Ω load in Figure 18–52, how much power will the external pass transistor have to dissipate? The maximum internal regulator current is set at 500 mA by R_{ext}.

27. Show how to include current limiting in the circuit of Figure 18–52. What should the value of the limiting resistor be if the external current is to be limited to 2 A?

28. Using an LM317, design a circuit that will provide a constant current of 500 mA to a load.

29. Repeat Problem 28 using a 7909.

30. If a 78S40 switching regulator is to be used to regulate a 12 V input down to a 6 V output, calculate the values of the external voltage-divider resistors.

■ **ANSWERS TO SECTION REVIEWS**

Section 18–1

1. The percentage change in the output voltage for a given change in input voltage.

2. The percentage change in output voltage for a given change in load current.

Section 18–2

1. Control element, error detector, sampling element, reference voltage

2. 2 V

Section 18–3

1. In a shunt regulator, the control element is in parallel with the load rather than in series.

2. A shunt regulator has inherent current limiting. A disadvantage is that a shunt regulator is less efficient than a series regulator.

Section 18–4

1. Step-down, step-up, inverting

2. Switching regulators operate at a higher efficiency.

3. The duty cycle varies to regulate the output.

Section 18–5

1. Input, output, and ground

2. A 7809 has a +9 V output; A 7915 has a −15 V output.

3. Input, output, adjustment

4. A two-resistor voltage divider

Section 18–6

1. A pass transistor increases the current that can be handled.
2. Current limiting prevents excessive current and prevents damage to the regulator.
3. Thermal overload occurs when the internal power dissipation becomes excessive.

■ **ANSWERS TO RELATED EXERCISES FOR EXAMPLES**	**18–1** 0.6%/V
	18–2 1.12%, 0.0224%/mA
	18–3 7.33 V
	18–4 0.7 A
	18–5 17.5 W
	18–6 12.7 V
	18–7 467 mA
	18–8 12 W
	18–9 12 Ω

TABLE OF STANDARD RESISTOR VALUES

Resistance Tolerance (± %)

0.1% 0.25% 0.5%	1%	2% 5%	10%	0.1% 0.25% 0.5%	1%	2% 5%	10%	0.1% 0.25% 0.5%	1%	2% 5%	10%	0.1% 0.25% 0.5%	1%	2% 5%	10%	0.1% 0.25% 0.5%	1%	2% 5%	10%	0.1% 0.25% 0.5%	1%	2% 5%	10%
10.0	10.0	10	10	14.7	14.7	—	—	21.5	21.5	—	—	31.6	31.6	—	—	46.4	46.4	—	—	68.1	68.1	68	68
10.1	—	—	—	14.9	—	—	—	21.8	—	—	—	32.0	—	—	—	47.0	—	47	47	69.0	—	—	—
10.2	10.2	—	—	15.0	15.0	15	15	22.1	22.1	22	22	32.4	32.4	—	—	47.5	47.5	—	—	69.8	69.8	—	—
10.4	—	—	—	15.2	—	—	—	22.3	—	—	—	32.8	—	—	—	48.1	—	—	—	70.6	—	—	—
10.5	10.5	—	—	15.4	15.4	—	—	22.6	22.6	—	—	33.2	33.2	33	33	48.7	48.7	—	—	71.5	71.5	—	—
10.6	—	—	—	15.6	—	—	—	22.9	—	—	—	33.6	—	—	—	49.3	—	—	—	72.3	—	—	—
10.7	10.7	—	—	15.8	15.8	—	—	23.2	23.2	—	—	34.0	34.0	—	—	49.9	49.9	—	—	73.2	73.2	—	—
10.9	—	—	—	16.0	—	16	—	23.4	—	—	—	34.4	—	—	—	50.5	—	—	—	74.1	—	—	—
11.0	11.0	11	—	16.2	16.2	—	—	23.7	23.7	—	—	34.8	34.8	—	—	51.1	51.1	51	—	75.0	75.0	75	—
11.1	—	—	—	16.4	—	—	—	24.0	—	24	—	35.2	—	—	—	51.7	—	—	—	75.9	—	—	—
11.3	11.3	—	—	16.5	16.5	—	—	24.3	24.3	—	—	35.7	35.7	—	—	52.3	52.3	—	—	76.8	76.8	—	—
11.4	—	—	—	16.7	—	—	—	24.6	—	—	—	36.1	—	36	—	53.0	—	—	—	77.7	—	—	—
11.5	11.5	—	—	16.9	16.9	—	—	24.9	24.9	—	—	36.5	36.5	—	—	53.6	53.6	—	—	78.7	78.7	—	—
11.7	—	—	—	17.2	—	—	—	25.2	—	—	—	37.0	—	—	—	54.2	—	—	—	79.6	—	—	—
11.8	11.8	—	—	17.4	17.4	—	—	25.5	25.5	—	—	37.4	37.4	—	—	54.9	54.9	—	—	80.6	80.6	—	—
12.0	—	12	12	17.6	—	—	—	25.8	—	—	—	37.9	—	—	—	56.2	—	—	—	81.6	—	—	—
12.1	12.1	—	—	17.8	17.8	—	—	26.1	26.1	—	—	38.3	38.3	—	—	56.6	56.6	56	56	82.5	82.5	82	82
12.3	—	—	—	18.0	—	18	18	26.4	—	—	—	38.8	—	—	—	56.9	—	—	—	83.5	—	—	—
12.4	12.4	—	—	18.2	18.2	—	—	26.7	26.7	—	—	39.2	39.2	39	39	57.6	57.6	—	—	84.5	84.5	—	—
12.6	—	—	—	18.4	—	—	—	27.1	—	27	27	39.7	—	—	—	58.3	—	—	—	85.6	—	—	—
12.7	12.7	—	—	18.7	18.7	—	—	27.4	27.4	—	—	40.2	40.2	—	—	59.0	59.0	—	—	86.6	86.6	—	—
12.9	—	—	—	18.9	—	—	—	27.7	—	—	—	40.7	—	—	—	59.7	—	—	—	87.6	—	—	—
13.0	13.0	13	—	19.1	19.1	—	—	28.0	28.0	—	—	41.2	41.2	—	—	60.4	60.4	—	—	88.7	88.7	—	—
13.2	—	—	—	19.3	—	—	—	28.4	—	—	—	41.7	—	—	—	61.2	—	—	—	89.8	—	—	—
13.3	13.3	—	—	19.6	19.6	—	—	28.7	28.7	—	—	42.2	42.2	—	—	61.9	61.9	62	—	90.9	90.9	91	—
13.5	—	—	—	19.8	—	—	—	29.1	—	—	—	42.7	—	—	—	62.6	—	—	—	92.0	—	—	—
13.7	13.7	—	—	20.0	20.0	20	—	29.4	29.4	—	—	43.2	43.2	43	—	63.4	63.4	—	—	93.1	93.1	—	—
13.8	—	—	—	20.3	—	—	—	29.8	—	—	—	43.7	—	—	—	64.2	—	—	—	94.2	—	—	—
14.0	14.0	—	—	20.5	20.5	—	—	30.1	30.1	30	—	44.2	44.2	—	—	64.9	64.9	—	—	95.3	95.3	—	—
14.2	—	—	—	20.8	—	—	—	30.5	—	—	—	44.8	—	—	—	65.7	—	—	—	96.5	—	—	—
14.3	14.3	—	—	21.0	21.0	—	—	30.9	30.9	—	—	45.3	45.3	—	—	66.5	66.5	—	—	97.6	97.6	—	—
14.5	—	—	—	21.3	—	—	—	31.2	—	—	—	45.9	—	—	—	67.3	—	—	—	98.8	—	—	—

NOTE: These values are generally available in multiples of 0.1, 1, 10, 100, 1 k, and 1 M.

B DERIVATIONS OF SELECTED EQUATIONS

EQUATION (2–1)

The average value of a half-wave rectified sine wave is the area under the curve divided by the period (2π). The equation for a sine wave is

$$v = V_p \sin \theta$$

$$V_{AVG} = \frac{\text{area}}{2\pi} = \frac{1}{2\pi} \int_0^\pi V_p \sin \theta \, d\theta = \frac{V_p}{2\pi} (-\cos \theta) \Omega \Big|_0^\pi$$

$$= \frac{V_p}{2\pi} [-\cos \pi - (-\cos 0)] = \frac{V_p}{2\pi} [-(-1) - (-1)] = \frac{V_p}{2\pi}(2)$$

$$V_{AVG} = \frac{V_p}{\pi}$$

EQUATION (2–11)

Refer to Figure B–1. When the filter capacitor discharges through R_L, the voltage is

$$v_C = V_{p(in)} e^{-t/R_L C}$$

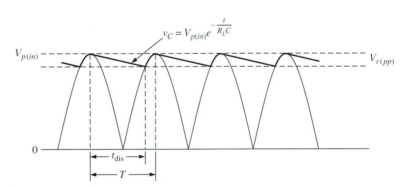

FIGURE B–1

Since the discharge time of the capacitor is from one peak to approximately the next peak, $t_{dis} \cong T$ when v_C reaches its minimum value.

$$v_{C(min)} = V_{p(in)} e^{-T/R_L C}$$

Since $RC \gg T$, T/R_LC becomes much less than 1 (which is usually the case); e^{-T/R_LC} approaches 1 and can be expressed as

$$e^{-T/RC} \cong 1 - \frac{T}{R_LC}$$

Therefore,

$$v_{C(min)} = V_{p(in)}\left(1 - \frac{T}{R_LC}\right)$$

The peak-to-peak ripple voltage is

$$V_{r(pp)} = V_{p(in)} - V_{C(min)} = V_{p(in)} - V_{p(in)} + \frac{V_{p(in)}T}{R_LC} = \frac{V_{p(in)}T}{R_LC}$$

$$V_{r(pp)} = \left(\frac{1}{fR_LC}\right)V_{p(in)}$$

EQUATION (2–12)

To obtain the dc value, one-half of the peak-to-peak ripple is subtracted from the peak value.

$$V_{DC} = V_{p(in)} - \frac{V_{r(pp)}}{2} = V_{p(in)} - \left(\frac{1}{2fR_LC}\right)V_{p(in)}$$

$$V_{DC} = \left(1 - \frac{1}{2fR_LC}\right)V_{p(in)}$$

DERIVATION OF THE RIPPLE VOLTAGE FOR A FULL-WAVE RECTIFIED SIGNAL AS USED IN EXAMPLE 2–8

The ac component of a full-wave rectified signal is the total voltage minus the dc value.

$$v = v_T - V_{DC}$$

The rms value of V_{ac} is

$$V_{r(rms)} = \left(\frac{1}{2\pi}\int_0^{2\pi}V_{ac}\,d\theta\right)^{1/2} = \left(\frac{1}{2\pi}\int_0^{2\pi}(v_T - V_{DC})^2\,d\theta\right)^{1/2}$$

$$= \left(\frac{1}{2\pi}\int_0^{2\pi}(v_T^2 - 2v_TV_{DC} + V_{DC}^2)\,d\theta\right)^{1/2}$$

$$= \left[\frac{1}{2\pi}\left(\int_0^{2\pi}v_T^2\,d\theta - \int_0^{2\pi}2v_TV_{DC}\,d\theta + \int_0^{2\pi}V_{DC}^2\,d\theta\right)\right]^{1/2}$$

$$= \left[\frac{1}{2\pi}\left(\int_0^{2\pi}v_T^2\,d\theta - 2V_{DC}\int_0^{2\pi}v_T\,d\theta + V_{DC}^2\int_0^{2\pi}d\theta\right)\right]^{1/2}$$

$$= (V_{T(rms)}^2 - 2V_{DC}^2 + V_{DC}^2)^{1/2}$$

$$V_{r(rms)} = (V_{T(rms)}^2 - V_{DC}^2)^{1/2}$$

For a full-wave rectified voltage:

$$V_{T(rms)} = \frac{V_p}{1.414}$$

$$V_{DC} = \frac{2V_p}{\pi}$$

$$V_{r(rms)} = \sqrt{\left(\frac{V_p}{1.414}\right)^2 - \left(\frac{2V_p}{\pi}\right)^2} = V_p \sqrt{\left(\frac{1}{1.414}\right)^2 - \left(\frac{2}{\pi}\right)^2}$$

$$V_{r(rms)} = 0.308V_p$$

EQUATION (6–10)

The Shockley equation for the base-emitter *pn* junction is

$$I_E = I_R(e^{VQ/kT} - 1)$$

where I_E = the total forward current across the base-emitter junction
  I_R = the reverse saturation current
  V = the voltage across the depletion layer
  Q = the charge on an electron
  k = a number known as Boltzmann's constant
  T = the absolute temperature

At ambient temperature, $Q/kT \cong 40$, so

$$I_E = I_R(e^{40V} - 1)$$

Differentiating, we get

$$\frac{dI_E}{dV} = 40I_R e^{40V}$$

Since $I_R e^{40V} = I_E + I_R$,

$$\frac{dI_E}{dV} = 40(I_E + I_R)$$

Assuming $I_R \ll I_E$,

$$\frac{dI_E}{dV} \cong 40I_E$$

The ac resistance r'_e of the base-emitter junction can be expressed as dV/dI_E.

$$r'_e = \frac{dV}{dI_E} \cong \frac{1}{40I_E} \cong \frac{25\text{ mV}}{I_E}$$

EQUATION (6–25)

The emitter-follower is represented by the r parameter ac equivalent circuit in Figure B–2(a). By Thevenizing from the base back to the source, the circuit is simplified to the form shown in Figure B–2(b).

$$V_{out} = V_e, I_{out} = I_e, \text{ and } I_{in} = I_b$$

$$R_{out} = \frac{V_e}{I_e}$$

$$I_e \cong \beta_{ac} I_b$$

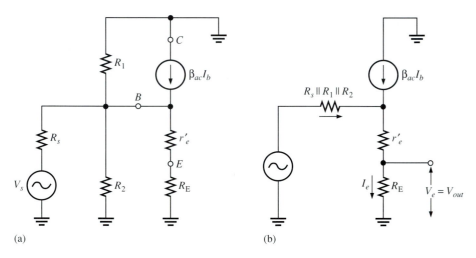

(a)   (b)

FIGURE B–2

With $V_s = 0$ and with I_b produced by V_{out}, and neglecting the base-to-emitter voltage drop (and therefore r'_e),

$$I_b \cong \frac{V_e}{R_1 \| R_2 \| R_s}$$

Assuming that $R_1 \gg R_s$ and $R_2 \gg R_s$,

$$I_b \cong \frac{V_e}{R_s}$$

$$I_{out} = I_e = \frac{\beta_{ac} V_e}{R_s}$$

$$\frac{V_{out}}{I_{out}} = \frac{V_e}{I_e} = \frac{V_e}{\beta_{ac} V_e / R_s} = \frac{R_s}{\beta_{ac}}$$

Looking from the emitter, R_E appears in parallel with R_s/β_{ac}. Therefore,

$$R_{out} = \left(\frac{R_s}{\beta_{ac}} \right) \| R_E$$

MIDPOINT BIAS: Proof that $I_D \cong 0.5 I_{DSS}$ when $V_{GS} = \dfrac{V_{GS(off)}}{3.4}$

Start with Equation (8–1):

$$I_D = I_{DSS}\left(1 - \frac{V_{GS}}{V_{GS(off)}}\right)^2$$

Let $I_D = 0.5 I_{DSS}$.

$$0.5 I_{DSS} = I_{DSS}\left(1 - \frac{V_{GS}}{V_{GS(off)}}\right)^2$$

Cancelling I_{DSS} on each side,

$$0.5 = \left(1 - \frac{V_{GS}}{V_{GS(off)}}\right)^2$$

We want a factor (call it F) by which $V_{GS(off)}$ can be divided to give a value of V_{GS} that will produce a drain current that is $0.5 I_{DSS}$.

$$0.5 = \left[1 - \frac{\left(\dfrac{V_{GS(off)}}{F}\right)}{V_{GS(off)}}\right]^2$$

Solving for F,

$$\sqrt{0.5} = 1 - \frac{\left(\dfrac{V_{GS(off)}}{F}\right)}{V_{GS(off)}} = 1 - \frac{1}{F}$$

$$\sqrt{0.5} - 1 = -\frac{1}{F}$$

$$\frac{1}{F} = 1 - \sqrt{0.5}$$

$$F = \frac{1}{1 - \sqrt{0.5}} \cong 3.4$$

Therefore, $I_D \cong 0.5 I_{DSS}$ when $V_{GS} = \dfrac{V_{GS(off)}}{3.4}$.

EQUATION (9–7)

$$I_D = I_{DSS}\left(1 - \frac{I_D R_S}{V_{GS(off)}}\right)^2 = I_{DSS}\left(1 - \frac{I_D R_S}{V_{GS(off)}}\right)\left(1 - \frac{I_D R_S}{V_{GS(off)}}\right)$$

$$= I_{DSS}\left(1 - \frac{2 I_D R_S}{V_{GS(off)}} + \frac{I_D^2 R_S^2}{V_{GS(off)}^2}\right) = I_{DSS} - \frac{2 I_{DSS} R_S}{V_{GS(off)}} I_D + \frac{I_{DSS} R_S^2}{V_{GS(off)}^2} I_D^2$$

Rearranging into a standard quadratic equation form,

$$\left(\frac{I_{DSS} R_S^2}{V_{GS(off)}^2}\right) I_D^2 - \left(1 + \frac{2 I_{DSS} R_S}{V_{GS(off)}}\right) I_D + I_{DSS} = 0$$

The coefficients and constant are

$$A = \frac{R_S I_{\text{DSS}}^2}{V_{\text{GS(off)}}^2}$$

$$B = -\left(1 + \frac{2R_S I_{\text{DSS}}}{V_{\text{GS(off)}}}\right)$$

$$C = I_{\text{DSS}}$$

In simplified notation, the equation is

$$A I_D^2 + B I_D + C = 0$$

The solutions to this quadratic equation are

$$I_D = \frac{-B \pm \sqrt{B^2 - 4AC}}{2A}$$

EQUATION (10–11)

An inverting amplifier with feedback capacitance is shown in Figure B–3. For the input,

$$I_1 = \frac{V_1 - V_2}{X_C}$$

Factoring V_1 out,

$$I_1 = \frac{V_1(1 - V_2/V_1)}{X_C}$$

The ratio V_2/V_1 is the voltage gain, $-A_v$.

$$I_1 = \frac{V_1(1 + A_v)}{X_C} = \frac{V_1}{X_C/(1 + A_v)}$$

FIGURE B–3

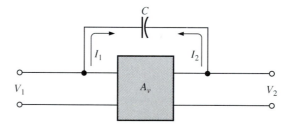

The effective reactance as seen from the input terminals is

$$X_{Cin(Miller)} = \frac{X_C}{1 + A_v}$$

or

$$\frac{1}{2\pi f C_{in(Miller)}} = \frac{1}{2\pi f C(1 + A_v)}$$

Cancelling and inverting, we get

$$C_{in(Miller)} = C(A_v + 1)$$

EQUATION (10–12)

For the output in Figure B–3,

$$I_2 = \frac{V_2 - V_1}{X_C} = \frac{V_2(1 - V_1/V_2)}{X_C}$$

Since $V_1/V_2 = -1/A_v$,

$$I_2 = \frac{V_2(1 + 1/A_v)}{X_C} = \frac{V_2}{X_C/(1 + 1/A_v)} = \frac{V_2}{X_C/[(A_v + 1)/A_v]}$$

The effective reactance as seen from the output is

$$X_{C_{out(Miller)}} = \frac{X_C}{(A_v + 1)/A_v}$$

$$\frac{1}{2\pi f C_{out(Miller)}} = \frac{1}{2\pi f C[(A_v + 1)/A_v]}$$

Cancelling and inverting, we get

$$C_{out(Miller)} = C\left(\frac{A_v + 1}{A_v}\right)$$

EQUATIONS (10–32) AND (10–33)

The total gain, $A_{v(tot)}$, of an individual amplifier stage at the lower critical frequency equals the midrange gain, $A_{v(mid)}$, times the attenuation of the high-pass RC circuit.

$$A_{v(tot)} = A_{v(mid)}\left(\frac{R}{\sqrt{R^2 + X_C^2}}\right) = A_{v(mid)}\left(\frac{1}{\sqrt{1 + X_C^2/R^2}}\right)$$

$$f_{cl} = \frac{1}{2\pi RC}$$

Dividing both sides by any frequency f,

$$\frac{f_{cl}}{f} = \frac{1}{(2\pi f C)R}$$

Since $X_C = 1/2\pi f C$,

$$\frac{f_{cl}}{f} = \frac{X_C}{R}$$

Substitution in the gain formula gives

$$A_{v(tot)} = A_{v(mid)}\left(\frac{1}{\sqrt{1 + (f_{cl}/f)^2}}\right)$$

The gain ratio is

$$\frac{A_{v(tot)}}{A_{v(mid)}} = \frac{1}{\sqrt{1 + (f_{cl}/f)^2}}$$

For a multistage amplifier with n stages, each with the same f_{cl} and gain ratio, the product of the gain ratios is

$$\left(\frac{1}{\sqrt{1 + (f_{cl}/f)^2}}\right)^n$$

The critical frequency f'_{cl} of the multistage amplifier is the frequency at which $A_{v(tot)} = 0.707A_{v(mid)}$, so the gain ratio at f'_{cl} is

$$\frac{A_{v(tot)}}{A_{v(mid)}} = 0.707 = \frac{1}{1.414} = \frac{1}{\sqrt{2}}$$

Therefore, for a multistage amplifier,

$$\frac{1}{\sqrt{2}} = \left[\frac{1}{\sqrt{1 + (f_{cl}/f'_{cl})^2}}\right]^n = \frac{1}{\left(\sqrt{1 + (f_{cl}/f'_{cl})^2}\right)^n}$$

So

$$2^{1/2} = \left(\sqrt{1 + f_{cl}/f'^2_{cl}}\right)^n$$

Squaring both sides,

$$2 = \left(1 + (f_{cl}/f'_{cl})^2\right)^n$$

Taking the nth root of both sides,

$$2^{1/n} = 1 + (f_{cl}/f'_{cl})^2$$

$$\left(\frac{f_{cl}}{f'_{cl}}\right)^2 = 2^{1/n} - 1$$

$$\left(\frac{f_{cl}}{f'_{cl}}\right) = \sqrt{2^{1/n} - 1}$$

$$f'_{cl} = \frac{f_{cl}}{\sqrt{2^{1/n} - 1}}$$

A similar process will give Equation (10–33):

$$f'_{cu} = f_{cu}\sqrt{2^{1/n} - 1}$$

EQUATIONS (10–34) AND (10–35)

The *rise time* is defined as the time required for the voltage to increase from 10 percent of its final value to 90 percent of its final value, as indicated in Figure B–4. Expressing the curve in its exponential form gives

$$v = V_{\text{final}}(1 - e^{-t/RC})$$

FIGURE B–4

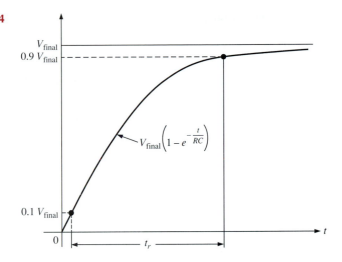

When $v = 0.1V_{final}$,

$$0.1V_{final} = V_{final}(1 - e^{-t/RC}) = V_{final} - V_{final}e^{-t/RC}$$
$$V_{final}e^{-t/RC} = 0.9V_{final}$$
$$e^{-t/RC} = 0.9$$
$$\ln e^{-t/RC} = \ln(0.9)$$
$$-\frac{t}{RC} = -0.1$$
$$t = 0.1RC$$

When $v = 0.9V_{final}$,

$$0.9V_{final} = V_{final}(1 - e^{-t/RC}) = V_{final} - V_{final}e^{-t/RC}$$
$$V_{final}e^{-t/RC} = 0.1V_{final}$$
$$\ln e^{-t/RC} = \ln(0.1)$$
$$-\frac{t}{RC} = -2.3$$
$$t = 2.3RC$$

The difference is the rise time.

$$t_r = 2.3RC - 0.1RC = 2.2RC$$

The critical frequency of an RC circuit is

$$f_c = \frac{1}{2\pi RC}$$

$$RC = \frac{1}{2\pi f_c}$$

Substituting,

$$t_r = \frac{2.2}{2\pi f_{cu}} = \frac{0.35}{f_{cu}}$$

$$f_{cu} = \frac{0.35}{t_r}$$

In a similar way, it can be shown that

$$f_{cl} = \frac{0.35}{t_f}$$

EQUATION (13–8)

The formula for open-loop gain in Equation (13–4) can be expressed in complex notation as

$$A_{ol} = \frac{A_{ol(mid)}}{1 + jf/f_{c(ol)}}$$

Substituting the above expression into the equation $A_{cl} = A_{ol}/(1 + BA_{ol})$, we get a formula for the total closed-loop gain.

$$A_{cl} = \frac{A_{ol(mid)}/(1 + jf/f_{c(ol)})}{1 + BA_{ol(mid)}/(1 + jf/f_{c(ol)})}$$

Multiplying the numerator and denominator by $1 + jf/f_{c(ol)}$ yields

$$A_{cl} = \frac{A_{ol(mid)}}{1 + BA_{ol(mid)} + jf/f_{c(ol)}}$$

Dividing the numerator and denominator by $1 + BA_{ol(mid)}$ gives

$$A_{cl} = \frac{A_{ol(mid)}/(1 + BA_{ol(mid)})}{1 + j[f/(f_{c(ol)}(1 + BA_{ol(mid)}))]}$$

The above expression is of the form of the first equation

$$A_{cl} = \frac{A_{cl(mid)}}{1 + jf/f_{c(cl)}}$$

where $f_{c(cl)}$ is the closed-loop critical frequency. Thus,

$$f_{c(cl)} = f_{c(ol)}(1 + BA_{ol(mid)})$$

EQUATION (17–1)

$$\frac{V_{out}}{V_{in}} = \frac{R(-jX)/(R - jX)}{(R - jX) + R(-jX)/(R - jX)} = \frac{R(-jX)}{(R - jX)^2 - jRX}$$

Multiplying the numerator and denominator by j,

$$\frac{V_{out}}{V_{in}} = \frac{RX}{j(R-jX)^2 + RX} = \frac{RX}{RX + j(R^2 - j2RX - X^2)}$$

$$= \frac{RX}{RX + jR^2 + 2RX - jX^2} = \frac{RX}{3RX + j(R^2 - X^2)}$$

For a $0°$ phase angle there can be no j term. Recall from complex numbers in ac theory that a *nonzero* angle is associated with a complex number having a j term. Therefore, at f_r the j term is 0.

$$R^2 - X^2 = 0$$

Thus,

$$\frac{V_{out}}{V_{in}} = \frac{RX}{3RX}$$

Cancelling, we get

$$\frac{V_{out}}{V_{in}} = \frac{1}{3}$$

EQUATION (17–2)

From the derivation of Equation (17–1),

$$R^2 - X^2 = 0$$
$$R^2 = X^2$$
$$R = X$$

Since $X = \dfrac{1}{2\pi f_r C}$,

$$R = \frac{1}{2\pi f_r C}$$

$$f_r = \frac{1}{2\pi RC}$$

EQUATIONS (17–3) AND (17–4)

The feedback circuit in the phase-shift oscillator consists of three RC stages, as shown in Figure B–5. An expression for the attenuation is derived using the mesh analysis method for the loop assignment shown. All Rs are equal in value, and all Cs are equal in value.

$$(R - j1/2\pi fC)I_1 - RI_2 + 0I_3 = V_{in}$$

$$-RI_1 + (2R - j1/2\pi fC)I_2 - RI_3 = 0$$

$$0I_1 - RI_2 + (2R - j1/2\pi fC)I_3 = 0$$

FIGURE B–5

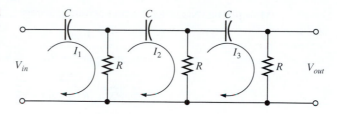

In order to get V_{out}, we must solve for I_3 using determinants:

$$I_3 = \frac{\begin{vmatrix} (R - j1/2\pi fC) & -R & V_{in} \\ -R & (2R - j1/2\pi fC) & 0 \\ 0 & -R & 0 \end{vmatrix}}{\begin{vmatrix} (R - j1/2\pi fC) & -R & 0 \\ -R & (2R - j1/2\pi fC) & -R \\ 0 & -R & (2R - j1/2\pi fC) \end{vmatrix}}$$

$$I_3 = \frac{R^2 V_{in}}{(R - j1/2\pi fC)(2R - j1/2\pi fC)^2 - R^2(2R - j1/2\pi fC) - R^2(R - 1/2\pi fC)}$$

$$\frac{V_{out}}{V_{in}} = \frac{RI_3}{V_{in}}$$

$$= \frac{R^3}{(R - j1/2\pi fC)(2R - j1/2\pi fC)^2 - R^3(2 - j1/2\pi fRC) - R^3(1 - 1/2\pi fRC)}$$

$$= \frac{R^3}{(R^3(1 - j1/2\pi fRC)(2 - j1/2\pi fRC)^2 - R^3[(2 - j1/2\pi fRC) - (1 - j1/2\pi RC)]}$$

$$= \frac{R^3}{R^3(1 - j1/2\pi fRC)(2 - j1/2\pi fRC)^2 - R^3(3 - j1/2\pi fRC)}$$

$$\frac{V_{out}}{V_{in}} = \frac{1}{(1 - j1/2\pi fRC)(2 - j1/2\pi fRC)^2 - (3 - j1/2\pi fRC)}$$

Expanding and combining the real terms and the j terms separately.

$$\frac{V_{out}}{V_{in}} = \frac{1}{\left(1 - \dfrac{5}{4\pi^2 f^2 R^2 C^2}\right) - j\left(\dfrac{6}{2\pi fRC} - \dfrac{1}{(2\pi f)^3 R^3 C^3}\right)}$$

For oscillation in the phase-shift amplifier, the phase shift through the RC circuit must equal 180°. For this condition to exist, the j term must be 0 at the frequency of oscillation f_0.

$$\frac{6}{2\pi f_0 RC} - \frac{1}{(2\pi f_0)^3 R^3 C^3} = 0$$

$$\frac{6(2\pi)^2 f_0^2 R^2 C^2 - 1}{(2\pi)^3 f_0^3 R^3 C^3} = 0$$

$$6(2\pi)^2 f_0^2 R^2 C^2 - 1 = 0$$

$$f_0^2 = \frac{1}{6(2\pi)^2 R^2 C^2}$$

$$f_0 = \frac{1}{2\pi\sqrt{6}RC}$$

Since the j term is 0,

$$\frac{V_{out}}{V_{in}} = \frac{1}{1 - \dfrac{5}{4\pi^2 f_0^2 R^2 C^2}} = \frac{1}{1 - \dfrac{5}{\left(\dfrac{1}{\sqrt{6}RC}\right)^2 R^2 C^2}} = \frac{1}{1 - 30} = -\frac{1}{29}$$

The negative sign results from the 180° inversion. Thus, the value of attenuation for the feedback circuit is

$$B = \frac{1}{29}$$

C

DATA SHEETS

All of the data sheets in this appendix are copyright of Motorola, Inc. Used by permission.

MOTOROLA

Designers' Data Sheet

"SURMETIC"▲ RECTIFIERS

. . . subminiature size, axial lead mounted rectifiers for general-purpose low-power applications.

Designers Data for "Worst Case" Conditions

The Designers▲ Data Sheets permit the design of most circuits entirely
from the information presented. Limit curves — representing boundaries on
device characteristics — are given to facilitate "worst case" design.

1N4001 thru 1N4007

**LEAD MOUNTED
SILICON RECTIFIERS**

50-1000 VOLTS
DIFFUSED JUNCTION

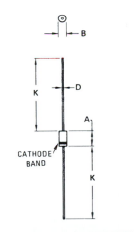

*MAXIMUM RATINGS

Rating	Symbol	1N4001	1N4002	1N4003	1N4004	1N4005	1N4006	1N4007	Unit
Peak Repetitive Reverse Voltage Working Peak Reverse Voltage DC Blocking Voltage	V_{RRM} V_{RWM} V_R	50	100	200	400	600	800	1000	Volts
Non-Repetitive Peak Reverse Voltage (halfwave, single phase, 60 Hz)	V_{RSM}	60	120	240	480	720	1000	1200	Volts
RMS Reverse Voltage	$V_{R(RMS)}$	35	70	140	280	420	560	700	Volts
Average Rectified Forward Current (single phase, resistive load, 60 Hz, see Figure 8, $T_A = 75^oC$)	I_O				1.0				Amp
Non-Repetitive Peak Surge Current (surge applied at rated load conditions, see Figure 2)	I_{FSM}				30 (for 1 cycle)				Amp
Operating and Storage Junction Temperature Range	T_J, T_{stg}				–65 to +175				^oC

*ELECTRICAL CHARACTERISTICS

Characteristic and Conditions	Symbol	Typ	Max	Unit
Maximum Instantaneous Forward Voltage Drop ($i_F = 1.0$ Amp, $T_J = 25^oC$) Figure 1	v_F	0.93	1.1	Volts
Maximum Full-Cycle Average Forward Voltage Drop ($I_O = 1.0$ Amp, $T_L = 75^oC$, 1 inch leads)	$V_{F(AV)}$	–	0.8	Volts
Maximum Reverse Current (rated dc voltage) $T_J = 25^oC$ $T_J = 100^oC$	I_R	0.05 1.0	10 50	μA
Maximum Full-Cycle Average Reverse Current ($I_O = 1.0$ Amp, $T_L = 75^oC$, 1 inch leads	$I_{R(AV)}$	–	30	μA

*Indicates JEDEC Registered Data.

CATHODE
BAND

DIM	MILLIMETERS		INCHES	
	MIN	MAX	MIN	MAX
A	5.97	6.60	0.235	0.260
B	2.79	3.05	0.110	0.120
D	0.76	0.86	0.030	0.034
K	27.94	–	1.100	–

CASE 59-04
Does Not Conform to DO-41 Outline.

MECHANICAL CHARACTERISTICS

CASE: Transfer Molded Plastic
MAXIMUM LEAD TEMPERATURE FOR SOLDERING PURPOSES: 350^oC, 3/8'' from
case for 10 seconds at 5 lbs. tension
FINISH: All external surfaces are corrosion-resistant, leads are readily solderable
POLARITY: Cathode indicated by color band
WEIGHT: 0.40 Grams (approximately)

▲Trademark of Motorola Inc.

FIGURE 1 – FORWARD VOLTAGE

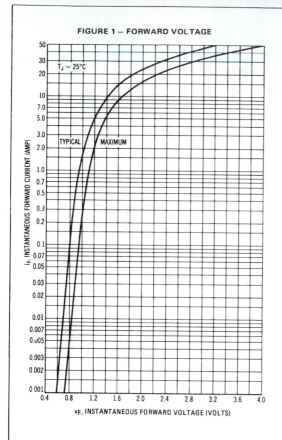

FIGURE 2 – NON-REPETITIVE SURGE CAPABILITY

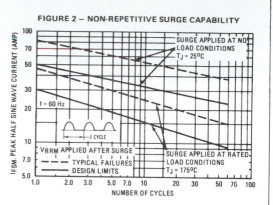

FIGURE 3 – FORWARD VOLTAGE TEMPERATURE COEFFICIENT

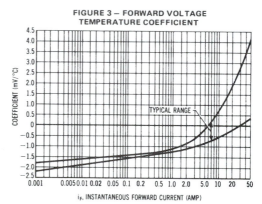

FIGURE 4 – TYPICAL TRANSIENT THERMAL RESISTANCE

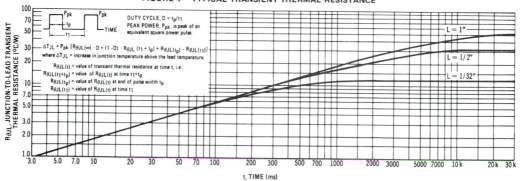

$$\Delta T_{JL} = P_{pk} \left[R_{\theta JL(\infty)} \cdot D + (1-D) \cdot R_{\theta JL}(t_1 + t_p) + R_{\theta JL}(t_p) - R_{\theta JL}(t_1) \right]$$

where ΔT_{JL} = increase in junction temperature above the lead temperature.

$R_{\theta JL}(t)$ = value of transient thermal resistance at time t, i.e.:

$R_{\theta JL}(t_1+t_p)$ = value of $R_{\theta JL}(t)$ at time t_1+t_p

$R_{\theta JL}(t_p)$ = value of $R_{\theta JL}(t)$ at end of pulse width t_p

$R_{\theta JL}(t_1)$ = value of $R_{\theta JL}(t)$ at time t_1

DUTY CYCLE, $D = t_p/t_1$
PEAK POWER, P_{pk}, is peak of an equivalent square power pulse.

The temperature of the lead should be measured using a thermocouple placed on the lead as close as possible to the tie point. The thermal mass connected to the tie point is normally large enough so that it will not significantly respond to heat surges generated in the diode as a result of pulsed operation once steady-state conditions are achieved. Using the measured value of T_L, the junction temperature may be determined by:

$$T_J = T_L + \Delta T_{JL}.$$

 MOTOROLA *Semiconductor Products Inc.*

MAXIMUM RATINGS

Rating	Symbol	2N2219 2N2222	2N2218A 2N2219A 2N2221A 2N2222A	Unit
Collector-Emitter Voltage	V_{CEO}	30	40	Vdc
Collector-Base Voltage	V_{CBO}	60	75	Vdc
Emitter-Base Voltage	V_{EBO}	5.0	6.0	Vdc
Collector Current — Continuous	I_C	800	800	mAdc
		2N2218A 2N2219,A	2N2221A 2N2222,A	
Total Device Dissipation @ T_A = 25°C Derate above 25°C	P_D	0.8 4.57	0.4 2.28	Watt mW/°C
Total Device Dissipation @ T_C = 25°C Derate above 25°C	P_D	3.0 17.1	1.2 6.85	Watts mW/°C
Operating and Storage Junction Temperature Range	T_J, T_{stg}	−65 to +200		°C

THERMAL CHARACTERISTICS

Characteristic	Symbol	2N2218A 2N2219,A	2N2221A 2N2222,A	Unit
Thermal Resistance, Junction to Ambient	$R_{\theta JA}$	219	145.8	°C/W
Thermal Resistance, Junction to Case	$R_{\theta JC}$	58	437.5	°C/W

2N2218A,2N2219,A
2N2221A,2N2222,A

JAN, JTX, JTXV AVAILABLE

2N2218, A/2N2219, A
CASE 79-04
TO-39 (TO-205AD)
STYLE 1

2N2221, A/2N2222, A
CASE 22-03
TO-18 (TO-206AA)
STYLE 1

3 Collector
2 Base
1 Emitter

GENERAL PURPOSE TRANSISTORS
NPN SILICON

ELECTRICAL CHARACTERISTICS (T_A = 25°C unless otherwise noted.)

Characteristic		Symbol	Min	Max	Unit
OFF CHARACTERISTICS					
Collector-Emitter Breakdown Voltage (I_C = 10 mAdc, I_B = 0)	Non-A Suffix A-Suffix	$V_{(BR)CEO}$	30 40	— —	Vdc
Collector-Base Breakdown Voltage (I_C = 10 μAdc, I_E = 0)	Non-A Suffix A-Suffix	$V_{(BR)CBO}$	60 75	— —	Vdc
Emitter-Base Breakdown Voltage (I_E = 10 μAdc, I_C = 0)	Non-A Suffix A-Suffix	$V_{(BR)EBO}$	5.0 6.0	— —	Vdc
Collector Cutoff Current (V_{CE} = 60 Vdc, $V_{EB(off)}$ = 3.0 Vdc)	A-Suffix	I_{CEX}	—	10	nAdc
Collector Cutoff Current (V_{CB} = 50 Vdc, I_E = 0) (V_{CB} = 60 Vdc, I_E = 0) (V_{CB} = 50 Vdc, I_E = 0, T_A = 150°C) (V_{CB} = 60 Vdc, I_E = 0, T_A = 150°C)	Non-A Suffix A-Suffix Non-A Suffix A-Suffix	I_{CBO}	— — — —	0.01 0.01 10 10	μAdc
Emitter Cutoff Current (V_{EB} = 3.0 Vdc, I_C = 0)	A-Suffix	I_{EBO}	—	10	nAdc
Base Cutoff Current (V_{CE} = 60 Vdc, $V_{EB(off)}$ = 3.0 Vdc)	A-Suffix	I_{BL}	—	20	nAdc
ON CHARACTERISTICS					
DC Current Gain (I_C = 0.1 mAdc, V_{CE} = 10 Vdc)	2N2218A, 2N2221A(1) 2N2219,A, 2N2222,A(1)	h_{FE}	20 35	— —	—
(I_C = 1.0 mAdc, V_{CE} = 10 Vdc)	2N2218A, 2N2221A 2N2219,A, 2N2222,A		25 50	— —	
(I_C = 10 mAdc, V_{CE} = 10 Vdc)	2N2218A, 2N2221A(1) 2N2219,A, 2N2222,A(1)		35 75	— —	
(I_C = 10 mAdc, V_{CE} = 10 Vdc, T_A = −55°C)	2N2218A, 2N2221A 2N2219,A, 2N2222,A		15 35	— —	
(I_C = 150 mAdc, V_{CE} = 10 Vdc)(1)	2N2218A, 2N2221A 2N2219,A, 2N2222,A		40 100	120 300	

2N2218A/19/19A/21A/22/22A

ELECTRICAL CHARACTERISTICS (continued) (T_A = 25°C unless otherwise noted.)

Characteristic		Symbol	Min	Max	Unit
(I_C = 150 mAdc, V_{CE} = 1.0 Vdc)(1) 2N2218A, 2N2221A 2N2219,A, 2N2222,A			20 50	— —	
(I_C = 500 mAdc, V_{CE} = 10 Vdc)(1) 2N2219, 2N2222 2N2218A, 2N2221A, 2N2219A, 2N2222A			30 25 40	— — —	
Collector-Emitter Saturation Voltage(1) (I_C = 150 mAdc, I_B = 15 mAdc) Non-A Suffix A-Suffix		$V_{CE(sat)}$	— —	0.4 0.3	Vdc
(I_C = 500 mAdc, I_B = 50 mAdc) Non-A Suffix A-Suffix			— —	1.6 1.0	
Base-Emitter Saturation Voltage(1) (I_C = 150 mAdc, I_B = 15 mAdc) Non-A Suffix A-Suffix		$V_{BE(sat)}$	0.6 0.6	1.3 1.2	Vdc
(I_C = 500 mAdc, I_B = 50 mAdc) Non-A Suffix A-Suffix			— —	2.6 2.0	

SMALL-SIGNAL CHARACTERISTICS

Characteristic		Symbol	Min	Max	Unit
Current Gain — Bandwidth Product(2) (I_C = 20 mAdc, V_{CE} = 20 Vdc, f = 100 MHz) All Types, Except 2N2219A, 2N2222A		f_T	250 300	— —	MHz
Output Capacitance(3) (V_{CB} = 10 Vdc, I_E = 0. f = 1.0 MHz)		C_{obo}	—	8.0	pF
Input Capacitance(3) (V_{EB} = 0.5 Vdc, I_C = 0, f = 1.0 MHz) Non-A Suffix A-Suffix		C_{ibo}	— —	30 25	pF
Input Impedance (I_C = 1.0 mAdc, V_{CE} = 10 Vdc, f = 1.0 kHz) 2N2218A, 2N2221A 2N2219A, 2N2222A		h_{ie}	1.0 2.0	3.5 8.0	kohms
(I_C = 10 mAdc, V_{CE} = 10 Vdc, f = 1.0 kHz) 2N2218A, 2N2221A 2N2219A, 2N2222A			0.2 0.25	1.0 1.25	
Voltage Feedback Ratio (I_C = 1.0 mAdc, V_{CE} = 10 Vdc, f = 1.0 kHz) 2N2218A, 2N2221A 2N2219A, 2N2222A		h_{re}	— —	5.0 8.0	X 10^{-4}
(I_C = 10 mAdc, V_{CE} = 10 Vdc, f = 1.0 kHz) 2N2218A, 2N2221A 2N2219A, 2N2222A			— —	2.5 4.0	
Small-Signal Current Gain (I_C = 1.0 mAdc, V_{CE} = 10 Vdc, f = 1.0 kHz) 2N2218A, 2N2221A 2N2219A, 2N2222A		h_{fe}	30 50	150 300	—
(I_C = 10 mAdc, V_{CE} = 10 Vdc, f = 1.0 kHz) 2N2218A, 2N2221A 2N2219A, 2N2222A			50 75	300 375	
Output Admittance (I_C = 1.0 mAdc, V_{CE} = 10 Vdc, f = 1.0 kHz) 2N2218A, 2N2221A 2N2219A, 2N2222A		h_{oe}	3.0 5.0	15 35	μmhos
(I_C = 10 mAdc, V_{CE} = 10 Vdc, f = 1.0 kHz) 2N2218A, 2N2221A 2N2219A, 2N2222A			10 15	100 200	
Collector Base Time Constant (I_E = 20 mAdc, V_{CB} = 20Vdc, f = 31.8 MHz) A-Suffix		$rb'C_c$	—	150	ps
Noise Figure (I_C = 100 μAdc, V_{CE} = 10 Vdc, R_S = 1.0 kohm, f = 1.0 kHz) 2N2222A		NF	—	4.0	dB
Real Part of Common-Emitter High Frequency Input Impedance (I_C = 20 mAdc, V_{CE} = 20 Vdc, f = 300 MHz) 2N2218A, 2N2219A 2N2221A, 2N2222A		$Re(h_{ie})$	—	60	Ohms

(1) Pulse Test: Pulse Width ≤ 300 μs, Duty Cycle ≤ 2.0%.
(2) f_T is defined as the frequency at which $|h_{fe}|$ extrapolates to unity.
(3) 2N5581 and 2N5582 are Listed C_{cb} and C_{eb} for these conditions and values.

2N2218A/19/19A/21A/22/22A

ELECTRICAL CHARACTERISTICS (continued) (T_A = 25°C unless otherwise noted.)

Characteristic		Symbol	Min	Max	Unit
SWITCHING CHARACTERISTICS					
Delay Time	(V_{CC} = 30 Vdc, $V_{BE(off)}$ = 0.5 Vdc, I_C = 150 mAdc, I_{B1} = 15 mAdc) (Figure 14)	t_d	—	10	ns
Rise Time		t_r	—	25	ns
Storage Time	(V_{CC} = 30 Vdc, I_C = 150 mAdc, I_{B1} = I_{B2} = 15 mAdc) (Figure15)	t_s	—	225	ns
Fall Time		t_f	—	60	ns
Active Region Time Constant (I_C = 150 mAdc, V_{CE} = 30 Vdc) (See Figure 12 for 2N2218A, 2N2219A, 2N2221A, 2N2222A)		T_A	—	2.5	ns

FIGURE 1 – NORMALIZED DC CURRENT GAIN

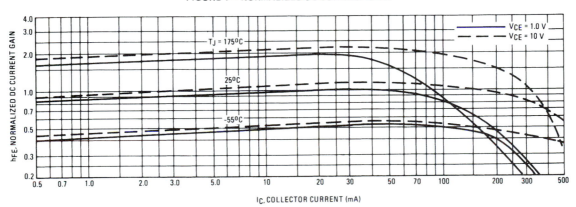

FIGURE 2 – COLLECTOR CHARACTERISTICS IN SATURATION REGION

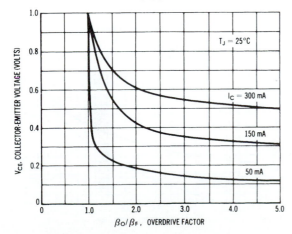

This graph shows the effect of base current on collector current. β_o (current gain at the edge of saturation) is the current gain of the transistor at 1 volt, and β_f (forced gain) is the ratio of I_c/I_{bf} in a circuit.

EXAMPLE: For type 2N2219, estimate a base current (I_{bf}) to insure saturation at a temperature of 25°C and a collector current of 150 mA.

Observe that at I_c = 150 mA an overdrive factor of at least 2.5 is required to drive the transistor well into the saturation region. From Figure 1, it is seen that h_{FE} @ 1 volt is approximately 0.62 of h_{FE} @ 10 volts. Using the guaranteed minimum gain of 100 @ 150 mA and 10 V, β_o = 62 and substituting values in the overdrive equation, we find:

$$\frac{\beta_o}{\beta_f} = \frac{h_{FE} @ 1.0 \, V}{I_c/I_{bf}} \qquad 2.5 = \frac{62}{150/I_{bf}} \qquad I_{bf} \approx 6.0 \, mA$$

2N2218A/19/19A/21A/22/22A

FIGURE 3 — "ON" VOLTAGES

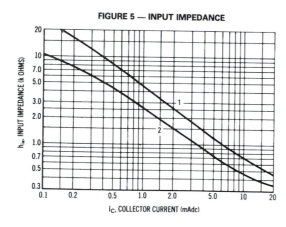

FIGURE 4 — TEMPERATURE COEFFICIENTS

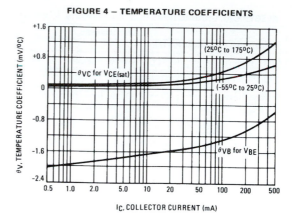

h PARAMETERS

V_{CE} = 10 Vdc, f = 1.0 kHz, T_A = 25°C

This group of graphs illustrates the relationship between h_{fe} and other "h" parameters for this series of transistors. To obtain these curves, a high-gain and a low-gain unit were selected and the same units were used to develop the correspondingly numbered curves on each graph.

FIGURE 5 — INPUT IMPEDANCE

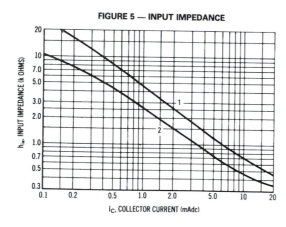

FIGURE 6 — VOLTAGE FEEDBACK RATIO

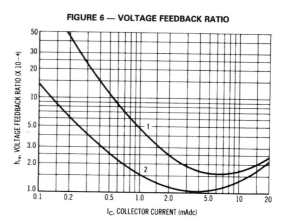

FIGURE 7 — CURRENT GAIN

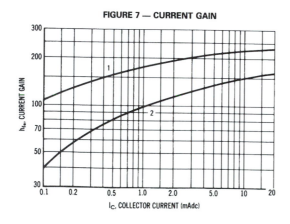

FIGURE 8 — OUTPUT ADMITTANCE

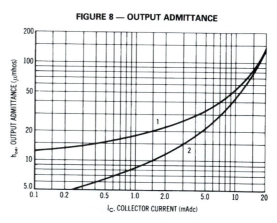

2N2218A/19/19A/21A/22/22A

SWITCHING TIME CHARACTERISTICS

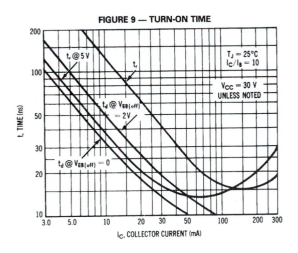

FIGURE 9 — TURN-ON TIME

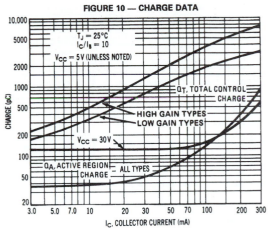

FIGURE 10 — CHARGE DATA

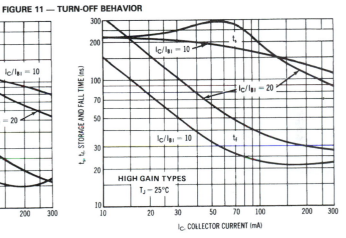

FIGURE 11 — TURN-OFF BEHAVIOR

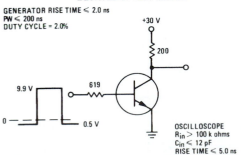

FIGURE 12 — DELAY AND RISE TIME
EQUIVALENT TEST CIRCUIT

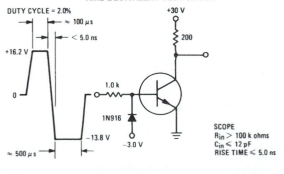

FIGURE 13 — STORAGE TIME AND FALL
TIME EQUIVALENT TEST CIRCUIT

MAXIMUM RATINGS

Rating	Symbol	Value	Unit
Collector-Emitter Voltage	V_{CEO}	40	Vdc
Collector-Base Voltage	V_{CBO}	60	Vdc
Emitter-Base Voltage	V_{EBO}	6.0	Vdc
Collector Current — Continuous	I_C	200	mAdc
Total Device Dissipation @ $T_A = 25°C$ Derate above 25°C	P_D	0.36 2.06	Watt mW/°C
Total Device Dissipation @ $T_C = 25°C$ Derate above 25°C	P_D	1.2 6.9	Watts mW/°C
Operating and Storage Junction Temperature Range	T_J, T_{stg}	−65 to +200	°C

THERMAL CHARACTERISTICS

Characteristic	Symbol	Max	Unit
Thermal Resistance, Junction to Case	$R_{\theta JC}$	0.15	°C/mW
Thermal Resistance, Junction to Ambient	$R_{\theta JA}$	0.49	°C/mW

2N3946
2N3947

CASE 22-03, STYLE 1
TO-18 (TO-206AA)

2 Base · 3 Collector · 1 Emitter · 3 2 1

GENERAL PURPOSE TRANSISTORS

NPN SILICON

ELECTRICAL CHARACTERISTICS ($T_A = 25°C$ unless otherwise noted.)

Characteristic		Symbol	Min	Max	Unit
OFF CHARACTERISTICS					
Collector-Emitter Breakdown Voltage(1) ($I_C = 10$ mAdc)		$V_{(BR)CEO}$	40	—	Vdc
Collector-Base Breakdown Voltage ($I_C = 10$ μAdc, $I_E = 0$)		$V_{(BR)CBO}$	60	—	Vdc
Emitter-Base Breakdown Voltage ($I_E = 10$ μAdc, $I_C = 0$)		$V_{(BR)EBO}$	6.0	—	Vdc
Collector Cutoff Current ($V_{CE} = 40$ Vdc, $V_{EB} = 3.0$ Vdc) ($V_{CE} = 40$ Vdc, $V_{EB} = 3.0$ Vdc, $T_A = 150°C$)		I_{CEX}	— —	0.010 15	μAdc
Base Cutoff Current ($V_{CE} = 40$ Vdc, $V_{EB} = 3.0$ Vdc)		I_{BL}	—	.025	μAdc
ON CHARACTERISTICS					
DC Current Gain(1) ($I_C = 0.1$ mAdc, $V_{CE} = 1.0$ Vdc)	2N3946 2N3947	h_{FE}	30 60	— 	—
($I_C = 1.0$ mAdc, $V_{CE} = 1.0$ Vdc)	2N3946 2N3947		45 90	— —	
($I_C = 10$ mAdc, $V_{CE} = 1.0$ Vdc)	2N3946 2N3947		50 100	150 300	
($I_C = 50$ mAdc, $V_{CE} = 1.0$ Vdc)	2N3946 2N3947		20 40	— —	
Collector-Emitter Saturation Voltage(1) ($I_C = 10$ mAdc, $I_B = 1.0$ mAdc) ($I_C = 50$ mAdc, $I_B = 5.0$ mAdc)		$V_{CE(sat)}$	— —	0.2 0.3	Vdc
Base-Emitter Saturation Voltage(1) ($I_C = 10$ mAdc, $I_B = 1.0$ mAdc) ($I_C = 50$ mAdc, $I_B = 5.0$ mAdc)		$V_{BE(sat)}$	0.6 	0.9 1.0	Vdc
SMALL-SIGNAL CHARACTERISTICS					
Current-Gain — Bandwidth Product ($I_C = 10$ mAdc, $V_{CE} = 20$ Vdc, $f = 100$ MHz)	2N3946 2N3947	f_T	250 300	— —	MHz
Output Capacitance ($V_{CB} = 10$ Vdc, $I_E = 0$, $f = 1.0$ MHz)		C_{obo}	—	4.0	pF

2N3946, 2N3947

ELECTRICAL CHARACTERISTICS (continued) (T_A = 25°C unless otherwise noted.)

Characteristic		Symbol	Min	Max	Unit
Input Capacitance (V_{EB} = 1.0 Vdc, I_C = 0, f = 1.0 MHz)		C_{ibo}	—	8.0	pF
Input Impedance (I_C = 1.0 mA, V_{CE} = 10 V, f = 1.0 kHz)	2N3946 2N3947	h_{ie}	0.5 2.0	6.0 12	kohms
Voltage Feedback Ratio (I_C = 1.0 mA, V_{CE} = 10 V, f = 1.0 kHz)	2N3946 2N3947	h_{re}	— —	10 20	X 10^{-4}
Small Signal Current Gain (I_C = 1.0 mA, V_{CE} = 10 V, f = 1.0 kHz)	2N3946 2N3947	h_{fe}	50 100	250 700	—
Output Admittance (I_C = 1.0 mA, V_{CE} = 10 V, f = 1.0 kHz)	2N3946 2N3947	h_{oe}	1.0 5.0	30 50	µmhos
Collector Base Time Constant (I_C = 10 mA, V_{CE} = 20 V, f = 31.8 MHz)		$rb'C_c$	—	200	ps
Noise Figure (I_C = 100 µA, V_{CE} = 5.0 V, R_g = 1.0 kΩ, f = 1.0 kHz)		NF	—	5.0	dB

SWITCHING CHARACTERISTICS

			Symbol	Min	Max	Unit
Delay Time	V_{CC} = 3.0 Vdc, V_{OB} = 0.5 Vdc, I_C = 10 mAdc, I_{B1} = 1.0 mA		t_d	—	35	ns
Rise Time			t_r	—	35	ns
Storage Time	V_{CC} = 3.0 V, I_C = 10 mA,	2N3946 2N3947	t_s	— —	300 375	ns
Fall Time	I_{B1} = I_{B2} = 1.0 mAdc		t_f	—	75	ns

(1) Pulse Test: PW ≤ 300 µs, Duty Cycle ≤ 2%.

TYPICAL SWITCHING CHARACTERISTICS
(T_A = 25°C unless otherwise noted)

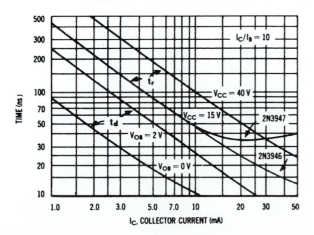

FIGURE 1 — DELAY AND RISE TIME

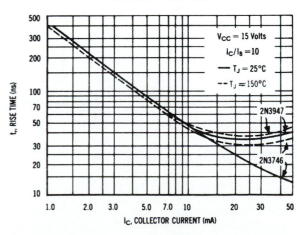

FIGURE 2 — RISE TIME

2N3946, 2N3947

FIGURE 3 — STORAGE AND FALL TIMES

2N3946

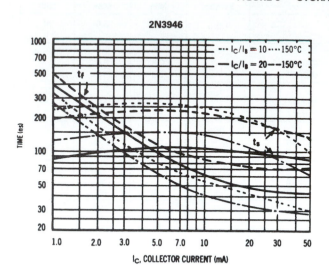

2N3947

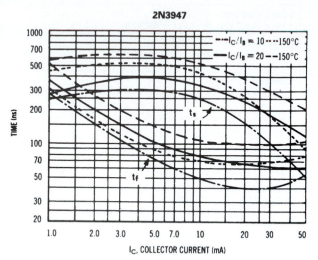

FIGURE 4 — TURN-ON TIME EQUIVALENT TEST CIRCUIT

FIGURE 5 — TURN-OFF TIME EQUIVALENT TEST CIRCUIT

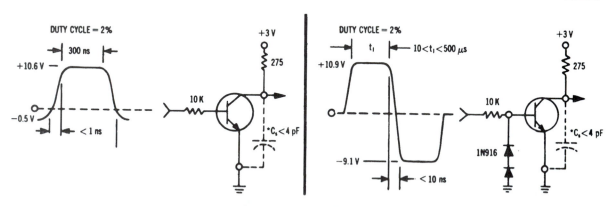

*TOTAL SHUNT CAPACITANCE OF TEST JIG AND CONNECTORS

2N3946, 2N3947

AUDIO SMALL-SIGNAL CHARACTERISTICS

FIGURE 6 — NOISE FIGURE VARIATIONS
V_{CE} = 5.0 V, T_A = 25°C

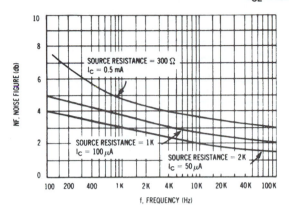

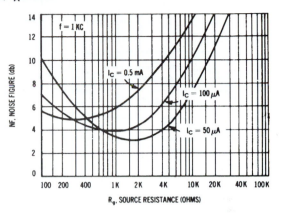

h PARAMETERS
V_{CE} = 10 V, T_A = 25°C, f = 1.0 kc

FIGURE 7 — CURRENT GAIN

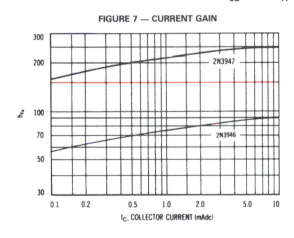

FIGURE 8 — OUTPUT CAPACITANCE

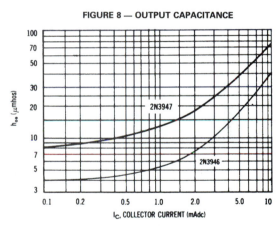

FIGURE 9 — INPUT IMPEDANCE

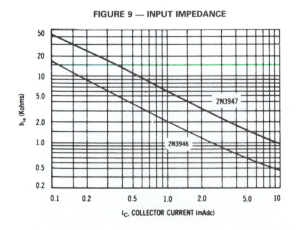

FIGURE 10 — VOLTAGE FEEDBACK RATIO

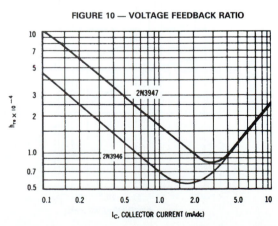

2N3946, 2N3947

FIGURE 11 — CURRENT GAIN CHARACTERISTICS
2N3946

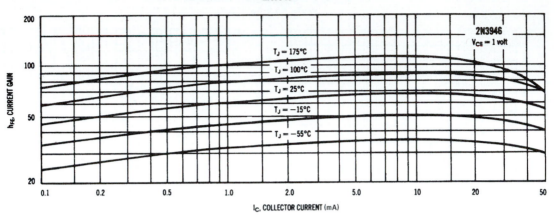

2N3947

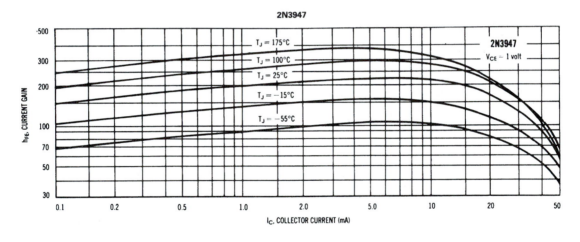

FIGURE 12 — CAPACITANCE

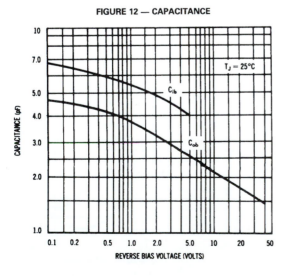

FIGURE 13 — CHARGE DATA

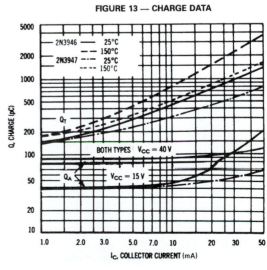

2N3946, 2N3947

FIGURE 14 — COLLECTOR SATURATION REGION
2N3946

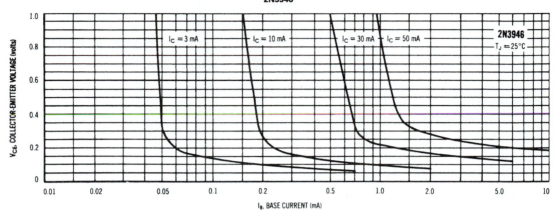

2N3947

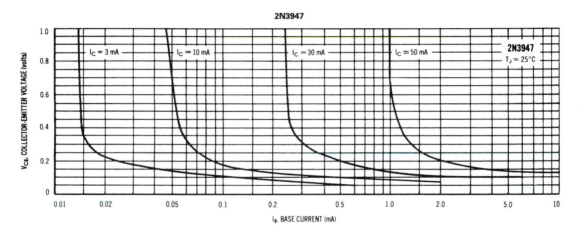

FIGURE 15 — "ON" VOLTAGES

FIGURE 16 — TEMPERATURE COEFFICIENTS

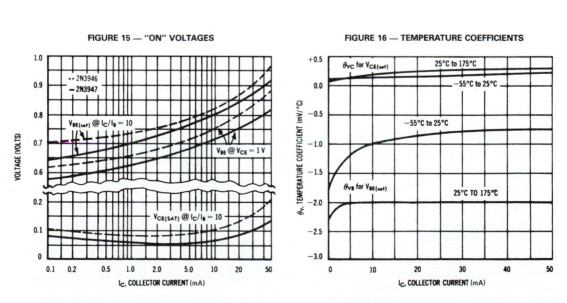

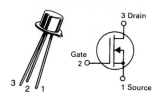

2N3796
2N3797

CASE 22-03, STYLE 2
TO-18 (TO-206AA)

3 Drain

Gate 2

3 2 1   1 Source

MOSFETs
LOW POWER AUDIO

N-CHANNEL — DEPLETION

MAXIMUM RATINGS

Rating	Symbol	Value	Unit

Drain-Source Voltage
  2N3796
  2N3797
	V_{DS}	
		25
		20
			Vdc

Gate-Source Voltage	V_{GS}	± 10	Vdc
Drain Current	I_D	20	mAdc
Total Device Dissipation @ T_A = 25°C Derate above 25°C	P_D	200 1.14	mW mW/°C
Junction Temperature Range	T_J	+ 175	°C
Storage Channel Temperature Range	T_{stg}	− 65 to + 200	°C

ELECTRICAL CHARACTERISTICS (T_A = 25°C unless otherwise noted.)

Characteristic	Symbol	Min	Typ	Max	Unit
OFF CHARACTERISTICS					

Drain-Source Breakdown Voltage
  (V_{GS} = −4.0 V, I_D = 5.0 μA)   2N3796
  (V_{GS} = −7.0 V, I_D = 5.0 μA)   2N3797
	$V_{(BR)DSX}$	
		25
		20
			
			30
			25
				
				—
				—
					Vdc

Gate Reverse Current(1)
  (V_{GS} = − 10 V, V_{DS} = 0)
  (V_{GS} = − 10 V, V_{DS} = 0, T_A = 150°C)
	I_{GSS}	
		—
		—
			
			—
			—
				
				1.0
				200
					pAdc

Gate Source Cutoff Voltage
  (I_D = 0.5 μA, V_{DS} = 10 V)   2N3796
  (I_D = 2.0 μA, V_{DS} = 10 V)   2N3797
	$V_{GS(off)}$	
		—
		—
			
			− 3.0
			− 5.0
				
				− 4.0
				− 7.0
					Vdc

Drain-Gate Reverse Current(1) (V_{DG} = 10 V, I_S = 0)	I_{DGO}	—	—	1.0	pAdc
ON CHARACTERISTICS					

Zero-Gate-Voltage Drain Current
  (V_{DS} = 10 V, V_{GS} = 0)   2N3796
                  2N3797
	I_{DSS}	
		0.5
		2.0
			
			1.5
			2.9
				
				3.0
				6.0
					mAdc

On-State Drain Current
  (V_{DS} = 10 V, V_{GS} = +3.5 V)   2N3796
                  2N3797
	$I_{D(on)}$	
		7.0
		9.0
			
			8.3
			14
				
				14
				18
					mAdc

SMALL-SIGNAL CHARACTERISTICS					

Forward Transfer Admittance
  (V_{DS} = 10 V, V_{GS} = 0, f = 1.0 kHz)   2N3796
                  2N3797
 	
  (V_{DS} = 10 V, V_{GS} = 0, f = 1.0 MHz)   2N3796
                  2N3797
	$\|y_{fs}\|$	
		900
		1500
		
		900
		1500
			
			1200
			2300
			
			—
			—
				
				1800
				3000
				
				—
				—
					μmhos

Output Admittance
  (V_{DS} = 10 V, V_{GS} = 0, f = 1.0 kHz)   2N3796
                  2N3797
	$\|y_{os}\|$	
		—
		—
			
			12
			27
				
				25
				60
					μmhos

Input Capacitance
  (V_{DS} = 10 V, V_{GS} = 0, f = 1.0 MHz)   2N3796
                  2N3797
	C_{iss}	
		—
		—
			
			5.0
			6.0
				
				7.0
				8.0
					pF

Reverse Transfer Capacitance (V_{DS} = 10 V, V_{GS} = 0, f = 1.0 MHz)	C_{rss}	—	0.5	0.8	pF
FUNCTIONAL CHARACTERISTICS					
Noise Figure (V_{DS} = 10 V, V_{GS} = 0, f = 1.0 kHz, R_S = 3 megohms)	NF	—	3.8	—	dB

(1) This value of current includes both the FET leakage current as well as the leakage current associated with the test socket and fixture when measured under best attainable conditions.

2N3796, 2N3797

TYPICAL DRAIN CHARACTERISTICS

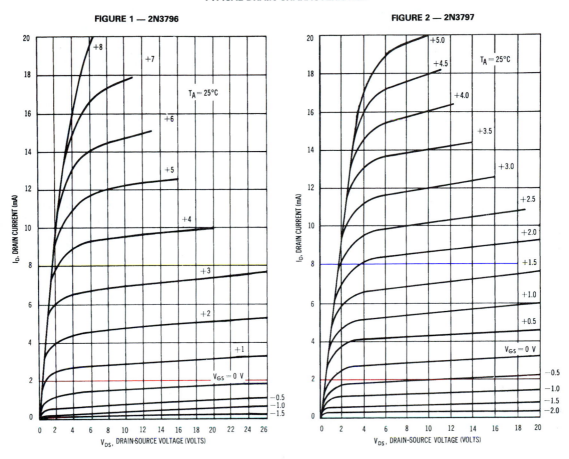

FIGURE 1 — 2N3796

FIGURE 2 — 2N3797

COMMON SOURCE TRANSFER CHARACTERISTICS

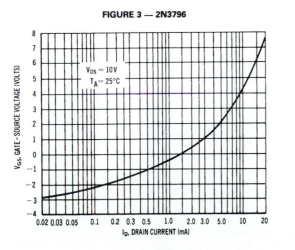

FIGURE 3 — 2N3796

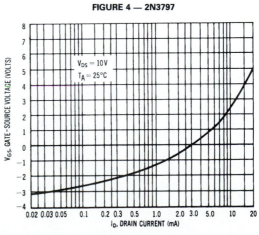

FIGURE 4 — 2N3797

2N3796, 2N3797

FIGURE 5 — FORWARD TRANSFER ADMITTANCE

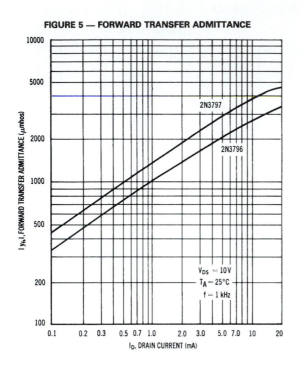

FIGURE 6 — OUTPUT ADMITTANCE

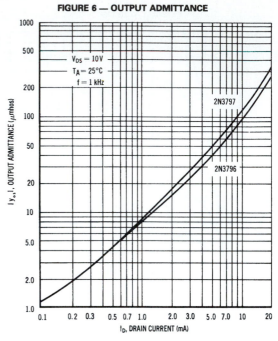

FIGURE 7 — NOISE FIGURE

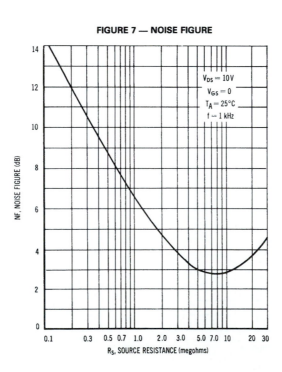

2N6394 MCR220-5
thru   MCR220-7
2N6399 MCR220-9

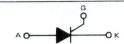

THYRISTORS

12 AMPERES RMS
50-800 VOLTS

SILICON CONTROLLED RECTIFIERS

. . . designed primarily for half-wave ac control applications, such as motor controls, heating controls and power supplies; or wherever half-wave silicon gate-controlled, solid-state devices are needed.

- Glass Passivated Junctions and Center Gate Fire for Greater Parameter Uniformity and Stability
- Small, Rugged, Thermowatt▲ Construction for Low Thermal Resistance, High Heat Dissipation and Durability
- Blocking Voltage to 800 Volts

*MAXIMUM RATINGS

Rating	Symbol	Value	Unit
Peak Reverse Voltage (1)	V_{RRM}		Volts
2N6394		50	
2N6395		100	
2N6396		200	
MCR220-5		300	
2N6397		400	
MCR220-7		500	
2N6398		600	
MCR220-9		700	
2N6399		800	
Forward Current RMS T_J = 125°C (All Conduction Angles)	$I_{T(RMS)}$	12	Amps
Peak Forward Surge Current (1/2 cycle, Sine Wave, 60 Hz, T_J = 125°C)	I_{TSM}	100	Amps
Circuit Fusing Considerations (T_J = –40 to +125°C, t = 1.0 to 8.3 ms)	I^2t	40	A^2s
Forward Peak Gate Power	P_{GM}	20	Watts
Forward Average Gate Power	$P_{G(AV)}$	0.5	Watt
Forward Peak Gate Current	I_{GM}	2.0	Amps
Operating Junction Temperature Range	T_J	–40 to +125	°C
Storage Temperature Range	T_{stg}	–40 to +150	°C

THERMAL CHARACTERISTICS

Characteristic	Symbol	Max	Unit
Thermal Resistance, Junction to Case	$R_{\theta JC}$	2.0	°C/W

(1) V_{RRM} for all types can be applied on a continuous dc basis without incurring damage. Ratings apply for zero or negative gate voltage. Devices should not be tested for blocking capability in a manner such that the voltage supplied exceeds the rated blocking voltage.

*Indicates JEDEC Registered Data.
▲Trademark of Motorola Inc.

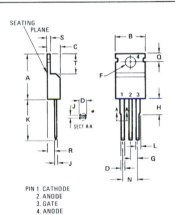

SEATING PLANE

PIN 1. CATHODE
2. ANODE
3. GATE
4. ANODE

All JEDEC dimensions and notes apply

DIM	MILLIMETERS		INCHES	
	MIN	MAX	MIN	MAX
A	14.23	15.87	0.560	0.625
B	9.66	10.66	0.380	0.420
C	3.56	4.82	0.140	0.190
D	0.51	1.14	0.020	0.045
F	3.531	3.733	0.139	0.147
G	2.29	2.79	0.090	0.110
H	–	6.35		0.250
J	0.31	1.14	0.012	0.045
K	12.70	14.27	0.500	0.562
L	1.14	1.77	0.045	0.070
N	4.83	5.33	0.190	0.210
Q	2.54	3.04	0.100	0.120
R	2.04	2.92	0.080	0.115
S	0.51	1.39	0.020	0.055
T	5.85	6.85	0.230	0.270

CASE 221-02
TO 220 AB

2N6394 thru 2N6399 ● MCR220-5 ● MCR220-7 ● MCR220-9

ELECTRICAL CHARACTERISTICS (T_C = 25°C unless otherwise noted.)

Characteristic		Symbol	Min	Typ	Max	Unit
*Peak Forward Blocking Voltage		V_{DRM}				Volts
(T_J = 125°C)	2N6394		50	–	–	
	2N6395		100	–	–	
	2N6396		200	–	–	
	MCR220-5		300	–	–	
	2N6397		400	–	–	
	MCR220-7		500	–	–	
	2N6398		600	–	–	
	MCR220-9		700	–	–	
	2N6399		800	–	–	
* Peak Forward Blocking Current (Rated V_{DRM} @ T_J = 125°C)		I_{DRM}	–	–	2.0	mA
* Peak Reverse Blocking Current (Rated V_{RRM} @ T_J = 125°C)		I_{RRM}	–	–	2.0	mA
* Forward "On" Voltage (I_{TM} = 24 A Peak)		V_{TM}	–	1.7	2.2	Volts
* Gate Trigger Current (Continuous dc) (Anode Voltage = 12 Vdc, R_L = 100 Ohms)		I_{GT}	–	5.0	30	mA
* Gate Trigger Voltage (Continuous dc) (Anode Voltage = 12 Vdc, R_L = 100 Ohms)		V_{GT}	–	0.7	1.5	Volts
* Gate Non-Trigger Voltage (Anode Voltage = Rated V_{DRM}, R_L = 100 Ohms, T_J = 125°C)		V_{GD}	0.2	–	–	Volts
* Holding Current (Anode Voltage = 12 Vdc)		I_H	–	6.0	40	mA
Turn-On Time (I_{TM} = 12 A, I_{GT} = 40 mAdc)		t_{gt}	–	1.0	2.0	μs
Turn-Off Time (V_{DRM} = rated voltage)		t_q				μs
(I_{TM} = 12 A, I_R = 12 A)			–	15	–	
(I_{TM} = 12 A, I_R = 12 A, T_J = 125°C)			–	35	–	
Forward Voltage Application Rate (T_J = 125°C)		dv/dt	–	50	–	V/μs

*Indicates JEDEC Registered Data.

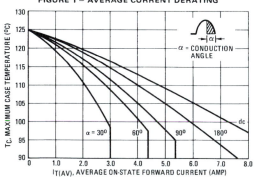

FIGURE 1 – AVERAGE CURRENT DERATING

T_C, MAXIMUM CASE TEMPERATURE (°C) vs $I_{T(AV)}$, AVERAGE ON-STATE FORWARD CURRENT (AMP)

α = CONDUCTION ANGLE

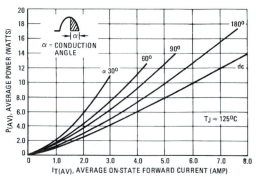

FIGURE 2 – MAXIMUM ON-STATE POWER DISSIPATION

$P_{(AV)}$, AVERAGE POWER (WATTS) vs $I_{T(AV)}$, AVERAGE ON-STATE FORWARD CURRENT (AMP)

α = CONDUCTION ANGLE

$T_J \approx 125°C$

 MOTOROLA *Semiconductor Products Inc.*

MOTOROLA Semiconductors
BOX 20912 ● PHOENIX, ARIZONA 85036

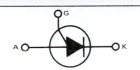

2N6027
2N6028

SILICON
PROGRAMMABLE UNIJUNCTION
TRANSISTORS

40 VOLTS
375 mW

DS 2520

SILICON PROGRAMMABLE
UNIJUNCTION TRANSISTORS

. . . designed to enable the engineer to "program" unijunction characteristics such as R_{BB}, η, I_V, and I_P by merely selecting two resistor values. Application includes thyristor-trigger, oscillator, pulse and timing circuits. These devices may also be used in special thyristor applications due to the availability of an anode gate. Supplied in an inexpensive TO-92 plastic package for high-volume requirements, this package is readily adaptable for use in automatic insertion equipment.

- Programmable — R_{BB}, η, I_V and I_P.

- Low On-State Voltage — 1.5 Volts Maximum @ I_F = 50 mA

- Low Gate to Anode Leakage Current — 10 nA Maximum

- High Peak Output Voltage — 11 Volts Typical

- Low Offset Voltage — 0.35 Volt Typical (R_G = 10 k ohms)

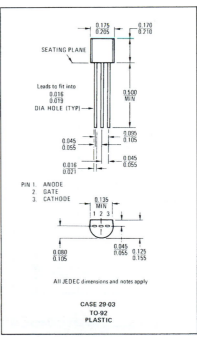

PIN 1. ANODE
    2. GATE
    3. CATHODE

CASE 29-03
TO-92
PLASTIC

All JEDEC dimensions and notes apply

MAXIMUM RATINGS

Rating	Symbol	Value	Unit
Power Dissipation(1)	P_F	375	mW
Derate Above 25°C	$1/\theta_{JA}$	5.0	mW/°C
DC Forward Anode Current(2)	I_T	200	mA
Derate Above 25°C		2.67	mA/°C
*DC Gate Current	I_G	±50	mA
Repetitive Peak Forward Current	I_{TRM}		
100 μs Pulse Width, 1.0% Duty Cycle		1.0	Amp
*20 μs Pulse Width, 1.0% Duty Cycle		2.0	Amp
Non-Repetitive Peak Forward Current	I_{TSM}	5.0	Amp
10 μs Pulse Width			
* Gate to Cathode Forward Voltage	V_{GKF}	40	Volt
* Gate to Cathode Reverse Voltage	V_{GKR}	–5.0	Volt
* Gate to Anode Reverse Voltage	V_{GAR}	40	Volt
* Anode to Cathode Voltage	V_{AK}	±40	Volt
Operating Junction Temperature Range	T_J	–50 to +100	°C
* Storage Temperature Range	T_{stg}	–55 to +150	°C

*Indicates JEDEC Registered Data
(1) JEDEC Registered Data is 300 mW, derating at 4.0 mW/°C.
(2) JEDEC Registered Data is 150 mA.

2N6027 ● 2N6028

ELECTRICAL CHARACTERISTICS (T_A = 25°C unless otherwise noted)

Characteristic	Figure	Symbol	Min	Typ	Max	Unit
*Peak Current	2,9,11	I_P				μA
(V_S = 10 Vdc, R_G = 1.0 MΩ) 2N6027			—	1.25	2.0	
2N6028			—	0.08	0.15	
(V_S = 10 Vdc, R_G = 10 k ohms) 2N6027			—	4.0	5.0	
2N6028			—	0.70	1.0	
*Offset Voltage	1	V_T				Volts
(V_S = 10 Vdc, R_G = 1.0 MΩ) 2N6027			0.2	0.70	1.6	
2N6028			0.2	0.50	0.6	
(V_S = 10 Vdc, R_G = 10 k ohms) (Both Types)			0.2	0.35	0.6	
*Valley Current	1,4,5,	I_V				μA
(V_S = 10 Vdc, R_G = 1.0 MΩ) 2N6027			—	18	50	
2N6028			—	18	25	
(V_S = 10 Vdc, R_G = 10 k ohms) 2N6027			70	270	—	
2N6028			25	270	—	
(V_S = 10 Vdc, R_G = 200 Ohms) 2N6027			1.5	—	—	mA
2N6028			1.0	—	—	
* Gate to Anode Leakage Current	—	I_{GAO}				nAdc
(V_S = 40 Vdc, T_A = 25°C, Cathode Open)			—	1.0	10	
(V_S = 40 Vdc, T_A = 75°C, Cathode Open)			—	3.0	—	
Gate to Cathode Leakage Current	—	I_{GKS}				nAdc
(V_S = 40 Vdc, Anode to Cathode Shorted)			—	5.0	50	
*Forward Voltage (I_F = 50 mA Peak)	1,6	V_F	—	0.8	1.5	Volts
*Peak Output Voltage	3,7	V_O	6.0	11	—	Volts
(V_B = 20 Vdc, C_C = 0.2 μF)						
Pulse Voltage Rise Time	3	t_r	—	40	80	ns
(V_B = 20 Vdc, C_C = 0.2 μF)						

*Indicates JEDEC Registered Data

FIGURE 1 – ELECTRICAL CHARACTERIZATION

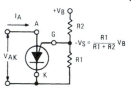 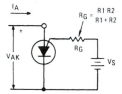 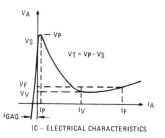

1A – PROGRAMMABLE UNIJUNCTION WITH "PROGRAM" RESISTORS R1 and R2

1B – EQUIVALENT TEST CIRCUIT FOR FIGURE 1A USED FOR ELECTRICAL CHARACTERISTICS TESTING (ALSO SEE FIGURE 2)

1C – ELECTRICAL CHARACTERISTICS

FIGURE 2 – PEAK CURRENT (Ip) TEST CIRCUIT

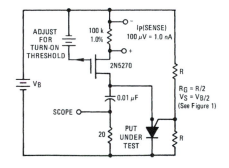

FIGURE 3 – V_O AND t_r TEST CIRCUIT

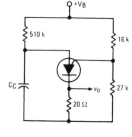 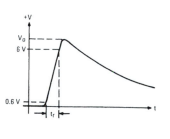

MOTOROLA *Semiconductor Products Inc.*

MOTOROLA
■ SEMICONDUCTOR ■
TECHNICAL DATA

MC1741
MC1741C

INTERNALLY COMPENSATED, HIGH PERFORMANCE OPERATIONAL AMPLIFIERS

. . . designed for use as a summing amplifier, integrator, or amplifier with operating characteristics as a function of the external feedback components.

- No Frequency Compensation Required
- Short-Circuit Protection
- Offset Voltage Null Capability
- Wide Common-Mode and Differential Voltage Ranges
- Low-Power Consumption
- No Latch Up

OPERATIONAL AMPLIFIER

SILICON MONOLITHIC
INTEGRATED CIRCUIT

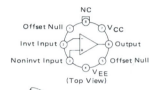

G SUFFIX
METAL PACKAGE
CASE 601

P1 SUFFIX
PLASTIC PACKAGE
CASE 626

U SUFFIX
CERAMIC PACKAGE
CASE 693

D SUFFIX
PLASTIC PACKAGE
CASE 751
(SO-8)

MAXIMUM RATINGS (T_A = +25°C unless otherwise noted)

Rating	Symbol	MC1741C	MC1741	Unit
Power Supply Voltage	V_{CC} V_{EE}	+18 −18	+22 −22	Vdc Vdc
Input Differential Voltage	V_{ID}	±30		Volts
Input Common Mode Voltage (Note 1)	V_{ICM}	±15		Volts
Output Short Circuit Duration (Note 2)	t_S	Continuous		
Operating Ambient Temperature Range	T_A	0 to +70	−55 to +125	°C
Storage Temperature Range Metal and Ceramic Packages Plastic Packages	T_{stg}	−65 to +150 −55 to +125		°C

NOTES:
1. For supply voltages less than +15 V, the absolute maximum input voltage is equal to the supply voltage.
2. Supply voltage equal to or less than 15 V.

PIN CONNECTIONS

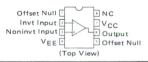

Offset Null	1		8	NC
Invt Input	2		7	V_{CC}
Noninvt Input	3		6	Output
V_{EE}	4		5	Offset Null

(Top View)

ORDERING INFORMATION

Device	Alternate	Temperature Range	Package
MC1741CD	—	0°C to +70°C	SO-8
MC1741CG	LM741CH, μA741HC		Metal Can
MC1741CP1	LM741CN, μA741TC		Plastic DIP
MC1741CU	—		Ceramic DIP
MC1741G	—	−55°C to +125°C	Metal Can
MC1741U	—		Ceramic DIP

EQUIVALENT CIRCUIT SCHEMATIC

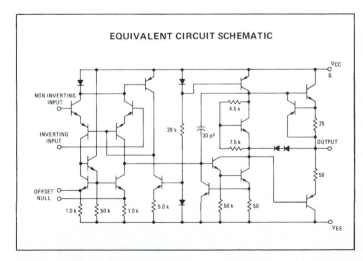

MC1741, MC1741C

ELECTRICAL CHARACTERISTICS (V_{CC} = + 15 V, V_{EE} = − 15 V, T_A = 25°C unless otherwise noted).

Characteristic	Symbol	MC1741 Min	MC1741 Typ	MC1741 Max	MC1741C Min	MC1741C Typ	MC1741C Max	Unit
Input Offset Voltage ($R_S \leqslant$ 10 k)	V_{IO}	—	1.0	5.0	—	2.0	6.0	mV
Input Offset Current	I_{IO}	—	20	200	—	20	200	nA
Input Bias Current	I_{IB}	—	80	500	—	80	500	nA
Input Resistance	r_i	0.3	2.0	—	0.3	2.0	—	MΩ
Input Capacitance	C_i	—	1.4	—	—	1.4	—	pF
Offset Voltage Adjustment Range	V_{IOR}	—	±15	—	—	±15	—	mV
Common Mode Input Voltage Range	V_{ICR}	±12	±13	—	±12	±13	—	V
Large Signal Voltage Gain (V_O = ±10 V, $R_L \geqslant$ 2.0 k)	A_v	50	200	—	20	200	—	V/mV
Output Resistance	r_o	—	75	—	—	75	—	Ω
Common Mode Rejection Ratio ($R_S \leqslant$ 10 k)	CMRR	70	90	—	70	90	—	dB
Supply Voltage Rejection Ratio ($R_S \leqslant$ 10 k)	PSRR	—	30	150	—	30	150	μV/V
Output Voltage Swing ($R_L \geqslant$ 10 k) ($R_L \geqslant$ 2 k)	V_O	±12 ±10	±14 ±13	— —	±12 ±10	±14 ±13	— —	V
Output Short-Circuit Current	I_{os}	—	20	—	—	20	—	mA
Supply Current	I_D	—	1.7	2.8	—	1.7	2.8	mA
Power Consumption	P_C	—	50	85	—	50	85	mW
Transient Response (Unity Gain — Non-Inverting) (V_I = 20 m V, $R_L \geqslant$ 2 k, $C_L \leqslant$ 100 pF) Rise Time	t_{TLH}	—	0.3	—	—	0.3	—	μs
(V_I = 20 m V, $R_L \geqslant$ 2 k, $C_L \leqslant$ 100 pF) Overshoot	os	—	15	—	—	15	—	%
(V_I = 10 V, $R_L \geqslant$ 2 k, $C_L \leqslant$ 100 pF) Slew Rate	SR	—	0.5	—	—	0.5	—	V/μs

ELECTRICAL CHARACTERISTICS (V_{CC} = + 15 V, V_{EE} = − 15 V, T_A = T_{low} to T_{high} unless otherwise noted).

Characteristic	Symbol	MC1741 Min	MC1741 Typ	MC1741 Max	MC1741C Min	MC1741C Typ	MC1741C Max	• Unit
Input Offset Voltage ($R_S \leqslant$ 10 kΩ)	V_{IO}	—	1.0	6.0	—	—	7.5	mV
Input Offset Current (T_A = 125°C) (T_A = -55°C) (T_A = 0°C to +70°C)	I_{IO}	— — —	7.0 85 —	200 500 —	— — —	— — —	— — 300	nA
Input Bias Current (T_A = 125°C) (T_A = -55°C) (T_A = 0°C to +70°C)	I_{IB}	— — —	30 300 —	500 1500 —	— — —	— — —	— — 800	nA
Common Mode Input Voltage Range	V_{ICR}	±12	±13	—	—	—	—	V
Common Mode Rejection Ratio ($R_S \leqslant$ 10 k)	CMRR	70	90	—	—	—	—	dB
Supply Voltage Rejection Ratio ($R_S \leqslant$ 10 k)	PSRR	—	30	150	—	—	—	μV/V
Output Voltage Swing ($R_L \geqslant$ 10 k) ($R_L \geqslant$ 2 k)	V_O	±12 ±10	±14 ±13	— —	— ±10	— ±13	— —	V
Large Signal Voltage Gain ($R_L \geqslant$ 2 k, V_{out} = ±10 V)	A_v	25	—	—	15	—	—	V/mV
Supply Currents (T_A = 125°C) (T_A = -55°C)	I_D	— —	1.5 2.0	2.5 3.3	— —	— —	— —	mA
Power Consumption (T_A = +125°C) (T_A = -55°C)	P_C	— —	45 60	75 100	— —	— —	— —	mW

*T_{high} = 125°C for MC1741 and 70°C for MC1741C
 T_{low} = -55°C for MC1741 and 0°C for MC1741C

MC1741, MC1741C

FIGURE 1 — BURST NOISE versus SOURCE RESISTANCE

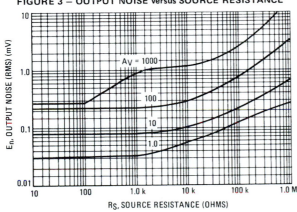

FIGURE 2 — RMS NOISE versus SOURCE RESISTANCE

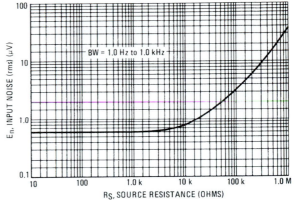

FIGURE 3 — OUTPUT NOISE versus SOURCE RESISTANCE

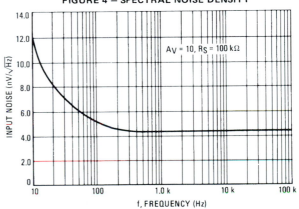

FIGURE 4 — SPECTRAL NOISE DENSITY

FIGURE 5 — BURST NOISE TEST CIRCUIT

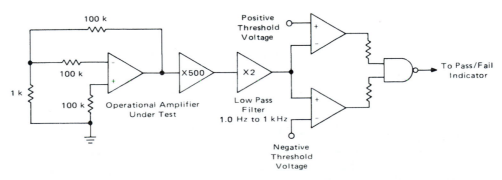

Unlike conventional peak reading or RMS meters, this system was especially designed to provide the quick response time essential to burst (popcorn) noise testing.

The test time employed is 10 seconds and the 20 μV peak limit refers to the operational amplifier input thus eliminating errors in the closed-loop gain factor of the operational amplifier under test.

MC1741, MC1741C

TYPICAL CHARACTERISTICS
(V$_{CC}$ = +15 Vdc, V$_{EE}$ = –15 Vdc, T$_A$ - +25°C unless otherwise noted)

FIGURE 6 – POWER BANDWIDTH
(LARGE SIGNAL SWING versus FREQUENCY)

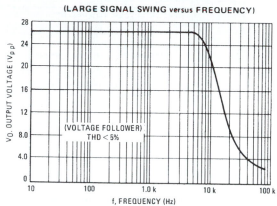

FIGURE 7 – OPEN LOOP FREQUENCY RESPONSE

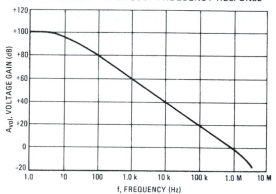

FIGURE 8 – POSITIVE OUTPUT VOLTAGE SWING
versus LOAD RESISTANCE

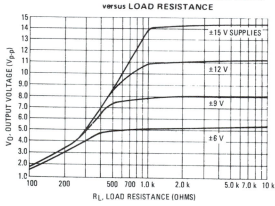

FIGURE 9 – NEGATIVE OUTPUT VOLTAGE SWING
versus LOAD RESISTANCE

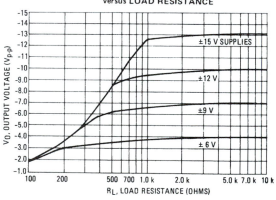

FIGURE 10 – OUTPUT VOLTAGE SWING versus
LOAD RESISTANCE (Single Supply Operation)

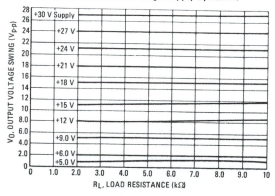

FIGURE 11 – SINGLE SUPPLY INVERTING AMPLIFIER

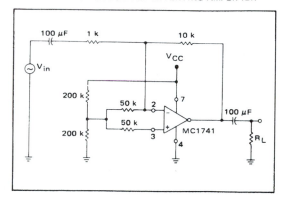

MOTOROLA
■■ SEMICONDUCTOR ■■
TECHNICAL DATA

THREE-TERMINAL POSITIVE VOLTAGE REGULATORS

These voltage regulators are monolithic integrated circuits designed as fixed-voltage regulators for a wide variety of applications including local, on-card regulation. These regulators employ internal current limiting, thermal shutdown, and safe-area compensation. With adequate heatsinking they can deliver output currents in excess of 1.0 ampere. Although designed primarily as a fixed voltage regulator, these devices can be used with external components to obtain adjustable voltages and currents.

- Output Current in Excess of 1.0 Ampere
- No External Components Required
- Internal Thermal Overload Protection
- Internal Short-Circuit Current Limiting
- Output Transistor Safe-Area Compensation
- Output Voltage Offered in 2% and 4% Tolerance

THREE-TERMINAL POSITIVE FIXED VOLTAGE REGULATORS

SILICON MONOLITHIC INTEGRATED CIRCUITS

K SUFFIX
METAL PACKAGE
CASE 1

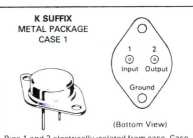

1 Input 2 Output

Ground

(Bottom View)

Pins 1 and 2 electrically isolated from case. Case is third electrical connection.

T SUFFIX
PLASTIC PACKAGE
CASE 221A

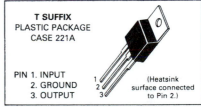

PIN 1. INPUT
2. GROUND
3. OUTPUT

(Heatsink surface connected to Pin 2.)

REPRESENTATIVE SCHEMATIC DIAGRAM

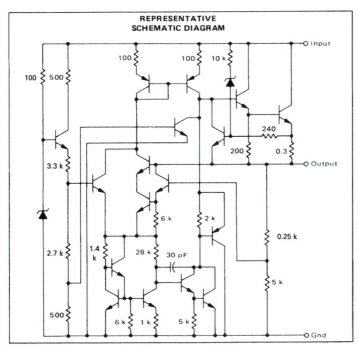

STANDARD APPLICATION

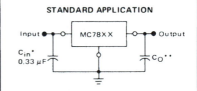

A common ground is required between the input and the output voltages. The input voltage must remain typically 2.0 V above the output voltage even during the low point on the input ripple voltage.

XX = these two digits of the type number indicate voltage.

* = C_{in} is required if regulator is located an appreciable distance from power supply filter.

** = C_O is not needed for stability; however, it does improve transient response.

XX indicates nominal voltage

ORDERING INFORMATION

Device	Output Voltage Tolerance	Tested Operating Junction Temp. Range	Package
MC78XXK	4%	−55 to +150°C	Metal Power
MC78XXAK*	2%		
MC78XXCK	4%	0 to +125°C	
MC78XXACK*	2%		
MC78XXCT	4%		Plastic Power
MC78XXACT	2%		
MC78XXBT	4%	−40 to +125°C	

*2% regulators in Metal Power packages are available in 5, 12 and 15 volt devices.

TYPE NO./VOLTAGE

MC7805	5.0 Volts	MC7812	12 Volts
MC7806	6.0 Volts	MC7815	15 Volts
MC7808	8.0 Volts	MC7818	18 Volts
MC7809	9.0 Volts	MC7824	24 Volts

MC7800 Series

MAXIMUM RATINGS (T_A = +25°C unless otherwise noted.)

Rating	Symbol	Value	Unit
Input Voltage (5.0 V − 18 V) (24 V)	V_{in}	35 40	Vdc

Power Dissipation and Thermal Characteristics
Plastic Package
  T_A = +25°C
  Derate above T_A = +25°C
  Thermal Resistance, Junction to Air
	P_D
	$1/\theta_{JA}$
	θ_{JA}
		Internally Limited
		15.4
		65
			Watts
			mW/°C
			°C/W

T_C = +25°C
  Derate above T_C = +75°C (See Figure 1)
  Thermal Resistance, Junction to Case
	P_D
	$1/\theta_{JC}$
	θ_{JC}
		Internally Limited
		200
		5.0
			Watts
			mW/°C
			°C/W

Metal Package
  T_A = +25°C
  Derate above T_A = +25°C
  Thermal Resistance, Junction to Air
	P_D
	$1/\theta_{JA}$
	θ_{JA}
		Internally Limited
		22.5
		45
			Watts
			mW/°C
			°C/W

T_C = +25°C
  Derate above T_C = +65°C (See Figure 2)
  Thermal Resistance, Junction to Case
	P_D
	$1/\theta_{JC}$
	θ_{JC}
		Internally Limited
		182
		5.5
			Watts
			mW/°C
			°C/W

Storage Junction Temperature Range	T_{stg}	− 65 to + 150	°C

Operating Junction Temperature Range
             MC7800,A
            MC7800C,AC
            MC7800B
	T_J	
		− 55 to + 150
		0 to + 150
		− 40 to + 150
			°C

DEFINITIONS

Line Regulation — The change in output voltage for a change in the input voltage. The measurement is made under conditions of low dissipation or by using pulse techniques such that the average chip temperature is not significantly affected.

Load Regulation — The change in output voltage for a change in load current at constant chip temperature.

Maximum Power Dissipation — The maximum total device dissipation for which the regulator will operate within specifications.

Quiescent Current — That part of the input current that is not delivered to the load.

Output Noise Voltage — The rms ac voltage at the output, with constant load and no input ripple, measured over a specified frequency range.

Long Term Stability — Output voltage stability under accelerated life test conditions with the maximum rated voltage listed in the devices' electrical characteristics and maximum power dissipation.

MC7800 Series

MC7805, B, C
ELECTRICAL CHARACTERISTICS (V_{in} = 10 V, I_O = 500 mA, T_J = T_{low} to T_{high} [Note 1] unless otherwise noted).

Characteristic	Symbol	MC7805 Min	Typ	Max	MC7805B Min	Typ	Max	MC7805C Min	Typ	Max	Unit
Output Voltage (T_J = +25°C)	V_O	4.8	5.0	5.2	4.8	5.0	5.2	4.8	5.0	5.2	Vdc
Output Voltage (5.0 mA ≤ I_O ≤ 1.0 A, P_O ≤ 15 W)	V_O										Vdc
7.0 Vdc ≤ V_{in} ≤ 20 Vdc		—	—	—	—	—	—	4.75	5.0	5.25	
8.0 Vdc ≤ V_{in} ≤ 20 Vdc		4.65	5.0	5.35	4.75	5.0	5.25	—	—	—	
Line Regulation (T_J = +25°C, Note 2)	Reg_{line}										mV
7.0 Vdc ≤ V_{in} ≤ 25 Vdc		—	2.0	50	—	7.0	100	—	7.0	100	
8.0 Vdc ≤ V_{in} ≤ 12 Vdc		—	1.0	25	—	2.0	50	—	2.0	50	
Load Regulation (T_J = +25°C, Note 2)	Reg_{load}										mV
5.0 mA ≤ I_O ≤ 1.5 A		—	25	100	—	40	100	—	40	100	
250 mA ≤ I_O ≤ 750 mA		—	8.0	25	—	15	50	—	15	50	
Quiescent Current (T_J = +25°C)	I_B	—	3.2	6.0	—	4.3	8.0	—	4.3	8.0	mA
Quiescent Current Change	ΔI_B										mA
7.0 Vdc ≤ V_{in} ≤ 25 Vdc		—	—	—	—	—	—	—	—	1.3	
8.0 Vdc ≤ V_{in} ≤ 25 Vdc		—	0.3	0.8	—	—	1.3	—	—	—	
5.0 mA ≤ I_O ≤ 1.0 A		—	0.04	0.5	—	—	0.5	—	—	0.5	
Ripple Rejection 8.0 Vdc ≤ V_{in} ≤ 18 Vdc, f = 120 Hz	RR	68	75	—	—	68	—	—	68	—	dB
Dropout Voltage (I_O = 1.0 A, T_J = +25°C)	$V_{in} - V_O$	—	2.0	2.5	—	2.0		—	2.0	—	Vdc
Output Noise Voltage (T_A = +25°C) 10 Hz ≤ f ≤ 100 kHz	V_n	—	10	40	—	10		—	10	—	μV/V_O
Output Resistance f = 1.0 kHz	r_O	—	17	—	—	17		—	17	—	mΩ
Short-Circuit Current Limit (T_A = +25°C) V_{in} = 35 Vdc	I_{sc}	—	0.2	1.2	—	0.2		—	0.2	—	A
Peak Output Current (T_J = +25°C)	I_{max}	1.3	2.5	3.3	—	2.2		—	2.2	—	A
Average Temperature Coefficient of Output Voltage	TCV_O	—	±0.6	—	—	-1.1		—	-1.1	—	mV/°C

MC7805A, AC
ELECTRICAL CHARACTERISTICS (V_{in} = 10 V, I_O = 1.0 A, T_J = T_{low} to T_{high} [Note 1] unless otherwise noted)

Characteristics	Symbol	MC7805A Min	Typ	Max	MC7805AC Min	Typ	Max	Unit
Output Voltage (T_J = +25°C)	V_O	4.9	5.0	5.1	4.9	5.0	5.1	Vdc
Output Voltage (5.0 mA ≤ I_O ≤ 1.0 A, P_O ≤ 15 W) 7.5 Vdc ≤ V_{in} ≤ 20 Vdc	V_O	4.8	5.0	5.2	4.8	5.0	5.2	Vdc
Line Regulation (Note 2)	Reg_{line}							mV
7.5 Vdc ≤ V_{in} ≤ 25 Vdc, I_O = 500 mA		—	2.0	10	—	7.0	50	
8.0 Vdc ≤ V_{in} ≤ 12 Vdc		—	3.0	10	—	10	50	
8.0 Vdc ≤ V_{in} ≤ 12 Vdc, T_J = +25°C		—	1.0	4.0	—	2.0	25	
7.3 Vdc ≤ V_{in} ≤ 20 Vdc, T_J = +25°C		—	2.0	10	—	7.0	50	
Load Regulation (Note 2)	Reg_{load}							mV
5.0 mA ≤ I_O ≤ 1.5 A, T_J = +25°C		—	2.0	25	—	25	100	
5.0 mA ≤ I_O ≤ 1.0 A		—	2.0	25	—	25	100	
250 mA ≤ I_O ≤ 750mA, T_J = +25°C		—	1.0	15	—	—	—	
250 mA ≤ I_O ≤ 750 mA		—	1.0	25	—	8.0	50	
Quiescent Current T_J = +25°C	I_B	—	—	5.0	—	—	6.0	mA
		—	3.2	4.0	—	4.3	6.0	
Quiescent Current Change	ΔI_B							mA
8.0 Vdc ≤ V_{in} ≤ 25 Vdc, I_O = 500 mA		—	0.3	0.5	—	—	0.8	
7.5 Vdc ≤ V_{in} ≤ 20 Vdc, T_J = +25°C		—	0.2	0.5	—	—	0.8	
5.0 mA ≤ I_O ≤ 1.0 A		—	0.04	0.2	—	—	0.5	
Ripple Rejection	RR							dB
8.0 Vdc ≤ V_{in} ≤ 18 Vdc, f = 120 Hz, T_J = +25°C		68	75	—	—	—	—	
8.0 Vdc ≤ V_{in} ≤ 18 Vdc, f = 120 Hz, I_O = 500 mA		68	75	—	—	68	—	
Dropout Voltage (I_O = 1.0 A, T_J = +25°C)	$V_{in} - V_O$	—	2.0	2.5	—	2.0	—	Vdc
Output Noise Voltage (T_A = +25°C) 10 Hz ≤ f ≤ 100 kHz	V_n	—	10	40	—	10	—	μV/V_O
Output Resistance (f = 1.0 kHz)	r_O	—	2.0	—	—	17	—	mΩ
Short-Circuit Current Limit (T_A = +25°C) V_{in} = 35 Vdc	I_{sc}	—	0.2	1.2	—	0.2	—	A
Peak Output Current (T_J = +25°C)	I_{max}	1.3	2.5	3.3	—	2.2	—	A
Average Temperature Coefficient of Output Voltage	TCV_O	—	±0.6	—	—	-1.1	—	mV/°C

NOTES: 1. T_{low} = –55°C for MC78XX, A T_{high} = +150°C for MC78XX, A
= 0° for MC78XXC, AC   = +125°C for MC78XXC, AC, B
= –40°C for MC78XXB

2. Load and line regulation are specified at constant junction temperature. Changes in V_O due to heating effects must be taken into account separately. Pulse testing with low duty cycle is used.

MOTOROLA
■■■ SEMICONDUCTOR ■■■
TECHNICAL DATA

MC7900 Series

THREE-TERMINAL
NEGATIVE VOLTAGE REGULATORS

The MC7900 Series of fixed output negative voltage regulators are intended as complements to the popular MC7800 Series devices. These negative regulators are available in the same seven-voltage options as the MC7800 devices. In addition, one extra voltage option commonly employed in MECL systems is also available in the negative MC7900 Series.

Available in fixed output voltage options from −5.0 to −24 volts, these regulators employ current limiting, thermal shutdown, and safe-area compensation — making them remarkably rugged under most operating conditions. With adequate heat-sinking they can deliver output currents in excess of 1.0 ampere.

- No External Components Required
- Internal Thermal Overload Protection
- Internal Short-Circuit Current Limiting
- Output Transistor Safe-Area Compensation
- Available in 2% Voltage Tolerance (See Ordering Information)

THREE-TERMINAL NEGATIVE FIXED VOLTAGE REGULATORS

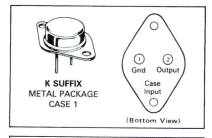

K SUFFIX
METAL PACKAGE
CASE 1

① Gnd   ② Output
Case Input
(Bottom View)

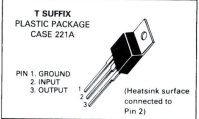

T SUFFIX
PLASTIC PACKAGE
CASE 221A

PIN 1. GROUND
2. INPUT
3. OUTPUT

(Heatsink surface connected to Pin 2)

SCHEMATIC DIAGRAM

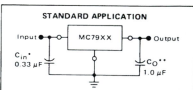

STANDARD APPLICATION

Input — MC79XX — Output

C_{in}^{*} 0.33 μF

C_O^{**} 1.0 μF

A common ground is required between the input and the output voltages. The input voltage must remain typically 2.0 V more negative even during the high point on the input ripple voltage.

XX = these two digits of the type number indicate voltage.

* = C_{in} is required if regulator is located an appreciable distance from power supply filter.

** = C_O improves stability and transient response.

ORDERING INFORMATION

Device	Output Voltage Tolerance	Tested Operating Junction Temp. Range	Package
MC79XXCK MC79XXACK*	4% 2%		Metal Power**
MC79XXCT MC79XXACT*	4% 2%	T_J = 0°C to +125°C	Plastic Power
MC79XXBT#	4%	T_J = −40°C to +125°C	

XX indicates nominal voltage.

*2% output voltage tolerance available in 5, 12 and 15 volt devices.

**Metal power package available in 5, 12 and 15 volt devices.

#Automotive temperature range selections are available with special test conditions and additional tests in 5, 12 and 15 volt devices. Contact your local Motorola sales office for information.

DEVICE TYPE/NOMINAL OUTPUT VOLTAGE

MC7905	5.0 Volts	MC7912	12 Volts
MC7905.2	5.2 Volts	MC7915	15 Volts
MC7906	6.0 Volts	MC7918	18 Volts
MC7908	8.0 Volts	MC7924	24 Volts

MC7900 Series

MAXIMUM RATINGS (T$_A$ = +25°C unless otherwise noted.)

Rating	Symbol	Value	Unit
Input Voltage (–5.0 V ⩾ V$_O$ ⩾ –18 V) (24 V)	V$_I$	–35 –40	Vdc

Power Dissipation
 Plastic Package
   T$_A$ = +25°C
   Derate above T$_A$ = +25°C
	P$_D$
	 1/R$_{\theta JA}$
		Internally Limited
		 15.4
			Watts
			 mW/°C

T$_C$ = +25°C Derate above T$_C$ = +95°C (See Figure 1)	P$_D$ 1/R$_{\theta JC}$	Internally Limited 200	Watts mW/°C

Metal Package
   T$_A$ = +25°C
   Derate above T$_A$ = +25°C
	P$_D$
	 1/R$_{\theta JA}$
		Internally Limited
		 22.2
			Watts
			 mW/°C

T$_C$ = +25°C Derate above T$_C$ = +65°C	P$_D$ 1/R$_{\theta JC}$	Internally Limited 182	Watts mW/°C
Storage Junction Temperature Range	T$_{stg}$	–65 to +150	°C
Junction Temperature Range	T$_J$	0 to +150	°C

THERMAL CHARACTERISTICS

Characteristic	Symbol	Max	Unit
Thermal Resistance, Junction to Ambient — Plastic Package — Metal Package	R$_{\theta JA}$	65 45	°C/W
Thermal Resistance, Junction to Case — Plastic Package — Metal Package	R$_{\theta JC}$	5.0 5.5	°C/W

MC7905C ELECTRICAL CHARACTERISTICS (V$_I$ = –10 V, I$_O$ = 500 mA, 0°C < T$_J$ < +125°C unless otherwise noted.)

Characteristic	Symbol	Min	Typ	Max	Unit
Output Voltage (T$_J$ = +25°C)	V$_O$	–4.8	–5.0	–5.2	Vdc
Line Regulation (Note 1)	Reg$_{line}$				mV

(T$_J$ = +25°C, I$_O$ = 100 mA)
   –7.0 Vdc ⩾ V$_I$ ⩾ –25 Vdc
   –8.0 Vdc ⩾ V$_I$ ⩾ –12 Vdc
		—
		 —
			7.0
			 2.0
				50
				 25	

(T$_J$ = +25°C, I$_O$ = 500 mA)
   –7.0 Vdc ⩾ V$_I$ ⩾ –25 Vdc
   –8.0 Vdc ⩾ V$_I$ ⩾ –12 Vdc
		—
		 —
			35
			 8.0
				100
				 50	

Load Regulation (T$_J$ = +25°C) (Note 1)
   5.0 mA ⩽ I$_O$ ⩽ 1.5 A
   250 mA ⩽ I$_O$ ⩽ 750 mA
	Reg$_{load}$	—
		 —
			11
			 4.0
				100
				 50
					mV

Output Voltage –7.0 Vdc ⩾ V$_I$ ⩾ –20 Vdc, 5.0 mA ⩽ I$_O$ ⩽ 1.0 A, P ⩽ 15 W	V$_O$	–4.75	—	–5.25	Vdc
Input Bias Current (T$_J$ = +25°C)	I$_{IB}$	—	4.3	8.0	mA

Input Bias Current Change
   –7.0 Vdc ⩾ V$_I$ ⩾ –25 Vdc
   5.0 mA ⩽ I$_O$ ⩽ 1.5 A
	ΔI$_{IB}$	—
		 —
			—
			 —
				1.3
				 0.5
					mA

Output Noise Voltage (T$_A$ = +25°C, 10 Hz ⩽ f ⩽ 100 kHz)	e$_{on}$	—	40	—	μV
Ripple Rejection (I$_O$ = 20 mA, f = 120 Hz)	RR	—	70	—	dB
Dropout Voltage I$_O$ = 1.0 A, T$_J$ = +25°C	V$_I$–V$_O$	—	2.0	—	Vdc
Average Temperature Coefficient of Output Voltage I$_O$ = 5.0 mA, 0°C ⩽ T$_J$ ⩽ +125°C	ΔV$_O$/ΔT	—	–1.0	—	mV/°C

Note:
1. Load and line regulation are specified at constant junction temperature. Changes in V$_O$ due to heating effects must be taken into account separately. Pulse testing with low duty cycle is used.

MC7900 Series

MC7912C ELECTRICAL CHARACTERISTICS (V_I = -19 V, I_O = 500 mA, 0°C < T_J < +125°C unless otherwise noted.)

Characteristic	Symbol	Min	Typ	Max	Unit
Output Voltage (T_J = +25°C)	V_O	-11.5	-12	-12.5	Vdc
Line Regulation (Note 1)	Reg_{line}				mV
(T_J = +25°C, I_O = 100 mA)					
-14.5 Vdc ≥ V_I ≥ -30 Vdc		—	13	120	
-16 Vdc ≥ V_I ≥ -22 Vdc		—	6.0	60	
(T_J = +25°C, I_O = 500 mA)					
-14.5 Vdc ≥ V_I ≥ -30 Vdc		—	55	240	
-16 Vdc ≥ V_I ≥ -22 Vdc		—	24	120	
Load Regulation (T_J = +25°C) (Note 1)	Reg_{load}				mV
5.0 mA ≤ I_O ≤ 1.5 A		—	46	240	
250 mA ≤ I_O ≤ 750 mA		—	17	120	
Output Voltage -14.5 Vdc ≥ V_I ≥ -27 Vdc, 5.0 mA ≤ I_O ≤ 1.0 A, P ≤ 15 W	V_O	-11.4	—	-12.6	Vdc
Input Bias Current (T_J = +25°C)	I_{IB}	—	4.4	8.0	mA
Input Bias Current Change	ΔI_{IB}				mA
-14.5 Vdc ≥ V_I ≥ -30 Vdc		—	—	1.0	
5.0 mA ≤ I_O ≤ 1.5 A		—	—	0.5	
Output Noise Voltage (T_A = +25°C, 10 Hz ≤ f ≤ 100 kHz)	e_{on}	—	75	—	μV
Ripple Rejection (I_O = 20 mA, f = 120 Hz)	RR	—	61	—	dB
Dropout Voltage I_O = 1.0 A, T_J = +25°C	V_I-V_O	—	2.0	—	Vdc
Average Temperature Coefficient of Output Voltage I_O = 5.0 mA, 0°C ≤ T_J ≤ +125°C	$\Delta V_O/\Delta T$	—	-1.0	—	mV/°C

MC7912AC ELECTRICAL CHARACTERISTICS (V_I = -19 V, I_O = 500 mA, 0°C < T_J < +125°C unless otherwise noted.)

Characteristic	Symbol	Min	Typ	Max	Unit
Output Voltage (T_J = +25°C)	V_O	-11.75	-12	-12.25	Vdc
Line Regulation (Note 1)	Reg_{line}				mV
-16 Vdc ≥ V_I ≥ -22 Vdc; I_O = 1.0 A, T_J = 25°C		—	6.0	60	
-16 Vdc ≥ V_I ≥ -22 Vdc; I_O = 1.0 A,		—	24	120	
-14.8 Vdc ≥ V_I ≥ -30 Vdc; I_O = 500 mA		—	24	120	
-14.5 Vdc ≥ V_I ≥ -27 Vdc; I_O = 1.0 A, T_J = 25°C		—	13	120	
Load Regulation (Note 1)	Reg_{load}				mV
5.0 mA ≤ I_O ≤ 1.5 A, T_J = 25°C		—	46	150	
250 mA ≤ I_O ≤ 750 mA		—	17	75	
5.0 mA ≤ I_O ≤ 1.0 A		—	35	150	
Output Voltage -14.8 Vdc ≥ V_I ≥ -27 Vdc, 5.0 mA ≤ I_O ≤ 1.0 A, P ≤ 15 W	V_O	-11.5	—	-12.5	Vdc
Input Bias Current	I_{IB}	—	4.4	8.0	mA
Input Bias Current Change	ΔI_{IB}				mA
-15 Vdc ≥ V_I ≥ -30 Vdc		—	—	0.8	
5.0 mA ≤ I_O ≤ 1.0 A		—	—	0.5	
5.0 mA ≤ I_O ≤ 1.5 A, T_J = 25°C		—	—	0.5	
Output Noise Voltage (T_A = +25°C, 10 Hz ≤ f ≤ 100 kHz)	e_{on}	—	75	—	μV
Ripple Rejection (I_O = 20 mA, f = 120 Hz)	RR	—	61	—	dB
Dropout Voltage I_O = 1.0 A, T_J = +25°C	V_I-V_O	—	2.0	—	Vdc
Average Temperature Coefficient of Output Voltage I_O = 5.0 mA, 0°C ≤ T_J ≤ +125°C	$\Delta V_O/\Delta T$	—	-1.0	—	mV/°C

Note:

1. Load and line regulation are specified at constant junction temperature. Changes in V_O due to heating effects must be taken into account separately. Pulse testing with low duty cycle is used.

ANSWERS TO SELF-TESTS

Chapter 1

1. (c)	**2.** (d)	**3.** (a)	**4.** (d)	**5.** (d)	**6.** (d)	**7.** (b)	**8.** (a)
9. (d)	**10.** (c)	**11.** (b)	**12.** (a)	**13.** (d)	**14.** (c)	**15.** (d)	**16.** (e)
17. (d)	**18.** (a)	**19.** (b)	**20.** (c)	**21.** (c)	**22.** (a)	**23.** (c)	**24.** (d)
25. (d)	**26.** (c)	**27.** (d)	**28.** (d)	**29.** (b)	**30.** (b)	**31.** (b)	**32.** (c)
33. (d)	**34.** (c)						

Chapter 2

1. (a)	**2.** (c)	**3.** (d)	**4.** (a)	**5.** (b)	**6.** (a)	**7.** (d)	**8.** (b)
9. (c)	**10.** (a)	**11.** (b)	**12.** (c)	**13.** (a)	**14.** (d)	**15.** (b)	**16.** (c)
17. (b)							

Chapter 3

1. (a)	**2.** (b)	**3.** (c)	**4.** (b)	**5.** (d)	**6.** (b)	**7.** (d)	**8.** (a)
9. (c)	**10.** (d)	**11.** (b)	**12.** (b)	**13.** (b)	**14.** (d)		

Chapter 4

1. (d)	**2.** (c)	**3.** (a)	**4.** (d)	**5.** (a)	**6.** (c)	**7.** (b)	**8.** (b)
9. (a)	**10.** (c)	**11.** (b)	**12.** (f)	**13.** (c)	**14.** (b)	**15.** (b)	**16.** (a)

Chapter 5

1. (b)	**2.** (c)	**3.** (d)	**4.** (d)	**5.** (c)	**6.** (d)	**7.** (a)	**8.** (d)
9. (c)	**10.** (a)	**11.** (b)	**12.** (c)	**13.** (a)	**14.** (c)	**15.** (f)	

Chapter 6

1. (a)	**2.** (b)	**3.** (c)	**4.** (b)	**5.** (d)	**6.** (d)	**7.** (a)	**8.** (b)
9. (d)	**10.** (c)	**11.** (a)	**12.** (b)	**13.** (d)	**14.** (c)	**15.** (a)	

Chapter 7

1. (b)	**2.** (d)	**3.** (a)	**4.** (c)	**5.** (b)	**6.** (c)	**7.** (a)	**8.** (d)
9. (b)	**10.** (c)	**11.** (a)	**12.** (d)				

Chapter 8

1. (e)	**2.** (b)	**3.** (a)	**4.** (c)	**5.** (d)	**6.** (c)	**7.** (a)	**8.** (c)
9. (b)	**10.** (d)	**11.** (a)	**12.** (c)	**13.** (d)	**14.** (c)	**15.** (c)	**16.** (b)
17. (a)	**18.** (c)						

Chapter 9

1. (f)	**2.** (b)	**3.** (c)	**4.** (d)	**5.** (a)	**6.** (c)	**7.** (a)	**8.** (c)
9. (b)	**10.** (a)	**11.** (d)	**12.** (a)	**13.** (c)	**14.** (a)	**15.** (b)	

Chapter 10

1. (d)	**2.** (c)	**3.** (b)	**4.** (a)	**5.** (d)	**6.** (b)	**7.** (c)	**8.** (c)
9. (a)	**10.** (c)	**11.** (b)	**12.** (d)	**13.** (c)	**14.** (a)	**15.** (b)	

Chapter 11

1. (b)	**2.** (d)	**3.** (c)	**4.** (c)	**5.** (a)	**6.** (e)	**7.** (b)	**8.** (b)
9. (d)	**10.** (d)	**11.** (c)	**12.** (d)	**13.** (a)	**14.** (d)	**15.** (c)	**16.** (b)

Chapter 12

1. (c)	**2.** (b)	**3.** (d)	**4.** (b)	**5.** (a)	**6.** (c)	**7.** (b)	**8.** (a)
9. (d)	**10.** (c)	**11.** (d)	**12.** (a)	**13.** (b)	**14.** (c)	**15.** (c)	**16.** (d)
17. (b)	**18.** (c)	**19.** (a)	**20.** (c)	**21.** (d)			

Chapter 13

1. (c)	**2.** (b)	**3.** (a)	**4.** (b)	**5.** (d)	**6.** (a)	**7.** (d)	**8.** (c)
9. (b)	**10.** (a) and (c)		**11.** (d)	**12.** (d)	**13.** (b)	**14.** (c)	**15.** (b)

Chapter 14

1. (c)	**2.** (a)	**3.** (c)	**4.** (e)	**5.** (b)	**6.** (d)	**7.** (c)	**8.** (a)
9. (c)	**10.** (a)	**11.** (b)	**12.** (c)	**13.** (b)	**14.** (d)	**15.** (d)	**16.** (a)
17. (d)	**18.** (c)						

Chapter 15

1. (d)	**2.** (b)	**3.** (a)	**4.** (e)	**5.** (c)	**6.** (b)	**7.** (a)	**8.** (c)
9. (d)	**10.** (c)	**11.** (a)	**12.** (b)	**13.** (b)	**14.** (b)	**15.** (b)	**16.** (c)

Chapter 16

1. (c)	**2.** (d)	**3.** (a)	**4.** (b)	**5.** (c)	**6.** (c)	**7.** (b)	**8.** (a)
9. (d)	**10.** (b)	**11.** (a)	**12.** (c)	**13.** (b)	**14.** (d)		

Chapter 17

1. (b)	**2.** (a)	**3.** (c)	**4.** (b)	**5.** (d)	**6.** (c)	**7.** (b)	**8.** (d)
9. (a)	**10.** (c)	**11.** (b)	**12.** (c)	**13.** (a)	**14.** (c)	**15.** (c)	

Chapter 18

1. (c)	**2.** (d)	**3.** (c)	**4.** (b)	**5.** (d)	**6.** (a)	**7.** (c)	**8.** (a)
9. (g)	**10.** (c)						

ANSWERS TO ODD-NUMBERED PROBLEMS

Chapter 1

1. 6 electrons; 6 protons
3. (a) insulator   (b) semiconductor   (c) conductor
5. Four
7. Conduction band and valence band
9. Antimony is a pentavalent material. Boron is a trivalent material. Both are used for doping.
11. No. The barrier potential is a voltage drop.
13. To prevent excessive forward current.
15. A temperature increase.
17. (a) −3 V   (b) 0.3 V   (c) 0.3 V   (d) 0.3 V
19. $V_A = 25$ V; $V_B = 24.3$ V; $V_C = 8.7$ V; $V_D = 8$ V

Chapter 2

1. See Figure ANS–1.

FIGURE ANS–1

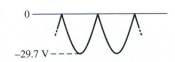

(a)

(b)

3. 23 V rms
5. (a) 1.59 V   (b) 63.7 V
   (c) 16.4 V   (d) 10.5 V
7. 173 V
9. 78.5 V
11. See Figure ANS–2.

FIGURE ANS–2

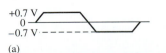

13. $V_r = 8.33$ V; $V_{DC} = 25.8$ V
15. 556 μF
17. $V_r = 1.47$ V; $V_{DC} = 30.6$ V
19. See Figure ANS–3.

FIGURE ANS–3

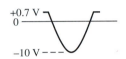

21. See Figure ANS–4.

FIGURE ANS–4

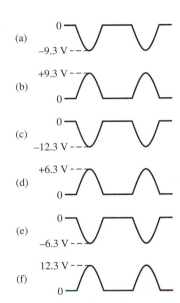

(a)
(b)
(c)
(d)
(e)
(f)

23. See Figure ANS–5.

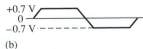

(a)   (b)

FIGURE ANS–5

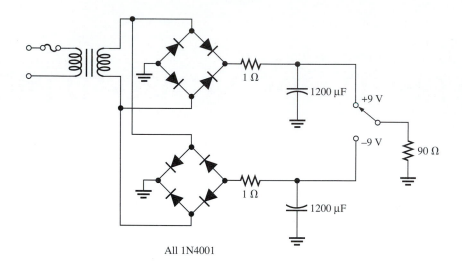

All 1N4001

FIGURE ANS–6

25. (a) A sine wave with a positive peak at +0.7 V, a
negative peak at −7.3 V, and a dc value of −3.3 V.
  (b) A sine wave with a positive peak at +29.3 V, a
negative peak at −0.7 V, and a dc value of +14.3 V.
  (c) A square wave varying from +0.7 V down to
−15.3 V, with a dc value of −7.3 V.
  (d) A square wave varying from +1.3 V down to −0.7 V,
with a dc value of +0.3 V.

27. 56.6 V

29. 50 V

31. 62.5 mΩ

33. Coil is open. Capacitor is shorted.

35. The circuit should not fail because the diode ratings
exceed the actual PIV and maximum current.

37. Excessive PIV or surge current causes diode to open
each time power is turned on. It could be a faulty
transformer or maybe the diodes do not have sufficient
PIV rating.

39. 177 μF

41. 651 mΩ (nearest standard 0.68 Ω)

43. See Figure ANS–6.

45. $V_{C1} = 155$ V; $V_{C2} = 310$ V

Chapter 3

1. See Figure ANS–7.

FIGURE ANS–7   Zener equivalent

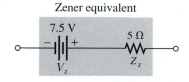

3. 5 Ω

5. 6.92 V

7. 14.3 V

9. See Figure ANS–8

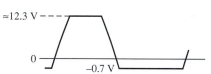

FIGURE ANS–8

11. 17.7%

13. 3.13%

15. 5.88%

17. 3 V

19. ≈ 2.5 V

21. See Figure ANS–9.

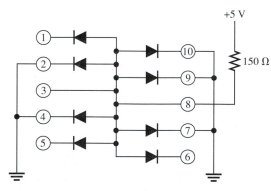

FIGURE ANS–9

23. (a) 30 kΩ   **(b)** 8.57 kΩ   **(c)** 5.88 kΩ

25. −750 Ω

27. The reflective ends cause the light to bounce back and forth, thus increasing the intensity of the light. The partially reflective end allows a portion of the reflected light to be emitted.

29. (a) ≈ 30 V dc

   (b) 0 V

   (c) Excessive 120 Hz ripple limited to 12 V by zener

   (d) Full-wave rectified waveform limited at 12 V by zener

   (e) 60 Hz ripple limited to 12 V

   (f) 60 Hz ripple limited to 12 V

   (g) 0 V

   (h) 0 V

31. Could be fuse, transformer, limiting resistor, or surge resistor open. Also, capacitor or zener could be shorted. Cannot isolate further with given measurements.

33. D_1 open, R_1 open, no dc voltage, short in threshold circuit

35. (a) 60 V   **(b)** 307 mW

   (c) 1.27 W   **(d)** 21 pF

   (e) 1N5139   **(f)** 4.82 pF

37. (a) Reverse-bias voltage   **(b)** 940 nm

   (c) 40 nA   **(d)** 940 nm

   (e) 40 µA/mW/cm^2 **(f)** 104 µA

39. $V_{OUT(1)} = 6.8$ V; $V_{OUT(2)} = 24$ V

41. See Figure ANS–10.

Chapter 4

1. Holes

3. The base is narrow and lightly doped so that a small recombination (base) current is generated compared to the collector current.

5. Positive, negative

7. 0.947

9. 24

11. 8.98 mA

13. 0.99

15. (a) $V_{BE} = 0.7$ V, $V_{CE} = 5.10$ V, $V_{CB} = 4.40$ V

   (b) $V_{BE} = -0.7$ V, $V_{CE} = -3.83$ V, $V_{CB} = 3.13$ V

17. $I_B = 26$ µA, $I_E = 1.3$ mA, $I_C = 1.27$ mA

19. 3 µA

21. 425 mW

23. 33.3

25. 500 µA, 3.33 µA, 4.03 V

27. See Figure ANS–12.

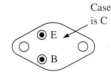

FIGURE ANS–12

29. Open, low resistance

31. (a) 27.8   **(b)** 109

33. Q_2 or Q_4 shorted collector to emitter, short from pin 2 of relay to ground, Q_1 or Q_3 open collector to emitter

35. (a) 40 V   **(b)** 200 mA dc   **(c)** 625 mW

   (d) 1.5 W   **(e)** 35

37. 1.26 W

39. (a) Saturated   **(b)** Not saturated

FIGURE ANS–10

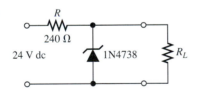

43. See Figure ANS–11.

FIGURE ANS–11

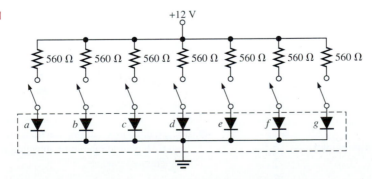

41. (a) No parameters are exceeded.
  (b) No parameters are exceeded.
43. Yes, marginally; $V_{CE} = 0.8$ V; $I_C = 75$ mA
45. See Figure ANS–13.

FIGURE ANS–13

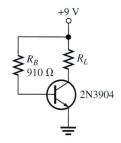

Chapter 5

1. Saturation
3. 18 mA
5. $V_{CE} = 20$ V; $I_{C(sat)} = 2$ mA
7. See Figure ANS–14.

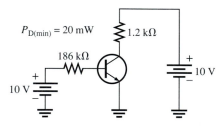

FIGURE ANS–14

9. $I_B = 514$ μA; $I_C = 46.3$ mA; $V_{CE} = 7.37$ V
11. I_C changes in the circuit using a common V_{CC} and V_{BB} supply, because a change in V_{CC} causes I_B to change, which in turn changes I_C.
13. 59.6 mA; 5.96 V
15. 639 Ω
17. When $R_E \gg R_B/\beta_{DC}$
19. 69.1
21. $I_C \cong 809$ μA; $V_{CE} = 13.2$ V
23. See Figure ANS–15.
25. (a) 1.41 mA, −8.67 V
  (b) 12.2 mW
27. 2.53 kΩ
29. 7.87 mA; 2.56 V
31. (a) Open collector or R_E open
  (b) No problems
  (c) Transistor shorted collector-to-emitter
  (d) Open emitter

FIGURE ANS–15

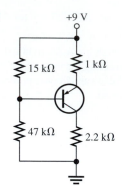

33. (a) 1: 10 V, 2: float,   3: 3.59 V, 4: 10 V
  (b) 1: 10 V, 2: 4.05 V, 3: 4.75 V, 4: 4.05 V
  (c) 1: 10 V, 2: 0 V,   3: 0 V,   4: 10 V
  (d) 1: 10 V, 2: 570 mV, 3: 1.27 V, 4: float
  (e) 1: 10 V, 2: 0 V,   3: 0.7 V,   4: 0 V
  (f) 1: 10 V, 2: 0 V,   3: 3.59 V, 4: 10 V
35. R_1 open, R_2 shorted, EB junction open
37. $V_C = V_{CC} = 9.1$ V, V_B normal, $V_E = 0$ V
39. None are exceeded.
41. 457 mW
43. See Figure ANS–16.

FIGURE ANS–16

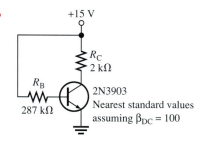

45. See Figure ANS–17.

FIGURE ANS–17

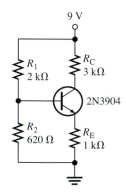

47. Yes
49. V_{CEQ} will be less, causing the transistor to saturate at a slightly higher temperature, thus limiting the low temperature response.

Chapter 6

1. Slightly greater than 1 mA min.

3. (a) $h_{ie} = 134\ \Omega$ (b) $h_{re} = 0.0001$

   (c) $h_{fe} = 147$ (d) $h_{oe} = 3.33$ mS

5. $r'_e \cong 19\ \Omega$

7. See Figure ANS–18.

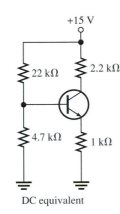

+15 V

22 kΩ   2.2 kΩ

4.7 kΩ   1 kΩ

DC equivalent

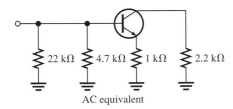

22 kΩ   4.7 kΩ   1 kΩ   2.2 kΩ

AC equivalent

FIGURE ANS–18

9. (a) 1.29 kΩ   (b) 968 Ω   (c) 171

11. (a) $V_B = 3.25$ V   (b) $V_E = 2.55$ V
   (c) $I_E = 2.55$ mA   (d) $I_C \cong 2.55$ mA
   (e) $V_C = 9.59$ V   (f) $V_{CE} = 7.04$ V

13. $A'_v = 131;\ \theta = 180°$

15. $A_{v(max)} = 65.5,\ A_{v(min)} = 2.06$

17. A_v is reduced to approximately 30. See Figure ANS–19.

FIGURE ANS–19

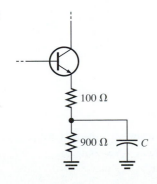

100 Ω

900 Ω   C

19. $R_{in} = 3.1$ kΩ; $V_{OUT} = 1.06$ V

21. 270 Ω

23. 8.8

25. $R_{in(emitter)} = 2.28\ \Omega$; $A_v = 526$; $A_i \cong 1$; $A_p = 526$

27. 400

29. (a) $A_{v1} = 93.6,\ A_{v2} = 302$
   (b) $A'_v = 28,267$
   (c) $A_{v1(dB)} = 39.4$ dB, $A_{v2(dB)} = 49.6$ dB, $A'_{v(dB)} = 89.0$ dB

31. $V_{B1} = 2.16$ V, $V_{E1} = 1.46$ V, $V_{C1} \cong 5.16$ V,
   $V_{B2} = 5.16$ V, $V_{E2} = 4.46$ V, $V_{C2} \cong 7.54$ V,
   $A_{v1} = 66,\ A_{v2} = 179,\ A'_v = 11,814$

33. (a) 1.41   (b) 2.00   (c) 3.16   (d) 10.0   (e) 100

35. Cutoff, 10 V

37.

Test Point	DC Volts	AC Volts (rms)
Input	0 V	25 μV
Q_1 base	2.99 V	20.8 μV
Q_1 emitter	2.29 V	0 V
Q_1 collector	7.44 V	1.95 mV
Q_2 base	2.99 V	1.95 mV
Q_2 emitter	2.29 V	0 V
Q_2 collector	7.44 V	589 mV
Output	0 V	589 mV

39. (a) $V_C = 5.87$ V, $V_c = 850$ mV
   (b) $V_C = 5.87$ V, $V_c = 0$ V
   (c) $V_C = 5.87$ V, $V_c = 0$ V
   (d) $V_C = 5.87$ V, $V_c = 203$ mV
   (e) $V_C = 5.87$ V, $V_c = 0$ V
   (f) $V_C \cong 0$ V, $V_c = 0$ V

41. (a) Q_1 is off   (b) 9 V   (c) 5.87 V

43. (a) 400   (b) 800 Ω   (c) 1 MΩ

45. A leaky coupling capacitor affects the bias voltages and attenuates the ac voltage.

47. See Figure ANS–20.

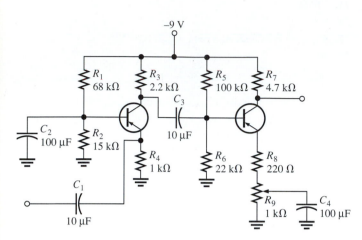

FIGURE ANS–20

49. See Figure ANS–21.

FIGURE ANS–21

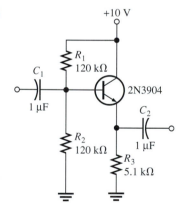

51. See Figure ANS–22.

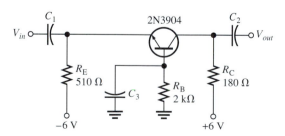

FIGURE ANS–22

53. $A_v = R_C/r'_e$

$A_v \cong (V_{R_C}/I_C)/(0.025 \text{ V}/I_C) = V_{R_C}/0.025 = 40V_{R_C}$

Chapter 7

1. 2.11 mA; 12 V

3. 8.47 V

5. (a) 2.45 mA, 4.41 V   (b) 6.40 mA, 1.69 V

7. (a) 3.46 mV rms   (b) 13.6 mV rms

9. (a) $P_{out} = 2.7$ mW, $\eta = 0.076$
   (b) $P_{out} = 14.1$ mW, $\eta = 0.154$

11. $V_{B1} = 10.7$ V; $V_{B2} = 9.3$ V; $V_{E1} = 10$ V;
   $V_{E2} = 10$ V; $V_{CEQ1} = 10$ V; $V_{CEQ2} = 10$ V

13. $P_{out} = 3.13$ W; $P_{DC} = 3.98$ W

15. $I_{CC} = 478$ mA; $P_{DC} = 11.5$ W; $P_{out} = 9$ W; $V_{CC} = 24$ V

17. 450 μW

19. 24 V

21. Negative half of input cycle

23. (a) No dc supply voltage or R_1 open
   (b) D_1 or D_2 open
   (c) No fault
   (d) Q_1 shorted C-to-E

25. 6 V dc, positive alternation of input signal

27. C_1 is connected backwards.

29. 51 W

31. Gain decreases.

33. T_C is much closer to the actual junction temperature than T_A. In a given operating environment, T_A is always less than T_C.

35. See Figure ANS–23.

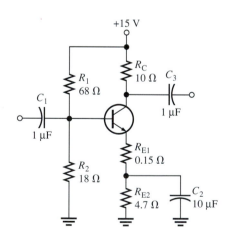

FIGURE ANS–23

Chapter 8

1. (a) Narrows   (b) Increases

3. See Figure ANS–24.

FIGURE ANS–24

n–channel   *p*–channel

5. 5 V

7. 10 mA

9. 4 V

11. −2.63 V

13. $g_m = 1429$ µS, $y_{fs} = 1429$ µS

15. $V_{GS} = 0$ V, $I_D = 8$ mA
$V_{GS} = -1$ V, $I_D = 5.12$ mA
$V_{GS} = -2$ V, $I_D = 2.88$ mA
$V_{GS} = -3$ V, $I_D = 1.28$ mA
$V_{GS} = -4$ V, $I_D = 0.320$ mA
$V_{GS} = -5$ V, $I_D = 0$ mA

17. 800 Ω

19. (a) 20 mA   (b) 0 A   (c) Increases

21. 211 Ω

23. 9.80 MΩ

25. $I_D \cong 5.3$ mA, $V_{GS} \cong 2.1$ V

27. $I_D \cong 1.9$ mA, $V_{GS} \cong -1.5$ V

29. The enhancement mode

31. The gate is insulated from the channel.

33. 4.69 mA

35. (a) Depletion   (b) Enhancement
(c) Zero bias   (d) Depletion

37. (a) 4 V   (b) 5.4 V   (c) −4.52 V

39. (a) 5 V, 3.18 mA   (b) 3.2 V, 1.02 mA

41. R_D or R_S open, JFET open D-to-S, $V_{DD} = 0$ V, or ground connection open.

43. No change

45. The 1 MΩ bias resistor is open.

47. $V_{OUT} = 300$ mV for pH = 5; $V_{OUT} = -400$ mV for pH = 9

49. $V_{OUT} = +12.1$ V assuming typical values.

51. (a) −0.5 V   (b) 25 V   (c) 310 mW
(d) −25 V

53. 2000 µS

55. 1 V

57. $I_D \cong 13$ mA when $V_{GS} = +3$ V, $I_D \cong 0.4$ mA when $V_{GS} = -2$ V.

59. −3.0 V

61. $I_D = 3.58$ mA; $V_{GS} = -4.21$ V

63. 6.01 V

65. See Figure ANS–25.

FIGURE ANS–25

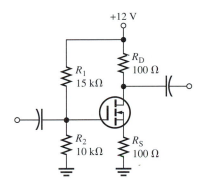

Chapter 9

1. (a) *n*-channel D-MOSFET with zero-bias; $V_{GS} = 0$
(b) *p*-channel JFET with self-bias; $V_{GS} = -0.99$ V
(c) *n*-channel E-MOSFET with voltage-divider bias;
$V_{GS} = 3.84$ V

3. (a) *n*-channel D-MOSFET   (b) *n*-channel JFET
(c) *p*-channel E-MOSFET

5. Figure 9–7(b): approximately 4 mA
Figure 9–7(c): approximately 3.2 mA

7. 5.71 kΩ

9. 2.73

11. 920 mV

13. (a) 4.32   (b) 9.92

15. ≈7.5 mA

17. 3.02

19. 33.6 mV rms

21. 9.84 MΩ

23. $V_{GS} = 9$ V; $I_D = 3.13$ mA; $V_{DS} = 13.3$ V; $V_{ds} = 675$ mV

25. $R_{in} \cong 10$ MΩ; $A_v = 0.783$

27. (a) 0.906   (b) 0.299

29. 250 Ω

31. (a) $V_{D1} = V_{DD}$; no Q_1 drain signal; no output signal
(b) $V_{D1} \cong 0$ V (floating); no Q_1 drain signal;
no output signal
(c) $V_{GS1} = 0$ V; $V_S = 0$ V; V_{D1} less than normal;
clipped output signal
(d) Correct signal at Q_1 drain; no Q_1 gate signal;
no output signal
(e) $V_{D2} = V_{DD}$; correct signal at Q_2 gate;
no Q_2 drain signal or output signal

33. The 10 µF capacitor between Q_1 drain and Q_2 gate is open.

35. $V_{DC} = 4.35$ V; $V_{ac} = 1.29$ V rms

37. (a) −3.0 V   (b) 20 V dc   (c) 200 mW   (d) ±10 V dc

39. 900 µS

41. 1.5 mA

43. 2.0; 6.82

45. See Figure ANS–26.

FIGURE ANS–26

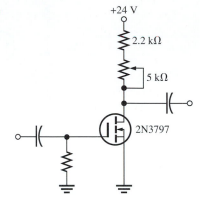

Chapter 10

1. If $C_1 = C_2$, the critical frequencies are equal, and they will both cause the gain to drop at 40 dB/decade below f_c.

3. Bipolar: C_{be}, C_{bc}, C_{ce}; FET: C_{gs}, C_{gd}, C_{ds}

5. 24 mV rms; 34 dB

7. (a) 3.01 dBm **(b)** 0 dBm **(c)** 6.02 dBm
(d) −6.02 dBm

9. (a) 318 Hz **(b)** 1.59 kHz

11. At $0.1f_c$: $A_v = 18.7$ dB
At f_c: $A_v = 35.7$ dB
At $10f_c$: $A_v = 38.7$ dB

13. Low-frequency response: C_1, C_2, and C_3
High-frequency response: C_{bc} and C_{be}

15. 4 pF

17. See Figure ANS–27.

19. Input circuit: $f_c = 4.32$ MHz
Output circuit: $f_c = 94.9$ MHz
Input f_c is dominant.

21. $f_{cl} = 136$ Hz, $f_{cu} = 8$ kHz

23. $BW = 5.26$ MHz, $f_{cu} \cong 5.26$ MHz

25. Input RC circuit: $f_c = 3.34$ Hz
Output RC circuit: $f_c = 3.01$ kHz
Output f_c is dominant.

27. Input circuit: $f_c = 12.9$ MHz
Output circuit: $f_c = 54.9$ MHz
Input f_c is dominant.

29. 230 Hz; 1.2 MHz

31. 514 kHz

33. ≈2.5 MHz

35. Increase the frequency until the output voltage drops to 3.54 V rms. This is f_{cu}.

37. 23.1 Hz

39. No effect

41. 112 pF

43. $C_{gd} = 1.3$ pF; $C_{gs} = 3.7$ pF; $C_{ds} = 3.7$ pF

45. ≈10.5 MHz

Chapter 11

1. $I_A = I_K = 646$ nA

3. See pages 598–600.

5. Add a transistor to provide inversion of negative half-cycle in order to obtain a positive gate trigger.

7. See Figure ANS–28.

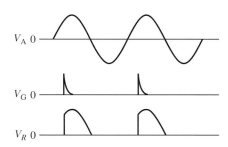

FIGURE ANS–28

9. Anode, cathode, anode gate, cathode gate

11. See Figure ANS–29.

13. 6.48 V

15. (a) 9.79 V **(b)** 5.2 V

17. See Figure ANS–30.

19. (a) 12 V **(b)** 0 V

FIGURE ANS–27

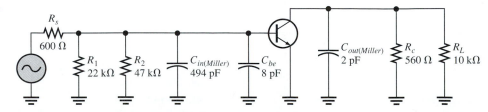

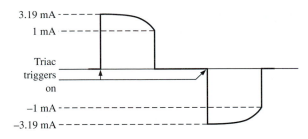

FIGURE ANS–29

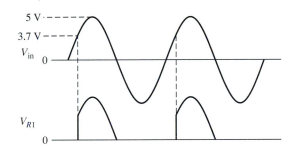

FIGURE ANS–30

21. When the switch is closed, the battery V_2 causes illumination of the lamp. The light energy causes the LASCR to conduct and thus energize the relay. When the relay is energized, the contacts close and 115 V ac are applied to the motor.

23. 30 mA

25. 0 V

27. As the PUT gate voltage increases, the PUT triggers on later to the ac cycle causing the SCR to fire later in the cycle, conduct for a shorter time, and decrease power to the motor.

29. See Figure ANS–31.

Chapter 12

1. *Practical op-amp:* High open-loop gain, high input impedance, low output impedance, high CMRR. *Ideal op-amp:* Infinite open-loop gain, infinite input impedance, zero output impedance, infinite CMRR.

3. **(a)** Single-ended input; differential output
   (b) Single-ended input; single-ended output
   (c) Differential input; single-ended output
   (d) Differential input; differential output

5. V_1: differential output voltage
   V_2: noninverting input voltage
   V_3: single-ended output voltage
   V_4: differential input voltage
   I_1: bias current

7. 8.1 μA

9. 108 dB

11. 0.30

13. 40 μs

15. $B = 9.90 \times 10^{-3}$, $V_f = 49.5$ mV

17. **(a)** 11   **(b)** 101   **(c)** 47.8   **(d)** 23

19. **(a)** 1   **(b)** −1   **(c)** 22   **(d)** −10

21. **(a)** 455 μA   **(b)** 455 μA   **(c)** −10 V   **(d)** −10

23. **(a)** $Z_{in(VF)} = 1.32$ TΩ
        $Z_{out(VF)} = 455$ μΩ
   (b) $Z_{in(VF)} = 500$ GΩ
        $Z_{out(VF)} = 600$ μΩ
   (c) $Z_{in(VF)} = 40$ GΩ
        $Z_{out(VF)} = 1.5$ mΩ

25. **(a)** 75 Ω placed in feedback path
   (b) 150 μV

27. 200 μV

29. **(a)** R_1 open or op-amp faulty
   (b) R_2 open
   (c) Nonzero output offset voltage; R_4 faulty or in need of adjustment

FIGURE ANS–31

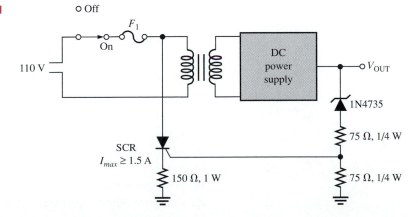

31. The voltage gain will be fixed at 100.

33. $Z_{in(NI)} = 3.96$ GΩ; $Z_{out(NI)} = 37.9$ mΩ

35. 50,000

37. See Figure ANS–32.

FIGURE ANS–32

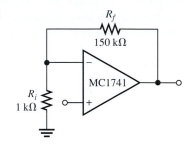

39. 6.32

Chapter 13

1. 70 dB

3. 1.67 kΩ

5. **(a)** 79,603 **(b)** 56,569 **(c)** 7960 **(d)** 80

7. **(a)** −0.674° **(b)** −2.69° **(c)** −5.71°
 (d) −45° **(e)** −71.2° **(f)** −84.3°

9. **(a)** 0 dB/decade **(b)** −20 dB/decade
 (c) −40 dB/decade **(d)** −60 dB/decade

11. 4.05 MHz

13. 21.1 MHz

15. Circuit (b) has smaller *BW* (97.5 kHz).

17. **(a)** 150° **(b)** 120° **(c)** 60°
 (d) 0° **(e)** −30°

19. **(a)** Unstable **(b)** Stable **(c)** Marginally stable

21. 25 Hz

23. D_1 or D_2 shorted, faulty Q_1 or Q_2, excessive gain from op-amp stage, R_6 open.

25. 0 V

27. 63,096

29. See Figure ANS–33.

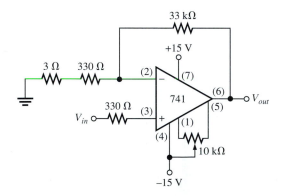

FIGURE ANS–33

31. See Figure ANS–34.

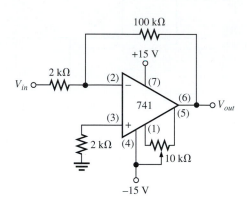

FIGURE ANS–34

Chapter 14

1. 24 V, with distortion

3. $V_{UTP} = +2.77$ V, $V_{LTP} = −2.77$ V

5. See Figure ANS–35.

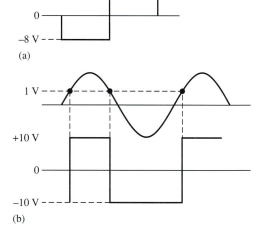

FIGURE ANS–35

7. +8.57 V and −0.968 V

9. **(a)** −2.5 V **(b)** −3.52 V

11. 110 kΩ

13. $V_{OUT} = −3.57$ V, $I_f = 357$ μA

15. −4.46 mV/μs

17. 1 mA

19. See Figure ANS–36.

21. Output not correct. Op-amp 2 or diode D_2 is faulty.

23. Output not correct. R_2 is open.

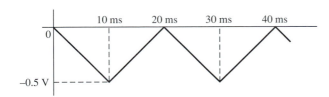

FIGURE ANS–36

25. Perform a visual check. The middle IC is installed backwards.

27. The output of IC1 will be 10 times normal, driving IC2 into saturation for the sample-and-hold operation.

29. +0.5 V

Chapter 15

1. $A_{v(1)} = A_{v(2)} = 101$

3. 1.005 V

5. 9.1

7. Change R_G to 1.8 kΩ.

9. 225

11. Change the 18 kΩ resistor to 270 kΩ.

13. Connect output (pin 22) directly to pin 23, and connect pin 38 directly to pin 40 to make $R_f = 0$.

15. 500 μA, 5 V

17. $A_v \cong 11.6$

19. See Figure ANS–37.

21. See Figure ANS–38.

23. **(a)** −0.301   **(b)** 0.301   **(c)** 1.70   **(d)** 2.11

25. The output of a log amplifier is limited to 0.7 V because of the transistor's *pn* junction.

27. −157 mV

29. $V_{out(max)} = -147$ mV, $V_{out(min)} = -89.2$ mV; the 1 V input peak is reduced 85% whereas the 100 mV input peak is reduced only 10%.

31. See Figure ANS–39.

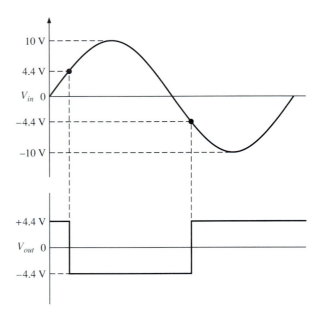

FIGURE ANS–38

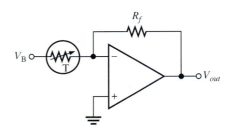

FIGURE ANS–39

33. TP1: ≈ 0 V
     TP2: ≈ 0 V
     TP3: 20 mV @ 1 kHz
     TP4: +12 V
     TP5: 0 V

FIGURE ANS–37

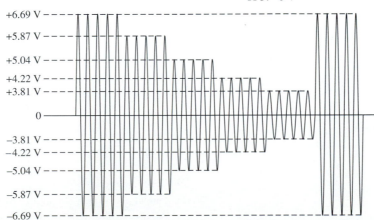

Chapter 16

1. **(a)** Band pass   **(b)** High pass   **(c)** Low pass
    (d) Band stop

3. 48.2 kHz, No

5. 700 Hz, 5.04

7. **(a)** 1, not Butterworth
    (b) 1.44, approximate Butterworth
    (c) 1st stage: 1.67; 2nd stage: 1.67; Not Butterworth

9. **(a)** Chebyshev   **(b)** Butterworth
    (c) Bessel   **(d)** Butterworth

11. 190 Hz

13. Add another identical stage and change R_1/R_2 ratio to 0.068 for first stage, 0.586 for second stage, and 1.482 for third stage.

15. Exchange positions of resistors and capacitors.

17. **(a)** Decrease R_1 and R_2 or C_1 and C_2.
    (b) Increase R_3 or decrease R_4.

19. **(a)** $f_0 = 4.95$ kHz, $BW = 3.84$ kHz
    (b) $f_0 = 449$ Hz, $BW = 96.5$ Hz
    (c) $f_0 = 15.9$ kHz, $BW = 838$ Hz

21. Sum the low-pass and high-pass outputs with a two-input adder.

Chapter 17

1. An oscillator requires no input (other than dc power).

3. $1/75 = 0.0133$

5. 733 mV

7. 50 kΩ

9. 7.5 V, 3.94

11. 136 kΩ, 691 Hz

13. 9.4

15. Change R_1 to 3.54 kΩ

17. $R_4 = 65.8$ kΩ, $R_5 = 47$ kΩ

19. 3.33 V, 6.67 V

21. 0.0076 µF

23. $f_{min} = 42.5$ kHz, $f_{max} = 57.5$ kHz

25. 25 kHz

Chapter 18

1. 0.025%

3. 1.01%

5. A: reference voltage, B: Control element, C: Error detector, D: Sampling circuit

7. 8.51 V

9. 9.57 V

11. 500 mA

13. 10 mA

15. $I_{L(max)} = 250$ mA, $P_{R1} = 6.25$ W

17. 40%

19. V_{out} increases

21. 14.3 V

23. 1.3 mA

25. 2.8 Ω

27. $R_{limit} = 0.35$ Ω

29. See Figure ANS–40.

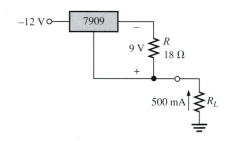

FIGURE ANS–40

GLOSSARY

Active filter   A frequency-selective circuit consisting of active devices such as transistors or op-amps coupled with reactive components.

A/D conversion   A process whereby information in analog form is converted into digital form.

Alpha (α)   The ratio of collector current to emitter current in a bipolar junction transistor.

Amplification   The process of increasing the power, voltage, or current by electronic means.

Amplifier   An electronic circuit having the capability to amplify power, voltage, or current.

Analog   Characterized by a linear process in which a variable takes on a continuous set of values.

Anode   The *p* region of a diode.

Atom   The smallest particle of an element that possesses the unique characteristics of that element.

Atomic number   The number of electrons in a neutral atom.

Atomic weight   Approximately the number of protons and neutrons in the nucleus of an atom.

Attenuation   The reduction in the level of power, current, or voltage.

Audio   Related to the frequency range of sound waves that can be heard by the human ear.

Avalanche   The rapid buildup of conduction electrons due to excessive reverse-bias voltage.

Avalanche breakdown   The higher voltage breakdown in a zener diode.

Band-pass filter   A type of filter that passes a range of frequencies lying between a certain lower frequency and a certain higher frequency.

Band-stop filter   A type of filter that blocks or rejects a range of frequencies lying between a certain lower frequency and a certain higher frequency.

Bandwidth   The characteristic of certain types of electronic circuits that specifies the usable range of frequencies that pass from input to output.

Barrier potential   The amount of energy required to produce full conduction across the *pn* junction in forward bias.

Base   One of the semiconductor regions in a bipolar junction transistor. The base is very thin and lightly doped compared to the other regions.

BASIC   A computer programming language: *B*eginner's *A*ll-purpose *S*ymbolic *I*nstruction *C*ode.

Bessel   A type of filter response having a linear phase characteristic and less than 20 dB/decade/pole roll-off.

Beta (β)   The ratio of collector current to base current in a bipolar junction transistor; current gain from base to collector.

Bias   The application of a dc voltage to a *pn* junction, transistor, or other device to produce a desired mode of operation.

Bipolar   Characterized by both free electrons and holes as current carriers.

Bipolar junction transistor (BJT)   A transistor constructed with three doped semiconductor regions separated by two *pn* junctions.

Bode plot   An idealized graph of the gain in dB versus frequency used to graphically illustrate the response of an amplifier or filter.

Bounding   The process of limiting the output range of an amplifier or other circuit.

Bridge rectifier   A type of full-wave rectifier consisting of diodes arranged in a four-cornered configuration.

Butterworth   A type of filter response characterized by flatness in the pass band and a 20 dB/decade/pole roll-off.

Carbon   A semiconductor material.

Carrier   The high frequency (RF) signal that carries modulated information in AM, FM, or other systems.

Cascade   An arrangement of circuits in which the output of one circuit becomes the input to the next.

Cathode   The *n* region of a diode.

Center-tapped rectifier   A type of full-wave rectifier consisting of a center-tapped transformer and two diodes.

Channel   The conductive path between the drain and source in an FET.

Chebyshev   A type of filter response characterized by ripples in the pass band and a greater than 20 dB/decade/pole roll-off.

Clamper   A circuit using a diode and a capacitor which adds a dc level to an ac voltage.

Clipper   See Limiter.

Closed-loop   An op-amp configuration in which the output is connected back to the input through a feedback circuit.

Closed-loop gain (A_{cl})   The voltage gain of an op-amp with external feedback.

Coherent light   Light having only one wavelength.

Collector   The largest of the three semiconductor regions of a BJT.

Common-base (CB)   A BJT amplifier configuration in which the base is the common terminal to an ac signal or ground.

Common-collector (CC)   A BJT amplifier configuration in which the collector is the common terminal to an ac signal or ground.

Common-drain (CD)   An FET amplifier configuration in which the drain is the grounded terminal.

Common-emitter (CE)   A BJT amplifier configuration in which the emitter is the common terminal to an ac signal or ground.

Common-gate (CG)   An FET amplifier configuration in which the gate is the grounded terminal.

Common mode   A condition characterized by the presence of the same signal on both op-amp inputs.

Common-mode rejection ratio (CMRR)   The ratio of open-loop gain to common-mode gain; a measure of an op-amp's ability to reject common-mode signals.

Common-source (CS)   An FET amplifier configuration in which the source is the grounded terminal.

Comparator   A circuit which compares two input voltages and produces an output in either of two states indicating the greater or less than relationship of the inputs.

Compensation   The process of modifying the roll-off rate of an amplifier to ensure stability.

Complementary pair   Two transistors, one *npn* and one *pnp,* having matched characteristics.

Conduction electron   A free electron.

Conductor   A material that conducts electrical current very well.

Core   The central part of an atom, includes the nucleus and all but the valence electrons.

Covalent   Related to the bonding of two or more atoms by the interaction of their valence electrons.

Critical frequency   The frequency at which the response of an amplifier or filter is 3 dB less than at midrange.

Crossover distortion   Distortion in the output of a class B push-pull amplifier at the point where each transistor changes from the cutoff state to the *on* state.

Crystal   The pattern or arrangement of atoms forming a solid material; a quartz device that operates on the piezoelectric effect and exhibits very stable resonant properties.

Current   The rate of flow of free electrons.

Cutoff   The nonconducting state of a transistor.

Cutoff frequency   Another term for critical frequency.

Cutoff voltage   The value of the gate-to-source voltage that makes the drain current approximately zero.

D/A conversion   The process of converting a sequence of digital codes to an analog form.

Damping factor   A filter characteristic that determines the type of response.

Dark current   The amount of thermally generated reverse current in a photodiode in the absence of light.

Darlington pair   A configuration of two transistors in which the collectors are connected and the emitter of the first drives the base of the second to achieve beta multiplication.

DC load line   A straight line plot of I_C and V_{CE} for a transistor circuit.

Decade   A ten-times increase or decrease in the value of a quantity such as frequency.

Decibel (dB)   The unit of the logarithmic expression of a ratio, such as power or voltage.

Depletion   In a MOSFET, the process of removing or depleting the channel of charge carriers and thus decreasing the channel conductivity.

Depletion region   The area near a *pn* junction on both sides that has no majority carriers.

Derivative   The instantaneous rate of change of a function, determined mathematically.

Diac   A two-terminal four-layer semiconductor device (thyristor) that can conduct current in either direction when properly activated.

Differential amplifier (diff-amp)   An amplifier that produces an output voltage proportional to the difference of the two input voltages.

Differentiator   A circuit that produces an output which approximates the instantaneous rate of change of the input function.

Digital   Characterized by a process in which a variable takes on either of two values.

Diode characteristic curve   A graph showing the relationship of current versus voltage.

Diode drop   The voltage across the diode when it is forward-biased; approximately the same as the barrier potential and typically 0.7 V for silicon.

Doping   The process of imparting impurities to an intrinsic semiconductor material in order to control its conduction characteristics.

Drain   One of the three terminals of an FET analogous to the collector of a BJT.

Dynamic resistance   The nonlinear internal resistance of a semiconductor material.

Electroluminescence   The process of releasing light energy by the recombination of electrons in a semiconductor.

Electron   The basic particle of negative electrical charge.

Electron-hole pair   The conduction electron and the hole created when the electron leaves the valence band.

Emitter   The most heavily doped of the three semiconductor regions of a BJT.

Emitter-follower   A popular term for a common-collector amplifier.

Enhancement   In a MOSFET, the process of creating a channel or increasing the conductivity of the channel by the addition of charge carriers.

Feedback The process of returning a portion of a circuit's output back to the input in such a way as to oppose a change in the output.

Feedforward A method of frequency compensation in op-amp circuits.

Field-effect transistor (FET) A type of unipolar, voltage-controlled transistor that uses an induced electric field to control current.

Figure of merit The ratio of energy stored and returned by a reactive component to the energy dissipated; also called quality factor, Q.

Filter A type of circuit that passes or blocks certain frequencies to the exclusion of all others.

Floating point A point in the circuit that is not electrically connected to ground or a "solid" voltage.

Fold-back current limiting A method of current limiting in voltage regulators.

Forced commutation A method of turning off an SCR.

Forward bias The condition in which a *pn* junction conducts current.

Forward-breakdown voltage The voltage at which a device enters the forward-blocking region.

Free electron An electron that has acquired enough energy to break away from the valance band of the parent atom; also called a conduction electron.

Frequency modulation (FM) A communication method in which a lower frequency intelligence-carrying signal modulates (varies) the frequency of a higher frequency signal.

Full-wave rectifier A circuit that converts an ac sinusoidal input voltage into a pulsating dc voltage with two output pulses occurring for each input cycle.

Fuse A protective device that burns open when the current exceeds a rated limit.

Gain The amount by which an electrical signal is increased or amplified.

Gain-bandwidth product A characteristic of amplifiers whereby the product of the gain and the bandwidth is always constant.

Gate One of the three terminals of an FET analogous to the base of a BJT.

Germanium A semiconductor material.

Half-wave rectifier A circuit that converts an ac sinusoidal wave input voltage into a pulsating dc voltage with one output pulse occurring for each input cycle.

Harmonics The frequencies contained in a composite waveform that are integer multiples of the repetition frequency (fundamental).

High-pass filter A type of filter that passes frequencies above a certain frequency while rejecting lower frequencies.

Holding current The value of the anode current below which a device switches from the forward-conduction region to the forward-blocking region.

Hole The absence of an electron in the valence band of an atom.

Hysteresis Characteristic of a circuit in which two different trigger levels create an offset or lag in the switching action.

Index of refraction A property of light-conducting materials that specifies how much a light ray will bend when passing from one material to another.

Infrared (IR) Light that has a range of wavelengths greater than visible light.

Input The terminal of a circuit to which an electrical signal is first applied.

Insulator A material that does not conduct current.

Integral The area under the curve of a function, determined mathematically.

Integrated circuit (IC) A type of circuit in which all the components are constructed on a single tiny chip of silicon.

Integrator   A circuit that produces an output which approximates the area under the curve of the input function.

Intrinsic   The pure or natural state of a material.

Inversion   The conversion of a quantity to its opposite value.

Inverting amplifier   An op-amp closed-loop configuration in which the input signal is applied to the inverting input.

Ionization   The removal or addition of an electron from or to a neutral atom so that the resulting atom (called an ion) has a net positive or negative charge.

Irradiance (H)   The power per unit area at a specified distance for the LED; the light intensity.

Junction field-effect transistor (JFET)   One of two major types of field-effect transistors.

Large-signal   A signal that operates an amplifier over a significant portion of its load line.

Laser   *L*ight *a*mplification by *s*timulated *e*mission of *r*adiation.

Light-activated silicon-controlled rectifier (LASCR)   A four-layer semiconductor device (thyristor) that conducts current in one direction when activated by a sufficient amount of light and continues to conduct until the current falls below a specified value.

Light-emitting diode (LED)   A type of diode that emits light when there is forward current.

Limiter   A diode circuit that clips off or removes part of a waveform above and/or below a specified level.

Linear   Characterized by a straight-line relationship.

Line regulation   The percentage change in output voltage for a given change in input (line) voltage.

Load   The amount of current drawn from the output of a circuit through a load impedance.

Load regulation   The percentage change in output voltage for a given change in load current.

Loop gain   An op-amp's open-loop gain times the attenuation.

Low-pass filter   A type of filter that passes frequencies below a certain frequency while rejecting higher frequencies.

Majority carrier   The most numerous charge carrier in a doped semiconductor material (either free electrons or holes).

Midrange   The frequency range of an amplifier lying between the lower and upper critical frequencies.

Minority carrier   The least numerous charge carrier in a doped semiconductor material (either free electrons or holes).

Monochromatic   Related to light of a single frequency; one color.

MOSFET   Metal oxide semiconductor field-effect transistor; one of two major types of FET; sometimes called IGFET for insulated-gate FET.

Multistage   Characterized by having more than one stage; a cascaded arrangement of two or more amplifiers.

Negative feedback   The process of returning a portion of the output signal to the input of an amplifier such that it is out of phase with the input signal.

Neutron   An uncharged particle found in the nucleus of an atomn.

Noise   An unwanted signal.

Noninverting amplifier   An op-amp closed-loop configuration in which the input signal is applied to the noninverting input.

Nucleus   The central part of an atom containing protons and neutrons.

Octave   A two-times increase or decrease in the value of a quantity such as frequency.

Open-loop gain (A_{ol}) The voltage gain of an op-amp without feedback.

Operational amplifier (op-amp)   A type of amplifier that has a very high voltage gain, very high input impedance, very low output impedance, and good rejection of common-mode signals.

Orbit   The path an electron takes as it circles around the nucleus of an atom.

Oscillator   An electronic circuit based on positive feedback that produces a time-varying output signal without an external input signal.

Output   The terminal of a circuit from which the final voltage is obtained.

Pentavalent atom   An atom with five valence electrons.

Percent regulation   A figure of merit used to specify the performance of a voltage regulator.

Phase   The relative angular displacement of a time-varying function relative to a reference.

Phase margin   The difference between the total phase shift through an amplifier and 180 degrees; the additional amount of phase shift that can be allowed before instability occurs.

Photodiode   A diode in which the reverse current varies directly with the amount of light.

Photon   A particle of light energy.

Phototransistor   A transistor in which base current is produced when light strikes the photosensitive semiconductor base region.

Piezoelectric effect   The property of a crystal whereby a changing mechanical stress produces a voltage across the crystal.

Pinch-off voltage   The value of the drain-to-source voltage of an FET at which the drain current becomes constant when the gate-to-source voltage is zero.

PN junction   The boundary between two different types of semiconductor materials.

Pole   A circuit containing one resistor and one capacitor that contributes 20 dB/decade to a filter's roll-off.

Positive feedback   The return of a portion of the output signal to the input such that it sustains the output. This output signal is in phase with the input signal.

Power supply   A circuit that converts ac line voltage to dc voltage and supplies constant power to operate a circuit or system.

Programmable unijunction transistor (PUT)   A type of three-terminal thyristor (more like an SCR than a UJT) that is triggered into conduction when the voltage at the anode exceeds the voltage at the gate.

Proton   The basic particle of positive charge.

Push-pull   A type of class B amplifier with two transistors in which one transistor conducts for one half-cycle and the other conducts for the other half-cycle.

Q-point   The dc operating (bias) point of an amplifier specified by voltage and current values.

Quality factor (Q) The ratio of a band-pass filter's center frequency to its bandwidth.

Radiant intensity (I_e) The output power of an LED per steradian in units of mW/sr.

Radiation   The process of emitting electromagnetic or light energy.

Recombination   The process of a free (conduction band) electron falling into a hole in the valence band of an atom.

Rectifier   An electronic circuit that converts ac into pulsating dc; one part of a power supply.

Regulator   An electronic device or circuit that maintains an essentially constant output voltage for a range of input voltage or load values; one part of a power supply.

Reverse bias   The condition in which a *pn* junction prevents current.

RF   Radio frequency.

Ripple factor   A measure of effectiveness of a power supply filter in reducing the ripple voltage; ratio of the ripple voltage to the dc output voltage.

Ripple voltage   The small variation in the dc output voltage of a filtered rectifier caused by the charging and discharging of the filter capacitor.

Roll-off   The decrease in the gain of an amplifier above or below the critical frequencies.

Saturation   The state of a BJT in which the collector current has reached a maximum and is independent of the base current.

Schematic   A symbolized diagram representing an electrical or electronic circuit.

Schmitt trigger   A comparator with hysteresis.

Schottky diode   A diode using only majority carriers and intended for high-frequency operation.

Semiconductor   A material that lies between conductors and insulators in its conductive properties.

Shell   An energy band in which electrons orbit the nucleus of an atom.

Shockley diode   The type of two-terminal thyristor that conducts current when the anode-to-cathode voltage reaches a specified "breakover" value.

Silicon   A semiconductor material.

Silicon-controlled rectifier (SCR)   A type of three-terminal thyristor that conducts current when triggered on by a voltage at the single gate terminal and remains on until the anode current falls below a specified value.

Silicon-controlled switch (SCS)   A type of four-terminal thyristor that has two gate terminals that are used to trigger the device on and off.

Slew rate   The rate of change of the output voltage of an op-amp in response to a step input.

Source   One of the three terminals of an FET analogous to the emitter of a BJT.

Source-follower   The common-drain amplifier.

Spectral   Pertaining to a range of frequencies.

Stability   A measure of how well an amplifier maintains its design values (Q-point, gain, etc.) over changes in beta and temperature; a condition in which an amplifier circuit does not oscillate.

Stage   One of the amplifier circuits in a multistage configuration.

Standoff ratio   The characteristic of a UJT that determines its turn-on point.

Step   A fast voltage transition from one level to another.

Switching current   The value of anode current at the point where the device switches from the forward-blocking region to the forward-conduction region.

Thermal overload   A condition in a rectifier where the internal power dissipation of the circuit exceeds a certain maximum due to excessive current.

Thermistor   A temperature-sensitive resistor with a negative temperature coefficient.

Thyristor   A class of four-layer (*pnpn*) semiconductor devices.

Transconductance   The ratio of a change in drain current for a change in gate-to-source voltage in an FET; in general, the ratio of the output current to the input current.

Transistor   A semiconductor device used for amplification and switching applications.

Triac   A three-terminal thyristor that can conduct current in either direction when properly activated.

Trigger   The activating input of some electronic devices and circuits.

Trivalent atom   An atom with three valence electrons.

Troubleshooting   The process and technique of identifying and locating faults in an electronic circuit or system.

Tuning ratio   The ratio of varactor capacitance at minimum and maximum reverse voltages.

Tunnel diode   A diode exhibiting a negative resistance characteristic.

Unijunction transistor (UJT)   A three-terminal single *pn* junction device that exhibits a negative resistance characteristic.

Valence   Related to the outer shell of an atom.

Varactor   A variable capacitance diode.

Voltage-follower   A closed-loop, noninverting op-amp with a voltage gain of one.

Voltage multiplier   A circuit using diodes and capacitors that increases the input voltage by two, three, or four times.

Wavelength   The distance in space occupied by one cycle of an electromagnetic or light wave.

Zener breakdown   The lower voltage breakdown in a zener diode.

Zener diode   A diode designed for limiting the voltage across its terminals in reverse bias.

INDEX